BIOPROCESSING OF RENEWABLE RESOURCES TO COMMODITY BIOPRODUCTS

BIOPROCESSING OF RENEWABLE RESOURCES TO COMMODITY BIOPRODUCTS

Edited by

Virendra S. Bisaria
Akihiko Kondo

WILEY

About the Cover: The pyramid represents successive and increasingly selective processing stages in bioconversion of plant biomass to industrial chemicals. The chemicals in white bubbles are the industrial commodity bioproducts pertaining to the realm of "white biotechnology".

Cover illustration/design by Ruchi Uppal.

Published by John Wiley & Sons, Inc., Hoboken, New Jersey.
Published simultaneously in Canada.

For general information on our other products and services or for technical support, please contact our Customer Care Department within the United States at (800) 762-2974, outside the United States at (317) 572-3993 or fax (317) 572-4002.

Wiley also publishes its books in a variety of electronic formats. Some content that appears in print may not be available in electronic formats. For more information about Wiley products, visit our web site at www.wiley.com.

Library of Congress Cataloging-in-Publication Data:

Bioprocessing of renewable resources to commodity bioproducts / edited by Virendra S. Bisaria, Akihiko Kondo.
pages cm
Includes bibliographical references and index.
ISBN 978-1-118-17583-5 (hardback)
1. Microbial biotechnology. 2. Biomass energy. I. Bisaria, Virendra S., editor of compilation. II. Kondo, Akihiko, 1959- editor of compilation.
TP248.27.M53B5626 2014
662′.88–dc23

2013046035

Printed in the United States of America.

10 9 8 7 6 5 4 3 2 1

CONTENTS

PREFACE

For the development of a sustainable, industrial society to meet our demands of energy and materials, it is being increasingly realized that we will have to shift from our dependence on petroleum to the use of renewable resources, such as starch- and cellulose-based plant materials. Historically till recently, petroleum-based resources were mainly targeted for research and development, and subsequent commercialization of the products derived therefrom. However, their rising costs and the anticipated threat to the earth's environment are providing the required incentive to find sustainable alternative resources. Biorefineries, based on renewable resources, shall enable the production of biofuels as well as commodity chemicals (those produced in excess of about 1 million tons per year). These processes which are based on carbohydrates (such as starch and cellulose) are also favorable from a chemical point of view because the functional groups that are introduced by costly oxidative process steps into naphta are already present in them. The commodity bioproducts can be produced by microbial processes. Most of them are natural products of microorganisms or can be produced by suitable pathway engineering of industrial organisms. As these bioproducts contain functional groups, they are extremely useful as starting materials for the chemical industry for synthesis of a wide variety of products such as polymers, surfactants, lubricants, and resins.

To avoid competition with starchy raw materials, which are largely used as food, as well as to realize the vision of a successful biorefinery, the renewable resource present in the form of abundant lignocellulosic biomass needs to be efficiently converted to its constituent monomers, comprising mainly of hexose (such as glucose, mannose, and galactose) and pentose sugars (such as xylose and arabinose). Accordingly, the Part I of the book deals with those enabling technologies that are crucial for the pretreatment (Chapter 3) and hydrolysis of biomass to give sugars in high yield (Chapter 4) by cellulolytic enzymes, primarily cellulase and xylanase (Chapter 5). This first part also covers the general aspect and the issues involved in the sustainability of a biorefinery (Chapter 1) and biomass feedstock logistics and the design of biomass feedstock supply systems (Chapter 2). Chapter 6 describes various bioprocessing technologies that in one form or the other will be required to be implemented for the development of biorefineries.

The Part II of the book contains state-of-the-art articles on a few chosen commodity bioproducts. These bioproducts represent most of those identified by the US Department of Energy for intensive investigation for their production from renewable resources. While covering these bioproducts, major emphasis has been given to the discipline of metabolic engineering for the development of suitable microbial biocatalysts/cell factories which shall enable their production from renewable resources. Ethanol which remains the most sought-after chemical and biofuel is covered in two chapters. While Chapter 7 describes the potential of recombinant bacteria for ethanol production, Chapter 8 is concerned mainly with strategies being developed to expand the genetic potential of the yeasts, already employed by the industry. Butanol, an excellent transportation fuel and a valuable chemical feedstock, is covered in Chapter 9 with respect to the advances that have taken place in recent years in the well-known ABE fermentation process for its production from renewable feedstock. Chapters 10 and 11 describe the recent advances being made for bio-based production of butanediols and propanediols, used extensively as solvent and for production of different types of chemicals, polymers, and so on. The feasibility of producing isobutanol, another higher alcohol besides butanol, possessing chemical features close to that of gasoline, through implementation of the Ehrlich pathway into several potential host microorganisms has been dealt with in Chapter 12. Lactic acid (LA), widely used in the food, pharmaceutical, and polymers industries, is already produced by microbial fermentations; Chapter 13, therefore, concentrates on production of LA and LA-based polymers from various genetically modified microorganisms from starchy and cellulosic materials. Chapter 14 describes the recent progress in biological production of 3-hydroxy propionic acid, used for the production of a wide range of commercially important chemicals such as acrylic acid, using different microorganisms and renewable substrates. Chapter 15 reviews the recent research and provides a critical analysis of future perspectives to develop an economically competitive bio-based process for producing fumaric acid, which is widely used in the food industry. Succinic acid with many applications including the production of important bulk chemicals, namely 1,4-butanediol (BDO), γ-butyrolactone (GBL), and tetrahydrofuran (THF), is covered in Chapter 16 with respect to its production from various substrates from natural and genetically modified organisms. Glutamic acid is the major amino acid produced by microbial fermentation on an industrial scale. Chapter 17 reviews the molecular mechanisms and metabolic engineering of glutamic acid production by *Corynebacterium glutamicum* and potential use of glutamic acid as a building block for producing several other chemicals. Xylitol, a natural sugar alcohol widely used as a sugar substitute in foods, toothpastes, and mouthwashes, is covered in Chapter 18 with respect to the application of recent approaches of genetic engineering, metabolic engineering, and cofactor engineering for its overproduction. New approaches for production of adipic acid, mainly used as an intermediate reactant for the production of nylon-6,6, are highlighted in Chapter 19 from several new feedstocks including lignin-rich streams.The commercial production of some of these commodity bioproducts in the near future will have a far reaching effect in catalyzing the realization of our goal of a sustainable biorefinery.

As research and development in this area has not yet achieved its full potential, the field of bioprocessing of renewable resources into commodity bioproducts will continue to expand to attain its commercial goal. Additionally, new bioproducts and fine chemicals will be added to the existing list of commodity bioproducts, as our capacity to produce sugars from cellulosic residues efficiently and economically increases and more efficient microbial biocatalysts are developed through application of modern biotechnology tools.

The book also provides a unique perspective to the industry about the scientific problems and their possible solutions in making a bioprocess work for commercial production of these commodity bioproducts. The book is suitable for researchers, practitioners, students, and consultants in metabolic engineering, bioprocess engineering, and biotechnology.

VIRENDRA S. BISARIA
AKIHIKO KONDO

CONTRIBUTORS

Somasundar Ashok Department of Chemical and Biomolecular Engineering, Pusan National University, Jangjeon-dong, Geumjeong-gu, Busan, Republic of Korea

Jozef Bernard Biochemical Engineering Institute, Technische Universität Braunschweig, Braunschweig, Germany

Virendra S. Bisaria Department of Biochemical Engineering and Biotechnology, Indian Institute of Technology Delhi, New Delhi, India

Ranjita Biswas BioEnergy Science Center, Biosciences Division, Oak Ridge National Laboratory, Oak Ridge, TN, USA

Hans P. Blaschek Department of Food Science and Human Nutrition, Institute for Genomic Biology, Center for Advanced Bioenergy Research (CABER), University of Illinois at Urbana-Champaign, Urbana, IL, USA

Bastian Blombach Institute of Biochemical Engineering, University of Stuttgart, Stuttgart, Germany

Ian J. Bonner Biofuels and Renewable Energy Technologies, Idaho National Laboratory, Idaho Falls, ID, USA

Michael Bott IBG-1: Biotechnology, Institute of Bio- and Geosciences, Jülich, Germany

Melanie Brocker IBG-1: Biotechnology, Institute of Bio- and Geosciences, Jülich, Germany

Roman Brunecky Biosciences Center, National Renewable Energy Laboratory, Golden, CO, USA

Gopal Chotani DuPont Industrial Biosciences, Palo Alto, CA, USA

Bernhard J. Eikmanns Institute of Microbiology and Biotechnology, University of Ulm, Ulm, Germany

Richard T. Elander National Bioenergy Center, National Renewable Energy Laboratory, Golden, CO, USA

Hongxin Fu School of Life Science and Biotechnology, Dalian University of Technology, Dalian, People's Republic of China

Israel Goldberg Department of Microbiology and Molecular Genetics, The Institute of Medical Research Israel-Canada, The Hebrew University—Hadassah Medical School, Jerusalem, Israel

Tomohisa Hasunuma Organization of Advanced Science and Technology, Graduate School of Engineering, Kobe University, Nada, Kobe, Japan

J. Richard Hess Energy Systems and Technologies, Idaho National Laboratory, Idaho Falls, ID, USA

Takashi Hirasawa Department of Bioengineering, Tokyo Institute of Technology, Mirodi-ku, Yokohama, Japan

He Huang State Key Laboratory of Materials-Oriented Chemical Engineering, College of Biotechnology and Pharmaceutical Engineering, Nanjing Tech University, Nanjing, People's Republic of China

Holger Janssen Department of Food Science and Human Nutrition, University of Illinois at Urbana-Champaign, Urbana, IL, USA

Xiao-Jun Ji State Key Laboratory of Materials-Oriented Chemical Engineering, College of Biotechnology and Pharmaceutical Engineering, Nanjing Tech University, Nanjing, People's Republic of China

Kevin L. Kenney Energy Systems and Technologies, Idaho National Laboratory, Idaho Falls, ID, USA

Sun-Ki Kim Department of Agricultural Biotechnology and Center for Food and Bioconvergence, Seoul National University, Seoul, Korea

Akihiko Kondo Department of Chemical Science and Engineering, Graduate School of Engineering, Kobe University, Nada, Kobe, Japan

Vinod Kumar Department of Chemical and Biomolecular Engineering, Pusan National University, Jangjeon-dong, Geumjeong-gu, Busan, Republic of Korea

Boris Litsanov IBG-1: Biotechnology, Institute of Bio- and Geosciences, Jülich, Germany

Current: Institute of Microbiology, Eidgenössische Technische Hochschule Zürich, Zürich, Switzerland

Chengwei Ma School of Life Science and Biotechnology, Dalian University of Technology, Dalian, People's Republic of China

Hans Marx Department of Biotechnology, BOKU-VIBT University of Natural Resources and Life Sciences, Vienna, Austria

Diethard Mattanovich Department of Biotechnology, BOKU-VIBT University of Natural Resources and Life Sciences, Vienna, Austria

Austrian Centre of Industrial Biotechnology (ACIB GmbH), Vienna, Austria

James D. McMillan National Bioenergy Center, National Renewable Energy Laboratory, Golden, CO, USA

Peyman Moslemy DuPont Industrial Biosciences, Palo Alto, CA, USA

Ying Mu School of Life Science and Biotechnology, Dalian University of Technology, People's Republic of China

David J. Muth Praxik, LLC, Ames, IA, USA

Kenji Okano Department of Biotechnology, Graduate School of Engineering, Osaka University, Osaka, Japan

Marco Oldiges IBG-1: Biotechnology, Institute of Bio- and Geosciences, Jülich, Germany

Sunghoon Park Department of Chemical and Biomolecular Engineering, Pusan National University, Jangjeon-dong, Geumjeong-gu, Busan, Republic of Korea

Yong-Cheol Park Department of Bio and Fermentation Convergence Technology, Kookmin University, Seoul, Korea

Caroline Peres DuPont Industrial Biosciences, Palo Alto, CA, USA

Abhishek Persad Department of Biochemical Engineering and Biotechnology, Indian Institute of Technology Delhi, New Delhi, India

Karthik Rajendran School of Engineering, University of Borås, Sweden

J. Stefan Rokem Department of Microbiology and Molecular Genetics, The Institute of Medical Research Israel-Canada, The Hebrew University—Hadassah Medical School, Jerusalem, Israel

Michael Sauer Department of Biotechnology, BOKU-VIBT University of Natural Resources and Life Sciences, Vienna, Austria
Austrian Centre of Industrial Biotechnology (ACIB GmbH), Vienna, Austria

Alexandra Schuler DuPont Industrial Biosciences, Palo Alto, CA, USA

Jin-Ho Seo Department of Agricultural Biotechnology and Center for Food and Bioconvergence, Seoul National University, Seoul, Korea

Hiroshi Shimizu Department of Bioinformatic Engineering, Graduate School of Information Science and Technology, Osaka University, Suita, Osaka, Japan

William A. Smith Biofuels and Renewable Energy Technologies, Idaho National Laboratory, Idaho Falls, ID, USA

Matthias Steiger Department of Biotechnology, BOKU-VIBT University of Natural Resources and Life Sciences, Vienna, Austria
Austrian Centre of Industrial Biotechnology (ACIB GmbH), Vienna, Austria

Nathan A. Stevens Materials and Physical Security, Idaho National Laboratory, Idaho Falls, ID, USA

Jonathan J. Stickel National Bioenergy Center, National Renewable Energy Laboratory, Golden, CO, USA

Yaqin Sun School of Life Science and Biotechnology, Dalian University of Technology, Dalian, People's Republic of China

Mohammad J. Taherzadeh School of Engineering, University of Borås, Borås, Sweden

Tsutomu Tanaka Department of Chemical Science and Engineering, Graduate School of Engineering, Kobe University, Nada-ku, Kobe, Japan

Jozef Bernhard Johann Henry van Duuren Biochemical Engineering Institute, Technische Universität Braunschweig, Braunschweig, Germany
Institute of Systems Biotechnology, Saarland University, Saarbrücken, Germany

Yi Wang Department of Food Science and Human Nutrition, Institute for Genomic Biology, University of Illinois at Urbana-Champaign, Urbana, IL, USA

Christoph Wittmann Institute of Systems Biotechnology, Saarland University, Saarbrücken, Germany

Zhilong Xiu School of Life Science and Biotechnology, Dalian University of Technology, Dalian, People's Republic of China

Ryosuke Yamada Organization of Advanced Science and Technology, Graduate School of Engineering, Kobe University, Nada, Kobe, Japan

Hideshi Yanase Department of Chemistry and Biotechnology, Graduate School of Engineering, Tottori University, Tottori, Japan

PART I

ENABLING PROCESSING TECHNOLOGIES

CHAPTER 1

Biorefineries—Concepts for Sustainability

MICHAEL SAUER, MATTHIAS STEIGER, and DIETHARD MATTANOVICH

Department of Biotechnology, University of Natural Resources and Life Sciences, Vienna, Austria
Austrian Centre of Industrial Biotechnology, Vienna, Austria

HANS MARX

Department of Biotechnology, University of Natural Resources and Life Sciences, Vienna, Austria

Bioprocessing of Renewable Resources to Commodity Bioproducts, First Edition.
Edited by Virendra S. Bisaria and Akihiko Kondo.

Abstract

Our society is highly dependent on fossil non-renewable resources. Therefore the main driving force in establishing a new industrial system is sustainability. Biorefineries, which are based on renewable biomass, can contribute to such a system. However, many current endeavors focus on single technologies and feedstocks such as starch or vegetable oils that could compete with food or feed. Nevertheless, in future it will be necessary to consider carefully for which purpose land is used to balance the needs of mankind for food and energy. We need to create flexible, zero-waste biorefineries that can accept a variety of low-value local feedstocks. The challenges are the development of efficient processes for the collection, handling, and pretreatment of biomass and for the selective conversion of biomass feedstocks into value-added products.

1.1 INTRODUCTION

Sustainability is the capacity to endure through renewal, maintenance, or sustenance. This is in contrast to durability, which is the capacity to endure through unchanging resistance to change. For humans in eco (and social) systems, sustainability is based on long-term maintenance of responsibility. In other words, as the Brundtland Commission of the United Nations (1987) has coined it: "sustainable development is development that meets the needs of the present without compromising the ability of future generations to meet their own needs."

For true sustainability this includes not only environmental, but also economic and social dimensions of resource use. All along the history access and availability of resources was a main driving force for the development of societies. But never in history was the demand for resources as high as it is now. Unfortunately, the development in the last 100 years was hardly sustainable because our energy demand is currently met mainly by fossil resources. While obviously based on natural carbon, the recycle time for the replenishment of fossil resources (estimated to be 280 million years; Liu et al., 2012a) is so high that they can be regarded as nonrenewable in context with human life. Use of this nonrenewable carbon sources is connected with two major problems: the first problem is the obvious limitation of these resources. Even avoiding any discussion about the time scale, the limitation of fossil resources is a fact due to the immense recycle time. This implies rising costs for energy and goods followed by increasing conflicts for access and distribution. The second problem inherently connected with prolonged use of fossil resources is the liberation of carbon dioxide—a greenhouse gas, which has been sequestered to the ground in ancient times. Its liberation is connected to various side effects, which shall not be discussed here in detail. However, we consider it commonly accepted that major pollution of the earth's atmosphere with greenhouse gasses is something unwanted, which we should strive to avoid for the sake of sustainability.

In this sense, CO_2 and compounds, which are obtainable from atmospheric CO_2, are the only renewable and thereby truly sustainable carbon sources. At present, this boils down to photosynthetic organisms as plants, algae, or cyanobacteria as basis for the production of carbon-containing goods, be it chemicals or fuels. In fact, the development of humanity started with the exclusive use of biomass as source for food, energy, and all goods. However, fossil resources have overtaken the role as a dominant energy and chemical source since the industrial revolution (Liu et al., 2012a).

Currently, biomass-derived energy sources supply about 50 EJ (exajoules) of the world's energy. The global energy demand was 463 EJ in 2005 and is supposed to increase to 691 EJ in 2030 (Lal, 2010). About 10% of the global primary energy consumption per year is based on biomass; this corresponds to 75% of the energy derived from alternative renewable energy sources (Haberl, 2010). Only 2% of the biomass-derived energy sources are utilized in the transportation sector. The rest is consumed for household uses predominantly as firewood (Srirangan et al., 2012). Model calculations suggest that a significant fraction of the energy demand could be met by the use of biomass. The World Energy Council and World Energy Assessment project estimates that bioenergy could supply a maximum of 250–450 EJ/year (probably a quarter of the global energy demand) by the year 2050 (Ragauskas et al., 2006a,b).

Biomass production by photosynthesis obtains its energy from the sun and its carbon from atmospheric CO_2. It is therefore a truly renewable source that can serve as food and feed, chemical and material, fuel and energy resource. A broad and valuable product mix can be created starting from biomass. The valorization possibilities in a biorefinery are at least as big as in fossil refineries. However, while the diversity of options allows a large range of configurations, this also implies different environmental and societal consequences or footprints. Decisions taken, regarding the valorization of biomass, should always take at least two critical points into account: the market impact (demands for products, possible displacements of products) and the ecology of the entire production chain (De Meester et al., 2011). Ideally, these decisions are well-thought-out and based on sound assessments to achieve an optimal sustainable development in the post fossil era.

1.2 THREE LEVELS FOR BIOMASS USE

Biomass can be used on three different levels:

- for food and feed,
- for (bio-based) products, and
- for (bio-) energy.

Food and feed inevitably rely on biomass. Materials and products are right now produced from fossil resources and biomass. If fossil resources are to be avoided or if they are depleted, biomass remains the only other basis for material production. For heat and power generation a variety of choices exist or are under development, such as solar power, wind energy, or geothermal energy. However, liquid fuels for transportation rely on fixed carbon, which again points to fossil resources or biomass.

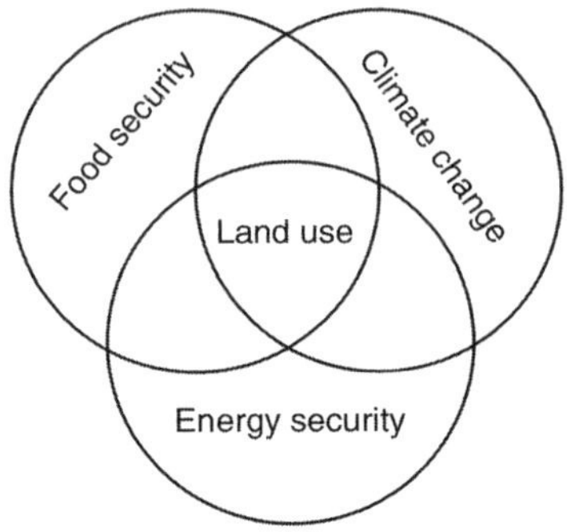

FIGURE 1.1 Food security, energy security, and climate change are centered around the limited availability of arable land. This constitutes a trilemma, which has to be addressed by our societies.

It is obvious that limited land availability makes it unlikely that biomass will be able to cover all of these demands fully (Ponton, 2009, De Meester et al., 2011). Figure 1.1 outlines the trilemma. Maximal valorization of biomass is therefore a key issue in the future.

This means that no biomass fraction should be considered as waste. From an economic perspective this is a major opportunity for biorefineries. From a societal perspective this appears as the major challenge of the future. However, focusing on the question of food *or* fuel appears not helpful (Karp and Richter, 2011). The pertinent challenge is how the increasing demands for food and energy can be met in the future, particularly when water and land availability will be limited and considering that food production requires significant fuel inputs, which are constantly increasing with intensification of agriculture. Food security has once again risen to the top of government agendas. Nevertheless, energy security is arguably an equally important challenge impacting food security and climate change.

Exemplary, in the United Kingdom, agriculture accounts for only 2% of energy use. However, almost 20% of United Kingdom's total energy consumption is used throughout the whole food production chain (Barling et al., 2008). Consequently, rising fuel prices or fuel shortages have a significant impact on the cost of food production. Decisions over land use should therefore be considered within the context of the bigger framework of all the challenges that lie ahead (Karp and Richter, 2011).

Carbon efficiency and energy efficiency are key parameters that should be taken into account for such decisions. As mentioned before—carbon can be obtained from fossil resources and biomass, while a variety of energy sources are conceivable including wind, photovoltaic, photothermic, geothermic, or hydro power. The energy mixture of the future will be diverse and will also contain a substantial amount of energy from biomass, but energy should be seen in this context more as a byproduct of the biorefinery than the main driving force. (This is different for the time being.)

One reason for this consideration is the inherent energy content of biomass: essentially the energy stored in biomass is a chemically captured form of solar energy. Energy from biomass can therefore be directly compared with energy obtained from photovoltaic systems. Blankenship et al. (2011) recently reviewed the energy efficiency of both technologies. They showed that a photovoltaic system coupled with hydrogen generation might capture up to 10–11% of the total solar energy per used area. In contrast, the solar energy conversion efficiencies of conventional crop plants usually do not exceed 1% (Blankenship et al., 2011). Only for microalgae grown in bioreactors, yields up to 3% are reported. These low efficiencies require a cautious consideration of biomass as a mere energy product, and strengthen the importance of material products obtained from biorefineries as a primary goal.

1.3 THE SUSTAINABLE REMOVAL OF BIOMASS FROM THE FIELD IS CRUCIAL FOR A SUCCESSFUL BIOREFINERY

The key factor for sustainability of currently cultivated biomass is their resource footprint. This concerns not only the direct land use and transformation and amongst others the related use of fertilizers, pesticides, fuels, and water for farming, but also the mineral balance and quality of the soil (Cherubini, 2010a).

This means that the production chain for biomass-derived goods is more demanding than its fossil equivalent. In fact, a variety of studies suggest that the agricultural phase is often the main contributor to the environmental impact of the production chain of bio-based products (Zah et al., 2007). Sustainable production of biomass is therefore of utmost importance. However, since the demand for bio-feedstock is increasing, while the arable land remains limited, much emphasis is given to higher agricultural yields. However, innovations for yield increase are often not focused on the simultaneous acceleration of environmental protection (Cassmann and Liska, 2007). Often, a blind striving for higher yields tends to cause severe damage to the natural environment. One example is the use of field crop residues. These residues are often seen as a renewable resource, which is freely available. In reality such residues are often required to enhance soil quality and to prevent erosion and nutrient depletion. Using this part of biomass might thus actually turn out to be a very bad choice in the long term (Reijnders, 2006).

Summing up, it is the production and supply of biomass rather than the demand for fuel or materials which limits the use of biomass as a renewable resource. In this context it is important to note that chemical production requires far lower amounts of carbon than fuel production. For example, in the United States, the chemical products segment consumed just over 3% of the total US petroleum consumption in 2007 (FitzPatrick et al., 2010). This opens an economic opportunity for the development of bio-sourced chemical products since the value of the chemical industry is comparable to the fuel industry, but requires only a fraction of the biomass (FitzPatrick et al., 2010).

So while the current industrial systems are split into three sectors namely food, bioenergy, and the chemical industry, these three sectors should come together and strive to valorize the used feedstock to the fullest to obtain the lowest resource footprint per (combined) output product(s).

The biorefinery approach is the promising concept, ideally combining the production of food, materials, and energy from biomass. Following the International Energy

Association's (IEA) definition, a biorefinery is the sustainable processing of biomass into a spectrum of marketable products (food, feed, materials, chemicals) and energy (fuels, power, heat) (De Jong et al., 2009).

The core technologies of a biorefinery are biochemical, microbial, and thermochemical processing. Energy and waste streams are internally recycled. Biochemical processes have the advantage of high selectivity at low processing temperatures. However, they normally require elaborate preprocessing stages and long processing times (FitzPatrick et al., 2010; Macrelli et al., 2012). Complementary, thermochemical routes include gasification, pyrolysis, and direct combustion to produce oils and gases. These are fast but nonspecific and generally require a high energy input. Biochemical/microbial and thermochemical processing complement each other. In an integrated system they can deliver significant advantages in terms of specificity of products, flexibility, and efficiency.

The biorefinery concept that has emerged is analogous to today's petroleum refineries. However, many current endeavors focus on single technologies and feedstock such as starch or vegetable oils that could compete with food or feed. We need to create flexible, zero-waste biorefineries that can accept a variety of low-value local feedstock. Biorefineries will then be able to compete with existing industries (Clark et al., 2012). Further down the value chain the development of green chemistry fills the gap between the sustainable resource and the product (Poliakoff and License, 2007).

1.4 MAKING ORDER: CLASSIFICATION OF BIOREFINERIES

A variety of classifications of biorefineries have been proposed: some of them consider feedstock, products, or processes. A very simple overview and classification of biorefineries subdivides them into three types: Phase I, Phase II, and Phase III (Clark et al., 2012).

Phase I biorefineries are integrated facilities limited to a single feedstock (e.g., corn or oils) which is converted into a single major product (e.g., ethanol or biodiesel).

Phase II biorefineries produce various end products from a single feedstock. They might be more flexible depending on product demand, prices, or others. An example is a biorefinery generating multiple products, ranging from sugar to ethanol, polymer precursors to animal feed, by utilizing sugar beet as the single feedstock. Exemplary, a biorefinery in Pomacle, France produces both ethanol and succinic acid in addition to beet sugar and glucose from a single facility with many processing streams (Hatti-Kaul et al., 2007; Le Henaff and Huc, 2008).

Finally, **Phase III** biorefineries are the most advanced, as they use a variety of biomass feedstock to yield a mix of products (Figure 1.2). Such biorefineries employ a combination of technologies, among them are chemical and/or biological transformations, extractions, and separations. Examples for Phase III biorefineries include whole-crop biorefineries encompassing an array of transformations of feedstock (e.g., corn, or rapeseed). The most promising type of Phase III biorefineries are

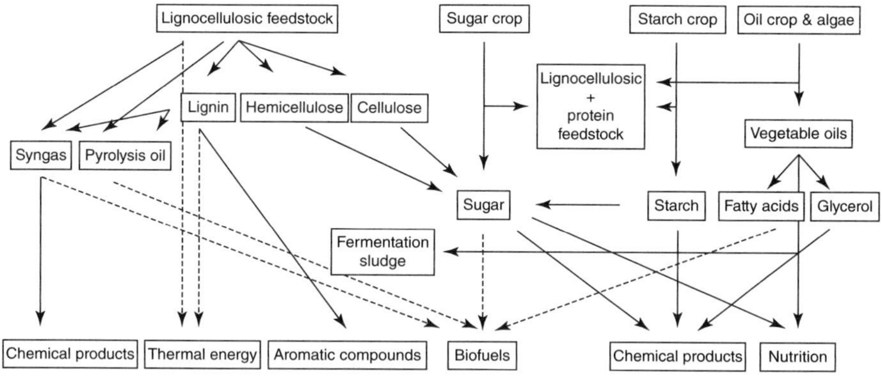

FIGURE 1.2 Schematic overview about the processing strategy of different feedstock used in biorefineries and the various products obtained. Processes leading to an energy product are shown as dashed lines. Fermentation sludge is the microbial biomass produced during the bioconversion processes.

based on lignocellulosic feedstock (e.g., wood, corn stover/cobs) to produce chemicals, fuels, energy, and other valuable outputs. Lignocellulosic biorefineries can be subclassified into various intermediate concepts based on the employed processes. This includes thermochemical biorefineries such as syngas platforms, or biochemical or microbial biorefineries such as sugar platforms. Sustainability is in the long run only obtainable with Phase III biorefineries. They are expected to expand the range and volume of bioproducts on the market as well as to improve the economics of biorefinery plants. At the same time the expectation is that they optimize the energy and environmental performance and enhance the cost competitiveness of bio-derived products. However, the development of such advanced integrated biorefineries is still ongoing.

More precise and detailed classifications of biorefineries rely on four main features: platforms, products, feedstock, and processes (Clark et al., 2012).

Platforms are defined as key intermediates between raw material and final products. They are considered as particularly relevant as these can be used to link different biorefinery concepts. Typical platforms are sugars (C_6 and/or C_5), lignin, syngas, or pyrolysis oils.

In terms of **products**, biorefineries can be broadly grouped into energy-driven and product-driven biorefineries. The main goal of energy-driven biorefineries is the production of one or more energy carriers (fuels, power, and/or heat) from biomass. The economic profitability of the plant is subsequently maximized by an upgrade and valorization of process residues. On the other hand, product-driven biorefineries are dedicated to the generation of one or more bio-based products. The economic profitability is maximized by production of bioenergy from process residues.

Biorefineries can be further classified based on their **feedstock**. For example, the feedstock can be subdivided into the main classes according to their origin, such as agriculture, forestry, industries, households, and aquaculture (Cherubini, 2010b).

Clearly, the feedstock for biorefineries will change further with the ongoing technical developments. **First-generation biorefineries** are based on edible feedstock from the agricultural sector (Srirangan et al., 2012), such as sugar and starch. **Second-generation biorefineries** are based on non-edible feedstock, comprising raw material derived from lignocellulosic biomass and crop waste residues from various agricultural and forestry processes (Nigam and Singh, 2011; Srirangan et al., 2012). **Third-generation biorefineries** make direct use of photosynthetic bacteria and algae, which can be cultivated in bioreactors and are thus independent of crop land (Nigam and Singh, 2011).

Processes employed in biorefinery concepts to convert biomass feedstock into marketable products include biochemical (e.g., anaerobic digestion, microbial fermentation, enzymatic conversion), chemical (e.g., hydrolysis, transesterification, hydrogenation, oxidation), and thermochemical (e.g., pyrolysis, gasification) processes.

1.5 QUANTITIES OF SUSTAINABLY AVAILABLE BIOMASS

The actual availability of the resources is the basic question to answer, when taking decisions about which resources to use. However, the sustainable and usable amount of biomass, which is present or possibly present in the future, is difficult to assess. Here we would like to give an overview about some numbers, particularly the orders of magnitude.

Nature produces over 200 billion tons of biomass by photosynthesis each year (Tschan et al., 2012). Other models estimate the global net primary production (NPP) to yield between 77.5 billion tons (Dhillon and von Wuehlisch, 2013) to over 100 billion tons (Field et al., 1998) of fixed carbon per year. The contribution of the marine environment to the global NPP is estimated to be between 31.8% up to about 50%. The standing biomass, however, is very distinct; more than 99% of the carbon fixed in biomass is estimated to be on land and less than 1% is found in marine environments (Dhillon and von Wuehlisch, 2013). Woody biomass is regarded the most abundant organic source on earth—around 90% of the carbon in standing biomass (Field et al., 1998; Liu, 2012b; Dhillon and von Wuehlisch, 2013).

Generally speaking, about 75% of the total biomass produced belongs to the class of carbohydrates. However, only 3.5% of these compounds are actually used by mankind (Tschan et al. 2012). Clearly, the theoretical availability of biomass does not mean that it is economically feasible or environmentally viable to collect it for industrial use. Parikka (2004) estimated the sustainable worldwide biomass energy potential to be about 100 EJ/year. Only 40% of this biomass is currently used according to Parikka (2004).

Current global bio-based chemical and polymer production (excluding biofuels) is estimated to be around 50 million tonnes (De Jong et al., 2012). The global petrochemical production of chemicals and polymers is estimated at around 330 million tonnes (De Jong et al., 2012).

Figure 1.3, taken from Vennestrøm et al. (2011) compares the total US oil consumption with harvested non-food biomass on a weight basis. On a weight basis

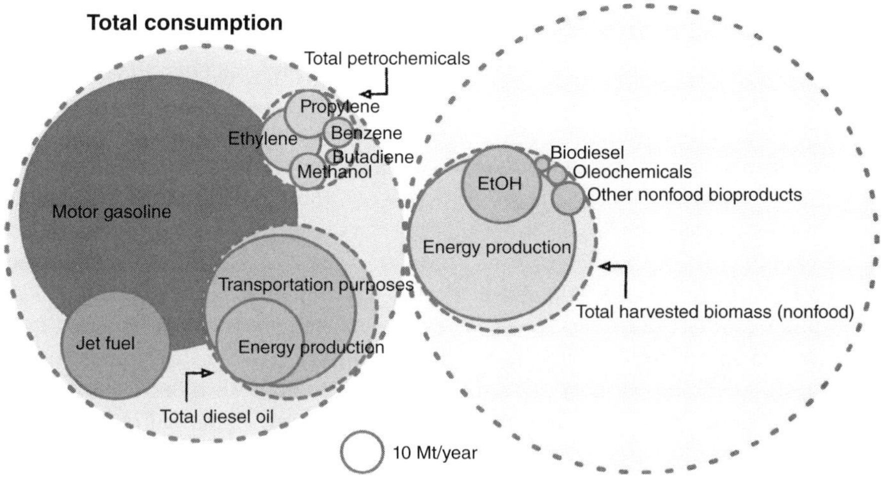

FIGURE 1.3 Total US oil consumption compared to potential and currently harvested nonfood biomass divided into its main uses. The area of each circle is proportional to the consumed amount. From Vennestrøm et al. (2011).

biomass has a lower energy and carbon density than crude oil. In fact, oil contains about twice the amount of carbon atoms and chemically stored energy as biomass. What becomes clear from this figure is that the orders of magnitude correspond to each other; however, with an increasing use of biomass in industry at some point biomass can become a scarce resource with increasing prizes. Feedstock which are very cheap at the time might become expensive when their industrial use is carried out on a scale that is comparable to that of current petrochemical processes. Large-scale use of biomass as feedstock will drastically alter the market. Thus, for long-term planning, the mature market must be considered instead of the current market, which is by no means a simple task. Anyhow, biomass has the potential to fully substitute petrol as carbon source for the chemical market, if the processes to produce and use them are sufficiently efficient.

1.6 QUANTIFICATION OF SUSTAINABILITY

In the development of sustainable industries, researchers are challenged to find innovative technical solutions without losing sight of the economic, societal, and environmental impacts of their work (Jenkins and Alles, 2011). Scientific and quantifiable methods are needed to guide research and industry into the right direction. Sustainability assessments aimed at quantifying the economic, environmental, and societal impacts can help to move debates and decision finding to a factual level. Science-based methods such as life-cycle analysis (LCA), defined as a holistic approach to quantify environmental impacts throughout the value chain of a product (International Organization for Standardization, 2006), can be applied as a decision support tool.

With LCA the feedstock selection for defined products and the decision of how to produce the feedstock and the product are set on a factual basis.

For the feedstock, there are logistical and sustainability concerns. Each potential biorefinery concept has specific coproduct and waste issues to consider. Transport is a general issue in this discussion. The biomass resource has to be transported to the refinery; subsequently the products have to be transported to the downstream industry and/or the consumer. Of interest is the approach of the company Nature-Works LLC that currently operates the largest biorefinery in the United States in Blair, Nebraska. The nameplate capacity of the polymer production plant is 140,000 tons of polymer per year. Corn is the basis for the production of the bioplastic polylactic acid (PLA) in a complex multistage process. Sixty percent of its corn feedstock is obtained from the local area (producers, located less than 40 kilometers from the plant). Several companies in an emergent network are now active on the Blair biorefinery campus (Wells and Zapata, 2012) reducing transportation from one industrial branch to the next one.

1.7 STARCH- AND SUGAR-BASED BIOREFINERY

Starch- and sugar-containing crops are quantitatively the most important products of today's agricultural system. Most of the existing biorefinery concepts are based on these plants and they are referred to as first-generation feedstock (*vide supra*), but they also constitute the backbone of human nutrition. General characteristics of this type of biorefinery are listed in Table 1.1.

The polysaccharide starch is found in most plants as a storage compound; however, only five plants, namely maize, rice, wheat, potatoes, and cassava account for the majority of worldwide produced starch-containing plants. Roughly 2.7 billion tons of these crops are annually harvested. The class of sugar-containing crops contains only two plants namely sugar cane and sugar beet of which about 2 billion tons are

TABLE 1.1 Characteristics of the Starch and Sugar Biorefinery

Advantages	Disadvantages
Starch and sugar crops are already cultivated worldwide today	Connection between the sugar/starch industry and classical chemical industry is still underdeveloped
Well-established industry and the handling of the different resources is developed	Direct competition with the food industry
Highly developed technology available for the primary raffination toward the platform chemicals saccharose and starch	Combined production of material and energetic products needs a better integration
A plethora of biochemical fermentations require fermentable sugars	Fertilizers required for high yield production

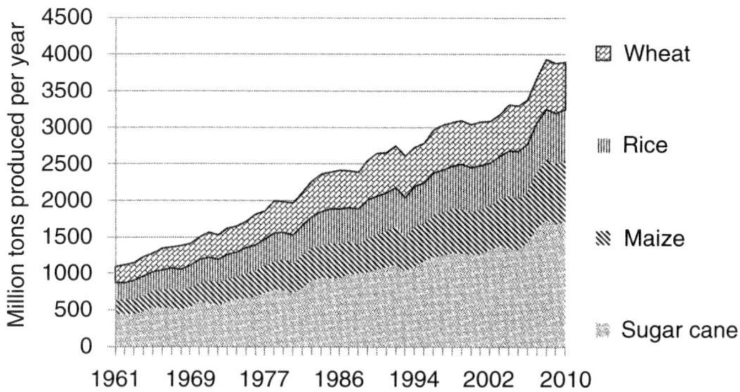

FIGURE 1.4 Worldwide production of the main sugar- and starch-containing crops. Data taken from the Food and Agriculture Organization of the United Nations (www.fao.org).

annually harvested. The shares of each plant in the overall worldwide production are shown in Figure 1.4. From this graph, it can be depicted that the overall annual production of these crops is rising. However, the future growth rate will depend on three main parameters: land, fertilizer, and plant productivity. The production of nitrogen fertilizers requires a high energy input. This directly connects energy price with crop price.

The starch and sugar processing industry is already highly developed and has the technology to readily deal with the conventional crops of today's agriculture. Its major limitation can be seen in the circumstance that only a part of the starch- or sugar-containing plant is implemented in the biorefinery approach. In the future, the whole crop including the stover has to be taken into account in order to develop a whole crop biorefinery concept (Kamm et al., 2006), keeping in mind that only a part can be used in a sustainable way.

The primary biorefinery products of the sugar and starch industry are glucose, fructose, gluconate, and bioethanol (Wagemann et al., 2012). At present, bioethanol is a fast-growing biorefinery energy product. Both in the United States and Brazil, bioethanol is produced in high quantities in order to substitute the dependence on fossil fuels in the transportation sector. In both cases, the biorefinery process can be structured in three steps: (1) obtainment of a solution of fermentable sugars; (2) bioconversion of sugars to ethanol; (3) ethanol separation and purification (Mussatto et al., 2010). Sugarcane is used as the main feedstock in Brazil whereas the majority of the bioethanol in the United States is produced from maize. More than 35% of harvested maize grain is used for bioethanol production in the United States (Perlack and Stokes, 2011) and by 2011, 52 million liters of ethanol were produced annually (Jerck et al., 2012). Two other important crops which may be used for biofuel production are cassava and sorghum. Cassava is grown as an annual crop in the tropical and subtropical countries and has the advantage that it is compatible with current corn ethanol technologies. Sorghum is a good alternative feedstock for dry regions, because of its lower water requirements compared to maize and sugarcane (Srirangan et al., 2012).

However, bioethanol is not the only biorefinery product which can be obtained from starch- and sugar-containing crops. Microorganisms can directly use sugars as a substrate and convert them to virtually any product. Today products like organic acids (e.g., citric acid, gluconic acid) and amino acids (e.g., glutamate, lysine) are already produced from sugars in high amounts and are not dependent on fossil resources as a substrate.

1.7.1 Sugar Crop Raffination

After harvesting, the main crop is further treated by a crushing and milling step that yields the sugar juice. As a by-product the insoluble lignocellulosic material of the plant is obtained. In case of sugar cane this residue is referred to as sugarcane bagasse. This feedstock is currently used (by burning) as an energy resource to drive the thermal requirements of the sugar plant. However, it is also considered a valuable feedstock for lignocellulosic biorefinery approaches (Dawson and Boopthay, 2008; Cherubini and Strømann 2011; Nigam and Singh, 2011; Macrelli et al., 2012). The sugar juice obtained can then either be directly used by the fermentation industry as a substrate or sugar is crystallized stepwise by water evaporation. The crystallized sugar can be used for various purposes including human nutrition and fermentation processes, if higher substrate purity is required.

1.7.2 Starch Crop Raffination

In a first milling step, the crop is broken up. In case of maize, a previous steeping step at 50°C enables a high yield and good starch quality. Starch is extracted with water and separated from the insoluble fibers, which are a potential feedstock for a lignocellulosic biorefinery. In successive steps the protein fraction is separated from the starch by either protein coagulation (heat or acid treatment) or centrifugation utilizing density differences. Starch can be readily hydrolyzed to fermentable glucose by means of amylases.

The processing of starch- and sugar-containing plants has a very long history and is directly connected to the main function of these crops as nutritional products. The highest potential for new biorefinery concepts based on sugar- and starch-containing crops can be expected in the various side products of this industry starting with plant residues, which are already separated from the crop on the field, and residues from milling and further processing steps. Those product streams often yield only a low value and are currently used as animal feed or for thermal processing (Nitayavardhana and Khanal, 2012). For example, current flour mills operate at 70–80% grain-to-flour yields. Various waste and by-product streams include bran, germ, and endosperm. These by-products contain a high proportion of starch (25–30%) that could be used for microbial bioconversion to produce valuable chemicals (Clark et al., 2012).

1.8 OILSEED CROPS

Plants like soybean, sunflower, rapeseed, peanut, oil palm, and coconut contain a high fraction of lipids and are referred to as oilseed crops. Vegetable oils have a long

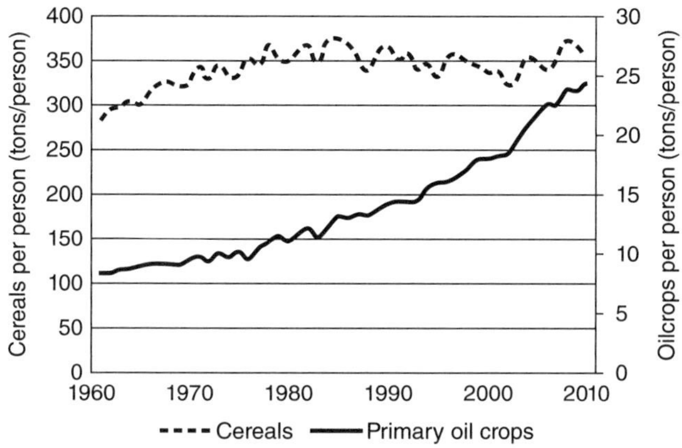

FIGURE 1.5 Development of the worldwide production of the major oil seed crops over the last 50 years normalized against the increase of the human population and compared to the development of cereals. Data taken from the Food and Agriculture Organization of the United Nations (www.fao.org).

tradition as edible oils and are of growing interest for the biofuel industry. Over the last 50 years the production of oilseed crops increased dramatically from around 100 million tons to over 800 million tons per year (Figure 1.5). Especially, the cultivation of oil palms and soybeans was significantly enlarged.

General characteristics of this type of biorefinery are listed in Table 1.2.

Besides the application as nutritional product, those plants have an important application for production of biofuels also. The triglycerides can be modified by a transesterification reaction with short-chain alcohols to produce alkyl esters, mainly methyl and ethyl esters. The product obtained is referred to as biodiesel. In that case, a fundamental concept of biorefineries was neglected, which requires a suitable application for all byproducts because during the transesterification reaction glycerol is obtained as a byproduct in high amounts. The production of 10 tons of biodiesel generates 1 ton of glycerol. Its price decreases with increased biodiesel production. Therefore, new biorefinery concepts need to be developed to convert glycerol to an

TABLE 1.2 Characteristics of the Oil Crop Biorefinery

Advantages	Disadvantages
Oilseed crops are already worldwide cultivated today	Economic processes for the utilization of glycerol and seed cakes are underdeveloped
Well-established industry and the handling of the different resources is developed	Fertilizers required for high-yield production
Fatty acids can be directly converted to a valuable-energy product (biodiesel)	

added-value product like 1,2- or 1,3-propanediol, acrolein, or lactic acid (Pflügl et al., 2012; Posada et al., 2012).

The situation for oil crop plants is comparable with the starch and sugar processing industry. The technology to obtain primary products like vegetable oils and biofuels is already established and is commercially applied. However, secondary refinery streams like plant residues, seed cakes of pressing and filtration steps, or process byproducts like the above-mentioned glycerol need to be considered in future. Processes need to be developed to add additional value to those products. Furthermore, waste streams like cooking oil can be fed back into the biodiesel production pipeline (Wang et al., 2007).

1.9 LIGNOCELLULOSIC FEEDSTOCK

The current use of lignocellulosic biomass is primarily the use of wood for combustion, construction, and furniture making; cellulose fibers are used for pulp and paper making or clothing (Möller et al., 2007). The global production of forest products has been estimated by the Food and Agriculture Organization of the United Nations (FAO) to be 3469 million m^3 of round wood in 2011, where roughly 50% are used as wood fuel and 50% for industrial use. Furthermore, 406 million m^3 are produced as sawn wood, 288 million m^3 are wood-based panels, wood pulp accounts for 173 million tons, paper and paperboard production has been 403 million tons, and 211 million tons are produced from recovered paper. However, not only wood can serve as feedstock for lignocellulose biorefineries but residues from agriculture can also be taken into account. The lignocellulose feedstock report from the EPOBIO project (Möller et al., 2007) states that the most abundant agricultural residue in Europe is wheat straw, rice straw in Asia, and corn stover in North America.

As the systematic cultivation of crops for biorefinery purposes is gaining more importance the question arises as to which crops are the most desirable ones. To select the best possible feedstock several criteria have to be considered. From an economic point of view the crop has to have a very high biomass yield, low requirement for fertilizers and pesticides, the ability to grow on marginal lands, and the cell wall structure should allow an easy access for bioconversion methods. In addition the crop should also cover environmental criteria such as low impact on biodiversity, water and soil quality, low greenhouse gas emission, and high carbon sequestration. Based on these criteria Möller et al. (2007) selected four candidates, namely poplar, willow, *Miscanthus*, and wheat straw for the region of the European Union.

Compared to the availability of fossil resources, areas for biomass cultivation are globally more evenly distributed, thereby enhancing the security of supply. Biorefineries utilizing lignocellulosic feedstock may even help to a certain extent to combat the unemployment status of rural areas (Menon and Rao, 2012).

1.9.1 Biochemical Biorefinery (Fractionation Biorefinery)

The term "lignocellulosic biomass" describes the material that constitutes the plant cell wall. This includes primarily cellulose (30–50%), hemicellulose (15–35%), and lignin (10–30%). As a result of the organization and interaction between these polymeric structures, the plant cell wall is naturally recalcitrant to biological degradation (Himmel et al., 2007).

The lignocellulose biorefinery is one of the most desirable forms of a biorefinery. As mentioned earlier, a Phase III type biorefinery uses virtually any lignocellulosic feedstock like wood, corn stover/cobs, straw, bagasse, and other lignocellulose-rich waste streams from agriculture, forestry, and municipal areas. The accessibility of the desired fermentable sugars is severely hindered by the assembly of the lignocellulosic biomass itself; therefore, a pretreatment by milling and grinding followed by a treatment with high temperature and pressure is required to access the fibers composed of fermentable mono- and oligosaccharides. The addition of mild or harsh acids, bases, or organic substances can further enhance the pulping. Depending on the particle size and the composition of the biomass that is delivered to the biorefinery, a suitable flow chart of pretreatment steps has to be established. Pretreatment techniques can be categorized into physical (milling, irradiation, and extrusion), physicochemical (steam explosion, ammonia fiber explosion, ammonia recycle percolation, microwave chemical, and liquid hot-water pretreatment), chemical (acid, alkaline, green solvents), and biological processes. The pretreatment gives rise to a solid or fluid stream of the three biomass main components: cellulose, hemicellulose, and lignin. These streams are further subjected to enzymatic hydrolysis to gain sugars for microbial fermentation. Menon and Rao (2012) conclude that the choice of pretreatment should consider the overall compatibility of feedstock, enzymes, and organisms to be applied. A more detailed description of pretreatment processes and lignocellulose hydrolysis is given in the following chapters of this book. The enzymatic conversion of cellulose and hemicellulose to fermentable sugars opens the possibility for the microbial conversion to chemical building blocks for the synthesis of bio-based materials or the conversion to biofuels for transport or energy purposes. Saccharification of cellulose and hemicellulose to platform compounds D-glucose (from cellulose), D-xylose, L-arabinose, D-mannose, D-glucose, D-galactose, and D-glucuronic acid (from hemicellulose) involve a series of hydrolytic enzymes or enzyme complexes to which Chapters 4 and 5 in this book are dedicated.

Platform compounds are further converted to products by microbial fermentation processes. Saccharification and fermentation can be accomplished in a sequential process by separate hydrolysis and fermentation (SHF), or in a consolidated one-pot process known as simultaneous saccharification and fermentation (SSF) of single sugars or simultaneous saccharification and co-fermentation (SSCF) of all monosaccharides. Future developments might even combine the production of saccharolytic enzymes, the hydrolysis of cellulose, and hemicellulose to monomeric sugars and the fermentation of hexose and pentose sugars in a single process, the so-called consolidated biomass processing (CBP) (Menon and Rao 2012).

The most prominent product in the biorefineries nowadays is bioethanol with a global production volume over 20 billion gallons (or 75 billion liters) in 2011 (Alternative Fuels Data Center, 2013). Huge efforts in research and development are on the way to broaden the product portfolio of the biorefineries in the future. Screening for new production strains in nature, or microorganisms accessible for metabolic engineering for directed biosynthesis of bioproducts, or a combination of

TABLE 1.3 Characteristics of the Biochemical Biorefinery

Advantages	Disadvantages
The availability of diverse lignocellulosic raw material is given	Sustainable usage of lignin for material production is still to be established
The use of these raw material does not interfere with food and feed production	Simultaneous fermentation of C6 and C5 sugars from the hemicellulose fraction is still to be optimized
Expertise of pretreatment technologies of milling and grinding as well as (thermo)-chemical pretreatment technologies is established in the pulp and paper industry sectors	Embedding product streams of the biorefinery into the chemical industry is still to be established
Fermentation technologies for the production of alcohols, organic acids, amino acids, and other chemical building blocks from sugars are well established	Supply of raw material at a "moderate" price has not been secured
Additional income for rural agricultural regions	

Source: Adapted from Wagemann et al. (2012).

both approaches will achieve this aim. Part II of this book is dedicated to specific biocommodity products from biorefineries and future aspects for the production of those.

The remaining lignin that is separated from the sugar stream during the pretreatment process can be used for the production of heat and power, which is done very frequently at the moment; nevertheless, it could be used in the future as a source of various aromatic compounds. The effective utilization of all the three components would play a significant role in the economic viability of an integrated biorefinery. General characteristics of this type of biorefinery are listed in Table 1.3.

To evaluate the sustainability concerning economic, environmental, and social aspects the Department of Energy (DOE) is funding integrated biorefinery projects in pilot, demonstration, and commercial scale. The importance of biochemical conversion is underlined by the fact that four out of five commercial scale plants funded by the DOE are based on biochemical conversion (Department of Energy, 2012).

1.9.2 Syngas Biorefinery (Gasification Biorefinery)

Thermochemical conversion of biomass can be accomplished in three ways that differ in the amount of oxygen that is supplied to the process. By the supply of excess air the biomass is combusted to generate heat and power. When oxygen is depleted from the process, the biomass undergoes pyrolysis or liquefaction. In the liquefaction biomass is decomposed into small molecules which then polymerize to oily compounds. Liquids gained from pyrolysis or hydrothermal liquefaction can be further refined to gasoline, diesel fuel, jet fuel, or chemicals. Syngas is produced by gasification by means of partial combustion of biomass, where only a limited amount of oxygen is supplied to the process (Demirbas, 2010). Depending on the type of gasifier and biomass used for the gasification (at temperatures of 750–800°C) a mixture of carbon monoxide (CO), hydrogen (H_2), methane (CH_4), nitrogen (N_2), carbon dioxide (CO_2), and some higher hydrocarbons in varying amounts are generated. The main constituents of syngas are H_2 (5–24%) and CO (14–67%) (Munasinghe and Khanal, 2010).

The H_2 and CO from the syngas can be converted via the Fischer–Tropsch process to long-chain hydrocarbons catalyzed by cobalt or iron. Before the Fischer–Tropsch synthesis (FTS) the syngas has to be cleaned and conditioned. The FTS can be operated at low temperatures to produce heavy, waxy hydrocarbons or at higher temperatures to produce olefins. By further product upgrading the array of products from FTS ranges from diesel, gasoline, methane, ethane, and to light and heavy waxes (Demirbas, 2010).

The microbial conversion of syngas offers some interesting opportunities for the future production of biofuels and biochemicals as well. According to Munasinghe and Khanal (2010), the merits over biochemical biorefinery approaches are the eliminations of costly pretreatment steps and enzymes, as well as the usage of all fractions from biomass including the lignin part. Compared to the Fischer–Tropsch process microbial catalysts have a much higher specificity and the ratio of H_2:CO is of minor importance.

Possible products from syngas fermentation can be ethanol, butanol, lactate, acetate, pyruvate, and butyrate. One of the most important factors in syngas fermentation is the microbial catalyst itself (anaerobic bacteria from the genera of *Clostridium*, *Acetobacterium*, *Butyribacterium*). The efficient conversion of syngas by the microbe can be negatively influenced by impurities like ethylene, ethane, sulfur, and nitrogen-containing gases as well as solid particles of tar, ash, and char. So these impurities have to be avoided by the appropriate choice of gasifier or the syngas has to be cleaned from these impurities before fermentation. The fermentation process is influenced by parameters such as pH (depending on the microorganism, optima are between 5.5 and 7.5), temperature (mesophilic organisms 37–40°C or thermophilic organisms 55–80°C), gas-to-liquid mass transfer in combination with reactor type (stirred tank, bubble column, membrane-based systems), and growth media (depending on the microorganism used). Despite the mass transfer limitations and the quality of the syngas Munasinghe and Khanal (2010) recommend, for future development of syngas fermentation, the genetic modification of existing syngas-fermenting microbes to high yield strains especially for solvent production, where the pathways to acid production have to be blocked. General characteristics of this type of biorefinery are listed in Table 1.4.

1.10 GREEN BIOREFINERY

In contrast to lignocellulose-feedstock biorefineries, where the composite of lignin, cellulose, and hemicellulose are very strong, the green biorefinery uses green biomass such as grasses, green crops like lucerne, clover, and green cereals. By wet fractionation, a fiber-rich press cake and a nutrient-rich green juice is obtained. The dried

TABLE 1.4 Characteristics of the Syngas Biorefinery

Advantages	Disadvantages
The gasification of charcoal is well established and works in large scales, the expertise of this process can be translated to biomass gasification	Large scale plants have a high raw material demand
Raw material for biomass gasification are available by using waste materials from agriculture and forestry, it does not interfere with food and feed production	Construction of industrial scale facilities is cost intensive
The chemical conversion and refining of syngas to Fischer–Tropsch diesel fuel, wax, methane, ethanol, and other specialty chemicals is established	The potential variety of products is not exploited yet
Complete (holistic) utilization of all biomass components	Biotechnological conversion of syngas to fermentation products is still in the research and development phase
	Supply of raw material at a moderate price has to be secured

Source: Adapted from Wagemann et al. (2012).

press cake can serve as fodder, as a raw material for hydrocarbons and chemicals or it can serve as a raw material for syngas production. By separation enzymes, dyes, flavorings, carbohydrates, and proteins can be recovered from the press juice. The press juice can also serve as a feedstock for fermentation where the fermentation broth is a source of lactic acid, amino acids, ethanol, and proteins (Kamm and Kamm, 2004).

1.11 MICROALGAE

Microalgae constitute a further source of industrially usable carbon fixed by photosynthesis. They offer a great potential for exploitation, such as biodiesel production, due to their oil content that can exceed 80% w/w (Amaro et al., 2011). Some possible advantages connected to microalgae as feedstock include that their cultivation is not linked to arable land (it is not linked to land at all, as off-shore cultivation is conceivable); they can grow in brackish or salty water and their efficiency in terms of energy use per hectare is potentially high. Following the classical concept of biorefineries, they are interesting because apart from their potential as oil for biodiesel producer, a variety of by-products are accumulated. These by-products include valuable omega-3-fatty acids, recombinant proteins, and algal meal containing high amounts of proteins (Subhadra and Grinson-George, 2011). The direct by-product of biodiesel production from oil is glycerol that can be used to grow more algae or which can be converted to higher-valued chemicals such as 1,3-propanediol.

The high oil productivity of microalgae cultures and the possible absence of competition for arable land and water resources justify the currently high investments into such projects. Nevertheless, care has to be taken with the evaluation of possible productivities. Various studies seem to claim unrealistically high numbers for the microalgal oil production potential. Decades of worldwide research have demonstrated that annual productivities beyond 100 tons of algal biomass per hectare appear not attainable at large scale, at least not with current strains and current technologies (Rodolfi et al., 2009). Thus, even under the best conditions a realistic oil yield will not exceed 40 tons per hectare per year. This compares to about 1000 liters of oil per hectare per year which can be typically produced by sunflower or rapeseed, and 6000 liters per hectare per year obtained with oil palms (Chisti, 2007).

Nevertheless, full commercialization of biodiesel from algae oil has not been realized yet. Till now the cost of algal biomass production of about US$5/kg is simply not compatible with the low costs required for biofuel production. One major problem connected to microalgae as oil resource is the large scale cultivation that has to be guaranteed the availability of light and CO_2 at high cell density. The energy demand for mixing and pumping is very high. Light does not penetrate more than a few centimeters into a dense culture of algal cells, so scale-up depends on an increase of surface area and not volume as is the case for heterotrophic fermentations (Scott et al., 2010). The costs of scale-up are much debated—estimates of production and capital costs vary widely. This points to the current painful lack of data from real-life demonstrations. There is a pressing need to conduct pilot studies at realistic scale to assess productivities likely to be achieved in practice.

Further drawbacks hindering the large-scale use of microalgae as resource is the enormous water content of the harvested biomass and the intracellular localization of the desired oil. Drying and oil extraction are very costly, particularly when environmentally benign technologies are applied and sustainability is an aim of the endeavor (Singh and Olsen, 2011). Lardon et al. (2009) exemplarily calculated that 1 MJ of energy in biodiesel from *Chlorella vulgaris* required an energy input of 1.66 MJ for production. Use of the algal biomass for energy generation could turn the balance to the positive side; however, this shows the significance of technological advancement before industrial exploitation of microalgae. An important step regarding the downstream processing is that it should allow the product generation without drying the biomass. For example, Levine et al. (2010) have developed a biodiesel production process starting from wet algal biomass with 80% moisture.

It is out of question that algal biomass can be utilized for the production of various bioproducts. However, significant improvements in the efficiency, cost structure and ability to scale-up algal growth, and lipid extraction are required to establish a commercially viable microalgae-based biorefinery.

1.12 FUTURE PROSPECTS—AIMING FOR HIGHER VALUE FROM BIOMASS

The concepts for biomass valorization are manifold. Some existing examples for biochemical products from bio-derived resources are summarized in Table 1.5

TABLE 1.5 Overview of Chemicals That Are Currently Produced, or Could Be Produced, From Biomass Together With Their Respective Market Type, Size of the Market, and Potential Biomass Feedstock

Chemical	Market type	Market size (Mt/y)[a]	Major player(s)	Feedstock
Acetic acid	Existing	9.0	–	Ethanol
Acrylic acid	Existing	4.2	Arkema, Cargill/Novozymes	Glycerol or glucose
C4 diacids	Emerging	(0.1–0.5)	BASF/Purac/CSM, Myriant	Glucose
Epichlorohydrin	Existing	1.0	Solvay, DOW	Glycerol
Ethanol	Existing	60	Cosan, Abengoa Bioenergy, ADM	Glucose
Ethylene	Existing	110	Braskem, DOW/Crystalsev, Borealis	Ethanol
Ethylene glycol	Existing	20	India Glycols, Dacheng Industrial	Glucose or xylitol
Glycerol	Existing	1.5	ADM, P&G, Cargill	Vegetable oil
5-Hydroxy-methylfurfural	Emerging	–	–	Glucose/fructose
3-Hydroxypropionic acid	Emerging	(≥0.5)	Novozymes/Cargill	Glucose
Isoprene	Existing/emerging	0.1 (0.1–0.5)	Danisco/Goodyear	Glucose
Lactic acid	Existing/emerging	0.3 (0.3–0.5)	Cargill, Purac/Arkema, ADM, Galactic	Glucose
Levulinic acid	Emerging	(≥0.5)	Segetis, Maine Bioproducts, Le Calorie	Glucose
Oleochemicals	Existing	10–15	Emery, Croda, BASF, Vantage Oleochemicals	Vegetable oil/fat
1,3-Propanediol	Emerging	(0.1–0.5)	Dupont/Tate & Lyle	Glucose
Propylene	Existing	80	Braskem/Novozymes	Glucose
Propylene glycol	Existing/emerging	1.4 (≥2.0)	ADM, Cargill/Ashland, Senergy, Dacheng Industrial	Glycerol or sorbitol
Polyhydroxyalkanoate	Emerging	(0.1–0.5)	Metabolix/ADM	Glucose

Source: Vennestrøm et al. (2011).

[a]Market size of an existing market is given as its current size including production from fossil resources; for emerging markets the expected market size is reported in parenthesis.

(Vennestrøm et al., 2011). In summary, an upgrading of biomass to higher-value products is a reasonable approach to replace crude oil. For electrical (on-grid) energy production alternative sources are simply conceivable. It is liquid transportation fuels where most problems occur for the judgment if they can or should be replaced by biomass-derived products or not.

The problems inherently connected with the production of liquid transportation fuels from biomass are, the amount of available biomass and the relatively low value of fuels. Fuel production from biomass should be limited to applications for which substitution is not a feasible alternative. This is, for example, the case for aviation and maybe marine traffic. Many other forms of traffic can be more and more shifted to electric or other energy forms, for example, to batteries or fuel cells. Clearly, this requires an overall modification of the current transportation infrastructure and a general reconsideration of transportation, which will take time.

The currently available biomass appears to be sufficient to replace fossil resources for the production of chemicals (Vennestrøm et al., 2011). The challenges in this context are the development of efficient processes for the collection, handling, and pretreatment of biomass and for the selective conversion of biomass feedstock into the value-added products.

The extensive current research into second-generation biofuels will significantly benefit the future renewable chemical industry. While products such as ethanol as fuel do not appear as perfectly sustainable solutions in the long run, the technologies currently developed to produce them are valuable for biomass use aiming at other products. Furthermore, many of the compounds at present produced by the biofuels industry might serve as interesting platform chemicals for a green chemical industry in the future. For example, ethanol is a possible starting point for acetic acid, ethylene, or ethylene glycol production (Christensen et al., 2008; Vennestrøm, 2011). An already existing example for this is the Brazilian company Braskem, producing biopolyethylene from sugarcane-derived ethanol. The polyethylene produced at Braskem is widely used for automobiles, cosmetics, packaging, and toys. In 2010, the company claimed to be the world leader as it opened a US$320 million sugarcane ethanol processing plant, which has the capacity to produce 200,000 tonnes of bio polyethylene per year (Wells and Zapata, 2012).

The production of transportation fuels is thus a good way of establishing processes and infrastructure needed for large-scale industrial utilization of biomass aiming at higher value. However, the assumption that biomass is available in excess, forming the basis for the current production of transportation fuels, will likely not hold true in the future. Careful evaluation is therefore needed when allocating these resources.

A further thought in this context is that current platform chemicals have been developed because they were convenient to produce from fossil resources. While it appears tempting to simply replace such fossil-based molecules by producing them from biomass, the inherent functionality of bio-derived molecules should be utilized as much as possible in the long run for the sake of sustainability.

REFERENCES

Alternative Fuels Data Center. (2013). Global ethanol production. Available at: http://www.afdc.energy.gov/data/tab/all/data_set/10331 [Accessed January 15, 2013].

Amaro HM, Guedes AC, Malcata FX. (2011). Advances and perspectives in using microalgae to produce biodiesel. *Appl Energy*, 88, 3402–3410.

Barling D, Sharpe R, Lang T. (2008). *Rethinking Britain's Food Security: A Research Report for the Soil Association*. London: Centre for Food Policy, City University.

Blankenship RE, Tiede DM, Barber J, Brudvig GW, Fleming G, Ghirardi M, Gunner MR, Junge W, Kramer DM, Melis A, Moore TA, Moser CC, Nocera DG, Nozik AJ, Ort DR, Parson WW, Prince RC, Sayre RT. (2011). Comparing photosynthetic and photovoltaic efficiencies and recognizing the potential for improvement. *Science*, 332, 805–809.

Cassman KG, Liska AJ. (2007). Food and fuel: realistic or foolish? *Biofuels Bioprod Bioref.* 1, 18–23.

Cherubini F. (2010a). GHG balances of bioenergy systems—overview of key steps in the production chain and methodological concerns. *Renew Energ*, 35, 1565–1573.

Cherubini F. (2010b). The biorefinery concept: using biomass instead of oil for producing energy and chemicals. *Energy Conversion Manage.*, 51, 1412–1421.

Cherubini F, Strømman AH. (2011). Chemicals from lignocellulosic biomass: opportunities, perspectives, and potential of biorefinery systems. *Biofuels Bioprod Bioref,* 5, 548–561.

Chisti Y. (2007). Biodiesel from microalgae, *Biotechnol Adv*, 25, 294–306.

Christensen CH, Rass-Hansen J, Marsden CC, Taarning E, Egeblad K. (2008). The renewable chemicals industry. *ChemSusChem*, 1, 283–289.

Clark JH, Luque R, Matharu A. (2012). Green chemistry, biofuels, and biorefinery, *Annu Rev Chem Biomol Eng*, 3, 183–207.

Dawson L, Boopathy R. (2008). Cellulosic ethanol production from sugarcane bagasse without enzymatic saccharification. *BioResources*, 3, 452–460.

Demirbas A. (2010). *Biorefineries for Biomass Upgrading Facilities*. ISBN 978-1-84882-720-2

De Jong E, Langeveld H, van Ree R. (2009). IEA Bioenergy Task 42 Biorefinery. Available at: http://www.biorefinery.nl/fileadmin/biorefinery/docs/Brochure_Totaal_definitief_HR_opt.pdf [Accessed January 24, 2014].

De Jong E, Higson A, Walsh P, Wellisch M. (2012). Bio-based Chemicals: Value Added Products from Biorefineries. Available at: http://www.ieabioenergy.com/wp-content/uploads/2013/10/Task-42-Biobased-Chemicals-value-added-products-from-biorefineries.pdf [24.01.2014].

De Meester S, Callewaert C, De Mol E, Van Langenhove H, Dewulf J. (2011). The resource footprint of biobased products: a key issue in the sustainable development of biorefineries. *Biofuels Bioprod Bioref*, 5, 570–580.

Department of Energy. (2012). IBR Portfolio Overview Fact Sheet. Available at: http://www1.eere.energy.gov/biomass/pdfs/ibr_portfolio_overview.pdf [Accessed January 15, 2013].

Dhillon RS, Wuehlisch G. (2013). Mitigation of global warming through renewable biomass. *Biomass Bioenergy*, 48, 75–89.

Field C, Behrenfeld M, Randerson J, Falkowski P. (1998). Primary production of the biosphere: integrating terrestrial and oceanic components. *Science*, 281, 237–240.

FitzPatrick M, Champagne P, Cunningham MF, Whitney RA. (2010). A biorefinery processing perspective: treatment of lignocellulosic materials for the production of value-added products. *Bioresour Technol*, 101, 8915–8922.

Haberl H, Beringer T, Bhattacharya SC, Erb K-H, Hoogwijk M. (2010). The global technical potential of bio-energy in 2050 considering sustainability constraints. Current Opinion in Environmental Sustainability, 2, 394–403.

Hatti-Kaul R, Tornvall U, Gustafsson L, Borjesson P. (2007). Industrial biotechnology for the production of bio-based chemicals—a cradle-to-grave perspective. *Trends Biotechnol*, 25, 119–124.

Himmel M, Ding S, Johnson D, Adney W, Nimlos M, Brady J, Foust T. (2007). Biomass recalcitrance: engineering plants and enzymes for biofuels production. *Science*, 315, 804–807.

International Organization for Standardization. (2006). Environmental Management, Life Cycle Assessment: Principles and Framework. ISO 14040 Ed. 2. International Organization for Standardization, Geneva.

Jenkins R, Alles C. (2011). Field to fuel: developing sustainable biorefineries. *Ecol Appl*, 21, 1096–1104.

Jerck M, Sather B, Henderson M, Gardner S. (2012). Accelerating Industry Innovation: 2012 Ethanol Industry Outlook. Renewable Fuels Foundation. Available at: http://ethanolrfa.3cdn.net/d4ad995ffb7ae8fbfe_1vm62ypzd.pdf [Accessed December 3, 2012].

Kamm B, Gruber PR, Kamm M. (2006). *Biorefineries—Industrial Processes and Products*. Weinheim: Wiley-VCH Verlag GmbH.

Kamm B, Kamm M. (2004). Principles of biorefineries. *Appl Microbiol Biotechnol*, 64, 137–145.

Karp A, Richter GM. (2011). Meeting the challenge of food and energy security. *J Exp Bot*, 62, 3263–3271.

Lal R. (2010). Managing soils for a warming earth in a food-insecure and energy-starved world. *J. Plant Nutr Soil Sci*, 173, 4–15.

Lardon L, Hélias A, Sialve B, Steyer JP, Bernard O. (2009). Life-cycle assessment of biodiesel production from microalgae. *Environ Sci Technol*, 43, 6475–6481.

Le Henaff Y, Huc JF. (2008). BIO-AMBER: from the patent to industrial demonstration. In: The Canadian Conference on Industrial Bioproduct Innovation, November 5–6, 2008. Montreal.

Levine RB, Pinnarat T, Savage PE. (2010). Biodiesel production from wet algal biomass through in situ lipid hydrolysis and supercritical transesterification. *Energy Fuels*, 24, 5235–5243.

Liu S, Lu H, Hu R, Shupe A, Lin L, Liang B. (2012a). A sustainable woody biomass biorefinery. *Biotechnol Adv*, 30, 785–810.

Liu S, Abrahamson LP, Scott GM. (2012b). Biorefinery: ensuring biomass as a sustainable renewable source of chemicals, materials, and energy. *Biomass Bioenergy*, 39, 1–4.

Macrelli S, Mogensen J, Zacchi G. (2012). Techno-economic evaluation of 2nd generation bioethanol production from sugar cane bagasse and leaves integrated with the sugar-based ethanol process. *Biotechnol Biof*, 5, 22.

Menon V, Rao M. (2012). Trends in bioconversion of lignocellulose: biofuels, platform chemicals & biorefinery concept. *Prog Energy Combust Sci*, 38, 522–550.

Möller R, Toonen M, van Beilen J, Salentijn E, Clayton D. (2007). Crop Platforms for Cell Wall Biorefining: Lignocellulose Feedstocks. CPL Press, Newbury, Berks, UK ISBN: 9781872691138.

Munasinghe PC, Khanal SK. (2010). Biomass-derived syngas fermentation into biofuels: opportunities and challenges. *Bioresour Technol*, 101, 5013–5022.

Mussatto SI, Dragone G, Guimarães PMR, Silva JP, Carneiro LM, Roberto IC, Vicente A, Dominguesa L, Teixeira JA. (2010). Technological trends, global market, and challenges of bio-ethanol production. *Biotechnol Adv*, 28, 817–830.

Nigam PS, Singh A. (2011). Production of liquid biofuels from renewable resources. *Prog Energy Combust Sci*, 37, 52–68.

Nitayavardhana S, Khanal SK. (2012). Biofuel residues/wastes: ban or boon? *Crit Rev Environ Sci Technol*, 42, 1–43.

Parikka M. (2004). Global biomass fuel resources. *Biomass Bioenergy*, 27, 613.

Perlack RD, Stokes BJ. (2011). US Billion-Ton Update: Biomass Supply for a Bioenergy and Bioproducts Industry. U.S. Department of Energy, August 2011.

Pflügl S, Marx H, Mattanovich D, Sauer M. (2012). 1,3-Propanediol production from glycerol with Lactobacillus diolivorans. *Bioresource Technol*, 119, 133–140.

Poliakoff M, License P. (2007). Green Chemistry. *Nature*, 450, 810–812.

Ponton JW. (2009). Biofuels: thermodynamic sense and nonsense. *J Clean Prod*, 17, 896–899.

Posada J, Rincón LE, Cardona C. (2012). Design and analysis of biorefineries based on raw glycerol: addressing the glycerol problem. *Bioresource Technol*, 111, 282–293.

Ragauskas AJ, Williams CK, Davison BH, Britovsek G, Cairney J, Eckert CA, Frederick W, Hallett JP, Leak DJ, Liotta CL, Mielenz JR, Murphy R, Templer R, Tschaplinski T. (2006a). The path forward for biofuels and biomaterials. *Science*, 311, 484–489.

Ragauskas AJ, Williams CK, Davison BH, Britovsek G, Cairney J, Eckert CA, Frederick WJ, Hallett JP, Leak DJ, Liotta CL, Mielenz JR, Murphy R, Templer R, Tschaplinski T. (2006b). Measuring the efficiency of biomass energy. *Science*, 312, 1744–1745.

Reijnders L. (2006). Conditions for the sustainability of biomass based fuel use. *Energ Policy*, 34, 863–876.

Rodolfi L, Chini Zittelli G, Bassi N, Padovani G, Biondi N, Bonini G, Tredici MR. (2009). Microalgae for oil: Strain selection, induction of lipid synthesis and outdoor mass cultivation in a low-cost photobioreactor. *Biotechnol Bioeng*, 102, 100–112.

Sacramento-Rivero JC. (2012). A methodology for evaluating the sustainability of biorefineries: framework and indicators. *Biof Bioprod Bioref*, 6, 32–44.

Scott SA, Davey MP, Dennis JS, Horst I, Howe CJ, Lea-Smith DJ, Smith AG. (2010). Biodiesel from algae: challenges and prospects. *Curr Opin Biotechnol*, 21, 277–286.

Singh A, Olsen SI. (2011). A critical review of biochemical conversion, sustainability and life cycle assessment of algal biofuels. *Appl Energy*, 88, 3548–3555.

Srirangan K, Akawi L, Moo-Young M, Chou CP. (2012). Towards sustainable production of clean energy carriers from biomass resources. *Appl Energy*, 100, 172–186.

Subhadra B. Grinson-George (2011). Algal biorefinery-based industry: an approach to address fuel and food insecurity for a carbon-smart world. *J Sci Food Agric*, 91, 2–13.

Tschan MJL, Brulé E, Haquette P, Thomas CM. (2012). Synthesis of biodegradable polymers from renewable resources. *Polym Chem*, 3, 836.

United Nations General Assembly. (1987). Report of the World Commission on Environment and Development: Our Common Future Chapter 2; Transmitted to the General Assembly as an Annex to document A/42/427. Available at: http://www.un-documents.net/ocf-02.htm.

Vennestrøm PNR, Osmundsen CM, Christensen CH, Taarning E. (2011). Beyond petrochemicals: the renewable chemicals industry. *Angew Chem Int Ed*, 50, 10502–10509.

Wagemann K, Benzig T, Böhland T, Fritsche UR, Fröhling M, Gröngröft A. (2012). Roadmap Bioraffinerien. Available at: http://www.bmbf.de/pub/roadmap_bioraffinerien.pdf.

Wang Y, Pengzhan Liu SO, Zhang Z. (2007). Preparation of biodiesel from waste cooking oil via two-step catalyzed process. *Energy Convers Manage*, 48, 184–188.

Wells P, Zapata C. (2012). Renewable eco-industrial development. *J Ind Ecol*, 16, 665–668.

Zah R, Böni H, Gauch M, Hischier R, Lehmann M, Wäger P. (2007). Life Cycle Assessment of Energy Products: Environmental Assessment of Biofuels. [Online]. Report written under contract from the Federal Office of Energy (BFE), the Federal Office for the Environment (CFE) and the Federal Office for Agriculture (BLW). Bern (2007). Available at: http://www.bioenergywiki.net/images/8/80/Empa_Bioenergie_ExecSumm_engl.pdf [Accessed January 4, 2013].

CHAPTER 2

Biomass Logistics

KEVIN L. KENNEY and J. RICHARD HESS
Energy Systems and Technologies, Idaho National Laboratory, Idaho Falls, ID, USA

NATHAN A. STEVENS
Materials and Physical Security, Idaho National Laboratory, Idaho Falls, ID, USA

WILLIAM A. SMITH and IAN J. BONNER
Biofuels and Renewable Energy Technologies, Idaho National Laboratory, Idaho Falls, ID, USA

DAVID J. MUTH
Praxik, LLC, Ames, IA, USA

Abstract

Understanding biomass feedstock logistics and the design of biomass feedstock supply systems requires that one understand how each logistics operation affects

Bioprocessing of Renewable Resources to Commodity Bioproducts, First Edition.
Edited by Virendra S. Bisaria and Akihiko Kondo.

feedstock cost, quality, and quantity metrics. With feedstock cost being a significant barrier to the economics of biofuel and bioproduct production, it tends to be the main focus of feedstock supply system design. In this case, feedstock quantity and quality are generally considered as constraints on the system rather than main objectives of feedstock supply system design. This chapter presents a methodology for quantifying the impact of logistics parameters on feedstock cost in order to identify and rank supply system barriers. These effects are presented in terms of statistical parameters of sensitivity, uncertainty, and influence. Discussion of these three analysis parameters will show that though feedstock cost is the primary outcome; feedstock cost and quality constraints are manifest in the form of uncertainty. Further, the sources of uncertainty are identified, and logistics solutions are discussed. Finally, we conclude that successful supply system designs and effective biomass logistics systems must control uncertainty to be successful.

2.1 INTRODUCTION

The emergent lignocellulosic biofuels industry has adapted agricultural forage and forest logging equipment and practices for the purpose of developing high-tonnage biomass supply systems for biorefining (Hess et al., 2009; Searcy and Hess, 2010; Perlack and Stokes, 2011; Shinners et al., 2011). High-tonnage biomass logistics development efforts have resulted in much progress with respect to fully understanding and demonstrating the capabilities and limitations of conventional agricultural forage and forestry supply equipment for both herbaceous and woody biomass resources.

Supply system modeling and analyses are instrumental in identifying and quantifying these limitations. Monte Carlo analysis is a common probabilistic analysis method for forecasting a model result based on the uncertainty of model inputs. These techniques have been used in technoeconomic assessments of both biochemical (Aden et al., 2002; Humbird et al., 2011) and thermochemical (Phillips et al., 2007; Dutta et al., 2011) conversion of lignocellulosic biomass to ethanol, as well as feedstock supply system designs (Hess et al., 2007). A Monte Carlo simulation represents the uncertainty of a result based on aggregated variability of inputs, but it does not distinguish the affects of multiple inputs such that they can be ranked and differentiated according to their impact on the model output.

This is a particular challenge for optimizing biomass supply systems, which, for economic viability must focus on cost reduction, but are also constrained by environmental sustainability and feedstock quality requirements, both of which introduce in-field variabilities that influence and are influenced by logistics parameters. Sustainability and quality constraints underpin the resource supply/demand balance fundamentally in such a way that methodologies are needed to identify not only a system's limitations, but also determine which variables have the greatest potential for cost savings within these constraints.

This chapter presents a methodology and approach to quantifying and ranking feedstock supply system variables that affect the delivered feedstock cost according to three statistically derived parameters: sensitivity, uncertainty, and influence. A discussion of supply chain uncertainty will show that significant variability is introduced when additional metrics—sustainability and quality in this case—impose requirements and limitations on supply chain variables that compete against feedstock cost. Examples of collection efficiency and storage dry matter losses will be presented to support both equipment development and best management practices solutions to reducing uncertainty. This chapter further demonstrates and concludes that supply systems that actively manage the uncertainty associated with biomass supply systems will, in the long run, be the most cost-effective and sustainable feedstock logistic systems.

2.2 METHOD OF ASSESSING UNCERTAINTY, SENSITIVITY, AND INFLUENCE OF FEEDSTOCK LOGISTIC SYSTEM PARAMETERS

2.2.1 Analysis Step 1—Defining the Model System

A lignocellulosic feedstock logistics system that supplies 800,000 dry matter tons of corn stover to a 60 Mgal/year biochemical ethanol biorefinery was modeled in Microsoft Excel using the methodology documented by Turhollow et al. (2009). The logistics system included raking and baling (large $4 \times 4 \times 8$ ft square bales) after grain harvest, collection of bales from the field to a roadside storage stack, transportation of bales to a biorefinery over a 350 day/year delivery schedule, and preprocessing of bales at the biorefinery to a hammer-milled bulk material ready for insertion into a pretreatment reactor (Hess et al., 2009).

2.2.2 Analysis Step 2—Defining Input Parameter Probability Distributions

In order to identify and rank the importance of model input parameters, an analysis was conducted using @RISK, a commercial simulation software package used to solve Excel spreadsheet models for a probable forecasted scenario (@RISK, Palisade Corp., Ithaca, NY; Excel, Microsoft Corp., Redmond, WA). Probability distributions were defined for each model input variable, including biomass resource availability parameters (e.g., grain yield, producer participation), biomass material property parameters (e.g., moisture content, bulk density), logistics system parameters (e.g., harvest window, transportation distance, fuel and electricity prices), machinery performance parameters (e.g., rates, capacities, efficiencies), and biomass loss parameters (e.g., collection efficiencies, storage losses). Most input variables were described according to a PERT distribution function, which is commonly used when data to define a distribution is sparse (@RISK, Palisade Corp., Ithaca, NY). A PERT distribution is defined by minimum, mode, and maximum values that are linked in a bell-shaped distribution, and to the extent that the minimum and maximum values are evenly distributed about the mode, this distribution mimics that of a normal distribution. Two input variables—biorefinery daily receiving hours and biorefinery feedstock inventory (i.e., tons of feedstock stored on site)—were defined by a uniform distribution that assumes all values between the defined minimum and maximum values are equally likely to occur. A uniform distribution was chosen for these two variables because all values within the distribution are equally likely to occur.

Minimum, mode, and maximum values of the PERT distributions were defined according to our own research data, where available (Hess et al., 2009); otherwise, literature data and input from experienced machinery operators, manufacturers, and vendors was used. American Society of Agricultural and Biological Engineers (ASABE) Agricultural Machinery Management Data (ASABE, 2006) was used to define machinery field speeds and efficiencies, repair and maintenance costs, annual operating hours, and estimated lifetime hours. The probability distribution represents either the inherent variability or the uncertainty of the input variables, as determined by the variability in collected field data, published data (e.g., field efficiency and field speed ranges published by ASABE), or range of operating parameters suggested by skilled operators of the equipment. The most likely value included in each distribution is the benchmark value input to the feedstock logistics model (Hess et al., 2009).

2.2.3 Analysis Step 3—Perform Deterministic Computations

A Latin Hypercube sampling method was used in the @Risk simulation to generate the input parameter values from the probability distribution functions. This method was chosen over the Monte Carlo technique, which samples randomly from the distribution function and causes clustering when low probability values are not sampled due to insufficient computational sampling iterations. In contrast, the Latin Hypercube stratified sampling technique systematically samples all segments (stratifications) of the distribution just once, resulting in fewer computational iterations required to produce a representative probability curve.

The analysis was conducted by incrementing each input parameter throughout the defined distribution while randomly varying the remaining parameters according to their own defined probability distributions. Thus, the impact of each parameter on delivered feedstock cost was determined individually, while also capturing the interdependence of the input parameters. Tens of thousands of scenarios were collected in this manner to generate the output shown in Figure 2.1.

This @Risk simulation was used to rank input variables based on the statistical relationship between each variable and the delivered feedstock cost. In order to resolve these rankings, the @Risk analysis results were further analyzed to isolate the individual influences with respect to three parameters: sensitivity, uncertainty, and influence (Figure 2.2). *Sensitivity*, a measure of how responsive a unit operation/process cost is to changes in a specific variable, was determined by approximating the slope using a linear regression of each response curve in units of $/ton per percent change from the variable's base value.

Sensitivity alone is not sufficient to rank input variables according to their impact on delivered feedstock cost because while feedstock cost may be highly sensitive to changes in a particular variable, the overall effect may be small if the range of that variable is small. This range or variability of an input variable, termed *Uncertainty*, was measured by the horizontal run of the response curve in units of percent change from the parameter base value. An additional parameter was included, referred to as *Influence*, to give preference to those variables that are more influential than others. A parameter's influence is represented by the curvature of the response curve, with greater curvature suggesting greater interdependence (or influence). A parameter's

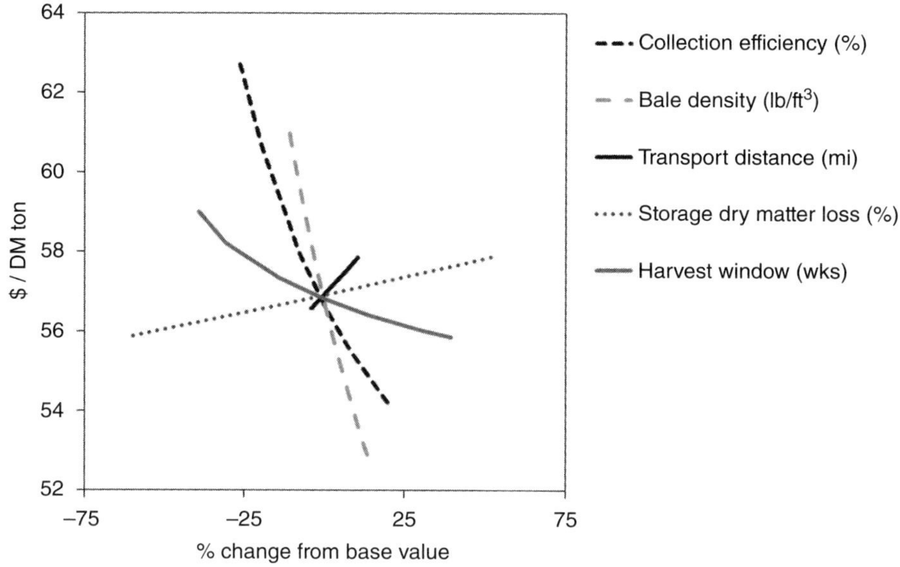

FIGURE 2.1 Response of delivered feedstock cost to changes in model input logistics parameters.

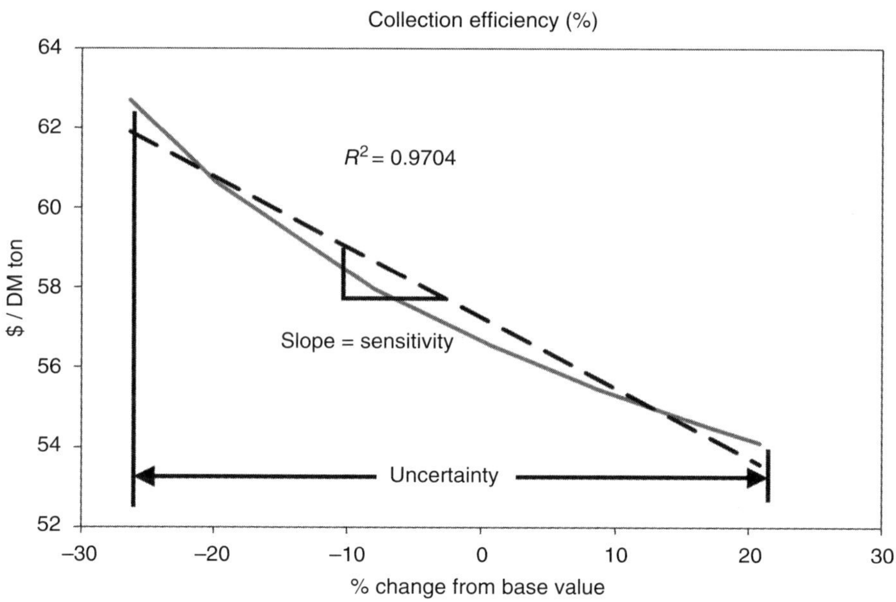

FIGURE 2.2 Illustrated definitions of sensitivity, influence, and uncertainty.

relative influence was estimated as the inverse of the R-squared value derived from a linear regression of the response curve.

2.2.4 Analysis Step 4—Deciphering the Results

Independent rankings of input variables according to sensitivity, uncertainty, and influence provided three disparate rankings (Figures 2.3–2.5). Bale bulk density, collection efficiency, and grain yield ranked highest in sensitivity; storage dry matter losses, harvest window, and bale moisture ranked the highest in uncertainty; and harvest window, collection efficiency, and stalk chopper field speed ranked highest in influence. A combined normalized ranking was determined by taking the product of sensitivity, uncertainty, and influence and dividing by the highest value (Figure 2.6). The resulting normalized values show the combined impact of the three parameters—sensitivity, uncertainty, and influence—relative to the input variable of greatest combined effect. The variables shown in Figure 2.6 were categorized as follows: (1) biomass yield (collection efficiency, grain yield, storage dry matter loss); (2) biomass material properties (bale bulk density, bale moisture); (3) machinery performance (shredder speed, baler capacity, baler field efficiency, semi speed, shredder field efficiency, loader capacity; and (4) system variables (harvest window, transportation distance winding factor—a multiplier applied to the transportation distance input, off-road diesel price).

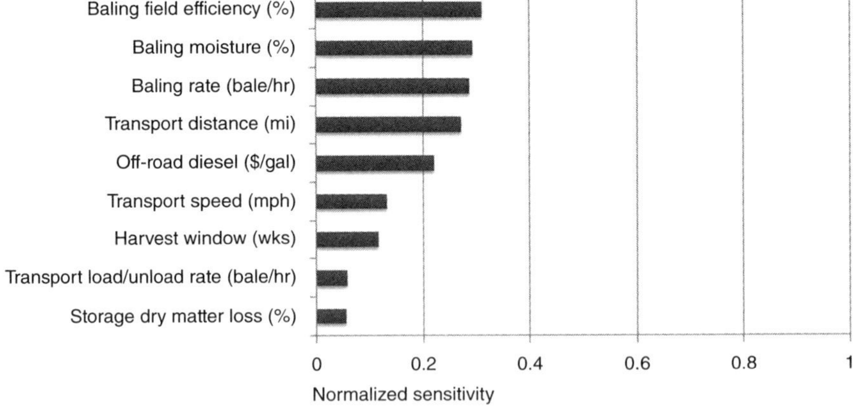

FIGURE 2.3 Normalized ranking of the sensitivity of feedstock cost to changes in input logistics parameters.

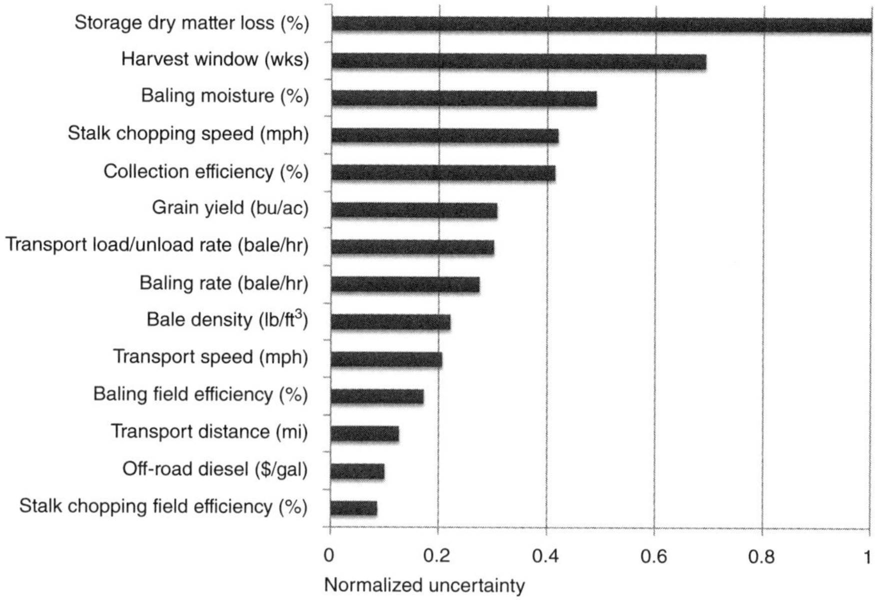

FIGURE 2.4 Normalized ranking of the uncertainty of feedstock cost related to the variability of input logistics parameters.

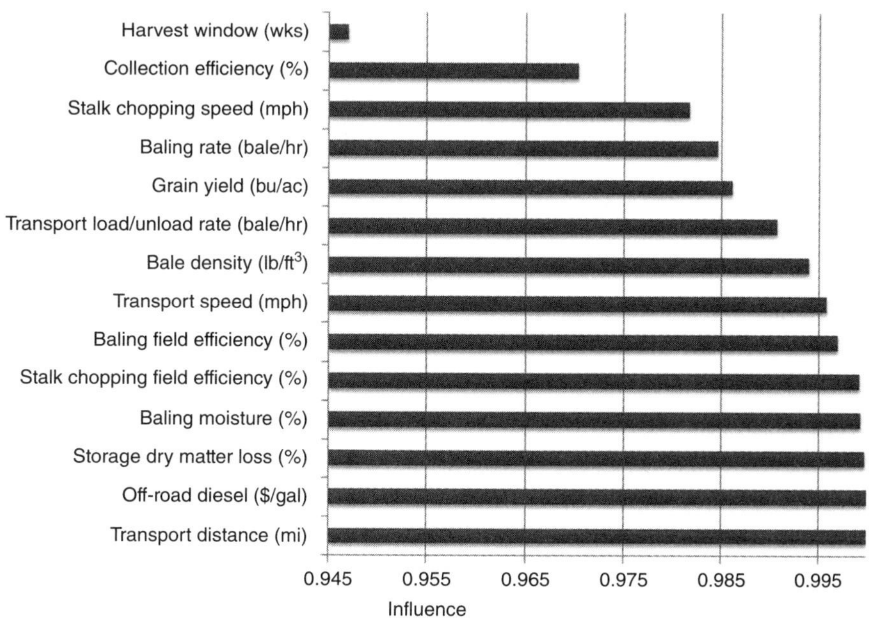

FIGURE 2.5 Ranking of the cumulative influence of input logistics parameters on feedstock cost.

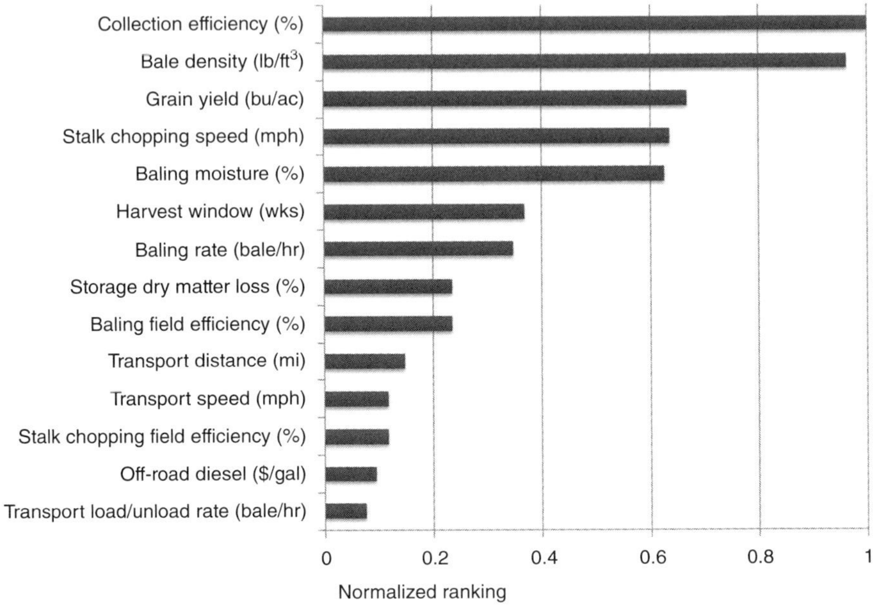

FIGURE 2.6 Normalized ranking of model input logistics parameters according to combined values of sensitivity, uncertainty, and influence.

Biomass yield variables—those that affect the mass per acre delivered to the biorefinery—are particularly important since they reside at the front end of the feedstock supply chain and thus have broad impacts that extend through the entire supply chain. In fact, biomass yield affects many of the other variable categories identified in the analysis, including machinery performance and system variables (harvest window and transportation distance). Accordingly, uncertainties of biomass yield variables—particularly collection efficiency and storage dry matter losses—are the focus of the discussion that follows.

2.3 UNDERSTANDING UNCERTAINTY IN THE CONTEXT OF FEEDSTOCK LOGISTICS

2.3.1 Increasing Biomass Collection Efficiency by Responding to In-Field Variability

Collection efficiency (the ratio of biomass collected to the total amount available in the field) ranked second in sensitivity, fifth in uncertainty, second in influence, and highest in overall combined ranking. The probability distribution for collection efficiency used in this analysis was based on a review of reported corn stover collection efficiencies (Richey et al., 1982; Shinners et al., 2003; Schechninger and Hettenhaus, 2004; Shinners and Binversie, 2004; Prewitt et al., 2007) from which we chose a most likely value of 43%, a minimum of 19%, and a maximum of 65%. The wide range of reported results show that current machinery itself is capable of high removal

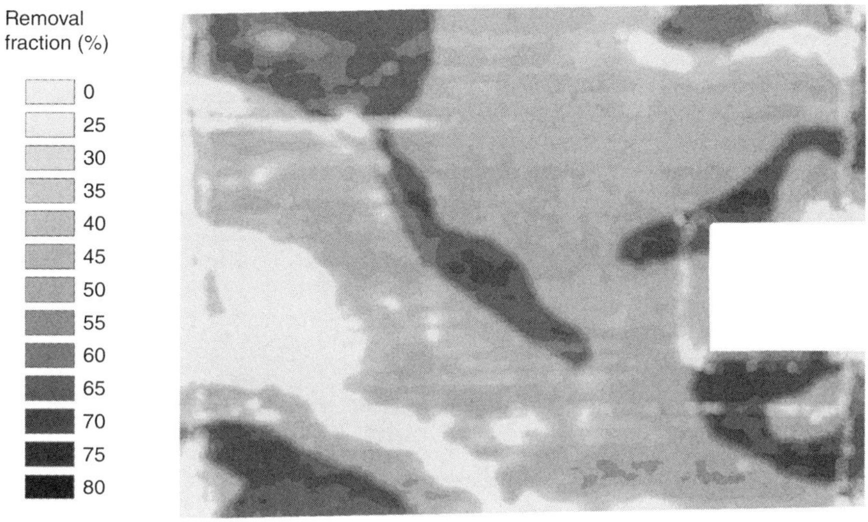

FIGURE 2.7 Sustainable subfield residue harvest plan that varies the removal rate between 0% and 80% (5.6 Mg/ha).

rates, but sustainability (Wilhelm et al., 2010) and quality (Prewitt et al., 2007) constraints often dictate deliberately conservative collection efficiencies. Therefore, sustainability and quality constraints are two main sources of uncertainty relating to corn stover collection efficiencies, and reducing feedstock cost involves solving uncertainty around these two issues.

It has generally been reported that corn stover removal rates of 30–40% could be sustainable over most corn acres (Nelson et al., 2004; Gregg and Izaurralde, 2010, Perlack et al., 2011). However, corn stover removal at a rate of 30–40% is often not economically viable. The emerging biorefining industry has estimated minimum removal rates of two dry short tons (DST) per acre for system economics to support corn stover removal operations.

Two management strategies have emerged to deal with this challenge and sustainably achieve a two DST/acre removal rate. The first is an equipment development strategy—variable rate residue harvesting—and the second is an agronomic strategy—implementing interval removal schemas (Muth and Bryden 2012). Multifactor sustainability analyses (Figure 2.7) have shown that advanced variable-rate harvesting systems capable of responding to subfield variability in topography, soil characteristics, and grain yield could achieve average removal rates as high as 75% (7.69 Mg/ha) without violating sustainability requirements (Muth and Bryden, 2012). Interval removal schemas allow conventional equipment to be used to collect corn stover one out of every 2 or 3 years, based on soil erosion and soil organic carbon constraints for an individual field.

Our own testing has shown that removal rates as high as 80% are attainable with wheel rake, flail shredder, or bar rake windrowers; however, increasing corn stover collection efficiencies with conventional harvest systems tends to reduce stover

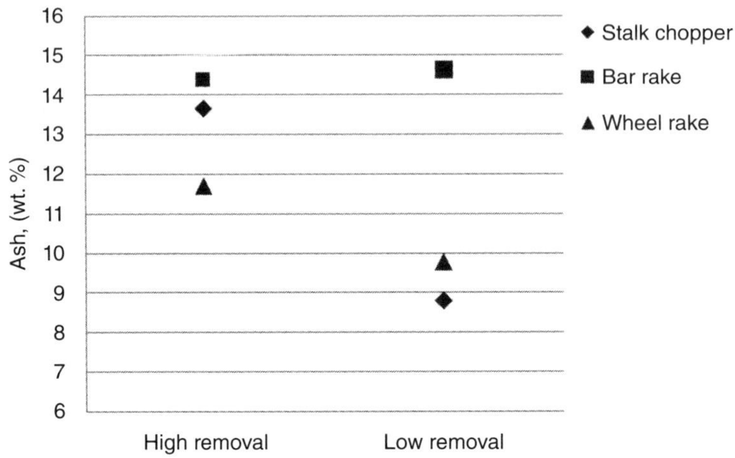

FIGURE 2.8 Corn stover large square bale ash content variability due to windrowing machinery and removal rate.

quality by increasing the ash content (Figure 2.8). The increase in stover ash content—both between equipment and between removal rates—is largely attributed to increased soil entrainment (Prewitt et al., 2007), but may also be attributed to differences in anatomical composition (Hoskinson et al., 2007). Though the dataset is too limited to support specific solutions, it shows that quality is a variable affecting collection efficiency uncertainty, and a priority on low ash content generally necessitates reduced collection efficiencies. The data also show that an equipment development solution that eliminates a variable from the uncertainty equation, such as a bar rake in this case, is beneficial. Like the sustainability problem, potential solutions exist that include both conventional and advanced harvesting systems. Understanding variable field factors that affect susceptibility of soil to disturbance and entrainment during windrowing may provide a solution for selection and configuration of conventional harvest equipment that can maintain ash content at acceptable levels regardless of removal rate. Ultimately, single-pass harvest systems that eliminate biomass/ground contact during harvest will provide the best opportunity to remove the quality variable affecting collection efficiencies.

2.3.2 Minimizing Storage Losses by Addressing Moisture Variability

Storage dry matter losses (loss of structural carbohydrates, water-soluble components, lignin, and ash resulting from biological deterioration and/or physical losses in storage) ranked fourteenth in sensitivity, first in uncertainty, twelfth in influence, and eighth in overall combined ranking. The probability distribution for dry matter loss used in this analysis was based on a review (Coble and Egg, 1987; Sanderson et al., 1997; Shinners et al., 2007; Shinners et al., 2010) of reported dry matter loss for dry (<20%, wet basis) aerobic storage, from which we chose a most likely value of 5%, a minimum of 1%, and a maximum of 8%. The range selected actually underestimates

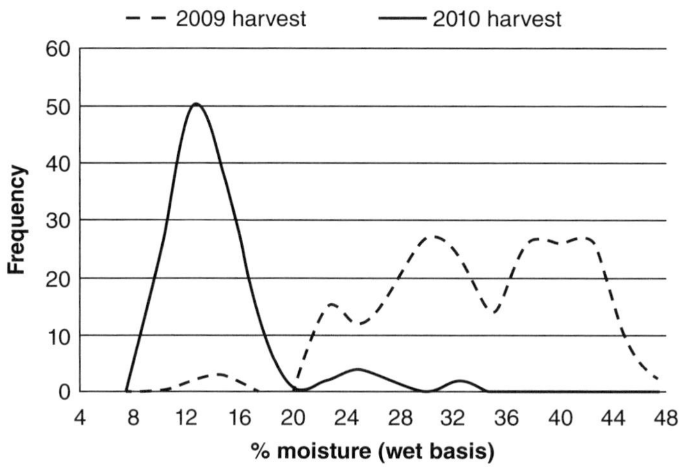

FIGURE 2.9 Midwest corn stover harvested in 2009 exceeded moisture limits for stable bale storage, while in 2010 nearly all harvested biomass could be stored stable in conventional stacks.

the true variability and uncertainty around storage dry matter losses. Storage losses can vary significantly depending on bale moisture, storage method, and environmental conditions during storage; these variables make control and prediction of dry matter loss in storage highly uncertain (Shinners et al., 2007; Darr and Shah, 2012).

Controlling dry matter losses and reducing uncertainty is highly dependent on moisture management. The ability to field-dry to safe aerobic storage conditions is compromised by often poor, always uncertain, field-drying conditions following the fall grain harvest. This is clearly demonstrated in Midwest corn stover moistures during the 2009 and 2010 feed corn harvest seasons (Figure 2.9). This annual variability in biomass moisture content presents significant challenges in implementing consistent and reliable storage practices.

Proper moisture management does not end in the field. Even with ideal agronomic conditions where the material dries down to safe aerobic storage levels before baling, other often subtle differences in storage methods can significantly affect dry matter losses. For example, the orientation of stacks as well as the characteristics of the ground surface on which bales are stacked can greatly affect storage conditions. In a corn stover storage study, we found bales to be more moist on the north-facing side than the south exhibiting up to 40% moisture (wb) differences from side to side (Figure 2.10). Bottom bales also accumulated more moisture, even when placed on crushed gravel, which was attributed to water running off the tarp and draining toward the stack (Smith et al., 2013). Thus, bale storage is a dynamic practice with no one-size-fits-all solution. Bales are capable of long-term storage with a proper understanding of the material's condition (moisture content at time of receipt and any past storage information) provided that they are properly managed. The key factors to control are the accumulation of moisture and the timing of this accumulation with respect to environmental conditions in storage.

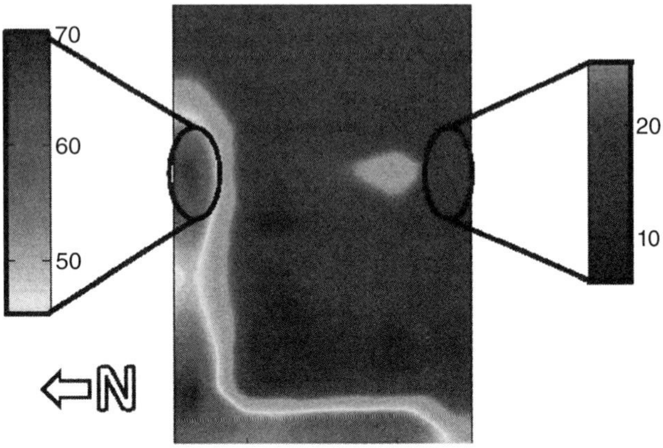

FIGURE 2.10 Isopleth moisture distribution of an end view of a stack of 3 × 4 × 8-ft large square bales (stacked 1-bale wide × 4 bales high) showing the influence of Northern exposure on bale moisture. (*For a color version, see the color plate section.*)

2.4 FUTURE PROSPECTS

Unlike engineered conversion processes operating within controlled (e.g., low uncertainty) environments of conversion reactors, feedstock logistics processes occur in distributed environments of high uncertainty. An analysis of feedstock cost variability, based on an assessment of sensitivity, uncertainty, and influence of feedstock logistics variables, showed that uncertainty dominates when sustainability and quality metrics compete with feedstock cost. This was demonstrated in the case of biomass collection efficiency and storage dry matter losses. Collection efficiency, though a measurement of machinery performance, is strongly dependent on sustainability and quality variables that often constrain collection efficiency to conservative levels. Reducing storage dry matter losses and maintaining feedstock quality through storage are possible through moisture management. Logistics solutions, whether through equipment development or improved methods and practices, must meet cost, sustainability, and quality requirements. Feedstock logistic systems that actively manage uncertainty by balancing these metrics will in the long run be the most cost-effective and sustainable logistic systems.

2.5 FINANCIAL DISCLOSURE/ACKNOWLEDGMENTS

Preparation of this manuscript was supported in part by the US Department of Energy, Office of Energy Efficiency and Renewable Energy under DOE Idaho Operations Office Contract DE-AC07-05ID14517. This manuscript has been authored by Battelle Energy Alliance, LLC under Contract No. DE-AC07-05ID14517 with the US Department of Energy. The US Government retains and the publisher, by accepting the article for publication, acknowledges that the US Government retains a nonexclusive, paid-up, irrevocable, world-wide license to publish or reproduce the published form of this manuscript, or allow others to do so, for US Government purposes.

REFERENCES

Aden A, Ruth M, Ibsen K, Jechura J, Neeves K, Sheehan J, Wallace B, Montague L, Slayton A, Lukas J. et al. (2002). Lignocellulosic biomass to ethanol process design and economics utilizing co-current dilute acid prehydrolysis and enzymatic hydrolysis for corn stover. National Renewable Energy Laboratory, Golden, CO. NREL/TP-510-32438.

ASABE. (2006). *ASAE D497.5 Agricultural Machinery Management Data*. St Joseph, MI: American Society of Agricultural and Biological Engineers.

Coble CG, Egg R. (1987). Dry matter losses during hay production and storage of sweet sorghum used for methane production. *Biomass*, 14, 209–217.

Darr MJ, Shah AJ. (2012). Biomass storage: An update on industrial solutions for baled biomass feedstocks. *Biofuels*, 3, 321–332.

Dutta A, Talmadge M, Hensley J, Worley M, Dudgeon D, Barton D, Groenendijk P, Ferrari D, Stears B, Searcy EM, Wright CT, Hess JR. (2011). Process design and economics for conversion of lignocellulosic biomass to ethanol-thermochemical pathway by indirect gasification and mixed alcohol synthesis. National Renewable Energy Laboratory, Golden, CO. NREL/TP-5100-51400.

Gregg JS, Izaurralde RC. (2010). Effect of crop residue harvest on long-term crop yield, soil erosion and nutrient balance: trade-offs for a sustainable bioenergy feedstock. *Biofuels*, 1, 69–83.

Hess JR, Kenney KL, Wright CT, Perlack R, Turhollow A. (2009). Corn stover availability for biomass conversion: situation analysis. *Cellulose*, 16, 599–619.

Hess JR, Wright CT, Kenney KL. (2007). Cellulosic biomass feedstocks and logistics for ethanol production. *Biofuels Bioprod Biorefin*, 1, 181–190.

Hoskinson RL, Karlen DL, Birrell SJ, Radtke CW, Wilhelm WW. (2007). Engineering, nutrient removal, and feedstock conversion evaluations of four corn stover harvest scenarios. *Biomass Bioenergy*, 31, 126–136.

Humbird DA, Davis R, Tao L, Kinchin C, Hsu D, Aden A, Schoen P, Lukas J, Olthof B, Worley M, Sexton D, Dudgeon D. (2011). Process design and economics for biochemical conversion of lignocellulosic biomass to ethanol–dilute-acid pretreatment and enzymatic hydrolysis of corn stover. National Renewable Energy Laboratory, Golden, CO. NREL/TP-5100-47764.

Muth DJ, Bryden KM. (2012). A conceptual evaluation of sustainable variable-rate agricultural residue removal. *J Environ Qual*, 41, 1796–1805.

Nelson RG, Walsh M, Sheehan JJ, Graham R. (2004). Methodology for estimating removable quantities of agricultural residues for bioenergy and bioproduct use. *Appl Biochem Biotechnol*, 113, 13–26.

Perlack RD, Stokes BJ. (2011). US Billion-ton update: Biomass supply for a bioenergy and bioproducts industry. Oak Ridge National Laboratory, Oak Ridge, TN. ORNL/TM-2011/224.

Phillips SD, Aden A, Jechura J, Dayton D, Eggeman T. (2007). Thermochemical ethanol via indirect gasification and mixed alcohol synthesis of lignocellulosic biomass. National Renewable Energy Laboratory, Golden, CO. NREL/TP-510-41168.

Prewitt RM, Montross MD, Shearer SA, Stombaugh TS, Higgins SF, McNeill SG, Sokhansanj S. (2007). Corn stover availability and collection efficiency using typical hay equipment. *Trans ASABE*, 50, 705–711.

Richey CB, Liljedahl JB, Lechtenberg VL. (1982). Corn stover harvest for energy production. *Trans ASABE*, 25, 834–839.

Sanderson MA, Egg RP, Wiselogel AE. (1997). Biomass losses during harvest and storage of switchgrass. *Biomass Bioenergy*, 12, 107–114.

Schechninger TM, Hettenhaus JR. (2004). Corn stover harvesting:grower, custom operator, and processor issues and answers. Report on corn stover harvest experiences in Iowa and Wisconsin for 1997–98 and 1998–99 crop years. US Department of Energy, Office of Energy Efficiency and Renewable Energy, Oak Ridge, TN. ORNL/SUB-04—4500008274-01.

Searcy EM, Hess JR. (2010). Uniform-format feedstock supply system: A Commodity-scale design to produce an infrastructure-compatible biocrude from lignocellulosic biomass. Idaho National Laboratory, Idaho Falls, ID. INL/EXT-10-20372.

Shinners KJ, Binversie BN. (2004). Harvest and storage of wet corn stover biomass. Paper Number: ASAE/CSAE 041159. 2004 ASAE/CSAE Annual International Meeting, Ottawa, Ontario, Canada, American Society of Agricultural and Biological Engineers.

Shinners KJ, Binversie BN, Muck RE, Weimer P. (2007). Comparison of wet and dry corn stover harvest and storage. *Biomass Bioenergy*, 31, 211–221.

Shinners KJ, Binversie BN, Savoie P. (2003). Harvest and storage of wet and dry corn stover as a biomass feedstock. ASAE Paper No. 03-6088. American Society of Agricultural Engineers Annual Meeting, Las Vegas, NV, American Society of Agricultural and Biological Engineers.

Shinners KJ, Boettcher GC, Muck RE, Weimer PJ, Casler MD. (2010). Harvest and storage of two perennial grasses as biomass feedstocks. *Trans ASABE*, 53, 359–370.

Shinners KJ, Digman MF, Runge TM. (2011). Biomass logistics—harvest and storage. In: Zhu J, Zhang X, Pan X, editors. *Sustainable Production of Fuels, Chemicals, and Fibers from Forest Biomass*. Washington, DC: American Chemical Society. 1067, pp. 65–86.

Smith WA, Bonner IJ, Kenney KL, Wendt LM. (2013). Practical considerations of moisture in baled biomass feedstocks. *Biofuels*, 4, 95–110.

Turhollow A, Wilkerson E, Sokhansanj S. (2009). Cost methodology for biomass feedstocks: herbaceous crops and agricultural residues. ORNL/TM-2008/105. Oak Ridge National Laboratory, Oak Ridge, TN.

Wilhelm WW, Hess JR, Karlen DL, Johnson JMF, Muth DJ, Baker JM, Gollany HT, Novak JM, Stott DE, Varvel GE. (2010). Review: Balancing limiting factors & economic drivers for sustainable Midwestern US agricultural residue feedstock supplies. *Ind Biotechnol*, 6, 271–287.

CHAPTER 3

Pretreatment of Lignocellulosic Materials

KARTHIK RAJENDRAN and MOHAMMAD J. TAHERZADEH
School of Engineering, University of Borås, Sweden

Bioprocessing of Renewable Resources to Commodity Bioproducts, First Edition.
Edited by Virendra S. Bisaria and Akihiko Kondo.

Abstract

Lignocelluloses are structural materials in plants expected to withstand physical, chemical, and biological degradation. Nonetheless, biomass, with their content of approximately 70% sugar polymers is attractive materials for production of bioproducts. The purpose of lignocelluloses in nature and their application as sugar sources is hence contrasting to each other. The structure of lignocelluloses and their recalcitrance are matters lifted early in this chapter in order to facilitate the understanding of the factors in need of attention, involving their pretreatment. This discussion is followed by brief descriptions of physical pretreatments, that is, milling, irradiation, steaming, etc., physicochemical methods, that is, explosions with steam, ammonia, CO_2, and SO_2, chemical pretreatments by alkali, acids, gases, oxidizing agents, and organosolvs, and biological pretreatments by fungi and their enzymes. The different methods for pretreatment of lignocelluloses are compared in terms of their efficiency and from an economic point of view.

3.1 INTRODUCTION

Lignocellulosic materials are composed of a variety of macromolecules, including cellulose, hemicelluloses, lignin, pectin, and minor amounts of inorganics. These materials represent a major fraction of agricultural residues, wood residues, and several waste streams from municipalities and companies (Taherzadeh and Karimi, 2008). In nature, lignocelluloses are the most abundant organic materials with an estimate of 10–50 billion tons in dry weight. This amount accounts to half the global biomass yield (Zhao et al., 2009). Lignocelluloses generally comprise 30–70% carbohydrates and are a potential source for production of various bioproducts, such as ethanol, butanol, and biogas (Lee et al., 2007a). Table 3.1 shows typical compositions of different lignocellulosic materials, including sugar polymers.

Due to the complex structure of lignocelluloses, the release of fermentable sugars is a hurdle for the industrialization of this renewable resource (Zhao et al., 2012a). The complex structure of plants functions as protection against microbial and enzymatic attacks and is called *biomass recalcitrance* (Himmel and Picataggio, 2009). Plant structures provide protection by many different strategies. The surface layer called "epidermis" comprises thick-walled cells, producing wax or oily substances. These substances make the lignocelluloses physically and chemically resistant to the environment. The arrangement of vascular tissues and cell walls constitute the next layer of protection (Taherzadeh and Jeihanipour, 2012). Cellulosic microfibrils in the plant cell walls are surrounded by polymers consisting of hemicelluloses, lignin, and pectin (Somerville et al., 2004). This complex arrangement might be the main reason for biomass recalcitrance in lignocelluloses (Himmel and Picataggio, 2009).

Understanding the chemistry and the morphological construction of biomass has a huge impact on the current pretreatment processes. Developing a deeper knowledge and more efficient technologies may be of great importance for future use. This

TABLE 3.1 Average Composition of Different Lignocellulosic Materials

Parameter	Unit (Dry)	Hardwood (Poplar)	Softwood (Pine)	Grass (Switchgrass)	Crop residue (Corn stover)
LHV	MJ/kg	19.5	19.6	17.1	16
Cellulose	%	42.9	44.5	32.0	37.7
Glucan	%	42.9	44.5	32.0	37.7
Hemicellulose	%	20.3	21.9	25.2	25.3
Xylan	%	17.0	6.30	21.1	21.6
Arabinan	%	1.20	1.60	2.84	2.42
Galactan	%	0.70	2.56	0.95	0.87
Mannan	%	1.42	11.4	0.30	0.38
Lignin	%	26.6	27.7	18.1	18.6
Acids	%	3.11	6.70	1.21	3.0
Extractives	%	4.70	2.88	17.5	5.61

Source: Adapted from Pandey et al. (2011).

chapter attempts at unveiling the chemistry (and its underlying mechanisms) of lignocellulosic biomass in relation to pretreatment methods.

3.2 COMPLEXITY OF LIGNOCELLULOSES

3.2.1 Anatomy of Lignocellulosic Biomass

Cell walls in plants have important functions including absorption, secretion, protection, food reserve, cell-shape maintenance, and control of cell expansion (Bidlack et al., 1992). In addition, plant cell walls with intercellular spaces provide mechanical support (Ding and Himmel, 2009). The architectural arrangement of the plant cell wall is complex. During cell wall division, two walls are formed from the existing wall (Mayer and Jürgens, 2004). These are called "primary cell walls." However, in some types of cells, for example, fiber cells in woods, an additional cell wall develops after the primary cell division. This cell wall is called a "secondary cell wall" and is located inside the primary cell wall (Somerville et al., 2004). In general, the primary cell wall is not lignified, while the secondary cell wall is extremely lignified. Some plant parts, for example, hairs of cotton seeds and stems of flax and hemp contain little or no lignin in the two cell walls (Gorshkova and Morvan, 2006).

The cell walls are composed of microfibrils interconnected by hemicelluloses. Other polymers such as pectin and lignin fill up the space between cellulosic microfibrils and hemicelluloses. Cells are connected to neighboring cells by middle lamellae, the thin lignin-rich outermost layer of the cell wall (Taherzadeh and Jeihanipour, 2012). The primary cell wall is the outer layer containing cellulose microfibrils arranged in random. The secondary cell wall is differentiated into outer (S1), middle (S2), and inner (S3) layers (Figure 3.1). The thickness of each layer differs, but the S2 layer is the thickest and constitutes the major part of the plant cell wall (Taherzadeh

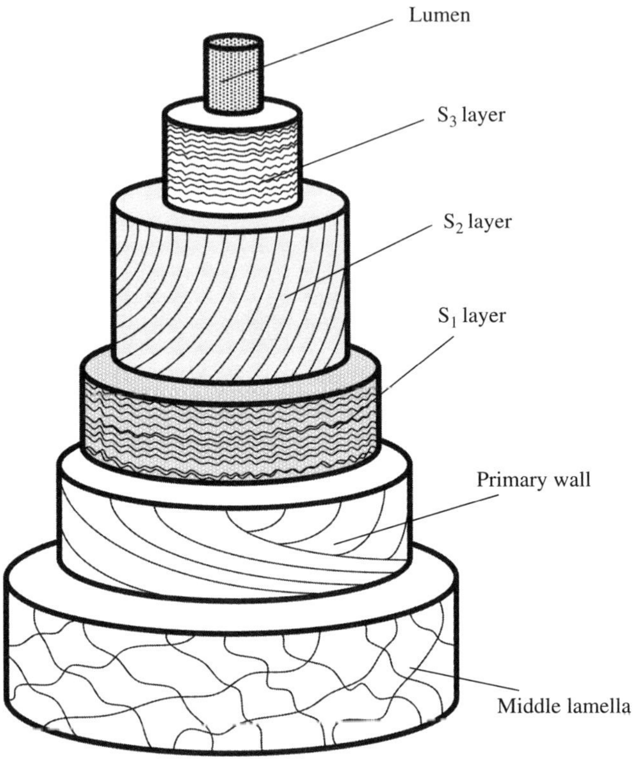

FIGURE 3.1 Cell wall of a wood fiber showing different layers.

and Jeihanipour, 2012). The cellulose microfibrils are arranged horizontally in the S1 layer, vertically in the S2 layer, and horizontally in the S3. The vertical orientation in the S2 layer provides tough mechanical and physical properties to the cell wall (Brett, 2000). Some cell walls have a tertiary wall (T) covering the secondary wall underneath the primary wall. Such a "T" layer is present in stem tissues of barley straw. The T-wall consists of randomly arranged cellulose microfibrils.

3.2.2 Proteins Present in the Plant Cell Wall

Proteins are present throughout the plant cell walls and comprise structural proteins, enzymes, signaling molecules, and wall-loosening proteins. These include hydroxyproline-rich glycoproteins, arabinogalactan proteins, glycine-rich proteins, proline-rich proteins, and chimeric proteins (Cassab, 1998; Harris and Stone, 2009). Chimeric proteins contain a high proportion of the amino acids hydroxyl proline, serine, and lysine (Brett and Waldron, 1996). Glycoproteins play a significant role in regulating the mechanochemical stability of the plant cell wall (Weng and Chapple, 2010).

3.2.3 Presence of Lignin in the Cell Wall of Plants

Lignin is an abundant phenolic macromolecule present in the secondary cell wall of the plants. Angiosperms, gymnosperms, monilophytes, isoetes, selaginella, lycopodium, zosterophylls, and rhyniophytes, all belong to tracheophytes, that is, higher plants containing lignin (Brett and Waldron, 1996; Weng and Chapple, 2010). Polymerization of lignin usually occurs through oxidative coupling, generated by three aromatic alcohols (Figure 3.2), including *p*-coumaryl alcohol, coniferyl alcohol, and sinapyl alcohol (Davin et al., 2009). The chemical composition of lignin varies between different taxa. For instance, in dicotyledonous gymnosperms the amount of guaiacyl and syringyl subunits are equal, while the guaiacyl subunits dominate in gymnosperms (Brett and Waldron, 1996).

Due to the nonenzymatic nature of polymerization in lignin, the linkage pattern is irregular (Brett and Waldron, 1996). Lignin continues to form as long as there is space in the cell wall. It fills all spaces in the cell wall and replaces water (Brett and Waldron, 1996). The lignin deposition in the cell wall is well planned and controlled. It starts at different rates in different cell corners (Davin et al., 2009). The effect

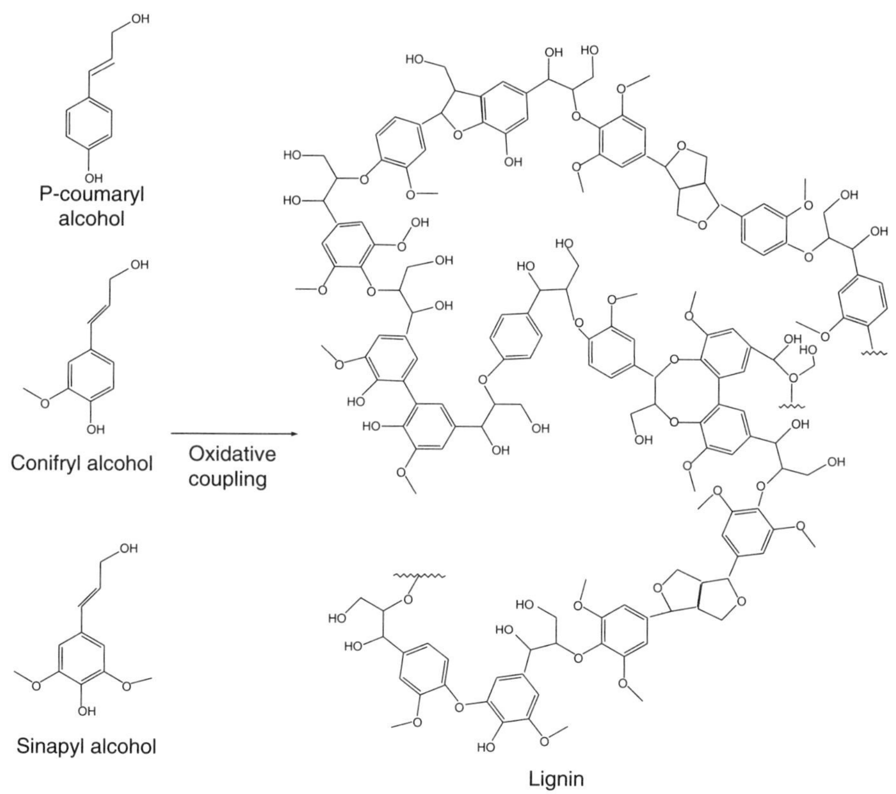

FIGURE 3.2 Three coupled aromatic alcohols form the lignin structure (Weng and Chapple, 2010).

of this deposition results in a strong hydrophobic structure packed with other cell wall components. Once the final structure is formed no further expansion is possible, thus stopping the growth (Brett and Waldron, 1996; Donaldson, 2001). Due to the incorporation of different components during polymerization, lignin functions as reinforcement in plants. It acts as a barricade in the cell wall blocking the penetration of nutrients and pathogens into the cell, which helps to protect against infection (Brett and Waldron, 1996; Davin et al., 2009; Kärkönen and Koutaniemi, 2010; Weng and Chapple, 2010).

3.2.4 Polymeric Interaction in the Plant Cell Wall

The cellulosic microfibrils and structural proteins are inserted into a gel matrix in the plant cell wall. These gel matrices are noncellulosic heteropolysaccharides, composed of pectin and hemicelluloses. Such heteropolysaccharides are capable of forming a polymeric gel, due to a linear backbone and a high solubility in water. Some regions in the primary cell wall are made of polysaccharides, containing intermolecular hydrogen bonding. Such non-covalent interactions in the gel matrix of the primary cell wall result in a thick three-dimensional layer being formed. The interaction between cellulosic microfibrils and noncellulosic polysaccharides is responsible for the cohesiveness of plant cell walls (Harris and Stone, 2009; Taherzadeh and Jeihanipour, 2012). Beside the non-covalent interactions between noncellulosic polysaccharides, matrix polymers, and cellulosic microfibrils many covalent bonds also exist between the cell wall components, that is, polysaccharide–polysaccharide, polysaccharide–lignin, polysaccharide–proteins, and lignin–protein. There are also indirect connections between polysaccharides, and between polysaccharides and lignin through covalent bridges (Cooper and Varner, 1984; Iiyama et al., 1994; Harris and Stone, 2009). Figure 3.3a displays the interactions between cellulose microfibrils, hemicelluloses, pectin, and structural proteins in plant cell walls.

Cross-linking between polysaccharide–polysaccharide units can occur in different forms such as ester-linked hydroxycinnamate dimers (Ralph et al., 2004), ferulic acid and/or *p*-coumaric acid monomers as cyclic dimers (Grabber et al., 2004), and direct linkage between polysaccharides and xyloglucans (Harris and Stone, 2009). Some important interactions between lignin and polysaccharides in the cell walls are the following:

(i) Ester–ether cross-links between hydroxycinnamate and heteroxylans etherified by hydroxyl groups of the monomeric lignin. This linkage is common in the secondary cell wall of grasses (Iiyama et al., 1994; Lam et al., 2001).
(ii) Direct ester linkage, formed when the hydroxyl group of the alcohols present in lignin are esterified by uronic acid on the polysaccharides (Iiyama et al., 1994).
(iii) Benzyl-ether linkages, formed between the hydroxyl groups present in polysaccharides and lignin monomers (Watanabe et al., 1989).

(iv) Phenyl glycoside linkages; the hydroxyl group on the side chain of lignin may be glycosylated by monosaccharides, oligosaccharides, or polysaccharides (Harris and Stone, 2009).

Cross-linking between polymers is the main reason for the limited accessibility of polysaccharides, while lignin replaces water, thereby increasing the occurrence of hydrogen bonds between hemicelluloses, polysaccharides, and cellulosic microfibrils (Donaldson, 2001).

3.2.5 Lignocellulosic Biomass Recalcitrance

Plant cell walls contain cellulose, which is synthesized in the plasma membrane and hemicelluloses, synthesized, and secreted from the Golgi apparatus (Ding and Himmel, 2006). The cellulose microfibril typically contains 36 hydrogen-bonded cellulose chains arranged parallel to one another (Somerville, 2006). However, recent studies revealed this to be a rough approximation, showing that the number of hydrogen-bonded cellulose chains may exceed 36 (Herth, 1983; Fernandes et al., 2011). In most plant cell walls, the synthesis of cellulosic microfibrils is enhanced

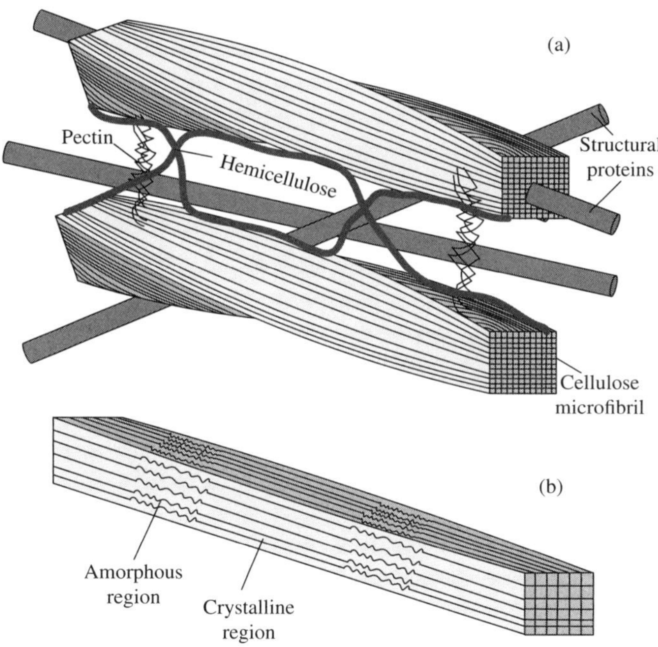

FIGURE 3.3 (a) Interactions between cellulose microfibrils, hemicelluloses, pectin, and structural proteins in plant cells, with (b) showing formation of an amorphous region from untwisted cellulose microfibrils, when completely dried. The crystalline region becomes fragile, turns amorphous, and further degradation occurs.

by a catalytic enzyme called cellulose synthase (Brown, 2004). This enzyme possesses many active sites catalyzing glucan polymerization resulting in aggregation of cellulose microfibrils (Mohnen et al., 2009).

3.2.5.1 *Cellulose morphologies* Cellulose can take various crystalline forms, due to its polymorphism (O'Sullivan, 1997). Crystalline cellulose occurs in four different allomorphs, namely Cellulose I, II, III, and IV (Kontturi, 2005). In nature, Cellulose I is formed comprising two crystal phases, namely monoclinic I_β and triclinic I_α (Atalla and Vanderhart, 1984). The crystallographic structure of the monoclinic and triclinic crystal phases is shown in Figure 3.4. The monoclinic cellulose is organized into two parallel chains packed as sheets in the upward direction, while the triclinic structure constitutes a single chain, while glucosyl linkage and hydroxymethyl groups occur in both structures. Although Cellulose I_β possesses a high crystallinity, highly unstructured intermolecular hydrogen bonds maintain the cellulose chains as sheets (Nishiyama et al., 2002, 2003). I_α and I_β form flat sheets, packed on top of each other. Due to this arrangement, an interaction occurs within the crystals, but without bonding (Nishiyama et al., 2002, 2003). There are three types of such interaction: (1) intrachain O–H–O bonds between glucose residues of the same chain and sheet; (2) interchain O–H–O bonds between glucose residues of the same sheet but different chains; (3) intersheet C–H–O "bonds" in terms of van der Waals interactions between neighboring sheets (Gross and Chu, 2010). The high resistance of cellulose toward biological and chemical hydrolysis is due to these complex interactions (Himmel et al., 2007).

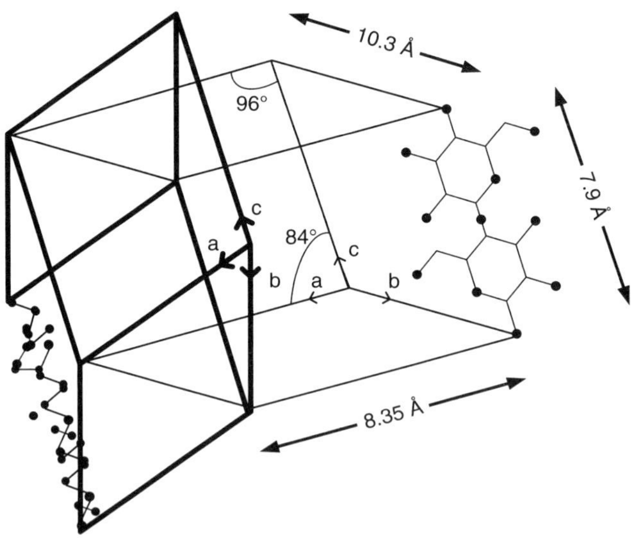

FIGURE 3.4 Arrangement of the monoclinic (thin lines) and triclinic (bold lines) structures in a single unit cell. The orientation of cellulose in the monoclinic phase, and its bond angle and bond distance (Sugiyama et al., 1991; El Oudiani et al., 2011).

Compared to triclinic I_α, monoclinic I_β is a more stable structure. The characterization of the interactions within cellulose microfibrils achieved by molecular simulations shows that the strongest and the most robust interactions are those between the sheets. Furthermore, intersheet hydrogen bonds are more abundant than intrachain and interchain hydrogen bonds per glucose molecule. The stability of monoclinic phases in comparison with triclinic phases, is due to the presence of these two extra intersheet hydrogen bonds in I_α (Gross and Chu, 2010). Heating at 260°C in the presence of alkali or heating at 280°C in the presence of inert atmosphere, for example, helium, transforms the less stable triclinic crystals into the more stable monoclinic phase (Hardy and Sarko, 1996; Wada, 2002). Furthermore, the monoclinic phase is the most abundant structure in the plant cell wall (Atalla and Vanderhart, 1984). The presence of carboxymethyl cellulose or xyloglucan decreases the proportion of the triclinic phase. However, at low temperatures, the amount of triclinic celluloses in plants increases. In the primary cell wall of apples and *Arabidopsis thaliana*, the amount of monoclinic and triclinic phases are equal (Yamamoto and Horn, 1994; Smith et al., 1998). The total amount of polysaccharides remains constant during cell division, increasing rapidly during the pre-enlargement period, while being constant during the enlargement, when partial interconversion, or partial break down of polysaccharides has been observed quite often (Yamaki et al., 1979; Labavitch, 1981).

The second allomorph of cellulose, "Cellulose II" is formed from Cellulose I by (a) the dissolution of Cellulose I in a Cellulose solvent followed by precipitation, using an antisolvent, (b) swelling Cellulose I fibers by using concentrated alkali-like sodium hydroxide, removing the excess swelling agent. Cellulose II is composed of antiparallel chains and the unit cell is monoclinic. The paired cellulose molecules can be rotated around their axes by 180 degrees (O'Sullivan, 1997; Kim et al., 2006). Cellulose III is obtained from Celluloses I and II by liquid ammonia treatment. Further, Cellulose IV is acquired from Cellulose III by heating; this transformation is however incomplete (Kontturi, 2005).

3.2.5.2 Twisting Cellulose Microfibrils There is evidence from studies by microscopy that cellulose microfibrils and nanofibrils are twisted in their native state (Hanley et al., 1997). In higher plants, a long helical twist is formed in the cellulose microfibrils (Figure 3.3a), and the distance between each twist depends on the lateral dimensions of the microfibrils. For instance, microfibrils in higher plants are only 3–5 nm, while they measure 6–7 nm in bacterial cellulose. An increase in the lateral dimension increases the resistance of the microfibril toward a twisted helix. This might be the reason for plants with load-bearing ability to have small dimensions, while large lateral dimensions are found in plants like cotton (Atalla et al., 2008).

Cellulose oligomers containing less than ten monomers are water-soluble (Atalla et al., 2008). It is expected that the cellulose chains are hydrated at the elementary nanofibril level and that cellulose processing at high temperature reduces the hydration. Thus, an increase in temperature changes the state of aggregation of native celluloses. Water aids in the relative motion of nanofibrils, while removal of water causes dryness and stimulates formation of hydrogen bonds between nanofibrils. This cellulose dehydration, whether it is by increasing the temperature or by drying, leads to formation of tight aggregates and more recalcitrant native cellulose. During the isolation of cellulose from plant cell walls, heating and drying are two methods that affect the final aggregation pattern of cellulose fibrils (Atalla et al., 2008; Taherzadeh and Jeihanipour, 2012).

3.3 CHALLENGES IN PRETREATMENT OF LIGNOCELLULOSES

Lignocelluloses are tough by nature and unlocking the structure by a pretreatment is obligatory to be able to use its carbohydrates for the production of various bioproducts. Pretreatment needs to be effective in opening up the structure of lignocelluloses. During the past decades, several methods of pretreatment have been developed, and are still developing. However, which are the challenges facing lignocelluloses during pretreatment? How is the composition of lignocelluloses changing during growth? Why and how does it change? This section focuses on unveiling the mystery of growth and development of the tough lignocelluloses.

Factors affecting the release of sugars during hydrolysis of cellulose are, for example, cellulose crystallinity, the degree of polymerization (DP), the surface area of the substrate, the content of lignin and hemicelluloses, particle size, porosity, and thickness of cell wall (Alvira et al., 2010). Understanding the effects of various pretreatments on these factors is significantly important. Lignocelluloses vary in this respect. It should be noticed that different plant parts possess different characteristics, and these may change during growth, or by harvesting, drying, grinding, etc. The factors listed above have mostly been ignored in research studies and there is a general presumption that the lignocellulose of a species respond similarly to pretreatment regardless of developmental stage.

The composition of the plant cell wall of cotton fiber during early stages of elongation has been compared with the cell wall composition of a full-grown plant (Meinert and Delmer, 1977). The thickness of the cell wall remains constant up to 12 days of photosynthesis. Between 12 and 16 days, all components of the cell wall grow steeply. The secondary cell wall starts to grow from 16 days and continues until 32 days after which time the growth stops. This is followed by a sharp decline in protein and uronic acid contents. The primary cell wall of cotton fibers has been reported to be similarly structured in other dicotyledons and gymnosperms. Even though the composition of the cell wall changes throughout the growth, the average composition of the primary cell wall remains unaltered. The DP in the primary cell wall is low and heterogeneous in the range of 2000–6000. However, after the secondary cell wall stops growing, the DP increases to 14,000 and turns homogenous. Similar studies were also performed on the maize cell wall (Carpita, 1984), showing that in the embryonic stage the cell wall mainly consists of glucuronoarabinoxylan (GAX), xyloglucan, and polymers enriched in 5-arbinosyl linkages. During the subsequent growth, GAX, and xyloglucan continue to increase, but not the arabinosyl polymers. This study further confirmed that the composition of the plant cell wall changes during growth.

Apart from the factors mentioned above formation of inhibitors during pretreatment, for example, furfural and 5-hydroxymethylfurfural (HMF), carboxylic acids, ketones, phenolics, and aldehydes, affects the subsequent processes. The chemistry and the effect of these inhibitors are well studied (Taylor et al., 2012). Effective pretreatment methods to reduce the inhibitor concentration, and to improve the sugar yields, are presently being researched.

3.4 PRETREATMENT METHODS AND MECHANISMS

Pretreatment is carried out with the purpose of modifying or altering the structural arrangements of the complex lignocellulosic materials, in order to improve the accessibility of the enzymes for cellulose digestion. The main purpose of the pretreatment is consequently to remove the coverage of the cellulose, for example, lignin, and to break the crystalline structure, opening it up for hydrolysis. Various pretreatment methods for improving enzyme accessibility have been developed in the last decades. Nevertheless, pretreatment is still conceived as the second most expensive process unit on an industrial scale (Alvira et al., 2010). A global summary of the different pretreatment methods developed is presented in Figure 3.5. In this section, the methods are briefly mentioned and compared.

3.4.1 Physical Pretreatment Methods

The general objective of this pretreatment category is to reduce particle size and crystallinity of lignocelluloses, thus increasing surface area accessibility, and decreasing DP (Alvira et al., 2010). Milling and irradiation are widely used physical pretreatments for lignocelluloses. Other physical methods are extrusion, pyrolysis, and high-pressure steaming. Extrusion is a combination of thermo-chemical and chemical pretreatment, which can be used at high throughput at high concentrations of lignocellulosic biomass (Choi and Oh, 2012). In pyrolysis, the materials are treated anaerobically at higher temperatures. The high-temperature process paves the way for other pretreatment methods such as dilute acid for further hydrolysis (Sun and Cheng, 2002).

3.4.1.1 Milling Milling disrupts the structure of the lignocelluloses by altering the degree of crystallinity and reducing the size of the material (Fan et al., 1980; Mais et al., 2002; Zeng et al., 2007). Milling can be applied prior to enzymatic hydrolysis or in combination with other pretreatment methods, such as dilute acid hydrolysis, steam, and ammonia (Tassinari and Macy, 1977; Zhang and Zhang, 1999; Mais et al., 2002; Muller et al., 2007; Taherzadeh and Karimi, 2008). Hammer milling is a favored milling process for grinding waste paper. Different milling processes are used based on the material and its nature. For instance, colloid mill, fibrillator, and dissolver are compatible with wet materials, while extruder, roller mill, cryogenic mill, and hammer mill are designed for dry materials. Ball milling can be used for both dry and wet materials (Walpot, 1986). Milling is an energy-demanding process. For instance,

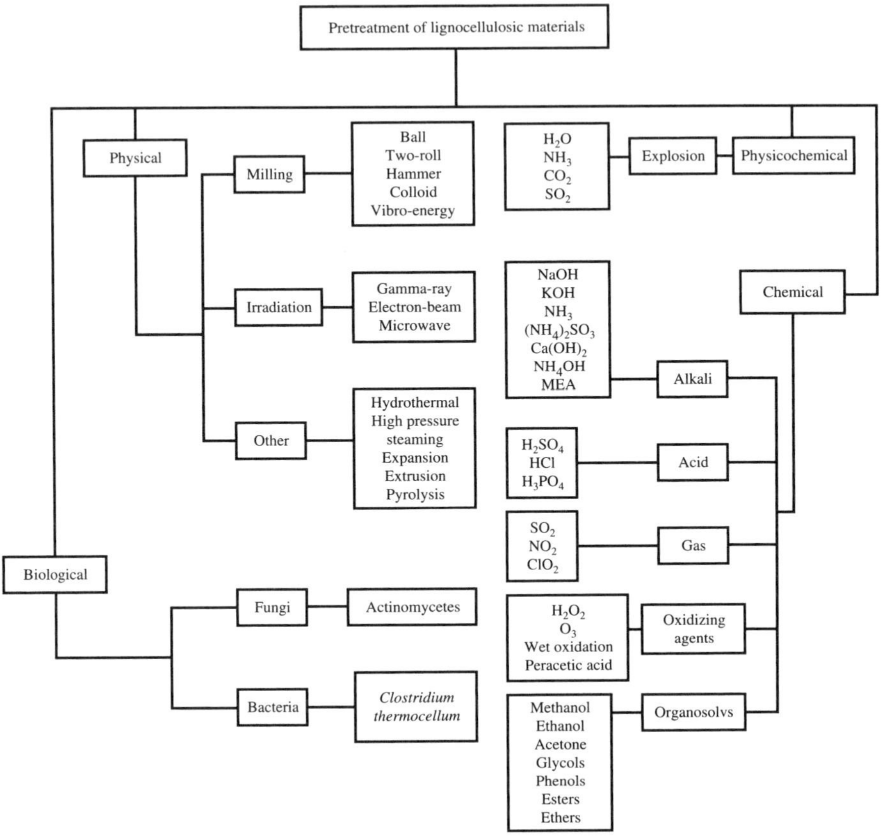

FIGURE 3.5 Schematic overview of different methods for pretreatment of lignocellulosic materials.

the power consumption of a 7 m long mill with a 5 m diameter would be in the range of 2.5–3.5 MW (Cleary, 2001). Milling does not change the composition of cellulose and hemicelluloses or the lignin content. It does, however, significantly increase the surface area with which the enzymes react during hydrolysis, while the degree of cellulose polymerization is also significantly reduced (Zhao et al., 2012b).

Acid hydrolysis of ball-mill-pretreated straw resulted in more than 50% of the cellulose recovery with the least loss of glucose, which was due to the crystallinity reduction by the ball mill (Sidiras and Koukios, 1989). Ball milling reduces the crystallinity index of cellulose by more than 70% (Fan et al., 1980). However, the number of ball beads in the milling process turns out to be an important factor in the efforts of improving enzymatic hydrolysis (Mais et al., 2002). In a study on superfine grinding (Jin and Chen, 2006), rice straw was cut into 5 to 8 cm pieces and steam exploded at 180–220°C for 4–5 minutes, then crushed until superfine using a fluidized-bed-opposed jet mill. The superfine material resulted in a high hydrolytic rate and a high yield of reducing sugars. The disadvantage of the milling pretreatment is its inability to remove lignin that restricts the enzymes' access to cellulose (Mooney et al., 1999; Berlin et al., 2006).

3.4.1.2 Irradiation Irradiation is another physical method for pretreatment of lignocelluloses. Irradiation is performed by means of gamma radiation, ultrasound, electron beam, and microwaves (Zhao et al., 2012b). Irradiation enhances the enzymatic degradation of cellulose to glucose. The cellulose fraction of lignocelluloses is degraded by irradiation to breakable fibers, low molecular oligosaccharides, or even to cellobiose (Kumakura and Kaetsu, 1983). The irradiation leads to dissociation of the glucoside bonds of cellulose resulting in low molecular compounds. Irradiation higher than 100 MR can result in glucose and small oligosaccharides being decomposed. Lignin is removed only to a limited extent by irradiation. Notwithstanding, the described methods are still expensive and are facing challenges in the application on an industrial scale (Kumakura and Kaetsu, 1978, 1983, 1984; Kumakura et al., 1982).

In ultrasound, one of the irradiation pretreatment types, the disintegration of lignocelluloses is dependent on the density and intensity of the ultrasonic waves (McDermott et al., 2001). This pretreatment might be the irradiation method closest to commercialization; ultrasounds for improving biogas production from grasses are now available in the market. Microwave pretreatment has thermal as well as nonthermal effects and is carried out by immersing biomass in dilute chemical agents, subsequently exposing it to microwaves for 5–20 minutes. Alkalis are the most suitable chemical agents, particularly sodium hydroxide (Alvira et al., 2010).

3.4.1.3 Others Other physical methods, well functioning as pretreatments of lignocelluloses are high-pressure steam, extrusion, pyrolysis, and electrical pulses (Lafitte-Trouque and Forster, 2002; Kim et al., 2003). Extrusion is a novel pretreatment method and is a combination of heating, shearing, and mixing. The combination of these different actions results in the lignocelluloses being physically as well as chemically modified. Ongoing research on pretreatment methods is exploring the potential of combining enzymatic hydrolysis with extrusion (Karunanithy et al., 2008; Choi and Oh, 2012). Pyrolysis may also be employed for pretreatment of lignocelluloses. In the pyrolysis pretreatment, the lignocelluloses are treated at high temperatures, up to 300°C, and if followed by dilute acid hydrolysis, this pretreatment yields 80–85% reducing sugars (Sun and Cheng, 2002).

In the high-pressure steam pretreatment, saturated steam in the temperature range 160–240°C and at a pressure of 6–34 bar is applied on lignocelluloses for several seconds, up to a few minutes. During this pretreatment, most of the hemicelluloses are in the liquid phase disintegrated into solubilized oligomeric and monomeric sugars. Cellulose present in the solid phase very easily becomes accessible to enzymes during hydrolysis. Addition of sulfuric acid or sulfur dioxide further stimulates the high-pressure steam pretreatment by catalysis. The addition of acids promotes the solubilization of lignocelluloses (Galbe and Zacchi, 2007).

3.4.2 Chemical and Physicochemical Methods

3.4.2.1 Steam and Other Explosions Explosions constitute combinations of mechanical and chemical pretreatment methods. Explosion pretreatments can be performed with steam (autohydrolysis), SO_2, NH_3, and CO_2. Fundamentally, the explosion pretreatment means heating up lignocellulose slurry under high pressure followed by sudden release of the pressure. This causes the fibers to separate, due to explosive decompression. No chemical needs to be added in the procedure, since acetic acid is released from hemicellulose at high temperature, facilitating the process (Taherzadeh and Karimi, 2008; Alvira et al., 2010). Steam explosion is therefore a very popular method for pretreatment. Steam explosion is usually carried out at 160–260°C and the operating time is 0.5–20 minutes (Taherzadeh and Karimi, 2008). Steam explosion pretreatment has nearly reached commercialization; pilot plants are presently operating in Sweden and Canada (Galbe and Zacchi, 2007). Steam explosion pretreatment of poplar wood at 210°C for 4 minutes resulted in 95% cellulose recovery (Taherzadeh and Karimi, 2008).

Supercritical CO_2 is used for carbon dioxide explosion pretreatment. CO_2 is cheap, nontoxic, inflammable, and easy to extract after explosion (Taherzadeh and Karimi, 2008). Due to the release of carbon dioxide at high pressure, lignocelluloses are disturbed, which increases the surface area for further hydrolysis. Glucose release was observed to increase with increasing pressure and temperature of the carbon dioxide was applied in supercritical carbon dioxide explosion. However, using subcritical carbon dioxide results in opposite scenario.

The ammonia fiber explosion (AFEX) process uses a combination of temperature (60–100°C), high pressure, and ammonia. A SEM analysis revealed that the principal target points during the AFEX process are the fibrous regions. However, some intermediates are formed during the process, for example, phenolic acids, aldehydes (from cleaving lignin–carbohydrates complexes), short-chain oligosaccharides, and organic acids. These compounds act as inhibitors in the subsequent processes. Figure 3.6 shows the schematic representation of an explosion process. Advantages of the AFEX process usually mentioned are high glucose yield upon enzymatic hydrolysis after the pretreatment, comprehensive recovery of the chemical used (ammonia), and no need to wash the materials after the pretreatment since the leftover ammonia can be used as a nutrient for microbial growth (Balan et al., 2009).

3.4.2.2 Dilute and Concentrated Acids Acid pretreatment is probably the most widely used method for pretreating lignocelluloses. It is aimed at dissolving the hemicellulose fraction, hence breaking the bonds between lignin, cellulose, and hemicellulose. The operating temperature usually varies between 30°C and 220°C, and the retention time is 5–90 minutes. Hydrochloric, sulfuric, nitric, and phosphorus acids are commonly used for acid-based pretreatments with sulfuric acid dominating (Taherzadeh and Karimi, 2007a). Apart from these acids, fumaric and maleic acids have been tested as well. Organic acids are more effective than sulfuric acid in pretreatment of wheat straw (Alvira et al., 2010). The acid pretreatment can be

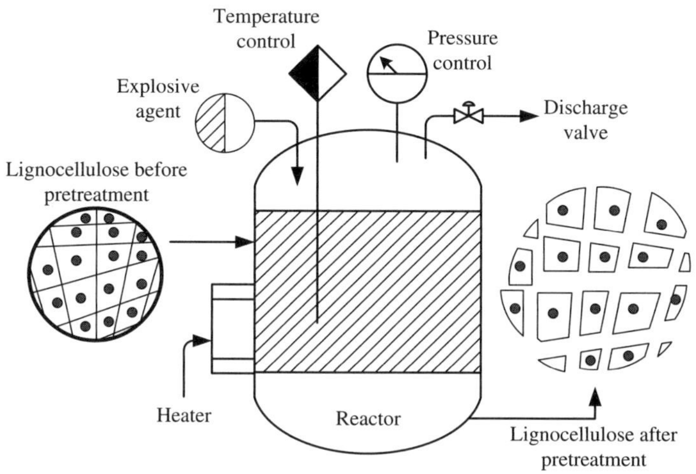

FIGURE 3.6 Schematic representation of an explosion pretreatment process, in which the explosive agents may be steam, ammonia, sulfur dioxide, and carbon dioxide.

performed in two modes: (i) low temperature and high acid concentration and (ii) high temperature and dilute acid.

Concentrated acid pretreatment is effective in opening up the structure of lignocelluloses at low temperature. In a study, combining acid and alkaline pretreatments (Chaudhary et al., 2012), lignocelluloses were treated with 30–70% sulfuric acid at 100°C for 1 hour followed by neutralization with calcium hydroxide resulting in up to 85% recovery of the monomeric sugars in the lignocellulosic materials. In another study (Jafari et al., 2011), pretreatment with 85.9% phosphoric acid resulted in 87–95% enzymatic hydrolysis of the cellulose. Phosphoric acid facilitates removal of most of the lignin. Phosphoric acid reacts with hemicelluloses and lignin, with significant reduction as a result. Concentrated phosphoric acid as pretreatment furthermore converts crystalline cellulose into amorphous form (Zhao et al., 2012b). However, the disadvantage of concentrated acid pretreatment is high investment and maintenance costs. With dilute acid pretreatment almost 100% of hemicelluloses are removed. Lignin is not dissolved as part of the acid pretreatment. However, dilute acid disrupts lignin, leading to increased susceptibility of the cellulose to enzymatic hydrolysis (Wyman, 1996; Yang and Wyman, 2004). During mild acid pretreatment, most of the hemicelluloses are removed (<120°C). Hemicelluloses are degraded to monosaccharides and the sugars ultimately to HMF and furfural (Zhao et al., 2012b).

Dilute acid pretreatment opens up the lignocelluloses structure as well as enhances the hydrolysis process, if right conditions are employed. It is a very popular method in which a very low concentration of acid (e.g., 0.1–1.0%) is used. Much research has explored the use of acid pretreatments of different lignocellulosic materials, focusing on chemical composition of the materials, operating time, concentration of acid, temperature, etc., searching for optimal conditions for recovering most of the sugars (Ma et al., 2010; Qi et al., 2010; Chen et al., 2011, 2012; Wei et al., 2011; Thirmal and Dahman, 2012). Two major drawbacks of acid pretreatment are its inability to reduce the crystallinity of cellulose and the formation of inhibitors, such as carboxylic acids, furans, and phenolic compounds, caused by the low pH (Taherzadeh, 1999; Taherzadeh and Karimi, 2007b). These shortcomings result in lower sugar yields, and a lower subsequent fermentation. Other challenges are equipment corrosion, and high operational and maintenance costs (Alvira et al., 2010).

Dilute acid pretreatment may be combined with other chemical pretreatments (Ma et al., 2010; Chen et al., 2011). For instance, Azzam pretreated bagasse with 0.5% HCl in a solution of zinc chloride for 10 minutes at 145°C. The pretreated bagasse was then cooled and precipitated with acetone, resulting in a yield of more than 93% (Azzam, 1987).

3.4.2.3 Alkali Sodium, potassium, calcium, and ammonium hydroxides are commonly used for alkaline pretreatment of lignocelluloses. NaOH causes swelling, which increases the internal surface, and decreases the DP of the cellulose. At concentrations above 10%, and at freezing temperature NaOH completely dissolves cellulose (Mirahmadi et al., 2010), while acting as an effective delignifier agent at a higher temperature. The intermolecular ester bonds, cross-linking xylan hemicelluloses, and lignin are saponified during alkali pretreatment, thus resulting in delignification of the biomass. Cleavage of the intramolecular α-aryl and β-aryl ether linkages is what essentially contributes to the lignin degradation. Alkali pretreatment also removes acetyl groups and uronic acid from the hemicelluloses. Generally, alkalis are good swelling agents, reducing cellulose crystallinity, which increases the accessibility of cellulose (Zhao et al., 2012b).

Calcium hydroxide or lime, removes amorphous substances such as lignin, from cellulose. The removal of lignin extends the adsorption site for enzymes, hence increasing the accessibility of cellulose. Addition of an oxidant agent (O_2 or H_2O_2) improves the lignin removal (Teghammar et al., 2010). Alkaline pretreatment is more effective on agricultural residues, than on woody substances (Taherzadeh and Karimi, 2008). Lime pretreatment can be performed as (i) short-term treatment (up to 6 hours) at 100–160°C at about 13 bar, with or without oxygen, (ii) long-term treatment (up to 8 weeks) at 55–65°C at atmospheric pressure, and (iii) simple pretreatment involving, for example, 1 hour in boiling water, with or without oxygen. Higher lignin contents are removed under oxidative conditions, at a pressure ranging from atmospheric to 20 bar (Sierra et al., 2009).

3.4.2.4 Oxidizing Agents Oxidizing agents such as ozone (Alvira et al., 2010), hydrogen peroxide, and oxygen, can be used for pretreatment of lignocelluloses. This pretreatment is usually combined with other chemical or hydrothermal treatments. In ozonolysis, degradation of aromatic and olefinic structures involves initial electrophilic attack by oxidants. In the oxidation by hydrogen peroxide, these structures are instead destroyed by a nucleophilic attack of hydrogen peroxide anions (Zhao et al., 2012b).

Ozonolysis pretreatment is an effective method and is specifically used for degradation of lignin and partial degradation of hemicelluloses. It is mostly carried out at room temperature. Prominent advantages of ozonolysis are that no inhibitory compounds are formed, and also no acids, alkali, or toxic compounds, making it a very attractive choice of pretreatment (Neely, 1984; Vidal and Molinier, 1988; Sun and Cheng, 2002). It is quite expensive to buy a large volume of ozone, which is a challenge to its feasibility. Important parameters to consider in ozonolysis pretreatment are moisture content of sample, particle size, and concentration of ozone.

In wet oxidation pretreatment, the material is treated with water using air or oxygen as oxidizing agents. The operating temperature is usually in the range of 150–200°C, with residence time, approximately 30 minutes. Temperature, reaction time, and the oxygen pressure are considered to be the crucial parameters in this pretreatment method (Schmidt and Thomsen, 1998). Wet oxidation attacks all fractions of the lignocelluloses. Hemicelluloses are decomposed into monomeric sugars. Lignin is both oxidized and cleaved, and cellulose is degraded to a certain extent. The main reactions in wet oxidation pretreatment comprise formation of acids and oxidative reactions.

3.4.2.5 Organosolvs Using organosolvs is another pretreatment means for lignocellulosic feedstock, in which biomass is treated with organic solvents or an aqueous solution of organosolv usually at 100–250°C and with or without a catalyst. A major function of organosolvs is complete removal of lignin and hemicelluloses, resulting in a highly accessible surface area, and large pore volumes (Zhao et al., 2012b). Alcohols, ketones, glycols, organic acids, phenols, esters, and ethers are the different organic solvents explored to date for organosolv pretreatments. Acetone is a widely used organic solvent for pretreatment of lignocelluloses. The pretreatment of D. Don chips with an acetone–water solution for 5 minutes at 195°C resulted in an ethanol yield of 99.5% of the theoretical (Araque et al., 2008).

Use of the organic solvent *N*-methylmorpholine-*N*-oxide (NMMO) for pretreatment of lignocelluloses has received increasing attention (Zhao et al., 2009). This chemical is able to open up the crystalline structure of cellulose, resulting in more than 99% sugar yield in the subsequent enzymatic hydrolysis (Jeihanipour and Taherzadeh, 2009). In the pretreatment with this organosolv, lignocelluloses are mixed with organic solvents and heated. Part of the lignin and the hemicelluloses are dissolved in the liquid phase, leaving cellulose in the solid phase (Taherzadeh and Karimi, 2008). A catalyst, for example dilute sulfuric acid, oxalic, salicylic, and acetylsalicylic acid, may be added to improve delignification or to decrease the operating temperature (Taherzadeh and Karimi, 2008; Huijgen et al., 2011). Removal of the solvent after the pretreatment is required as the solvent might inhibit the enzymatic hydrolysis (Sun and Cheng, 2002). Recovery and recycling of the solvent is also necessary when using this pretreatment, or the process will be too expensive (Zhao et al., 2009).

Peracetic acid (PAA) is a powerful oxidizing agent which can be used for pretreating lignocelluloses. This pretreatment can be accomplished at low temperature

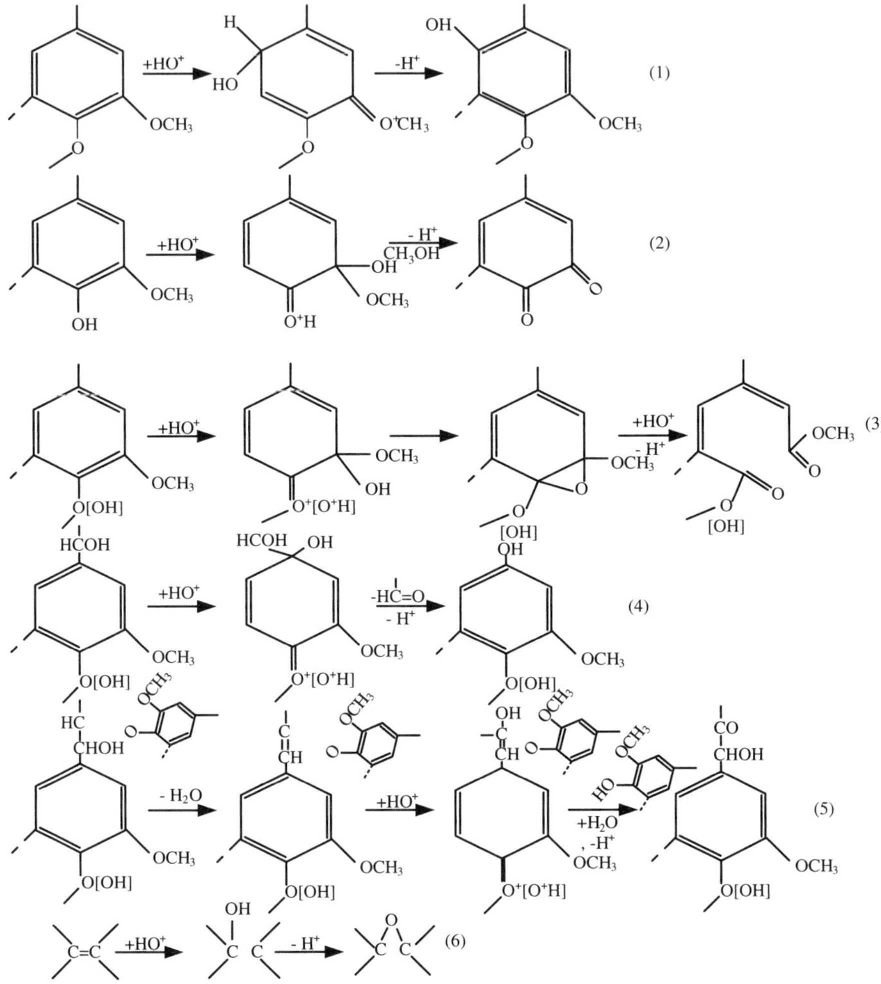

FIGURE 3.7 Mechanisms of organosolv peracetic acid pretreatment (Zhao et al., 2009).

and atmospheric pressure. The mechanism of the PAA pretreatment is most likely that hydroxonium ions (H_3O^+), present in PAA, attack the lignin electron sites electrophilicly, resulting in the breakdown of lignin. The main reactions of PAA pretreatment are shown in Figure 3.7, and include (i) ring hydroxylation, (ii) oxidative demethylation, (iii) oxidative ring opening, (iv) displacement of side chains, (v) cleavage of β-aryl ether bonds, and (vi) epoxidation. Lignin is then disordered and dissolved in the liquid phase. Delignification increases the surface area of cellulose and, as a consequence, its accessibility (Suchy and Argyropoulos, 2001; Zhao et al., 2009). The effectiveness of various pretreatment methods described above on enzymatic hydrolysis of different lignocellulosic substrates is shown in Table 3.2.

TABLE 3.2 Comparison of Enzymatic Hydrolysis for Different Biomass With and Without Pretreatment

Lignocellulose biomass	Type of pretreatment	Enzymatic hydrolysis without pretreatment (%)	Enzymatic hydrolysis with pretreatment (%)	References
Softwood spruce	Alkali (NaOH)	14.1	35.7	Mirahmadi et al. (2010)
Harwood birch	Alkali (NaOH)	6.9	82.3	Mirahmadi et al. (2010)
Corn stover	Dilute acid	15.1	60.6	Lloyd and Wyman (2005)
Corn stover	Steam explosion	60	80	Öhgren et al. (2007)
Poplar	Organosolv	8	36–41	Chum et al. (1988)
White pine	Sulfuric acid	4	40	Grethlein et al. (2004)

3.4.2.6 Ionic Liquids Ionic liquids (ILs) are green solvents, since they contain no toxics or explosives. They are generally salts, existing as liquids below 100°C, and contain large organic cations and small inorganic anions (Zhu et al., 2006; Alvira et al., 2010). Cations and anions play different significant roles during pretreatment. Cations interact with lignin by hydrogen bonding and π–π interactions, while anions act as hydrogen bond receptors, interacting with the hydroxyl group of the cellulose. As a consequence, the crystalline structure of the lignocelluloses is disrupted (Fukaya et al., 2008; Janesko, 2011; Wu et al., 2011), resulting in a very effective pretreatment. ILs convert the crystalline form of cellulose into an amorphous form partially removing hemicelluloses and lignin (Zhao et al., 2012b). After the pretreatment, the ILs can be recovered and reused, and different methods such as evaporation, salting out, pervaporation, ion exchange, and reverse osmosis, can be employed for this purpose (Zhu et al., 2006). However, using ILs for pretreatment is still expensive that provides a hurdle for its industrialization (Zhu et al., 2006; FitzPatrick, 2011; Mora-Pale et al., 2011; Wu et al., 2011). Different ILs and their specific targets for pretreatment action are summarized in Table 3.3.

3.4.3 Biological Methods

Although biological degradation of wood has been known for a considerable time, biological pretreatment methods have been put in focus only in recent years. These methods work at ambient temperature, demand no chemicals, and are environmentally friendly. Biological pretreatment can be performed by fungi, and delignification occurs by means of ligninolytic enzymes such as lignin peroxidase, manganese peroxidase, and laccase (Isroi et al., 2011).

TABLE 3.3 Ionic Liquids (ILs) used for Pretreatment of Lignocelluloses, Showing the General Formula, With the R and X for Each IL (Mora-Pale et al., 2011)

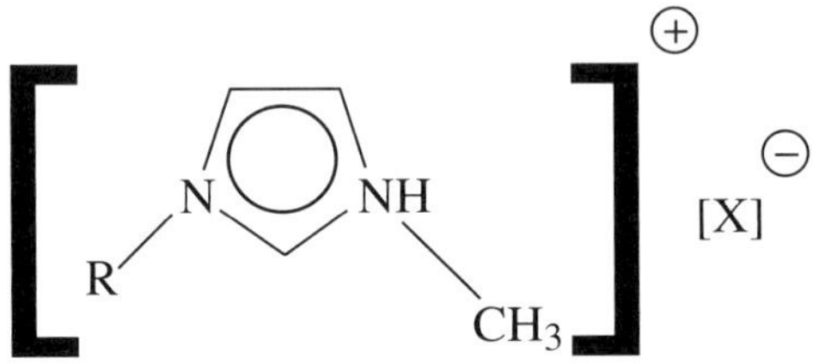

Ionic liquids	R	X
1-Ethyl-3-methylimidazolium acetate [EMIM]OAc	Ethyl	Acetate
1-Ethyl-3-methylimidazolium chloride [EMIM]Cl	Ethyl	Chloride
1-Allyl-3-methylimidazolium chloride [AMIM]Cl	Allyl	Chloride
1-Butyl-3-methylimidazolim chloride [BMIM]Cl	Butyl	Chloride
1-Butyl-3-methylimidazolium bromide [BMIM]Br	Butyl	Bromide
1-Butyl-3-methylimidazolium acetate [BMIM]OAc	Butyl	Acetate
1-Butyl-3-methylimidazolium tetrafluoroborate [BMIM] BF_4	Butyl	Tetrafluoroborate
1-Butyl-3-methylimidazolium hexafluorophosphate [BMIM] PF_6	Butyl	Hexafluorophosphate
1-Butyl-3-methylimidazolium methylsulfate [BMIM] $MESO_4$	Butyl	Methylsulfate
1-Octyl-3-methylimidazolium chloride [OMIM] Cl	Octyl	Chloride

3.4.3.1 Fungi White- and soft-rot fungi are able to degrade lignin and hemicelluloses, leaving cellulose almost intact. Some white-rot fungi, such as *Phanerochaete chrysosporium, Ceriporia lacerata, Cyathus stercolerus, Ceriporiopsis subvermispora, Pycnoporus cinnabarinus*, and *Pleurotus ostreatus* show high efficiency in terms of removing lignin (Alvira et al., 2010; Isroi et al., 2011). These fungi mainly produce ligninolytic enzymes and H_2O_2, when entering the Fenton reaction results in release of free radicals. The radicals attack the lignin walls, and they split open, allowing the enzymes to perform. White-rot fungi degrade from the lumen and inwards, toward the secondary wall, and finally at the middle lamella. Oxidative reactions cause depolymerization of lignin. The final products would be carbon dioxide and water (Yu et al., 2009; Ma et al., 2010; Yunqin et al., 2010; Isroi et al., 2011; Zhao et al., 2012b).

The biodegradation of lignocelluloses by fungi is associated with their mycelial growth. This habit allows fungi to transfer rare nutrients such as nitrogen and iron to a nutrient-poor lignocellulosic substrate that constitutes carbon source. Fungal degradation occurs extracellularly by association with outer cell envelope layer. There are two types of extracellular enzymatic systems: (a) hydrolytic system and (b) ligninolytic system. In hydrolytic system, hydrolases are produced for degrading polysaccharides, while ligninolytic system opens phenyl rings and degrades lignin. *Trichoderma reesei* is one of the widely studied fungi used in commercial production of cellulases and hemicellulases (Sánchez, 2009).

Hatakka (1983) compared 19 white-rot fungi for the pretreatment of wheat straw. The results show that *Pleurotus ostreatus* could convert 35% of straw to reducing sugars in 5 weeks. Similar results were also obtained by other white-rot fungi including *Phanerochaete sordid* and *Pycnoporus cinnabarinus* in 4 weeks (Okano et al., 2005). Zhang et al. (2007) compared 34 isolates of white-rot fungi for the pretreatment of bamboo for improving the enzymatic hydrolysis. *Echinodontium taxodii 2538* and *Trametes versicolor G20* were selected for further pretreatment due to high lignin loss (>20%) and high selectivity value of lignin degradation (>2) after 4 weeks. After 120 days of pretreatment, the sugar yield from bamboo culms increased 5.15- and 8.76-fold for *T. versicolor G20* and *E. taxodii 2538*, respectively.

Loss of cellulose during pretreatment by fungi is always a concern. In order to prevent this loss, a mutant strain was developed to degrade lignin in wood chips. The strain was cellulase-deficient mutant from *Sporotrichum pulverulentum* (Ander and Eriksson, 2006). Akin et al. (1995) studied the delignification by white-rot fungi on bermuda grass. The results show biodegradation improvement by 29–32% and 63–77% after 6 weeks, when *Ceriporiopsis subvermispora* or *Cyathus stercoreus*, respectively, were used. Japanese red pine was studied for biological pretreatment by three white-rot fungi, viz., *Ceriporia lacerata, Stereum hirsutum,* and *Polyporus brumalis*; it revealed that *S. hirsutum* degraded lignin and not cellulose or hemicelluloses (Lee et al., 2007b).

Biological pretreatment by fungi is environmentally friendly, energy saving, and no chemicals are used. This makes it economically feasible. However, rate of the pretreatments is quite low compared to the other pretreatment methods. An additional challenge is that some cellulose and hemicellulose are consumed during the pretreatment which reduces the total content of reducing sugars after the pretreatment (Wyman, 1996; Galbe and Zacchi, 2007).

3.4.3.2 Enzymes Many of the white- and brown-rot fungi produce enzymes which are competent to degrade lignin, such as laccase, manganese peroxidase, versatile peroxidase, glyoxal peroxidase, and aryl alcohol peroxidase. Nonetheless, lignin peroxidase, another enzyme essential for lignin degradation, is not produced by many white-rot fungi (Isroi et al., 2011). Laccase and peroxidase were directly used to pretreat lignocelluloses, resulting in 30% increased productivity in ethanol yield compared to biological pretreatment by *T. reesei.* The reason may be the difference in the detoxification mechanism for the same enzyme when applied *in vitro* and *in vivo* (Palmqvist and Hahn-Hägerdal, 2000). Laccase is widely applied in the forest

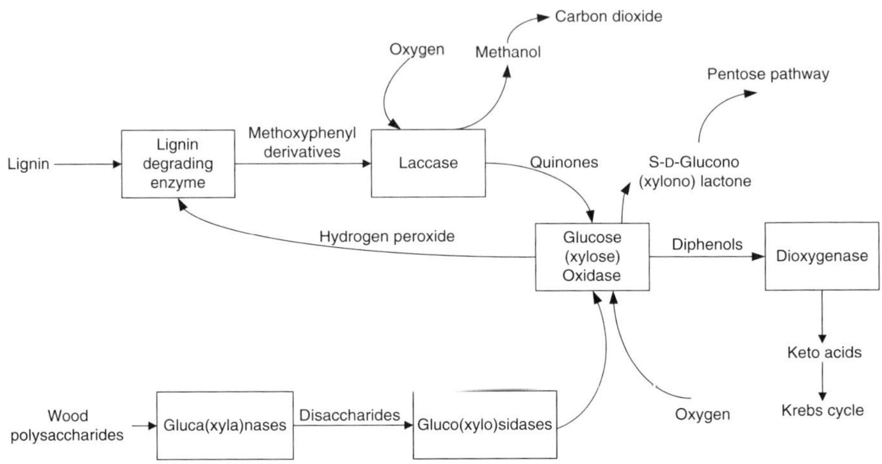

FIGURE 3.8 Proposed mechanisms of lignocellulose transformation by enzymes in white-rot fungi (adapted from Leonowicz et al., 1986).

industry for the bioleaching of lignocellulosic pulps (Widsten and Kandelbauer, 2008). Other enzymes include pectinases and pectin lyases produced from solid-state fermentation of agricultural residues. Ligninolytic enzymes principally target at lignin by the oxidative reaction in the presence of mediators. These enzymatic reactions are carried out at 30–45°C with low enzyme loading rate for about 6–26 hours. More research is happening recently in the purification of these enzymes and improving the activity of enzymes by mutagenesis and rDNA technology (Hao et al., 2006; Mtui and Nakamura, 2007, 2010; Mtui, 2010).

Figure 3.8 shows the transformation of lignocelluloses by the enzymes present in the white-rot fungi. According to this feedback mechanism, laccase oxidizes lignin radicals to quinones, which is the oxygen source of glucose oxidases. Another enzyme namely alcohol oxidase, which produces hydrogen peroxide, prevents the polymerization of laccase. When oxygen is consumed in glucose reaction, it is supplied from quinones or methoxy phenyl derivatives. Glucose oxidase counteracts the excess quinones present, leaving the laccase to proceed with its function (Leonowicz et al., 1999). An outline of advantages and drawbacks of various pretreatment processes is provided in Table 3.4.

3.5 ECONOMIC OUTLOOK

Profitability is always a major concern when selecting a pretreatment method. As mentioned previously, pretreatment is the second most expensive procedure in the processing of lignocelluloses into various commodity bioproducts. A careful economic analysis is essential in order to choose the best pretreatment method for a particular biomass and a specific product. The economic analysis is usually made after experimental data has been gathered from the laboratory or from a pilot study.

TABLE 3.4 Summary of the Advantages and Disadvantages of Different Methods for Pretreatment of Lignocellulosic Biomass (Alvira et al., 2010)

Pretreatment method	Advantages	Disadvantages
Biological	Degrades lignin and hemicellulose Low energy consumption	Low rate of hydrolysis
Milling	Reduces cellulose crystallinity	High power and energy consumption
Steam explosion	Causes lignin transformation and hemicellulose solubilization Cost-effective Higher yield of glucose and hemicellulose in the two-step method	Generation of toxic compounds Partial hemicellulose degradation
AFEX	Increases accessible surface area Low formation of inhibitors	Not efficient for raw materials with high lignin content High cost of large amounts of ammonia
CO_2 explosion	Increases accessible surface area Cost-effective Does not imply generation of toxic compounds	Does not affect lignin and hemicelluloses Requires very high pressure
Wet oxidation	Efficient removal of lignin Low formation of inhibitors Minimizes the energy demand (exothermic)	High cost of oxygen and alkaline catalyst
Ozonolysis	Reduces lignin content Does not imply generation of toxic compounds	High cost of large amounts of ozone
Organosolv	Causes lignin and hemicellulose hydrolysis	High cost Solvents need to be drained and recycled
Concentrated acid	High glucose yield Ambient temperatures	High cost of acid; needs to be recovered Reactor corrosion problems Formation of inhibitors
Diluted acid	Less corrosion problems than concentrated acid Less formation of inhibitors	Generation of degradation products Low sugar concentration in exit stream
Enzymatic	Higher hydrolysis rate	High cost of enzymes

The parameters in the economic analysis can be used as a benchmark for comparison of pretreatment methods which gives a clear picture of the most economically profitable pretreatment. The impact of pretreatment on other unit operations, such as fermentation and downstream processing, also need to be studied. Process economic analysis will grant a precise estimate of the production costs, and other necessary costs, which can be compared with the costs of existing processes. Economic modeling also offers an opportunity to forecast the amount of product obtained with a specific method using various feedstock. Process modeling and cost estimation facilitate the final optimization, realizing the best possible output, in terms of technicality and economical viability (Wooley et al., 1999b; Mosier et al., 2005).

The competition for developing optimal pretreatment methods, with process economy being considered as well, is fierce. Several factors are of vital importance to the economy of the pretreatment.

(a) **Efficiency of pretreatment:** It is probably the most important factor affecting the economy of the process. Some pretreatment methods, such as ionic liquids, are quite effective in opening up the crystalline structure of cellulose, while other methods, such as steam explosion or biological pretreatments, are not as effective in this sense. Effectiveness is a major point to consider when selecting pretreatment. On the other hand, the weight this factor carries varies in different processes, depending on the value of the residuals.

(b) **Energy consumption of pretreatment**: Most of the pretreatments are for processes where energy is the final product, such as ethanol, butanol, or biogas. Consequently, the energy consumption of the pretreatment is very important to the energy balance of the whole process. Some processes, such as milling or irradiation, have high energy demands in the form of electricity, while other processes, such as steam explosion or dilute acid hydrolysis, use high-pressure steam, which might provide a challenge to the process.

(c) **Materials recovery:** Several chemical and physicochemical methods use chemicals for the pretreatments in which the materials are recovered. Ionic liquids or concentrated acid processes are two examples in which full recovery of the pretreating agents are crucial or the process would not be economically feasible.

(d) **Duration of pretreatment:** Some of the pretreatment methods, such as explosions, are quite fast, while other methods, like the environmentally friendly fungal pretreatments, require weeks. Handling vast amounts of raw materials for a long period of time, might present a challenge.

(e) **Risks and environmental aspects of pretreatment**: Pretreatment methods involve various technical and environmental concerns leading to higher costs of the process. These include the materials of the constructions used in the process (e.g., the expensive Hastelloy for dilute acid hydrolysis), or the environmental effects of chemicals, or genetically improved organisms, escaping to the environment. A life cycle assessment is necessary to consider all aspects and risks.

The conclusion is that no pretreatment method is as yet proven to be more advantageous than any other method. An optimal pretreatment method includes consideration of the entire process, that is, optimization of energy sources and energy consumption, environmental aspects, and also the biorefinery concept.

National Renewable Energy Laboratory (NREL) in the United States has developed several processes and economic models for various bioproducts, using, for example, the Aspen Plus® process simulator for fully integrated material balances, and the Aspen Process Economic Analyzer for the economic models. Other types of software, such as Intelligen SuperPro Designer®, were also used to predict economical viability of a process. Models were accepted after meticulous thermodynamics calculations, and analyses of physical properties, developed in a large database (Wooley et al., 1996, 1999a,b; Mosier et al., 2005). Some prominent reports concerning techno-economic analyses are noted here for further studies (Galbe and Zacchi, 1992; Wyman, 1994, 1995; Avellar and Glasser, 1998; Wooley et al., 1999b; Hamelinck et al., 2005).

3.6 FUTURE PROSPECTS

The cellulosic biofuel industry presently finds itself in a "chicken and egg" situation; with fewer industries starting up, farmers produce biomass at a slower pace, leading to industries worrying about continuous feedstock availability.

This situation needs to be changed (Dale and Ong, 2012). To achieve this, a strong focus needs to be put on improving the understanding of lignocelluloses in terms of the cell's structural components, that is, their synthesis, their development into fibrils and further into cellulose, up to a mature cell wall is accomplished. This foundation will facilitate significant changes of current pretreatment methods for future appropriateness. Pretreatments comprise physical, chemical, physicochemical, and biological methods. Combining these methods and optimizing them is a necessary approach for the future. New progress has been made concerning the handling of high solids. Furthermore, fermentation technology is developing; media optimization with the aid of genetic engineering improves the efficiency of cellulases' activity. Synthetic biology approaches to metabolic engineering, optimizing the microbial conversion of substrates to desired bioproducts, should come in handy for the development of robust and efficient microbes (Menon and Rao, 2012).

In the past century, several research groups have investigated lignocellulose pretreatments in an attempt to develop a method that is environmentally friendly as well as economically viable. However, the industries are having problems with selecting the right pretreatment method, due to the pretreatment methods not being comparable. A conclusive picture of the techno-economic analysis is still devoid. A uniform research reporting system is required for comparison of results. A novel method with less environmental impact needs to be developed that also reduces the processing water, since waste and wastewater are being discharged in huge amounts at the end of pretreatments. Furthermore, the use of extreme temperatures, pressures, and hazardous chemicals increase the costs of pretreatments. The choice of pretreatment should be technically feasible, economically affordable, with minimal release of inhibitor compounds. A more industrial viewpoint is imperative for process development.

REFERENCES

Akin DE, Rigsby LL, Sethuraman A, Morrison WH, Gamble GR, Eriksson KE. (1995). Alterations in structure, chemistry, and biodegradability of grass lignocellulose treated with the white rot fungi *Ceriporiopsis subvermispora* and *Cyathus stercoreus*. *Appl Environ Microbiol*, 61, 1591–1598.

Alvira P, Tomás-Pejó E, Ballesteros M, Negro MJ. (2010). Pretreatment technologies for an efficient bioethanol production process based on enzymatic hydrolysis: a review. *Bioresour Technol*, 101, 4851–4861.

Ander P, Eriksson KE. (2006). Selective degradation of wood components by white-rot fungi. *Physiol Plant*, 41, 239–248.

Araque E, Parra C, Freer J, Contreras D, Rodríguez J, Mendonça R, Baeza J. (2008). Evaluation of organosolv pretreatment for the conversion of *Pinus radiata* D. Don to ethanol. *Enzyme Microb Technol*, 43, 214–219.

Atalla RH, Brady JW, Matthews JF, Ding SY, Himmel ME. (2008). Structures of plant cell wall celluloses. In: Himmel ME, editor. *Biomass Recalcitrance*. Blackwell Publishing, pp. 188–212.

Atalla RH, Vanderhart DL. (1984). Native cellulose: a composite of two distinct crystalline forms. *Science*, 223, 283–285.

Avellar BK, Glasser WG. (1998). Steam-assisted biomass fractionation. I. Process considerations and economic evaluation. *Biomass Bioenergy*, 14, 205–218.

Azzam A. (1987). Saccharification of bagasse cellulose pretreated with ZnCl2 and HCl. *Biomass*, 12, 71–77.

Balan V, Bals B, Chundawat SPS, Marshall D, Dale BE, Mielenz JR. (2009). Lignocellulosic biomass pretreatment using AFEX. *Methods Mol Biol*, 581, 61–77.

Berlin A, Balakshin M, Gilkes N, Kadla J, Maximenko V, Kubo S, Saddler J. (2006). Inhibition of cellulase, xylanase and β-glucosidase activities by softwood lignin preparations. *J Biotechnol*, 125, 198–209.

Bidlack J, Malone M, Benson R. (1992). Molecular structure and component integration of secondary cell walls in plants. *Proc Okla Acad Sci*, 72, 51–56.

Brett CT. (2000). Cellulose microfibrils in plants: biosynthesis, deposition, and integration into the cell wall. *Int Rev Cytol*, 199, 161–199.

Brett CT, Waldron KW. (1996). The molecular components of the wall. In: Brett, C., Waldron, K., Eds. *Physiology and Biochemistry of Plant Cell Walls*. Chapman and Hall: Great Britain, pp. 4–43.

Brown RM Jr. (2004). Cellulose structure and biosynthesis: what is in store for the 21st century? *J Polym Sci A Polym Chem*, 42, 487–495.

Carpita NC. (1984). Cell wall development in maize coleoptiles. *Plant Physiol*, 76, 205.

Cassab GI. (1998). Plant cell wall proteins. *Annu Rev Plant Biol*, 49, 281–309.

Chaudhary G, Singh LK, Ghosh S. (2012). Alkaline pretreatment methods followed by acid hydrolysis of *saccharum spontaneum* for bioethanol production. *Bioresour Technol*, 124, 111–118.

Chen WH, Tu YJ, Sheen HK. (2011). Disruption of sugarcane bagasse lignocellulosic structure by means of dilute sulfuric acid pretreatment with microwave-assisted heating. *Appl Energy*, 88, 2726–2734.

Chen X, Shekiro J, Franden MA, Wang W, Zhang M, Kuhn E, Johnson DK, Tucker MP. (2012). The impacts of deacetylation prior to dilute acid pretreatment on the bioethanol process. *Biotechnol Biofuels*, 5, 8–22.

Choi CH, Oh KK. (2012). Application of a continuous twin screw-driven process for dilute acid pretreatment of rape straw. *Bioresour Technol*, 110, 349–354.

Chum HL, Johnson DK, Black S, Baker J, Grohmann K, Sarkanen KV, Wallace K, Schroeder HA. (1988). Organosolv pretreatment for enzymatic hydrolysis of poplars: I. Enzyme hydrolysis of cellulosic residues. *Biotechnol Bioeng*, 31, 643–649.

Cleary PW. (2001). Charge behaviour and power consumption in ball mills: sensitivity to mill operating conditions, liner geometry and charge composition. *Int J Miner Process*, 63, 79–114.

Cooper JB, Varner JE. (1984). Cross-linking of soluble extensin in isolated cell walls. *Plant Physiol*, 76, 414–417.

Dale BE, Ong RG. (2012). Energy, wealth, and human development: why and how biomass pretreatment research must improve. *Biotechnol Prog*, 28(4), 893–898.

Davin LB, Patten AM, Jourdes M, Lewis NG. (2009). Lignins: a twenty-first century challenge. In: Himmel ME, editor. *Biomass Recalcitrance*. Blackwell Publishing, pp. 213–305.

Ding SY, Himmel ME. (2006). The maize primary cell wall microfibril: a new model derived from direct visualization. *J Agric Food Chem*, 54, 597–606.

Ding SY, Himmel ME. (2009). Anatomy and ultrastructure of maize cell walls: an example of energy plants. In: Himmel ME, editor. *Biomass Recalcitrance*. Blackwell Publishing, pp. 38–60.

Donaldson LA. (2001). Lignification and lignin topochemistry—an ultrastructural view. *Phytochemistry*, 57, 859–873.

El Oudiani A, Chaabouni Y, Msahli S, Sakli F. (2011). Mercerization of *Agave americana* L. fibers. *Text Res J*, 103, 565–574.

Fan L, Lee YH, Beardmore DH. (1980). Mechanism of the enzymatic hydrolysis of cellulose: effects of major structural features of cellulose on enzymatic hydrolysis. *Biotechnol Bioeng*, 22, 177–199.

Fernandes AN, Thomas LH, Altaner CM, Callow P, Forsyth VT, Apperley DC, Kennedy CJ, Jarvis MC. (2011). Nanostructure of cellulose microfibrils in spruce wood. *Proc Natl Acad Sci U S A*, 108, E1195–E1203.

FitzPatrick M. (2011). Characterization and processing of lignocellulosic biomass in ionic liquids. Chem Eng. Ontario, Canada Queen's University, pp. 178.

Fukaya Y, Hayashi K, Wada M, Ohno H. (2008). Cellulose dissolution with polar ionic liquids under mild conditions: required factors for anions. *Green Chem*, 10, 44–46.

Galbe M, Zacchi G. (1992). Simulation of ethanol production processes based on enzymatic hydrolysis of lignocellulosic materials using ASPEN PLUS. *Appl Biochem Biotechnol*, 34, 93–104.

Galbe M, Zacchi G. (2007). Pretreatment of lignocellulosic materials for efficient bioethanol production. *Biofuels*, 108, 41–65.

Gorshkova T, Morvan C. (2006). Secondary cell-wall assembly in flax phloem fibres: role of galactans. *Planta*, 223, 149–158.

Grabber JH, Ralph J, Lapierre C, Barrière Y. (2004). Genetic and molecular basis of grass cell-wall degradability. I. Lignin–cell wall matrix interactions. *C R Biol*, 327, 455–465.

Grethlein HE, Allen DC, Converse AO. (2004). A comparative study of the enzymatic hydrolysis of acid-pretreated white pine and mixed hardwood. *Biotechnol Bioeng*, 26, 1498–1505.

Gross AS, Chu JW. (2010). On the molecular origins of biomass recalcitrance: the interaction network and solvation structures of cellulose microfibrils. *J Phys Chem B*, 114, 13333–13341.

Hamelinck CN, Hooijdonk G, Faaij APC. (2005). Ethanol from lignocellulosic biomass: techno-economic performance in short-, middle-, and long-term. *Biomass Bioenergy*, 28, 384–410.

Hanley SJ, Revol JF, Godbout L, Gray DG. (1997). Atomic force microscopy and transmission electron microscopy of cellulose from *Micrasterias denticulata*; evidence for a chiral helical microfibril twist. *Cellulose*, 4, 209–220.

Hao JI, Tian X, Song F, He XB, Zhang ZH, Zhang P. (2006). Involvement of lignocellulolytic enzymes in the decomposition of leaf litter in a subtropical forest. *J Eukaryot Microbiol*, 53, 193–198.

Hardy BJ, Sarko A. (1996). Molecular dynamics simulations and diffraction-based analysis of the native cellulose fibre: structural modelling of the I-α and I-β phases and their interconversion. *Polymer*, 37, 1833–1839.

Harris PJ, Stone BA. (2009). Chemistry and molecular organization of plant cell walls. In: Himmel ME, editor. *Biomass Recalcitrance*. Blackwell Publishing, pp. 61–93.

Hatakka AI. (1983). Pretreatment of wheat straw by white-rot fungi for enzymic saccharification of cellulose. *Appl Microbiol Biotechnol*, 18, 350–357.

Herth W. (1983). Arrays of plasma-membrane "rosettes" involved in cellulose microfibril formation of *Spirogyra*. *Planta*, 159, 347–356.

Himmel ME, Ding SY, Johnson DK, Adney WS, Nimlos MR, Brady JW, Foust TD. (2007). Biomass recalcitrance: engineering plants and enzymes for biofuels production. *Science*, 315, 804–807.

Himmel ME, Picataggio SK. (2009). Our challenge is to acquire deeper understanding of biomass recalcitrance and conversion. In: Himmel ME, editor. *Biomass Recalcitrance*. Blackwell Publishing, pp. 1–6.

Huijgen WJJ, Smit AT, Reith JH, Uil Hd. (2011). Catalytic organosolv fractionation of willow wood and wheat straw as pretreatment for enzymatic cellulose hydrolysis. *J Chem Technol Biotechnol*, 86, 1428–1438.

Iiyama K, Lam TBT, Stone BA. (1994). Covalent cross-links in the cell wall. *Plant Physiol*, 104, 315–320.

Isroi RM, Syamsiah S, Niklasson C, Cahyanto MN, Ludquist K, Taherzadeh MJ. (2011). Biological pretreatment of lignocelluloses with white-rot fungi and its applications: a review. *Bioresources*, 6, 5224–5259.

Jafari V, Labafzadeh SR, Jeihanipour A, Karimi K, Taherzadeh MJ. (2011). Construction and demolition lignocellulosic wastes to bioethanol. *Renew Energy*, 36, 2771–2775.

Janesko BG. (2011). Modeling interactions between lignocellulose and ionic liquids using DFT-D. *Phys Chem Chem Phys*, 13, 11393–11401.

Jeihanipour A, Taherzadeh MJ. (2009). Ethanol production from cotton-based waste textiles. *Bioresour Technol*, 100, 1007–1010.

Jin S, Chen H. (2006). Superfine grinding of steam-exploded rice straw and its enzymatic hydrolysis. *Biochem Eng J*, 30, 225–230.

Karunanithy C, Muthukumarappan K, Julson JL. (2008). Influence of high shear bioreactor parameters on carbohydrate release from different biomasses, *Am Soc Agric Biol Eng*, ASABE 084114. ASABE, St. Joseph, Mich.

Kim J, Park C, Kim TH, Lee M, Kim S, Kim SW, Lee J. (2003). Effects of various pretreatments for enhanced anaerobic digestion with waste activated sludge. *J Biosci Bioeng*, 95, 271–275.

Kim NH, Imai T, Wada M, Sugiyama J. (2006). Molecular directionality in cellulose polymorphs. *Biomacromolecules*, 7, 274–280.

Kontturi EJ. (2005). Surface chemistry of cellulose: from natural fibres to model surfaces, Eindhoven University of Technology.

Kumakura M, Kaetsu I. (1978). Radiation-induced decomposition and enzymatic hydrolysis of cellulose. *Biotechnol Bioeng*, 20, 1309–1315.

Kumakura M, Kaetsu I. (1983). Effect of radiation pretreatment of bagasse on enzymatic and acid hydrolysis. *Biomass*, 3, 199–208.

Kumakura M, Kaetsu I. (1984). Pretreatment by radiation and acids of chaff and its effect on enzymatic hydrolysis of cellulose. *Agricultural Wastes*, 9, 279–287.

Kumakura M, Kojima T, Kaetsu I. (1982). Pretreatment of lignocellulosic wastes by combination of irradiation and mechanical crushing. *Biomass*, 2, 299–308.

Kärkönen A, Koutaniemi S. (2010). Lignin biosynthesis studies in plant tissue cultures. *J Integr Plant Biol*, 52, 176–185.

Labavitch JM. (1981). Cell wall turnover in plant development. *Annu Rev Plant Physiol*, 32, 385–406.

Lafitte-Trouque S, Forster C. (2002). The use of ultrasound and γ-irradiation as pre-treatments for the anaerobic digestion of waste activated sludge at mesophilic and thermophilic temperatures. *Bioresour Technol*, 84, 113–118.

Lam TBT, Kadoya K, Iiyama K. (2001). Bonding of hydroxycinnamic acids to lignin: ferulic and p-coumaric acids are predominantly linked at the benzyl position of lignin, not the [beta]-position, in grass cell walls. *Phytochemistry*, 57, 987–992.

Lee DK, South Dakota State U, Initiative SG. (2007a). Composition of herbaceous biomass feedstocks, North Central Sun Grant Center, South Dakota State University.

Lee J, Gwak K, Park J, Park M, Choi D, Kwon M, Choi I. (2007b). Biological pretreatment of softwood *Pinus densiflora* by three white rot fungi. *J Microbiol*, 45, 485–491.

Leonowicz A, Matuszewska A, Luterek J, Ziegenhagen D, Wojtaś-Wasilewska M, Cho NS, Hofrichter M, Rogalski J. (1999). Biodegradation of lignin by white rot fungi. *Fungal Genet Biol*, 27, 175–185.

Leonowicz A, Rogalski J, Szczodrak J, Piedurek J, Curie-Sklodowska M. (1986). The possible key role of glucose oxidase in transformation of lignocellulose. Biotechnology in the Pulp and Paper Industry, p. 159.

Lloyd TA, Wyman CE. (2005). Combined sugar yields for dilute sulfuric acid pretreatment of corn stover followed by enzymatic hydrolysis of the remaining solids. *Bioresour Technol*, 96, 1967–1977.

Ma F, Yang N, Xu C, Yu H, Wu J, Zhang X. (2010). Combination of biological pretreatment with mild acid pretreatment for enzymatic hydrolysis and ethanol production from water hyacinth. *Bioresour Technol*, 101, 9600–9604.

Mais U, Esteghlalian AR, Saddler JN, Mansfield SD. (2002). Enhancing the enzymatic hydrolysis of cellulosic materials using simultaneous ball milling. *Appl Biochem Biotechnol*, 98, 815–832.

Mayer U, Jürgens G. (2004). Cytokinesis: lines of division taking shape. *Curr Opin Plant Biol*, 7, 599–604.

McDermott B, Chalmers A, Goodwin J. (2001). Ultrasonication as a pre-treatment method for the enhancement of the psychrophilic anaerobic digestion of aquaculture effluents. *Environ Technol*, 22, 823–830.

Meinert MC, Delmer DP. (1977). Changes in biochemical composition of the cell wall of the cotton fiber during development. *Plant Physiol*, 59, 1088–1097.

Menon V, Rao M. (2012). Trends in bioconversion of lignocellulose: biofuels, platform chemicals & biorefinery concept. *Prog Energy Combust Sci*, 38, 522–550.

Mirahmadi K, Kabir MM, Jeihanipour A, Karimi K, Taherzadeh M. (2010). Alkaline pretreatment of spruce and birch to improve bioethanol and biogas production. *Bioresources*, 5, 928–938.

Mohnen D, Bar-Peled M, Somerville C. (2009). Cell wall polysaccharide synthesis. In: Himmel ME, editor. *Biomass Recalcitrance*. Blackwell Publishing, pp. 94–187.

Mooney CA, Mansfield SD, Beatson RP, Saddler JN. (1999). The effect of fiber characteristics on hydrolysis and cellulase accessibility to softwood substrates. *Enzyme Microb Technol*, 25, 644–650.

Mora-Pale M, Meli L, Doherty TV, Linhardt RJ, Dordick JS. (2011). Room temperature ionic liquids as emerging solvents for the pretreatment of lignocellulosic biomass. *Biotechnol Bioeng*, 108, 1229–1245.

Mosier N, Wyman C, Dale B, Elander R, Lee YY, Holtzapple M, Ladisch M. (2005). Features of promising technologies for pretreatment of lignocellulosic biomass. *Bioresour Technol*, 96, 673–686.

Mtui G, Nakamura Y. (2007). Characterization of lignocellulosic enzymes from white-rot fungus *Phlebia crysocreas* isolated from a marine habitat. *J Eng Appl Sci*, 2, 1501–1508.

Mtui G, Nakamura Y. (2010). Lignocellulosic enzymes from Flavodon flavus, a fungus isolated from western Indian Ocean off the Coast of Dar es Salaam, Tanzania. *Afr J Biotechnol*, 7, 3066–3072.

Mtui GYS. (2010). Recent advances in pretreatment of lignocellulosic wastes and production of value added products. *Afr J Biotechnol*, 8, 1398–1415.

Muller CD, Abu-Orf M, Novak JT. (2007). Application of mechanical shear in an internal-recycle for the enhancement of mesophilic anaerobic digestion. *Water Environ Res*, 79, 297–304.

Neely W. (1984). Factors affecting the pretreatment of biomass with gaseous ozone. *Biotechnol Bioeng*, 26, 59–65.

Nishiyama Y, Langan P, Chanzy H. (2002). Crystal structure and hydrogen-bonding system in cellulose Iβ from synchrotron X-ray and neutron fiber diffraction. *J Am Chem Soc*, 124, 9074–9082.

Nishiyama Y, Sugiyama J, Chanzy H, Langan P. (2003). Crystal structure and hydrogen bonding system in cellulose Iα from synchrotron x-ray and neutron fiber diffraction. *J Am Chem Soc*, 125, 14300–14306.

Öhgren K, Bura R, Saddler J, Zacchi G. (2007). Effect of hemicellulose and lignin removal on enzymatic hydrolysis of steam pretreated corn stover. *Bioresour Technol*, 98, 2503–2510.

O'Sullivan AC. (1997). Cellulose: the structure slowly unravels. *Cellulose*, 4, 173–207.

Okano K, Kitagawa M, Sasaki Y, Watanabe T. (2005). Conversion of Japanese red cedar *Cryptomeria japonica* into a feed for ruminants by white-rot basidiomycetes. *Anim Feed Sci Technol*, 120, 235–243.

Palmqvist E, Hahn-Hägerdal B. (2000). Fermentation of lignocellulosic hydrolysates. I: inhibition and detoxification. *Bioresour Technol*, 74, 17–24.

Pandey A, Larroche C, Ricke SC, Dussap CG, Gnansounou E. (2011). *Biofuels: Alternative Feedstocks and Conversion Processes*: Academic Press.

Qi B, Chen X, Wan Y. (2010). Pretreatment of wheat straw by nonionic surfactant-assisted dilute acid for enhancing enzymatic hydrolysis and ethanol production. *Bioresour Technol*, 101, 4875–4883.

Ralph J, Bunzel M, Marita JM, Hatfield RD, Lu F, Kim H, Schatz PF, Grabber JH, Steinhart H. (2004). Peroxidase-dependent cross-linking reactions of p-hydroxycinnamates in plant cell walls. *Phytochem Rev*, 3, 79–96.

Sánchez C. (2009). Lignocellulosic residues: biodegradation and bioconversion by fungi. *Biotechnol Adv*, 27, 185–194.

Schmidt AS, Thomsen AB. (1998). Optimization of wet oxidation pretreatment of wheat straw. *Bioresour Technol*, 64, 139–151.

Sidiras D, Koukios E. (1989). Acid saccharification of ball-milled straw. *Biomass*, 19, 289–306.

Sierra R, Granda CB, Holtzapple MT. (2009). Lime pretreatment. *Methods Mol Biol*, 581, 115–124.

Smith BG, Harris PJ, Melton LD, Newman RH. (1998). Crystalline cellulose in hydrated primary cell walls of three monocotyledons and one dicotyledon. *Plant Cell Physiol*, 39, 711–720.

Somerville C. (2006). Cellulose synthesis in higher plants. *Annu Rev Cell Dev Biol*, 22, 53–78.

Somerville C, Bauer S, Brininstool G, Facette M, Hamann T, Milne J, Osborne E, Paredez A, Persson S, Raab T. (2004). Toward a systems approach to understanding plant cell walls. *Science*, 306, 2206.

Suchy M, Argyropoulos DS. (2001). Catalysis and activation of oxygen and peroxide delignification of chemical pulps: a review. *Oxidative Delignification Chemistry*, 785, 2–43.

Sugiyama J, Vuong R, Chanzy H. (1991). Electron diffraction study on the two crystalline phases occurring in native cellulose from an algal cell wall. *Macromolecules*, 24, 4168–4175.

Sun Y, Cheng J. (2002). Hydrolysis of lignocellulosic materials for ethanol production: a review. *Bioresour Technol*, 83, 1–11.

Taherzadeh MJ. (1999). Ethanol from lignocellulose: physiological effects of inhibitors and fermentation strategies. Department of Chemical Reaction Engineering. Goteborg, Sweden, Chalmers University of Technology.

Taherzadeh MJ, Jeihanipour A. (2012). Recalcitrance of lignocellulosic biomass to anaerobic digestion. In: Mudhoo A, editor. *Biogas Production: Pretreatment Methods in Anaerobic Digestion*. Wiley Online Library, pp. 27–54.

Taherzadeh MJ, Karimi K. (2007a). Acid-based hydrolysis processes for ethanol from lignocellulosic materials: a review. *Bioresources*, 2, 472–499.

Taherzadeh MJ, Karimi K. (2007b). Enzymatic-based hydrolysis processes for ethanol from lignocellulosic materials: a review. *Bioresources*, 2, 707–738.

Taherzadeh MJ, Karimi K. (2008). Pretreatment of lignocellulosic wastes to improve ethanol and biogas production: a review. *Int J Mol Sci*, 9, 1621–1651.

Tassinari T, Macy C. (1977). Differential speed two roll mill pretreatment of cellulosic materials for enzymatic hydrolysis. *Biotechnol Bioeng*, 19, 1321–1330.

Taylor MP, Mulako I, Tuffin M, Cowan D. (2012). Understanding physiological responses to pre-treatment inhibitors in ethanologenic fermentations. *Biotechnol J*, 7, 1169–1181.

Teghammar A, Yngvesson J, Lundin M, Taherzadeh MJ, Horváth IS. (2010). Pretreatment of paper tube residuals for improved biogas production. *Bioresour Technol*, 101, 1206–1212.

Thirmal C, Dahman Y. (2012). Comparisons of existing pretreatment, saccharification, and fermentation processes for butanol production from agricultural residues. *Can J Chem Eng*, 90, 745–761.

Wada M. (2002). Lateral thermal expansion of cellulose Iβ and IIII polymorphs. *J Polym Sci B Polym Phys*, 40, 1095–1102.

Walpot J. (1986). Enzymatic hydrolysis of waste paper. *Conserv Recycl*, 9, 127–136.

Watanabe T, Ohnishi J, Yamasaki Y, Kaizu S, Koshijima T. (1989). Binding-site analysis of the ether linkages between lignin and hemicelluloses in lignin-carbohydrate complexes by DDQ-oxidation. *Agric Biol Chem*, 53, 2233–2252.

Wei H, Donohoe BS, Vinzant TB, Ciesielski PN, Wang W, Gedvilas LM, Zeng Y, Johnson DK, Ding SY, Himmel ME. (2011). Elucidating the role of ferrous ion cocatalyst in enhancing dilute acid pretreatment of lignocellulosic biomass. *Biotechnol Biofuels*, 4, 48–65.

Weng JK, Chapple C. (2010). The origin and evolution of lignin biosynthesis. *New Phytol*, 187, 273–285.

Vidal PF, Molinier J. (1988). Ozonolysis of lignin—improvement of in vitro digestibility of poplar sawdust. *Biomass*, 16, 1–17.

Widsten P, Kandelbauer A. (2008). Laccase applications in the forest products industry: a review. *Enzyme Microb Technol*, 42, 293–307.

Wooley R, Ruth M, Glassner D, Sheehan J. (1999a). Process design and costing of bioethanol technology: a tool for determining the status and direction of research and development. *Biotechnol Prog*, 15, 794–803.

Wooley R, Ruth M, Sheehan J, Ibsen K, Majdeski H, Galvez A. (1999b). Lignocellulosic biomass to ethanol process design and economics utilizing co-current dilute acid prehydrolysis and enzymatic hydrolysis current and futuristic scenarios, DTIC Document.

Wooley RJ, Putsche V, National Renewable Energy L. (1996). Development of an ASPEN PLUS physical property database for biofuels components, National Renewable Energy Laboratory.

Wu H, Mora-Pale M, Miao J, Doherty TV, Linhardt RJ, Dordick JS. (2011). Facile pretreatment of lignocellulosic biomass at high loadings in room temperature ionic liquids. *Biotechnol Bioeng*, 108, 2865–2875.

Wyman C. (1996). *Handbook on Bioethanol: Production and Utilization*. CRC.

Wyman CE. (1994). Ethanol from lignocellulosic biomass: technology, economics, and opportunities. *Bioresour Technol*, 50, 3–15.

Wyman CE. (1995). Economic fundamentals of ethanol production from lignocellulosic biomass. In: Saddler JN, Penner MH, editors. *Enzymatic, Degradation of Insoluble Carbohydrates*. Washington, DC: American Chemical Society, pp. 272–290.

Yamaki S, Machida Y, Kakiuchi N. (1979). Changes in cell wall polysaccharides and monosaccharides during development and ripening of Japanese pear fruit. *Plant Cell Physiol*, 20, 311–321.

Yamamoto H, Horn F. (1994). In situ crystallization of bacterial cellulose I. Influences of polymeric additives, stirring and temperature on the formation celluloses I α and I β as revealed by cross polarization/magic angle spinning (CP/MAS) 13 C NMR spectroscopy. *Cellulose*, 1, 57–66.

Yang B, Wyman CE. (2004). Effect of xylan and lignin removal by batch and flowthrough pretreatment on the enzymatic digestibility of corn stover cellulose. Biotechnol Bioeng, 86, 88–98.

Yu J, Zhang J, He J, Liu Z, Yu Z. (2009). Combinations of mild physical or chemical pretreatment with biological pretreatment for enzymatic hydrolysis of rice hull. *Bioresour Technol*, 100, 903–908.

Yunqin L, Dehan W, Lishang W. (2010). Biological pretreatment enhances biogas production in the anaerobic digestion of pulp and paper sludge. *Waste Manage Res*, 28, 800–810.

Zeng M, Mosier NS, Huang CP, Sherman DM, Ladisch MR. (2007). Microscopic examination of changes of plant cell structure in corn stover due to hot water pretreatment and enzymatic hydrolysis. *Biotechnol Bioeng*, 97, 265–278.

Zhang R, Zhang Z. (1999). Biogasification of rice straw with an anaerobic-phased solids digester system. *Bioresour Technol*, 68, 235–245.

Zhang X, Yu H, Huang H, Liu Y. (2007). Evaluation of biological pretreatment with white rot fungi for the enzymatic hydrolysis of bamboo culms. *Int Biodeterior Biodegrad*, 60, 159–164.

Zhao X, Cheng K, Liu D. (2009). Organosolv pretreatment of lignocellulosic biomass for enzymatic hydrolysis. *Appl Microbiol Biotechnol*, 82, 815–827.

Zhao X, Zhang L, Liu D. (2012a). Biomass recalcitrance. Part I: the chemical compositions and physical structures affecting the enzymatic hydrolysis of lignocellulose. *Biofuels, Bioprod Biorefin*, 6, 465–482.

Zhao X, Zhang L, Liu D. (2012b). Biomass recalcitrance. Part II: fundamentals of different pre-treatments to increase the enzymatic digestibility of lignocellulose. *Biofuels, Bioprod Biorefin*, 6, 561–579.

Zhu S, Wu Y, Chen Q, Yu Z, Wang C, Jin S, Ding Y, Wu G. (2006). Dissolution of cellulose with ionic liquids and its application: a mini-review. *Green Chem*, 8, 325–327.

CHAPTER 4

Enzymatic Hydrolysis of Lignocellulosic Biomass

JONATHAN J. STICKEL, RICHARD T. ELANDER, and JAMES D. McMILLAN

National Bioenergy Center, National Renewable Energy Laboratory, Golden, CO, USA

ROMAN BRUNECKY

Biosciences Center, National Renewable Energy Laboratory, Golden, CO, USA

Bioprocessing of Renewable Resources to Commodity Bioproducts, First Edition.
Edited by Virendra S. Bisaria and Akihiko Kondo.

Abstract

Enzymatic hydrolysis is the central technology in a prototypical biochemical conversion process. Preceded by pretreatment and followed by microbial or inorganic catalyst conversion, enzymatic hydrolysis is the process that liberates monomeric sugars from the structural carbohydrates, cellulose, and hemicellulose in lignocellulosic biomass. In order for a biochemical conversion process to be economically viable, enzymatic hydrolysis must produce sugars at high yields and at reasonable conversion rates using a minimum of input energy and chemicals (including hydrolytic enzymes). This chapter reviews research and development efforts to improve enzymatic hydrolysis technology, with particular emphasis on enzyme systems, operation at high solids concentrations, process modeling, and process integration. The development of more active enzyme systems, improved biomass conversion process technology and integration, and more efficient enzyme production systems (as reviewed in Chapter 5) have dramatically improved the economics of enzymatic hydrolysis over the past decade, and commercial biorefineries incorporating enzymatic hydrolysis of lignocellulosic biomass are anticipated within the next few years.

4.1 INTRODUCTION

Enzymatic hydrolysis lies at the heart of "biochemical platform" or "sugar platform" approaches to utilizing renewable lignocellulosic biomass to produce fuels and chemicals. The processing steps of pretreatment followed by enzymatic hydrolysis can be used to depolymerize biomass cell wall carbohydrates, cellulose, and hemicellulose to simple sugars. Biomass sugars produced by (pretreatment and) enzymatic hydrolysis can be converted into a wide range of fuels or chemical products using a variety of biological (fermentation or microbial conversion) or inorganic catalyst-based technologies.

Lignocellulosic biomass is primarily composed of three biopolymers: lignin, hemicellulose, and cellulose. While cellulose is a homogeneous polymer of glucose, hemicellulose is a heterogeneous branched polymer primarily composed of several sugars including xylose, mannose, galactose, rhamnose, and arabinose. There are a variety of approaches to deconstructing lignocellulose carbohydrates (cellulose, hemicellulose) into sugars, spanning so-called biochemical and thermochemical approaches, including methods using acids, enzymes, or other hydrolytic catalysts. Most approaches use a combination of biological, chemical, mechanical, and thermal "pretreatment and hydrolysis" steps to saccharify lignocellulosic materials. Regardless of the approach, the key is to be able to produce the sugars at a cost and quality sufficient for them to be profitably used for downstream conversion to targeted fuel and chemical products.

This chapter focuses on production of sugars from cellulosic or fibrous biomass feedstocks using conversion routes based on enzymatic hydrolysis. Section 4.2 describes the nature of "cellulase" and "hemicellulase" enzyme systems capable of depolymerizing cellulose and hemicellulose to soluble sugars, and reviews the prominent concepts of enzyme synergy and biomass recalcitrance. Performance of enzymatic hydrolysis under anticipated real-world processing conditions involving high solids concentrations, where factors such as slurry rheology, water activity, product feedback inhibition, and mixing become especially important, is discussed in section 4.3, along with methods for accurately calculating conversion yields in high solids systems. Considerations of heat and mass transfer within such slurries are also briefly introduced. This background provides a foundation in section 4.4 for reviewing nascent efforts to mechanistically model and simulate enzymatic hydrolysis processes and ultimately to develop predictive models to aid process optimization and scale up. Final discussion in section 4.5 focuses on the key role of integration in enzymatic hydrolysis process development, emphasizing the centrality of the enzymatic hydrolysis step within an overall biorefining process and the many opportunities to improve technical and economic performance by better integrating the enzymatic hydrolysis step with both upstream and downstream conversion unit operations.

4.2 CELLULASE, HEMICELLULASE, AND ACCESSORY ENZYME SYSTEMS AND THEIR SYNERGISTIC ACTION ON LIGNOCELLULOSIC BIOMASS

The enzyme systems capable of digesting cellulose and hemicellulose are reviewed in this section. Although hemicellulose can be partly hydrolyzed by some thermochemical pretreatments, digestion of residual hemicellulose by enzymes can significantly improve overall conversion yields. Lignin, an aromatic-based biopolymer, is generally resistant to enzymatic digestion. While a few lignin-degrading enzymes have been discovered (Chen et al., 2011), lignin-degrading enzymes have not yet been sufficiently developed for use in a biorefinery, and are not included in the discussion below.

4.2.1 Biomass Recalcitrance

As our understanding of how the complex physiochemical structure of biomass contributes to recalcitrance has evolved, so has our understanding of how the assorted biomass-degrading enzymes work together to overcome biomass recalcitrance. It is now clear that the classical viewpoint of recalcitrance, which states that recalcitrance is primarily associated with the "physical presence of lignin and hemicellulose and the form of cellulose, with crystalline cellulose being more recalcitrant than the amorphous form," is overly simplistic (Kohlmann et al., 1996). Himmel et al. (2007) emphasize the importance of the phrase "biomass recalcitrance" and expand upon the traditionally terse definition of this term by discussing several barriers that biological systems face when performing degradation of lignocellulosic structural carbohydrates. They describe a number of anatomical features in plant tissues that contribute to recalcitrance. These include "the arrangement and density of the vascular bundles" and "sclerenchymatous" tissues, lignifications within the plant cell wall, and "the structural heterogeneity and complexity of cell wall constituents, such as microfibrils and matrix polymers."

The major enzymes in all biomass conversion cocktails today are cellulases and hemicellulases. However, it is also important to recognize and consider the effects

TABLE 4.1 Enzyme Families and Their Activities on Lignocellulosic Substrates

Substrate	Enzyme activity	CAZy families
Cellulose	β-1,4-Endoglucanase	GH5,7,12,45,9,48
	Cellobiohydrolase	GH6,7,4,48,9
	β-1,4-Glucosidase	GH1,3,9
	β-1,4-Endoglucanase/polysaccharide mono-oxygenase	GH61, CBM33
Xylan	β-1,4-Endoxylanase	GH10,11,9,8
	β-1,4-Xylosidase	GH3,43
	α-Glucuronidase	GH67,115
	α-Arabinofuranosidase	GH51,54
	Arabinoxylan arabinofuranohydrolase	GH62
	β-1,4-Galactosidase	GH2,35
	Acetyl xylan/feruloyl esterase	CE1,4,5,16
Xyloglucan	Xyloglucan β-1,4-endoglucanase	GH12,74
	α-Arabinofuranosidase	GH51,54
	α-Xylosidase	GH31
	α-Fucosidase	GH29,95
	α-1,4-Galactosidase	GH27,36
	β-1,4-Galactosidase	GH2,35

Source: Adapted from van den Brink and de Vries (2011).

of secondary accessory enzymes that aid the major cellulases and hemicellulases in performing enzymatic digestion. Also, when considering the carbohydrate-active enzymes (CAZy) glycoside hydrolase (GH) classification system, it is important to keep in mind that although enzymes within the same CAZy GH family share sequence similarity, many families can contain multiple activities. One example of this is the GH5 family, which contains many catalytic activities, including exoglucanases, endoglucanases, and endomannanases. Table 4.1 lists groups of several biomass-degrading enzyme families by function and substrate.

4.2.2 Cellulases

The primary target product of most traditional conversion processes is the fermentable monomer, glucose. Glucose *in planta* largely exists in the form of crystalline cellulose, which consists of chains of β-1,4 linked glucosyl units that are stabilized by hydrogen bonds, van der Waals forces, and stacking of hydrophobic regions. These chains of glucose are the primary targets of processive cellobiohydrolases such as fungal GH6s (active on the nonreducing end) and fungal GH7s (active on the reducing end), but are aided by enzymes such as fungal family 5 endo-glucanases that create new reducing ends for the enzymes to function on (Nutt et al., 1998; van den Brink and de Vries, 2011). While fungal GH5,6,7's are the most common cellulose-active enzymes in fungal cellulase mixtures, there also exist unique analogous families in bacteria, for example GH48's are exo-acting cellulases (Zverlov et al., 1998; Irwin et al., 2000). GH9's are also unique in that some are processive endo-cellulases (Table 4.1) (Li et al., 2007).

The crystalline nature of the cellulose chains in plant cell walls has always been regarded as one barrier to conversion, though a variety of cellulose-degrading enzyme systems have evolved to digest the crystalline form of cellulose at various rates (Fan et al., 1981; Kohlmann et al., 1996; Himmel et al., 2007). However, native cellulose in plants is composed of both crystalline and amorphous components. Some researchers have reported increases in the binding of a cellobiohydrolase (Cel7A) to the amorphous form of cellulose, and resulting 24 hour extents of conversion, compared with the highly crystalline form (Schroeder et al., 1986; Jeoh et al., 2007). One possible benefit of an acidic thermo-chemical pretreatment process may be to modify cellulose to contain more extensive regions of disordered and reduced crystallinity. If the disordered cellulose is also amorphous-like, this would explain the enhanced cellulase action observed on dilute acid pretreated cellulose (Matthews et al., 2010).

The classical endo–exo deconstruction model of pure cellulose is undergoing revision in light of the recent discovery of a family of copper-dependent polysaccharide mono-oxygenases from the GH61 family (Table 4.1) (Harris et al., 2010; Quinlan et al., 2011). These enzymes represent a novel mechanism to deconstruct cellulose in that they cleave cellulose oxidatively, unlike traditional acid/base catalyzed enzyme systems. Family GH61 enzymes can also provide powerful synergistic benefits to cocktails of traditional GH5, GH6, and GH7 enzyme mixtures (Harris et al., 2010).

4.2.3 Hemicellulases

The second most abundant sugar found in lignocellulosic feedstocks (excepting softwoods) is xylose that comprises the xylan backbone of hemicellulose. In general, it is thought that cellulose microfibrils are held together loosely by a network of various cross-linking glycans (hemicelluloses) using hydrogen bonding and carbohydrate–carbohydrate interactions (Gorshkova et al., 2010). The primary enzymes required for removal of the xylan backbone, which comprises the bulk of xylan found in plant materials, are endo-xylanases from diverse families (GH5, GH8, GH10, and GH11) that act on β-D-(1,4) linkages. In particular, GH10 enzymes degrade linear chains of β-D-(1,4)-linkages, xylan backbones with high degrees of substitutions, and smaller xylo-oligosaccharides, while GH11 xylanases function primarily on β-D-(1,4) linkages (Collins et al., 2005). A smaller body of work to date has focused on the impact of other less abundant hemicelluloses, particularly those associated with primary cell walls, including xyloglucans, glucomannans, and mixed linkage glucans, that is, β-D-(1,3-1,4) glucan (Scheller and Ulvskov, 2010). All of these carbohydrate linkages are known to exist at low levels in cell walls. Finally, the released xylo-oligosaccharides, typically xylobiose, must also be degraded by β-xylosidases. Most fungal β-xylosidases belong to the GH3 family, but several putative β-xylosidases are assigned to GH43 (Table 4.1) (van den Brink and de Vries, 2011).

4.2.4 Accessory Enzymes

While cellulose and xylan comprise the vast majority of sugars available in plant cell walls, other polysaccharides are also present at much lower levels. Depending on the plant species, these secondary polysaccharides include xyloglucans, glucomannans, and mixed linkage glucans such as β-1,3 glucans or mixed linkage β-(1,3);β-(1,4) glucans, (Matthews et al., 2006; Scheller and Ulvskov, 2010). Thus, these small pools of secondary sugars are largely inaccessible to traditional cellulases and require β-(1,3) debranching activities provided by family GH55 enzymes. One example of a glucose releasing accessory enzyme is a xyloglucanase obtained from *Trichoderma reesei*. When added to simple cellulase mixtures, this enzyme improves the extent of conversion, though the modest nature of these improvements makes it difficult to discern whether the benefit is synergistic or merely additive (Benko et al., 2008).

When considering the conversion of the xylan backbone, one feature that is notable is its decoration with additional sugars and other moieties (e.g., acetyl), adding to its complexity. Accessory enzymes hydrolyze some of the linkages of these complex xylooligomers. Xylose can also be present in other hemicellulose forms, such as in xyloglucans. Substitutions on the individual xylose units comprising these materials can include *acetyl*, *arabinofuranosyl*, and *methyl glucuronosyl* groups that vary in placement and abundance widely depending on the plant material. Additionally, *arabinofuranosyl* units often cross-link the xylan backbones to lignins in the plant cell wall via *feruloyl ester* units, which are thought to be ether-bonded to the lignin (Jeffries, 1990; Scheller and Ulvskov, 2010). In corn stover specifically, chemically isolated xylans have been shown to be primarily decorated with 2-*O*- and 3-*O*-monoacetyl, [MeGlcA-α-(1,2)][3-OAc], and arabinofuranosyl units (Naran et al., 2009). It is primarily these "decorations" on xylose and various cross-links that are the targets of accessory enzymes. To target the acetyl decorations, acetyl xylan esterase (AXE) enzymes are employed. AXEs inhabit several carbohydrate esterase (CE) families, the key differences being the specific O-linked acetyl linkage to be hydrolyzed. CE families 1, 4, and 5 have a strong preference for 2-*O*-linked residues, the most common linkage in hemicellulose, while CE16 prefers 3-*O*- and 4-*O*-linked residues (Margolles-Clark et al., 1996; van den Brink and de Vries, 2011; Zhang et al., 2011). Cleavage of arabinofuranosyl linkages typically requires fungal α-arabinofuranosidases mainly found in GH families 51 and 54, although some bifunctional enzymes from GH3 and GH43 have also been shown to have α-arabinofuranosidase activity (Numan and Bhosle, 2006; van den Brink and de Vries, 2011). For the deconstruction of methyl-glucuronosyl linkages there are two primary families for α-glucuronidases: the family GH67 α-glucuronidases are active on short oligosaccharides, while some of the GH115 α-glucuronidases are active on polymeric xylan (Table 4.1) (Nagy et al., 2002, 2003; van den Brink and de Vries, 2011).

4.2.5 Synergy with Xylan Removal and Cellulases

The synergy between cellulases, hemicellulases, and accessory enzymes has been explored in some depth but is still not fully understood. Studies utilizing purified cellulases and hemicellulases have shown that when some of the hemicellulose component in lignocellulose remains intact following pretreatment, the enzymatic hydrolysis of the remaining xylan and substituted xylan moieties improves cellulose hydrolysis by cellulases (Selig et al., 2008, 2009). This improvement in cellulose conversion by hemicellulose removal has been shown to be a linear relationship. One proposed hypothesis for this improvement is that hemicellulose provides a steric hindrance for cellulases, but there is not yet any conclusive evidence for this. To improve both rates and total extents of conversion of hemicellulose to monomers, one needs to effectively cleave side-chain sugars from the xylan backbone using enzymes such as AXE, α-glucuronidase, and α-L-arabinofuransidase (Naran et al., 2009). The impact on xylan hydrolysis by removing such side groups was initially observed by Biely et al. (1986), who in their early studies on AXEs noted that using them with xylanases improved the overall rate of conversion. A similar improvement using arabinofuranosidases was also reported that indicated arabinofuranosyl side groups impeded hydrolysis of the xylan backbone in arabinoxylans (Sørensen et al., 2003). Furthermore, acetyl side groups, and feruloyl esters linked to arabinofuranosyl side chains, also have a direct impact on the hydrolysis of residual xylan remaining after chemical pretreatment (Selig et al., 2008, 2009). It has also been reported that commercially available hemicellulase preparations have been shown to be effective in providing adequate supplementation of debranching enzyme activities (Sørensen et al., 2007).

One possible alternative to the steric hindrance hypothesis of synergistic action is the recent work by Qing et al. (2010), in which they have demonstrated that complex xylooligomers, much more so than monomeric xylose and insoluble xylan, are strongly inhibitory toward hydrolysis of pretreated lignocelluloses when using commercial cellulase complexes. This may be another explanation for the improvements seen when using cellulase–xylanase mixtures. Furthermore, Qing et al. (2010) also noted that pre-hydrolysis of xylan and xylooligomers prior to cellulase addition was more effective than simultaneous hydrolysis of these carbohydrates, which may suggest that the hemicellulosic carbohydrates may somehow irreversibly bind to and inactivate the cellulases.

Our understanding of the interaction of biomass-degrading enzymes on lignocellulosic biomass has advanced considerably in the last decade. Nonetheless, further work in this area will help point the way toward optimized systems of enzymes tailored for specific feedstocks and pretreatment methods, as further discussed in section 4.5.

4.3 ENZYMATIC HYDROLYSIS AT HIGH CONCENTRATIONS OF BIOMASS SOLIDS

Although the biochemical conversion of lignocellulosic biomass to biofuels and bioproducts requires the use of water as a carrier fluid and reaction solvent, there are substantial economic and environmental benefits to reducing the amount of water used throughout the conversion process. Performing the conversion process with less water results in relatively smaller process volumes, enabling smaller equipment sizes, improved fermentation yields due to higher sugar concentrations, and a reduction in the energy needed to separate water from the product (Humbird et al., 2010). However, operating at low water conditions, and consequently at high solids concentrations

($\gtrsim$10% insoluble solids), gives rise to several chemical and physical phenomena that can limit product yields and increase some operating costs. For the unit operation of enzymatic hydrolysis, these phenomena include inhibition of the enzyme catalysts, inefficient slurry transport and mixing, and ineffective heat and mass transfer. Implementation of appropriate process technologies can reduce the impact of some of these phenomena, but ultimately there is an economic trade-off between reducing water content and obtaining high product yields (Humbird et al., 2010).

4.3.1 Conversion Yield Calculations

Scientific investigation and development of conversion processes with slurries at high solids concentrations requires careful definition and mathematical treatment of quantities of interest. Several investigators have addressed the need to use detailed mass-balance relationships when calculating the conversion yield of enzymatic hydrolysis at high solids loading (Hodge et al., 2009; Kristensen et al., 2009; Roche et al., 2009a,b; Zhu et al., 2011). Calculating yields from liquid-phase concentrations alone, without accounting for the presence of the solid phase, can result in overestimation of conversion yield by as much as 30% (Kristensen et al., 2009). Instead, a mass-based fractional conversion yield equation is recommended, given by

$$\xi = \frac{\sum_i r_i (f_i - f_{i,0})}{f_{\mathrm{is},0} \sum_j x_{j,0}}, \tag{4.1}$$

where ξ is the mass fraction of insoluble polysaccharides that have been hydrolyzed to soluble sugars, f_i is the mass fraction of soluble sugar i (per unit mass of slurry), f_{is} is the mass fraction of insoluble solids, x_j is the mass fraction of polysaccharide j in the solid phase (per unit mass of insoluble solids), and r_i is the molecular weight ratio of a polysaccharide unit to its corresponding hydrated (soluble) unit i. The additional "0" in a subscript denotes the value at a reference point, typically the start of a batch reaction. Soluble sugars that are usually measured and accounted for are glucose, cellobiose, and xylose, and polysaccharide components that are usually included are glucan and xylan. Additional sugars, such as arabinose and xylooligomers, and polysaccharides, such as arabinan, may be included if they are present in significant amounts. A conversion yield calculation accounting only for the conversion of glucan to glucose would be given as

$$\xi_G = \frac{r_g (f_g - f_{g,0})}{f_{\mathrm{is},0} x_{G,0}}. \tag{4.2}$$

The mass fraction of a sugar in the slurry (g/g) may be calculated from its liquid-phase concentration c_i (g/L) using the relationship

$$f_i - f_{i,0} = \frac{c_i}{\rho_{\mathrm{liq}}} f_{\mathrm{liq}} - \frac{c_{i,0}}{\rho_{\mathrm{liq},0}} f_{\mathrm{liq},0}, \tag{4.3}$$

where $f_{\text{liq}} = 1 - f_{\text{is}}$ is the mass fraction of the liquid phase. The liquid density ρ_{liq} depends on the concentration of dissolved sugars, where a simple linear relationship has been shown to agree well with measured values (Zhu et al., 2011). Finally, the mass fraction of liquid in the slurry is related to sugar concentrations by the equation

$$f_{\text{liq}} = \frac{f_{\text{liq},0}(1 - \sum_i r_i c_{i,0}/\rho_{\text{liq},0})}{1 - \sum_i r_i c_i/\rho_{\text{liq}}}. \tag{4.4}$$

To determine the extent of conversion of an enzymatic hydrolysis reaction using this approach, measured liquid concentration values are first used with Equations 4.3 and 4.4 to obtain values for the mass fractions of liquid and soluble sugars in the slurry. These values are then used with Equation 4.1 to calculate the conversion yield.

4.3.2 Product Inhibition of Enzymes

It is well known that fungal hemicellulases and cellulases are inhibited by their reaction products. Cross inhibition, for example, inhibition of cellobiohydrolase by xylooligomers, has also been demonstrated to contribute significantly to reduced reaction rates and conversion yields (Qing et al., 2010). Reducing the water content of hydrolyzing biomass slurries results in an inversely proportional increase in the concentrations of soluble oligosaccharides and monomeric sugar products. Although product inhibition is inherent to fungal enzymes and cannot be avoided, a few approaches can be used to alleviate it. Through targeted genetic engineering, for example, it may be possible to modify the enzymes themselves to exhibit less inhibition (Bu et al., 2011). Commercial enzyme cocktails now include many different cellulases and hemicellulases that act synergistically and prevent a buildup of any intermediate sugar or oligosaccharide family. Enzyme mixtures can also be tailored to the feedstock and pretreatment conditions (Banerjee et al., 2010). Nevertheless, product inhibition of β-glucosidase and β-1,4-xylosidase by the monomeric sugars glucose and xylose will still be problematic unless these enzymes are overloaded, which is generally not economically feasible to do.

In order to truly mitigate product inhibition, the sugar products must be removed from the reaction as the reaction is taking place. One means of doing so is to use a process configuration based on simultaneous saccharification and (co-)fermentation (SSF or SSCF), a practice that has a long history in the laboratory (Ghosh et al., 1982; Wang et al., 2013) and has been scaled to pilot-plant operations (Dale and Moelhman, 2009). In SSF, the fermentative microorganism continuously metabolizes the sugars to product (e.g., ethanol), hence preventing the accumulation of high sugar concentrations that cause inhibition of the enzymes. However, the SSF must be operated at temperatures that are favorable for fermentation (typically 30–37°C) rather than the optimum for enzymatic saccharification (typically 50°C). This slows the rate of enzymatic conversion such that the system is typically in a sugar-depleted state, starving the fermentation. SSF may still be limited by fermentation product inhibition, both to the enzymes and the microorganism.

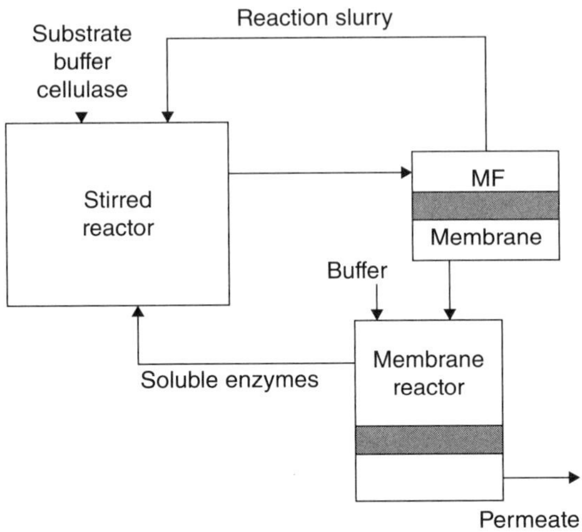

FIGURE 4.1 Block diagram of combined enzymatic hydrolysis and two-stage membrane separation (microfiltration (MF) and ultrafiltration) to remove sugars while retaining undigested biomass and enzymes. Reprinted from Andric et al. (2010b) with permission from Elsevier.

Membrane filtration is another potential way to remove sugars during enzymatic hydrolysis while retaining undigested solid biomass and enzymes (Knutsen and Davis, 2004; Gan et al., 2005). Andric et al. (2010b) provide a review of several combined reaction and separation processes for enzymatic hydrolysis that were studied in the laboratory, and a diagram of one such process is shown in Figure 4.1. In addition to reducing enzyme inhibition, these so-called "membrane reactors" have the potential for continuous operation with the retention or recycle of enzymes. However, they have not yet been sufficiently developed for scale-up and routine use in pilot-scale operations. Few studies have used lignocellulosic substrates, where the buildup of lignin is likely to be problematic.

There have been numerous empirical and semi-empirical models proposed for the enzymatic hydrolysis of cellulose that include inhibition terms (Andric et al., 2010a,b). The use of these models can be helpful with process design and economic calculations where it is necessary to account for the limitations imposed by inhibition. More rigorous mechanistic models for enzymatic hydrolysis have also included terms for product inhibition, as described in section 4.4.

4.3.3 Slurry Transport and Mixing

Pretreated biomass slurries at high solids concentrations have highly non-Newtonian rheology (Knutsen and Liberatore, 2009; Stickel et al., 2009; Ehrhardt et al., 2010). Dilute-acid pretreated corn stover slurries exhibit a yield stress at concentrations above 5% insoluble solids. The slurries are shear-thinning after yielding, although measuring shear-flow profiles of sufficient quality for fitting to constitutive models has

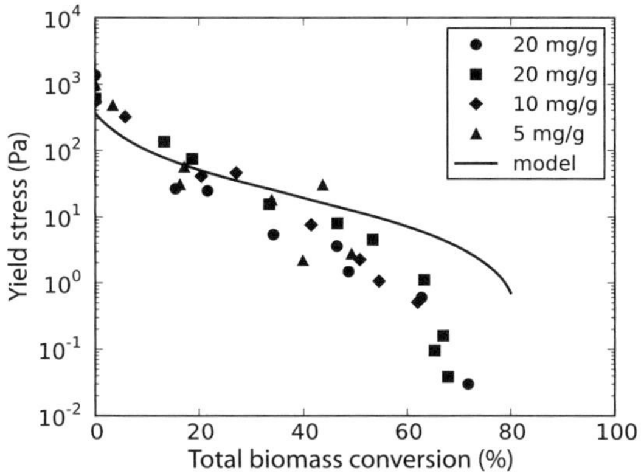

FIGURE 4.2 Yield stress evolution through the course of enzymatic hydrolysis of pretreated corn stover, starting at 20% insoluble solids content and varying enzyme loading (mass enzyme/mass cellulose), as a function of biomass conversion. The predictive model is described by Roche et al. (2009a).

proved challenging. The yield stress increases by orders of magnitude as the insoluble solids concentration increases. Above 20% insoluble solids, air voids appear in the slurry as most of the water resides inside the pores of the biomass. These three-phase (air–liquid–solid) materials obey the principles of wet granular materials rather than hydrodynamics. During enzymatic hydrolysis, the insoluble cellulose is hydrolyzed to soluble species, effectively transferring solid-phase material to the liquid phase. Consequently, as enzymatic hydrolysis proceeds, the slurry thins out considerably, and the yield stress decreases by orders of magnitude as illustrated in Figure 4.2 (Roche et al., 2009a). Transporting high solids slurries by pump (or other means) and mixing such slurries can be highly energy intensive. Mixing of pretreated biomass that has a high yield stress (initially) is ineffective using traditional impellers in vertical stirred tanks. Mixing in horizontal reaction vessels by axial rotation has been proposed by a few investigators and has been shown to scale well (Jørgensen et al., 2007; Roche et al., 2009a,b). Some types of water-soluble polymers and surfactants have been shown to act as rheology modifiers for biomass slurries, effectively reducing their yield stress (Knutsen and Liberatore, 2010; Samaniuk et al., 2012).

4.3.4 Heat and Mass Transport

As recently reviewed by Viamajala et al. (2010), the inherently multi-scale structure of lignocellulosic biomass can limit heat and mass transport in reacting biomass slurries. Although heat transfer is of more concern for pretreatment, it should not be neglected when designing large-scale enzymatic hydrolysis reactors where heating or cooling the slurries could take considerably more time than for more typical dilute aqueous streams. Mass transport, however, is a significant issue for enzymatic hydrolysis, especially at high solids concentrations. A few experimental studies have demonstrated the need for effective bulk mixing of enzymes with the pretreated biomass, especially at the start of the reaction (Mais et al., 2002; Roche et al., 2009b; Lavenson et al., 2012). Slow continuous mixing or moderate intermittent mixing has been shown to be sufficient to achieve good conversion yields, provided mixing occurs throughout the reactor with few dead zones. Mixing with overly high shear rates has been shown to deactivate cellulases (Cao and Tan, 2004).

Diffusive transport of enzymes and sugars to the surface of and within the porous structure of biomass particles is also an important consideration. Grethlein (1985) demonstrated experimentally that the initial rate of enzymatic hydrolysis correlates strongly with the volume of pores that are larger than the size of cellulases. Luterbacher et al. (2012) used confocal fluorescent microscopy to observe that digesting biomass particles do not shrink but rather fade, indicating that the pore structure enlarges during enzymatic digestion. Roberts et al. (2011) used NMR to probe water mobility and diffusivity of solutes in cellulose suspensions. They found that the amount of water associated with the cellulose, supposed to be water in the pores, decreased with increasing solids concentration. They also found that the diffusivity of sugar and enzyme surrogate (bovine serum albumin) through the cellulose suspension decreased with increasing solids concentration. It has been hypothesized that product inhibition is amplified when there is poor mass transfer in the reacting slurry (Hodge et al., 2008; Viamajala et al., 2010). As enzymes hydrolyze the polysaccharides to sugars, locally high concentrations of the sugars may accumulate near the enzymes unless they are transported away by diffusive and convective transport. Because mixing becomes more difficult at high solids loading, product inhibition and ineffective transport are antagonistic to the overall reaction rates and yields.

4.4 MECHANISTIC PROCESS MODELING AND SIMULATION

The use of mathematical modeling and computer simulation has become prevalent throughout the petrochemical process industries. Models that are based on scientifically demonstrated chemical and physical mechanisms, that is, "mechanistic models," allow engineers to design and test unit operations and their integration *in silico* with only limited supporting laboratory and pilot-scale experimentation. This has not yet been possible for biomass conversion processes, excepting techno-economic analyses that rely on empirical relationships. Mechanistic modeling of biochemical conversion of biomass to chemicals and fuels generally, including enzymatic hydrolysis specifically, is challenging for several reasons: the presence of multiple phases, non-Newtonian rheology, multiple reacting species, and phenomena occurring on multiple length and time scales. Over the last few years, progress has been made to address some of these challenges for enzymatic hydrolysis.

There has been a substantial effort to model the kinetics of enzymatic hydrolysis of cellulose and (pretreated) lignocellulosic substrates. Bansal et al. (2009) provide a comprehensive review of many of the models that have been developed. Most of these models are strictly empirical or based on highly simplified Michaelis–Menten concepts. Unfortunately, the assumptions commonly used with Michaelis–Menten kinetics models, namely, that reaction takes place in solution and there is a single substrate species that is completely accessible, are not valid for insoluble, polymeric cellulose. Hence, these models are limited to the parameter space from which they were developed and provide few insights into the underlying mechanisms of the reaction. A couple of models treat cellulose as polymers with a distribution of chain lengths (Okazaki and Mooyoung, 1978; Zhang and Lynd, 2006), but do not address the insoluble structure of crystalline cellulose.

Recently, a few groups have independently developed mechanistic models for the enzymatic hydrolysis of cellulose that account for the distribution of cellulose degree of polymerization (DP), the changing morphology and accessibility of insoluble cellulose, and the differing functionality of the component cellulase enzymes (Zhou et al., 2009a,b; Levine et al., 2010; Zhou et al., 2010; Levine et al., 2011; Griggs et al., 2012a,b). The inhibition of enzymes by their products have also been included in some of the models (Griggs et al., 2012b). Although the specific features and mathematical implementation for these three models are different, all predict that surface accessibility to enzymes is the rate-limiting phenomenon. These mechanistic models are also able to predict synergism between endo- and exo-acting cellulases and show that the synergism changes with the DP of the cellulose substrate. Figure 4.3

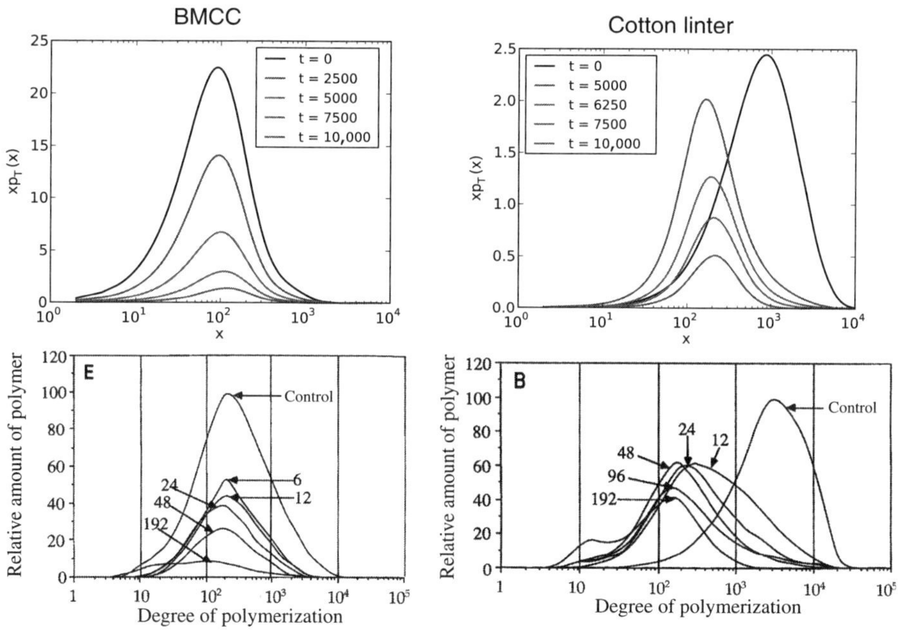

FIGURE 4.3 Qualitative comparison of mechanistic model predictions (top row, from Griggs et al. (2012b)) with experimental results (bottom row, from Srisodsuk et al. (1998)) for the changing DP distribution of cellulose during enzymatic hydrolysis by EG_I and CBH_I. The left column compares results for bacterial microcrystalline cellulose (BMCC) with a relatively low initial DP, and the right column compares results for cotton linter with a relatively high initial DP. (*For a color version, see the color plate section.*)

shows simulated and experimentally measured DP distributions during enzymatic hydrolysis for two different substrates. For an initially low DP substrate (bacterial microcrystalline cellulose), there is little-to-no right-to-left shift in the DP distribution. However, for an initially high DP substrate (cotton linter), there is a shift in the DP distribution to lower DP at early times. The mechanistic model predictions qualitatively match the experimental results. Griggs et al. (2012b) suggest that for high DP substrates, endoglucanases (EG_I) hydrolyze β-(1,4)-glycosidic bonds in the middle of the cellulose polymers, making chain ends available to exoglucanases (CBH_I). Once the DP of the cellulose is sufficiently reduced, the activity of exoglucanases dominates the digestion, and the total mass of the cellulose population is reduced.

Although these kinetic mechanistic models represent a significant step forward in our ability to predict transient behavior and process yields for the enzymatic hydrolysis of cellulose, they still fall short in several ways. Further model development will be needed to account for the presence of lignin and hemicellulose in lignocellulosic substrates. Furthermore, pretreated lignocellulosic biomass can vary considerably depending on the feedstock and pretreatment conditions. It may prove difficult to develop a universal mechanistic model that applies for all industrially relevant biomass substrates, but some progress can be made by incorporating measurable chemical and physical properties of the substrates, such as chemical composition and particle morphology. Recently, Luterbacher et al. (2012) developed a kinetic model that accounts for the pore-size distribution of biomass particles. Some experimental studies have suggested that processive enzymes such as cellobiohydrolase may get "stuck" (Jalak and Valjamae, 2010), and others have suggested that cellulases may be spatially confined on the cellulose surface (Xu and Ding, 2007). Further work is needed to verify these phenomena and effectively incorporate them into mechanistic models, if necessary. As discussed in section 4.2, there exist many different cellulase, hemicellulase, and accessory enzymes with varying modes of action. As commercial enzyme preparations include more of these enzymes in appreciable concentrations, it may be necessary to explicitly include their action in mechanistic models. However, as additional features are added to the mechanistic models, their complexity increases proportionally. Model developers and users should consider carefully the need to balance the level of detail needed to describe the phenomena of interest and the mathematical complexity and computational time needed for performing simulations.

In addition to models for the kinetics of enzymatic hydrolysis reactions, models for the transport of biomass slurries within reactors are needed. Computational fluid dynamics (CFD) can be used simulate convective and diffusive transport in process vessels, but only a few researchers have applied CFD to the enzymatic hydrolysis of lignocellulosic biomass (Um and Hanley, 2008; Shao et al., 2010; Carvajal et al., 2012), perhaps because it has been challenging to determine appropriate constitutive equations for use in CFD software. The CFD modeling approach may differ depending on the concentration of biomass solids. For low solids concentrations, settling suspension models with a Newtonian suspending liquid may be used. The rheology will be non-Newtonian at higher solids concentrations, and simple yield stress and power law models, for example, the Herschel–Bulkley model, may be used.

In order to conduct comprehensive computer simulations of enzymatic hydrolysis for the purposes of performing reactor design and evaluating scale-up, it will be necessary to couple a mechanistic kinetics model with an appropriate CFD model. To the authors' knowledge, this has not yet been accomplished, and it could prove to be quite challenging. The number of independent equations in a CFD model for a reasonably complex vessel can be on the order of 10^6. If the mechanistic kinetics model has roughly 10^3 equations, then the total system size will be on the order of 10^9. Solving such a system will require a high performance computer utilizing a parallel architecture. In addition, reaction and mixing may occur at different timescales, further complicating the implementation. Despite these challenges, the potential benefits that can be obtained from coupled transport and kinetics simulations, namely, accelerated technology development and reduced commercialization risk, should motivate further R&D in this area.

4.5 CONSIDERATIONS FOR PROCESS INTEGRATION AND ECONOMIC VIABILITY

To be effectively practiced in a commercially relevant (i.e., economical) process context, enzymatic hydrolysis must be conducted with consideration of an entire lignocellulosic biomass conversion process. The choice of biomass feedstock and its specific compositional and structural attributes define the various types and amounts of structural carbohydrates that must be hydrolyzed, while the amounts and properties of noncarbohydrate components (lignin, ash, protein, etc.) also determine the overall recalcitrance that must be overcome to effectively deconstruct the biomass feedstock and produce sugars in high yields. Biomass pretreatment processes have been developed to help overcome the natural recalcitrance of lignocellulosic biomass by rendering it much more susceptible to rapid and complete enzymatic deconstruction. Additionally, the process context in which the resulting soluble carbohydrates are used to produce a chemical and/or fuel product of interest can also define the specific enzyme system and process configuration in which enzymatic hydrolysis is performed.

4.5.1 Feedstock

There is a wide variety of biomass feedstocks that are potentially viable for utilization in commercial-scale biochemical conversion processes. While biomass feedstocks all consist of three primary compounds (cellulose, hemicellulose, lignin), the relative amounts of these compounds can substantially vary across different feedstock classes and the structural features and associated cross-linking cause wide differences in biomass recalcitrance across feedstock classes (McMillan, 1997; Perez et al., 2010; Zhao et al., 2012). In general, woody feedstocks (particularly softwoods) contain more lignin than herbaceous agricultural residues and perennial energy crops. The types of covalent linkages between lignin and hemicellulose also vary across feedstock classes and can impact the requirement for accessory enzyme activities if such linkages survive pretreatment in significant amounts. The levels of the minor sugars (arabinose, galactose, and mannose) also vary considerably with biomass type. Softwoods typically contain more galactose and mannose than hardwoods, whereas hardwoods, herbaceous plants, and agricultural residues generally contain higher levels of arabinose and xylose. In some herbaceous energy crops and agricultural residues, arabinose levels are high enough that conversion of arabinose (in addition to glucose and xylose) is required to achieve overall economic viability, which may impact the required amounts of additional accessory enzyme activities if levels of residual arabinan (or arabinose that has not fully been depolymerized) are still substantial after pretreatment of these feedstock types.

The presence of other relatively minor components in feedstocks (or extraneous substances harvested with the feedstock) can either directly influence the feedstock reactivity and resulting enzyme activity and processing requirements or require inclusion of feedstock "pre-processing" steps to mitigate the effects of these components. While ash components (either from extraneous soil collected during the feedstock harvesting process or from intrinsic ash components within the feedstock itself (Vassilev et al., 2010)) are not widely believed to directly impact enzymatic hydrolysis performance, their presence can impact pretreatment performance by causing a buffering effect on added pretreatment chemicals, which can indirectly impact enzymatic hydrolysis performance. Acetyl groups, which are often present on side chains of the primary hemicellulosic xylan backbone, have also been shown to affect pretreatment and enzymatic hydrolysis performance (Kong et al., 1992; Chang and Holtzapple, 2000). Research is being conducted to alter plant physiological processes to reduce the acetyl content in potential biomass feedstocks (Lee et al., 2011). Chen et al. (2012) have recently developed a feedstock deacetylation process for a lignocellulosic ethanol process using corn stover that has been shown to lower the optimal pretreatment severity, improve overall enzymatic saccharification, reduce the requirement for AXE enzyme activities, and improve the fermentability of high concentration sugar hydrolysates. As shown in Figure 4.4, a simple feedstock soaking deacetylation process can significantly improve the enzymatic hydrolysis of cellulose in pretreated corn stover, even when using whole pretreatment slurry at an enzymatic hydrolysis total solids loading of 25% (Chen et al., 2012). In this case, the combined (glucose + xylose) liquid-phase sugar concentration exceeds 150 g/L, although little additional xylose is enzymatically generated due to apparent end-product inhibition at such high sugar concentrations.

4.5.2 Pretreatment

Naturally occurring cellulolytic bacteria and fungi produce distinct enzyme components that work synergistically to degrade lignocellulosic biomass structural carbohydrates to sugars, as discussed in section 4.2. The enzymatic hydrolysis of native biomass is typically very slow, requiring that a pretreatment of the biomass be conducted in order to increase the rate and extent of enzymatic hydrolysis reactions to be economically viable for producing a commodity chemical and/or fuel product. The pretreatment operation can directly influence the amount and types of enzymes

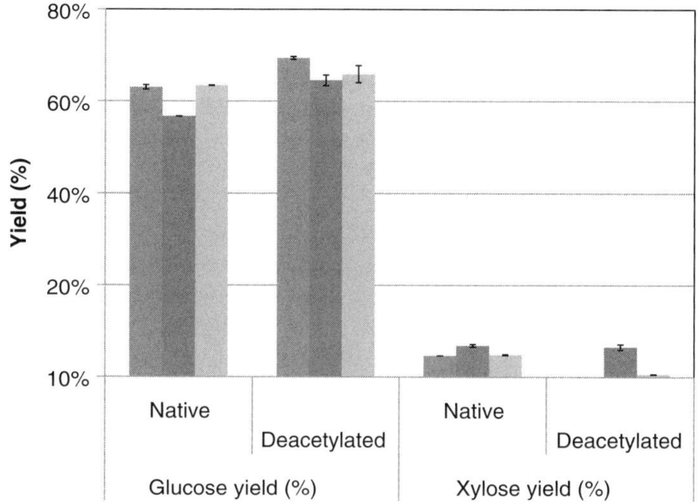

FIGURE 4.4 Whole slurry enzymatic hydrolysis yields of dilute acid pretreated corn stover, with and without feedstock deacetylation. Pretreatment conditions: 150°C, 10 minutes, 8 mg H_2SO_4/g dry biomass, 45% solids in pretreatment reactor prior to steam injection. Enzymatic hydrolysis conditions: 25% total solids, no detoxification/conditioning (other than pH adjustment to 4.8), 50°C, 168 hours, Novozymes Cellic Ctec2 (20 mg/g cellulose in pretreated slurry) plus Novozymes Cellic Htec2 (20 mg/g cellulose in pretreated slurry). The three bars in each group represent different corn stover varieties as described in Chen et al. (2012).

needed to saccharify all plant cell wall structural carbohydrates to monomeric sugars, as some pretreatment approaches can hydrolyze virtually all of the hemicellulose directly to monomeric xylose, while other pretreatment approaches largely leave the hemicellulose intact or only achieve partial hydrolysis to oligomeric sugars (Sun and Cheng, 2002; Mosier et al., 2005b; Elander et al., 2009). Therefore, the pretreatment process directly impacts the types and relative amounts of enzyme activities that are required in the subsequent enzymatic hydrolysis step, including hemicellulases and other accessory enzymes needed to deconstruct and hydrolyze hemicellulose that was not saccharified during pretreatment, in addition to hydrolyzing cellulose that is not typically converted at appreciable extents in leading pretreatment approaches. Many pretreatment processes that do solubilize hemicellulose to a significant extent often do not fully convert soluble oligosaccharides to monomeric sugars, thus requiring oligomer-hydrolyzing enzyme activities if high yields of monomer sugars are ultimately desired (Qing et al., 2010). In order to achieve high yields of soluble sugars from hemicellulose, there are also covalent linkages between hemicellulose and lignin residues that require cleavage by appropriate enzymes if such linkages survive pretreatment (Kong et al., 1992; Chang and Holtzapple, 2000). In general, acidic (Lloyd and Wyman, 2005; Weiss et al., 2009; Humbird et al., 2011) and hot water/steam (Brownell and Saddler, 1987; Mok and Antal, 1992; Liu and Wyman, 2003; Mosier et al., 2005a) pretreatments solubilize greater amounts of hemicellulose directly in the pretreatment step, while alkaline (Iyer et al., 1996; Chang et al., 2001; Kim and Lee, 2005; Chundawat et al., 2010) and oxidative (Klinke et al., 2002; Varga et al., 2003; Saha and Cotta, 2006) pretreatments primarily depolymerize and disrupt the lignin structure and/or the cellulose crystalline structure to enhance enzymatic digestibility. Pretreatment approaches that solubilize hemicellulose to a large extent but leave much of the solubilized hemicellulose in oligomeric form, including less severe dilute-acid pretreatment conditions, may require additional mild thermochemical and/or enzymatic steps to convert oligosaccharides to monomeric sugars. The enzymatic hydrolysis process configuration, along with the concentration of solubilized pretreatment products and the extent to which the pretreated slurry has been separated or washed, can impact the ability of enzyme systems to effectively hydrolyze hemicellulose-derived oligosaccharides (Shekiro et al., 2012).

While various pretreatment approaches have been extensively studied in laboratory-scale reactor systems, economic viability can only be achieved when pretreatment is conducted in a commercially relevant manner. For large-scale processes, this generally involves the use of high solids reactor systems designed for continuous operation. Pretreatment performance can change significantly when scaling from batch, low solids, laboratory-scale systems to continuous, high solids, pilot-scale (or larger) systems (Schell et al., 2003; Shekiro et al., 2012). It is important to characterize enzymatic hydrolysis of pretreated solids from commercially relevant pretreatment systems, as factors such as grinding/compression and residence time distribution will cause changes in enzymatic hydrolysis performance as compared with similarly pretreated biomass generated in small laboratory-scale reactor systems.

4.5.3 Downstream Conversion

The simplest downstream conversion process configuration would utilize all biomass sugars derived from pretreatment and enzymatic hydrolysis in a combined, high concentration hydrolysate. The 2011 National Renewable Energy Laboratory corn stover lignocellulosic ethanol design report (Humbird et al., 2011) describes such a scenario. In this case, enzymatic hydrolysis is performed using the whole slurry obtained from a high solids dilute acid pretreatment, where all soluble compounds generated during pretreatment (including monomeric and oligomeric sugars, acetic acid, soluble extractives, and some solubilized lignin) are present in relatively high concentrations. Such a configuration is possible for lignocellulosic ethanol, as cofermenting microorganisms capable of utilizing five- and six-carbon sugars exist and have been shown to perform well under process-relevant conditions. Other enzymatic hydrolysis process configurations, including SSF and washed-solids enzymatic hydrolysis, both of which reduce the background and in-process sugar concentrations in order to alleviate end-product sugar inhibition of hydrolytic enzymes, are no longer found to be necessary, as improved commercially ready enzyme preparations can generate combined sugar concentrations greater than 150 g/L at reasonable enzyme loadings and reaction times. However, the presence of high concentrations of sugars and other compounds generated or liberated in the pretreatment step (primarily hemicellulosic sugars, acetate, and sugar degradation products) can negatively impact enzyme performance compared with the situation where the background concentration of such compounds is lower. Background concentrations of xylose from high solids dilute-acid pretreatment of corn stover can cause inhibition of enzymatic release of additional xylose from remaining xylan and/or oligomeric xylose (Shekiro et al., 2012). Additionally, acetyl groups present in the feedstock or when released as acetic acid into hydrolysates can impact performance of various cellulolytic enzymes (Kong et al., 1992; Chang and Holtzapple, 2000; Selig et al., 2009; Agger et al., 2010), unless selectively removed as described in section 4.5.1.

Enzymatic hydrolysis of pretreated biomass has generally been studied within the context of a process that requires released sugars to be in monomer form for subsequent microbial conversion to fuel and chemical products. Enzyme preparations with sufficient beta-glucosidase (and possibly beta-xylosidase) activities to produce high yields of monomeric sugars are typically required to prevent accumulation of dimeric sugars. However, there are some microorganisms that can utilize some sugar oligomers like cellobiose and would therefore not require an enzyme preparation rich in beta-glucosidase activity (Weimer and Zeikus, 1977; Yang et al., 2010; Ha et al., 2011). Additionally, processes are now being developed that utilize chemical catalysts to upgrade sugars (monomeric and oligomeric) and other forms of soluble carbon, such as sugar degradation products from pretreatment, acetic acid, and certain soluble lignin compounds. Such processes would not require all of the enzyme activities needed for hydrolysis processes optimized for producing monomeric sugars.

4.6 ECONOMIC OUTLOOK

The cost of enzymes for saccharifying (pretreated) lignocellulosic biomass is estimated to have decreased over 20-fold since 1999, largely facilitated by US Department of Energy (DOE) cost-shared subcontracts to industrial enzyme producers (Teter et al., 2006; Merino and Cherry, 2007; Aden and Foust, 2009; McMillan et al., 2011). Owing to this progress, while the cost of enzymes for biomass saccharification still remains a significant expense, it is no longer viewed as a major barrier to demonstrating cost-competitive cellulose feedstock to biomass sugars technology. Techno-economic cost modeling at the National Renewable Energy Laboratory, based on a commercial scale "sugar platform" cellulosic ethanol process design, estimates that enzymes cost approximately USD 0.30–0.50 per gallon of ethanol (or roughly USD 0.04–0.08 per pound sugar). Thus, while enzymes still represent the second largest contributor to operating cost in the process, after feedstock, they no longer dominate processing costs (Humbird and Aden, 2009; Humbird et al., 2010).

Quantitative data on enzyme and biomass sugar production processing costs are largely not available in the public domain. Regardless, enzyme production and enzymatic saccharification technologies have advanced sufficiently for sugar production from lignocellulosic feedstocks to become economically viable for some particular process concepts, or at least quite close to economic viability. Reflecting this, many enzymatic hydrolysis processes are being tested at bench, pilot, and demonstration scales, and several larger demonstration- and commercial-scale plants for cellulosic biofuels production that incorporate enzymatic hydrolysis technology are operating or under construction around the world. Companies pursuing commercialization of fuels and chemicals productions from biomass-derived sugars produced using an enzymatic hydrolysis approach include Abengoa, Borregaard, Chemtex, DuPont, Poet-DSM, and Mascoma (Anonymous, 2013).

4.7 FUTURE PROSPECTS

Tremendous advances have been made over the past several decades in understanding biomass composition and structure and the diversity in function and structure of hydrolytic enzymes and their mechanisms of overcoming lignocellulosic plant cell wall recalcitrance to deconstruction. Many cellulase-based enzyme systems (i.e., multicomponent, multi-activity enzyme cocktails) are now known to be capable of deconstructing or hydrolyzing the major biomass carbohydrates cellulose and xylan to glucose and xylose, respectively, following a suitable pretreatment. As described in section 4.2, required enzyme activities can vary widely depending upon biomass feedstock type and pretreatment reaction conditions.

Enzymatic hydrolysis can be performed in a stand-alone manner (sequential hydrolysis and fermentation or SHF), or combined with fermentation (SSF), or carried out in a hybrid fashion that starts off as SHF and finishes as SSF. As discussed in section 4.3, process economics and commercial viability favor operation at higher solids (lower water) conditions where material handling, mixing, and mass transfer issues are present that must be properly addressed to achieve good enzymatic hydrolysis performance. Significant progress has been made in developing predictive mechanistic models of enzymatic hydrolysis kinetics and at characterizing and modeling the fluid dynamics of high solids enzymatic hydrolysis processes, although, as described in section 4.4, further work is required to develop and validate models that couple mechanistic kinetic models with appropriate fluid dynamics models.

The economics of enzymatic hydrolysis have dramatically improved over the past decade, largely as a result of three areas where substantial progress has been achieved. First, newly discovered or catalytically enhanced enzyme components are being incorporated into multienzyme cellulase preparations to improve their efficacy. Second, enzyme production has improved such that host production microorganisms are producing enzymes more efficiently and economically than has been historically possible. Third, process technology for enzymatic hydrolysis and its integration with the upstream and downstream process steps that it interacts with, that is, pretreatment and sugar upgrading, respectively, has improved substantially, enabling high yields to be achieved using cost-effective enzyme loadings.

Many well-financed technology development companies are actively pursuing scale-up and commercialization of variations of enzymatic hydrolysis technology for specific pretreated biomass feedstocks. Within the next several years, it is anticipated that enzymatic hydrolysis technologies, as practiced by these companies, will be contributing to commercial production of a range of biofuels and chemical products from lignocellulosic feedstocks.

ACKNOWLEDGMENTS

This work was funded by the US Department of Energy under Contract No. DE-AC36-08-GO28308 with the National Renewable Energy Laboratory and through the BioEnergy Technologies Office.

REFERENCES

Aden A, Foust T. (2009). Technoeconomic analysis of the dilute sulfuric acid and enzymatic hydrolysis process for the conversion of corn stover to ethanol. *Cellulose*, 16(4), 535–545.

Agger J, Vikso-Nielsen A, Meyer AS. (2010). Enzymatic xylose release from pre-treated corn bran arabinoxylan: differential effects of deacetylation and deferuloylation on insoluble and soluble substrate fractions. *J Agric Food Chem*, 58(10), 6141–6148.

Andric P, Meyer AS, Jensen PA, Dam-Johansen K. (2010a). Reactor design for minimizing product inhibition during enzymatic lignocellulose hydrolysis: I. significance and mechanism of cellobiose and glucose inhibition on cellulolytic enzymes. *Biotechnol Adv*, 28(3), 308–324.

Andric P, Meyer AS, Jensen PA, Dam-Johansen K. (2010b). Reactor design for minimizing product inhibition during enzymatic lignocellulose hydrolysis: II. quantification of inhibition and suitability of membrane reactors. *Biotechnol Adv*, 28(3), 407–425.

Anonymous. (2013). Analysis: Advanced Biofuel Companies Now in More Than Half of U.S. States; Clean Fuel Standards Fueling Growth in Top States. PR Newswire (Feb 6).

Banerjee G, Scott-Craig JS, Walton JD. (2010). Improving enzymes for biomass conversion: a basic research perspective. *BioEnergy Res*, 3(1), 82–92.

Bansal P, Hall M, Realff MJ, Lee JH, Bommarius AS. (2009). Modeling cellulase kinetics on lignocellulosic substrates. *Biotechnol Adv*, 27(6), 833–848.

Benko Z, Siika-aho M, Viikari L, Reczey K. (2008). Evaluation of the role of xyloglucanase in the enzymatic hydrolysis of lignocellulosic substrates. *Enzyme Microb Technol*, 43(2), 109–114.

Biely P, MacKenzie CR, Puls J, Schneider H. (1986). Cooperativity of esterases and xylanases in the enzymatic degradation of acetyl xylan. *Nat Biotechnol*, 4(8), 731–733.

Brownell HH, Saddler JN. (1987). Steam pretreatment of lignocellulosic material for enhanced enzymatic hydrolysis. *Biotechnol Bioeng*, 29(2), 228–235.

Bu LT, Beckham GT, Shirts MR, Nimlos MR, Adney WS, Himmel ME, Crowley MF. (2011). Probing carbohydrate product expulsion from a processive cellulase with multiple absolute binding free energy methods. *J Biol Chem*, 286(20), 18161–18169.

Cao Y, Tan HM. (2004). The effect of shear field on the hydrolysis of cellulose. *J Macromol Sci Phys*, B43(6), 1115–1121.

Carvajal D, Marchisio DL, Bensaid S, Fino D. (2012). Enzymatic hydrolysis of lignocellulosic biomasses via CFD and experiments. *Ind Eng Chem Res*, 51(22), 7518–7525.

Chang VS, Holtzapple MT. (2000). Fundamental factors affecting biomass enzymatic reactivity. *Appl Biochem Biotechnol*, 84–86, 5–37.

Chang VS, Nagwani M, Kim CH, Holtzapple MT. (2001). Oxidative lime pretreatment of high-lignin biomass—poplar wood and newspaper. *Appl Biochem Biotechnol*, 94(1), 1–28.

Chen M, Zeng GM, Tan ZY, Jiang M, Li H, Liu LF, Zhu Y, Yu Z, Wei Z, Liu YY, Xie GX. (2011). Understanding lignin-degrading reactions of ligninolytic enzymes: binding affinity and interactional profile. *PLoS One*, 6(9), e25647.

Chen XW, Shekiro J, Franden MA, Wang W, Zhang M, Kuhn E, Johnson DK, Tucker MP. (2012). The impacts of deacetylation prior to dilute acid pretreatment on the bioethanol process. *Biotechnol Biofuels*, 5, 8.

Chundawat SPS, Vismeh R, Sharma LN, Humpula JF, Sousa LD, Chambliss CK, Jones AD, Balan V, Dale BE. (2010). Multifaceted characterization of cell wall decomposition products formed during ammonia fiber expansion (AFEX) and dilute acid based pretreatments. *Bioresour Technol*, 101(21), 8429–8438.

Collins T, Gerday C, Feller G. (2005). Xylanases, xylanase families and extremophilic xylanases. *FEMS Microbiol Rev*, 29(1), 3–23.

Dale MC, Moelhman M. (2009). Enzymatic Simultaneous Saccharification and Fermentation (SSF) of Biomass to Ethanol in a Pilot 130 Liter Multistage Continuous Reactor Separator. Tech. Rep. Bio-Process Innovation, Inc., W. Lafayette, IN.

Ehrhardt MR, Monz TO, Root TW, Connelly RK, Scott CT, Klingenberg DJ. (2010). Rheology of dilute acid hydrolyzed corn stover at high solids concentration. *Appl Biochem Biotechnol*, 160(4), 1102–1115.

Elander RT, Dale BE, Holtzapple M, Ladisch MR, Lee YY, Mitchinson C, Saddler JN, Wyman CE. (2009). Summary of findings from the biomass refining consortium for applied fundamentals and innovation (CAFI): corn stover pretreatment. *Cellulose*, 16(4), 649–659.

Fan LT, Lee YH, Beardmore DR. (1981). The influence of major structural features of cellulose on rate of enzymatic hydrolysis. *Biotechnol Bioeng*, 23(2), 419–424.

Gan Q, Allen SJ, Taylor G. (2005). Analysis of process integration and intensification of enzymatic cellulose hydrolysis in a membrane bioreactor. *J Chem Technol Biotechnol*, 80(6), 688–698.

Ghosh P, Pamment NB, Martin WRB. (1982). Simultaneous saccharification and fermentation of cellulose—effect of beta-d-glucosidase activity and ethanol inhibition of cellulases. *Enzyme Microb Technol*, 4(6), 425–430.

Gorshkova T, Mikshina P, Gurjanov O, Chemikosova S. (2010). Formation of plant cell wall supramolecular structure. *Biochemistry(Moscow)*, 75(2), 159–172.

Grethlein HE. (1985). The effect of pore size distribution on the rate of enzymatic hydrolysis of cellulosic substrates. *Nat Biotechnol*, 3(2), 155–160.

Griggs AJ, Stickel JJ, Lischeske JJ. (2012a). A mechanistic model for enzymatic saccharification of cellulose using continuous distribution kinetics I: depolymerization by EGI and CBHI. *Biotechnol Bioeng*, 109(3), 665–675.

Griggs AJ, Stickel JJ, Lischeske JJ. (2012b). A mechanistic model for enzymatic saccharification of cellulose using continuous distribution kinetics II: cooperative enzyme action, solution kinetics, and product inhibition. *Biotechnol Bioeng*, 109(3), 676–685.

Ha SJ, Galazka JM, Kim SR, Choi JH, Yang XM, Seo JH, Glass NL, Cate JHD, Jin YS. (2011). Engineered *Saccharomyces cerevisiae* capable of simultaneous cellobiose and xylose fermentation. *Proc Natl Acad Sci USA*, 108(2), 504–509.

Harris PV, Welner D, McFarland KC, Re E, Navarro Poulsen JC, Brown K, Salbo R, Ding H, Vlasenko E, Merino S, Xu F, Cherry J, Larsen S, Lo Leggio L. (2010). Stimulation of lignocellulosic biomass hydrolysis by proteins of glycoside hydrolase family 61: structure and function of a large, enigmatic family. *Biochemistry*, 49(15), 3305–3316.

Himmel ME, Ding SY, Johnson DK, Adney WS, Nimlos MR, Brady JW, Foust TD. (2007). Biomass recalcitrance: engineering plants and enzymes for biofuels production. *Science*, 315(5813), 804–807.

Hodge DB, Karim MN, Schell DJ, McMillan JD. (2008). Soluble and insoluble solids contributions to high-solids enzymatic hydrolysis of lignocellulose. *Bioresour Technol*, 99(18), 8940–8948.

Hodge DB, Karim MN, Schell DJ, McMillan JD. (2009). Model-based fed-batch for high-solids enzymatic cellulose hydrolysis. *Appl Biochem Biotechnol*, 152(1), 88–107.

Humbird D, Aden A. (2009). Biochemical production of ethanol from corn stover: 2008 state of technology model. Tech. Rep. *NREL/TP-5100–47764*, National Renewable Energy Laboratory, Golden, CO.

Humbird D, Mohagheghi A, Dowe N, Schell DJ. (2010). Economic impact of total solids loading on enzymatic hydrolysis of dilute acid pretreated corn stover. *Biotechnol Prog*, 26(5), 1245–1251.

Humbird D, Davis R, Tao L, Kinchin C, Hsu D, Aden A, Schoen P, Lukas J, Olthof B, Worley M, Sexton D, Dudgeon D. (2011). Process design and economics for biochemical conversion of lignocellulosic biomass to ethanol: dilute-acid pretreatment and enzymatic hydrolysis of corn stover. Tech. Rep. *NREL/TP-5100-47764*, National Renewable Energy Laboratory, Golden, CO.

Irwin DC, Zhang S, Wilson DB. (2000). Cloning, expression and characterization of a family 48 exocellulase, Cel48a, from *Thermobifida fusca*. *Eur J Biochem*, 267(16), 4988–4997.

Iyer PV, Wu ZW, Kim SB, Lee YY. (1996). Ammonia recycled percolation process for pretreatment of herbaceous biomass. *Appl Biochem Biotechnol*, 57–58, 121–132.

Jalak J, Valjamae P. (2010). Mechanism of initial rapid rate retardation in cellobiohydrolase catalyzed cellulose hydrolysis. *Biotechnol Bioeng*, 106(6), 871–883.

Jeffries TW. (1990). Biodegradation of lignin-carbohydrate complexes. *Biodegradation*, 1(2), 163–176.

Jeoh T, Ishizawa CI, Davis MF, Himmel ME, Adney WS, Johnson DK. (2007). Cellulase digestibility of pretreated biomass is limited by cellulose accessibility. *Biotechnol Bioeng*, 98(1), 112–122.

Jørgensen H, Vibe-Pedersen J, Larsen J, Felby C. (2007). Liquefaction of lignocellulose at high-solids concentrations. *Biotechnol Bioeng*, 96(5), 862–870.

Kim TH, Lee YY. (2005). Pretreatment of corn stover by soaking in aqueous ammonia. *Appl Biochem Biotechnol*, 121, 1119–1131.

Klinke HB, Ahring BK, Schmidt AS, Thomsen AB. (2002). Characterization of degradation products from alkaline wet oxidation of wheat straw. *Bioresour Technol*, 82(1), 15–26.

Knutsen JS, Davis RH. (2004). Cellulase retention and sugar removal by membrane ultrafiltration during lignocellulosic biomass hydrolysis. *Appl Biochem Biotechnol*, 114, 585–599.

Knutsen JS, Liberatore MW. (2009). Rheology of high-solids biomass slurries for biorefinery applications. *J Rheol*, 53(4), 877–892.

Knutsen JS, Liberatore MW. (2010). Rheology modification and enzyme kinetics of high solids cellulosic slurries. *Energy Fuels*, 24, 3267–3274.

Kohlmann KL, Westgate P, Velayudhan A, Weil J, Sarikaya A, Brewer MA, Hendrickson RL, Ladisch MR. (1996). Enzyme conversion of lignocellulosic plant materials for resource recovery in a controlled ecological life support system. *Adv Space Res*, 18(1–2), 251–265.

Kong FR, Engler CR, Soltes EJ. (1992). Effects of cell-wall acetate, xylan backbone, and lignin on enzymatic hydrolysis of aspen wood. *Appl Biochem Biotechnol*, 34–35, 23–35.

Kristensen JB, Felby C, Jorgensen H. (2009). Determining yields in high solids enzymatic hydrolysis of biomass. *Appl Biochem Biotechnol*, 156(1–3), 557–562.

Lavenson DM, Tozzi EJ, Karuna N, Jeoh T, Powell RL, McCarthy MJ. (2012). The effect of mixing on the liquefaction and saccharification of cellulosic fibers. *Bioresour Technol*, 111, 240–247.

Lee CH, Teng Q, Zhong RQ, Ye ZH. (2011). The four *Arabidopsis* reduced wall acetylation genes are expressed in secondary wall-containing cells and required for the acetylation of xylan. *Plant Cell Physiol*, 52(8), 1289–1301.

Levine SE, Fox JM, Blanch HW, Clark DS. (2010). A mechanistic model of the enzymatic hydrolysis of cellulose. *Biotechnol Bioeng*, 107(1), 37–51.

Levine SE, Fox JM, Clark DS, Blanch HW. (2011). A mechanistic model for rational design of optimal cellulase mixtures. *Biotechnol Bioeng*, 108(11), 2561–2570.

Li Y, Irwin DC, Wilson DB. (2007). Processivity, substrate binding, and mechanism of cellulose hydrolysis by *Thermobifida fusca* Cel9a. *Appl Environ Microbiol*, 73(10), 3165–3172.

Liu CG, Wyman CE. (2003). The effect of flow rate of compressed hot water on xylan, lignin, and total mass removal from corn stover. *Ind Eng Chem Res*, 42(21), 5409–5416.

Lloyd TA, Wyman CE. (2005). Combined sugar yields for dilute sulfuric acid pretreatment of corn stover followed by enzymatic hydrolysis of the remaining solids. *Bioresour Technol*, 96(18), 1967–1977.

Luterbacher JS, Parlange JY, Walker LP. (2012). A pore-hindered diffusion and reaction model can help explain the importance of pore size distribution in enzymatic hydrolysis of biomass. *Biotechnol Bioeng*, 110, 127–136. doi:10.1002/bit.24614

Mais U, Esteghlalian AR, Saddler JN. (2002). Influence of mixing regime on enzymatic saccharification of steam-exploded softwood chips. *Appl Biochem Biotechnol*, 98, 463–472.

Margolles-Clark E, Tenkanen M, Söderlund H, Penttilä M. (1996). Acetyl xylan esterase from *Trichoderma reesei* contains an active-site serine residue and a cellulose-binding domain. *Eur J Biochem*, 237(3), 553–560.

Matthews JF, Skopec CE, Mason PE, Zuccato P, Torget RW, Sugiyama J, Himmel ME, Brady JW. (2006). Computer simulation studies of microcrystalline cellulose Iß. *Carbohydr Res*, 341(1), 138–152.

Matthews JF, Himmel ME, Brady JW. (2010). Simulations of the structure of cellulose. In: Nimlos MR, Crowley MF, editors. *Computational Modeling in Lignocellulosic Biofuel Production*, Vol. 1052: ACS Symposium Series. American Chemical Society, pp. 17–53.

McMillan JD. (1997). Bioethanol production: status and prospects. *Renew Energy*, 10(2–3), 295–302.

McMillan JD, Jennings EW, Mohagheghi A, Zuccarello M. (2011). Comparative performance of precommercial cellulases hydrolyzing pretreated corn stover. *Biotechnol Biofuels*, 4, 29.

Merino ST, Cherry J. (2007). Progress and challenges in enzyme development for biomass utilization. *Adv Biochem Eng Biotechnol*, 108, 95–120.

Mok WSL, Antal MJ. (1992). Uncatalyzed solvolysis of whole biomass hemicellulose by hot compressed liquid water. *Ind Eng Chem Res*, 31(4), 1157–1161.

Mosier N, Hendrickson R, Ho N, Sedlak M, Ladisch MR. (2005a). Optimization of pH controlled liquid hot water pretreatment of corn stover. *Bioresour Technol*, 96(18), 1986–1993.

Mosier N, Wyman C, Dale B, Elander R, Lee YY, Holtzapple M, Ladisch M. (2005b). Features of promising technologies for pretreatment of lignocellulosic biomass. *Bioresour Technol*, 96(6), 673–686.

Nagy T, Emami K, Fontes CMGA, Ferreira LMA, Humphry DR, Gilbert HJ. (2002). The membrane-bound alpha-glucuronidase from *Pseudomonas cellulosa* hydrolyzes 4-O-methyl-D-glucuronoxylooligosaccharides but not 4-O-methyl-D-glucuronoxylan. *J Bacteriol*, 184(17), 4925–4929.

Nagy T, Nurizzo D, Davies GJ, Biely P, Lakey JH, Bolam DN, Gilbert HJ. (2003). The alpha-glucuronidase, Glca67a, of *Cellvibrio japonicus* utilizes the carboxylate and methyl groups of aldobiouronic acid as important substrate recognition determinants. *J Biol Chem*, 278(22), 20286–20292.

Naran R, Black S, Decker S, Azadi P. (2009). Extraction and characterization of native heteroxylans from delignified corn stover and aspen. *Cellulose*, 16(4), 661–675.

Numan M, Bhosle N. (2006). Alpha-L-arabinofuranosidases: the potential applications in biotechnology. *J Ind Microbiol Biotechnol*, 33(4), 247–260.

Nutt A, Sild V, Pettersson G, Johansson G. (1998). Progress curves. *Eur J Biochem*, 258(1), 200–206.

Okazaki M, Mooyoung M. (1978). Kinetics of enzymatic-hydrolysis of cellulose: analytical description of a mechanistic model. *Biotechnol Bioeng*, 20(5), 637–663.

Perez DD, Guillemain A, Berthelot A, N'Guyen-The N, De Morogues F, Gomes C. (2010). Evaluation of forestry biomass quality for the production of second-generation biofuels. *Cell Chem Technol*, 44(1–3), 1–14.

Qing Q, Yang B, Wyman CE. (2010). Xylooligomers are strong inhibitors of cellulose hydrolysis by enzymes. *Bioresour Technol*, 101(24), 9624–9630.

Quinlan RJ, Sweeney MD, Lo Leggio L, Otten H, Poulsen JCN, Johansen KS, Krogh KBRM, Jørgensen CI, Tovborg M, Anthonsen A, Tryfona T, Walter CP, Dupree P, Xu F, Davies GJ, Walton PH. (2011). Insights into the oxidative degradation of cellulose by a copper metalloenzyme that exploits biomass components. *Proc Natl Acad Sci USA*, 108(37), 15079–15084.

Roberts KM, Lavenson DM, Tozzi EJ, McCarthy MJ, Jeoh T. (2011). The effects of water interactions in cellulose suspensions on mass transfer and saccharification efficiency at high solids loadings. *Cellulose*, 18(3), 759–773.

Roche CM, Dibble CJ, Knutsen JS, Stickel JJ, Liberatore MW. (2009a). Particle concentration and yield stress of biomass slurries during enzymatic hydrolysis at high-solids loadings. *Biotechnol Bioeng*, 104(2), 290–300.

Roche CM, Dibble CJ, Stickel JJ. (2009b). Laboratory-scale method for enzymatic saccharification of lignocellulosic biomass at high-solids loadings. *Biotechnol Biofuels*, 2, 28.

Saha BC, Cotta MA. (2006). Ethanol production from alkaline peroxide pretreated enzymatically saccharified wheat straw. *Biotechnol Prog*, 22(2), 449–453.

Samaniuk JR, Scott CT, Root TW, Klingenberg DJ. (2012). Rheological modification of corn stover biomass at high solids concentrations. *J Rheol*, 56(3), 649–665.

Schell DJ, Farmer J, Newman M, McMillan JD. (2003). Dilute-sulfuric acid pretreatment of corn stover in pilot-scale reactor—investigation of yields, kinetics, and enzymatic digestibilities of solids. *Appl Biochem Biotechnol*, 105, 69–85.

Scheller HV, Ulvskov P. (2010). Hemicelluloses. *Annu Rev Plant Biol*, 61(1), 263–289.

Schroeder LR, Gentile VM, Atalla RH. (1986). Nondegradative preparation of amorphous cellulose. *J Wood Chem Technol*, 6(1), 1–14.

Selig MJ, Knoshaug EP, Adney WS, Himmel ME, Decker SR. (2008). Synergistic enhancement of cellobiohydrolase performance on pretreated corn stover by addition of xylanase and esterase activities. *Bioresour Technol*, 99(11), 4997–5005.

Selig M, Adney W, Himmel M, Decker S. (2009). The impact of cell wall acetylation on corn stover hydrolysis by cellulolytic and xylanolytic enzymes. *Cellulose*, 16(4), 711–722.

Shao XJ, Lynd L, Bakker A, LaRoche R, Wyman C. (2010). Reactor scale up for biological conversion of cellulosic biomass to ethanol. *Bioprocess Biosyst Eng*, 33(4), 485–493.

Shekiro J, Kuhn EM, Selig MJ, Nagle NJ, Decker SR, Elander RT. (2012). Enzymatic conversion of xylan residues from dilute acid-pretreated corn stover. *Appl Biochem Biotechnol*, 168(2), 421–433.

Sørensen HR, Meyer AS, Pedersen S. (2003). Enzymatic hydrolysis of water-soluble wheat arabinoxylan. 1. Synergy between α-l-arabinofuranosidases, endo-1,4-ß-xylanases, and ß-xylosidase activities. *Biotechnol Bioeng*, 81(6), 726–731.

Sørensen HR, Pedersen S, Meyer AS. (2007). Synergistic enzyme mechanisms and effects of sequential enzyme additions on degradation of water insoluble wheat arabinoxylan. *Enzyme Microb Technol*, 40(4), 908–918.

Srisodsuk M, Kleman-Leyer K, Keranen S, Kirk TK, Teeri TT. (1998). Modes of action on cotton and bacterial cellulose of a homologous endoglucanase–exoglucanase pair from *Trichoderma reesei*. *Eur J Biochem*, 251(3), 885–892.

Stickel JJ, Knutsen JS, Liberatore MW, Luu W, Bousfield DW, Klingenberg DJ, Scott CT, Root TW, Ehrhardt MR, Monz TO. (2009). Rheology measurements of a biomass slurry: an inter-laboratory study. *Rheol Acta*, 48(9), 1005–1015.

Sun Y, Cheng JY. (2002). Hydrolysis of lignocellulosic materials for ethanol production: a review. *Bioresour Technol*, 83(1), 1–11.

Teter SA, Xu F, Nedwin GE, Cherry JR. (2006). Enzymes for biorefineries. In: Kamm B, Gruber P, Michael K, editors. *Biorefineries—Industrial Processes and Products*, vol. 1. Weinheim, Germany: Wiley-VCH Verlag GmbH, pp. 357–383.

Um BH, Hanley TR. (2008). A comparison of simple rheological parameters and simulation data for *Zymomonas mobilis* fermentation broths with high substrate loading in a 3-l bioreactor. *Appl Biochem Biotechnol*, 145(1–3), 29–38.

van den Brink J, de Vries R. (2011). Fungal enzyme sets for plant polysaccharide degradation. *Appl Microbiol Biotechnol*, 91(6), 1477–1492.

Varga E, Schmidt AS, Reczey K, Thomsen AB. (2003). Pretreatment of corn stover using wet oxidation to enhance enzymatic digestibility. *Appl Biochem Biotechnol*, 104(1), 37–50.

Vassilev SV, Baxter D, Andersen LK, Vassileva CG. (2010). An overview of the chemical composition of biomass. *Fuel*, 89(5), 913–933.

Viamajala S, Donohoe BS, Decker SR, Vinzant TB, Selig MJ, Himmel ME, Tucker MP. (2010). Heat and mass transport in processing of lignocellulosic biomass for fuels and chemicals. In:

Singh OV, Harvey SP, editors. *Sustainable Biotechnology: Sources of Renewable Energy*. Springer, pp. 1–18.

Wang LJ, Luo ZL, Shahbazi A. (2013). Optimization of simultaneous saccharification and fermentation for the production of ethanol from sweet sorghum (*Sorghum bicolor*) bagasse using response surface methodology. *Ind Crop Prod*, 42, 280–291.

Weimer PJ, Zeikus JG. (1977). Fermentation of cellulose and cellobiose by *Clostridium thermocellum* in absence and presence of *Methanobacterium thermoautotrophicum*. *Appl Environ Microbiol*, 33(2), 289–297.

Weiss ND, Nagle NJ, Tucker MP, Elander RT. (2009). High xylose yields from dilute acid pretreatment of corn stover under process-relevant conditions. *Appl Biochem Biotechnol*, 155(1–3), 418–428.

Xu F, Ding HS. (2007). A new kinetic model for heterogeneous (or spatially confined) enzymatic catalysis: contributions from the fractal and jamming (overcrowding) effects. *Appl Catal A Gen*, 317(1), 70–81.

Yang SJ, Kataeva I, Wiegel J, Yin YB, Dam P, Xu Y, Westpheling J, Adams MWW. (2010). Classification of '*Anaerocellum thermophilum*' strain DSM 6725 as *Caldicellulosiruptor bescii* sp. nov. *Int J Syst Evol Microbiol*, 60, 2011–2015.

Zhang YHP, Lynd LR. (2006). A functionally based model for hydrolysis of cellulose by fungal cellulase. *Biotechnol Bioeng*, 94(5), 888–898.

Zhang J, Siika-aho M, Tenkanen M, Viikari L. (2011). The role of acetyl xylan esterase in the solubilization of xylan and enzymatic hydrolysis of wheat straw and giant reed. *Biotechnol Biofuels*, 4(1), 60.

Zhao XB, Zhang LH, Liu DH. (2012). Biomass recalcitrance. Part I: the chemical compositions and physical structures affecting the enzymatic hydrolysis of lignocellulose. *Biofuels Bioprod Biorefin*, 6(4), 465–482.

Zhou W, Hao ZQ, Xu Y, Schuttler HB. (2009a). Cellulose hydrolysis in evolving substrate morphologies II: numerical results and analysis. *Biotechnol Bioeng*, 104(2), 275–289.

Zhou W, Schuttler HB, Hao ZQ, Xu Y. (2009b). Cellulose hydrolysis in evolving substrate morphologies I: a general modeling formalism. *Biotechnol Bioeng*, 104(2), 261–274.

Zhou W, Xu Y, Schuttler HB. (2010). Cellulose hydrolysis in evolving substrate morphologies III: time-scale analysis. *Biotechnol Bioeng*, 107(2), 224–234.

Zhu YM, Malten M, Torry-Smith M, McMillan JD, Stickel JJ. (2011). Calculating sugar yields in high solids hydrolysis of biomass. *Bioresour Technol*, 102(3), 2897–2903.

Zverlov V, Mahr S, Riedel K, Bronnenmeier K. (1998). Properties and gene structure of a bifunctional cellulolytic enzyme (CelA) from the extreme thermophile '*Anaerocellum thermophilum*' with separate glycosyl hydrolase family 9 and 48 catalytic domains. *Microbiology*, 144(2), 457–465.

CHAPTER 5

Production of Cellulolytic Enzymes

RANJITA BISWAS

BioEnergy Science Center, Biosciences Division, Oak Ridge National Laboratory, Oak Ridge, TN, USA

ABHISHEK PERSAD and VIRENDRA S. BISARIA

Department of Biochemical Engineering and Biotechnology, Indian Institute of Technology Delhi, New Delhi, India

Abstract

For the conversion of plant biomass into various bioproducts, a significant bottleneck is enzymatic hydrolysis of lignocelluloses to soluble sugars. These sugars are then metabolized through various natural or engineered pathways toward products of interest. The success of projected biorefinery processes depend to a large extent on the economics of hydrolytic enzyme production. Presently, mesophilic fungal strains

Bioprocessing of Renewable Resources to Commodity Bioproducts, First Edition.
Edited by Virendra S. Bisaria and Akihiko Kondo.

like *Trichoderma reesei* and *Aspergillus niger* produce cellulases at an industrial scale. Various recombinant cellulases have been successfully expressed in industrial strains which can improve the economics due to their high specificities for targeted bioproducts. In addition, designer cellulosomes and xylanosomes are expected to make the hydrolytic enzymes more effective because of efficient surface binding and direct action on lignocelluloses. This chapter describes various genetic and process approaches which have been used to increase the production of cellulase and xylanase enzymes.

5.1 INTRODUCTION

With the growing human population, the world is facing tremendous pressure to meet its needs of food, feed, chemicals, and energy, and also to balance the demand and supply in keeping with environmental safeguards. The natural gas and oil fields are shrinking fast to meet the demands of our progress. Today, alternate energy sources from environmental wastes and energy crops are high priority research interests worldwide. For production of biocommodities from biomass, one of the major bottlenecks has been the efficient hydrolysis of lignocelluloses to sugars. Primarily, lignocelluloses comprise cellulose (~30 to 50%), hemicellulose (~20% to 35%), and lignin (~15% to 25%). This composition can vary amongst plant species and plant tissues. Even in the same species of corn, there is considerable variability in the composition of corn stover due to genetic background and environmental conditions (Templeton et al., 2009). The plant kingdom produces lignocellulosic biomass to the tune of 10–50 billion tons per annum worldwide (Sticklen, 2006). This renewable biomass can be hydrolyzed to pentose and hexose sugars that serve as building blocks for various industrial products (Chandrakant and Bisaria, 1998; Lynd et al., 2005; Bevan and Franssen, 2006; Bayer et al., 2007; Hatti-Kaul et al., 2007; Madhavan et al., 2012; Seiboth et al., 2012). For this, enzymatic methods to hydrolyze the plant biomass are preferred due to non-inhibitory by-products and nontoxic effluents.

Hydrolytic enzymes like cellulases and xylanases convert lignocelluloses to sugars that can be fermented by various microbes to biofuels and other value-added products. The relatively high cost of these enzymes remains a major barrier to their commercial application in any bioindustry, although significant reduction in the cost of these enzymes has been made in recent years. The focus areas of research have been to improve the efficiency of known enzymes, identify new and more active enzymes, find optimized enzyme mixes for pretreated lignocelluloses and reduce the cost of enzyme production (Merino and Cherry, 2007). Also, for an industrial process to be economically viable, enzymatic breakdown of lignocellulose to fermentable sugars must occur as quickly as possible, preferably in hours. This chapter describes various promising microbial sources of hydrolytic enzymes, and the genetic and process methods employed to produce the enzymes.

5.2 HYDROLYTIC ENZYMES FOR DIGESTION OF LIGNOCELLULOSES

5.2.1 Cellulases

Cellulase is the most prominent group of hydrolytic enzymes that catalyze the hydrolysis of β-1,4 linkages present in cellulose to give glucose. This hexose is fed into the central metabolic pathways of various microorganisms to produce various bioproducts. The cellulases are primarily produced in nature by plants, fungi, bacteria, and even some protozoa, mollusks, and nematodes (Watanabe and Tokuda, 2001). Presently, the common microbial sources of cellulases for industrial-scale production are mesophiles grown in the temperature range of 30–35°C, such as *Trichoderma reesei* (also known as *Hypocrea jecorina*) and *Aspergillus niger*. The enzymes from thermophiles grown in the temperature range of 50–90°C are advantageous for harsh industrial processes, have faster reaction rates and are less prone to contamination. A few examples of such promising microorganisms are *Clostridium thermocellum*, *Thermoanaerobacter* sp, and *Thermotoga maritima* (Gomes and Steiner, 2004). The enzymes from psychrophiles such as *Clostridium* strain PXYL1 (Akila and Chandra, 2003) and *Pseudoalteromonas haloplanktis* (Violot et al., 2003) which grow in the temperature range of 5–20°C have advantages in applications where temperature is detrimental to the product and also reduce the energy consumption. Hence, depending on the suitability of the desired product, an appropriate microbial producer of hydrolytic enzymes can be selected from a wide array available in nature.

Cellulases are multienzyme complexes that consist of three major components: (1) endo-β1-4-glucanases (endo-β1-4-D-glucan 4-glucanohydrolase, EC 3.2.1.4), (2) exo-β1-4-glucanase or cellobiohydrolase (exo-β1-4-D-glucan 4-cellobiohydrolase, EC 3.2.1.91), and (3) β-glucosidase (EC 3.2.1.21); some cellulase systems also contain exo-β1-4-D-glucan 4-glucohydrolase (EC 3.2.1.74) and exo-β1-4-cellobiosidase (EC 3.2.1.176). These three components act synergistically to hydrolyze cellulose to glucose. Endo-glucanase hydrolyzes internal β-1,4 linkages of cellulose chains and creates new reducing and nonreducing ends. Thereupon exoglucanase cleaves disaccharide cellobiose from the nonreducing end (cellobiohydrolase) and in some cases from the reducing end (cellobiosidase) of the cellulose chain. These cellobiose units and short-chain cellodextrins are hydrolyzed by β-glycosidase into individual monomeric units of glucose (Beguin and Aubert, 1994). Various forms of these enzymes are present in different species to digest diverse types of celluloses present in nature. Therefore, the cellulase activities are expressed in terms of the substrates used to quantify them; for example, hydrolysis of carboxymethyl cellulose is measured as carboxymethyl cellulase (CMCase), filter paper hydrolysis as filter paper cellulase units (FPU), Avicel cellulose digestion as Avicelase, cotton fiber hydrolysis as cotton cellulase, etc.

The best known producer of cellulases, *T. reesei*, has two exoglucanases, Cel6A and Cel7A. Cel6A digests cellulose chain from the nonreducing end and Cel7A from the reducing end. It is supported by 8 endoglucanses and 7 β-glycosidases in the multienzyme cellulase system (Martinez et al., 2008). The genome sequence of hyperthermophile, *Caldicellulosiruptor bescii* and *Caldicellulosiruptor saccharolyticus*, reveals diverse glycoside hydrolase families. It has at least four endo- or exo-acting cellulases and a β-glycosidase with the ability to hydrolyze crystalline cellulose (Blumer-Schuette et al., 2010; Kanafusa-Shinkai et al., 2013). The diversity of cellulases in nature secreted by various microbes makes them versatile and capable of digesting almost all kinds of plant material made of cellulose.

Different mechanisms for digestion of cellulose are employed by various cellulase-producing microorganisms. Most aerobes have cellulases made up of carbohydrate-binding module (CBM), a flexible linker protein and catalytic domain. CBMs are responsible for increasing the interaction between cellulase and cellulose surface. Higher binding affinity of CBM has been shown to be responsible for higher cellulose conversion (Linder et al., 1995; Takashima et al., 2007). Soluble forms such as carboxymethyl cellulose and amorphous forms of cellulose do not require CBM for hydrolysis. The crystalline form of cellulose requires CBM of cellulases for its hydrolysis (Rabinovich, 2002). The cellulosomal cellulases consist of a dockerin domain that binds to cohesion modules of the scaffolding subunit. The dockerin–cohesin complex binds to enzyme by a flexible linker peptide to display catalytic domain. The scaffolding also has a CBM to recognize and bind to cellulosic substrates (Bayer et al., 2004).

5.2.2 Xylanases

For maximizing hydrolysis of lignocellulosic residues by cellulase enzymes, synergistic action of xylanase is required (Ghose and Bisaria, 1979). The cellulose fibers in plant biomass are held together by lignin and xylan. The major enzymes responsible for the hydrolysis of xylan are xylanase (endo-1,4-β-D-xylanohydrolase, EC 3.2.1.8) and β-xylosidase (1,4-β-D-xylohydrolase, EC 3.2.1.37). Xylanase cleaves internal β-1,4 linkages of the xylan backbone. β-Xylosidase hydrolyzes xylobiose and small xylooligosaccharides to xylose and facilitates the hydrolysis of xylan. A few additional debranching enzymes are also required for complete hydrolysis of xylans by removal of the side chain substituents from the xylan backbone. Often xylans are in partially acetylated form, which are cleaved by acetylxylan esterase (EC 3.1.1.72) (Tenkanen and Poutanen, 1992). Arabinosyl substituents are removed by α-L-arabinofuranosidase (EC 3.2.1.55), which results in an increase in the number of potential access points of xylanase to the xylan backbone (Dekker, 1985). 4-*O*-Methyl glucuronic acid side groups, which are among the major substituents of wood xylans, form ester linkages between uronic acid residues and lignin. The debranching enzyme α-glucuronidase acts synergistically with xylanases to digest glucuronoxylan and releases 4-*O*-methyl glucuronic acid (Puls et al., 1987). Thus, xylanase and various debranching enzymes act together to breakdown xylan to primarily pentose sugars while improving accessibility of cellulases to hydrolyze cellulose fibers.

Xylanases are synthesized by fungi, bacteria, marine algae, insects, seeds, etc. but the principal source of commercial xylanases is filamentous fungi. The fungi secrete xylanase into the extracellular medium along with several accessory xylanolytic enzymes for debranching substituted xylans. There is considerable interest in the enzymes of xylan hydrolysis for their use as a supplement in animal feed, manufacture of bread, food, textiles, pulp bleaching and in the production of ethanol and xylitol (Polizeli et al., 2005). For commercial production of these enzymes, the focus is on utilization of agro-residual wastes along with development of efficient bioprocess strategies to obtain high-enzyme titers.

5.3 DESIRABLE ATTRIBUTES OF CELLULASE FOR HYDROLYSIS OF CELLULOSE

The hydrolytic enzymes which are to be produced should be of desirable characteristics for their application in the hydrolysis of lignocelluloses. Some of the desirable characteristics of the enzymes are (1) catalytic efficiency, (2) thermal stability, (3) adsorption, (4) end-product inhibition resistance, and (5) shear inactivation (Kloysov, 1988; Bisaria, 1998). The catalytic efficiency, for example, of one of the most active endoglucanases from *T. reesei* on crystalline cellulose is on the order of 0.051 per second. Glucoamylase, on the other hand, has catalytic efficiency of 58 per second which is nearly the same as shown by endoglucanase against soluble cellulose, CMC. This indicates that the catalytic efficiency is low for crystalline cellulose owing to the inherent features of cellulose. Therefore, crystalline cellulose is normally subjected to a pretreatment process for efficient cellulose hydrolysis. The desirable attributes of cellulase for efficient hydrolysis of cellulose and the impact of these attributes on cellulose hydrolysis are listed in Table 5.1. A review of cellulase properties produced by *T. reesei* indicates that they also have the following disadvantages under the actual operating conditions of cellulose hydrolysis.

(a) Low thermostability which requires replenishment of lost activity at prolonged usage.
(b) High inhibition which results in cessation of hydrolysis at increased concentration of glucose and cellobiose.
(c) Inactivation by shear due to mixing by impellers in cellulose hydrolysis reactors.
(d) Low adsorption of some cellulase components on crystalline cellulose, which may result in their washing away in a flow-through reactor during continuous hydrolysis.

In view of the projected large-scale usage of cellulase enzymes in bioconversion of lignocellulosic residues, there is a continuous renewed interest in search of novel sources of cellulases which produce these enzymes with desired novel properties that cause rapid hydrolysis of cellulose. Several research groups, including those from National Renewable Energy Laboratory (NREL), Genencor, Novozymes, etc., have reported improvements in one or more of these properties, especially with respect to thermal stability (Mousdale, 2010). It is pertinent to mention here that such an

TABLE 5.1 Desirable Attributes of Cellulase for Efficient Hydrolysis of Cellulose

Attributes	Impact
1. High specific activity	Enzymes with higher specific activity (units/mg protein) will be required in lower quantities to achieve the same degree of hydrolysis.
2. High catalytic efficiency against crystalline cellulose	Cellulases with high catalytic efficiency will reduce the time required to hydrolyze crystalline cellulose.
3. High thermostability	Thermostable enzymes can be used at higher temperatures to accelerate the rates of cellulose hydrolysis.
4. Resistance to end-product inhibition	Most cellulases are inhibited by the products of cellulose hydrolysis (i.e, glucose and cellobiose). End-product inhibition resistant cellulases can produce more concentrated mixtures of sugars, which will obviate the need to concentrate them before bioconversion into bioproducts such as ethanol.
5. Stability against shear forces	Shear-resistant cellulases will be useful where agitation must be provided to suspend the solid cellulose in a hydrolysis reactor.

improvement is desirable in view of the fact that hydrolysis rate doubles with 10°C increase in reaction temperature.

5.4 STRATEGIES USED FOR ENHANCED ENZYME PRODUCTION

5.4.1 Genetic Methods

Strain improvement by random mutagenesis is often limited by the microbe's genetic makeup. To overcome such barriers toward developing a workhorse strain, various genetic tools are used with good success. Thus, a tailored enzyme for a specific need of a bioprocess can be designed through genetic engineering of the strain. Various industrial microbial strains like *T. reesei*, *A. niger*, *Pichia pastoris*, *Saccharomyces cerevisae*, *Escherichia coli,* and *Bacillus subtilis* have been genetically modified to express heterologous genes of commercial significance. A number of genetic approaches have been used over the last two decades to overexpress cellulase and xylanase genes. We have not made an attempt to cover all those procedures in this review. Nevertheless, a few recent approaches have been described in the following paragraphs which are indicative of the trend of research to overexpress the enzymes in suitable hosts under appropriate genetic controls.

In case of cellulase enzymes, endoglucanases from *Aureobasidium pullulans* ApCel5A, *Gloeophyllum trabeum* GtCel2A, and *Sporotrichum thermophile* StCel5A were expressed at relatively high levels ($\geq$0.3 g/L) in *A. niger*. ApCel5A and StCel5A hydrolyzed CMC five- and two-times faster, respectively, than *T. reesei* endoglucanase Cel5A. These recombinant strains can, therefore, be promising candidates for developing new cellulase system (Tambor et al. 2012). Again, a Cel7A exoglucanase from *T. reesei* expressed in *A. niger* was able to grow on spent bagasse hydrolysate and spent spruce hydrolysate with higher endoglucanase activity than *T. reesei*. This was done to overcome glucose repression in *T. reesei* strain; the cellulase gene in *A. niger* was under constitutive promoter (Alriksson et al., 2009). The post-translation glycosylation of proteins is known to influence the stability, conformation, secretion, and activity of enzymes (Lis and Sharon, 1993). *Trichoderma reesei* cellulase proteins are known to undergo hyper-glycosylation when expressed in strains such as *A. niger*, *S. cerevisiae,* and *Yarrowia lipolitica*. Recent studies revealed that removal of *N*-glycosylation site via alanine mutation from asparagine-270 slightly increased its activity on crystalline cellulose. In another fungus *Penicillium funiculosum*, removal of three *N*-glycan sites from rCel7A increased enzyme activity by 35%, whereas addition of *N*-glycan at asparagine-194 via mutation of alanine-196 to serine (to create an N-X-S motif) enhanced activity by 85% in *A. niger* (Adney et al., 2009).

Genome sequence of *T. reesei* is now available (Martinez et al., 2008). The cellulase genes are known to be regulated in a coordinated manner. The presence of at least three transcriptional activators, viz. XYR1, ACE 2, and HAH2/3/5 complex; and two repressors, viz. CRE1 and ACE1 has been reported (Seiboth et al., 2012). Up- and down-regulation of certain genes has been shown to result in overexpression of cellulase and xylanase genes. In one of the highest producers of cellulases, viz. *T. reesei* RUT C30, upregulation of *xyr1* gene under a strong *pdc* promoter and downregulation of negative regulator *ace1* increased total protein secretion, filter paper activity, and CMCase activity by 103%, 114%, and 134% respectively (Wang et al., 2013). The *ace1* repressor gene was knocked out in cellulase expression system of *Trichoderma koningii* YC01by ribonucleic acid interference (RNAi) method. The *ace1*-silenced strains had improved cellulase and xylanase production under inducing condition. After 5 days of cultivation, the expression levels of *cbh1*, *cbh2*, *egl1*, *egl2*, and *xyn2* in strain A553 of *T. koningii* were 1.9-, 2.0-, 0.6-, 1.8-, and 1.5-times higher, respectively, than the control strain (Wang et al., 2012). The subject of cellulase regulation in *T. reesei* and its manipulation through genetic engineering approaches have been elaborately discussed recently (Seiboth et al., 2012).

Suitable codon optimization of target genes can also be used to enhance the expression of heterologous protein production. Thus, cellobiohydrolase 1, cellobiohydrolase 2, and β-mannanase from *T. reesei* and xylanase A from *Thermomyces lanuginosus* were individually expressed after codon optimization and under synthetic alcohol oxidase 1 promoter in *P. pastoris*. Such recombinant *P. pastoris* expressed 1.142 g/L of β-mannanase, 6.55 g/L of cellobiohydrolase 2, and 1.2 g/L of xylanase A in fed-batch cultures (Mellitzer et al., 2012). A hyperthermophilic β-glycosidase (CelB) from *Pyrococcus furiosus* was expressed in *Lactobacillus plantarum* NC8 and *Lactobacillus casei* using pSIP system. The specific activity of CelB was 44% higher in *L. plantarum* than in *L. casei*. On cultivation of *L. plantarum* in a bioreactor using whey medium, 33.4 U/mg protein of CelB activity was obtained (Böhmer et al., 2012).

Typically, in aerobic bacteria, cellulolytic enzymes are produced in "free" form and work synergistically during cellulose hydrolysis. In contrast, several anaerobic bacteria have been found to produce a cell-associated enzyme complex called the cellulosome (Bayer et al., 2007). In addition, a few anaerobic fungi, such as *Orpinomyces* and *Piromyces*, have been reported to produce cellulosome (Doi and Kosugi, 2004). Recently, a facultative bacterium *Paenibacillus curdlanolyticus* B-6 was found to produce cellulosome-like multienzyme complexes under aerobic conditions (Waeonukul et al., 2009). Whether cellulase is in monomeric form or displayed on scaffolding protein in cellulosomes, it acts synergistically to attack complex polymeric substrates. This synergy can be achieved between cellulases from different microbial systems, between cellulosomal and noncellulosomal enzymes, between different types of enzymes from different families, and between enzymes that have different modes of action. For example, β-glucosidase (Gluc1C) and β-1,4-endoglucanase (Endo5A) from *Paenibacillus* sp. (isolated from the gut of cotton bollworm) were used to construct and express a bifunctional chimeric protein in *E. coli*. It was found that one of the several constructs, namely, EG5 generated 24% more sugar from alkali-treated rice straw than Endo5A or a mixture of Endo5A–Gluc1C enzymes. The chimeric protein EG5 is reported to have enhanced thermostability, specific activity, substrate affinity, and catalytic efficiency (Adlakha et al., 2012). Similarly, a synergy between different glycoside hydrolases of *Thermobifida fusca* was found. A designer cellulosome was constructed using two *T. fusca* cellulases, Cel48A exoglucanase and Cel5A endoglucanase, and two *T. fusca* xylanases, endoxylanases Xyn10B, and Xyn11A. The final tetravalent cellulosome assembly had ~2.4-fold enhanced activity on wheat straw, a complex cellulosic substrate (Moraïs et al., 2010). An earlier work showed that desired endoglucanase along with scaffolding protein can be expressed in a host of choice and produce designer mini cellulosomes with specific functions. Endoglucanase Eng B and scaffolding protein mini-CbpA1 from *Clostridium cellulovorans* were expressed and produced as designer cellulosomes in *B. subtilis* WB800 (Cho et al., 2004). Similarly, noncellulosomal cell surface display of hydrolytic enzymes using anchoring proteins can act as whole cell biocatalysts in biomass saccharification. For example, CBH2 (cellobiohydrolase 2) and EG2 (endoglucanase 2) from *T. reesei* and BGL1 (β-glucosidase 1) from *Aspergillus aculeatus* were integrated to C-terminal of cell-surface protein α-agglutinin of *S. cerevisae*. It was observed that increasing the copy number of *CBH2* increased saccharification efficiency by 1.7-fold and increasing the copy number of both *CBH2* and *EG2* improved saccharification efficiency by 2.5-fold of the resulting yeast strain (Matano et al., 2012). Such designer cellulosomes and whole cell biocatalysts may change complicated bioprocesses to cost-effective consolidated bioprocesses in future.

Efforts to overexpress xylanase in the industrial strain of *S. cerevisae* have been made by various groups. For example, endo-β-1,4 xylanase of *A. niger* IME-216 was expressed with 1.5-fold higher yield in *S. cerevisiae* strain YS2-2 compared to the parent (Tian et al., 2013). Endoxylanase from *Geobacillus thermodentrificans* TSAA1 was expressed in *E. coli* BL21 (DE3). It was characterized as thermostable at 70°C and alkalistable at pH 9.0 suggesting suitable application in various industrial processes (Verma et al., 2013). As thermostable enzymes have improved hydrolytic performance, many potential candidates are being studied for their expression and production of xylanases. Thermophilic and hyperthermophilic xylan-degrading organisms possess a homolog of Xyn10A, a multi-modular endoxylanase. The Xyn10A module is flanked on the N-terminus by a tandem repeat of carbohydrate binding module 22 (CBM 22) and on C-terminus by another tandem repeat of CBM 9 and repeats of surface layer homology (SLH) modules. It was found that N-terminal CBM 22 was critical for imparting thermostability and thermophilicity to xylanases like TsXynA from *Thermoanaerobacterium saccharolyticum* (Lee et al., 1993), TmXynA from *T. maritima* (Meissner et al., 2000) and PbXynC from *Paenibacillus barcinonensis* (Blanco et al., 1999). Partial genome analysis of thermophilic bacterium *Caldanaerobius polysaccharolyticus* predicted that Xyn10A is anchored to the surface of the bacterium. It had two families of CBM 22 at N-terminus and two families of CBM 9 and three SLH modules at C-terminus (Han et al., 2012).

The alcohol oxidase, AOX1, in *P. pastoris* has a strong inducible promoter and therefore is a promoter of choice by most researchers for the expression of xylanase genes in *P. pastoris*. It gives high levels of xylanase expression under methanol induction. Ruanglek et al. (2007) reported an enzyme activity of 3676 U/mL and volumetric productivity of 22,832 U/L · h with methanol induction of the gene product of *xyl*B from *A. niger* when expressed under AOX1 promoter in *P. pastoris*. Similarly, a recombinant thermostable xylanase A from *Bacillus halodurans* C-125 was expressed in *P. pastoris*. On induction by methanol, a very high xylanase activity of 3361 U/mL was achieved in 132 hours (Lin et al., 2013).

Marine microorganisms are potential sources of enzymes which possess unique characteristics such as salt tolerance, hyperthermostability, and cold adaptation (Trincone, 2011). Ribosome engineering was used to introduce mutation in gene *rpsL* (encoding ribosomal protein S12) in marine *Streptomyces viridochromogenes* which produces thermostable xylanase. A mutant M11-1(10) of *S. viridochromogenes* strain M11 selected on streptomycin-containing plates showed 14% higher xylanase activities than that of the wild-type strain. Ribosomal protein S12 plays an essential role in the decoding function and is involved in both inspection of codon–anticodon pairings and response to streptomycin (Liu et al., 2013). An interesting recombinant strain of *Chlamydomonas reinhardtii*, a freshwater, green microalga, was constructed to hyper express xylanase gene *xyn1* from industrial strain of *T. reesei*. The ble-2A is a nuclear expression vector, resistant for zeocin/bleomycin, of *C. reinhardtii*. Linking expression of *xyn1* with *ble2A* expression on the same open reading frame led to a dramatic (100-fold) increase in xylanase activity in cells lysates compared to the unlinked construct. Further, a secretion signal sequence placed between *xyn1* and *ble2A* led to the secretion of xylanase (Rasala et al., 2012). In another approach, the pEXPYR shuttle vector was used for the heterologous overexpression in *Aspergillus nidulans* of the *P. funiculosum* endo-1,4-xylanase (XynC) and the *A. niger* α-L-arabinofuranosidase (AbfB) together. There was synergistic improvement of xylooligosaccharides production from pretreated sugar cane bagasse when both enzymes were combined (Gonçalves et al., 2012).

The strong cbh1 (cellobiohydrolase 1) promoter has been used frequently for heterologous and homologous protein expression in *T. reesei*. However, this promoter needs induction and is partly regulated by catabolite repression. On the other hand, the mRNA levels of *pdc* (encoding pyruvate decarboxylase) and *eno* (encoding enolase) genes are much higher than other genes under high glucose conditions. Recombinant *T. reesei* strains that homologously expressed xylanase II were constructed by using the promoters of the *pdc* and *eno* whereby they were found to produce 9266 IU/mL and 8866 IU/mL of xylanase, respectively, in the medium containing high glucose concentration (Li et al., 2012). Similarly, other promoters of cellulase system derived from cellobiohydrolase 2 (*cbh2*) and endoglucanase 2 (*egl2*), were used to create an expression platform. A thermophilic xylanase (XynB) from *Dictyoglomus thermophilum* was then expressed synergistically in *T. reesei* RUT-C30. Transformants with the combination of *cbh2* promoter, secretion signal, cellobiohydrolase 2 CBM, and linker as well as *XynB* secreted multiple forms of XynB containing various *N*- and *O*-glycans (Miyauchi et al., 2013). The fusion proteins can also be successfully expressed on the surface of the whole cells like the cell surface display of Cex protein, encoding xylanase, and exoglucanase in *Cellulomonas fimi*, using anchor protein PgsA on *E. coli* surface. The xylanase-surface display achieved the highest activity of 54 U/g dry cell weight at 60 hours and 37°C. The study showed that cell surface display of xylanase embedded in the cell membrane was more thermostable and had a broader pH range than that of the purified enzyme. The optimum temperature shifted from 40°C to 60°C in membrane anchored protein, probably due to the presence of more hydrophobic amino acids (Chen et al., 2012).

Like cellulosomes, designer nanostructured, multifunctional, self-assembling protein complex termed as xylanosomes, are specifically targeted for hydrolysis of complex hemicelluloses. It was reported by McClendon et al. (2012) that a designer xylanosome composed of a xylanase and a bifunctional arabinofuranosidase/xylosidase released 30% more soluble sugars than the corresponding free and unstructured enzymes. Another xylanosome composed of xylanase and a ferulic acid esterase removed 20% more ferulic acid from wheat arabinoxylan than free enzymes. In another study, xylanosome constructed with three endoxylanases of *T. fusca* and a β-xylosidase succeeded in degrading 25% of the total xylan component of wheat straw (Moraïs et al., 2011).

5.4.2 Process Methods

Although efficient and novel genetic methods have been employed by various researchers to overexpress a few components of cellulase and xylanase enzymes in several hosts, information on modification of the properties of the recombinant enzymes is rather scanty (cf. section 5.3). Furthermore, the enzyme components which have been overexpressed are unlikely to hydrolyze cellulose efficiently unless mixed with other components to have a complete enzyme system, which is required for efficient hydrolysis. This is perhaps the reason as to why such recombinant organisms have not been widely studied for production of the enzymes in bench or pilot-scale bioreactors, which is required to assess the economics of their large-scale production. The production of hydrolytic enzymes economically is of prime concern for their subsequent commercial application in bioproduct manufacture. For successful industrial production of enzymes, the choice of microorganism and the process strategies are the key. Microorganisms that can use agro and industrial wastes as carbon source for cellulolytic enzyme production are potential candidates for industrial-scale production. The following paragraphs describe a few important approaches that have been used for production of cellulase and xylanase enzymes with high titer and productivity in submerged and in solid-state fermentation (SSF) systems.

The microbial production systems are multivariable controlled processes. Media composition and environmental factors can enhance the titer and productivity of enzymes in microbial cell cultures. It is, therefore, essential to study and quantify the effects of key medium components on growth and enzyme production and strike a balance between the two to enhance the enzyme titer and productivity. It is well known that *T. reesei* strains are used for industrial production of cellulases. Its mutant strain *T. reesei* RUT C30 is amongst the top cellulase producers (Kubicek, 2013). Mandels' group was one of the pioneering groups that worked extensively on cellulase system of *T. reesei* during the 1960s and 70s. The group showed that C30 when cultivated on cellulose produced 15 FPU/mL with productivity of 80 U/L · h (Ryu and Mandels, 1980). A six-step mutation in one of its earliest mutants QM9414 increased productivity by two-fold in comparison to RUT C30 (Durand and Clanet, 1988). In another microbe *Acremonium cellulolyticus*, classical mutation technique was used to isolate hyper-cellulase producer. After UV irradiation and NTG treatments, *A. cellulolyticus* CF-2612 mutant produced increased FPase activity of 17.8 IU/mL. Further, in fed-batch mode *A. cellulolyticus* improved cellulase productivity to 140 FPU/L · h (Fang et al., 2009). Several other microbes have also been studied by various laboratories for production of cellulases in bench- and pilot-scale submerged fermenters as summarized in Table 5.2.

Cellulase is an inducible enzyme system for which several carbon sources have been tested to find the best inducer (Mandels, 1975; Seiboth et al., 2012). Cellulose itself has been recognized as one of the best inducers for synthesizing the complete cellulase system; other important inducers include sophorose and lactose. Cellulase biosynthesis is repressed by glucose; however, *T. reesei* grows rapidly on simple sugars like glucose but more slowly with a lag phase on lactose or cellulose. *T. reesei* does not normally produce cellulases when grown in media containing glucose. In *T. reesei* RUT-C30 strain, the mutation in *cre1* (carbon catabolite repressor) released glucose repression and expressed cellulases (Ilmen et al., 1996). Other factors which influence the enzyme production include concentration and quality of the carbon source, growth conditions, aeration, temperature, and pH. The effect of medium composition in *T. reesei* RUT-C30 cultures on cell growth and enzyme production was investigated recently by Ahamed and Vermette (2008a) with the aim of reducing production costs by using cellulose as a main substrate and lactose plus lactobionic acid in the fed-batch phase. Four different media compositions were used to test the production level of cellulase enzyme in *T. reesei* during fed-batch cultures in a 7-L stirred tank bioreactor (STR). *T. reesei* grew very well in cellulose–yeast extract medium with lactose and lactobionic acid as inducers and produced a maximum FPA of 5.02 U/mL, which was two to three times higher than those obtained with other media, that is, corn steep–glucose (1.4 U/mL), cellulose–yeast extract–peptone (2.3 U/mL),

TABLE 5.2 Production of Cellulases by a Few Prominent Microorganisms in Laboratory and Pilot-Scale Bioreactors Under Submerged Fermentation

		Conditions			Enzyme activities			
Strain	Substrate	Temp (°C)	pH	Scale (L)	Cellulase (FPU/mL)	CMCase (U/mL)	β-Glucosidase (U/mL)	Reference
Acremonium cellulolyticus CF-2612	Solka-Floc	30	NA	2	17.8			Fang et al. (2009)
Acremonium cellulolyticus C-1	Milk pack	28	4.5	3	16			Park et al. (2011)
Acremonium cellulolyticus C-1	Solka-Floc	30		50	13.08			Ikeda et al. (2007)
Aspergillus niger A12	Glucose and sugarcane bagasse	30	5	5		0.72		Cunha et al. (2012)
Penicillum funiculosum	Pretreated sugarcane bagasse	30	5	10	1.13	10.25		Maeda et al. (2013)
Streptomyces sp.	CMC	50	6.5	50		148	137	Jang and Chang (2005)
Trichoderma pseudokoingii S-38	Glucose and cellulose powder	28	3.6	2.5		10.86		Duan et al. (2004)
Trichoderma reesei 2U-02	Corn cob	30	4.8	30		5.48		Liming and Xueliang (2004)
Trichoderma viride	Sugarcane bagasse	28		7	0.6	0.6		Lan et al. (2013)
Trichoderma harzianum P49P11	Pretreated sugarcane bagasse and sucrose	29	5	3	1.21		17.32	Delabona et al. (2012)

CMC, carboxymethyl cellulose; NA, not Available.

and cellulose–yeast nitrogen base–CMC (1.4 U/mL). The volumetric enzyme productivity in cellulose–yeast extract medium (69.8 U/L · h) was also approximately two to three times higher than those in the other media (Ahamed and Vermette, 2008a).

STRs are the most common reactors employed for cellulase production. However, it is sometimes difficult to modulate the shear stress levels within these reactors, which often damage filamentous microorganisms such as *T. reesei* and lower enzyme production. Many reactor designs such as air-lift reactors have been suggested to increase the oxygen transfer rate while limiting the deleterious effects of shear stresses on filamentous microorganisms and minimizing power consumption. The absence of impellers, moving parts, and bearings allows extended aseptic operation of these reactors. They have lower power consumption as compared to stirred tanks, good mixing capacity for media containing suspended solids, efficient mass and heat transfer, and better defined flow patterns (Chisti, 1989). Ahamed and Vermette (2010) used a draft tube air-lift bioreactor to produce cellulases by *T. reesei* RUT-C30 in a cellulose medium with lactose and lactobionic acid in fed-batch mode. The airlift reactor was operated without agitation and with agitation where shear was provided by an impeller running at 400 rpm. Significant increases in filter paper activity (from 12 to 17 U/mL) and volumetric enzyme productivity (from 147 to 200 U/mL · h) were obtained in cultures carried out with no mechanical agitation. These results could be explained on the basis of *T. reesei*'s morphology, as mechanical agitation in STR caused more fragmentation of its hyphae. The lowered enzyme activity could also be attributed to the shear sensitive nature of the *T. reesei* cellulase enzyme system.

The association of filamentous fungi such as *Trichoderma* and *Aspergillus*, which are amongst the most important microorganisms used in industry, allows higher production of cellulolytic enzymes by synergistic action. The synergistic reaction occurs as a result of synthesis of some metabolites by one organism which has a positive effect on the other in terms of increased production of cellulolytic enzymes. It was shown by Ghose's group (Ghose et al., 1985; Panda et al., 1987) that there was increased synthesis of both cellulase and xylanase enzymes when the addition of *Aspergillus wentii* culture was phased by 15 hours to a growing culture of *T. reesei*. More recently, Ahamed and Vermette (2008b) have shown that a mixed culture of *T. reesei* and *A. niger* cultured in media containing cellulose–yeast extract in a 3 L fed-batch bioreactor, with lactose and lactobionic acid feeding during fed-batch phase, produced 2.1-fold improved filter paper activity (7.1 U/mL) and enzyme productivity (98.4 FPU/L · h) compared to the monocultures. The enzyme system derived from the co-fermentation of the two species had higher cellulose hydrolysis potential than the enzymes derived from the monocultures.

Recently Lan et al. (2013) studied the production of cellulolytic enzymes by *Trichoderma viride* in a rotating fibrous-bed bioreactor (RFBB) and compared its performance with that of an STR. It was envisaged that the immobilization of *T. viride* mycelia on solid support of polypropylene cloth would eliminate some of the difficulties associated with the use of STRs (such as disintegration of the mycelia by the impeller, fouling of the pH and temperature probes, and clogging of the sample ports by the mycelia). It was found that the *T. viride* mycelia formed biofilm in the RFBB, which was operated under a relatively low-shear environment, and promoted the enhanced production of enzymes. Under the operating conditions employed, the immobilized mycelia in RFBB produced 35.5% higher FPase activity and gave 69.7% higher saccharification yield of sugarcane bagasse compared to the enzyme produced in STR.

Application of statistical optimization tools like response surface methodology where several parameters are considered simultaneously for their influence and interaction during the bioprocess has resulted in getting increased enzyme titers in xylanase-producing organisms. For example, statistical optimization of xylanase production parameters such as pH, temperature, substrate concentration, cultivation time, and aeration improved enzyme activity by 315% (Bocchini et al., 2002). There was a significant increase of 20-fold xylanase activity to 174 IU/mL by optimizing process conditions in shake flask using response surface methodology in a halophillic eubacterium strain SX15 (Wejse et al., 2003). The xylanase activity of 210 IU/mL was obtained at the end of cultivation in a bioreactor, which was higher than in the shake flask when carbon source was replaced by soluble oat husk extract in *T. lanuginosus* DSM 10635 (Xiong et al., 2004a). A white rot fungus, *Armillaria gemina* SKU2114, secreted lignocellulose-degrading enzymes which showed maximum endoglucanase, cellobiohydrolase, and β-glucosidase activities of 146, 34, and 15 U/mL respectively, and also xylanase, laccase, mannanase, and lignin peroxidase with activities of 1270, 0.16, 57, and 0.31 U/mL, respectively, when grown on rice straw as a carbon source. Application of response surface methodology enhanced the hydrolysis of wood, *Populus tomentiglandulosa*, to an extent of 62% by this enzyme (Jagtap et al., 2013).

In submerged fermentation, the biomass remains evenly dispersed in liquid medium with equal access to the dissolved nutrients and oxygen. Efficient heat and mass transfer also take place. The fungus *Melanocarpus albomyces* which is filamentous, multinucleate, thermophilic, and non-sporulating has been studied with respect to xylanase production in submerged fermentation as it produces thermostable xylanase along with xylosidase and xylan-debranching enzymes. A mutant of *M. albomyces* IITD3A produced 415 IU/mL xylanase on soluble alkaline lignocellulose extract in a 14 L bioreactor with volumetric productivity of 11,530 IU/L · h, which was 8-fold higher than that of the wild-type strain (Biswas et al., 2010a). It was found by response surface methodology that the pH of the medium was critical for the production of xylanase by *M. albomyces* IITD3A. Cyclic maintenance of pH of fermentation medium between 7.8 and 8.2 increased the productivity to 16,670 IU/L · h (Figure 5.1). Cycling of pH in systems such as this one, in which different components of xylanase have different pH optima, seems to be a simple strategy to improve enzyme titer and productivity. Further optimization of process parameters like aeration and agitation in 14 L bioreactor, changed the fungal morphology to a pellet form with an increase in productivity to 22,000 IU/L · h, which is amongst one of the highest reported in filamentous fungi (Biswas et al., 2010b). Maintenance of a particular fungal morphology has been extensively reported to be essential for the production of several secondary metabolites such as antibiotics also.

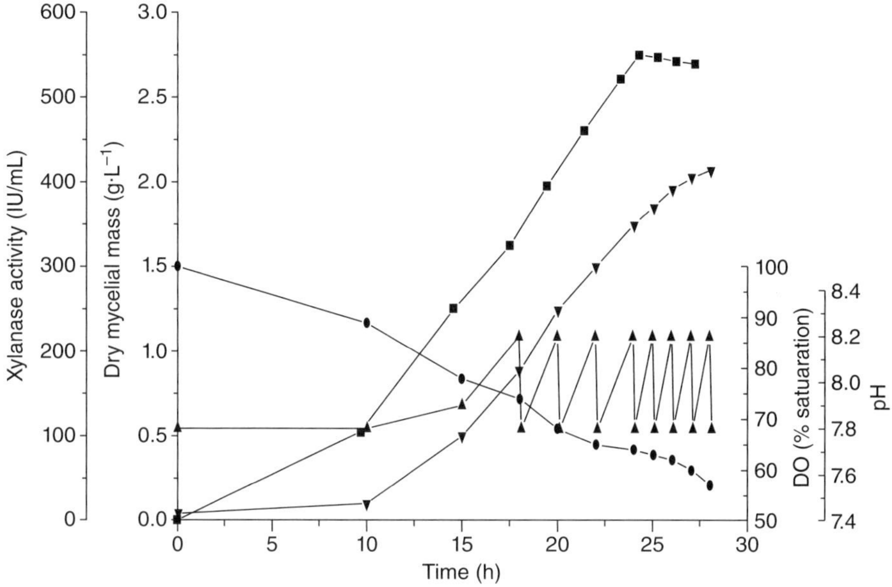

FIGURE 5.1 Production of xylanase by *Melanocarpus albomyces* IITD3A in 14 L bioreactor with pH cycling. Filled square, xylanase activity; filled triangle, pH; filled circle, dissolved oxygen; and inverted filled triangle, dry mycelia mass (Biswas et al., 2010b).

Some microbes have been reported to give better productivity of enzymes in SSF. In SSF, the solid waste residues from agriculture, forest, mills, etc. are used as the support substrates on whose surface the dispersed microbial biomass grows under controlled temperature, moisture, and aeration. It offers several advantages over liquid cultivation, especially for the fungal cultures, as there is higher productivity per unit volume, reduced energy requirements, lower capital investment, low wastewater output, higher concentrations of metabolites obtained, and low downstream processing cost (Narang et al., 2001). During SSF of wheat straw by *M. albomyces*, process conditions were optimized as 600–850 μm particle size of wheat straw, 43 hours inoculum age, 1.37% Tween 80, 86% initial moisture content, 5.1% urea, and 0.74% yeast extract. Under these optimized conditions, xylanase activity of 7760 U/g initial dry substrate was produced (Narang et al., 2001). Production of xylanase by *Fusarium oxysporum* under solid-state culture on corn stover was enhanced by optimization of the type of nitrogen source, initial moisture level, growth temperature, and initial pH (Panagiotou et al., 2003). A marine strain of *A. niger* used water weed, *Eichchornia crassipes*, as main substrate and produced 17.80 U/g dry substrate in 6 days in SSF indicating the potential for economic cellulase production (Xue et al., 2012). Another industrial strain of *A. niger* produced cellulase under different cultivation conditions like biofilms submerged and SSF, and produced 1.77, 1.17, and 1.17 FPU/mL, respectively. Although biofilm cultures had less growth rate than the submerged and solid-state cultures, they gave higher cellulase yields (370, 212, and

217 U/galactose, respectively) and volumetric productivities (24, 16, and 16 U/L · h, respectively). The biofilm cultures also produced higher endoglucanase and xylanase activities (Gamarra et al., 2010).

The SSF in general faces major problems with respect to design and operation of large-scale bioreactors (Ashley et al., 1999). However, various types of bioreactors have been studied for solid-state fermentation processes, including packed beds, rotating drums, gas–solid fluidized beds, and stirred bioreactors but were unable to completely overcome the mass and heat transfer limitations (Mitchell et al., 2000). A novel intermittent agitation rotating drum type bioreactor for thermophilic microorganism *Thermoascus aurantiacus* was used for xylanase production. Important fermentation variables like initial moisture content of the medium, growth temperature, and airflow rate were effectively controlled at the set values through this design and xylanase activity of 4490 U/g of dry wheat straw was achieved (Kalogeris et al., 2003). The operational conditions in SSF like air flow rate, inlet air relative humidity, and initial substrate moisture content influenced the production of biomass-degrading multienzyme in *A. niger* strain 12. An instrumented lab-scale bioreactor equipped with an online automated monitoring and control system maintained initial substrate moisture content of 84%, an inlet air humidity of 70%, and a flow rate of 24 mL/min for *A. niger* fermentation on soya bean meal; it produced 0.55 IU/g FPase, 35.1 IU/g endoglucanase, and 47.7 IU/g xylanase (Vitcosque et al., 2012). The production of xylanases by a few microorganisms in submerged and SSF systems is shown in Table 5.3.

Some microbes such as *A. niger* KK2, *A. pullulans*, *F. oxysporum*, and *T. aurantiacus* simultaneously secrete cellulases and xylanases to digest lignocelluloses. The production of these cellulolytic and xylanolytic enzymes by such organisms has been studied under submerged and solid-state cultivations where they were found to produce these enzymes (Table 5.4). These can be helpful in extracting more sugars from the lignocelluloses and decrease the cost of production of bioproducts. Also, they can be potential candidates for consolidated bioprocesses.

Thermophiles that have faster metabolic rates can result in faster growth lignocellulosic substrates and high productivity of hydrolytic enzymes. Thus, thermophiles producing these enzymes are of great interest for the biocommodity industry. At high temperatures, the reaction rates are faster, the risk of contamination is low, and the process is more energy efficient for pretreatment of biomass. A newly isolated *Geobacillus* sp T1 has been reported to grow on barley and wheat straw to produce cellulase units of 143.5 U/mL after 24 hours of fermentation (Assareh et al., 2012). *Streptomyces* sp CS802, isolated from soil in the Republic of Korea, produced 1482 U/mL of xylanase on corn cob medium at 60°C (Simkhada et al., 2012). An interesting thermophile, *Thermobacillus xylanolyticus*, was able to use glucose and xylans as primary carbon source. Its growth on lignocellulosic biomass like wheat straw and wheat bran induced hemicellulolytic enzyme production, and xylanase was the primary enzyme secreted. Debranching enzyme esterase was produced by the bacterial cultures grown on wheat straw whereas arabinofuranosidase production was significantly higher in bacterial cultures grown on wheat bran. Thus *T. xylanolyticus* seems to have adapted its enzymatic profile to better address the composition of various lignocellulosic substrates (Rakotoarivonina et al., 2012). Multiple acidophilic

TABLE 5.3 Microbial Xylanase Production in Submerged and Solid-State Fermentation

	Organism	Carbon source	Fermentation conditions				Xylanase activity		References
			Temp. (°C)	pH	Type	Scale (L)	(IU/mL or IU/g)	Time (h)	
1.	*Armillaria gemina*	Wheat bran	–		SmF	7.0	1270.0	216	Dhiman et al. (2013)
2.	*Aspergillus oryzae*	Spent sulphite liquor	30	7.5	SmF	15	199.0	48	Chipeta et al. (2008)
3.	*Aspergillus niger*	Corn cob	28	3.5	SmF	20.0	290.0	80	Qi-peng et al. (2005)
4.	*Aspergillus sulphureus*	Wheat bran	32	6.5	SSF	75.0	7.5	72	Wenqing et al. (2003)
5.	*Bacillus coagulans* BL69	Fibrous soy residue	37	–	SSF	0.5	1.5	72	Heck et al. (2005b)
6.	*Bacillus circulans* BL53	Fibrous soy residue	37	–	SSF	12.0	8.25	96	Heck et al. (2005a)
7.	*Bacillus* sp.	Oat spelt xylan	40	8.0	SmF	1.5	65.0	24	Chivero et al. (2001)
8.	*Bacillus subtilis*	Oat spelt xylan	50	6.0	SmF	2.0	1.73	18	Sá-Pereira et al. (2002)
9.	*Melanocarpus albomyces* IIS 68	Wheat straw	45	7.0	SSF	0.5	7760.0	120	Narang et al. (2001)
10.	*Melanocarpus albomyces* IITD3A	Wheat straw extract	45	7.0	SmF	14	550.0	24	Biswas et al. (2010a)
12.	*Neocallimastix frontalis*	Cellulose	39	6.8	SmF	1.2	3.68	144	Srinivasan et al. (2001)
13.	*Penicillium* sp. *WX-Z1*	Wheat bran	30	–	SmF	30	46.5	144	Cui and Zhao (2012)
14.	*Rhizopus oryzae*	Corn cob	35	5.0	SmF	2.0	260.0	120	Bakir et al. (2001)
15.	*Rhodothermus marinus*	Birchwood xylan	61	7.5	SmF	5.0	7.5	72	Gomes et al. (2000a)
16.	*Streptomyces* sp. Ab106	Sugarcane bagasse	50	7.0	SmF	5.0	32.0	144	Techapun et al. (2003)
17.	*Thermoascus aurantiacus*	Sugarcane bagasse	45	–	SSF	0.2	1597.0	240	Milagres et al. (2004)
18.	*Thermoascus aurantiacus*	Solka-Floc	47	4.5	SmF	5.0	208.0	264	Gomes et al. (2000b)
19.	*Thermomyces lanuginosus* SSBP	Xylose	50	6.5	SmF	30.0	405.0	48	Reddy et al. (2002)
20.	*Trichoderma reesei Rut* C-30	Lactose	28	6.0	SmF	2.0	94.7	120	Xiong et al. (2004b)

SmF, Submerged Fermentation (enzyme activity in IU/mL); SSF, Solid-State Fermentation (enzyme activity in IU/g solid subtrate).

TABLE 5.4 Microbial Strains Coproducing Cellulases and Xylanases on Lignocellulosic Residues

Strains	Substrate	SSF/SmF	FPase (U/g or U/mL)	CMCase (U/g or U/mL)	β-Glucosidase (U/g or U/mL)	Xylanase (U/g or U/mL)	β-Xylosidase (U/g or U/mL)	References
Acremonium cellulolyticus CF-2612	Wheat bran	SmF	10.5	356.5	22.6	580.6	1.1	Hideno et al. (2011)
Aspergillus niger KK2	Rice straw	SSF	19.5	129.0	100	5070	193	Kang et al. (2004)
Aureobasidium pullulans	Corn bran	SSF	–	1.05	1.3	5	–	Leite et al. (2007)
Cellulomonas flavigena	Sugarcane bagasse	SmF	–	7.56	–	75.6	–	Rojas-Rejón et al. (2011)
Fusarium oxysporum	Corn stover	SSF	–	211	0.088	1216	0.052	Panagiotou et al. (2003)
Myceliophthora sp IMI 387099	Rice straw	SSF	2.44	32.9	7.48	900.2	–	Badhan et al. (2007)
Penicillium echinulatum 9A02S1	Sugarcane bagasse and wheat straw	SSF	32.89	282.36	58.95	10	–	Camassola and Dhillon (2007)
Thermoascus aurantiacus	Wheat straw	SSF	5.5	1709	79	4490	45	Kalogeris et al. (2003)

SmF, submerged fermentation (enzyme activity in U/mL); SSF, solid-state fermentation (enzyme activity in U/g solid subtrate).

xylanolytic enzymes were produced by *Penicillium oxalicum* GZ-2, and wheat straw was the best inducer, resulting in the highest xylanase (115.2 U/mL) and β-xylosidase (89 mU/mL) activities at 50°C (Liao et al., 2012). *Paenibacillus macerans* IIPSP3 isolated from the gut of the wood-feeding termite produced thermostable xylanase that was active over a broad range of temperatures (40–90°C) with xylanase activity of 100 U/mL on corn cobs. The strain showed maximum xylanase production at its optimum growth temperature, that is, 50°C (Dheeran et al., 2012). Two potential strains of marine fungi obtained from degrading mangrove leaves, *Fusarium* sp KAWIT-A, and *Aureobasidium* sp 2LIPA-M, produced thermophilic xylanase in submerged fermentation supplemented with 1.5% marine salts (Torres and Cruz, 2013). Another thermophilic microorganism, *Streptomyces thermovulgaris* TISTR1948 produced 274.5 U/mL xylanase on rice straw (Chaiyaso et al., 2011). Moreover, several thermophilic cellulase producers like *Paenibacillus barcinoneusis* (Asha et al., 2012), *B. subtilis* strain LFS3 (Rawat and Tewari, 2012), *C. thermocellum* strains CS7 and CS8 (Lv and Yu, 2012), and *Phiahophora* sp G5 (Zhao et al., 2012) have recently been studied for enzyme production with a view to utilize them for production of bioproducts on a large scale.

5.5 ECONOMIC OUTLOOK

Providing cost-effective enzymes is a big challenge for industrial application of enzymes. Serious efforts have been made in this direction by application of both genetic as well as process engineering tools, by the use of waste materials as substrates, and by using the minimum steps in downstream recovery of enzymes. Although significant cost reductions in cellulase production by 20- to 30-fold have been made (Lynd et al., 2005), it is still far to compete with chemical industries for production of ethanol. Many commercial enzyme manufacturers like Novozymes, Genencor, Danisco, Armano Enzyme, etc. are striving to develop highly efficient and less expensive cellulolytic enzyme cocktails (Chandel et al., 2012). Current estimates suggest that the cost of producing cellulosic ethanol is $1.80 per gallon or higher, which is almost twice as high as the cost of producing ethanol from starch. A major emphasis on research is, therefore, on reduction of the cost of hydrolytic enzymes. Current research is being aimed at bringing down the enzyme cost from $0.50 to $0.10 per gallon of ethanol (EERE, DOE, 2010). The economics of production of hydrolytic enzymes will remain an integral part of any successful biocommodity industry. With combined knowledge of protein engineering, metabolic engineering, genomics, and bioprocess engineering, it is expected that cellulases and xylanases can be produced in copious amounts on inexpensive substrates that will make biorefinery industry a reality.

5.6 FUTURE PROSPECTS

The role of cellulase and xylanase enzymes is crucial in realizing the aim of producing ethanol (and other biofuels) and commodity bioproducts from cellulosic feedstocks. The cost of producing cellulase has been brought down significantly over the last decade, but it still seems to be about two to three times higher than the threshold value projected for economic production of cellulosic ethanol. We believe that the development of more effective cellulases with desirable characteristics such as high catalytic efficiency on crystalline cellulose, end-product inhibition resistance and higher thermostability shall be able to bring out a faster and more complete hydrolysis of celluloses to reducing sugars. To achieve these goals, modern tools of protein and metabolic engineering, together with process engineering and scale-up in suitable bioreactors need to be applied. This will result in the usage of decreased amounts of cellulolytic enzymes, thereby lowering their contribution to the overall production cost of ethanol and other bioproducts. Further, information on genome of the best cellulase producer, *T. reesei*, and its several transcriptional regulators which are involved in cellulase regulation is now available. As our understanding to modulate these regulators in response to various external stimuli increases, it is expected that it shall be possible to overproduce the desired mix of biomass-degrading cellulolytic enzymes with high yield and productivity.

REFERENCES

Adlakha N, Sawant S, Anil A, Lali A, Yazdani SS. (2012). Specific fusion of β-1,4-endoglucanase and β-1,4-glucosidase enhances cellulolytic activity and helps in channeling of intermediates. *Appl Environ Microbiol*, 78, 7447–7454.

Adney WS, Jeoh T, Beckham GT, Chou YC, Baker JO, Michener W, Brunecky R, Himmel ME. (2009). Probing the role of N linked glycans in the stability and activity of fungal cellobiohydrolases by mutational analysis. *Cellulose*, 16, 699–709.

Ahamed A, Vermette P. (2008a). Culture-based strategies to enhance cellulase enzyme production from *Trichoderma reesei* RUT-C30 in bioreactor culture conditions. *Biochem Eng J*, 40, 399–407.

Ahamed A, Vermette P. (2008b). Enhanced enzyme production from mixed cultures of *Trichoderma reesei* RUT-C30 and *Aspergillus niger* LMA grown as fed-batch in a stirred tank bioreactor. *Biochem Eng J*, 42, 41–46.

Ahamed A, Vermette P. (2010). Effect of mechanical agitation on the production of cellulases by *Trichoderma reesei* RUT-C30 in a draft-tube airlift bioreactor. *Biochem Eng J*, 49, 379–387.

Akila G, Chandra TS. (2003). A novel cold-tolerant *Clostridium* strain PXYL1 isolated from a psychrophilic cattle manure digester that secretes thermolabile xylanase and cellulase. *FEMS Microbiol Lett*, 219, 63–67.

Alriksson B, Rose SH, van Zyl WH, Sjöde A, Nilvebrant NO, Jönsson LJ. (2009). Cellulase production from spent lignocellulose hydrolysates by recombinant *Aspergillus niger*. *Appl Environ Microbiol*, 75(8), 2366–2374.

Asha BM, Malini B, Revathi M, Yadav A, Sakthivel N. (2012). Purification and characterization of a thermophilic cellulase from a novel cellulolytic strain, *Paenibacillus barcinonensis*. *J Microbiol Biotechnol*, 22, 1501–1509.

Ashley VM, Mitchell DA, Howes T. (1999). Evaluating strategies for overcoming overheating problems during solid-state fermentation in packed bed bioreactors. *Biochem Eng J*, 3, 141–150.

Assareh R, Zahiri HS, Noghabi KA, Aminzadeh S, Khaniki GB. (2012). Characterization of the newly isolated *Geobacillus* sp. T1, the efficient cellulase-producer on untreated barley and wheat straws. *Bioresour Technol*, 120, 99–105.

Badhan AK, Chadha BS, Kaur J, Saini HS, Bhat MK. (2007).Production of multiple xylanolytic and cellulolytic enzymes by thermophilic fungus *Myceliophthora* sp. IMI 387099. *Bioresour Technol*, 98, 504–510.

Bakir U, Yavascaoglu S, Guvenc F, Ersayin A. (2001). An endo-β-1, 4-xylanase from *Rhizopus oryzae*: production, partial purification and biochemical characterization. *Enzyme Microb Technol*, 29, 328–334.

Bayer EA, Belaich JP, Shoham Y, Lamed R. (2004).The cellulosomes: multienzyme machines for degradation of plant cell wall polysaccharides. *Annu Rev Microbiol*, 58, 521–554.

Bayer EA, Lamed R, Himmel ME. (2007). The potential of cellulases and cellulosomes for cellulosic waste management. *Curr Opin Biotechnol*, 18, 237–245.

Béguin P, Aubert JP. (1994).The biological degradation of cellulose. *FEMS Microbiol Rev*, 13, 25–58.

Bevan MW, Franssen MCR. (2006). Investing in green and white biotech. *Nat Biotechnol*, 24, 765–767.

Bisaria VS. (1998). Bioprocessing of agro-residues to value added products. In: Martin AM, editor. *Bioconversion of Waste Materials to Industrial Products*, 2nd ed. UK: Chapman & Hall, pp. 197–246.

Biswas R, Sahai V, Mishra S, Bisaria VS. (2010a). Development of mutants of *Melanocarpus albomyces* for hyperproduction of xylanase. *Biotechnol Bioprocess Eng*, 15, 800–809.

Biswas R, Sahai V, Mishra S, Bisaria VS. (2010b). Bioprocess strategies for enhanced production of xylanase by *Melanocarpus albomyces* IITD3A on agro-residual extract. *J Biosci Bioeng*, 110, 702–708.

Blanco A, Díaz P, Zueco J, Parascandola P, Javier-Pastor FI. (1999). A multidomain xylanase from a *Bacillus* sp. with a region homologous to thermostabilizing domains of thermophilic enzymes. *Microbiology*, 145, 2163–2170.

Blumer-Schuette SE, Lewis DL, Kelly RM. (2010). Phylogenetic, microbiological and glycoside hydrolase diversities within the extremely thermophilic, plant biomass-degrading genus *Caldicellulosiruptor*. *Appl Environ Microbiol*, 76, 8084–8092.

Bocchini DA, Alves-Prado HF, Baida LC, Roberto IC, Gomes E, da Silva R. (2002). Optimization of xylanase production by *Bacillus circulans* D1 in submerged fermentation using response surface methodology. *Process Biochem*, 38, 727–731.

Böhmer N, Lutz-Wahl S, Fischer L. (2012). Recombinant production of hyperthermostable CelB from *Pyrococcus furiosus* in *Lactobacillus* sp. *Appl Microbiol Biotechnol*, 96, 903–912.

Camassola M, Dillon APJ. (2007). Production of cellulases and hemicellulases by *Penicillium echinulatum* grown on pretreated sugar cane bagasse and wheat bran in solid-state fermentation. *J Appl Microbiol*, 103, 2196–2204.

Chaiyaso T, Kuntiya A, Techapun C, Leksawasdi N, Seesuriyachan P, Hanmoungjai P. (2011). Optimization of cellulase-free xylanase production by thermophillic *Streptomyces thermovulgaris* TISTR1948 through Plackett-Burman and response surface methodological approaches. *Biosci Biotechnol Biochem*, 75, 531–537.

Chandel AK, Chandrasekhar G, Silva MB, da Silva SS. (2012).The realm of cellulases in biorefinery development. *Crit Rev Biotechnol*, 32, 187–202.

Chandrakant P, Bisaria VS. (1998). Simultaneous bioconversion of cellulose and hemicellulose to ethanol. *Crit Rev Biotechnol*, 18, 295–331.

Chen YP, Hwang IE, Lin CJ, Wang HJ, Tseng CP. (2012). Enhancing the stability of xylanase from *Cellulomonas fimi* by cell-surface display on *Escherichia coli*. *J Appl Microbiol*, 112, 455–463.

Chipeta, ZA, Preez JC, Christopher L. (2008). Effect of cultivation pH and agitation rate on growth and xylanase production by *Aspergillus oryzae* in spent sulphite liquor. *J Ind Microbiol Biotechnol*, 35, 587–594.

Chisti MY. (1989). *Airlift Bioreactors*. London: Elsevier, pp. 5–6.

Chivero ET, Mutukumira AN, Zvauyu R. (2001). Partial purification and characterization of a xylanase enzyme produced by a microorganism isolated from selected indigenous fruits of Zimbabwe. *Food Chem*, 72, 179–185.

Cho HY, Yukawa H, Inui M, Doi RH, Wong SL. (2004). Production of mini cellulosomes from *Clostridium cellulovorans* in *Bacillus subtilis* WB800. *Appl Environ Microbiol*, 70, 5704–5707.

Cui F, Zhao L. (2012). Optimization of xylanase production from *Penicillium* sp.WX-Z1 by a two-step statistical strategy: Plackett-Burman and Box-Behnken experimental design. *Int J Mol Sci*, 13, 10630–10646.

Cunha FM, Esperança MN, Zangirolami TC, Badino AC, Farinas CS. (2012). Sequential solid-state and submerged cultivation of *Aspergillus niger* on sugarcane bagasse for the production of cellulase. *Bioresour Technol*, 112, 270–274.

Dekker RFH. (1985). Biodegradation of the hemicelluloses. In Higuchi T, editor. *Biosynthesis and Biodegradation of Wood Components*. London: Academic Press, pp. 505–533.

Delabona PDS, Farinas CS, da Silva MR, Azzoni SF, Pradella JGDC. (2012). Use of a new *Trichoderma harzianum* strain isolated from the Amazon rainforest with pretreated sugar cane bagasse for on-site cellulase production. *Bioresour Technol*, 107, 517–521.

Dheeran P, Nandhagopal N, Kumar S, Jaiswal YK, Adhikari DK. (2012). A novel thermostable xylanase of *Paenibacillus macerans* IIPSP3 isolated from the termite gut. *J Ind Microbiol Biotechnol*, 39, 851–860.

Dhiman SS, Kalyani D, Jagtap SS, Haw JR, Kang YC, Lee JK. (2013). Characterization of a novel xylanase from *Armillaria gemina* and its immobilization onto SiO2 nanoparticles. *Appl Microbiol Biotechnol*, 97, 1081–1091.

Doi RH, Kosugi A. (2004). Cellulosomes: plant cell-wall-degrading enzyme complexes. *Nat Rev Microbiol*, 2, 541–551.

Duan XY, Liu SY, Zhang WC, Zhang QX, Gao PJ. (2004). Volumetric productivity improvement for endoglucanase of *Trichoderma pseudokoingii* S-38. *J Appl Microbiol*, 96, 772–776.

Durand H, Clanet M. (1988). Genetic improvement of *Trichoderma reesei* for large-scale cellulose production. *Enzyme MicrobTechnol*, 10, 341–346.

Energy Efficiency & Renewable Energy (EERE), Department of Energy (DOE). (2010). United States of America. Available at: http://www1.eere.energy.gov/bioenergy/pdfs/current_state_of_the_us_ethanol_industry.pdf Last accessed on Jan 20, 2014.

Fang X, Yano S, Inoue H, Sawayama S. (2009). Strain improvement of *Acremonium cellulolyticus* for cellulose production by mutation. *J Biosci Bioeng*, 107, 256–261.

Gamarra NN, Villena GK, Gutiérrez-Correa M. (2010). Cellulase production by *Aspergillus niger* in biofilm, solid-state and submerged fermentations. *Appl Microbiol Biotechnol*, 87, 545–551.

Ghose TK, Bisaria VS. (1979). Studies on the mechanism of enzymatic hydrolysis of cellulosic substances. *Biotechnol Bioeng*, 21, 131–146.

Ghose TK, Panda T, Bisaria VS. (1985). Effect of culture phasing and mannanase on production of cellulase and hemicellulase by *Trichoderma reesei* D1-6 and *Aspergillus wentii* Pt 2804, *Biotechnol Bioeng*, 27, 1353–1361.

Gomes I, Gomes J, Gomes DJ, Steiner W. (2000b). Simultaneous production of high activities of thermostable endoglucanase and β-glucosidase by the wild thermophilic fungus *Thermoascus aurantiacus*. *Appl Microbiol Biotechnol*, 53, 461–468.

Gomes J, Gomes I, Terler K, Gubala N, Ditzelmüller G, Steiner W. (2000a). Optimization of culture medium and conditions for α-L-arabinosidase production by the extreme thermophilic eubacterium *Rhodothermus marinus*. *Enzyme MicrobTechnol*, 27, 414–422.

Gomes J, Steiner W. (2004). The biocatalytic potential of extremophiles and extremozymes. *Food Technol Biotechnol*, 42, 223–235.

Gonçalves TA, Damásio ARL, Segato F, Alvarez TM, Bragatto J, Brenelli LB, Citadini APS, Murakami MT, Ruller R, Leme AFP, Prade RA, Squina FM. (2012). Functional characterization and synergic action of fungal xylanase and arabino-furanosidase for production of xylooligosaccharides. *Bioresour Technol*, 119, 293–299.

Han Y, Agarwal V, Dodd D, Kim J, Bae B, Mackie RI, Nair SK, Cann IKO. (2012). Biochemical and structural insights into xylan utilization by the thermophilic bacterium *Caldanaerobius polysaccharolyticus*. *J Biol Chem*, 42, 34946–34960.

Hatti-Kaul R, Ulrika T, Linda G, Börjesson P. (2007). Industrial biotechnology for the production of bio-based chemicals acradle to grave perspective. *Trends Biotechnol*, 25, 119–24.

Heck JX, Flôres SH, Hertz PF, Ayub MAZ. (2005a). Optimization of cellulase-free xylanase activity produced by *Bacillus coagulans* BL 69. *Process Biochem*, 40, 107–112.

Heck JX, Hertz PF, Ayub MAZ. (2005b). Extraction optimization of xylanases obtained by solid-state cultivation of *Bacillus circulans* BL53. *Process Biochem*, 40, 2891–2895.

Hideno A, Inoue H, Tsukahara K, Yano S, Fang X, Endo T, Sawayama S. (2011). Production and characterization of cellulases and hemicellulases by *Acremonium cellulolyticus* using rice straw subjected to various pretreatments as the carbon source. *Enzyme Microb Technol*, 48, 162–168.

Ikeda Y, Hayashi H, Okuda N, Park EY. (2007). Efficient cellulase production by the filamentous fungus *Acremonium cellulolyticus*. *Biotechnol Prog*, 23, 333–338.

Ilmen M, Thrane C, Penttila M. (1996). The glucose repressor gene *crel* of *Trichoderma:* isolation and expression of a full-length and a truncated mutant form. *Mol Gen Genet*, 251, 451–460.

Jagtap SS, Dhiman SS, Kim TS, Li J, Lee JK, Kang YC. (2013). Enzymatic hydrolysis of aspen biomass into fermentable sugars by using lignocellulases from *Armillaria gemina*. *Bioresour Technol*, 133, 307–314.

Jang HD, Chang KS. (2005). Thermostable cellulases from *Streptomyces* sp.: scale-up production in a 50-L fermenter. *Biotechnol Lett*, 27, 239–242.

Kalogeris E, Iniotaki F, Topakas E, Christakopoulos P, Kekos D, Macris BJ. (2003). Performance of an intermittent agitation rotating drum type bioreactor for solid-state fermentation of wheat straw. *Bioresour Technol*, 86, 207–213.

Kanafusa-Shinkai S, Wakayama J, Tsukamoto K, Hayashi N, Miyazaki Y, Ohmori H, Tajima K, Yokoyama H. (2013). Degradation of microcrystalline cellulose and non-pretreated plant biomass by a cell-free extracellular cellulase/hemicellulase system from the extreme thermophilic bacterium. *J Biosci Bioeng*, 115, 64–70.

Kang SW, Park YS, Lee JS, Hong SI, Kim SW. (2004).Production of cellulases and hemicellulases by *Aspergillus niger* KK2 from lignocellulosic biomass. *Bioresour Technol*, 91, 153–156.

Klyosov AA. (1988). Cellulases of the third generation. In: Aubert, JP, Beguin, P and Millet, J, editors. *Biochemistry and Genetics of Cellulose Degradation*. London: Academic Press, pp. 87–99.

Kubicek CP. (2013). Systems biological approaches towards understanding cellulase production by *Trichoderma reesei*. *J Biotechnol*, 163, 133–142.

Lan TQ, Wei D, Yang ST, Liu X. (2013). Enhanced cellulase production by *Trichoderma viride* in a rotating fibrous bed bioreactor. *Bioresour Technol*, 133, 175–182.

Lee YE, Lowe SE, Henrissat B, Zeikus JG. (1993). Characterization of the active site and thermostability regions of endoxylanase from *Thermoanaerobacterium saccharolyticum* B6A-RI. *J Bacteriol*, 175, 5890–5898.

Leite RSR, Bocchini DA, Martins EDS, Silva D, Gomes E, Silva RD. (2007). Production of cellulolytic and hemicellulolytic enzymes from *Aureobasidium pulluans* on solid state fermentation. *Appl Biochem Biotechnol*, 137–140(1–12), 281–288.

Li J, Wang J, Wang S, Xing M, Yu S, Liu G. (2012). Achieving efficient protein expression in *Trichoderma reesei* by using strong constitutive promoters. *Microb Cell Fact*, 11, 84.

Liao H, Xu C, Tan S, Wei Z, Ling N, Yu G, Raza W, Zhang R, Shen Q, Xu Y. (2012). Production and characterization of acidophilic xylanolytic enzymes from *Penicillium oxalicum* GZ-2. *Bioresour Technol*, 123, 117–124.

Liming X, Xueliang S. (2004). High-yield cellulase production by *Trichoderma reesei* ZU-02 on corncob residue. *Bioresour Technol*, 91, 259–262.

Lin XQ, Han SY, Zhang N, Hu H, Zheng SP, Ye YR, Lin Y. (2013). Bleach boosting effect of xylanase A from *Bacillus halodurans* C-125 in ECF bleaching of wheat straw pulp. *Enzyme Microb Technol*, 52, 91–98.

Linder M, Lindeberg G, Reinikainen T, Teeri TT, Pettersson G. (1995). The difference in affinity between two fungal cellulose-binding domains is dominated by a single amino acid substitution. *FEBS Lett*, 372, 96–98.

Lis H, Sharon N. (1993). Protein glycosylation: structural and functional aspects. *Eur J Biochem*, 218, 1–27.

Liu Z, Zhao X, Bai F. (2013). Production of xylanase by an alkaline-tolerant marine-derived Streptomyces viridochromogenes strain and improvement by ribosome engineering. *Appl Microbiol Biotechnol*, 97(10), 4361–4368.

Lv W, Yu Z. (2012). Isolation and characterization of two thermophilic cellulolytic strains of *Clostridium thermocellum* from a compost sample. *J App Microbiol*, 114, 1001–1007.

Lynd LR, van Zyl WH, McBride JE, Laser M. (2005). Consolidated bioprocessing of cellulosic biomass: an update. *Curr Opin Biotechnol*, 16, 577–583.

Madhavan A, Srivastava A, Kondo A, Bisaria VS. (2012). Bioconversion of lignocellulose-derived sugars to ethanol by engineered *Saccharomyces cerevisiae*. *Crit Rev Biotechnol*, 32, 22–48.

Maeda RN, Barcelos CA, Anna LMMS, Pereira N Jr. (2013). Cellulase production by *Penicillium funiculosum* and its application in the hydrolysis of sugar cane bagasse for second generation ethanol production by fed batch operation. *J Biotechnol*, 163, 38–44.

Mandels M. (1975). Microbial sources of cellulases. In: Wilkie CR, editor. *Cellulose as a Chemical and Energy Resource*. New York: John Wiley and Sons, pp. 81–105.

Martinez D, Berka RM, Henrissat B, Saloheimo M, Arvas M, Baker SE, Chapman J, Chertkov O, Coutinho PM, Cullen D, et al. (2008). Genome sequencing and analysis of the biomass-degrading fungus *Trichoderma reesei* (syn. *Hypocrea jecorina*). *Nat Biotechnol*, 26, 553–560.

Matano Y, Hasunuma T, Kondo A. (2012). Display of cellulases on the cell surface of *Saccharomyces cerevisiae* for high yield ethanol production from high-solid lignocellulosic biomass. *Bioresour Technol*, 108, 128–133.

McClendon SD, Mao Z, Shin HD, Wagschal K, Chen RR. (2012). Designer xylanosomes: protein nanostructures for enhanced xylan hydrolysis. *Appl Biochem Biotechnol*, 167, 395–411.

Meissner K, Wassenberg D, Liebl W. (2000). The thermostabilizing domain of the modular xylanase XynA of *Thermotoga maritima* represents a novel type of binding domain with affinity for soluble xylan and mixed-linkage β-1,3/β-1,4-glucan. *Mol Microbiol*, 36, 898–912.

Mellitzer A, Weis R, Glieder A, Flicker K. (2012). Expression of lignocellulolytic enzymes in *Pichia pastoris*. *Microb Cell Fact*, 11, 61.

Menkhaus TJ, Bai Y, Zhang C, Nikolov ZL, Glatz CE. (2004). Considerations for the recovery of recombinant proteins from plants. *Biotechnol Prog*, 20, 1001–1014.

Merino ST, Cherry J. (2007). Progress and challenges in enzyme development for biomass utilization. *Adv Biochem Eng Biotechnol*, 108, 95–120.

Milagres AMF, Santos E, Piovan T, Roberto IC. (2004). Production of xylanases by *Thermomyces aurantiacus* from sugarcane bagasse in an aerated growth fermenter. *Process Biochem*, 39, 1387–1391.

Mitchell DA, Krieger N, Stuart DM, Pandey A. (2000). New developments in solid-state fermentation. II. Rational approaches to the design operation and scale-up of bioreactors. *Process Biochem*, 3, 1211–1225.

Miyauchi S, Te'o VS Jr, Bergquist PL, Nevalainen KM. (2013). Expression of a bacterial xylanase in *Trichoderma reesei* under the egl2 and cbh2 glycosyl hydrolase gene promoters. *N Biotechnol*, 30(5), 523–530.

Moraïs S, Barak Y, Caspi J, Hadar Y, Lamed R, Shoham Y, Wilson DB, Bayer EA. (2010). Cellulase-xylanase synergy in designer cellulosomes for enhanced degradation of a complex cellulosic substrate. *MBio*,1(5), pii: e00285.

Moraïs S, Barak Y, Hadar Y, Wilson DB, Shoham Y, Lamed R, Bayer EA. (2011). Assembly of xylanases into designer cellulosomes promotes efficient hydrolysis of the xylan component of a natural recalcitrant cellulosic substrate. *MBio*, 2(6), pii: e00233-11.

Mousdale DM. (2010). *Introduction to Biofuels*. Boca Raton, FL: CRC Press, pp. 45–90.

Narang S, Sahai V, Bisaria VS. (2001). Optimization of xylanase production by *Melanocarpus albomyces* IIS 68 in solid-state fermentation using response surface methodology. *J Biosci Bioeng* 91, 425–427.

Panagiotou G, Kekos D, Macris BJ, Christakopoulos P. (2003). Production of cellulolytic and xylanolytic enzymes by *Fusarium oxysporum* grown on corn stover in solid state fermentation. *Ind Crops Prod*, 18, 37–45.

Panda T, Bisaria VS, Ghose TK. (1987). Effect of culture phasing and polysaccharide on production of xylanase by mixed culture of *Trichoderma reesei* D1-6 and *Aspergillus wentii* Pt 2804. *Biotechnol Bioeng*, 30, 868–874.

Park EY, Naruse K, Kato T. (2011). Improvement of cellulase production in cultures of *Acremonium cellulolyticus* using pretreated waste milk pack with cellulase targeting for biorefinery. *Bioresour Technol*, 102, 6120–6127.

Polizeli MLTM, Rizzatti ACS, Monti R, Terenzi HF, Jorge JA, Amorim DS. (2005). Xylanases from fungi: properties and applications. *Appl Microbiol Biotechnol*, 67, 577–591.

Puls J, Schmidt O, Granzow C. (1987). α-Glucuronidase in two microbial xylanolytic systems. *Enzyme Microb Technol*, 9, 83–88.

Qi-peng Y, Jian-dong W, Huai Z, Zhong-ming Q. (2005).Effect of temperature shift on production of xylanase by *Aspergillus niger*. *Process Biochem*, 40, 3255–3257.

Rabinovich ML, Melnick MS, Bolobova AV. (2002). The structure and mechanism of action of cellulolytic enzymes. *Biochemistry (Mosc)*, 67, 850–871.

Rakotoarivonina H, Hermant B, Monthe N, Rémond C. (2012). The hemicellulolytic enzyme arsenal of *Thermobacillus xylanolyticus* depends on the composition of biomass used for growth. *Microb Cell Fact*, 11, 159.

Rasala BA, Lee PA, Shen Z, Briggs SP, Mendez M, Mayfield SP. (2012). Robust expression and secretion of xylanase1 in *Chlamydomonas reinhardtii* by fusion to a selection gene and processing with the FMDV 2A Peptide. *PLoS One*, 7(8), e43349.

Rawat R, Tewari L. (2012). Purification and characterization of an acidothermophilic cellulase enzyme produced by *Bacillus subtilis* strain LFS3. *Extremophiles*, 16, 637–644.

Reddy V, Reddy P, Pillay B, Singh S. (2002). Effect of aeration on the production of hemicellulases by *Thermomyces lanuginosus* SSBP in a 30 L bioreactor. *Process Biochem*, 37, 1221–1228.

Rojas-Rejón OA, Poggi-Varaldo HM, Ramos-Valdivia AC, Martínez-Jiménez A, Cristiani-Urbina E, Martínez MDLT, Ponce-Noyola T. (2011). Production of cellulases and xylanases under catabolic repression conditions from mutant PR-22 of *Cellulomonas flavigena*. *J Ind Microbiol Biotechnol*, 38, 257–264.

Ruanglek V, Sriprang R, Ratanaphan N, Tirawongsaroj P, Chantasigh D, Tanapongpipat S, Pootanakit K, Eurwilaichitr L. (2007). Cloning expression, characterization and high cell density production of recombinant endo 1,4-β-xylanases from *Aspergillus niger* in *Pichia pastoris*. *Enzyme Microb Technol*, 41, 19–25.

Ryu DDY, Mandels M. (1980). Cellulases: biosynthesis and applications. *Enzyme Microb Technol*, 2, 91–102.

Sá-Pereira P, Mesquita A, Duarte JC, Barros MRA, Costa-Ferreira M. (2002). Rapid production of thermostable cellulase-free xylanase by a strain of *Bacillus subtilis* and its properties. *Enzyme Microb Technol*, 30, 924–933.

Seiboth B, Herold S, Kubicek CP. (2012). Metabolic engineering of inducer formation for cellulase and hemicellulase gene expression in *Trichoderma reesei*. In: Wang X, Chen J, Quinn P, editors. *Reprogramming Microbial Metabolic Pathways*. New York: Springer, pp. 367–390.

Simkhada JR, Yoo HY, Choi YH, Kim SW, Yoo JC. (2012). An extremely alkaline novel xylanase from a newly isolated *Streptomyces* strain cultivated in corncob medium. *Appl Biochem Biotechnol*, 168, 2017–2027.

Srinivasan K, Murakami M, Nakashimada Y, Nishio N. (2001). Efficient production of cellulolytic and xylanolytic enzymes by the rumen anaerobic fungus, *Neocallimastix frontalis*, in a repeated batch culture. *J Biosci Bioeng*, 91, 153–158.

Sticklen M. (2006). Plant genetic engineering to improve biomass characteristics for biofuels. *Curr Opin Biotechnol*, 17, 315–319.

Takashima S, Ohno M, Hidaka M, Nakamura A, Masaki H. (2007). Correlation between cellulose binding and activity of cellulose-binding domain mutants of *Humicola grisea* cellobiohydrolase1. *FEBS Lett*, 581, 5891–5896.

Tambor JH, Ren H, Ushinsky S, Zheng Y, Riemens A, St-Francois C, Tsang A, Powlowski J, Storms R. (2012). Recombinant expression, activity screening and functional characterization identifies three novel endo-1,4-β-glucanases that efficiently hydrolyze cellulosic substrates. *Appl Microbiol Biotechnol*, 93, 203–214.

Techapun C, Poosaran N, Watanabe M, Sasaki K. (2003). Optimization of aeration and agitation rates to improve cellulase-free xylanase production by thermotolerant *Streptomyces* sp. Ab106. *J Biosci Bioeng*, 95, 398–301.

Templeton DW, Sluiter AD, Hayward TK, Hames BR, Thomas SR. (2009). Assessing corn stover composition and sources of variability via NIRS. *Cellulose*, 16, 621–639.

Tenkanen M, Poutanen K. (1992). Significance of esterases in degradation of xylans. In: Visser J, Kusters-Van Someran MA, Beldman G, Voragen AGJ, editors. *Xvlans and Xvlanases*. Amsterdam: Elsevier Science Publishers, pp. 203–212.

Tian B, Xu Y, Cai W, Huang Q, Gao Y, Li X, Huang J. (2013). Molecular cloning and overexpression of an endo-β-1,4-xylanase gene from *Aspergillus niger* in industrial *Saccharomyces cerevisiae* YS2 strain. *Appl Biochem Biotechnol*, 170(2), 320–328.

Torres JMO, dela Cruz TEE. (2013). Production of xylanases by mangrove fungi from the Philippines and their application in enzymatic pretreatment of recycled paper pulps. *World J Microbiol Biotechnol*, 29, 645–655.

Trincone A. (2011) Marine biocatalysts: enzymatic features and applications. *Mar Drugs*, 9, 478–499.

Verma D, Anand A, Satyanarayana T. (2013). Thermostable and alkalistable endoxylanase of the extremely thermophilic bacterium *Geobacillus thermodenitrificans* TSAA1: cloning, expression, characteristics and its applicability in generating xylooligo-saccharides and fermentable sugars. *Appl Biochem Biotechnol*, 170(1), 119–130.

Violot S, Haser R, Sonan G, Georlette D, Feller G, Aghajari N. (2003). Expression, purification, crystallization and preliminary X-ray crystallographic studies of a psychrophilic cellulase from *Pseudoalteromonas haloplanktis*. *Acta Crystallogr D-Biol Crystallogr*, 59, 1256–1258.

Vitcosque GL, Fonseca RF, Rodríguez-Źũniga UF, Neto VB, Couri S, Farinas CS. (2012). Production of biomass-degrading multienzyme complexes under solid-state fermentation of soybean meal using a bioreactor. *Enzyme Res*, 2012, 248983.

Waeonukul R, Kyu KL, Sakka K, Ratanakhanokchai K. (2009). Isolation and characterization of a multienzyme complex (cellulosome) of the *Paeniöacilluscurdlanolyticus* B-6 grown on Avicel under aerobic conditions. *J Biosci Bioeng*, 107, 610–614.

Wang S, Liu G, Wang J, Yu J, Huang B, Xing M. (2013). Enhancing cellulase production in *Trichoderma reesei* RUT C30 through combined manipulation of activating and repressing genes. *J Ind Microbiol Biotechnol*, 40(6), 633–641.

Wang SW, Xing M, Liu G, Yun SW, Wang J, Tian SL. (2012). Improving cellulase production in *Trichoderma koningii* through RNA interference on *ace1*gene expression. *J Microbiol Biotechnol*, 22, 1133–1140.

Watanabe H, Tokuda G. (2001). Animal cellulases. *Cell Mol Life Sci*, 58, 1167–1178.

Wejse PL, Ingversen K, Mortensen KK. (2003). Xylanase production by a novel halophilic bacterium increased 20-fold by response surface methodology. *Enzyme Microb Technol*, 32, 721–727.

Wenqing L, Defa L, Wu Y. (2003). Influence of water activity and temperature on xylanase biosynthsis in pilot-scale solid-state fermenter by *Aspergillus sulphureus*. *Enzyme Microb Technol*, 32, 305–311.

Xiong H, Nyyssölä A, Jänis J, Pastinen O, von Wemarn N, Leisola M, Turumen O. (2004a). Characterization of the xylanase produced by submerged cultivation of *Thermomyces lanuginosus* DSM 10635. *Enzyme Microb Technol*, 35, 93–99.

Xiong H, von Weymarn N, Leisola M, Turunen O. (2004b). Influence of pH on the production of xylanases by *Trichoderma reesai* Rut C-30. *Process Biochem*, 39, 729–734.

Xue DS, Chen HY, Lin DQ, Guan YX, Yao SJ. (2012). Optimization of a natural medium for cellulase by a marine *Aspergillus niger* using response surface methodology. *Appl Biochem Biotechnol*, 167, 1963–1972.

Zhao J, Shi P, Bai Y, Huang H, Luo H, Zhang H, Xu D, Wang Y, Yao B. (2012). A thermophilic cellulase complex from *Phialophora* sp. G5 showing high capacity in cellulose hydrolysis. *Appl Biochem Biotechnol*, 166, 952–960.

CHAPTER 6

Bioprocessing Technologies

GOPAL CHOTANI, CAROLINE PERES, ALEXANDRA SCHULER, and PEYMAN MOSLEMY

DuPont Industrial Biosciences, Palo Alto, CA, USA

Bioprocessing of Renewable Resources to Commodity Bioproducts, First Edition.
Edited by Virendra S. Bisaria and Akihiko Kondo.

Abstract

The primary objective of bioprocess development is the integration of feedstock conversion and cell factory operation to produce the needed enzymes and biochemicals. Bioprocessing technologies are enabling the development of biorefineries through decades of experience combined with advances in metabolic engineering and synthetic biology. In this chapter, we present the evidence of the interdependence of cell factory, fermentation, downstream processing, and formulation with product development. Through examples of biocatalysts and biochemicals produced by state-of-the-art bioprocesses, we demonstrate how the biorefineries of the future will be shaped.

6.1 INTRODUCTION

Advances in cell factory design and integrated bioprocessing are promoting phenomenal applications of the growing concept of biorefineries. Cell factories (biocatalysts) are being used in two related but distinct production processes. The first is focused on enzyme production, while the second targets the production of biochemicals, biomaterials, and biofuels. By developing more efficient enzymes and microbes, bio-based industry leaders are creating new processes, products, and applications, which require sophisticated enzyme cocktails that combine many distinct activities. Likewise, synthetic biology is allowing the engineering of microorganisms so that the biofuels produced are adaptable to the currently used fuels in transportation. Production of 1,3-propanediol and lactic acid by engineered microorganisms from sugar are recent examples of the biorefinery concept progressing toward reality. The commercial production of 1,3-propanediol for Sorona® is based on the technology developed in a unique collaboration between DuPont and Genencor. This process has become the hallmark of an emerging commercially viable application of the biorefinery concept for the production of basic chemical building blocks competing with petrochemically derived materials (Nakamura and Whited, 2003).

The concept of a biorefinery is not new. Apart from coproduct streams that result from the processing of grain or biomass, the released sugars are used as a source of carbon to produce an ever-growing list of products such as fuel and the building blocks for the synthesis of chemicals and polymers. In its simplistic form, several pieces of the biorefinery have been in operation for thousands of years and have evolved depending upon market needs, for example, driven by the rising demand for biofuel like ethanol. Biorefineries will continue to evolve in a manner similar to the growth of petroleum refineries in the 1900s. As technological developments expand

and the spectrum of products that can be produced from biorefineries grows, a wide variety of feedstocks and bioprocessing technologies will come into use.

Grains, trees, grasses, and crop residues have been used to produce a range of products like biofuels, bioplastics, pulp and paper, solvents, and building materials. The agriculture industry produces large varieties and quantities of sugary crops and grains, which are used to produce a host of materials. For example, starch from grain crops and sucrose from beets and sugar cane are being converted to a large number of products. Some of these products include: high fructose corn syrups (HFCS), ethanol, 1,3-propanediol, lactic acid, ascorbic acid, citric acid, acetic acid, glycerol, amino acids, and gluconic acid. As additional advances in pathway engineering are made, one can expect many more biochemicals and biopolymers to be produced from sugars (Chotani et al., 2000). As more biochemicals and biomaterials are produced from fermentable carbon sources, favorable changes in the agriculture and chemical sectors can be expected. For the last 30 years, Genencor and DuPont have been active in developing bioprocessing technologies, and by developing state-of-the-art technologies for biocatalyst design, integrated bioprocessing, and industrial applications, they have focused on the production and conversion of fermentable sugars.

Advances in pathway engineering have enabled the commercialization of a number of competitive bioprocesses and will continue to provide a low-cost route to biochemical, biomaterial, and biofuel production. For producing commodity products, the cost of carbon is often more than 50% of the total direct cost of production, and therefore efforts to lower this cost will play an important role in the overall development of a biorefinery. The challenge then is to develop and optimize a portfolio of technologies that will allow growth of the bio-industry. The two main scientific developments to focus on will be (a) enzymes required for the conversion of starch and cellulosic biomass to fermentable sugars within an enabling cost structure and (b) engineered organisms, the cell factories, to produce chemicals competitively from these sugars.

The essential elements of an integrated bioprocess are as shown in Figure 6.1: (a) feedstock processing capability for the conversion of a wide variety of feedstocks to fermentable carbon and coproduct streams, preferably using enzymes, (b) engineered biocatalysts (cell factories) for the conversion of a variety of carbon substrates

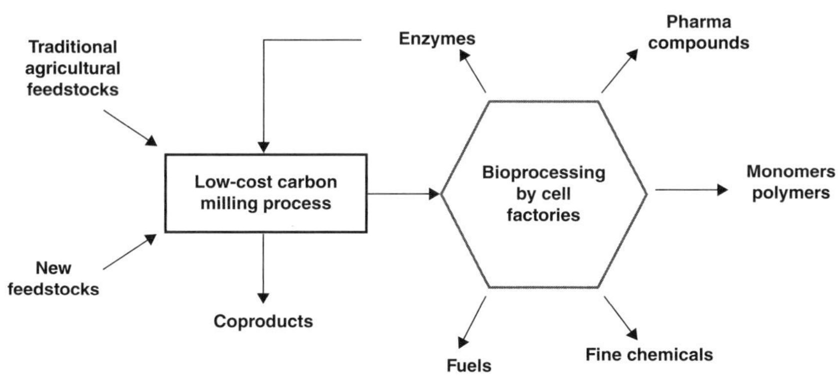

FIGURE 6.1 Integrated bioprocessing.

to a plethora of products, and (c) bioprocessing capabilities for the production of enzymes and biochemicals to meet multitude of customer applications and market economics. This chapter presents the issues to consider for state-of-the-art bioprocess development, with focus on integration with feedstock conversion and biocatalyst engineering.

6.2 CELL FACTORY PLATFORM

The production of enzymes and chemicals has been around for centuries. One of the first examples for use of enzymes was the extraction of naturally occurring enzymes from the source substrate, for example, β-amylases from grains used in brewing. Enzymes have been extracted from plant and animal sources. Then came the era of microbially produced enzymes, where the enzymes were initially natural products of the microbial culture. Today, most enzymes are produced microbially; however, many enzymes are no longer native to the host but rather engineered versions. Protein engineering has allowed optimizing enzyme properties for specific use (Estell, 1993). A number of different approaches have been taken to improve enzyme performance including increasing thermal stability and/or specific activity, and recruitment of novel activities. Random mutagenesis and screening has been traditionally utilized to improve enzyme properties, where an important part of this effort was the development of a small-scale screen to quickly and accurately measure the effect of the changes made.

Along with the ability to engineer proteins with improved properties, it is imperative that the engineered proteins be expressed at high levels in production strains approved by the regulatory agencies for no adverse impact on the living systems (Ferrari and Peres, 2009). Microbial production strains have also been engineered to excrete molecules during fermentation, and often a conventional mutagenesis and screening approach has been applied to this end. Through successive rounds of mutagenesis and selection with different screens, an improved production host strain is often developed. Several new traits can be recruited in the strain to positively impact fermentation, for example, specific growth rate to reduce the growth time needed for the fermentation, which positively impacts productivity; or product per cell, which impacts both product yield from the carbon source and volumetric productivity.

Future efforts to develop commercial fermentation processes will require development of production strains different from traditional organisms like yeast, *Bacillus*, *Escherichia coli* (*E. coli*), and other fungi and bacteria. There is a need to develop hosts capable of fermenting at higher or lower pH, temperatures, concentrations of substrates and products. Furthermore, these new microorganisms would also need to be able to resist by-products generated during feedstock conversion. In the last few years, the knowledge from genomes of hundreds of microorganisms has exponentially increased. In this collection of newly characterized microorganisms, one can find many different desired physiological traits and learn the strategies that nature has imposed for the selection of strains able to deal with high temperatures, high concentration of toxic compounds, sugars, salts, pH, etc. Furthermore, some of these genomes may harbor new enzymes and pathways that would allow the use of non-conventional substrates for fermentation.

6.2.1 Properties of a Biocatalyst

Every bioprocess starts with the best possible cell factory. A better performing cell will allow for less capital, lower variable cost for the fermentation, and a simpler recovery process as the product will be more concentrated. More than one property makes for a good cell factory and none of the commonly used cell factories have all the needed attributes. Case-by-case development of a biocatalyst is still the norm, even though a platform for a given organism would help accelerate process development. Described below are elements from a process point of view to consider in choosing a biocatalyst for a bioprocess.

6.2.1.1 Fixed Cost The biocatalyst has a strong influence on what capital is needed and hence affects the fixed cost. The oxygen supply to the fermentation is an important determinant of cost (lower the oxygen needed, lower are the costs for agitation, pressure, and aeration). The heat generated by a process is also tightly linked to the oxygen uptake rate (OUR), and therefore cooling costs increase when oxygen demand increases.

The oxygen required can be a function of the organism used. For example, *Bacillus* or *Pseudomonas* are non-fermentative organisms and thus perform better with excess oxygen. Their growth rates can reach $\geq$0.7/h which can lead to significant OUR. Other organisms, like *E. coli* or *Saccharomyces cerevisiae*, can operate under fermentative metabolism and require low or no oxygen. Slow-growing organisms, like filamentous fungi, are able to avoid the need for high oxygen transfer rate (OTR) fermenters because of their high specific productivity. These organisms are useful for the production of industrially popular hydrolytic enzymes.

In addition, production of the same biochemical by the same microorganism could have a different oxygen requirement depending on the pathway used. A good example is the production of Bioisoprene^{TM} by DuPont and Goodyear. The precursor for isoprene, dimethylallylpyrophosphate (DMAPP), is essential for all living organisms. It is a precursor to several essential molecules such as cholesterol, ubiquinone, and chlorophyll (Kuzuyama and Seto, 2012). Most prokaryotes produce DMAPP via the DXP (1-deoxy-D-xylulose 5-phosphate) or MEP (methylerythritol) pathway, while eukaryotes use the mevalonate pathway (MVA) (Kuzuyama and Seto, 2012). Plants harbor the MVA pathway in their cytosol and the DXP pathway in their plastid. Based on the reactions to produce DMAPP from glucose via each pathway, theoretical yield and oxygen amount can be estimated. Figure 6.2 shows that those two parameters vary depending on the pathway used. By engineering the carbon fluxes through both pathways at the right ratio, the process would result in the best yield and the lowest oxygen demand. Bioprocessing capital could possibly be reduced by exploring the richness of bacterial diversity.

Microorganisms have been isolated which can perform at a wide range of pH or temperature (Zambare et al., 2011), use synthesis gas as a carbon source (Köpke et al.,

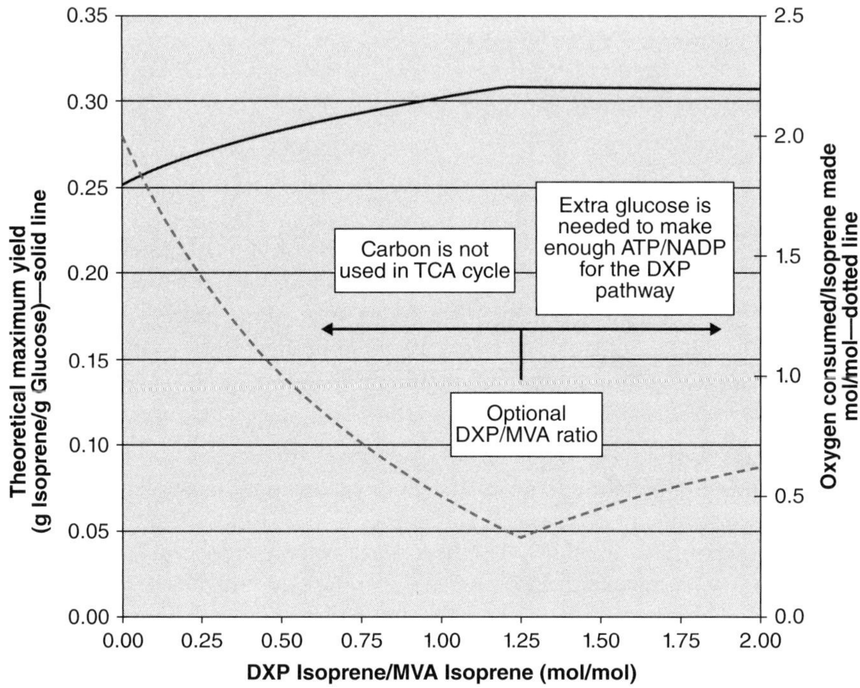

FIGURE 6.2 Theoretical maximum yield of isoprene from glucose (solid line), and oxygen consumed per mol of isoprene made (dotted line), as a function of the ratio of the two dimethylallylpyrophosphate (DMAPP)-producing pathways.

2010), or transfer electrons to and from an electrode (Ross et al., 2011). Currently most of these organisms are not tractable, but methods are being developed that would allow for expression of exogenous proteins in those very attractive chassis (Leang et al., 2013; Ye and Kelly, 2012). Recent sequencing endeavors have generated millions of sequences, and diving into those sequences could help in the understanding of this diversity. For example, through thermo-tolerance, running a bioprocess at higher temperature generates considerable savings from cooling system; or, through wide pH survival, lowering the pH of a fermentation decreases chances of contamination.

Capital cost can also be reduced by creatively "recycling" production plants aimed at a different production process. One should consider green-, gray- and brown-field designing options for a plant (http://sustainabilityworkshop.autodesk.com/fundamentals/building-site-location). For example, to produce isobutanol as a better biofuel option than ethanol, companies like Butamax and Gevo have focused on retrofitting existing ethanol production plants rather than building new ones (Fountain, 2012). Retrofitting a plant offers benefits of infrastructure already in place.

6.2.1.2 Variable Cost Although several factors impact direct or variable cost, the cost of carbon to feed cell factories is often more than 50% of the total direct cost of production. In other words, the yield of a process (g product/g substrate) is a strong driver of cost. Industrial R&D efforts are often aimed at optimizing yield of product made by a certain pathway, or new pathways/enzymes are added to increase theoretical yield (Burgard et al., 2012; Figure 6.2). Product yields are less than theoretical as some carbon is devoted toward growth of the biocatalyst. Commercial targets, however, are often more than 80% of theoretical yield. A well-known case is the conversion of glucose into ethanol by *S. cerevisiae*, where high commercial yields >0.46 g ethanol/g glucose (>90% of 0.51 theoretical yield) are commonly observed (Erdei et al., 2010).

Production of an enzyme by a bacterial or a fungal strain is more complicated since a protein is composed of 20 amino acids as building blocks. Theoretical protein mass yield on glucose is then best estimated from the stoichiometry of the building blocks and is close to 50%, according to Equation 6.1, based on the elemental composition (Nielsen et al., 2002):

$$C_6H_{12}O_6 + 1.05\,NH_3 + 1.63\,O_2 \rightarrow 0.7\,C_6H_{8.9}O_{1.7}N_{1.5} + 1.8\,CO_2 + 4.46\,H_2O \qquad (6.1)$$

where the stoichiometric coefficient 0.7 is calculated from metabolic flux analysis.

Almost equally important is the ability of the biocatalyst to utilize varieties of carbon sources. While corn syrup, for example, dextrose, is still the major carbon source for industrial processes in the United States, carbon sources elsewhere are sucrose from sugar cane, sugar beet, or molasses. The cheaper and less refined feedstocks also contain C_5 sugars such as xylose or arabinose, and it is advantageous if the biocatalyst is able to use such sugars. While *E. coli* is a natural C_5 metabolizing organism, *S. cerevisiae* has to be engineered to utilize such substrates (Gosh et al., 2011). Likewise, not all *E. coli* strains are able to utilize sucrose, but they can also be engineered to process C_{12} sugar substrates (Bruschi et al., 2012).

Further efforts to lower carbon source cost play an important role in the overall development of the biorefinery concept. Several new initiatives are focused on cellulosic biomass (e.g., agricultural waste), one of the lowest cost source of fermentable sugar in the biorefinery. Much knowledge has been gained, and significant progress has been made on the saccharification of cellulosic biomass to fermentable sugars for their subsequent bioconversion to chemicals. As a result, such processes are coming closer to commercial reality. Since 2008, DuPont Cellulosic Ethanol has been operating a demonstration-scale facility in Vonore, Tennessee, to generate the data necessary to finalize the integrated scale-up technology for commercial production. The next step is the construction and operation of a commercial-scale cellulosic ethanol biorefinery in Nevada, Iowa. DuPont broke ground for that plant in December 2012, with a 12–18 month build out. Once completed, this plant will be one of the first commercial-scale cellulosic biorefineries in the world and will generate 28 million gallons of cellulosic biofuels from corn stover (corn residues, including cobs, stalks, and leaves).

Conversion of cellulosic biomass into sugars has traditionally been achieved enzymatically (Dean et al., 2006). With the rise of Cleantech technologies, several companies, such as Renmatix or Virdia, are now developing chemical extraction processes for conversion of cellulosic biomass to highly refined fermentable sugars (Bullis,

2012). Further developments of enzymatic and/or chemical processes for bioconversion of cellulosic biomass will help reduce dependency on food resources.

Efforts to navigate upstream of the carbon cycle by working with microorganisms which can use syngas (Köpke et al., 2010, 2011) or natural gas (Jiang et al., 2010) as a carbon source are also in place. Such promising technologies would allow harnessing a cheap carbon source, avoiding the "food versus fuel" issue, as well as reducing greenhouse gases in the atmosphere.

6.2.1.3 *Robustness* Other properties of a biocatalyst can also have a very significant impact on the bioprocess. For example, substrate and product tolerances are important considerations (Nicolaou et al., 2010). Tolerance can be improved by strain evolution (Miller et al., 2009) or by targeted genetic modification (Zingaro and Papoutsakis, 2013).

The abilities to resist phage contamination or to grow at a pH other than neutral are attributes which can make a bioprocess more robust. Those are among reasons why yeast is sometimes preferred to *E. coli*. In some cases, even if a particular organism would be preferred, a less desirable catalyst might have to be used if a crucial enzyme cannot be expressed. An example is the poor expression of plant P450 enzymes in *E. coli*, compared to *Saccharomyces* (Duan and Schuler, 2006). Naturally, genetic tractability is also important. Genetic tools for all workhorses used in bioprocessing such as *E. coli*, *Bacillus*, *Aspergillus*, *Trichoderma*, *Saccharomyces* are available, and progress is being constantly made on understanding the genetics of more out-of-the-box organisms.

Additional costs related to the cell factory could include the use of an external inducer. Indeed, most bioprocesses are comprised of a growth phase and a production phase. A cell factory can in some cases be switched from growth to production phase using nutritional parameters only (production of most hydrolytic enzymes follows this pattern) but could sometimes need an inducer to go into production phase. A well-known inducer for *E. coli* is isopropyl β-D-1-thiogalactopyranoside (IPTG). If an inducer must be used, then it needs to be affordable and acceptable for wastewater treatment (Table 6.1).

6.2.2 Recent Trends in Cell Factory Construction for Bioprocessing

Recent technical advances in strain/pathway construction have moved the limiting step from strain construction to screening. Those advances are, among others, affordable gene synthesis, synthetic construction of gene libraries (Collier et al., 2009; Bott et al., 2012), and novel techniques for DNA assembly (e.g., Engler et al., 2008; Gibson et al., 2009).

An alternative design being tested is an organism which can produce enzymes to release fermentable sugars from complex carbohydrate feedstock and convert the released sugars to products. Such integration has been called consolidated bioprocessing (CBP). The CBP strategy is already being put to work in the industry for production of ethanol from starch by yeast (Brevnova et al., 2011), and there are proposals for developing CBP for biomass conversion using the cellulolytic fungus,

TABLE 6.1 Microorganisms Commonly Utilized in Bioprocessing, and Examples of Their Pros and Cons for Bioprocessing

Organism	Advantages	Drawbacks
Yeast such as *Saccharomyces*, *Pichia*, *Yarrowia*	• No phage contamination • Easy to use in existing plant retrofit	• Less tolerance • Less tractable than some bacteria • No C5 utilization
Lactobacillus	• High tolerance to solvent	• Less tractable
Bacillus	• Secretes enzymes • No inducer needed	• High OUR
Escherichia coli	• Highest tractability	• Sensitive to phage contamination
Trichoderma, *Aspergillus* (fungi)	• High specific productivity • Secretes enzymes	• Less tractable for metabolic engineering

OUR, oxygen uptake rate.

Trichoderma reesei (Xu et al., 2009). CBP combines enzyme production and simultaneous saccharification of feedstock with fermentation of the resulting sugars into a single microorganism, that is, CBP if successful will not involve a separate enzyme production process.

Another platform worth mentioning in bioprocessing is cell-free catalysis, which is used not only in academia (Hodgman and Jewett, 2012) but also for the commercial production of protein therapeutics (Zawada et al., 2011).

All the knowledge gathered in bioprocessing has enabled the biological production of chemicals and enzymes to be a reality (Table 6.2).

At any time during bioprocess development, it should be kept in mind that changes upstream can lead to major changes downstream. The cell factory determines not only the fermentation attributes but also the recovery and the formulation parts of the process. For example, for better performance in the application of a detergent-additive, the enzyme had to be produced by *Bacillus* containing an inactivated *bglC* gene (Cervin, 2011). An integrated bioprocess design of cell factory, fermentation, recovery, and formulation can result in considerable savings of time and money.

TABLE 6.2 Some Examples of High-Volume Commercial Bioprocessing

Product	Organism	Company
Enzymes	*Bacillus*, *Trichoderma*	DuPont, Novozymes
1,3-Propanediol	*Escherichia coli*	DuPont/Tate & Lyle Joint Venture
Oil	Microalgae	Solazyme
Polylactic acid	Yeast	Nature Works
Succinic acid	*Escherichia coli*	Myriant
Succinic acid	Yeast	Reverdia
Ethanol	*Saccharomyces*	Poet
Ethanol	*Zymomonas*	DuPont Cellulosic Ethanol

6.3 FERMENTATION PROCESS

A fermentation process involves consideration of a number of factors to achieve optimal cell factory performance. Such factors can be grouped into two main categories: (a) process recipe and (b) equipment capabilities. For example, the production microorganism may prefer aerobic or anaerobic conditions, the product may be extracellular or intracellular (or the cell mass itself), and it may be produced during growth or after growth. Therefore, the fermentation process design needs to be considered in the context of the integrated production process since fermentation can significantly affect downstream processing. In other words, the manufacturing process can best be described as a product recipe for successful fermentation (growth of organism and production of enzyme or metabolite), recovery (separation of molecule(s) of interest by processing of fermentation broth), and formulation (preparation of a stable product, suitable for use in various applications).

Fermentation process can be divided into three stages:

- Pre-seed culture
- Seed propagation
- Main fermentation

In all stages, media or raw materials are batched and then sterilized, in-line or in the fermenter. The raw materials for the fermentation medium are typically carbohydrates such as corn syrup and glucose, proteins such as corn extracts, soy meal and yeast extracts, various salts, antifoam, and ammonia, sodium hydroxide, sulfuric acid, or phosphoric acid for pH control. The inoculum buildup starts in a laboratory, by revival of generally frozen cells or spores; propagation of cells in shake flasks or wave bags. The culture is aseptically inoculated into shake flasks containing media, followed by growth under aseptic conditions. The grown inoculum is then transferred to a sterilized container which allows for aseptic transfer to seed fermenter in the manufacturing area. After the required time period, following appreciable growth of microorganisms in the seed fermenter, the seed is then transferred to the main fermenter via direct piping. All stages of seed buildup and media preparation are done under strict sterility to prevent contaminating microorganisms from entering the manufacturing process. The entire process is carried out in sealed vessels designed to prevent both the release of the production organism and the entry of other microorganisms.

The fermenter operation can typically be categorized as one of the four types. The *batch* process is a constant volume fermentation where all the contents of the fermenter are added at the beginning (though air for dissolved oxygen (DO) supply, and alkali, e.g., ammonia, or acid, e.g., sulfuric for pH control may be added throughout the fermentation). In a *fed-batch* process, the essential nutrients are initially charged into the fermenter and then more are added during the fermentation, resulting in a volume that is not constant. A *continuous* fermentation is one where the fermenter contents are continuously added and removed while the volume is typically held constant. Finally, in a continuous system with a *partial* or *complete cell recycle*, a cell mass concentration/separation apparatus is used to return the concentrated cell mass to the fermenter while the media is removed. The detailed kinetics of these processes has been described in the literature (Chotani et al., 2007).

The production organism is grown in a submerged (or deep) tank in pure culture form. Most enzyme production processes are aerobic; oxygen is required to supply metabolic energy to the culture and therefore require continuous addition of air to the fermenter. This oxygen requirement results in two specific issues for the fermentation equipment: oxygen mass transfer and heat removal. Oxygen is only sparingly soluble in water, so to supply the required oxygen, a pressurized fermenting vessel is used and high pressure air is introduced at the bottom of the fermenter. Also, sometimes mixing is enhanced in the fermenting liquid through impellers driven by a high-power motor. All of these requirements (pressure, air flow, and power) increase the complexity of the fermentation equipment. In an aerobic fermentation process, heat is generated by the organism through the respiration of oxygen (metabolic heat) and also through the transfer of power from the agitator motor to the liquid (mechanical heat). This heat must be removed from the fermenter through cooling surfaces and finally transferred to cooling water that is circulated through the jacket and cooling pipes. A fermentation process with high oxygen demand requires sophisticated and expensive equipment and results in high utility usages.

Stoichiometry, that is, chemical compositions of the production microorganism and the product molecule(s), can be used for the development of the fermentation process. For example, chemical balances can be used to determine the media composition, amounts of carbon and nitrogen sources for growing cells and making products, estimate respiration (oxygen, carbon dioxide) rates, and other fermenter parameters. Based on the typical composition of a microbial cell (Roels, 1983), cell mass molecular formula used is $CH_{1.8}O_{0.5}N_{0.2}$, MW = 24.6. The microbial composition can vary depending on the cell type and its physiological state.

Product formation in a fermenter can be characterized using the most common metric(s) such as titer (product concentration), yield (product per substrate), and rate (productivity). Equations 6.1–6.3 can be explicitly written for product formation, cell growth, and maintenance. These equations assume that the only fermentation products containing carbon are product, cell mass, and carbon dioxide, though there are often additional side products that can be taken into account to improve the accuracy of the model. For production of enzyme(s), Equation 6.1 can be used:

$$C_6H_{12}O_6 + 1.05\,NH_3 + 1.63\,O_2 \rightarrow 0.7\,C_6H_{8.9}O_{1.7}N_{1.5} + 1.8\,CO_2 + 4.46\,H_2O \tag{6.1}$$

For cell mass, of typical elemental composition $C_6H_{10.8}O_3N_{1.2}$, formation from glucose in defined media, the equation is

$$C_6H_{12}O_6 + 0.76\,NH_3 + 2.03\,O_2 \rightarrow 0.63\,C_6H_{10.8}O_3N_{1.2} + 2.22\,CO_2 + 3.73\,H_2O \tag{6.2}$$

where the stoichiometric coefficient 0.63 is calculated from metabolic flux analysis.

Cell maintenance reactions in case of aerobic fermentation can be followed by equation:

$$C_6H_{12}O_6 + 6\,O_2 \rightarrow 6\,CO_2 + 6\,H_2O \tag{6.3}$$

Additional equations can be added to the existing system of Equations 6.1–6.3 to account for these side products when appropriate. The theoretical product yield and cell mass coefficients derived from Equations 6.1 and 6.2 are respectively:

$$Y_{\mathrm{p/s}}^{\max} = \frac{M_{\mathrm{p}}}{M_{\mathrm{s}}} \tag{6.4}$$

$$Y_{x/\mathrm{s}}^{\max} = \frac{M_x}{M_{\mathrm{s}}} \tag{6.5}$$

where M_{p} is the total product made and M_x is the total cell mass (dry cell weight) from total substrate (carbon source). $Y_{x/\mathrm{s}}$ and $Y_{\mathrm{p/s}}$ values on average are in 0.5 range.

These equations have many uses throughout fermentation development, such as an estimation of maximum practical product yield, biocatalyst's (cell) performance (ratio of product to dry cell weight), fermentation cooling requirements, or the cumulative heat of reaction.

Fed-batch fermentation process is in extensive use in industry for its ability to control key variables such as substrate, nutrient concentration, pH, and DO. A typical total mass balance around a *fed-batch* fermenter is

$$\frac{d}{dt}(\rho \cdot V) = \rho_{\mathrm{F}} \cdot F_{\mathrm{s}}(t) - CO_2\ \text{loss} + O_2\ \text{gain} - \text{Vapor loss} + NH_3,\ \text{etc. gain} \tag{6.6}$$

where ρ and V are the density and volume, respectively, of the fermentation broth. ρ_{F} is the density of the feed solution, and $F_{\mathrm{s}}(t)$ is the rate of substrate via feed addition, as a function of fermentation time.

The cell mass balance provides Equation 6.7:

$$\frac{d}{dt}(X \cdot V) = \mu \cdot X \cdot V \tag{6.7}$$

where$X$ is the cell mass concentration in the fermenter broth, and μ is the specific growth rate.

The substrate balance equation is

$$\frac{d}{dt}(S \cdot V) = F_{\mathrm{s}}(t) \cdot S_{\mathrm{f}} - V \cdot q_{\mathrm{s}} \cdot X \tag{6.8}$$

where S is the substrate concentration in the fermenter broth, S_{f} is the substrate concentration in the feed solution, and q_{s} is the specific substrate uptake rate.

Product accumulation follows the equation:

$$\frac{d}{dt}\left(C_{\mathrm{p}} \cdot V\right) = q_{\mathrm{p}} \cdot X \cdot V \tag{6.9}$$

where C_{p} is the product concentration in the fermenter broth, and q_{p} is the specific production rate.

The specific substrate uptake rate q_{s} is dependent on growth, production, and maintenance rates according to:

$$q_{\mathrm{s}} = \frac{\mu}{Y_{x/\mathrm{s}}^{\max}} + \frac{q_{\mathrm{p}}}{Y_{\mathrm{p/s}}^{\max}} + m \tag{6.10}$$

where $Y_{x/\mathrm{s}}^{\max}$ is the maximum cell yield, $Y_{\mathrm{p/s}}^{\max}$ is the maximum product yield and m is the maintenance rate. Empirically, the specific product formation rate q_{p} can be expressed by the Leudeking–Piret equation for many production systems:

$$q_{\mathrm{p}} = \alpha \cdot \mu + \beta \tag{6.11}$$

where α is growth-associated product to cell mass ratio and μ is a nongrowth associated rate of production.

The product concentration (titer) in the fermenter C_{p}, varies with time as:

$$C_{\mathrm{p}} = \frac{\int_0^t \left(q_{\mathrm{p}} \cdot X \cdot V\right) dt}{V_{\mathrm{t}}} = \frac{\text{cumulative product mass}}{\text{cumulative volume}} \tag{6.12}$$

The volumetric productivity, P_{v} depends on fermenter volume and fermentation time, for example, average productivity:

$$P_{\mathrm{v}} = \frac{\int_0^t \left(q_{\mathrm{p}} \cdot X \cdot V\right) dt}{\overline{V} \cdot t} = \frac{\text{cumulative product mass}}{\text{average working volume} \cdot \text{time}} \tag{6.13}$$

where $\overline{V}$ is the average working volume. Likewise, final productivity can be estimated by replacing $\overline{V}$ with V_{f}, final working volume; or nominal productivity by replacing $\overline{V}$ with nominal fermenter volume.

The overall cell mass yield, $Y_{x/\mathrm{s}}^{\mathrm{obs}}$, with time:

$$Y_{x/\mathrm{s}}^{\mathrm{obs}} = \frac{X_{\mathrm{t}} \cdot V_{\mathrm{t}}}{\left[S_0 \cdot V_0 + S_{\mathrm{f}} \int_0^t F(t)\, dt\right]} = \frac{\text{biomass produced}}{\text{substrate mass consumed}} \tag{6.14}$$

The overall product yield, $Y_{p/s}^{obs}$, with time:

$$Y_{p/s}^{obs} = \frac{C_p \cdot V_t}{\left[S_0 \cdot V_0 + S_f \int_0^t F(t)\,dt\right]} = \frac{\text{cumulative product mass}}{\text{substrate mass consumed}} \tag{6.15}$$

where S_0 is the substrate concentration in the batch volume V_0 at time zero.

A fermenter is cooled through the heat exchange area in order to remove the metabolic heat and maintain the set point temperature. The total heat of reaction, $\Delta H_{total}^{\Theta}$, can be calculated from heats of reactions (Equations 6.1–6.3):

$$\Delta H_{total}^{\Theta} = \sum_{rxn} n \Delta H_{rxn}^{\Theta} = \text{heat produced} \tag{6.16}$$

where n is the fractional contribution of each reaction to $\Delta H_{total}^{\Theta}$. Heats of reactions can be estimated using standard values of heat of combustion or formation, for substrates, cells, and products. Therefore, in order to calculate the heat of reaction, the molar stoichiometry of the reaction and the standard heat of combustions or formations of the reactants and products must be known (Doran, 1995). In case of reactions with oxygen as electron acceptor (aerobic fermentation), heat of combustion, for a variety of substrates, is on average $-460\frac{\text{kJ}}{\text{gmol O}_2}$ (Roels, 1983). Interestingly, this result had been observed for a variety of organisms including *Bacillus*, *E. coli*, and *Aspergillus* (Cooney et al., 1968).

Experimentally, precise value of $\Delta H_{total}^{\Theta}$ can be measured only after accounting for heats of mixing, evaporative cooling, agitation, airflow, etc. However, most of these contributions can be small compared to the heat of respiration, in commercial-scale large-size aerobic fermenters.

The oxygen uptake rate (OUR) provides a metabolic activity metric for aerobic fermentations:

$$\text{OUR} = \frac{F_i(O_2) - F_o(O_2)}{\overline{V}} = \frac{\text{rate of moles of oxygen consumed}}{\text{working volume}} \tag{6.17}$$

And, the carbon dioxide evolution rate (CER) provides a metabolic activity metric for aerobic and anaerobic fermentations:

$$\text{CER} = \frac{F_i(CO_2) - F_o(CO_2)}{\overline{V}} = \frac{\text{rate of moles of carbon dioxide produced}}{\text{working volume}} \tag{6.18}$$

where F_i (O_2 or CO_2) and F_o (O_2 or CO_2) are the flow rates of O_2 or CO_2, in and out of the fermenter, respectively.

The ability of the biocatalyst in terms of production rate can be estimated by the specific productivity metric, $\overline{q}_{\mathrm{p}}$:

$$\overline{q}_{\mathrm{p}} = \frac{P_{\mathrm{v}}}{\overline{X}} = \frac{\text{volumetric productivity}}{\text{average dry cell weight}} \tag{6.19}$$

$\overline{q}_{\mathrm{p}}$ can also be quantified on cumulative cell mass basis or instantaneous product formation rate basis by changing numerator and denominator of Equation 6.19.

Overall, a robust industrial process must include the characteristics listed below:

Essential

- High product yield
- Minimal oxygen demand; low installed capital cost
- Minimal product inhibition; high titer
- Low pH; low sterilization cost
- Able to utilize broad range of feedstock
- Generally regarded as safe (GRAS) host; low cell disposal cost
- Insensitive to scale-up; tolerant to shear, gradients of concentration, pH, temperature, DO in fermenter

Desirable

- High specific productivity; high substrate uptake rate
- Minimal nutrients media
- Minimal salts media
- High temperature
- No spore formation

After days or weeks, depending on the bacterium- or fungus-type production organism, when the product titer (concentration), yield, or rate, sometimes combined, are maximal, the fermenter is ready for harvest. Before harvest, in some case, stabilizing salts are added, or temperature and pH parameters are adjusted.

6.4 RECOVERY PROCESS

Recovery or downstream processing refers to the isolation and purification of fermentation products. The central themes in downstream processing are separation techniques. The purpose is to get the product out of the fermentation broth into a form that can be used in the target application. As the name implies, downstream processing occurs following fermentation. The fermentation process is the center point, which uses cell factories to manufacture the target molecule. Recovery needs to generate the product by separating the target molecule from the waste. The extent to which processing needs to be done depends on the final bioproduct. Purity requirements for commercial bioproducts range from essentially as is in fermentation broth to highly purified forms, depending upon the application. The goal of downstream processing is to recover the product of interest at high yield and in a form that is safe, cost-efficient, stable, and suitable for the target application. Commodity bioproducts usually are recovered at much lower cost and lower purity compared to pharmaceutical bioproducts.

A wide variety of products can be produced by fermentation. In some cases the microorganism itself is the product, for example, in the manufacturing of active dry yeast (ADY). Well-known pharmaceutical fermentation products are insulin and penicillin G. Fermentation processes are also used to produce various commodity bioproducts like organic and amino acids, polysaccharides, lipids, chemical compounds like isoprene (Whited et al., 2010), 1,3-propanediol (Nakamura and Whited, 2003), RNA, DNA, enzymes, and other proteins. The large variety of commodity bioproducts produced by fermentation requires an equally large variety of different methods to separate and purify them. Compared to fermentation processes, where usually one unit, the fermenter vessel, is used, several different steps and unit operations are necessary to achieve purification and formulation of bioproducts.

To emphasize the large variety of downstream processes and the importance of integrated process development between fermentation, recovery, and formulation, four bioprocesses are briefly discussed below.

6.4.1 Active Dry Yeast

An example of a commodity bioproduct everybody has probably encountered is ADY. In this case, the cells are the product. The downstream processing for ADY can consist of several washing steps and concentration by centrifugation. The washing steps remove the fermentation media from the yeast cells. The yeast cream obtained by centrifugation is further concentrated using a filter-press to produce a yeast cake. Extrusion and drying of the yeast cake yields the final ADY product.

All the processing steps that are done after the yeast cream is generated serve only one purpose: water removal. The water removal steps are necessary to stabilize the active yeast cells. Additional processing steps are costly, but the yeast cream has a short shelf life compared to the dried product.

One important factor when drying yeast cells is intracellular trehalose. The intracellular trehalose content is considered a critical determinant of stress tolerance in yeast (Nishida et al., 2004). Trehalose is a nonreducing α-linked disaccharide commonly found in any hydrobiotic organisms. When phospholipid membranes are dried, the temperature at which the gel to liquid crystal phase transition occurs increases (Crowe et al., 1984). It has been shown that trehalose interacts with model membranes during drying and lowers the dry membrane phase transition temperature (Crowe et al., 1986).

To facilitate the drying of yeast cells, the trehalose has to be accumulated in the yeast cells already during the fermentation. Therefore, integration of the fermentation

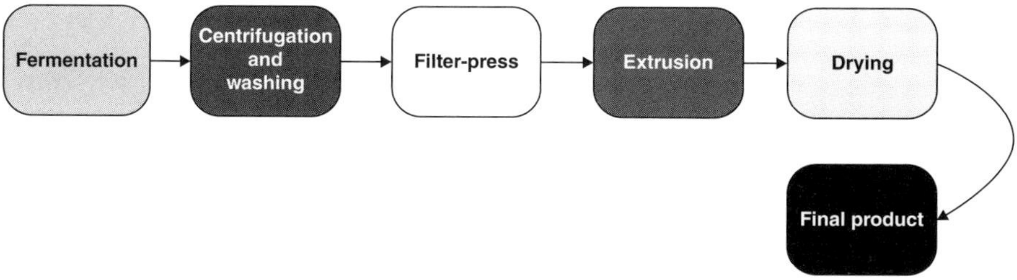

FIGURE 6.3 Simplified recovery schematic for an ADY product.

process with the downstream processing is critical in the manufacturing process for ADY (Figure 6.3).

6.4.2 Unclarified Enzyme Product

Another simple example is an unclarified enzyme product. In this case, the cells in the fermenter produce a target molecule, for example, a cellulase enzyme mixture. The enzyme mixture is the product, but the cells remain in the final product. This approach is preferable when a product has very low purity requirements and a separation process is cost prohibitive. Cellulase enzyme mix is used to break down biomass into glucose which enables the use of cellulosic raw materials as fermentation feedstocks (Himmel et al., 1999). This application requires a very inexpensive enzyme mix at low purity. Cellulases are also used in textile applications, where they are required in a purified form.

The challenge with preparing an unclarified broth product is to stabilize the fermentation broth containing cells. One of the best cellulase producers is the filamentous fungus *T. reesei* (Durand et al., 1988; Foreman et al., 2003). These cells need to be inactivated at the end of fermentation since they are not removed from the product. For *T. reesei*, this is often done through a combination of pH adjustment and addition of chemicals. For other production organisms, mechanical cell disruption or lysis is very commonly used. The integrated bioprocess considerations are that the selected method to inactivate the cells needs to be suitable for the specific organism, compatible with the target molecule and cost-efficient.

Cell disruption methods are often categorized as mechanical and nonmechanical. Common mechanical disruption methods are homogenization and wet grinding. A homogenizer pumps the feed slurry with a positive displacement pump through a restricted nozzle or orifice. Cell disruption occurs through the shear forces and simultaneous decompression. Alternatively, bead mill consists of a grinding chamber filled with grinding media, such as glass beads. The beads are moved by a shaft with multiple disks. The disruption results through a combination of high shear, cavitation, grinding, and by direct collision with the beads (Goldberg, 1997). Common nonmechanical methods for cell disruption are physical and chemical or enzymatic lysis. Physical cell lysis can be achieved by freezing or osmotic shock. Chemical

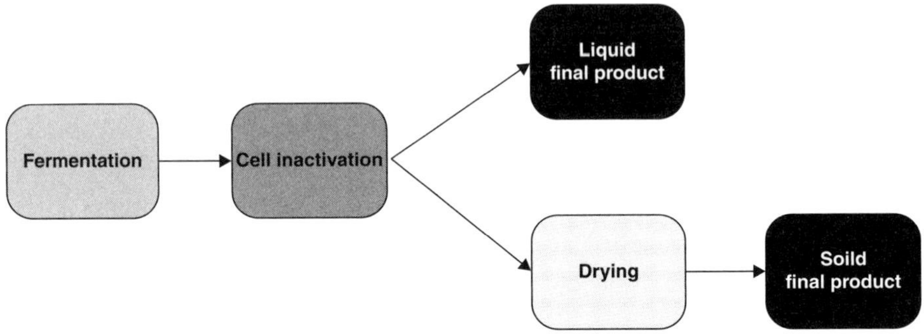

FIGURE 6.4 Simplified recovery process schematic for an unclarified enzyme product.

or enzymatic lysis can be achieved with detergents, antibiotics, or enzymes like lysozyme.

The inactivated fermentation broth is stabilized further either by drying or through the addition of chemicals. The key stability aspects are the enzymatic activity and microbial stability. Antimicrobials and polyols are commonly added to the inactivated fermentation broth for stabilization, but the choice of chemicals is very restricted by the target application.

Overall, as in the previous example, the entire process and the final product application need to be considered in order to develop the most competitive downstream processing (Figure 6.4).

6.4.3 Clarified Enzyme Product

Enzymes can be expressed as either intracellular or extracellular products. Isolation of intracellular molecules often involves the separation of complex biological mixtures. On the other hand, extracellular molecules are generally released into the medium containing fewer other components.

One of the major steps in bioprocess development is the removal of cells. Prior to the cell removal, additives are often used to ensure stability of the molecule during processing. The range of additives is as wide as the range of products. Proteins are often kept stable during processing simply by adjusting pH and keeping process temperatures low. Additives are often avoided, or used only in small quantities. In the first processing steps, quantities treated are large and therefore additives can quickly become cost prohibitive. The second reason for avoiding additives is that large parts of the initial process streams go into the waste system. If the added chemical is not compatible with regular waste water treatment permits, expensive waste treatment steps need to be included in the process. In the final product formulation, a wider range of additives can be used than during processing since the quantities that require treatment are usually smaller and waste disposal is normally not an issue.

Depending on the cell separation equipment used, the cell removal can be facilitated by inducing self-flocculation of cells (Stratford, 1989), or by adding flocculation polymers to the fermentation broth. Flocculants are added to create larger particles

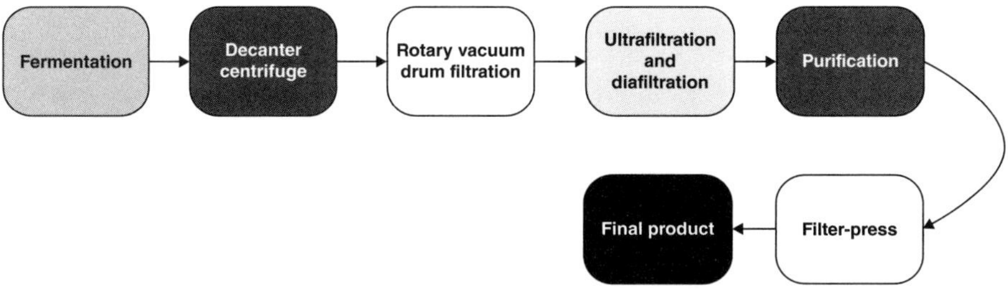

FIGURE 6.5 Simplified recovery schematic for a clarified enzyme recovery process.

since individual cells are too small to be removed by the cell removal equipment. Water can also be added to the fermentation broth which can be beneficial for several reasons: it enhances the flocculation since it lowers the ionic strength of the liquid; it increases the process yield since the water entrapped in the cell cake or sludge contains a lower product concentration due to the dilution; and it lowers the density of the liquid medium. In addition to the particle size and viscosity, the density difference between the cells and the liquid medium is one of the governing factors for the sedimentation rate as described by Stoke's law.

Cell removal by rotary vacuum drum filtration (RVDF), centrifugation, or filter-press often requires flocculation to obtain good separation. On the other hand, cell removal by microfiltration does not require flocculation. Centrifugation often does not remove all the cells and has to be combined with a polish filtration step to achieve complete cell removal. A process that is used to recover an extracellular enzyme for a liquid product can have the following steps: the fermentation broth is treated with additives to ensure the stability of the molecule during processing; the broth is then diluted and flocculated; the flocculated fermentation broth is processed using a decanter centrifuge process, followed by RVDF; the clarified liquid is concentrated using ultrafiltration; to remove small impurities, the ultrafiltration concentrate is diafiltered, the diafiltered concentrate is treated with activated carbon and other additives to further remove impurities; the carbon is removed by filtration and finally, formulation chemicals are added to stabilize the product. In this process, the formulation and purification are combined. The need for the purification steps is determined and developed by evaluating the final product application. Steps are added to the downstream process to remove by-products which interfere with the final application (Figure 6.5).

6.4.4 Bioisoprene™

Isoprene is a commodity chemical used in the production of synthetic rubber, adhesives, and specialty elastomers. Manufacturing of isoprene is currently based on petrochemical sources, but significant progress has been made to develop an alternative fermentation-based isoprene source (Whited et al., 2010).

Fermentation-based isoprene production uses a continuous gas-phase product recovery. This is possible due to the high vapor pressure of isoprene (400 mm Hg

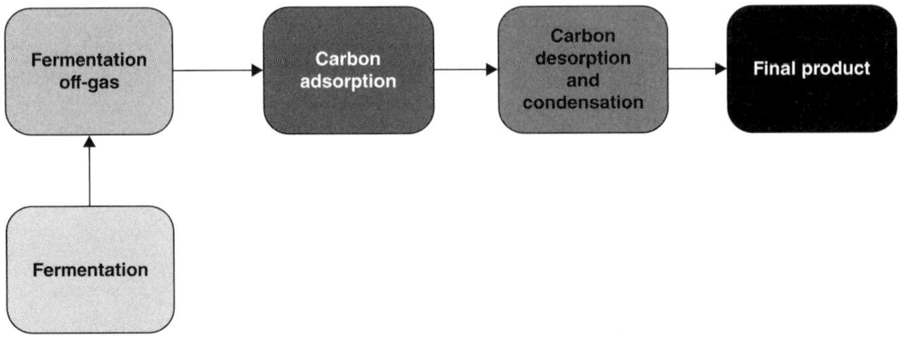

FIGURE 6.6 Simplified schematic of bioisoprene recovery process.

at 20°C; in vapor form at 37°C). The properties of isoprene are both beneficial and challenging for the product recovery. The high vapor pressure allows continuous and easy removal of the product from the fermentation medium through the off-gas, but the molecule is also extremely flammable and requires special safety considerations to ensure proper handling and containment.

Isoprene can be recovered from the off-gas by various means. The method depends on the scale, the percentage of isoprene in the off-gas, and the level of impurities. On laboratory scale, adsorption to activated carbon can be used conveniently to capture isoprene from the off-gas. The desorption of isoprene from activated carbon can be achieved using steam or nitrogen followed by condensation to obtain concentrated liquid isoprene. The purity of the concentrated isoprene liquid from this simple process is usually over 99.5%.

The gas-phase recovery is important for the overall process economics since it allows the use of inexpensive, crude feed stocks in fermentation. The use of very crude feedstocks for aqueous products often requires additional purification steps which can negate any economic advantage gained from their use (Figure 6.6).

As can be seen from the four different examples, the choice of purification steps and procedures depends on the product location and properties and ultimately very much on the target application. Considerations are if the molecule is intracellular, cell associated, or extracellular and if it is soluble in the aqueous fermentation liquid or not. Other considerations are the fermentation medium itself and the fermentation organism. Large-scale industrial fermentation media often relies on complex, crude feedstocks as nutrient sources, as using defined nutrient sources is usually cost prohibitive. One of the major challenges is that complex fermentation media can lead to more variability in the fermentation output. To consistently deliver a robust product, downstream processing and formulation need to be able to normalize the upstream variability.

Another important consideration is the waste stream generated during the downstream processing. Waste costs can significantly contribute to the overall product cost. Due to this, various material and water recycle steps are usually utilized in a manufacturing plant. Waste treatment considerations also heavily restrict the choice of process additives.

6.5 FORMULATION PROCESS

After downstream processing, liquid forms of product, especially enzymes, may not satisfy product quality requirements of biorefinery application or long-term storage. The purpose of formulation is to develop commercially viable products that meet safety, stability, and performance requirements. For instance, robust stable formulations are required in animal feed applications to protect enzymes such as phytase from elevated temperature and moisture during in-feed pelleting processes.

Formulations enhance protein stability by counteracting the primary forces of deactivation, that is, protein denaturation, catalytic site deactivation, and proteolysis (Becker et al., 1997). Enzyme denaturation occurs by physical unfolding of a protein's tertiary structure under thermal and or chemical stress. Once a protein begins to unfold, it becomes dramatically more vulnerable to deactivation and proteolysis. To minimize unfolding, the formulator can alter the protein's (micro)environment so as to induce a compact protein structure; this is done most effectively by "preferential exclusion" of water from the protein surface by adding water-associating compounds such as sugars, polyhydric alcohols, and lyotropic salts (DePaz et al., 2000). Catalytic site inactivation can be best prevented by introducing sufficient levels of cofactors, by addition of reversible inhibitors, and by exclusion of oxidizing or reactive species from the formulation.

Besides enzymatic stability, a formulation should meet several key secondary requirements including preservation against microbial contamination, prevention of physical precipitation or haze formation, minimizing potentially sensitizing dust or aerosol formation, and optimization of esthetic criteria such as color and odor. Many of these problems are best addressed by integrated product development strategy, for example, focusing as far "upstream" as strain development, choice of raw materials in the fermentation or recovery processes. Downstream operations such as diafiltration, adsorption, chromatography, crystallization, and extraction can be used to remove impurities responsible for color, odor, haze, and precipitation (Becker, 1995).

The risk of physical precipitation can be minimized by formulating near the isoelectric point of enzyme with hydrophilic solvents such as glycerol or propylene glycol. Solvating salts can be added at moderate levels to avoid either salting-out or "reverse salting-in." A combination of microfiltration, acidification, and minimization of free (unbound) water can be used to prevent microbial contamination; biocides can be effective, but the range of acceptable chemicals for controlling or killing microbes is increasingly circumscribed by health and safety regulations.

It is important that each ingredient in enzyme formulation is compatible with enzyme and other inactive components of the formulation and yet is qualified for final product. Product safety is an important consideration as enzymes can cause immune responses in humans, leading to sensitization and allergic reactions. Inhalation is the principal route of sensitization. Therefore, a formulation is needed that prevents enzyme intrusion in lung tissues through nasal and mucal routes.

Although few publications on novel formulation technologies are available in the literature (Becker et al., 1997; Hasan et al., 2010), the intellectual property landscape is filled with composition of matter patents and patent applications

TABLE 6.3 Considerations in Selection of Enzyme Product Form

Protein physical and chemical properties:
Molecular weight Solubility Isoelectric point (pI) Protein inherent stability
Compatibility with:
Inactive ingredients Final product physical form
Method of addition to final product Final product stability

(US7,300,779; US7,157,416; US6,602,841; 2011/0300201; 2011/0065163; 2010/0124586). Enzyme formulations are marketed in solid, paste, slurry, or liquid form. Some considerations in selection of enzyme product form are listed in Table 6.3. Common product forms of enzymes are discussed below.

6.5.1 Solid Forms

Granulation is a generic term for particle size enlargement through agglomeration or microencapsulation. Multiparticulates are often the preferred form of solid enzyme products for various industrial applications. In general terms, such enzyme formulations are referred to as granule or granulate. Large volumes of enzyme granules may be fabricated in the form of microparticles, microspheres, or microcapsules, utilizing a variety of microencapsulation technologies such as spray drying, freeze drying, spray congealing, high shear granulation, granulation–extrusion–spheronization, fluid-bed granulation, fluid-bed spray coating, and tableting. In practice, a combination of these processes may be used to manufacture enzyme granules with desired specifications. For instance, in production of enzyme granules through drying processes, the liquid enzyme concentrate is mixed with stabilizing agents and or diluents and then converted to a dry form through spray drying or freeze drying.

Spray drying has been utilized for decades and is still one of the common production methods for enzyme products. Other modes of drying such as freeze drying are typically costly and are not economically justified for the production of large volume, low cost enzyme product.

Enzyme particle characteristics such as size, shape, porosity, and morphology can be altered by adapting robust formulation and manufacturing processes to satisfy performance (aqueous solubility, immediate release, modified release, bioavailability, etc) and stability (chemical, biochemical, thermal) requirements. Granules may be engineered to have a matrix type or core and shell, the so-called "onion-type," morphology (Figures 6.7a–6.7c). Enzymes can be encapsulated in a uniform matrix core without a coating (Figure 6.7a) or within a core coated with a protective layer (Figure 6.7b). Alternatively, they can be embedded within a layer over inactive cores and overcoated with multiple stabilizer and protective layers (Figure 6.7c). Industrial enzyme granules in the size range of 425–850 μm (40–20 US standard mesh) are

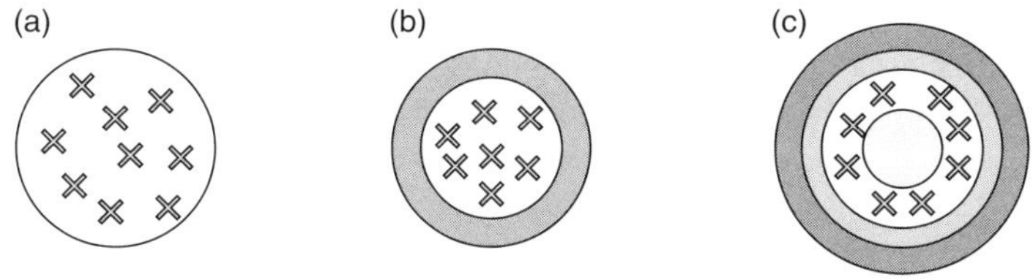

FIGURE 6.7 Enzyme granules with matrix (a) and core/shell (b and c) morphologies.

preferred for both utility and safety reasons. The various manufacturing processes for the production of enzyme granules are discussed below.

6.5.1.1 Spray Drying Spray drying provides a fast and cost-effective method for encapsulating enzymes and for maintaining enzyme activity over a product's shelf life. A simplified box diagram of spray drying process is shown in Figure 6.8. A liquid enzyme concentrate is mixed with stabilizing agents and or diluents so that the resulting product has the desired activity for targeted application as well as good stability during storage and application process. The stabilizing agents may be selected from a wide range of excipients such as salts, sugars, sugar alcohols, natural and synthetic polymers. After the ingredients are mixed, the resulting solution or suspension is atomized into small droplets through a stationary nozzle or a rotary nozzle with pressurized air. The atomized liquid is introduced at the top of the spray tower and is exposed to hot air at temperatures as high as 200°C. Due to the evaporative cooling, however, the exhaust air temperatures can be as low as 100°C. The forming enzyme powder may experience even lower temperatures. The resulting powders are homogeneous and puffy but tend to have small particle sizes which could generate dust during collection and transport. To control the dust, the spray dried powders can be blended with low levels of oil to effectively reduce any materials that may become airborne. Uncoated spray dried granules, however, could lead to allergic reactions in manufacturing plant personnel. As such, formulation technologies used to produce modern dry products have evolved far beyond simple spray- or freeze drying in order to meet people's workplace safety requirements.

The spray dried powder can also be incorporated into other product forms. These can be as simple as agglomerated or more sophisticated structured particles. The powder can be agglomerated to form larger particles and to reduce dust. This is done by filling the powder into mixing equipment such as a low or high shear granulator and adding a binder solution to the powder while mixing. This process results in binding the fine particles together and forming larger particles. The agglomeration process can

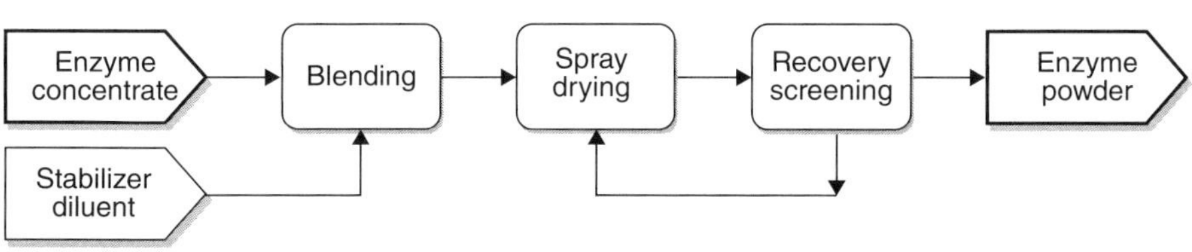

FIGURE 6.8 Manufacture of enzyme granules (microparticles) by spray drying process.

also be carried out in a fluid-bed granulator. In this process, the powder is suspended in an air stream and then an atomized spray of binder is applied through a nozzle. As the particles pass through the atomized liquid, they become wet, agglomerate, and then dry in agglomerated form. This process is carried out multiple times thus resulting in larger and larger particles. Both of these methods are somewhat uncontrolled, and the resulting product has a broad particle size distribution.

In specialized equipment, such as multistage driers, the spray drying and agglomeration processes can be carried out simultaneously. In this process, the enzyme concentrate and any added stabilizers are atomized though a nozzle similar to a standard spray drier. However, any fine powders are evacuated from the drying chamber and subsequently reintroduced into the drying chamber near the point of liquid atomization. This results in the dry particle rewetting and agglomerating with other nearby particles. By controlling the airflows in the system, it is possible to target a narrow size distribution of the final agglomerated particles.

The spray dried powders can also be incorporated into more structured particles such as extrudates or high shear granules. These particles can offer other advantages including improved stability and particle integrity. They also have the possibility of including other functional ingredients and customized enzyme release mechanisms.

6.5.1.2 High Shear Granulation

In the high shear process, dry ingredients including the enzyme powder are added to the product bowl of granulation equipment. While being mixed with a plough shear mixer and a high-speed chopper, a binder solution is added to the powder. The shearing action generates small particles which can then be dried in a tray dryer or a fluidized-bed dryer. A simplified box diagram of granulation process is shown in Figure 6.9.

Various binders and stabilizers can be used in high shear granulation to produce non-friable particles with different morphologies, which protect enzymes during storage but allow for their ready release in solution during application.

6.5.1.3 Granulation–Extrusion–Spheronization

In the granulation–extrusion–spheronization process, enzyme concentrate or spray dried enzyme

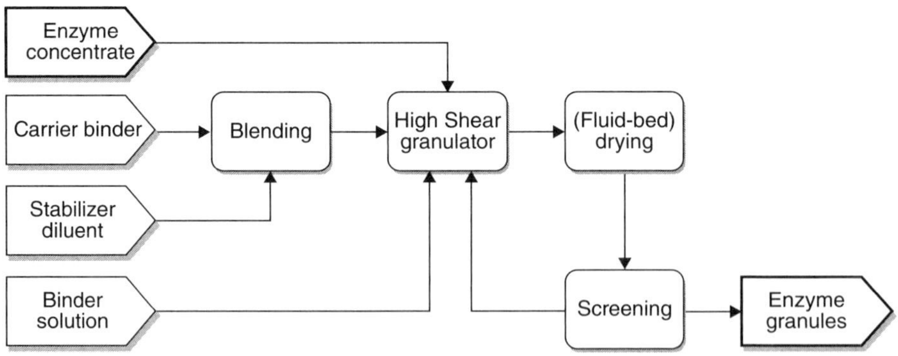

FIGURE 6.9 Manufacture of enzyme granules with high shear granulation-drying processes.

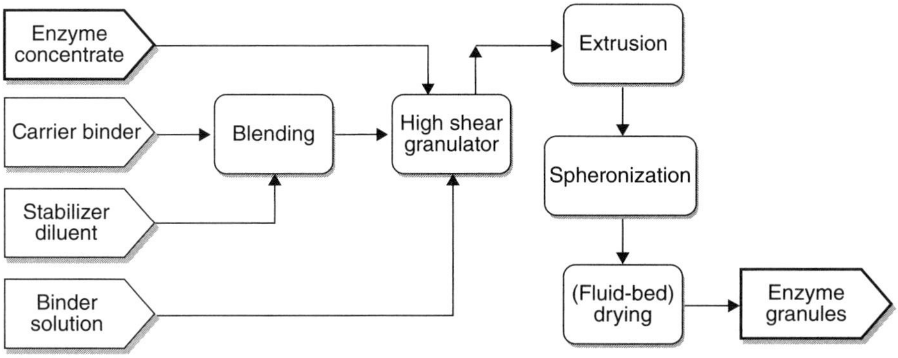

FIGURE 6.10 Manufacture of enzyme granules with high shear granulation–extrusion–spheronization and drying processes.

powder is mixed with inactive ingredients that function as diluent, stabilizer, binder, extrusion and spheronization aid, solubilizer, and other performance or stability enhancing agents. The materials are blended to create a dough-like mass. This mass is then pressed through a perforated plate or a mesh screen, and the resulting "noodles" are either cut off at regular intervals or allowed to fall off on their own. The wet noodles can be dried in a tray dryer or a fluidized-bed dryer. Alternatively, the noodles can be converted into spherical particles by using a device known as a spheronizer (or marumerizer) which consists of a spheronization plate with crosshatched surface. Wet noodles break into smaller pieces by spinning off the plate at high speeds. Finally, the particles are dried in a fluid-bed drier. A simplified box diagram is presented in Figure 6.10.

6.5.1.4 Spray Congealing Spray dried enzyme powder can be incorporated into a molten fat or wax. This mixture is atomized through a rotary nozzle into a cooled chamber essentially equivalent to a spray drier tower. The atomized droplets cool, solidify into particles, and harden as they fall, resulting in round particles. A simplified box diagram of spray congealing process is shown in Figure 6.11.

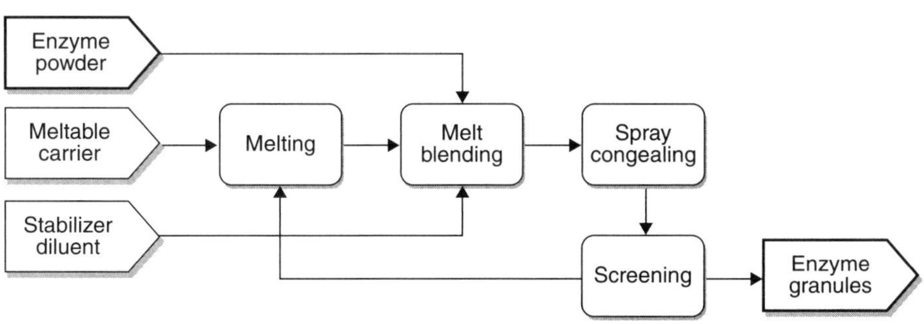

FIGURE 6.11 Manufacture of enzyme granules with spray congealing process.

Spray congealing is less commonly used due to its various practical limitations. This method is straightforward and inexpensive but has the drawback of being limited by melting properties of ingredients. In addition, high melt temperatures can lead to deactivation of enzyme. Likewise, moderate to high temperatures during the shipment or use of the product may cause agglomeration of granules.

6.5.1.5 *Fluidized-Bed Granulation/Coating* The advantage of the fluidized-bed method is that the entire granulation process can be carried out within a single, contained piece of equipment. The spray-coating process allows the sequential application of multiple layers of coating materials of different thickness and composition with almost infinite flexibility. A simplified box diagram is shown in Figure 6.12.

Enzyme release can be customized to meet final product application requirements. By creating core-shell particle morphology where the enzyme is included in the core or inner layers, a functionalized coating can be applied as outer layer. These coatings can not only further improve the stability of enzyme product but can also be triggered in response to environmental conditions to release the enzyme under desired conditions. These triggers may include pH, temperature, ion strength, water activity, or shear. The enzyme core units can be manufactured separately as agglomerates, extrudates, high shear granules, or spray chilled particles.

Additionally the enzyme core may be generated with fluidized-bed spray coating. In the fluid-bed process, an inert carrier (seed) is suspended in a hot air stream within a process chamber. While the particles are suspended, an air atomized enzyme liquid is introduced into the chamber. The atomized droplets are captured on the carrier and subsequently dried. The carrier/enzyme particles pass through the atomized liquid multiple times until the desired enzyme concentration is obtained. The fluid-bed process can also be utilized to apply the functional coating to any of the enzyme core units described simply by changing from the enzyme solution to a solution of the desired coating material.

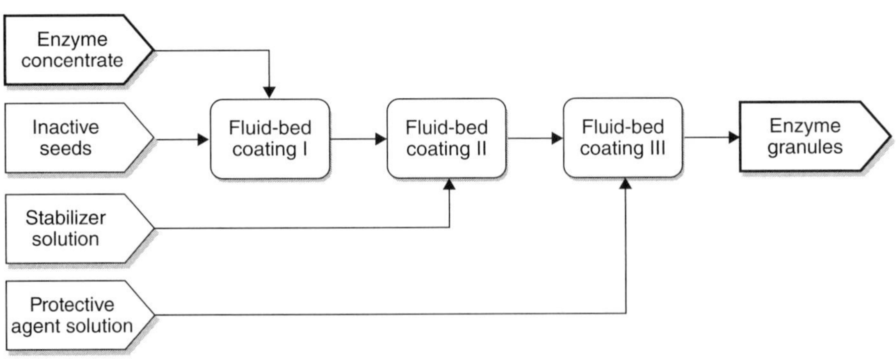

FIGURE 6.12 Manufacture of enzyme granules with fluid-bed granulation process.

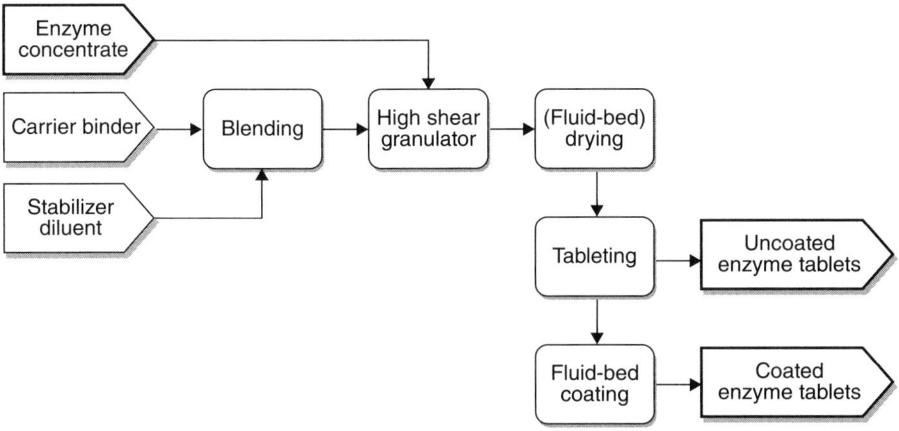

FIGURE 6.13 Manufacture of enzyme tablets with high shear granulation, drying, and tableting processes.

Various binders, stabilizers, and coatings can be used in a fluid-bed process to create different morphologies of non-friable particles with good storage stability and performance functionality.

6.5.1.6 Tableting Tablets (5–20 mm dia.) and mini-tablets (1–5 mm dia.) are another useful format for delivery of enzymes and can be advantageous in several ways. A simplified box diagram in Figure 6.13 shows the processes used in the manufacture of tablets. The enzyme can be incorporated into the tablet from any of the solid forms previously discussed including spray dried powders, granules, and coated particles. Similar to enzyme blends, tablets can incorporate multiple enzyme activities as well as other components that can benefit the particular application. However, tablets eliminate the possibility of segregation that blends can encounter. The tablets thus provide a consistent dose of enzymes and other components compared to blends. Finally handling and dosing of tablets into the application is far easier, less dusty, and cleaner compared to a powder blend.

6.5.2 Slurry or Paste Forms

Enzymes produced by fermentation processes are recovered and refined through downstream processing into the slurry or paste form which can directly be used for industrial applications. For example, the concentrate from a polyethylene glycol (PEG) extraction step may directly be used if the enzyme is sufficiently pure and free of microbial contamination. In this case, the formulator must work closely with the recovery engineer to produce a clean and sterile formulation. The addition of a sterile filtration step prior to a concentration step, in a clean, sterile tank, might be all that is necessary to obtain the final product. Another option is to force the precipitation or crystallization of the enzyme and then disperse it into a formulation that is compatible with targeted industrial products. The final step is a preservative addition to the slurry or paste to ensure good shelf stability.

6.5.3 Liquid Forms

Liquid formulations of enzymes are prepared in a variety of ways including aqueous high or low solids systems, dispersions, and emulsions. Many enzymes exhibit a solubility limit in aqueous media. Therefore, producing a liquid product with a high active payload may be challenging.

Liquid enzyme products have various applications such as additives in dishwasher detergents and animal feed (PPLA, post-pelleting liquid application). The formulation of enzymes in liquid form can be challenging as the liquid forms are more dynamic systems compared to solids because of the presence of water, and for this reason additional stabilizing agents are required in liquid formulations.

Stabilizers function by maintaining the protein structure, preventing aggregation, and therefore avoiding denaturation. Stabilizers that are used in liquids include common carbohydrates such as sucrose and dextrose, and sugar alcohols such as mannitol, sorbitol, and xylitol. These chemicals help reduce the water activity which plays an important role in controlling microbial growth.

The most common preservatives to control microbial contamination are sodium benzoate and potassium sorbate. They are both effective at low pH and can control a broad spectrum of microorganisms. However, one of the major food trends today is consumer's desire to reduce or remove the level of preservatives in the products. Subsequently, all the materials that go into the production of food products, such as enzymes, also need to be preservative-free, and this can be a challenge for enzyme products, especially in liquid form. As long as the solid products are protected from moisture, the likelihood of any microbial or fungal growth during storage is extremely low. In the case of liquids, the situation is more challenging. To control microbial growth during storage, the environment of the solution has to be such that microbial growth is not favored. This can be done by controlling the water activity, pH, or osmolality. Also, more and more natural preservatives are being developed that are effective at controlling microbial growth. These natural preservatives include certain plant extracts or peptides that inhibit microbial growth.

The liquid product formulation process is usually integrated with the downstream production process. After the enzyme has been concentrated, the enzyme stabilizers are added followed by a filtration process. The filtration is done to remove any undissolved solids and/or to reduce the level of microbial burden in the formulated product. The filtration process can be done using standard filtration equipment such as an RVDF, plate and frame, or depth filtration. The resulting products are transparent and commonly range from clear to amber in color.

6.5.3.1 Aqueous Systems

The formulation of an aqueous system begins with characterizing the solubility of enzyme. Determination of the maximum solubility concentration will be helpful in establishing the upper concentration limit for the formulated product. A solubility profile in a variety of carriers such as water, glycerol, polyethylene glycol (PEG), propylene glycol, or butylene glycol should be created for the evaluation of aqueous systems. Glycols have the advantage of both stabilizing the three-dimensional structure of the protein and reducing the need for preservation. Additionally, the attachment or association of PEG to the protein may improve its stability. It may be necessary to solubilize the enzyme in water first and then dilute with the solvent.

6.5.3.2 Dispersions Dispersions can be prepared from spray dried enzyme powder, or from an enzyme precipitate or crystal paste. The dried or partially dried powder is suspended in a suitable carrier such as glycols or glycol/water mixtures and preserved if needed. The dispersion may be visually clear or cloudy in appearance. As salts are commonly used to force the precipitation or crystallization of proteins, it may be difficult to remove or reduce the salt concentration prior to formulation.

6.5.3.3 Emulsions Enzymes may be incorporated into emulsions. Proteins are typically hydrophilic in nature and would most likely reside in the aqueous phase of the emulsion. Proteins may be covalently attached to fatty acids such as palmitic acid, that is, palmitoylated, which may partition them into the oil phase.

6.6 FINAL PRODUCT BLENDS

No matter what delivery form is chosen, the components of the carrier should be evaluated for their compatibility with the final product formulation in which they will be incorporated. Likewise is true for their safety profiles. Moreover, components that are already approved globally should be considered, especially if the product is to be used worldwide. It is possible that certain components may be more suitable for particular applications, but higher concentrations of certain excipients may not be desirable. Having chosen a desired delivery system, all enzyme formulations should be evaluated for their shelf stability as stand-alone products. Stability evaluations should encompass activity and phase stability after several freeze/thaw cycles, at elevated temperatures and in humid environments, and long-term stability at ambient and cold temperatures.

Many of the products that are used in feed and other industrial applications contain more than one enzyme activity. Enzyme manufacturers can produce these blends through combining already formulated intermediates, such as spray dried or microencapsulated solids, or liquids in defined ratios. For the solid product blends, it is important that the intermediate products are similar in size and density. This allows for good mixing and content uniformity while also reducing the likelihood of segregation during shipping and handling. The blending operation can also be utilized to reduce the activity of a product by blending with a nonactive diluent. Liquid blends are also possible, but the compatibility of the intermediate products is an important consideration so that unstable products are not created.

The blending process can be carried out in a variety of equipment in batch or continuous mode. For liquid blends a filtration step can be done post-mixing to remove any residual particulates if necessary. Solid product blends can be made in a variety of blending equipment, such as ribbon blenders, plough shear mixers, Nauta mixers, or fluid-bed processors.

6.7 ECONOMIC OUTLOOK AND FUTURE PROSPECTS

Positive economical and environmental impacts of using alternative sustainable feedstocks will ultimately favor biorefineries. To meet this challenge, development of efficient and robust enzymes to process a large variety of feedstocks will remain on the critical path. Likewise, advances in the use of synthetic biology and metabolic pathway engineering will expand the development of cell factories leading to growth in the development of innovative products. A thorough knowledge of systems biology and integrated process engineering will expand the use of biosynthesis, by both enzymes and whole cells. Hybrid processes through combination of biochemical and chemical processes will enhance competitiveness of bio-based products. A bio-based industry would build and progress through development of bioprocesses for unmet needs, rather than by competing with existing chemical processes. It may be difficult for a bio-based process to displace highly efficient, environmentally favorable and sustainable chemical processes used widely for the production of certain established products. For example, bio-based polymers and plastics industries will grow their market share by synergizing and collaborating with the chemical process industry that generates these products.

A 2009 report on bioeconomy published by the Organization for Economic Cooperation and Development (OECD) predicted that the value of biochemicals other than active pharmaceutical ingredients could reach between 12% and 20% of all chemical production by 2015, a tenfold increase since 2005. The report also forecasted that biofuel production could partly shift from starch-based bioethanol to higher energy density fuels manufactured from sugar cane or to cellulosic ethanol from lignocellulosic feedstock (OECD, 2009). Sales of products made by biotechnological processes in 2007 were reported to be around 48 billion Euros, matching 3.5% of total chemical sales in that year, and following a similar trend as chemical sales in regional markets, that is, North America (NAFTA), Europe (EU-27), Asia (including China and Japan), and rest of the world (OECD, 2010a). The report projected that the total sales of bio-based products could reach 340 billion Euros in 2017, leveling at 15.4% of total chemical sales in four major segments of base chemicals (33.3%), consumer chemicals (24.8%), specialty chemicals (21.4%), and active pharmaceutical ingredients (20.5%) (OECD, 2010b). A recent report by the World Economic Forum (WEF) concluded that converting biomass into fuels, energy, and chemicals has the potential to generate upward of 230 billion USD to the global economy by 2020 (World Economic Forum, 2010).

A bio-based industry of critical size will cope with a cycle time of 6–8 years product and process development, with a corresponding high cost of new investment for pre-commercial demonstration, first and *N*th production plant. Economy of scale of development programs and of operation will establish production platforms for chemical intermediates from renewable carbon sources. The growth of a bio-based industry will continue to rely on streamlined government approvals of genetically modified feedstocks, cell factories, products, and their impact on environment (Sherpa Group, 2011). The development of bioprocesses based on the use of alternative feedstock will ultimately mitigate the debate between food and fuel.

ACKNOWLEDGMENT

We thank Dr Roopa Ghirnikar and Dr Anthony Calabria, DuPont Industrial Biosciences, for helpful suggestions in preparation of the manuscript.

NOMENCLATURE

ADY	Active dry yeast
CBP	Consolidated bioprocessing
DMAPP	Dimethylallylpyrophosphate
DXP	1-Deoxy-D-xylulose 5-phosphate
GRAS	Generally regarded as safe
HFCS	High fructose corn syrups
MEP	Methylerythritol
MVA	Mevalonate
OECD	Organization for Economic Cooperation and Development
OTR	Oxygen transfer rate
OUR	Oxygen uptake rate
PEG	Polyethylene glycol
PPLA	Post-pelleting liquid application
RVDF	Rotary vacuum drum filtration

REFERENCES

Becker NT. (1995) Separation and purification processes for recovery of industrial enzymes. In: Singh RK, Rizvi SSH, editors. *Bioseparation Processes in Foods*. New York: Marcel Dekker, pp. 427–445.

Becker T, Park G, Gaertner AL. (1997) Formulation of detergent enzymes. In: Misset EJH, Baas EJ, editors. *Enzymes in Detergency*. New York: Marcel Dekker, pp. 299–325.

Bott R, Cervin M, Kellis J, McAuliffe J, Miasnikov A, Peres C, Rife C, Wells D, Weyler W, Whited G. (2012) Isoprene synthase variants for improved microbial production of isoprene. US Patent 8,173,410.

Brevnova E, McBride JE, Wiswall E, et al. (2011) Yeast expressing saccharolytic enzymes for consolidated bioprocessing using starch and cellulose. WIPO Patent Application WO/2011/153516.

Bruschi M, Boyes SJ, Sugiarto H, Nielsen LK Vickers C. (2012) A transferable sucrose utilization approach for non-sucrose-utilizing *Escherichia coli* strains. *Biotechnol Adv*, 30, 1001–1010.

Bullis K. (2012) Cellulosic ethanol gets a $100 million boost. *MIT Technology Review*. Available at: http://www.technologyreview.com/news/427145/cellulosic-ethanol-gets-a-100-million-boost/ (last accessed 1/15/2014)

Burgard A, Osterhout R, Sun J, Pharkya P. (2012) Methods for increasing product yields. WIPO Patent Application WO/2012/099621.

Cervin M. (2011) Cellulase-free enzyme compositions and host cells for producing the same. US Patent Application Publication 2011/0098207.

Chotani G, Dodge T, Hsu A, Kumar M, LaDuca R, Trimbur D, Weyler W, Sanford K. (2000) Commercial production of chemicals using pathway engineering. *Biochem Biophys Acta*, 1543, 434–455.

Chotani GK, Dodge TC, Gaertner AL, Arbige MV. (2007) Industrial biotechnology: discovery to delivery. In: Kent JA, editor. *Kent and Riegel's Handbook of Industrial Chemistry and Biotechnology*. New York: Springer Science & Business Media, LLC, pp. 1311–1374.

Collier K, Vogtentanz G, Amin A, Estabrook M, Estell D, Fox B, Power S, Rao R, Schmidt B. (2009) Generation and identification of variants with improved purification yield of Bowman-Birk protease inhibitors carrying protein binding loops. *Protein Expr Purif*, 68, 146–160.

Cooney CL, Wang DIC, Mateles RI. (1968) Measurement of heat evolution and correlation with oxygen consumption during microbial growth. *Biotechnol Bioeng*, 11, 269–281.

Crowe JH, Crowe LM, Chapman D. (1984) Preservation of membranes in anhydrobiotic organisms: the role of trehalose. *Science*, 223, 701–703.

Crowe LM, Womersley C, Crowe JH, Reid D, Appel L, Rudolph A. (1986) Prevention of fusion and leakage in freeze-dried liposomes by carbohydrates. *BBA Biomembranes*, 861, 131–140.

Dean D, Dodge T, Valle F, Chotani G. (2006) Development of biorefineries—technical and economic considerations. In: Kamm B. et al., editor. *Biorefineries—Industrial Processes and Products. Status Quo and Future Directions*. Weinheim, Germany: Wiley-VCH, pp. 67–83.

DePaz R, Barnett CC, Dale DA, Carpenter JF, Gaertner AL, Randolph TW. (2000) The excluding effects of sucrose on a protein chemical degradation pathway: methionine oxidation in subtilisin. *Arch Biochem Biophys*, 384, 123–132.

Doran PM. (1995) *Bioprocess Engineering Principles*. San Diego: Academic Press Limited.

Duan H, Schuler M. (2006) Heterologous expression and strategies for encapsulation of membrane-localized plant P450s. *Phytochem Rev*, 5, 507–523.

Durand H, Clanet M, Tiraby G. (1988) Genetic improvement of *Trichoderma reesei* for large scale cellulase production. *Enzyme Microb Technol*, 10, 341–346.

Engler C, Kandzia R, Marillonnet S. (2008) A one pot, one step, precision cloning method with high throughput capability. *PLOS One*, 3, e3647.

Erdei B, Barta Z, Sipos B, Reczey K, Galbe M, Zacchi G. (2010) Ethanol production from mixtures of wheat straw and wheat meal. *Biotechnol Biofuels*, 3, 16.

Estell D. (1993) Engineering enzymes for improved performance in industrial applications. *J Biotechnol*, 28, 25–30.

Ferrari E, Peres C. (2009) Method for amplifying locus in bacterial cell. WIPO Patent Application WO/2009/120929.

Foreman PK, Brown D, Dankmeyer L, Dean R, Diener S, Dunn-Coleman NS, Goedegebuur F, Houfek TD, England GJ, Kelley AS, Meerman HJ, Mitchell T, Mitchinson C, Olivares HA, Teunissen PJ, Yao J, Ward M. (2003) Transcriptional regulation of biomass-degrading enzymes in the filamentous fungus *Trichoderma reesei*. *J Biol Chem*, 278, 31988–31997.

Fountain H. (2012) Corn Ethanol Makers Weigh Switch to Butanol. *New York Times*, October 23.

Gibson D, Young L, Chuang RY, Venter J, Hutchison C, Smith H. (2009) Enzymatic assembly of DNA molecules up to several hundred kilobases. *Nat Methods*, 6, 343–345.

Goldberg E. (1997) *Handbook of Downstream Processing*. London, UK: Blackie Academic & Professional.

Gosh A, Zhao H, Price N. (2011). Genome-scale consequences of cofactor balancing in engineered pentose utilization pathways in *Saccharomyces cerevisiae*. *PLOS One*, 6, e27316.

Hasan F, Shah AA, Javed S, Hameed A. (2010). Enzymes used in detergents: lipases. *Afr J Biotechnol*, 9, 4836–4844.

Himmel ME, Ruth MF, Wyman CE. (1999) Cellulase for commodity products from cellulosic biomass. *Curr Opin Biotechnol*, 10, 358–364.

Hodgman C, Jewett M. (2012) Cell-free synthetic biology: thinking outside the cell. *Metab Eng*, 14, 261–269.

Jiang H, Chen Y, Jiang P, Zhang C, Smith T, Murrell C, Xing XH. (2010) Methanotrophs: multifunctional bacteria with promising applications in environmental bioengineering. *Biochem Eng J*, 49, 277–288.

Köpke M, Held C, Hujer S, Liesegang H, Wiezer A, Wollherr A, Ehrenreich A, Liebel W, Gottschalk G, Durre P. (2010) *Clostridium ljungdahlii* represents a microbial production platform based on syngas. *Proc Natl Acad Sci U S A*, 107, 13087–13092.

Köpke M, Mihalcea C, Liew F, Tizard J, Ali M, Conolly J, Al-Sinawi B, Simpson S. (2011) 2, 3-Butanediol production by acetogenic bacteria, an alternative route to chemical synthesis, using industrial waste gas. *Appl Environ Microbiol*, 77, 5467.

Kuzuyama T, Seto H. (2012) Two distinct pathways for essential metabolic precursors for isoprenoid biosynthesis. *Proc Jpn Acad Ser B Phys Biol Sci*, 88, 41–52.

Leang C, Ueki T, Nevin K, Lovley D. (2013) A genetic system for *Clostridium ljungdahlii*: a chassis for autotrophic production of biocommodities and a model homoacetogen. *Appl Environ Microbiol*. 79, 1102–1109.

Miller E, Jarboe L, Yomano L, York S, Shanmugam K, Ingram L. (2009). Silencing of NADPH-dependent oxidoreductase genes (yqhD and dgkA) in furfural-resistant ethanologenic *Escherichia coli*. *Appl Environ Microbiol*, 75, 4315–4323.

Nakamura C, Whited G. (2003) Metabolic engineering for the microbial production of 1,3 propanediol. *Curr Opin Biotechnol*, 14, 454–459.

Nicolaou S, Gaida S, Papoutsakis E. (2010) A comparative view of metabolite and substrate stress and tolerance in microbial bioprocessing: from biofuels and chemicals, to biocatalysis and bioremediation. *Metab Eng*, 12, 307–331.

Nielsen J, Villadsen J, Liden G. (2002) *Bioreaction Engineering Principles*. New York: Kluwer Academics/Plenum Publishers.

Nishida O, Kuwazaki S, Suzuki C, Shima J. (2004) Superior molasses assimilation, stress tolerance, and trehalose accumulation of baker's yeast isolated from dried sweet potatoes (hosi-imo). *Biosci Biotechnol Biochem*, 68, 1442–1448.

OECD (2009) The Bioeconomy to 2030: Designing a Policy Agenda—Main Findings and Policy Conclusions. Available at: http://www.oecd.org/futures/long-termtechnological societalchallenges/42837897.pdf. (last accessed 1/21/2014)

OECD (2010a) Industry Structure and Business Models for Industrial Biotechnology Research Methodology and Results for Discussion. Available at: http://www.oecd.org/sti/biotech/44776744.pdf. (last accessed 1/21/2014)

OECD (2010b) The Bioeconomy in the OECD Countries and Beyond. Available at: http://www.oecd.org/sti/biotech/46394481.pdf. (last accessed 1/21/2014)

Roels JA. (1983) *Energetics and Kinetics in Biotechnology*. Amsterdam, The Netherlands: Elsevier Biomedical Press.

Ross D, Flynn J, Baron D, Gralnick J, Bond D. (2011) Towards electrosynthesis in Shewanella: energetics of reversing the Mtr pathway for reductive metabolism. *PLOS One*, 6, e16649.

Sherpa Group (2011) KET Industrial Biotechnology—Working Group Report. Available at: http://ec.europa.eu/enterprise/sectors/ict/files/kets/4_industrial_biotechnology-final_report_en.pdf. (last accessed 1/21/2014)

Stratford M. (1989) Yeast flocculation: calcium specificity. *Yeast*, 5, 487–496.

Whited GM, Feher FJ, Benko AD, Cervin MA, Chotani CK, McAuliffe JC, La Duca RJ, Ben-Shoshan EA, Sanford KJ. (2010) Development of a gas-phase bioprocess for isoprene-monomer production using metabolic pathway engineering. *Ind Biotechnol*, 6, 152–163.

World Economic Forum (2010) The Future of Industrial Biorefineries. Available at: http://www3.weforum.org/docs/WEF_FutureIndustrialBiorefineries_Report_2010.pdf. (last accessed 1/21/2014)

Xu Q, Singh A, Himmel ME. (2009) Perspectives and new directions for the production of bioethanol using consolidated bioprocessing of lignocellulose. *Curr Opin Biotechnol*, 20, 364–371.

Ye R, Kelly K. (2012) Construction of carotenoid biosynthetic pathways through chromosomal integration in methane-utilizing bacterium *Methylomonas* sp. strain 16a. *Methods Mol Biol*, 892, 185–195.

Zambare V, Bhalla A, Muthukumarappan K, Sani R, Christopher L. (2011) Bioprocessing of agricultural residues to ethanol utilizing a cellulolytic extremophile. *Extremophiles*, 15, 611–618.

Zawada J, Yin G, Steiner A, Yang J, Nresh A, Roy S, Gold D, Heinsohn H, Murray C. (2011) Microscale to manufacturing scale-up of cell-free cytokine production—a new approach for shortening protein production development timelines. *Biotech Bioeng*, 108, 1570–1578.

Zingaro KA, Papoutsakis E. (2013) GroESL overexpression imparts *Escherichia coli* tolerance to i-, n-, and 2-butanol, 1,2,4-butanetriol and ethanol with complex and unpredictable patterns. *Metab Eng*, 15, 196–205.

PART II

SPECIFIC COMMODITY BIOPRODUCTS

CHAPTER 7

Ethanol from Bacteria

HIDESHI YANASE

Department of Chemistry and Biotechnology, Graduate School of Engineering, Tottori University, Tottori, Japan

Bioprocessing of Renewable Resources to Commodity Bioproducts, First Edition.
Edited by Virendra S. Bisaria and Akihiko Kondo.

Abstract

For the commercial production of second-generation bioethanol, an efficient microbial platform to achieve high yield, productivity, and final titer of ethanol is needed. The powerful tools of metabolic engineering and pathway engineering have allowed the generation of engineered strains of *Zymomonas mobilis* capable of cofermenting major sugar components of lignocellulosic biomass, such as glucose, xylose, and arabinose, and ethanologenic *Escherichia coli* capable of producing almost the theoretical yield of ethanol. By using similar approaches, a number of ethanologenic bacteria with different abilities such as cofermentation of hexoses and pentoses and direct fermentation of cellulose and hemicellulose have also been genetically engineered. These studies have shown that in addition to *Zm. mobilis* and *E. coli*, *Zymobacter palmae*, *Klebsiella oxytoca*, *Erwinia chrysanthemi*, *Enterobacter asburiae*, and *Corynebacterium glutamicum* are potential biocatalysts for second-generation bioethanol processes such as separate hydrolysis and fermentation (SHF), liquefaction followed by simultaneous saccharification and cofermentation (L+SSCF), and simultaneous saccharification and cofermentation (SSCF). Recently, thermophilic bacteria have begun to attract attention as potential new biocatalysts for consolidated bioprocessing (CBP). In this chapter, the metabolic engineering of ethanologenic bacteria and their specific biocatalytic properties for second-generation bioethanol production are described.

7.1 INTRODUCTION

The need to develop alternative sources of liquid fuels has stimulated much interest in bioethanol production from renewable resources. Today's biofuel industry produces ethanol primarily from feedstocks such as cereals and sugar cane, using traditional brewing methods with yeast. However, industries producing these first-generation bioethanols, especially from corn starch, are in competition with the food and animal feed industries. In contrast, lignocellulosic biomass including crop waste, forestry residues, and municipal solid waste have great potential to be important sources of bioethanol. This is in large part because they are the most abundant and sustainable raw materials worldwide and occur as byproducts without competing uses. Consequently, many researchers and companies around the world have been developing industrial processes for producing lignocellulosic ethanol as a second-generation bioethanol. In that context, researchers have been searching for ethanol-producing microorganisms that are superior to the traditional yeast. In 1982, the research group of Rogers et al. (1997) demonstrated that the rate of ethanol productivity of *Zymomonas mobilis* is three to five times faster than that of brewing yeast and also reported that the ethanol productivity from glucose by using the cell-recycle method with *Zm. mobilis* reached a maximum productivity of 200 g/L·h (Lee et al., 1980). Because of this higher ethanol productivity of *Zymomonas*, bacterial ethanologenesis has attracted interest in the field of bioethanol production. Initially, *Zm. mobilis* was investigated for application in the production of first-generation bioethanol from cornstarch and sugar cane. However, *Zm. mobilis* could not replace traditional yeast as a biocatalyst for bioethanol production for several reasons. One problem is the salt sensitivity of *Zm. mobilis* (Sahm et al., 1991). *Zymomonas mobilis* is unable to grow on molasses from sugar cane, which is one of the least expensive materials available for bioethanol production, because it has a high concentration of potassium. Subsequently, ethanologenesis by bacteria underwent a renaissance due to the demand for second-generation bioethanol production from non-edible parts of crops, including corncobs, cornstalks, sugarcane bagasse, and wheat straw, because first-generation bioethanol production competes with global food production.

Cellulose and hemicellulose in lignocellulosic biomass are heavily coated with lignin. There are several steps in the complex process for efficient recovery of bioethanol from lignocellulosic biomass. It is necessary to remove the lignin coating to expose the cellulose and hemicellulose by using thermochemical or physical treatment before enzymatic saccharification and fermentation of lignocellulosic biomass (Esteghlalian et al., 1997; Taherzadeh and Karimi, 2007; Sun and Chen, 2008; Zhang and Cai, 2008; Chen et al., 2009; Kumar et al., 2009; Rabelo et al., 2009). The hard coating is loosened by the pretreatment, and then the lignocellulosic biomass is enzymatically saccharified to release glucose, xylose, arabinose, and mannose. Finally, all of the released sugars must be efficiently converted to ethanol. The hydrolysate of herbaceous biomass contains glucose, xylose, and arabinose as the major sugar components, and the hydrolysate of woody biomass contains glucose, mannose, and xylose. Bacteria must be improved to allow co-fermentation of xylose, arabinose, and mannose along with glucose. Brewing yeasts are unable to ferment the major sugar components from lignocellulosic biomass. Therefore, bacteria capable of utilizing a broad range of carbohydrate substrates are of interest for second-generation bioethanol production. As a result, metabolic engineering of *Escherichia coli*, for which there are highly developed strategies for genetic manipulation, was initiated. The research group of Ingram et al. (1987) has been actively breeding ethanologenic *E. coli* and closely related gram-negative bacteria. This strategy has been adopted to develop a non-ethanologenic bacterium with the ability to degrade and assimilate a broad range of substrate sugars. As an alternative strategy, the National Renewable Energy Laboratory (NREL) research group of Zhang broadened the range of carbohydrate substrates utilized by *Zm. mobilis* by conferring the ability to ferment xylose to this ethanologenic bacterium with a naturally limited range of utilizable carbohydrates (Zhang et al., 1995).

To accelerate commercial production of second-generation bioethanol, the cost of the cellulase cocktail utilized in the saccharification process has to be significantly reduced. One effective solution to reduce the amount of cellulase cocktails utilized is to breed an ethanologenic bacterium capable of fermenting cellulose directly by expressing cellulase genes and secreting the gene products. Therefore, strains belonging to the genera *Klebsiella* and *Erwinia* (Tolan and Finn, 1987; Ohta et al., 1991b), which have various inherent extracellulases, are of interest, and conferring the ability to produce ethanol on these bacteria was attempted. Moreover, a similar concept has been applied to *Zm. mobilis* and *Zymobacter palmae*. The ideal process for second-generation bioethanol production is assumed to be via consolidated bioprocessing (CBP) (Lynd et al., 2005); in which saccharification of holocellulose and co-fermentation of all of the released sugars are simultaneously occurring in one vessel by a single ethanologenic microorganism. To date, no information is available concerning bacteria capable of CBP. However, potential biocatalysts for CBP are thermophilic bacteria, including strains belonging to the genera *Thermoanaerobium*, *Thermoanaerobacter*, and *Geobacillus*, which are capable of cellulose saccharification and assimilation of a broad range of sugars at high temperature (Payton, 1984; Rainey et al., 1993).

In this chapter, the development of ethanologenic bacteria for the efficient production of second-generation ethanol is reviewed. Recent research on yeast is not included, as it is reviewed in the following chapter. Although the specifics of how each of these bacteria was engineered for ethanol fermentation vary, there are many similar features in the breeding strategies. The development of heteroethanologenic bacteria capable of producing the second-generation bioethanol is discussed first and then that of homoethanologenic bacteria is described.

7.2 HETEROETHANOLOGENIC BACTERIA

Almost all heteroethanologenic bacteria capable of assimilating various sugars from lignocellulosic biomass accumulate several metabolites from carbohydrates under anaerobic conditions. In these bacteria, the highly polymerized sugars such as cellulose and hemicellulose are degraded to release mono- and oligosaccharides, and the released sugars are transported into cells by active transport systems accompanied by phosphorylation or passive transport systems mediated by specific permeases. Generally, once taken up, sugars are metabolized to form pyruvate as the end product via the Embden–Meyerhof–Parnas (EMP) pathway (Figure 7.1). Then, under anaerobic conditions, the liberated pyruvate is converted to several end products, such as formate, lactate, acetate, butyrate, acetone, ethanol, isopropanol, and butanol, whose production is dependent on the oxidation–reduction conditions and the state of the cellular energy balance (Figure 7.2).

Unfortunately, these strains produce a small amount of ethanol relative to other metabolites under anaerobic conditions. Therefore, to improve ethanol productivity of the bacteria with a broad substrate range from polysaccharides to pentose monosaccharides, it is necessary to introduce genes encoding ethanologenesis from homoethanologenic bacteria. The research group of Ingram has been very successful in metabolically engineering *E. coli*, and these efforts have invaluable merit. However, the introduction of genes encoding ethanologenesis did not result in the production of the theoretical yield of ethanol in the genetically engineered strains. To increase the recovery of ethanol, maintaining the redox conditions and energy state of the cells is critical, and the carbon flow of pyruvate must be diverted from side pathways to ethanol production by controlling and blocking branched pathways from pyruvate. The metabolic engineering of heteroethanologenic bacteria is outlined below.

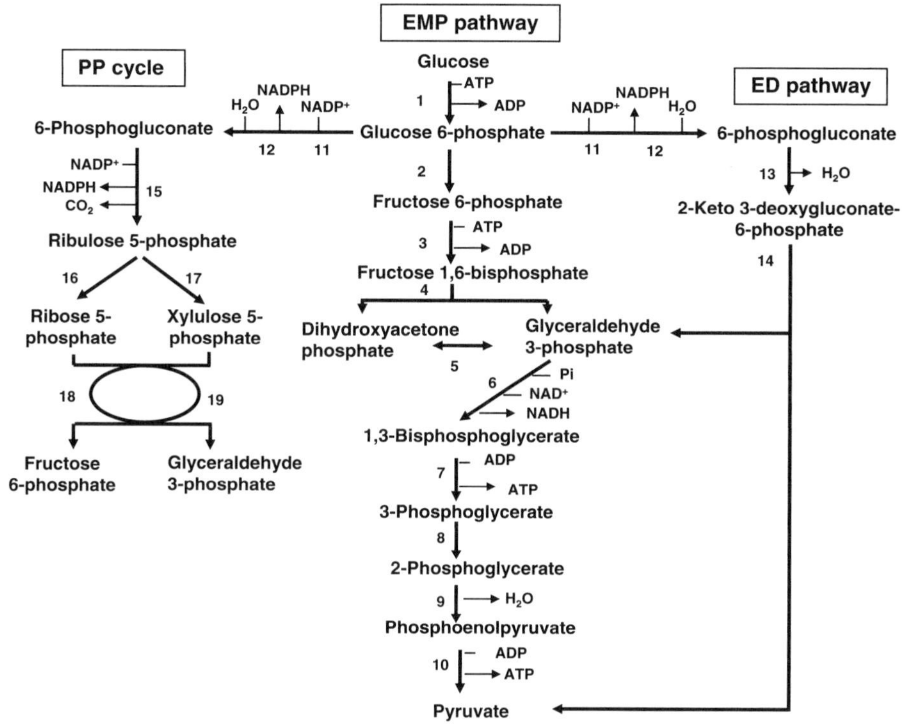

FIGURE 7.1 Glycolysis pathways in bacteria: 1, hexokinase; 2, glucose-6-phosphate isomerase; 3, phosphofructokinase; 4, fructose bisphosphate aldolase; 5, triosephosphate isomerase; 6, glyceraldehyde-3-dehydrogenase; 7, phosphoglycerate kinase; 8, phosphoglycerate mutase; 9, 2-phosphoglycerate dehydrogenase; 10, pyruvate kinase; 11, glucose-6-phosphate dehydrogenase; 12, 6-phosphogluconolactonase; 13, 6-phosphogluconate dehydratase; 14, 2-keto-3-deoxygluconate aldolase; 15, 6-phosphogluconate dehydrogenase; 16, ribose-5-phosphate isomerase; 17, ribulose phosphate 3-epimerase; 18, transaldolase; 19, transketolase.

7.2.1 *Escherichia coli*

7.2.1.1 Ethanologenesis *Escherichia coli* produces lactate as a major metabolite under anaerobic conditions, but only produces a small amount of ethanol. However, *E. coli* has advantages as a biocatalyst for ethanol production, including the ability to ferment a broad range of utilizable sugars. When *E. coli* ferments sugars to a mixture of ethanol and organic acids (Figure 7.2), ethanol is produced from pyruvate using pyruvate formate lyase (PFL). This fermentation pathway is unbalanced because one NADH is generated for each pyruvate, and two NADHs are required for converting pyruvate into ethanol. *Escherichia coli* balances the fermentation by producing acetic and succinic acids. Therefore, in native *E. coli*, only half of the available pyruvate is converted to ethanol. To develop an ethanologenic *E. coli*, the

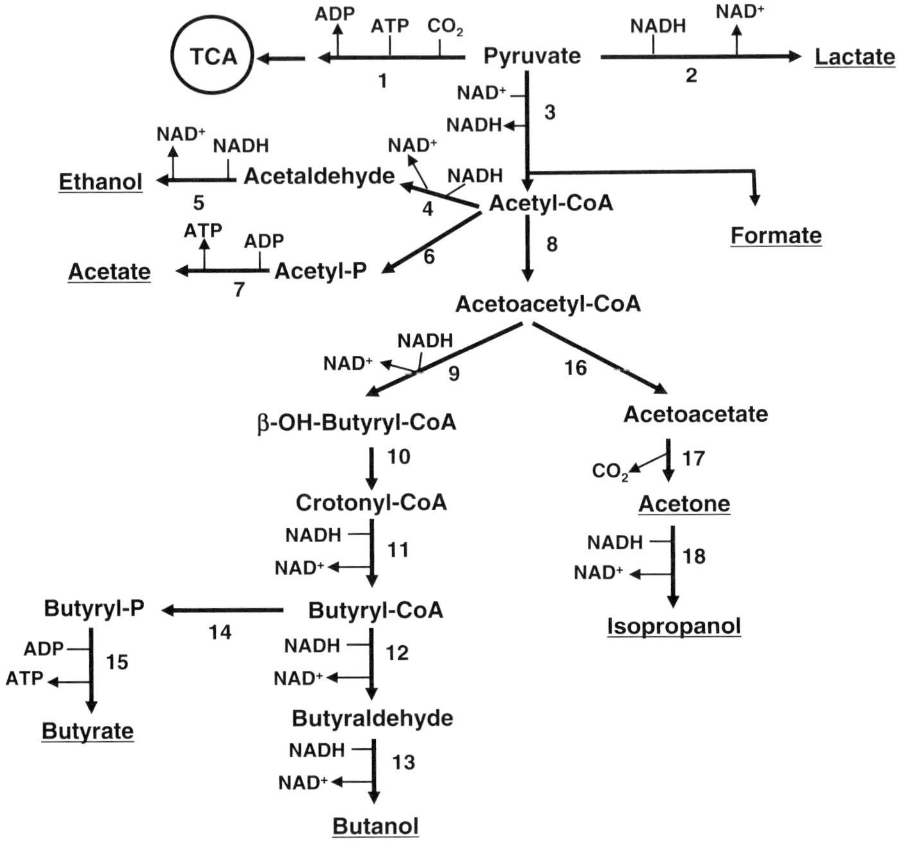

FIGURE 7.2 Fermentation in heteroethanologenic bacteria: 1, pyruvate dehydrogenase; 2, lactate dehydrogenase; 3, pyruvate formate lyase; 4, acetaldehyde dehydrogenase; 5, alcohol dehydrogenase; 6, phosphotransacetylase; 7, acetate kinase; 8, acetyl-CoA-acetyl transferase; 9, L(+)-β-hydroxybutyryl-CoA dehydrogenase; 10, 1,3-hydroxy-acyl-CoA hydrolase; 11, butyryl-CoA dehydrogenase; 12, butyraldehyde dehydrogenase; 13, butanol dehydrogenase; 14, phosphotransbutyrylase; 15, butyrate kinase; 16, CoA transferase; 17, acetoacetate decarboxylase; 18, isopropanol dehydrogenase.

reinforcement of the pathway from pyruvate to ethanol is essential. Homoethanologenic microorganisms like *Saccharomyces cerevisiae* and *Zm. mobilis* convert pyruvate directly to acetaldehyde by means of a pyruvate decarboxylase (PDC), allowing them to produce 1 mole of ethanol for every mole of pyruvate. Bräu and Sahm (1986) successfully expressed the *Zm. mobilis* PDC at high levels in *E. coli*, but the expression of only PDC in *E. coli* was not sufficient to confer ethanologenesis. Subsequently, Ingram et al. (1987) cloned the alcohol dehydrogenase II (ADH II) gene from a genomic library of *Zm. mobilis* using a unique assay system for ADH activity. The ADH II is primarily involved in the reduction of acetaldehyde to ethanol in *Zm. mobilis*. *Escherichia coli* expressing *pdc* and *adh II* produced ethanol with the theoretical yield. Both *adh II* and *pdc* were co-expressed under the control of the *lac* promoter, and the construct was named the production of ethanol (PET) operon. The research group of Ingram evaluated several *E. coli* strains as hosts (Alterthum and Ingram, 1989). The major criteria for an appropriate host included tolerance to ethanol, plasmid stability in nonselective media, and superior ethanol yield on glucose. *Escherichia coli* ATCC 11303 (pLOI297) was identified as the most promising strain.

7.2.1.2 Pentose Fermentation One of the useful properties of *E. coli* is its ability to utilize a broad spectrum of carbohydrates. In the commercial production of cellulosic bioethanol, ethanol production from the pentoses present in hemicellulose is an important issue. Basically, the ethanologenic *E. coli* can take up xylose through the specific permease encoded by *xylE*, and produce ethanol. Xylose fermentations with *E. coli* ATCC 11303 (pLOI297) were evaluated under various culture conditions to fully characterize the strain (Beall et al., 1991). Optimal culture conditions were determined to be below 42°C and above pH 6.0. Under these growth conditions, the recombinant *E. coli* had a maximum ethanol tolerance of 53–56 g/L and a final average productivity (for 80 g/L xylose) of 0.72 g/L·h. To improve the genetic stability of the strain, the PET operon was integrated into the chromosome of ATCC 11303 (Attfield and Bell, 2006). However, the transformant produced only low levels of ethanol due to the low expression level of the integrated PET operon. To improve the expression of the PET operon, a chloramphenicol (Cm)-tolerant strain was isolated, because the chloramphenicol acetyltransferase gene (*cat*) was integrated as a drug marker gene in the PET operon. Spontaneous mutant KO4, which was called a hyper ethanol-producing strain, produced ethanol efficiently. This strain was further modified by deletion of the fumarate reductase gene (*frd*) to prevent the formation of succinate, a major by-product of *E. coli* metabolism via the TCA cycle. Ultimately, strain KO11, which was able to ferment glucose and xylose to ethanol at theoretical yields of 100% in rich media, was generated (Ohta et al., 1991a; Lindsay et al., 1995). To improve xylose fermentation, the glycolytic flux and the growth rate of recombinant strain were evaluated (Gonzalez et al., 2002). Using a different approach, Dien et al. (2000) developed ethanologenic strains of *E. coli* such as FBR3, FBR4, and FBR5. These strains were constructed by transforming a xylose-utilizing isolate of strain *E. coli* FMJ39 with plasmid pLOI297. In another study, *E. coli* SE2738 was developed from wild-type *E. coli* K-12 W3110; this mutant strain exhibited 82% theoretical ethanol yield when grown on xylose under anaerobic conditions (Kim et al., 2007).

To achieve efficient ethanol production from hydrolysates of lignocellulosic biomass, it is essential to develop an ethanologenic microorganism that can simultaneously ferment a mixture of pentoses and glucose. In *E. coli* the utilization of the pentose sugars, xylose and arabinose, is normally repressed in the presence of glucose, due to carbon catabolite repression (CCR) (Görke and Stülke, 2008). With this system, glucose blocks the expression of sugar-specific transporters and key enzymes needed for the metabolism of alternative sugars. Enabling simultaneous fermentation is an important step for increasing the fermentation rate, and glucose-inhibited uptake of pentoses is primarily responsible for the lack of co-fermentation

(Nichols et al., 2001). Although Nichols et al. demonstrated the simultaneous fermentation of glucose, xylose, and arabinose by inactivating *ptsG*, which encodes the glucose-specific transporter of the phosphotransferase system (PTS), the rate of glucose fermentation of the mutant was decreased to a level similar to that of xylose fermentation. Furthermore, a glucose facilitator protein (GFL) from *Zm. mobilis*, which also transports xylose, was evolved to eliminate glucose inhibition on xylose uptake. Through random mutagenesis and partial deletion, the evolved transporter showed improved xylose transport activity in the presence of glucose. Nevertheless, no obvious advantage in xylose consumption was observed when xylose was co-fermented with glucose, which might be attributed to an inhibition of the intracellular xylose metabolism by glucose (Ren et al., 2009). In another investigation, Yomano and coworkers deleted the methylglyoxal synthase gene (*mgsA*) in an ethanologenic *E. coli* strain, which enabled this strain to co-utilize a 10% mixture of mannose, glucose, arabinose, xylose, and galactose (2% each) within 72 hours (Yomano et al., 2009). If *E. coli* is to be used in the commercial production of second-generation bioethanol, the synchronous fermentation of pentoses and hexoses has to be improved further.

An additional problem of using engineered *E. coli* in bioethanol production is ethanol sensitivity. Yomano et al. (1998) used long-term adaptation on the medium supplemented with ethanol to increase the ethanol tolerance of *E. coli* KO11 by approximately 10%. A number of isolates from the adapted culture were tested for ethanol tolerance, and one strain (LY01) was chosen for further study. Adaptation successfully reduced the time required to ferment 140 g/L xylose from 120 hours (for strain KO11) to 96 hours. Interestingly, LY01 also tolerated hydrolysate-associated inhibitors, including aldehydes (Zaldivar et al., 1999a,b), alcohols (Zaldivar et al., 2001), and organic acids (Zaldivar and Ingram, 1999a), better than KO11. Cultures of LY01 were able to tolerate up to 25 g/L acetic acid, 3.5 g/L hydroxymethylfurfural (HMF), and 3.5 g/L furfural. A novel area of research focuses on increasing the unsaturated fatty acid levels in the lipid membrane to improve ethanol tolerance. Overexpression of β-hydroxydecanoyl thioester dehydratase, which is involved in fatty acid synthesis, resulted in increased tolerance to ethanol (Luo et al., 2009).

7.2.1.3 Cellobiose and Cellulose Fermentation Although some wild-type strains of *E. coli* can utilize cellobiose as a sole carbon source, its metabolism has not been carefully examined (Moniruzzaman et al., 1997). To confer the ability to ferment cellobiose, the *casAB* operon from *Klebsiella oxytoca* was introduced into *E. coli*. *Klebsiella oxytoca* contains a PTS, which enables it to utilize cellobiose. The *K. oxytoca casAB* operon, which encodes an enzyme II cellobiose and a phospho-β-glucosidase, was expressed in *E. coli* KO11. Unfortunately, expression was very poor. However, spontaneous mutants with 15-fold higher specific activities for cellobiose utilization were isolated. The best mutant produced 45 g/L ethanol with 94% of the theoretical yield. Two of the mutants were tested for their ability to produce ethanol from mixed waste office paper. The engineered strain produced 32.7 g/L ethanol with 72% of the theoretical yield.

Numerous endoglucanases have been expressed in *E. coli*, allowing it to hydrolyze amorphous and soluble cellulose to shorter cellooligosaccharides (Srivastava et al., 1995; Wood et al., 1997; Da Silva et al., 2005). Among the enzymes that were successfully produced are Cel5Z and Cel8Y from *Erwinia chrysanthemi*. Unfortunately *E. coli* has limited ability to secrete these cellulases into the extracellular medium and as a result most of the recombinant Cel5Z accumulated in the periplasmic space, while 90% of Cel8Y was secreted as an extracellular product. Zhou et al. successfully reconstructed the type II secretion system, encoded by the *out* genes from *Er. chrysanthemi* in *E. coli* (Zhou and Ingram, 2000). The expression of *out* genes enabled *E. coli* to secrete more than 50% of the recombinant Cel5Z.

7.2.2 *Klebsiella oxytoca*

7.2.2.1 Ethanologenesis *Klebsiella oxytoca* is an enteric bacterium found growing in paper and pulp streams as well as around other sources of wood. *Klebsiella oxytoca* is capable of growing at a pH at least as low as 5.0 and temperatures as warm as 35°C. *Klebsiella oxytoca* can grow on a wide variety of sugars including hexoses and pentoses, as well as on cellobiose and cellotriose. This latter characteristic makes the strain especially appealing for cellulose fermentations. However, *K. oxytoca* has a complex fermentative metabolism, leading to various different metabolic products, such as lactate, formate, acetate, and butanediol in addition to ethanol. *K. oxytoca* metabolizes glucose to pyruvate through the EMP pathway, and then pyruvate is converted to ethanol, acetate, and formate in a ratio of 1:1:2, respectively. Although *K. oxytoca* can produce ethanol from pyruvate via the PFL pathway, the amount of ethanol that accumulates is small. Therefore, using a strategy similar to that used for the breeding of *E. coli*, a plasmid containing the *Zymomonas* PET operon (pLOI555) was introduced into *K. oxytoca*. In the resulting strain, *K. oxytoca* M5A1, expression of the PET operon enabled more than a 90% increase in fermentation products (Ohta et al., 1991a). The ethanol yield for *K. oxytoca* M5A1 (pLOI555) was 94–98% of the theoretical based upon initial glucose. The strain fermented 100 g/L glucose to produce 43 g/L ethanol within 30 hours, and after further 48 hours, the added glucose was completely consumed to produce 48 g/L ethanol with maximal ethanol productivity of 2.1 g/L·h.

7.2.2.2 Pentose and Hemicellulose Fermentation Strain M5A1 (pLOI555) appears to be particularly well suited for fermenting xylose. Under anaerobic conditions at pH 6.0, 30°C, the recombinant *K. oxytoca* rapidly fermented 100 g/L xylose to produce 46 g/L ethanol within 78 hours with maximal ethanol productivity of 2.0 g/L·h, indicating that its ethanol productivity was twice as fast as *E. coli* strain KO11. To stabilize the PET operon in *K. oxytoca*, the operon along with the *cat* gene marker was integrated into the pyruvate formate lyase gene (*pfl*) on the chromosome (Wood and Ingram, 1992). By screening integrated strains exhibiting high resistant to Cm, enhanced ethanol production was achieved in *K. oxytoca*. Then, to confer the ability to ferment hemicellulose, a xylanase gene (*xynZ*) from *Clostridium thermocellum* and a thermotolerant xylosidase gene (*xylB*) from *Butyrivibrio fibrisolvens* were introduced into ethanologenic *K. oxytoca* (Burchhardt and Ingram, 1992). However, most of the xylanase and xylosidase were localized to the cytoplasm of the transformant. An attempt was made to ferment xylan in a two-stage saccharification and fermentation process using the recombinant *K. oxytoca*. In the first (saccharification) step, the harvested cells were incubated in the medium containing 4% xylan at 60°C for 65 hours. Then, in the second (fermentation) step, the recombinant *K. oxytoca* was reinoculated into the hydrolyzed solution to ferment released xylose. As a result, 7.7–7.9 g/L ethanol was produced with about 30% of the theoretical yield.

7.2.2.3 *Cellobiose and Cellulose Fermentation* Some strains of *K. oxytoca* are capable of fermenting cellobiose. Brook and Ingram (1995) reported cellobiose fermentation by ethanologenic *K. oxytoca* P2, which readily fermented 100 g/L cellobiose to produce 45.2 g/L of ethanol with 96% of the theoretical yield. Strain P2 was tested successfully on various feedstocks including mixed office paper, sugarcane bagasse (Doran et al., 1994), corn fiber (Moniruzzaman et al., 1996), and sugar beet pulp (Doran et al., 2000). Golias et al. (2002) compared strain P2 and cellobiose fermenting yeasts for fermenting microcrystalline cellulose (Sigmacell 50). Ethanol production by *K. oxytoca* was 25–50% faster than with yeast, but the final ethanol concentration was limited to 37 g/L. Furthermore, to confer the ability to ferment cellulose directly, Wood and Ingram (1992) introduced four genes encoding thermotolerant cellulases, *celA*, *celB*, *celC*, and *celD* from *C. thermocellum*. Under two-stage saccharification and fermentation conditions, the obtained transformant expressing *celD* efficiently produced 5% ethanol from 36 g/L alkaline-treated cellulose. A combination of a commercial cellulase cocktail enhanced the ethanol production from crystalline cellulose, producing 17.4 g/L ethanol with 65% of the theoretical yield. Zou and Ingram (2000, 2001) integrated two genes encoding extracellular endoglucanases (*celZ* and *celY*) from *Er. chrysanthemi* into the chromosome of strain P2 and introduced the required auxiliary transporter genes (*out*) on a plasmid (pCPP2006); the resulting cellulolytic strain was named SZ21. The strain fermented cellulose (Sigmacell 50) poorly when no additional cellulase was added. Encouragingly, when commercial cellulase was added, SZ21 (pCPP2006) produced a 7–16% higher ethanol yield than the parental strain.

7.2.3 *Erwinia* spp. and *Enterobacter asburiae*

Erwinia spp. are gram-negative facultative anaerobes that are capable of utilizing a broad range of carbohydrates, especially pentose sugars. Generally, under anaerobic conditions, *Erwinia* assimilates xylose and arabinose and accumulates formate, acetate, lactate, and ethanol as the fermentation products. The first attempt to genetically engineer an ethanologenic *Erwinia* strain was reported by Tolan and Finn (1987). After screening for strains capable of growing on 5 g/L xylose in the presence of 3% ethanol, *Er. chrysanthemi* B374 was selected as the host strain for genetic engineering. To increase ethanol production from xylose and arabinose, *pdc* from *Zm. mobilis* was introduced into strain B375. Ethanol production was increased and lower amounts of organic acids were produced by the recombinant strain. The yield of ethanol from xylose and arabinose reached 1.45 mol/L of xylose and 1.1 mol/mol of arabinose, respectively. However, the growth rate and ethanol tolerance of the transformant were declined by dissonance of intracellular energy balance. Subsequently, Beall and Ingram (1993) reported the metabolic engineering of *Er. carotovora* SR38 and *Er. chrysanthemi* EC16, which were capable of producing ethanol efficiently from xylose and cellobiose in addition to glucose. The PET operon from *Zm. mobilis* was introduced and expressed in both strains. Both of the resultant ethanologenic *Erwinia* strains produced almost the theoretical yield of ethanol from 100 g/L glucose with 2 g ethanol/L·h, 80 g/L xylose with 1.1 g ethanol/L·h, and 100 g/L cellobiose with 1.5 g ethanol/L·h, respectively.

Recently, Bi et al. (2009) isolated *En. asburiae* JDR-1, a strain capable of fermenting both xylose and methylglucuronoxylose, which is released from hemicellulose upon acid treatment. The strain produced predominantly ethanol and acetate from hydrolysates of sweet gum xylan. To enhance ethanol production, the PET operon was introduced into *En. asburiae* JDR-1. Introduction of the PET operon led to homoethanol fermentation with acetate as a by-product. In addition, deletion of the native *pfl* gene enhanced ethanol production from xylan hydrolysates with 99% of theoretical yield and a rate of 0.11 g ethanol/L·h, which was 1.57 times the yield and 1.48 times the rate obtained with the ethanologenic strain *E. coli* KO11.

7.2.4 *Corynebacterium glutamicum*

7.2.4.1 Ethanologenesis *Corynebacterium glutamicum*, a nonpathogenic, Gram-positive soil bacterium, has been widely used for the industrial production of various amino acids and nucleic acids. Generally the industrial production of amino acids and organic acids has been performed under aerobic conditions. Interestingly, Inui et al. (2004) found that the major glycolysis pathways function to produce various organic acids, such as succinate and lactate under oxygen-deprived conditions. In oxygen-deprived conditions, after cultivation of *C. glutamicum* R cells, harvested cells were packed in a reactor at very high concentration, and then an industrial raw material was added to start the production of target compounds. From transcriptome analysis of cells under oxygen-deprived conditions, the metabolic balance was shifted according to the oxidized or reduced condition. In oxygen-deprived conditions, accumulation of lactate and succinate was enhanced by stimulation of the expression level of the glycolysis pathway in *C. glutamicum* R. Then, Inui et al., introduced the PET operon from *Zm. mobilis* into *C. glutamicum* R under the control of the native promoter of the lactose dehydrogenase gene (*ldhA*). The obtained transformant produced ethanol with acetate, lactate, and succinate as by-products. A glucose conversion yield of 79% was obtained for ethanol and organic acids, with ethanol accounting for 20%. By deleting the endogenous *ldh* gene, a final ethanol productivity of 29.5 g/L·h was achieved, corresponding to a glucose conversion yield of 53% for ethanol in the presence of small amount of pyruvate. The addition of small quantities of pyruvate and acetaldehyde has been shown to increase the sugar consumption and ethanol production in strains deficient in genes required for the synthesis of organic acids under oxygen-deprived conditions (Inui et al., 2004). Moreover, under growth-arrested conditions ethanol-producing *C. glutamicum* showed tolerance toward inhibitors like furan, organic acids, and phenolics, which are present in various pretreated lignocellulosic hydrolysates. The *C. glutamicum* strain (R-*ldhA*-pCRA723) retained 62–100% ethanol productivity in the presence of a model cocktail of inhibitors (Sakai et al., 2007).

7.2.4.2 *Pentose and Cellobiose Fermentation* Like *S. cerevisiae* and *Zm. mobilis*, most wild-type *Corynebacterium* strains are not naturally able to utilize pentose. When the *E. coli xylA* and *xylB* genes were integrated into the genome of *C. glutamicum* R (Kawaguchi, 2009), the resulting strain assimilated xylose as sole carbon source, and diauxic consumption of xylose by the engineered strain was observed in the presence of glucose under aerobic conditions. In contrast, under oxygen-deprived conditions, growth-arrested cells of the transformant could ferment both xylose and glucose simultaneously. Furthermore, when genes for xylose and arabinose catabolism from *E. coli* were introduced, the obtained transformant was able to utilize both sugars under aerobic and oxygen-deprived conditions (Kawaguchi et al., 2006, 2008). Overexpression of an arabinose transporter gene (*araE*) enhanced the consumption rate of both sugars (Sasaki et al., 2009). In the presence of glucose, the consumption rate of xylose or arabinose was not inhibited under growth-arrested conditions, which is an advantage compared with *E. coli* and *Zm. mobilis*.

The ability to utilize cellobiose was also conferred on *C. glutamicum* R. The strain utilizes methyl β-glucoside and the natural aryl β-glucosides, for example, salicin and arbutin through the PTS. To enhance transport of cellobiose, *bglF* encoding β-glucoside PTS permease was subjected to site-directed mutagenesis to construct BglF317A. Residue 317 of BglF is near the putative PTS active site H313 in the membrane spanning IIC domain. The introduction of BglF317A enabled cells to utilize cellobiose and produce organic acids as metabolites. Recently, Sasaki et al. (2008) conferred the ability to co-ferment cellobiose, glucose, xylose, and arabinose to ethanologenic *Corynebacterium*. The mutant *bglF* (317A) and *bglA* for β-glucosidase were integrated into the chromosome of the recombinant *C. glutamicum* R capable of utilizing glucose, xylose, and arabinose. All of the integrated genes were expressed efficiently in the transformant, and a mixture of cellobiose, glucose, xylose, and arabinose was consumed simultaneously and completely to produce the organic acids succinate, lactate, and acetate under oxygen-deprived conditions. However, the production of ethanol from mixed sugars by the engineered *Corynebacterium* was not demonstrated.

Based on the abilities discussed above, including the simultaneous utilization of a sugar mixture comprised of glucose, xylose, arabinose, and cellobiose and tolerance toward inhibitory by-products generated during pretreatment of lignocellulosic biomass, *Corynebacterium* has the potential to be developed into a useful ethanologenic bacterium.

7.2.5 Thermophilic Bacteria

7.2.5.1 *Ethanologenesis* Bacteria with the ability to ferment and produce ethanol at high temperatures are ideal candidates for developing economically feasible ethanologenic processes because of the reduced energy requirements for cooling reactors as well as the reduced risk of contamination. Thermophilic anaerobic bacteria that have been isolated from geothermal sources such as hot springs can grow at above 60°C (Lynd et al., 1991, 1996). Thermophiles are commonly able to ferment both pentoses and hexoses and, in some cases, produce enzymes to degrade cellulose and hemicellulose in lignocellulosic biomass (Lynd et al., 2002, 2008). Generally, however, thermophilic bacteria have a complex fermentative metabolism, resulting in the production of a variety of metabolic products, such as lactate, formate, and acetate in addition to ethanol (Figure 7.2). Most thermophilic bacteria take up carbohydrates via ATP-binding cassette (ABC) transporters or energy-independent facilitated diffusion proteins, and the carbohydrates are metabolized through the EMP pathway (Lamed and Zeikus 1980a,b). In contrast to homoethanologenic microorganisms like *Zm. mobilis* and *S. cerevisiae*, thermophilic ethanologenic bacteria do not possess PDC for the direct conversion of pyruvate to acetaldehyde. Instead, thermophilic bacteria produce ethanol from pyruvate using acetyl-CoA as an intermediate. Under fermentative conditions, two major enzymes, PFL and pyruvate ferredoxin oxidoreductase (PFOR), are present in thermophilic bacteria to direct pyruvate to acetyl-CoA. Ethanol is a key metabolite in ethanologenic thermophiles and is a result of a series of reactions, in which the final steps to form ethanol are catalyzed by acetaldehyde dehydrogenase (ALDH) and ADH, which reduce acetyl-CoA to acetaldehyde and further to ethanol, respectively. Members of the genus *Geobacillus* convert pyruvate to acetyl-CoA using pyruvate dehydrogenase (PDH) under aerobic conditions. PDH is inactive under anaerobic conditions, and *Geobacillus* spp. carry out mixed acid fermentation whereby pyruvate is converted to formate, acetate, and ethanol via PFL (Payton, 1984; Tang et al., 2009).

7.2.5.2 Pentose and Cellulosic Materials Fermentation Desai et al. (2004) reported the first site-directed gene disruption in *Thermoanaerobacterium saccharolyticum* by knocking out the *ldh* gene. The Δ*ldh* mutant (TD1) strain produced no detectable lactic acid during fermentation on either glucose or xylose, resulting in proportionately increased yields of acetate and ethanol. Later on, a further developed strain was reported by Shaw et al. (2008), in which both lactic acid and acetic acid formation pathways were deleted. The knockout strain (Δ*ldh*, Δ*ak*, and Δ*pta*) ALK2 produced ethanol as the only significant organic end product on xylose with an ethanol yield increased by nearly 160% compared with the wild type. Long-term adaptation of strain ALK2 in a continuous culture with progressively increasing feed xylose (up to 70 g/L) further improved its xylose consumption capacity. The resulting strain produced a final concentration of ethanol of 33.1 g/L with an ethanol yield of 0.46 g per gram xylose.

Another example of a thermophilic bacterium that is amenable to metabolic engineering is *Thermoanaerobacter mathranii*. An early constructed mutant, *T. mathranii* BG1L1, in which the *ldh* gene was knocked out, produced no detectable lactate and showed significantly increased ethanol yields (between 0.39 and 0.42 g ethanol per gram glucose/xylose) compared with the wild type (Georgieva et al., 2007). Strain BG1L1 was also tested on undetoxified pretreated lignocellulosic biomass hydrolysate, such as corn stover and wheat straw, and produced stable and high ethanol yields (0.39–0.42 g per gram sugars) in long-term continuous fermentations (Georgieva and Ahring, 2007; Georgieva et al., 2008). Tripathi et al. established a targeted gene knockout system for *C. thermocellum* and knocked out the phosphotransacetylase gene (*pta*), which catalyzes the formation of acetyl-P from acetyl-CoA (Tripathi et al., 2010). The *C. thermocellum* Δ*pta* mutant produced no detectable acetate when growing on cellobiose or Avicel as sole carbon sources. The Δ*pta* strain showed 29% and 27% increases in ethanol yields on cellobiose and Avicel, respectively, compared with the wild type.

Following metabolic engineering of a *Geobacillus* strain, Cripps et al. (2009) reported upregulated expression of *pdh* under anaerobic conditions in an *ldh*-inactivated *Geobacillus thermoglucosidasius* strain, in which the native promoter region upstream of *pdhA* was replaced with the *ldh* promoter from *Geobacillus stearothermophilus* NCA1503. The upregulation of *pdh* in strain TM180 (Δ*ldh* and *pdh* upregulated) resulted in a significantly enhanced ethanol yield (0.37 g ethanol per gram glucose) compared with both the wild-type strain (0.10 g ethanol per gram glucose) and the *ldh* deletion strain TM89 (0.24 g ethanol per gram glucose). However, fermentation by strain TM180 required moderate aeration conditions to achieve high ethanol yield. Subsequent inactivation of the *pfl* gene in strain TM180 created the triple mutant TM242 (Δ*ldh*, *pdh* upregulated, and Δ*pfl*) that showed an ethanol yield of 0.42 g/g on glucose, which was greater than 80% of the theoretical yield. In addition, the recombinant strain TM242 rapidly metabolized cellobiose with an ethanol productivity of 3.2 g/L·h. Metabolism on xylose by strain TM242 was slower than with hexoses, and an ethanol yield of 0.34 g/g was produced from xylose. These studies represent the engineering strategies applied to facultative anaerobic thermophilic bacteria.

An engineered thermophilic bacterium has the potential to produce bioethanol directly from lignocellulosic biomass, because most thermophilic bacteria can utilize the carbohydrate polymers cellulose and hemicellulose. Unlike cellulolytic fungi and anaerobic bacteria, anaerobic clostridia lack the ability to release cellulose-degrading enzymes into culture medium. They construct a complex cellulase system called cellulosome that is tightly bound to the cell surface (Schwarz, 2001). Georgieva et al. (2007, 2008) reported the fermentation of wheat straw hydrolysate that had undergone wet-exploded treatment (3–12% dry matter) and dilute acid corn stover hydrolysate (2.5–15% dry matter). The fermentations were conducted in a continuous upflow anaerobic sludge bed (UASB) reactor with a residence time of 2 days. The ethanol yields based on consumed sugars (mixed C5 and C6 sugars) were in the range of 0.39–0.42 g/g, and the maximum ethanol concentrations were in the range of 10.4–14.4 g/L in both of the fermentations. Shaw et al. (2008) reported a simultaneous saccharification and fermentation (SSF) process based on Avicel by applying a commercial cellulase (derived from *Trichoderma reesei*) together with a thermophilic bacterial fermentation utilizing the above engineered *T. saccharolyticum* strain ALK2 (Δ*ldh* Δ*ak* Δ*pta* mutant) at 50°C. They reported a 2.5-fold reduction in the cellulase requirement in a thermophilic SSF compared with a mesophilic SSF using *S. cerevisiae* as the fermentation organism at 37°C.

Efficient performance of thermophilic bacteria is also expected in CBP, which includes cellulase production, enzymatic hydrolysis, and ethanol fermentation in one single step.

7.3 HOMOETHANOLOGENIC BACTERIA

Zymomonas mobilis is a homoethanologenic bacterium that produces almost the theoretical yield of ethanol from glucose under anaerobic conditions, but not under strict anaerobic conditions. Another homoethanologenic bacterium is *Zb. palmae*. The two organisms have characteristics in common, but while the range of utilizable carbohydrates is limited to glucose, fructose, and sucrose for *Zm. mobilis*, the spectrum of carbohydrates utilized by *Zb. palmae* is broader and includes not only hexoses but also α-linked di- and tri-saccharides and sugar alcohols, from which it produces nearly the theoretical yield of ethanol. However, neither organism is able to ferment other major sugar components of lignocellulosic biomass such as xylose, arabinose, and mannose, or undecomposed cellooligosaccharide residues from cellulose. In these homoethanologenic bacteria, glucose is metabolized to form pyruvate through the Entner–Doudoroff (ED) pathway (Figure 7.1). The ED pathway is usually found in aerobic bacteria belonging to the genera *Acetobacter* and *Pseudomonas*; anaerobic bacteria having the ED pathway for glycolysis are unusual. When ATP formation via the EMP and the ED pathway was compared, it was found that the EMP yields 2 moles of ATP per mole of glucose, whereas the ED pathway yields 1 mole of ATP per mole of glucose. Thus the cell yield of ATP from glucose is less in *Zm. mobilis* and *Zb. palmae* than heteroethanologenic bacteria, resulting in superior ethanol productivity per cell mass. Furthermore, the pathway from pyruvate to ethanol is the same as that in *S. cerevisiae* (Figure 7.3). The liberated pyruvate is decarboxylated

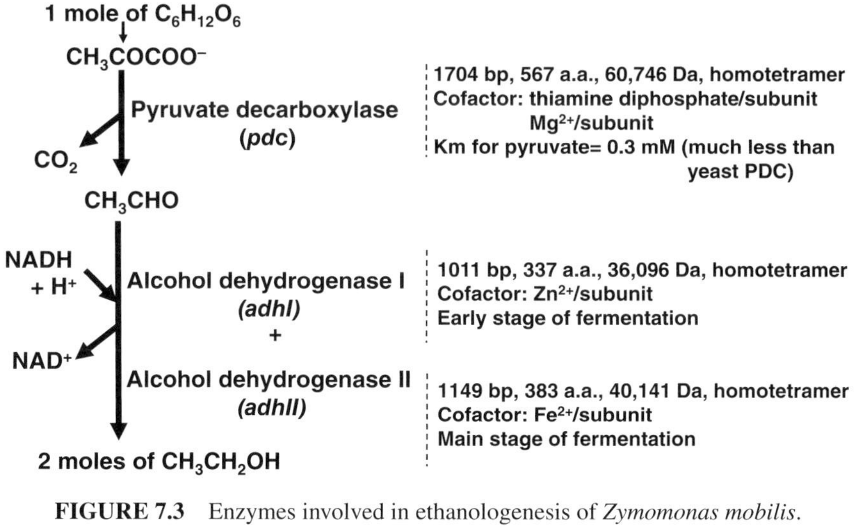

FIGURE 7.3 Enzymes involved in ethanologenesis of *Zymomonas mobilis*.

by PDC, yielding acetaldehyde and CO_2, after which the acetaldehyde is reduced to ethanol by ADH. Like heteroethanologenic bacteria, a branched pathway from pyruvate functions in both strains, but the combined effect of high expression levels and low K_m of the enzyme is to divert carbon flow to ethanol. Therefore, the breeding strategy in homoethanologenic bacteria is to broaden the range of utilizable sugars while retaining the higher ethanol productivity and the ethanol tolerance. The metabolic engineering of homoethanologenic bacteria is outlined as follows.

7.3.1 *Zymomonas mobilis*

7.3.1.1 Ethanologenesis *Zymomonas mobilis* has been found in fermenting tropical plant juices and is a fermenting agent in the production of alcoholic beverages, such as pulque and palm wines (Swings and De Ley, 1977). In Europe, *Zm. mobilis* has been isolated as a spoilage agent from beer, cider, and perry. *Zymomonas mobilis* performs a highly productive ethanol fermentation and offers a number of advantages over traditional yeast fermentation. This is suggestive of the potential utility of *Zm. mobilis* in the second-generation bioethanol production. The microorganism has a homoethanologenic fermentation pathway and tolerates up to 120 g/L ethanol. It has a higher ethanol yield (5–10% more ethanol per gram fermented glucose) and a much higher specific ethanol productivity (2.5×) than *Saccharomyces* sp. (Rogers et al., 1982). Furthermore, *Zm. mobilis* is generally regarded as safe (GRAS). *Saccharomyces* metabolizes glucose to pyruvate via the EMP pathway; ethanol is then formed from pyruvate. In contrast, *Zymomonas* anaerobically ferments sugars via the ED pathway, forming pyruvate from gluconate (Figure 7.1). As in *Saccharomyces*, the liberated pyruvate is decarboxylated, yielding acetaldehyde and CO_2, after which the acetaldehyde is reduced to produce ethanol (Sahm et al., 1991). In *Zm. mobilis*, glycolytic enzymes account for 30–50% of the soluble protein. Among these enzymes, glyceraldehyde 3-phosphate dehydrogenase and phosphoglycerate kinase are key regulators in the ED pathway. Moreover, the presence of PDC and ADH isozymes that are tolerant to high ethanol concentrations enables *Zymomonas* to perform a pure ethanol fermentation (Figure 7.3). The kinetic parameters for ethanol fermentation by *Zymomonas* and *Saccharomyces carlsbergensis* have been compared in 25% glucose under anaerobic conditions (Table 7.1) (Rogers et al., 1982). Ethanol production and glucose uptake by *Zymomonas* are three to four times faster than by *S. carlsbergensis*. The higher production results in a greater ethanol yield than is traditionally seen

TABLE 7.1 Ethanol Production by *Zymomonas mobilis* and *Saccharomyces carlsbergensis*

Kinetic parameters	*Zm. mobilis*	*S. carlbergensis*	*S. cerevisiae*
Specific ethanol productivity, $q_{p/x}$ (g/g·h)	5.44	0.82	0.57
Cell yield, $Y_{x/s}$ (g/g)	0.028	0.043	0.038
Ethanol yield, $Y_{p/s}$ (%), 100% = 0.511 (g/g)	95	90	86

Source: Adapted from Rogers et al. (1982) and data from Yanase laboratory.

with yeast, despite the fact that the cell yield is smaller for *Zymomonas* than *S. carlsbergensis*.

7.3.1.2 Pentose Fermentation Despite its advantages as an ethanologen, *Zm. mobilis* is not well suited for biomass conversion because it ferments only glucose, fructose, and sucrose. *Zymomonas* strains can take up D-xylose via its D-glucose transport system, but xylose is not utilized (DiMarco and Romano, 1985; Parker et al., 1995). However, Zhang et al. (1995) have successfully engineered strains capable of fermenting xylose (Figure 7.4). To confer xylose-fermenting ability, four genes (*xylA*, *xylB*, *tktA*, *talB*), which encode the xylose catabolic enzymes xylose isomerase, xylulose kinase, transketolase, and transaldolase, respectively, were introduced from the *E. coli* genome. Xylose isomerase and xylulose kinase convert xylose into xylulose-5-phosphate, and then this intermediate is converted to glyceraldehyde 3-phosphate and fructose 6-phosphate through the reinforced pentose phosphate pathway. The genes were expressed on a plasmid using either the enolase or glyceraldehyde-3-phosphate dehydrogenase promoters from *Zm. mobilis*; both of which are strong constitutive promoters. The transformed strain CP4 (pZB5) grew on xylose, and the ethanol yield was 86%. By using a similar strategy, Deanda et al. (1996) conferred the ability to ferment arabinose on *Zm. mobilis*. Five genes from *E. coli* (*araA, araB, araD, tktA, talB*), encoding L-arabinose isomerase, L-ribulokinase, L-ribulose-5-phosphate-4-epimerase, transketolase, and transaldolase, respectively, were inserted into a shuttle vector to construct an arabinose fermentation plasmid (Figure 7.4). The first three

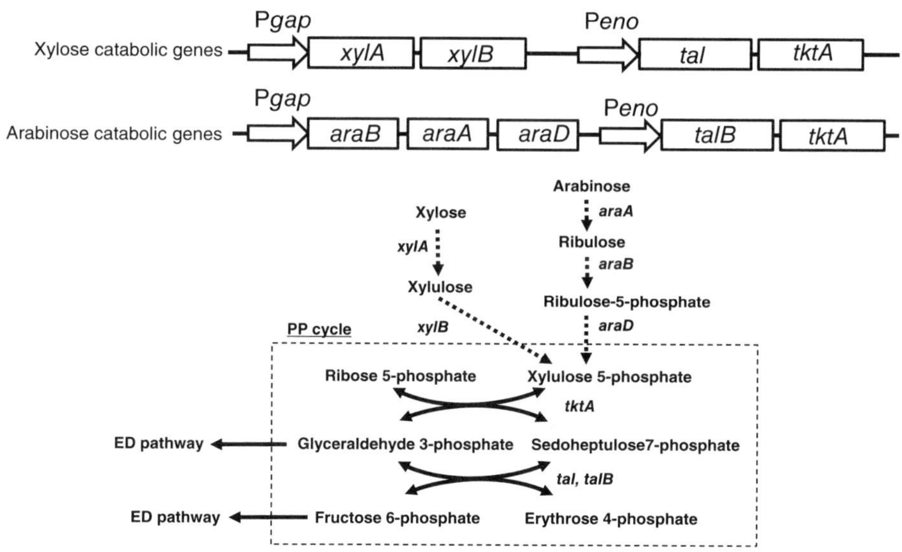

FIGURE 7.4 Ethanologenesis from pentose sugars in metabolically engineered *Zymomonas mobilis*: P*gap*, glyceraldehyde dehydrogenase promoter; P*eno*, enolase promoter; *xylA*, xylose isomerase; *xylB*, xylulokinase; *araA*, arabinose isomerase; *araB*, ribulokinase; *araD*, ribulose-5-phosphate-4-epimerase; *tal/talB*, transaldolase; *tktA*, transketolase.

enzymes are responsible for converting arabinose to xylulose-5-phosphate. And, as before, xylulose-5-phosphate was converted to ED pathway intermediates via transketolase and transaldolase. The resultant transformant, ATCC 39676 (pZB206), fermented arabinose (25 g/L) to ethanol with 84% theoretical yield. But the rate of arabinose fermentation was much lower compared to the rate of xylose fermentation by the xylose-fermenting strain, and small amount of arabinose remained in the culture medium after 100-hour cultivation. This slower fermentation rate was attributed to the affinity of the glucose transporter Glf for arabinose (Parker et al., 1995).

Rogers' group has been improving the xylose fermentation of *Zm. mobilis* and selected a highly ethanol tolerant strain, ZM4 (ATCC 31821) (Joachimsthal and Rogers, 2000). They transformed strain ZM4 with the xylose pathway plasmid pZB5, which was constructed by Zhang et al., and the obtained transformant demonstrated much higher ethanol production than strain CP4 (pZB5). When both strains were used to ferment an equal mixture of glucose (65 g/L) and xylose (65 g/L), strain ZM4 (pZB5) produced 62 g/L ethanol in 48 hours, while CP4 (pZB5) produced only 52 g/L after 60 hours. However, when the sugar concentration was further increased to 75 g/L each, the ZM4 (pZB5) fermentation stalled after the ethanol concentration rose to 67 g/L. Mohagheghi et al. (2002) genetically engineered *Zm. mobilis* to enable co-fermentation of glucose, xylose, and arabinose by genomic integration of the xylose and arabinose catabolic genes. The co-fermentation process yield from *Zm. mobilis* AX101 was about 84%, with a preferential order of sugar utilization: glucose > xylose and arabinose. However, these strains showed acetic acid sensitivity. To address the problem of sensitivity to toxic fermentation inhibitors, a new strain of *Zm. mobilis*, ZM4/AcR (pZB5), which had increased acetate resistance, was developed. This strain had enhanced performance in the presence of 12 g/L sodium acetate (Jeon et al., 2005). *Zymomonas mobilis* ZM4 produced nearly theoretical yields of ethanol with high specific productivity and was able to ferment both C5 and C6 sugars. The transformed *Zm. mobilis* ZM4 performed best under anaerobic conditions but also exhibited tolerance to aerobic conditions.

Yanase et al. (2012) genetically engineered a strain of *Zm. mobilis* capable of co-fermentation of a wood hydrolysate containing glucose, mannose, and xylose as major sugar components. This was accomplished by introducing genes encoding mannose and xylose catabolic enzymes from *E. coli* (Figure 7.5). Integration of *E. coli* phosphomannose isomerase gene (*manA*) into the *Zm. mobilis* chromosome conferred the ability to co-ferment mannose and glucose and resulted in the production of 91% of the theoretical yield of ethanol within 36 hours. Then, by introducing a recombinant plasmid harboring the genes encoding *E. coli (xylA*, *xylB*, *tal*, and *tktA)*, the obtained transformant *Zm. mobilis sucZE2::manA* (pZA22-xt) could co-ferment a mixture of 20 g/L glucose, 20 g/L mannose, and 20 g/L xylose within 72 hours, producing 89.8% of the theoretical yield. The recombinant *Zm. mobilis* also efficiently fermented an acid hydrolysate of a cellulosic feedstock containing glucose, mannose, and xylose. Moreover, a reactor packed with the strain continuously produced ethanol from an acid hydrolysate of wood biomass from coniferous trees for 10 days without accumulation of residual sugars. Ethanol productivity was 10.27 g/L·h at a dilution rate of 0.25/h (Figure 7.6). Yamada et al. (2002) reported fermentation of a dilute acid

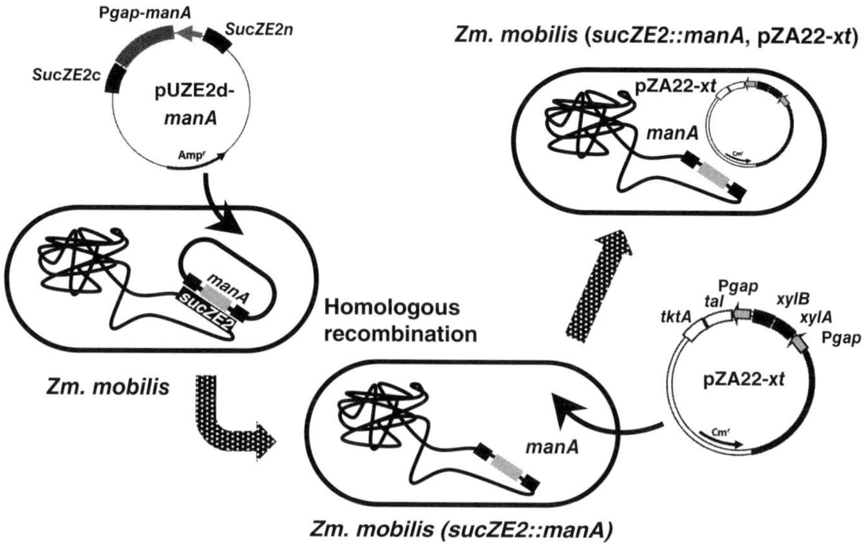

FIGURE 7.5 Metabolic engineering of *Zm. mobilis* capable of fermenting major sugar components of woody biomass. *Zymomonas mobilis* that can produce ethanol from major sugar components of wood hydrolysates was metabolically engineered by induction of genes encoding xylose- and mannose-catabolic enzymes from *Escherichia coli*. *manA*, phosphomannose isomerase.

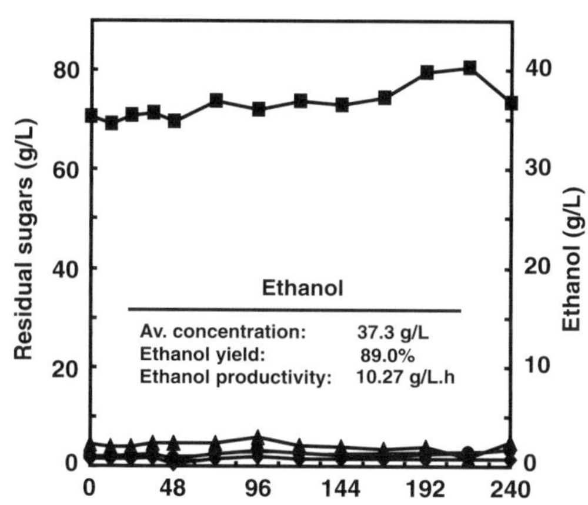

FIGURE 7.6 Time courses of 25-mL-scale continuous fermentation packed with the engineered *Zm. mobilis* using 9:1 acid hydrolysate medium at 30°C and pH 6.0 with a dilution rate of 0.25/h. After 48 hours of continuous feeding of 9:1 acid hydrolysate medium at a dilution rate of 0.01/h, feeding of 9:1 acid hydrolysate medium at a dilution rate of 0.25/h was started. Closed squares indicate the accumulation of ethanol, closed diamonds the residual glucose, closed triangles the residual mannose, and closed circles the residual xylose. From Yanase et al. (2012).

hydrolysate (41.2 g/L glucose and 11.2 g/L xylose) prepared from Japanese Cedar by immobilized *Zm. mobilis* 31821 carrying pZB5. The immobilized *Zm. mobilis* could ferment both sugars completely to produce 20 g/L ethanol within 50 hours. In addition, Jeon et al. (2010) reported a comparative evaluation of ethanol production from eight different hydrolysates of cellulosic raw materials by *Zm. mobilis* CP4 carrying pZB5. The highest ethanol yields and productivities were obtained with wheat straw and bagasse hydrolysates; fermentation of wood hydrolysate, such as pine and eucalyptus, resulted in relatively low ethanol production; the yields from pine and eucalyptus hydrolysates were 25% and 71%, respectively. *Zymomonas mobilis sucZE2*::*manA* (pZA22-xt) thus appears to be effective in producing ethanol from wood biomass.

7.3.1.3 Cellobiose and Cellulose Fermentation To breed *Zymomonas* strains able of fermenting cellulose and producing ethanol directly, it was necessary to introduce and express both the endoglucanase and β-glucosidase genes. Bacterial cellulase genes, including endoglucanase from *Cellulomonas uda*, *Acetobacter xylinum*, *Er. chrysanthemi*, *Pseudomonas fluorescens*, or *Bacillus subtilis*, have been introduced into and expressed in *Zm. mobilis*. Brestic-Goachet et al. (1989) expressed *Er. chrysanthemi* Cel5Z in *Zm. mobilis*. The endoglucanase gene was expressed efficiently at a level close to that of the parent strain, and most of the enzyme activity was released into the periplasmic space of *Zm. mobilis*. About 35% of the total enzyme activity was released into the culture medium, although cell lysis was not detected. To breed *Zymomonas* strains that can convert cellooligosaccharides to ethanol, the β-glucosidase gene (*bgl*) from the cellulosic bacterium *Xanthomonas albilineans* was cloned in *E. coli*, and then the gene was subcloned onto the small broad-host-range plasmid, pRK404 and introduced into *Zm. mobilis* (Su et al., 1989). The *bgl* gene was expressed in *Zm. mobilis*, and intact cells of the recombinant strain produced 132 mM ethanol from 110 mM cellobiose in 11 days. Yanase et al. expressed *bgl* from the rumen bacterium *Ruminococcus albus* in *Zm. mobilis* (Yanase et al., 2005a). To enhance the secretion of BGL expressed in *Zm. mobilis*, *bgl* was fused at the N-terminus with the 53-amino acid Tat signal peptide from the native periplasmic enzyme glucose–fructose oxidoreductase and introduced into *Zm. mobilis*. The tag enabled 61% of the β-glucosidase activity to be transported through the cytoplasmic membrane of the recombinant strain. The obtained transformant produced 10.7 g/L ethanol from 22 g/L cellobiose during 2 days of cultivation, which is more than 95% of the theoretical yield of ethanol.

Many approaches aimed at broadening the spectrum of utilizable substrates through transfer of appropriate hydrolase genes have been attempted. However, the recombinant strains were unable to produce ethanol directly from cellulosic biomass. Although there are obstacles that remain to be overcome, the company DuPont Danisco Cellulosic Ethanol (DDCE) has established a cellulosic ethanol demonstration facility based on genetically modified *Zm. mobilis* (http://www.ddce.com). The facility in Vonore, Tennessee, has an annual production capacity of 250,000 gallons of ethanol. The plant produces ethanol from agricultural residue and bioenergy crops, including corncobs and switchgrass.

7.3.2 *Zymobacter palmae*

7.3.2.1 *Ethanologenesis* The gram-negative strain *Zb. palmae*, which was isolated as a homoethanologenic fermentative bacterium from palm sap, is a facultatively anaerobic organism that ferments hexoses, α-linked di- and tri-saccharides, and sugar alcohols (fructose, galactose, glucose, mannose, maltose, melibiose, sucrose, raffinose, mannitol, and sorbitol) (Okamoto et al., 1993, 1994). This strain produces approximately 2 mole of ethanol per mole of glucose without accumulation of by-products and shows productivity similar to that of *Zm. mobilis*. However, the strain does not utilize cellulose, hemicellulose, cellooligosaccharides produced from degradation of cellulose or the pentoses, xylose and arabinose. If *Zymobacter* is to be used for bioethanol production, the number of substrates that can be utilized has to be broadened by metabolic engineering. As the efficiency of genetic engineering is generally dependent on the amount of genomic DNA information available for the target strain, Yanase et al. analyzed the entire sequence of the 3,023,645 base pair *Zb. palmae* genome and found it to contain 2475 open reading frames. The encoded metabolic pathways were predicted based on the genome annotation of *Zb. palmae*. Like glycolysis in *Zm. mobilis*, *Zb. palmae* ferments sugars via the ED pathway, forming pyruvate from gluconate. The liberated pyruvate is decarboxylated, yielding acetaldehyde and carbon dioxide. Then, the acetaldehyde is reduced to produce ethanol. This glycolytic pathway is present in both *Zb. palmae* and *Zm. mobilis*, and the cell yield of ATP from glucose is lower in both homoethanologenic bacteria than in yeast, suggesting that almost all of the carbohydrate utilized is efficiently converted to ethanol. However, this strain does not have genes encoding xylose or mannose catabolic enzymes.

7.3.2.2 Pentose Fermentation Yanase et al. (2007) engineered *Zb. palmae* in order to broaden the range of fermentable sugar substrates to include the pentose sugar xylose. The *E. coli* genes *xylA*, *xylB*, *tal*, and *tktA*, which encode the xylose catabolic enzymes, were introduced into *Zb. palmae*, where their expression was driven by the *Zm. mobilis* P*gap*. When cultured with 40 g/L xylose, the recombinant *Zb. palmae* strain was able to ferment 16.4 g/L xylose within 5 days, producing 91% of the theoretical yield of ethanol with no accumulation of xylitol or organic acids as metabolic byproducts. Notably, xylose acclimation enhanced both the expression of xylose catabolic enzymes and the rate of xylose uptake into recombinant *Zb. palmae*, which enabled the acclimated organism to completely and simultaneously ferment a mixture of 40 g/L glucose and 40 g/L xylose within 8 hours, producing 95% of the theoretical yield of ethanol (Figure 7.7). Yanase et al. speculated that xylose might be incorporated via a different transport system with a broad range of substrate sugars other than glucose. In addition, a specific sugar transport system in the xylose acclimated strain might be upregulated by xylose. Thus, efficient fermentation of a mixture of glucose and xylose to ethanol can be accomplished by using *Zb. palmae* expressing *E. coli* xylose catabolic genes. This property of the acclimated *Zb. palmae* strain appears to be suitable for the continuous fermentation of acid hydrolysates prepared from cellulosic feedstocks.

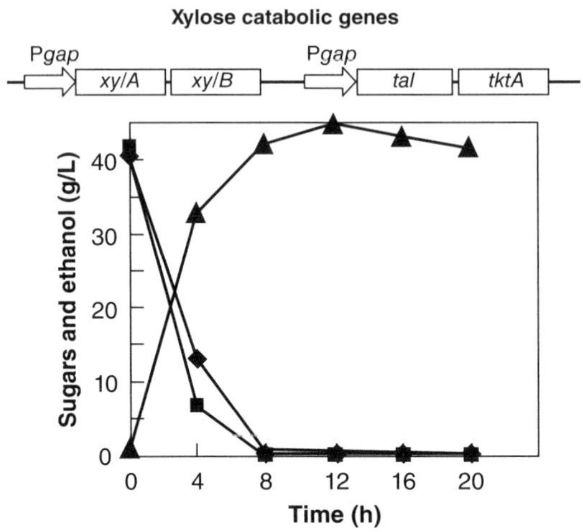

FIGURE 7.7 Simultaneous fermentation of xylose by acclimated *Zymobacter palmae* [pMFY31-P*gap-xylA/xylB*-P*gap-tal/tktA*] under resting conditions. The acclimated strain was precultured statically in T medium containing 40 g/L xylose for 48 hours at 30°C. Harvested cells were resuspended in T medium containing a mixture of 40 g/L glucose and 40 g/L xylose to give a concentration of cells at 8.2 mg dry weight cells/mL, and incubated statically at 30°C. Closed squares indicate the residual glucose, closed diamonds the residual xylose, and closed triangles the accumulation of ethanol. From Yanase et al. (2007).

Yanase (2011) also conferred the ability to co-ferment major sugar components of woody biomass such as glucose, mannose, and xylose. The artificial mannose operon was constructed by inserting the *E. coli* phosphomannose isomerase gene (*manA*) and *Zm. mobilis* fructokinase gene (*frk*) under the strong *Zm. mobilis* promoter P*gap* in the shuttle vector pMFY31, and the constructed plasmid was introduced into *Zb. palmae*. The obtained transformant fermented glucose and mannose completely to ethanol in amounts nearing the theoretical yield. Note that simultaneous fermentation of 30 g/L mannose occurred in the presence of 60 g/L glucose, and that the rate of mannose consumption was similar to that for glucose. Furthermore, to stabilize the strain, Yanase et al. integrated the genes encoding the catabolic enzymes for both xylose and mannose into the genome of *Zb. palmae*. The target genes selected for integration were the levansucrase (*lvs*) and the endoglucanase (*eg*) genes. Finally, a doubly integrated strain could ferment glucose, xylose, mannose, or their mixture to produce ethanol with almost the theoretical yield. This enabled the efficient recovery of bioethanol from all three major sugar components of woody biomass.

7.3.2.3 Cellobiose and Cellulose Fermentation To confer the ability to ferment cellobiose to *Zb. palmae*, the *bgl* gene from *R. albus* was introduced into *Zb. palmae*, where its expression was driven by its endogenous promoter (Yanase et al., 2005b). Interestingly, in *Zb. palmae* approximately 50% of the enzyme activity was found in the washing solution, which corresponds to the cell-surface fraction, and osmotic-shock solution, which corresponds to the periplasmic fraction. This situation is favorable for the fermentation of cellobiose by *Zb. palmae*, and the recombinant strain was able to assimilate cellobiose to produce ethanol. The recovery of ethanol from cellobiose by the recombinant strain reached approximately 95% of the theoretical yield from glucose; moreover, no glucose liberated from cellobiose or other common metabolites, such as pyruvate, lactate, acetate, or glycerol were detected in the culture fluid. In the presence of a mixture of glucose and cellobiose, the recombinant *Zb. palmae* fermented both sugars to ethanol during growth of the organism, but glucose was preferentially utilized and fermented at a faster rate than cellobiose.

To confer the ability to ferment cellulose on *Zb. palmae*, Kojima et al. (2012) developed a tool for cell-surface display of heterologous cellulolytic enzymes on *Zb. palmae*. They focused on the cell-surface display motif of the *Pseudomonas* ice nucleation protein (Ina). The gene encoding Ina from *Pseudomonas syringae* IFO3310 was cloned, and its product was composed of three functional domains: an N-terminal domain, a central domain with repeated amino acid residues, and a C-terminal domain. The N-terminal domain of Ina was shown to function as the anchoring motif for a green fluorescence protein fusion protein in *E. coli*. Then, the N-terminal anchoring motif of Ina was fused to the coding sequence of an N-terminally truncated *Cellulomonas* endoglucanase. *Zymobacter palmae* cells carrying the fusion endoglucanase gene were shown to degrade carboxymethyl cellulose as judged by the large halo on CMC plates stained with Congo red. Most of the fused endoglucanase was localized to the culture medium, and a portion was confirmed by immunofluorescence microscopy to be displayed on the cell surface. Tools for cell-surface display on *Zb. palmae* should therefore be effective for improving direct fermentation of pretreated lignocellulosic biomass by fusing the anchoring motif of Ina to the N-termini of additional bacterial cellulolytic enzymes, including endoglucanases, cellobiohydrolases, and BGL.

7.4 ECONOMIC OUTLOOK

In nature, there is a vast array of lignocellulosic biomass available for fermentation, which has encouraged many researchers to continue to develop new processes for second-generation bioethanol production. To accelerate the commercial production of second-generation bioethanol, it is essential to produce a continuous supply of low-cost bioethanol by developing a simple and inexpensive process. For the past two decades, the process of bioethanol production has gradually evolved from separate hydrolysis and fermentation (SHF) to simultaneous saccharification and fermentation (SSF), liquefaction followed by simultaneous saccharification and co-fermentation (L+SSCF), and finally SSCF. These improvements have reduced the cost of cellulosic ethanol, but the cost of production from lignocellulosic biomass is not yet competitive with those from grain-based ethanol (Table 7.2) (Hinman et al., 1992; Wooley et al., 1999; Wingren et al., 2003; Lynd et al., 2005; Arifeen et al., 2009; Vega-Sanchez and Ronald, 2010). Bioethanol from sugarcane in Brazil costs US$0.23–0.29/L, while in the European Union and the United States sugar beet and corn-derived bioethanol cost

TABLE 7.2 Comparison of the Cost Economics of Ethanol Production from Various Feedstocks

Feedstock	Production cost (US$ per liter)	Reference
Softwood	0.42	Lynd et al. (2005)
Yellow poplar	0.38	Wingren et al. (2003)
Hardwood	0.36	Wooley et al. (1999)
Willow	0.126	Arifeen et al. (2009)
Corn stover	0.28	Vega-Sanchez and Ronald (2010)
Sugarcane bagasse	1.20	Hinman et al. (1992)

TABLE 7.3 Summary of the NREL Process for Production of Ethanol From Corn Stover

Feedstock	Corn stover
Pretreatment	Dilute acid (total solid loading 30 wt%) Sulfuric acid loading 18 mg/g dry biomass at 158°C, for 5 min, at 5.5 atm
Hydrolysis	Whole slurry adjusted with ammonia (total solid loading 20 wt%) Fungal cellulase cocktail (20 mg/g cellulose) at 48°C, for 3.5 days
Fermentation	Hydrolysate (total solid loading 19.8%) Recombinant *Zm. mobilis* (co-fermentation of glucose, xylose, and arabinose) at 32°C, 1.5 days
Ethanol yield	76% theoretical yield (calculated from feedstock)
Selling cost	US$2.15/gal (US$0.57/L)

Source: Adapted from Humbird et al. (Mar. 2011)

US$0.29/L and US$0.53/L, respectively (Balat, 2011). In 2011, NREL (Colorado, USA) published the detailed report "Process Design and Economics for Biochemical Conversion of Lignocellulosic Biomass to Ethanol" (Humbird et al., Mar. 2011). The NREL process design converts corn stover to ethanol by dilute-acid pretreatment, enzymatic saccharification, and co-fermentation, and with a minimum ethanol selling price (MESP) of US$2.15/gal (US$0.57/L calculated) by 2012 conversion targets (Table 7.3). In the report, the biomass amount processed is 2205 dry ton/day at 76% theoretical ethanol yield (79 gal/dry ton). It is expected that this MESP will become the standard for the cost of cellulosic bioethanol.

7.5 FUTURE PROSPECTS

To reduce the cost of second-generation bioethanol, CBP is proposed as an ultimate process. In response to the updated processes, a number of ethanologenic bacteria have been bred, but to date no single microorganism suitable for CBP has been developed. The essential and desirable traits of microorganisms for efficient lignocellulosic ethanol production are summarized as follows: (1) broad substrate utilization range including the ability to utilize hexoses and pentoses simultaneously; (2) high ethanol titers, yields, and productivity with minimal byproduct formation; (3) increased tolerance to ethanol and inhibitors; (4) ability to degrade cellulose/hemicellulose; (5) minimal nutrient requirements; and (6) low risk of contamination. The genetically engineered *Zm. mobilis* which is the most promising ethanologenic bacterium is used in the NREL process. *Zymomonas mobilis* is able to co-ferment the mixture of glucose, xylose, and arabinose in hydrolysates of corn stover generated by pretreatment with dilute acid and enzymatic hydrolysis. Although the engineered *Zm. mobilis* has the above listed traits (1), (2), (3: ethanol tolerance), and (5), the strain cannot ferment cellulose and hemicellulose directly at higher temperature. Another homoethanologenic bacterium, *Zb. Palmae*, has traits (1), (2), (4: cellooligosaccharide fermentation), and (5), but tolerance to inhibitors and thermotolerance need to be improved. Ethanologenic *E. coli* have traits (1), (2), (3: inhibitor tolerance), and (5), but biphasic fermentation of C5- and C6-sugars, the tolerance to ethanol, saccharification of cellulose, and thermotolerance will need to be improved in the future. *Klebsiella oxytoca*, *Er. chrysanthemi*, and *Enterobacter* sp. exhibit broad substrate range of fermentable sugars for (1), saccharification of cellulose and hemicellulose (4), and tolerance to inhibitors (3), but ethanol productivity needs to be improved. The ethanologenic *C. glutamicum* has similar weak points to *E. coli*, and ethanol fermentation by *C. glutamicum* required unusual conditions, involving oxygen-deprived conditions in concentrated cell suspensions. Recently thermophilic bacteria have been investigated as potential biocatalysts for CBP, because the bacteria ferment cellulose and hemicellulose directly at temperatures over 65°C. However, because the thermophilic bacteria produce by-products besides ethanol, ethanol productivity needs to be improved to the level of *Zm. mobilis*.

None of the above-mentioned ethanologenic bacteria are able to directly ferment highly polymerized cellulose and hemicellulose at high temperature (over 45°C), indicating that they are not appropriate biocatalysts for CBP. Judging from the traits of each ethanologenic bacterium, the L+SSCF and SSCF processes are suitable for commercial production of second-generation bioethanol. However, to use the current ethanologenic bacteria as biocatalysts for L+SSCF and SSCF in commercial production, more robust ethanologenic bacteria must be developed. First, to eliminate the cost of detoxification of the process, increased tolerance to inhibitors such as furfural, hydroxymethyl furfural, phenolic compounds, and acetate is essential. Second, to synchronize both steps of the enzymatic saccharification and fermentation, the ability to ferment at temperatures over 45°C is essential, because the fungal cellulases perform optimally at temperatures over 45°C. In the future, the development of robust biocatalysts capable of saccharification and co-fermentation of cellulose and hemicellulose at high temperature will allow the production of second-generation bioethanol via CBP.

REFERENCES

Alterthum F, Ingram LO. (1989). Efficient ethanol-production from glucose, lactose, and xylose by recombinant *Escherichia coli*. *Appl Environ Microbiol*, 55, 1943–1948.

Arifeen N, Wang R, Kookos I, Webb C, Koutinas AA. (2009). Optimization and cost estimation of novel wheat bio-refining for continuous production of fermentation feedstock. *Biotechnol Prog*, 23, 872–880.

Attfield PV, Bell PJ. (2006). Use of population genetics to driven on recombinant *Saccharomyces cerevisiae* strains that grow using xylose as a sole carbon source. *FEMS Yeast Res*, 6, 862–868.

Balat M. (2011). Production of bioethanol from lignocellulosic materials via the biochemical pathway: A review. *Energ Convers Manage*, 52, 858–875.

Beall DS, Ingram LO. (1993). Genetic engineering of soft-rot bacteria for ethanol production from lignocellulose. *J Ind Microbiol*, 11, 151–155.

Beall DS, Ohta K, Ingram LO. (1991). Parametric studies of ethanolproduction from xylose and other sugars by recombinant *Escherichia coli*. *Biotechnol Bioeng*, 38, 296–303.

Bi C, Zhang X, Ingram LO, Preston JF. (2009). Genetic engineering of *Enterobacter asburiae* strain JDR-1 for efficient production of ethanol from hemicellulose hydrolysates. *Appl Environ Microbiol*, 75, 5743–5749.

Bräu B, Sahm H. (1986). Cloning and expression of the structural gene for pyruvate decarboxylase of *Zymomonas mobilis* in *Escherichia coli*. *Arch Microbiol*, 144, 296–301.

Brestic-Goachet N, Gunasekaran P, Cami B, Baratti JC. (1989). Transfer and expression of an *Erwinia chrysanthemi* cellulase gene in *Zymomonas mobilis*. *J Gen Microbiol*, 135, 893–902.

Brooks TA, Ingram LO. (1995). Conversion of mixed waste office paper to ethanol by genetically engineered *Klebsiella oxytoca* strain P2. *Biotechnol Prog*, 11, 619–625.

Burchhardt G, Ingram LO. (1992). Conversion of xylan to ethanol by ethanologenic strains of *Escherichia coli* and *Klebsiella oxytoca*. *Appl Environ Microbiol*, 58, 1128–1133.

Chen M, Zhao J, Xia L. (2009). Comparison of four different chemical pretreatments of corn stover for enhancing enzymatic digestibility. *Biomass Bioenergy*, 33, 1381–1385.

Cripps RE, Eley K, Leak DJ, Rudd B, Taylor M, Todd M, Boakes S, Martin S, Atkinson T. (2009). Metabolic engineering of *Geobacillus thermoglucosidasius* for high yield ethanol production. *Metab Eng*, 11, 398–408.

Da Silva GP, De Araujo EF, Silva D, Guimaraes WV. (2005). Ethanolic fermentation of sucrose, sugarcane juice and molasses by *Escherichia coli* strain KO11 and *Klebsiella oxytoca* strain P2. *Braz J Microbiol*, 36, 395–404.

Deanda K, Zhang M, Eddy C, Picataggio S. (1996). Development of an arabinose-fermenting *Zymomonas mobilis* strain by metabolic pathway engineering. *Appl Environ Microbiol*, 62, 4465–4470.

Desai SG, Guerinot ML, Lynd LR. (2004). Cloning of L-lactate dehydrogenase and elimination of lactic acid production via gene knockout in *Thermoanaerobacterium saccharolyticum* JW/SLYS485. *Appl Microbiol Biotechnol*, 65, 600–605.

Dien BS, Nichols NN, O'Bryan PJ, Bothast RJ. (2000). Development of new ethanologenic *Escherichia coli* strains for fermentation of lignocellulosic biomass. *Appl Biochem Biotechnol*, 84, 181–196.

DiMarco AA, Romano AH. (1985). D-Glucose transport system of *Zymomonas mobilis*. *Appl Environ Microbiol*, 49, 151–157.

Doran JB, Aldrich HC, Ingram LO. (1994). Saccharification and fermentation of sugar-cane bagasse by *Klebsiella oxytoca* P2 containing chromosomally integrated genes encoding the *Zymomonas mobilis* ethanol pathway. *Biotechnol Bioeng*, 44, 240–247.

Doran JB, Cripe J, Sutton M, Foster B. (2000). Fermentations of pectin-rich biomass with recombinant bacteria to produce fuel ethanol. *Appl Biochem Biotechnol*, 84–86, 141–152.

Esteghlalian A, Hashimoto AG, Fenske JJ, Penner MH. (1997). Modeling and optimization of the dilute-sulfuric-acid pretreatment of corn stover, poplar and switchgrass. *Bioresour Technol*, 59, 129–136.

Georgieva TI, Ahring BK. (2007). Evaluation of continuous ethanol fermentation of dilute-acid corn stover hydrolysate using thermophilic anaerobic bacterium *Thermoanaerobacter* BG1L1. *Appl Microbiol Biotechnol*, 77, 61–68.

Georgieva TI, Mikkelsen MJ, Ahring BK. (2007). High ethanol tolerance of the thermophilic anaerobic ethanol producer *Thermoanaerobacter* BG1L1. *Cent Eur J Biol*, 2, 364–377.

Georgieva TI, Mikkelsen MJ, Ahring BK. (2008). Ethanol production from wet-exploded wheat straw hydrolysate by thermophilic anaerobic bacterium *Thermoanaerobacter* BG1L1 in a continuous immobilized reactor. *Appl Biochem Biotechnol*, 145, 99–110.

Golias H, Dumsday GJ, Stanley GA, Pamment NB. (2002). Evaluation of a recombinant *Klebsiella oxytoca* strain for ethanol production from cellulose by simultaneous saccharification and fermentation: comparison with native cellobiose utilizing yeast strains and performance in co-culture with thermotolerant yeast and *Zymomonas mobilis*. *J Biotechnol*, 96, 155–168.

Gonzalez R, Tao H, Shanmugam KT, York SW, Ingram LO. (2002). Global gene expression differences associated with changes in glycolytic flux and growth rate in *Escherichia coli* during the fermentation of glucose and xylose. *Biotechnol Prog*, 18, 6–20.

Görke B, Stülke J. (2008). Carbon catabolite repression in bacteria: many ways to make the most out of nutrients. *Nat Rev Microbiol*, 6, 613–624.

Hinman ND, Schell DJ, Riley CJ, Bergeron PW, Walter PJ. (1992). Preliminary estimate of the cost of ethanol production for SSF technology. *Appl Biochem Biotechnol*, 34, 639–649.

Humbird D, Davis R, Tao L, Kinchin C, Hsu D, Aden A, Schoen P, Lukas J, Olthof B, Worley M, Sexton D, Dudgeon D. (Mar. 2011). Process design and economics for biochemical conversion of lignocellulosic biomass to ethanol dilute-acid pretreatment and enzymatic hydrolysis of corn stover. Technical Report, NREL/TP-5100-47764. Available at: http://www.osti.gov/bridge

Ingram LO, Conway T, Clark DP, Sewell GW, Preston JF. (1987). Genetic engineering of ethanol production in *Escherichia coli*. *Appl Environ Microbiol*, 53, 2420–2425.

Inui M, Kawaguchi H, Murakami S, Vertès AA, Yukawa H. (2004). Metabolic engineering of *Corynebacterium glutamicum* for fuel ethanol production under oxygen-deprivation conditions. *J Mol Microbiol Biotechnol*, 8, 243–254.

Jeon YJ, Svenson CJ, Rogers PL. (2005). Over-expression of xylulokinase in a xylose-metabolising recombinant strain of *Zymomonas mobilis*. *FEMS Microbiol Lett*, 244, 85–92.

Jeon YJ, Xun Z, Rogers PL. (2010). Comparative evaluations of cellulosic raw materials for second generation bioethanol production. *Lett Appl Microbiol*, 51, 518–524.

Joachimsthal EL, Rogers PL. (2000). Characterization of a high productivity recombinant strain of *Zymomonas mobilis* for ethanol production from glucose/xylose mixtures. *Appl Biochem Biotechnol*, 84–86, 343–356.

Kawaguchi H, Vertès AA, Okino S, Inui M, Yukawa H. (2006). Engineering of a xylose metabolic pathway in *Corynebacterium glutamicum*. *Appl Environ Microbiol*, 72, 3418–3428.

Kawaguchi H, Sasaki M, Vertès AA, Inui M, Yukawa H. (2008). Engineering of an L-arabinose metabolic pathway in *Corynebacterium glutamicum*. *Appl Microbiol Biotechnol*, 77, 1053–1062.

Kawaguchi H, Sasaki M, Vertes AA, Inui M, Yukawa H. (2009). Identification and functional analysis of the gene cluster for L-arabinose utilization in *Corynebacterium glutamicum*. *Appl Environ Microbiol*, 75(11), 3419–3429.

Kim Y, Ingram IO, Shanmugam KT. (2007). Construction of an *Escherichia coli* k-12 mutant for homoethanologenic fermentation of glucose or xylose without foreign genes. *Appl Environ Microbiol*, 73, 1766–1771.

Kojima M, Akahoshi T, Yanase H. (2012). Expression and surface display of *Cellulomonas* endoglucanase in the ethanologenic bacterium *Zymobacter palmae*. *Appl Microbiol Biotechnol*, 96, 1093–1104.

Kumar P, Barrett DM, Delwiche MJ, Stroeve P. (2009). Methods for pretreatment of lignocellulosic biomass for efficient hydrolysis and biofuel production. *Ind Eng Chem Res*, 48, 3713–3729.

Lamed R, Zeikus JG. (1980a). Ethanol production by thermophilic bacteria—relationship between fermentation product yields of and catabolic enzyme activities in *Clostridium thermocellum* and *Thermoanaerobium brockii*. *J Bacteriol*, 144, 569–578.

Lamed R, Zeikus JG. (1980b). Glucose fermentation pathway of *Thermoanaerobium brockii*. *J Bacteriol*, 141, 1251–1257.

Lee KJ, Lefebvre M, Tribe DE, Rogers PL. (1980). High productivity ethanol fermentations with *Zymomonas mobilis* using continuous cell recycle. *Biotechnol Lett*, 2, 487–492.

Lindsay SE, Bothast RJ, Ingram LO. (1995). Improved strains of recombinant *Escherichia coli* for production of ethanol from sugar mixtures. *Appl Microbiol Biotechnol*, 43, 70–75.

Luo LH, Seo P, Seo J, Heo S, Kim D. (2009). Improved ethanol tolerance in *Escherichia coli* by changing the cellular fatty acids composition through genetic manipulation. *Biotechnol Lett*, 31, 1867–1871.

Lynd LR. (1996). Overview and evaluation of fuel ethanol from cellulosic biomass: technology, economics, the environment, and policy. *Annu Rev Energy Environ*, 21, 403–465.

Lynd LR, Cushman JH, Nichols RJ, Wyman CE. (1991). Fuel ethanol from cellulosic biomass. *Science*, 251, 1318–1323.

Lynd LR, Weimer PJ, van Zyl WH, Pretorius IS. (2002). Microbial cellulose utilization: fundamentals and biotechnology. *Microbiol Mol Biol Rev*, 66, 506–577.

Lynd LR, van Zyl WH, McBride JE, Laser M. (2005). Consolidated bioprocessing of cellulosic biomass: an update. *Curr Opin Biotechnol*, 16, 577–583.

Lynd R, Currie D, Ciazza N, Herring C, Orem N. (2008). Consolidated bioprocessing of cellulosic biomass to ethanol using thermophilic bacteria. *Bioenergy*, Washington: ASM, pp. 55–73.

Mohagheghi A, Evans K, Chou YC, Zhang M. (2002). Cofermentation of glucose, xylose, and arabinose by genomic DNA-integrated xylose/arabinose fermenting strain of *Zymomonas mobilis* AX101. *Appl Biochem Biotechnol*, 98–100, 885–898.

Moniruzzaman M, Dien BS, Ferrer B, Hespell RB, Dale BE, Ingram LO, Bothast RJ. (1996). Ethanol production from AFEX pretreated corn fiber by recombinant bacteria. *Biotechnol Lett*, 18, 985–990.

Moniruzzaman M, Lai X, York SW, Ingram LO. (1997). Isolation and molecular characterization of high-performance cellobiosefermenting spontaneous mutants of ethanologenic *Escherichia coli* KO11 containing the *Klebsiella oxytoca casAB* Operon. *Appl Environ Microbiol*, 63, 4633–4637.

Nichols NN, Dien BS, Bothast RJ. (2001). Use of catabolite repression mutants for fermentation of sugar mixtures to ethanol. *Appl Microbiol Biotechnol*, 56, 120–125.

Ohta K, Beall DS, Mejia JP, Shanmugam KT, Ingram LO. (1991a). Genetic improvement of *Escherichia coli* for ethanol production: chromosomal integration of *Zymomonas mobilis* genes encoding pyruvate decarboxylase and alcohol dehydrogenase II. *Appl Environ Microbiol*, 57, 893–900.

Ohta K, Beall DS, Mejia JP, Shanmugam KT, Ingram LO. (1991b). Metabolic engineering of *Klebsiella oxytoca* M5A1 for ethanol production from xylose and glucose. *Appl Environ Microbiol*, 57, 2810–2815.

Okamoto T, Taguchi H, Nakamura K, Ikenaga H, Kuraishi H, Yamasato K. (1993). *Zymobacter palmae* gen. nov., sp. nov., a new ethanol-fermenting peritrichous bacterium isolated from palm sap. *Arch Microbiol*, 21, 333–337.

Okamoto H, Taguchi H, Nakamura K, Ikenaga H. (1994). Production of ethanol from maltose by *Zymobacter palmae* fermentation. *Agric Biol Chem*, 58, 1328–1329.

Parker C, Barnell WO, Snoep JL, Ingram LO, Conway T. (1995). Characterization of the *Zymomonas-mobilis* glucose facilitator gene-product (Glf) in recombinant *Escherichia coli*: examination of transport mechanism, kinetics and the role of glucokinase in glucose transport. *Mol Microbiol*, 15, 795–802.

Payton MA. (1984). Production of ethanol by thermophilic bacteria. *Trends Biotechnol*, 2, 153–158.

Rabelo SC, Filho RM, Costa AC. (2009). Lime pretreatment of sugarcane bagasse for bioethanol production. *Appl Biochem Biotechnol*, 153, 139–153.

Rainey FA, Ward NL, Morgan HW, Toalster R, Stackebrandt E. (1993). Phylogenetic analysis of anaerobic thermophilic bacteria—aid for their reclassification. *J Bacteriol*, 175, 4772–4779.

Ren C, Chen T, Zhang J, Liang L, Lin Z. (2009). An evolved xylose transporter from *Zymomonas mobilis* enhances sugar transport in *Escherichia coli*. *Microb Cell Fact*, 8, 66. doi:10.1186/1475-2859-8-66

Rogers PL, Lee KJ, Skotnicki ML, Tribe DE. (1982). Ethanol production by *Zymomonas mobilis*. *Advances Biochem Eng*, 23, 37–84.

Rogers PL, Lee KJ, Tribe DE. (1997). Kinetics of alcohol production by *Zymomonas mobilis* at high sugar concentrations. *Biotechnol Lett*, 1, 165–170.

Sahm H, Brinber-Meyer S, Sprenger G. (1991). The Prokaryotes. *A Handbook on the Biology of Bacteria: Ecophysiology, Isolation, Identification, Applications*, 2nd ed. New York: Springer-Verlag, pp. 2287–2298.

Sakai S, Tsuchida Y, Nakamoto H, Okino S, Ichihashi O, Kawaguchi H, Watanabe T, Inui M, Yukawa H. (2007). Effect of lignocellulose-derived inhibitors on growth of and ethanol production by growth-arrested *Corynebacterium glutamicum* R. *Appl Environ Microbiol*, 73, 2349–2353.

Sasaki M, Jojima T, Kawaguchi H, Inui M, Yukawa H. (2008). Simultaneous utilization of D-cellobiose, D-glucose, and D-xylose by recombinant *Corynebacterium glutamicum* under oxygen-deprived conditions. *Appl Microbiol Biotechnol*, 81, 691–699.

Sasaki M, Jojima T, Kawaguchi H, Inui M, Yukawa H. (2009). Engineering of pentose transport in *Corynebacterium glutamicum* to improve simultaneous utilization of mixed sugars. *Appl Microbiol Biotechnol*, 85, 105–115.

Schwarz WH. (2001). The cellulosome and cellulose degradation by anaerobic bacteria. *Appl Microbiol Biotechnol*, 56, 634–649.

Shaw AJ, Podkaminer KK, Desai SG, Bardsley JS, Rogers SR, Thorne PG, Hogsett DA, Lynd LR. (2008). Metabolic engineering of a thermophilic bacterium to produce ethanol at high yield. *Proc Natl Acad Sci USA*, 105, 13769–13774.

Srivastava R, Kumar GP, Srivastava KK. (1995). Construction of a recombinant cellulolytic *Escherichia coli*. *Gene*, 164, 185–186.

Su P, Delaney SF, Rogers PL. (1989). Cloning and expression of a β-glucosidase gene from *Xanthomonas albilineans* in *Escherichia coli* and *Zymomonas mobilis*. *J Biotechnol*, 9, 139–152.

Sun F, Chen H. (2008). Enhanced enzymatic hydrolysis of wheat straw by aqueous glycerol pretreatment. *Bioresour Technol*, 99, 6156–6161.

Swings J, De Ley J. (1977). Biology of *Zymomonas*. *Bacteriol Rev*, 41, 1–46.

Taherzadeh MJ, Karimi K. (2007). Enzymatic-based hydrolysis processes for ethanol from lignocellulosic materials. *BioResources*, 2, 707–738.

Tang YJ, Sapra R, Joyner D, Hazen TC, Myers S, Reichmuth D, Blanch H, Keasling JD. (2009). Analysis of metabolic pathways and fluxes in a newly discovered thermophilic and ethanol tolerant *Geobacillus* strain. *Biotechnol Bioeng*, 102, 1377–1386.

Tolan JS, Finn RK. (1987). Fermentation of D-xylose and L-arabinose to ethanol by *Erwinia chrysanthemi*. *Appl Environ Microbiol*, 53, 2033–2038.

Tripathi SA, Olson DG, Argyros DA, Miller BB, Barrett TF, Murphy DM, Mccool JD, Warner AK, Rajgarhia VB, Lynd LR, Hogsett DA, Caiazza NC. (2010). Development of *pyrF*-based genetic system for targeted gene deletion in *Clostridium thermocellum* and creation of a *pta* mutant. *Appl Environ Microbiol*, 76, 6591–6599.

Vega-Sanchez ME, Ronald PC. (2010). Genetic and biotechnological approaches for biofuel crop improvement. *Curr Opin Biotechnol*, 21, 218–224.

Wingren A, Galbe M, Zacchi G. (2003). Techno-economic evaluation of producing ethanol from softwood: comparison of SSF and SHF and identification of bottlenecks. *Biotechnol Prog*, 9, 1109–1117.

Wood BE, Ingram LO. (1992). Ethanol-production from cellobiose, amorphous cellulose, and crystalline cellulose by recombinant *Klebsiella oxytoca* containing chromosomally integrated *Zymomonas mobilis* genes for ethanol-production and plasmids expressing thermostable cellulase genes from *Clostridium thermocellum*. *Appl Environ Microbiol*, 58, 2103–2110.

Wood BE, Beall DS, Ingram LO. (1997). Production of recombinant bacterial endoglucanase as a co-product with ethanol during fermentation using derivatives of *Escherichia coli* KO11. *Biotech Bioeng*, 55, 547–555.

Wooley R, Ruth M, Glassner D, Sheehan J. (1999). Process design and costing of bioethanol technology: a tool for determining the status and direction of research and development. *Biotechnol Prog*, 15, 794–803.

Yamada T, Fatigati MA, Zhang M. (2002). Performance of immobilized *Zymomonas mobilis* 31821 (pZB5) on actual hydrolysates produced by Arkenol technology. *Appl Biochem Biotechnol*, 98–100, 899–907.

Yanase H. (2011). Metabolic engineering of ethanologenic bacteria capable of producing cellulosic biomass. International Union of Microbiological Societies 2011 Congress, BR35-(2).

Yanase H, Nozaki K, Okamoto K. (2005a). Ethanol production from cellulosic materials by genetically engineered *Zymomonas mobilis*. *Biotechnol Lett*, 27, 259–263.

Yanase H, Yamamoto K, Sato D, Okamoto K. (2005b). Ethanol production from cellobiose by *Zymobacter palmae* carrying the *Ruminococcus albus* β-glucosidase gene. *J Biotechnol*, 118, 35–43.

Yanase H, Sato D, Yamamoto K, Matsuda S, Yamamoto S, Okamoto K. (2007). Genetic engineering of *Zymobacter palmae* for production of ethanol from xylose. *Appl Environ Microbiol*, 73, 2592–2599.

Yanase H, Miyawaki H, Sakurai M, Kawakami A, Matsumoto M, Haga K, Kojima M, Okamoto K. (2012). Ethanol production from wood hydrolysate using genetically engineered *Zymomonas mobilis*. *Appl Microbiol Biotechnol*, 94, 1667–1678.

Yomano LP, York SW, Ingram LO. (1998). Isolation and characterization of ethanol-tolerant mutants of *Escherichia coli* KO11 for fuel ethanol production. *J Ind Microbiol Biotechnol*, 20, 132–138.

Yomano LP, York SW, Shanmugam KT, Ingram LO. (2009). Deletion of methylglyoxal synthase gene (*mgsA*) increased sugar cometabolism in ethanol-producing *Escherichia coli*. *Biotechnol Lett*, 31, 1389–1398.

Zaldivar J, Ingram LO. (1999a). Effect of organic acids on the growth and fermentation of ethanologenic *Escherichia coli* LY01. *Biotechnol Bioeng*, 66, 203–210.

Zaldivar J, Martinez A, Ingram LO. (1999b). Effect of selected aldehydes on the growth and fermentation of ethanologenic *Escherichia coli*. *Biotechnol Bioeng*, 65, 24–33.

Zaldivar J, Nielsen J, Olsson L. (2001). Fuel ethanol production from lignocellulose: a challenge for metabolic engineering and process integration. *Appl Microbiol Biotechnol*, 56, 17–34.

Zhang M, Eddy C, Deanda K, Finkestein M, Picataggio S. (1995). Metabolic engineering of a pentose metabolism pathway in ethanologenic *Zymomonas mobilis*. *Science*, 267, 240–243.

Zhang Q, Cai W. (2008). Enzymatic hydrolysis of alkali-pretreated rice straw by *Trichoderma reesei* ZM4-F3. *Biomass Bioenergy*, 32, 1130–1135.

Zhou S, Ingram LO. (2000). Synergistic hydrolysis of carboxymethyl cellulose and acid-swollen cellulose by two endoglucananses (CelZ and CelY) from *Erwinia chrysanthemi*. *J Bacteriol*, 182, 5676–5682.

Zhou S, Ingram LO. (2001). Simultaneous saccharification and fermentation of amorphous cellulose to ethanol by recombinant *Klebsiella oxytoca* SZ21 without supplemental cellulase. *Biotechnol Lett*, 23, 1455–1462.

CHAPTER 8

Ethanol Production from Yeasts

TOMOHISA HASUNUMA and RYOSUKE YAMADA

Organization of Advanced Science and Technology, Kobe University, Nada, Kobe, Japan

AKIHIKO KONDO

Department of Chemical Science and Engineering, Graduate School of Engineering, Kobe University, Nada, Kobe, Japan

Abstract

Numerous environmental and social benefits could result from the replacement of petroleum-based transport fuels with bioethanol converted from renewable biomass.

Bioprocessing of Renewable Resources to Commodity Bioproducts, First Edition.
Edited by Virendra S. Bisaria and Akihiko Kondo.

One of the key elements for the development of environmentally benign ethanol production is the construction of biomass-hydrolyzing yeast strains. The commonly used yeast *Saccharomyces cerevisiae* is a superior ethanol producer with demonstrated fast sugar consumption, high ethanol yield from glucose, and high resistance to ethanol. Hence, many researchers have targeted the heterologous expression of biomass-degrading enzymes in yeast to utilize glucose released from biomass for the production of ethanol by itself. In particular, cell-surface engineering is a powerful tool for yeast engineering. The display of amylolytic and cellulolytic enzymes on the yeast cell surface has accomplished direct ethanol production from starchy and cellulosic biomass. Moreover, reutilization of the cell surface–engineered yeast has the advantage of reducing enzyme cost, enabling reuse of enzymes on the cell surface by collecting the cells. For the efficient production of ethanol from biomass, improved assimilation of a wide variety of substrates could be achieved by overexpressing or deleting genes encoding traits responsible for yeast fermentability. Establishing economically feasible fermentation processes requires a marked increase in ethanol product titers due to the high energy demands of product recovery steps, as well as the high capital and production costs associated with bioethanol production equipment. A combination of biomass-degrading enzyme capacity and metabolic engineering in yeast strains could be an effective approach to developing cells with novel fermentation ability for industrial applications, and development of functional consolidated bioprocessing.

8.1 INTRODUCTION

Environmental concerns and the depletion of oil reserves have resulted in governmental actions and incentives to establish greater energy independence by promoting research on environmentally benign and sustainable fuels. Utilization of biomass as the starting material for the production of fuels has received considerable interest in recent years (Ragauskas et al., 2006). Bioethanol is currently one of the most promising alternatives to conventional transport fuels because of desirable characteristics such as high octane value and good combustion efficiency (Mussatto et al., 2010). The use of ethanol produced from biomass as a transport fuel could help to reduce CO_2 buildup by recycling CO_2 that is released when bioethanol is combusted as fuel. Also, burning ethanol instead of gasoline could eliminate the release of acid rain–causing sulfur dioxide.

Bioethanol can be obtained directly from sucrose or from starchy and lignocellulosic biomass via several steps (Figure 8.1). Although the process complexity depends on the type of feedstock, the spectrum of designed and implemented technologies ranges from simple conversion of sugars by fermentation, to multistage conversion of lignocellulosic biomass into ethanol. Sugar crops such as sugarcane and sugar beet, need only a milling process for the extraction of sugars for fermentation, which is a relatively simple process of sugar transformation into ethanol. In particular, molasses, a main by-product of the final runoff syrup from sugar

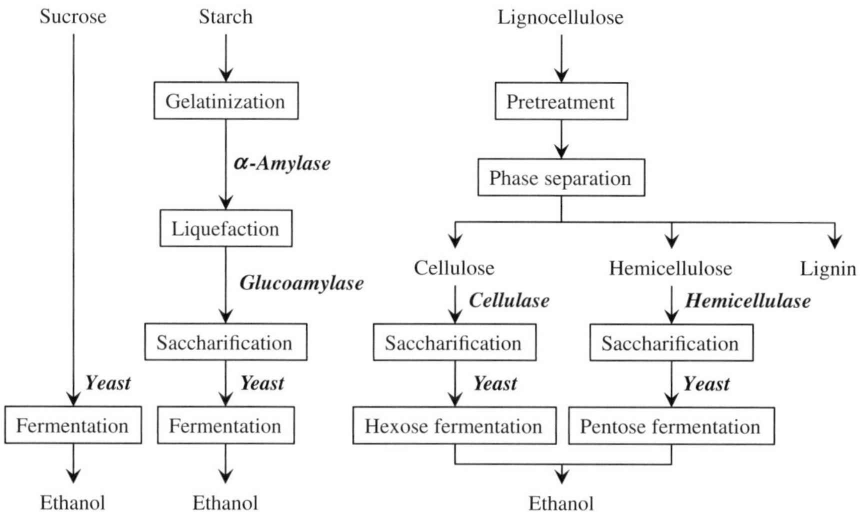

FIGURE 8.1 Schematic diagram of ethanol production from sucrose, starch, and lignocellulose.

manufacture, represents a major fermentation feedstock for commercial ethanol production (Senthilkumar and Gunasekaran, 2009). The total sugar in molasses can amount to 50–60% (w/v), of which about 60% is sucrose, which makes molasses a suitable substrate for industrial-scale ethanol production. The commercial production of ethanol from molasses mostly utilizes yeast strains belonging to the *Saccharomyces* spp. for the fermentation. The commonly used yeast *Saccharomyces cerevisiae* has many advantages as an ethanol producer, including fast sugar consumption, high ethanol yield from glucose, and high resistance to ethanol. For the efficient production of ethanol, three aspects have been developed: the supplementation of molasses, cell recycle fermentation, and the use of thermotolerant strains for improving both the rate of alcohol production and the final ethanol concentration.

Starch and lignocellulose of plant origin are the most abundant utilizable biomass resources. On the other hand, in processes that use these polysaccharide materials, saccharification is necessary before fermentation (Figure 8.1) because yeast strains cannot utilize these materials due to a lack of polysaccharide hydrolysis activities. So far, many researchers have reported attempts to address this issue using recombinant yeast strains capable of hydrolyzing starch and lignocellulose to glucose by expressed biomass-degrading enzymes (Figure 8.2). Recombinant strains can promote biomass utilization by the implementation of simultaneous saccharification and fermentation (SSF), which simplifies the bioethanol production process by the integration of biomass conversion steps. This chapter reviews the work that has been done on the development of yeast strains that demonstrate direct ethanol production from starchy and lignocellulosic biomass.

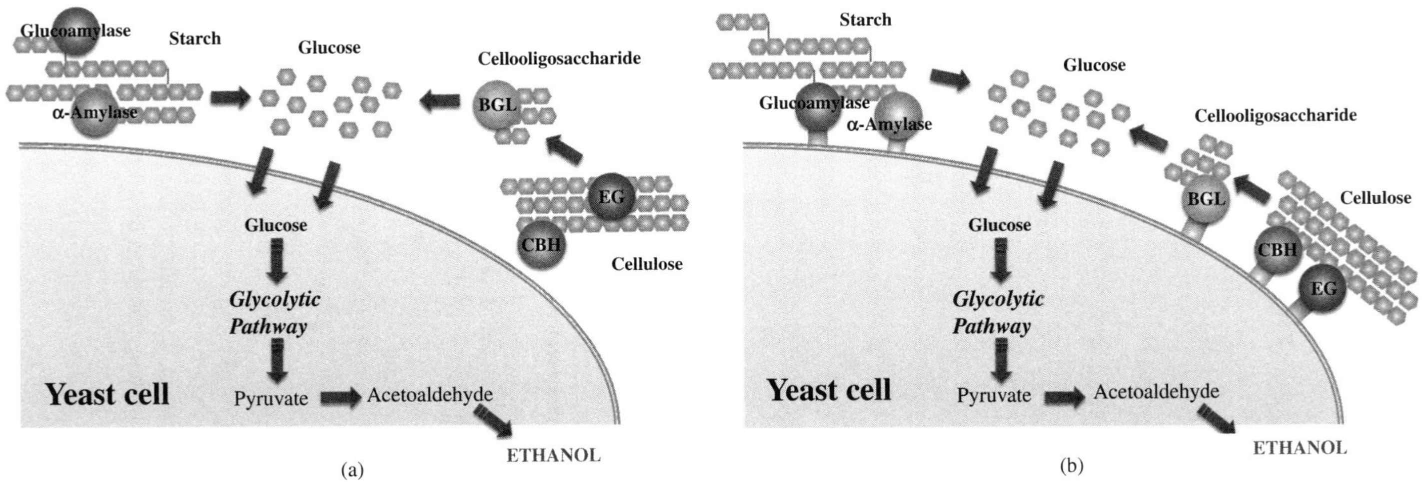

FIGURE 8.2 Direct conversion of starchy and lignocellulosic biomass to ethanol using enzyme-secreting (a) or enzyme surface–displaying (b) yeast cells. (*For a color version, see the color plate section.*)

8.2 ETHANOL PRODUCTION FROM STARCHY BIOMASS

8.2.1 Starch Utilization Process

Practical ethanol production from starchy biomass such as cassava, rice, sweet sorghum, and sweet potato have been reported; of these, corn is the most commonly used starchy feedstock for bioethanol production. This process uses starch from grains resembling corn; it is gelatinized by cooking and hydrolyzed to form glucose, which can be fermented by the microorganisms. Conventional ethanol production from raw starch requires the following three steps: (1) cooking at high temperature (140–180°C), (2) addition of α-amylase and glucoamylase for enzymatic saccharification of the cooked materials to glucose, and (3) fermentation of glucose to ethanol. The liquefaction process, which accounts for 30–40% of the total energy used for ethanol production, combined with the large quantities of enzymes that are required to convert the raw starch into glucose, both contribute to making conventional ethanol production an expensive and complex process. Matsumoto et al. (1982) reported non-cooking and low temperature cooking fermentation systems to reduce energy consumption by approximately 50%, though it is still necessary to add large amounts of amylolytic enzymes to hydrolyze the starchy materials to glucose.

Many researchers have attempted to resolve the problem by using recombinant amylase-expressing yeasts with the ability to directly ferment starch to ethanol (Tamalampudi et al., 2009). These amylases can be generally defined as the enzymes that hydrolyze the *O*-glycosyl linkage of starch (Nair et al., 2009). α-Amylases that cleave the internal α-1,4-glucosidic linkages in the starch are one of the most popular and important forms of industrial amylases, which are chiefly required for the thinning of starch in the liquefaction process. Glucoamylases are capable of cleaving both α-1,6- and α-1,4-glucoside linkages in the low molecular maltodextrins, removing one glucose unit at a time from the nonreducing end of the carbohydrate molecule. Recombinant yeast strains that coproduce α-amylase and glucoamylase have been developed to improve the efficiency of starch fermentation (Eksteen et al., 2003). Recent advances in yeast cell–surface engineering have provided the tools for the display of amylolytic enzymes that allow the utilization of yeast as a whole-cell biocatalyst for direct ethanol production from starch (Tamalampudi et al., 2009; Yamada et al., 2010a). Moreover, the integration of hydrolysis and fermentation steps by surface engineering of yeast cells can reduce the unit operation costs compared to separate hydrolysis and fermentation (SHF) process.

8.2.2 Yeast Cell–Surface Engineering System for Biomass Utilization

Yeast cell–surface engineering has been established to display enzymes, functional proteins, antibodies, and combinatorial protein libraries (Kondo and Ueda, 2004). The cell surface is a functional interface between the inside and outside of the cell allowing some surface proteins to extend across the plasma membrane, while others are bound by non-covalent or covalent interactions to the cell surface components. For anchoring surface-specific proteins, yeast cells have molecular systems to confine proteins to particular domains on the cell surface (Pittet and Conzelmann, 2007). For instance, agglutinin (Agα1 and Aga1), flocculin Flo1, Sed1, Cwp1, Cwp2, Tip1, and Tir1/Srp1 all have glycosylphosphatidylinositol (GPI) anchor moieties that are covalently attached to the carboxy termini of proteins (Lesage and Bussey, 2006). The GPI-anchored proteins contain hydrophobic peptides at their carboxy termini. After the completion of protein synthesis, the precursor protein remains anchored in the endoplasmic reticulum (ER) membrane by the hydrophobic carboxy-terminal sequence, with the rest of the protein in the ER lumen. Within less than a minute, the hydrophobic carboxy-terminal sequence is cleaved at the ω site and concomitantly replaced with a GPI anchor, presumably by the action of a transamidase (Ueda and Tanaka, 2000).

Biotechnology allows the cell surface to be exploited using yeast's natural mechanisms for anchoring proteins onto the cell surface (Figure 8.2). The use of the GPI-anchoring system has enabled display of various kinds of functional proteins on the cell surface without the loss of their activity through genetic engineering (Kondo and Ueda, 2004). Among the GPI anchor proteins, α-agglutinin and flocculin have been mainly used for the display of biomass-degrading enzymes. In the α-agglutinin system, the carboxy-terminal half of the α-agglutinin containing the GPI anchor attachment signal is connected to a target enzyme to anchor it as a fusion protein on the yeast cell surface. In the flocculin system, two types of cell surface display methods have been developed. In one system, the carboxy-terminal region of Flo1p, which contains a GPI attachment signal, is used, while the second system attempts to utilize the ability of the flocculation functional domain of Flo1p to create a novel surface display apparatus (Kondo and Ueda, 2004).

An advantage of the cell-surface engineering system is that biomass-degrading enzymes are genetically self-immobilized on the yeast cell surface so that the activities of the enzymes are retained as long as the yeast continues growing, whereas it is not easy to maintain the activities for a long reaction period in the conventional direct fermentation system in which the enzymes are secreted into the medium (Ueda and Tanaka, 2000). Another advantage of this system is the easy separation of the biocatalyst from the product. Reutilization of the yeast cells enables reuse of the enzymes displayed on the cell surface without reproduction of the yeast cells, which would reduce the cost of yeast propagation as well as enzyme addition (Kondo et al., 2002; Matano et al., 2013a). *Saccharomyces cerevisiae* is useful as a host for genetic engineering, since it allows the folding and glycosylation of expressed heterologous eukaryotic proteins and can be genetically manipulated. Moreover, the yeast can be cultivated to a high density in an inexpensive medium, so that the display of enzymes on the cell surface can have several applications in bioconversion processes.

8.2.3 Ethanol Production from Starchy Biomass Using Amylase-Expressing Yeast

Surface expression of amylolytic enzymes was initiated by the pioneering work of Murai et al. (1997) who reported a strategy for development of recombinant *Sa.*

cerevisiae strains displaying amylolytic enzymes on the cell surface. Glucoamylase derived from *Rhizopus oryzae* glucoamylase was displayed as a fusion protein with α-agglutinin using the secretion signal peptide of *Rh. oryzae* glucoamylase on the cell surface of a laboratory *Sa. cerevisiae* strain, MT8-1. The anchoring on the cell wall was verified by immunofluorescence labeling with anti-glucoamylase IgG. Kondo et al. (2002) used a flocculating yeast strain, YF207, for the surface display of glucoamylase because such a yeast strain offers the advantage of easy harvest of cells with displayed enzymes by flocculation after the batch fermentation. In this experiment, the display of glucoamylase did not negatively affect the growth and flocculation ability of the yeast during the ethanol production phase. Moreover, the activity of the glucoamylase displayed on the surface of flocculent yeast was similar to that displayed on non-flocculent yeast. The recombinant glucoamylase-displaying flocculent yeast strain maintained a high ethanol production rate of 0.6–0.7 g/L·h during repeated batch fermentation of soluble starch over 300 hours. In the fermentation by glucoamylase-displaying yeast cells, glucose was maintained at a very low concentration, which might be because the recombinant yeast cells metabolize glucose as soon as it is released from soluble starch by the glucoamylase. This low concentration of glucose in the fermentation is advantageous in minimizing the risk of contamination. On the other hand, the display of only glucoamylase led to the accumulation of an insoluble starch fraction during fed-batch fermentation because of the lack of a liquefying enzyme, α-amylase. In order to overcome this problem, Shigechi et al. (2002) developed two recombinant yeast strains: one co-displaying glucoamylase and α-amylase on the cell surface and the other displaying glucoamylase and secreting α-amylase into the culture medium. In fed-batch fermentations with soluble potato starch as a substrate, these two recombinant strains produced more than 60 g/L ethanol after 100 hours under anaerobic conditions. However, in using the α-amylase-secreting strain, glucose concentration in the culture medium was slightly higher, which was probably due to higher accessibility of starch to the secreted α-amylase in the medium.

Ethanol production from low temperature cooked starch has several advantages over the conventional high temperature cooking process because the high temperature cooking requires high energy and the addition of large amounts of amylolytic enzymes. Shigechi et al. (2004a) directly produced ethanol from corn starch cooked at 80°C in a single step using a recombinant yeast strain co-displaying both glucoamylase from *Rh. oryzae* and α-amylase from *Bacillus stearothermophilus*. α-Amylase hydrolysis of α-1,4-linkages of starch in a random fashion played an important role in the cooperative and sequential decomposition of the starch, which led to the efficient production of ethanol. The maximum ethanol titer, ethanol production rate, and substrate consumption rate from the fermentation of low-temperature cooked starch were almost the same as those from high-temperature cooked starch. Using the co-displaying strain, the yield of ethanol produced was 0.50 g per gram of carbohydrate consumed, which corresponds to 97.2% of the theoretical yield (0.51 g of ethanol per gram of glucose). The isolation of amylase from the lactic acid bacterium *Streptococcus bovis* opened new horizons for the hydrolysis of raw starch. Shigechi et al. (2004b) developed a novel non-cooking fermentation system for direct ethanol production with a yeast strain co-displaying *Rh. oryzae* glucoamylase and *St. bovis* α-amylase using α-agglutinin and Flo1p, respectively, as anchor proteins. They reported that α-amylase activity depends on the anchor protein. Glucoamylase activity is anchor independent while the use of the Flo1 anchor led to 40 times higher α-amylase activity than with the α-agglutinin anchor. Since several α-amylases have a starch-binding domain at their carboxy-terminal region, the display of α-amylase using Flo1p as anchor would be effective for ethanol production from raw corn starch. The yield, in terms of grams, of ethanol produced per gram of sugar consumed was 0.44 g/g, which corresponds to 86.5% of the theoretical yield (Shigechi et al., 2004b). Recently, Yamada et al. (2010a) constructed a high-performance, starch-degrading yeast capable of direct ethanol production from purified raw corn starch by combining the δ-integration technique and polyploidization with high yields. In addition, brown rice was directly converted to ethanol by the polyploid amylase-expressing yeast strain without any pretreatment or addition of enzymes or nutrients (Yamada et al., 2011b). Yamakawa et al. (2012) reported repeated batch fermentation of raw starch using a recombinant *Sa. cerevisiae* strain co-displaying α-amylase and glucoamylase. Fermentation from 100 g/L of raw starch was repeated for 23 cycles without the loss of enzyme activity. By displaying amylolytic enzymes on yeast cell surfaces, efficient whole-cell biocatalysts for SSF have been constructed.

8.3 ETHANOL PRODUCTION FROM LIGNOCELLULOSIC BIOMASS

8.3.1 Lignocellulose Utilization Process

Lignocellulosic materials such as sugarcane bagasse, corn stover, rice and wheat straw, grasses, wood chip, and other agricultural residues are of great interest as feedstock for second-generation biofuel production because they represent abundant, inexpensive, and renewable sources of sugars. However, the processes involved are more complicated than from sugar/starch crops, which mean the costs of ethanol production from lignocellulosic materials are higher. Ethanol production from lignocellulosic materials includes the following steps: (1) chemical and physicochemical pretreatment to swell the biomass, (2) hydrolysis of cellulose and hemicellulose to fermentable sugars by enzymes such as cellulases and hemicellulases, (3) microbial fermentation for the production of ethanol, (4) separation and concentration of ethanol by distillation and dehydration (Sánchez and Cardona, 2008).

Lignocellulosic materials are composed mainly of cellulose, hemicellulose, and lignin (Saha, 2003; Gnansounou, 2009). Cellulose, a homopolymer of β-1,4-linked glucose units, is the major component of lignocelluloses and constitutes nearly 40–60% of its total dry weight. The linear chains of cellulose are highly crystalline microfibrils. Hemicellulose, a substituted polysaccharide with a xylan backbone consisting of β-1,4-linked xylose monomers to which substituents and saccharides such as glucose, galactose, glucuronic acid, and arabinose are attached, constitutes 20–35% of lignocellulosic biomass. The hemicellulose molecules are linked to the cellulose microfibrils by hydrogen bonds. Lignin is a phenolic compound formed by polymerization of three types of monomers such as *p*-coumaryl, coniferyl, and synapyl alcohols and is present as an amorphous mass surrounding the cellulose and hemicellulose fibers. The recalcitrant structure makes the lignocellulose resistant to saccharification by hydrolytic enzymes. In general, the chemical and physicochemical pretreatment of lignocellulose causes cellulose to swell, thereby increasing its accessibility to saccharification enzymes (Chandel et al., 2012). However, the hydrolysis of cellulose remains a major limiting factor for the efficient utilization of lignocellulosic materials (Matano et al., 2012).

To release soluble sugars from cellulose, the activities of multiple enzymes, including endoglucanase (EG), exoglucanases such as cellobiohydrolase (CBH) and cellodextrinase, and β-glucosidase (BGL) are required (Chandel et al., 2012). Since cellulase reactions are inhibited by their intermediary and final products such as cellooligosaccharides and glucose, SSF processes that combine enzymatic hydrolysis with sugar consumption are preferential for the alleviation of cellulase activity inhibition (van Zyl et al., 2007). However, the difference in optimum temperatures between saccharification and fermentation during SSF is a drawback of the bioethanol production (Hasunuma and Kondo, 2012a). To overcome this limitation, large amounts of saccharification enzymes are required, which severely impacts the cost effectiveness of bioethanol production from lignocellulosic materials.

The recent development of microorganisms capable of efficient cellulose hydrolysis and fermentation represents a significant step toward reducing the requirement for enzyme addition in SSF processes (Matano et al., 2012). The consolidation of enzyme production, saccharification, and fermentation into a single process, which is referred to as consolidated bioprocessing (CBP), is becoming increasingly recognized as having potential for the low-cost production of bioethanol, as the high costs of capital investment, raw materials, and equipment associated with microbial enzyme production can be reduced (Olson et al., 2011; Menon and Rao, 2012). To develop recombinant cellulolytic microorganisms, heterologous expression of cellulase genes has been developed in yeast hosts (Hasunuma and Kondo, 2012b).

8.3.2 Fermentation of Cellulosic Materials

8.3.2.1 Ethanol Production from Cellulase-Secreting Yeast

For the construction of a cellulose fermentation system, multiple molecular components of a cellulolytic network have been expressed in recombinant *Sa. cerevisiae* strains. van Rensburg et al. (1998) first reported the expression of multiple cellulolytic genes in *Sa. cerevisiae*. They developed a recombinant cellulolytic *Sa. cerevisiae* strain simultaneously expressing EG from *Butyrivibrio fibrisolvens*, CBH from *Phanerochaete chrysosporium*, cellodextrinase from *Ruminococcus flavefaciens*, and cellobiase from *Endomycopsis fibuliger*. The resulting strain could grow with cellobiose as the sole carbon source. den Haan et al. (2007) reported co-expression of EG from *Trichoderma reesei* and BGL from *Saccharomycopsis fibuligera* in a recombinant *Sa. cerevisiae* strain, which enabled growth on a pure cellulose, phosphoric acid-swollen cellulose (PASC), with direct production of ethanol. Anaerobic growth of up to 0.27 g-dry-cell-weight/L was observed on medium containing 10 g/L PASC as the sole carbon source with concomitant ethanol production of up to 1.0 g/L. Jeon et al. (2009) expressed a bacterial cellulose–binding domain-containing EG from *Clostridium cellulovorans* and BGL from *Sa. fibuligera* in a recombinant *Sa. cerevisiae* strain. The resultant yeast strain efficiently hydrolyzed cellulosic materials such as carboxymethylcellulose (CMC) and β-glucan to smaller fragments for efficient fermentation to ethanol through the cellulose-binding domain's affinity for the substrate. Recently, a high-affinity cellodextrin transport system was isolated from the cellulolytic fungus, *Neurospora crassa*. A recombinant *Sa. cerevisiae* strain expressing the cellodextrin transporter together with intracellular BGL grew on not only cellobiose but also on cellodextrin for longer than on cellobiose (Galazka et al., 2010).

8.3.2.2 Ethanol Production from Cellulase-Displaying Yeast As mentioned above, many attempts have been made to engineer *Sa. cerevisiae* strains to secrete cellulases into the fermentation medium. An alternative approach is cell-surface engineering to display cellulolytic enzymes on the yeast cell surface, affording whole-cell biocatalysis for SSF and CBP (Kotaka et al., 2008; Yamada et al., 2011a; Yanase et al., 2010b). Fujita et al. (2002) first attempted to convert barley β-glucan into ethanol by constructing cellulose-degrading yeast cells that co-display *Aspergillus aculeatus* BGL1 and *T. reesei* EGII on the cell surface. β-Glucan is a linear, soluble polysaccharide composed of an average of 1200 glucose residues joined by approximately 70% β-1,4-glycosidic linkages and approximately 30% β-1,3-glycosidic linkages. The recombinant yeast strain showed growth in synthetic medium containing β-glucan as the sole carbon source and was able to directly ferment 45 g/L β-glucan to produce 16.5 g/L ethanol within 50 hours. The yield was 0.48 g ethanol produced per gram of carbohydrate utilized. Subsequently, PASC was fermented to ethanol with a recombinant yeast strain co-displaying *T. reesei* EGII and CBHII, and *A. aculeatus* BGL1 on the cell surface (Fujita et al., 2004). The yield of ethanol produced from PASC was 0.45 g/g, which corresponds to 88.5% of the theoretical yield. By displaying three cellulolytic enzymes on the yeast cell surface, the cellooligosaccharides produced by the reaction of EGII and CBHII were converted to glucose by BGL1 and the glucose was immediately taken up by the cells. This study reported that CBHII plays a crucial role in the hydrolysis of cellulose and that synergism between EGII and CBHII is successfully induced on the yeast cell surface. Although surface-displayed CBHII had only relatively low activity with respect to PASC, the yeast strain co-displaying EGII and CBHII showed significantly higher cellulolytic activity than the yeast displaying only EGII (Fujita et al., 2004). Although efficient degradation of lignocellulosic biomass requires an appropriate expression ratio of cellulolytic enzymes such as EG, CBH, and BGL in recombinant *Sa. cerevisiae*, it is difficult to simultaneously control multiple enzymes. Yamada et al. (2010b) developed a combinatorial method to optimize the cellulase expression ratio in yeast using a cocktail δ-integration system through an evolutionary engineering approach, which reveals the importance of EG expression for efficient PASC degradation. A diploid recombinant yeast strain with an optimized ratio of cellulases displayed on the yeast cell surface successfully produced ethanol from agricultural waste biomass without the need for addition of exogenous enzymes (Yamada et al., 2011a).

The strategy aimed at regulating cellulase ratios on the microbial cell surface is referred as the minicellulosome strategy (Ito et al., 2009; Lilly et al., 2009; Tsai et al., 2009, 2010; Wen et al., 2010). Ito et al. (2009) developed a yeast cell–surface display system by assembling (i) the Z domain of *Staphylococcus aureus* protein A and the Fc domain of human immunoglobulin G and (ii) the cohesin and dockerin domains of *C. cellulovorans* to construct chimeric scaffolding at both the Z and cohesion domains. Heterologous *T. reesei* EGII and *A. aculeatus* BGL1 fused with the Fc and dockerin domains were expressed, secreted, and assembled into functional complexes on the surface of the recombinant *Sa. cerevisiae* cells via interactions with complementary domains on scaffolding, yielding yeast strains capable of hydrolyzing β-glucan. Wen et al. (2010) reported recombinant *Sa. cerevisiae* strains displaying a trifunctional minicellulosome consisting of a mini scaffolding containing a cellulose-binding domain and three cohesin modules, which was tethered to the cell surface through the yeast α-agglutinin adhesion receptor and three types of cellulases, EGII, CBHII, and BGL1, each bearing a carboxy-terminal dockerin. Surface display of the minicellulosomes gave the yeast cells the ability to simultaneously break down and ferment PASC to ethanol with a titer of 1.8 g/L. The close proximity and ordering of EGII and CBHII on the mini scaffolding enhanced the hydrolysis of PASC. Tsai et al. (2010) constructed yeast strains capable of either displaying a trifunctional scaffolding carrying three divergent cohesin domains from *Clostridium thermocellum*, *Clostridium cellulolyticum*, and *Ru. flavefaciens*, or secreting one of the three corresponding dockerin-tagged cellulases (EG from *C. thermocellum*, exoglucanase from *C. cellulolyticum*, or BGL from *Ru. flavefaciens*). By using a yeast consortium composed of one strain displaying the mini scaffolding and three strains secreting dockerin-tagged cellulases, the secreted cellulases were docked onto the displayed mini scaffolding in a highly organized manner. Furthermore, by adjusting the ratio of different populations in the consortium, cellulose hydrolysis and ethanol production could be successfully fine-tuned.

In order to establish an economically feasible ethanol production process, ethanol titers markedly increased from those presently achievable are required because of the high energy demands of the subsequent ethanol distillation process (Galbe and Zacchi, 2007). Increasing the final ethanol titer would have a significant effect on lowering capital and production costs due to the reduced size of the necessary equipment, such as fermentation tanks and distillation columns (Wingren et al., 2003). On the other hand, achieving higher titers inevitably requires increased loading of lignocellulose in the SSF and CBP processes. Generally, pretreatment of raw lignocellulosic biomass with diluted-acid, alkaline, liquid hot water, or steam explosions has yielded cellulosic materials in solid form that contain approximately 20–40% (w/w) dry matter (Tucker et al., 2003; Pérez et al., 2008; Zhang et al., 2010). Therefore, the fermentation of lignocellulose must be initiated with the solid form. However, operating the SSF at solid concentrations exceeding approximately 10% (w/w) poses a number of technical problems, as the water content of substrates is directly correlated to the rheology of the fermentation mixture, which is important for the interaction between lignocellulosic materials and cellulolytic enzymes (Pimenova and Hanley, 2004). Upon increasing the solid lignocellulose content in the bioreactor, solids and enzymes become insufficiently mixed. In addition, a low water content increases the viscosity of the substrate material, which also makes mixing within the bioreactor more difficult and leads to decreased ethanol yields. To overcome the problem, Matano et al. (2012) performed short-term liquefaction and subsequent fermentation in a novel drum-type rotary fermentation system using a yeast strain developed for cell-surface display of fungal cellulases to achieve high ethanol production from hydrothermally pretreated rice straw. In the recombinant yeast strain, three copies of *T. reesei* EG and CBHII genes and one copy of the *A. aculeatus BGL* gene were expressed. In the presence of 10 FPU/g-biomass added cellulase, the recombinant cellulolytic strain produced 1.4-fold higher ethanol (89% of the theoretical yield) from high-solid (200 g-dry-weight/L) rice straw within 72 hours of fermentation than the wild-type strain. Notably, cellulases displayed on the yeast cell surface were capable of hydrolyzing cellulose that was not hydrolyzed by commercial cellulases, leading to increased sugar utilization for improved ethanol production (Figure 8.3).

Surface-engineered yeast strains displaying cellulases have the following advantages: (1) close proximity of multiple cellulases on the cell surface enables synergistic hydrolysis of cellulose, which leads to increased sugar availability for ethanol production (Fujita et al., 2004; Matano et al., 2012); (2) glucose liberated from cellulose is concurrently taken up at the yeast cell surface so that the glucose concentration is maintained at low levels, which reduces both the risk of contamination by other glucose-dependent organisms and product inhibition of cellulases (Fujita et al., 2004); (3) since the steady-state concentration of glucose in the medium can be maintained near zero, glucose repression, which prevents the uptake, catabolism or both of non-glucose sugar is alleviated to facilitate consumption of xylose (Nakamura et al., 2008); (4) cellulolytic enzymes are genetically self-immobilized on the yeast cell surface so that the activities of the enzymes are retained as long as the yeast

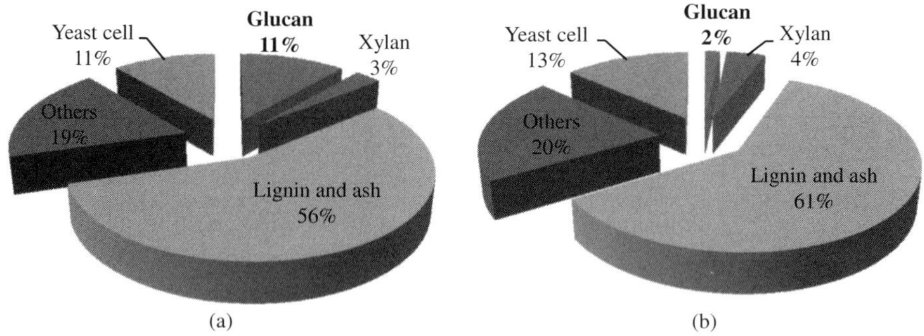

FIGURE 8.3 Composition of residual matter obtained after the fermentation of 200 g-dry-weight/L rice straw, hydrothermally pretreated (Matano et al., 2012), in the presence of 10 FPU/g-biomass cellulase with wild-type (a) and cellulase-displaying *Saccharomyces cerevisiae* strain, NBRC1440/B-EC3 (b). (*For a color version, see the color plate section.*)

continues to grow, while the activity of enzymes secreted into the medium is poorly maintained over prolonged reaction periods (Ueda and Tanaka, 2000; Matano et al., 2013b); (5) reutilization of the yeast cells enables reuse of the enzymes displayed on the cell surface without a need to reproduce the cells, which would reduce the cost of yeast propagation as well as enzyme addition (Kondo et al., 2002; Matano et al., 2013a).

For effective ethanol production, cell recycle batch fermentation (CRBF) is a promising strategy where cell recycling significantly reduces time and costs associated with inoculum preparation. However, in the case of the fermentation of lignocellulosic materials, it is impossible to retrieve microbial cells by conventional cell collection procedures because lignocellulosic residue, which is not utilized by microbes, is retained in the fermentation medium with yeast cells in the solid fraction after fermentation. Also, lignin left behind as solid residue in the fermentation medium inhibits the activity of cellulase, which should be removed for the efficient saccharification of cellulose in the subsequent batch. So far, CRBF for lignocellulosic materials has not yet been realized in the lignocellulose fermentation processes. Recently, Matano et al. (2013a) carried out a two-stage separation consisting of rough removal of lignocellulosic residue under low-speed centrifugation, and solid–liquid separation under high-speed centrifugation to collect yeast cells, enabling cell recycling for the repeated batch fermentation of lignocellulosic materials. Through the optimization of the first-stage centrifugal force, the fermentation capacity of *Sa. cerevisiae* was maintained for more than five cycles (Figure 8.4). Additionally, the display of cellulase on the yeast cell surface improved ethanol production. During the five consecutive batches, the average ethanol production from 200 g/L high-solid

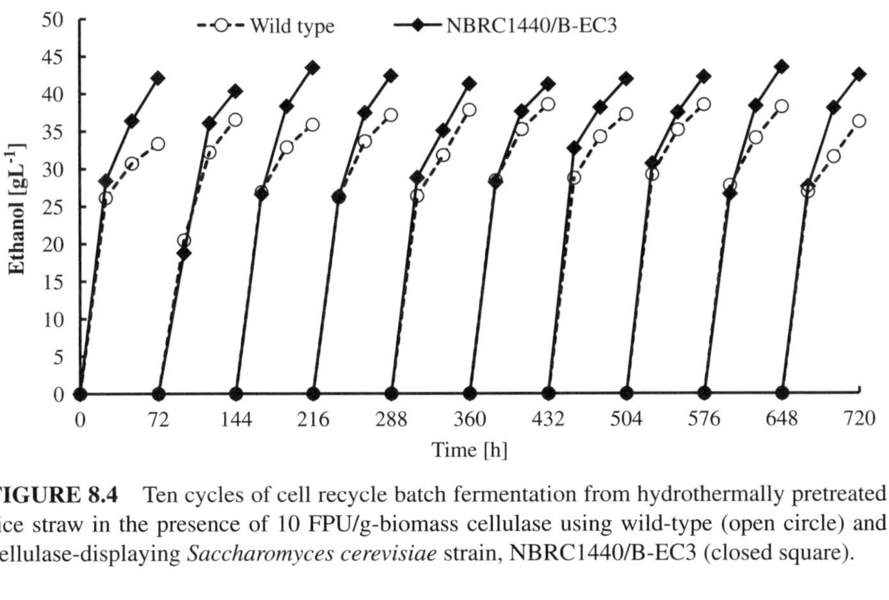

FIGURE 8.4 Ten cycles of cell recycle batch fermentation from hydrothermally pretreated rice straw in the presence of 10 FPU/g-biomass cellulase using wild-type (open circle) and cellulase-displaying *Saccharomyces cerevisiae* strain, NBRC1440/B-EC3 (closed square).

lignocellulosic material was 42.2 g/L, and the total ethanol yield was 86.3% of the theoretical yield based on biomass input.

8.3.2.3 Application of Cell-Surface Engineering to Thermotolerant Yeast

The difference in optimum temperature between saccharification and fermentation poses a drawback for efficient ethanol production by the SSF process. Saccharification with fungal cellulolytic enzymes is best done at around 50°C, while most fermenting microbes have an optimum temperature for ethanol fermentation of between 28°C and 37°C. Thus, heterologous production of cellulases has been pursued using thermotolerant yeast strains as hosts (Hasunuma and Kondo, 2012a). The application of thermotolerant yeast strains engineered according to the SSF process would overcome this drawback by allowing hydrolysis and fermentation to occur at elevated temperatures. Furthermore, ethanol production at elevated temperature has received much attention because of advantages such as reduction in the costs associated with cooling, continuous evaporation of ethanol from broth under reduced pressure, reduced risk of contamination, and suitability for application in tropical countries as well as the improvement of SSF efficiency.

Thermotolerant yeast strains of *Kluyveromyces*, *Saccharomyces*, and *Fabospora* genera that can produce more than 5% (w/v) ethanol at elevated temperature (>40°C) have been identified (Szczodrak and Targoński, 1988; D'Amore et al., 1989). *Candida glabrata*, *Hansenula polymorpha* (*Ogataea angusta*), and *Pichia kudriavzevii* (*Issatchenkia orientalis*) are also expected to be useful for the development of high-temperature SSF processes (reviewed in Hasunuma and Kondo, 2012a). In the presence of a small quantity of saccharification enzymes, a recombinant *P. kudriavzevii* expressing *A. aculeatus* BGL produced 29 g/L ethanol from 100 g/L Avicel microcrystalline cellulose within 72 hours at 40°C, without addition of BGL (Kitagawa et al., 2010). *Kluyveromyces marxianus* appears to be particularly promising in this regard. Many strains of *K. marxianus* grow well at temperatures as high as 45–52°C and can produce ethanol at temperatures between 38°C and 45°C. Moreover, *K. marxianus* offers additional benefits that include a high growth rate and the ability to utilize a wide variety of sugar substrates including arabinose, galactose, mannose, and xylose. Because of these advantages, *K. marxianus* has been used for ethanol production from industrially relevant substrates such as sugarcane juice, corn silage juice, molasses, and whey powder. Yanase et al. (2010a) engineered a *K. marxianus* strain for the display of cellulolytic enzymes on the cell surface. *Trichoderma reesei* EGII and *A. aculeatus* BGL1 were co-displayed on the cell surface of the recombinant *K. marxianus*, which produced 20.4 g/L ethanol from 53.4 g/L cellobiose at 45°C. The recombinant strains successfully converted β-glucan at 48°C to a titer of 4.24 g/L from 10 g/L within 12 hours in the absence of cellulase addition. The yield (in grams of ethanol produced per gram of β-glucan consumed) was 0.47 g/g. This study indicates that high-temperature cellulose fermentation to ethanol is efficiently accomplished using recombinant thermotolerant *K. marxianus* strains displaying thermostable cellulolytic enzymes on the cell surface. Engineering of *K. marxianus* would make a significant contribution to the improvement of CBP for bioethanol production.

8.3.3 Fermentation of Hemicellulosic Materials

8.3.3.1 Utilization of Hemicellulosic Materials Hemicellulose is the second most abundant constituent of lignocellulosic biomass and, if economic success of lignocellulosic ethanol is to be realized, is required to be utilized for ethanol production, along with cellulose. In order to directly convert hemicellulosic materials to ethanol, xylanolytic enzymes should be expressed using ethanol-producing yeast strains through genetic engineering. Furthermore, since *Sa. cerevisiae* cannot naturally utilize xylose, the most common pentose sugar in hemicellulosic materials, considerable efforts to engineer xylose metabolic pathways in *Sa. cerevisiae* have been demanded for hemicellulose utilization.

8.3.3.2 Xylose Fermentation The construction of an efficient xylose assimilation pathway in *Sa. cerevisiae* has been approached via heterologous expression of genes for xylose reductase (XR) and xylitol dehydrogenase (XDH) derived from *Scheffersomyces stipitis*, along with overexpression of *Sa. cerevisiae* xylulokinase (XK) to produce ethanol through xylose fermentation (Nevoigt, 2008; van Vleet and Jeffries, 2009). Xylose is first reduced to xylitol by XR, and then xylitol is oxidized to xylulose by XDH. Xylulose is phosphorylated by XK to xylulose-5-phosphate (X5P), which is then metabolized through the nonoxidative pentose phosphate pathway (PPP) and the glycolysis pathway. On the other hand, a xylose isomerase (XI) gene derived from anaerobic fungi such as *Piromyces* and *Orpinomyces* has been introduced into *Sa. cerevisiae* (van Maris et al., 2007; Madhavan et al., 2009a,b; Tanino et al., 2010). In the XI-based strain, heterologous expression of the bacterial XI gene in the recombinant *Sa. cerevisiae* has proved to be a challenging task marked by low XI activity in yeast cells (Hahn-Hägerdal et al., 2007). Heterologous expression of XR and XDH gave higher specific ethanol productivity as a result of a higher xylose consumption rate. However, xylitol formation arising from the cofactor imbalance between XR and XDH was higher for the XR/XDH-based strain than for the XI-based strain, where ethanol yield close to the theoretical value were achieved because of the avoidance of this cofactor imbalance (Karhumaa et al., 2007). Thus, the mere introduction of xylose utilization pathways into *Sa. cerevisiae* was not enough to confer upon the recombinant strains, the necessary traits for efficient ethanol fermentation. A number of engineering strategies to enhance ethanolic xylose fermentation ability by recombinant *Sa. cerevisiae* strains have been advanced; they include (1) introducing an efficient xylose transporter to bring more xylose into cell; (2) relieving the intracellular redox imbalance; and (3) boosting the PPP through genetic engineering (Kondo et al., 2013). In a recent metabolic modification, deletion of the *pho13* gene encoding *p*-nitrophenyl phosphatase increased the capacities of the recombinant XR/XDH-based *Sa. cerevisiae* to grow on and ferment xylose (van Vleet and Jeffries, 2009; Fujitomi et al., 2012). Despite substrate and intracellular functions of the PHO13 protein, DNA microarray analysis demonstrated that central metabolism, including *ZWF1*, *SOL3*, and *ADH1*, was increased by deletion of the *PHO13* gene. In contrast, the expression levels of respiratory chain-related genes such as *COX2* and *CYC1*, and mitochondrial proton-transporting ATP synthase genes in the Δ*PHO13* mutant were lower than those in the control strain (Fujitomi et al., 2012). While there have been numerous efforts to improve ethanolic xylose fermentation, comprehensive analysis based on biological profiling would yield insight into the key requirements for metabolic engineering.

8.3.3.3 Fermentation of Hemicellulosic Materials to Ethanol Hemicellulose consists of a xylan backbone of β-1,4-linked xylopyranoside partially substituted with acetyl, glucuronosyl, and arabinosyl side chains. Xylan is hydrolyzed to xylooligosaccharides by endoxylanase, after which β-xylosidase hydrolyzes xylooligosaccharides to release xylose. A number of bacterial and fungal species are able to utilize xylan as a carbon source (Jeffries, 1983); however, *Sa. cerevisiae* cannot. Many researchers have studied the production of xylanolytic enzymes in the yeasts such as *Sa. cerevisiae* and *Sc. stipitis*.

Katahira et al. (2004) displayed both *T. reesei* xylanase, XYNII, and *A. oryzae* β-xylosidase on the cell surface of *Sa. cerevisiae*, which enabled hydrolysis of xylan to xylose by *Sa. cerevisiae*. They also introduced XR, XDH, and XK into a xylan-degrading yeast strain (Figure 8.5). The resultant strain produced 7.1 g/L ethanol from birchwood xylan with a yield of 0.30 g ethanol per gram carbohydrate. Recently, Sakamoto et al. (2011) successfully produced ethanol from a rice straw hydrolysate obtained by liquid hot water treatment. The treatment solubilizes the hemicellulosic component of rice straw to yield liquid-soluble fractions containing not only cellooligosaccharide but also xylan and xyloglucan. Accordingly, recombinant XR/XDH/XK-based *Sa. cerevisiae* strains co-displaying *A. aculeatus* BGL, *A. oryzae* β-xylosidase, and *T. reesei* XYNII on the cell surface were developed, which

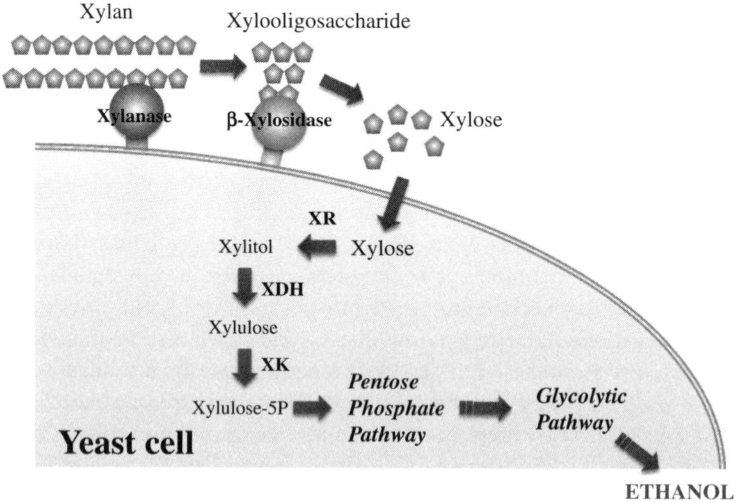

FIGURE 8.5 A recombinant yeast strain displaying hemicellulolytic enzymes such as xylanase and β-xylosidase on the cell surface and expressing xylose-assimilating enzymes such as XR, XDH, and XK. (*For a color version, see the color plate section.*)

directly produce ethanol from the hemicellulosic materials without a requirement for addition of sugar-hydrolyzing enzymes or detoxification of the hydrolysate. The ethanol yield was 0.41 g ethanol per gram sugar consumed, which corresponded to 82% of the theoretical yield.

Xylanolytic enzymes have also been expressed in *Sc. stipitis*, a species with the ability to ferment xylose. The direct production of ethanol from xylan was performed using a recombinant *Sc. stipitis* strain secreting *Cryptococcus albidus* xylanase; however, the ethanol productivity and yield were not reported (Morosoli et al., 1993). Voronovsky et al. (2009) engineered a thermotolerant yeast, *H. polymorpha*, coexpressing *T. reesei* endoxylanase and *Aspergillus niger* β-xylosidase by integrating these genes into the genome of *H. polymorpha* with the promoter of the *H. polymorpha* glyceraldehyde-3-phosphate dehydrogenase gene. The resulting strains were capable of growth and ethanol production at 48°C on a minimal medium supplemented with birchwood xylan as the sole carbon source.

8.3.4 Ethanol Production in the Presence of Fermentation Inhibitors

To exploit lignocellulosic biomass for ethanol production, the fermenting yeast must utilize hexose and pentose sugars in the presence of toxic compounds such as acetic acid, formic acid, furfural, and 5-hydroxymethylfurfural released during the process of biomass pretreatment that can decrease ethanol yield and productivity and also disturb cell growth (Hasunuma and Kondo, 2012b; Madhavan et al., 2012). In particular, xylose utilization by recombinant *Sa. cerevisiae* strains was severely affected in the presence of such fermentation inhibitors. Hence, in order to improve fermentation capacity of yeast strains in the presence of toxic compounds, several strategies have been applied to overcome the inhibitory effect. These approaches include evolutionary strategies such as the improvement of natural tolerance by controlling inhibitor concentrations during the fermentation (Martín et al., 2007) and a mutagenesis and genome-shuffling approach (Liu et al., 2009; Zheng et al., 2011). On the other hand, overexpression of genes encoding enzymes that confer resistance towards specific inhibitors has been rare to date due to lack of knowledge about the physiological basis of yeast tolerance to toxic compounds. Therefore, the mechanisms of yeast tolerance and adaptation toward fermentation inhibitors have been the subject of numerous studies with a view to improvement of inhibitor tolerance through metabolic engineering.

Multi-omics analysis methods such as genome-wide approaches, proteomics, and metabolomics have been recently exploited to gain insight into the molecular and genetic traits involved in tolerance and adaptation to fermentation inhibitors (reviewed in Hasunuma and Kondo, 2012b). The functional basis of furfural tolerance was elucidated by comparing a furfural-adapted *Sa. cerevisiae* strain and its parent by means of a metabolic flux and global transcript analysis (Heer et al., 2009), the results of which suggested that the NADPH-dependent oxidoreductases are central to the major resistance mechanism with increased NADPH-generating flux through the PPP. The effect of acetic acid on xylose fermentation by the recombinant xylose-fermenting *Sa. cerevisiae* has been analyzed by a metabolomic approach (Hasunuma et al., 2011a).

Metabolic profiling in *Sa. cerevisiae* cells revealed metabolites involved in nonoxidative PPP significantly accumulated upon addition of acetic acid, indicating that acetic acid may slow down the flux of the pathway. Indeed, overexpression of the *TAL1* gene, coding for transaldolase, reduced the inhibition of the fermentation capacity by weak acids such as acetic and formic acids. DNA microarray analysis has been applied to survey the unpredictable transcriptional response of *Sa. cerevisiae* to stress (Li and Yuan, 2010). For example, expression of the *FDH1*, *FDH2*, *ALD5*, *HXK2*, *VCX1*, *GPD2*, and *HXT4* genes in *Sa. cerevisiae* cells was upregulated in response to increased formic acid concentration (Hasunuma et al., 2011b), which might be related to the defense mechanism of *Sa. cerevisiae*. Since formic acid breakdown converts NADH to NAD^+, NAD^+-dependent reductases such as *ALD5* and *GPD2* might be induced to maintain the intracellular redox balance. *HXT4* and *HXK2*, encoding a hexose transporter and hexokinase, respectively, might be upregulated to produce more ATP. To reinforce the capability for formic acid breakdown, *FDH1* was overexpressed in the xylose-fermenting *Sa. cerevisiae* strain. This modification allowed the yeast to rapidly decompose excess formic acid in order to effectively produce ethanol from xylose (Hasunuma et al., 2011b). Thus, multi-omics is a powerful tool to gain insight into the effects of environmental perturbation on microbial metabolism and to then develop rational strategies for conferring stress tolerance through genetic engineering. On the other hand, the information available from multi-omics represents only a snapshot in time taken at the moment of sampling the metabolites. To further our understanding, dynamic metabolic flux should be determined to address this issue by means of kinetic modeling and stable isotope labeling.

8.4 ECONOMIC OUTLOOK

Ethanol could be produced from ethylene derived from petroleum via a chemical synthesis route, or from sugars derived from biomass by biosynthesis. To produce ethanol by chemical synthesis, petroleum-derived ethylene is hydrolyzed using sulfuric acid as catalyst. On the other hand, using biosynthesis, sugars are first produced from starchy biomass such as corn or lignocellulosic biomass such as switchgrass by enzymatic saccharification, or are directly extracted from sugar-containing biomass such as sugar cane. The resultant sugars are then fermented into ethanol by ethanol-producing yeasts such as *Sa. cerevisiae*, *Sc. stipitis*, and *Candida shehatae*. There are many concerns, such as exhaustion of petroleum and environmental issues, associated with producing petroleum-based ethanol. By contrast, because ethanol can be produced under moderate conditions with a relatively high yield using ethanol-producing yeast as a biocatalyst, biosynthesized ethanol should be more compatible with the environment. Before the fermentation industry was well established, petroleum-based ethanol synthesized from ethylene was the predominant source of industrial ethanol. However, biomass-based ethanol has become predominant in recent times. The amount of biomass-based ethanol production was 4 million tons per year, and for petroleum-based ethanol 1 million tons per year, in 2000 in the United States (Demain, 2000). The production cost of petroleum-based ethanol has remained relatively stable, but

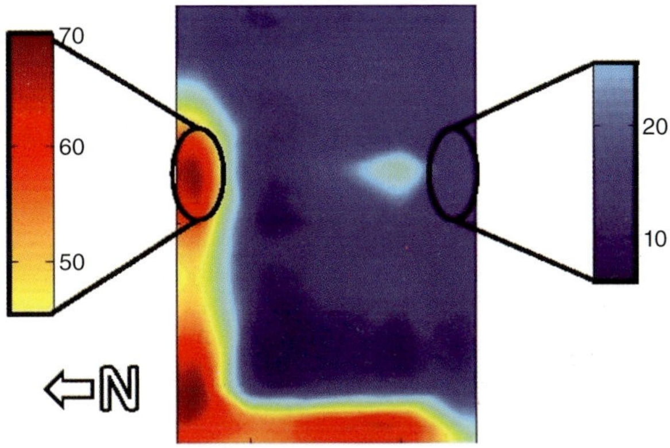

FIGURE 2.10 Isopleth moisture distribution of an end view of a stack of 3 × 4 × 8-ft large square bales (stacked 1-bale wide × 4 bales high) showing the influence of Northern exposure on bale moisture.

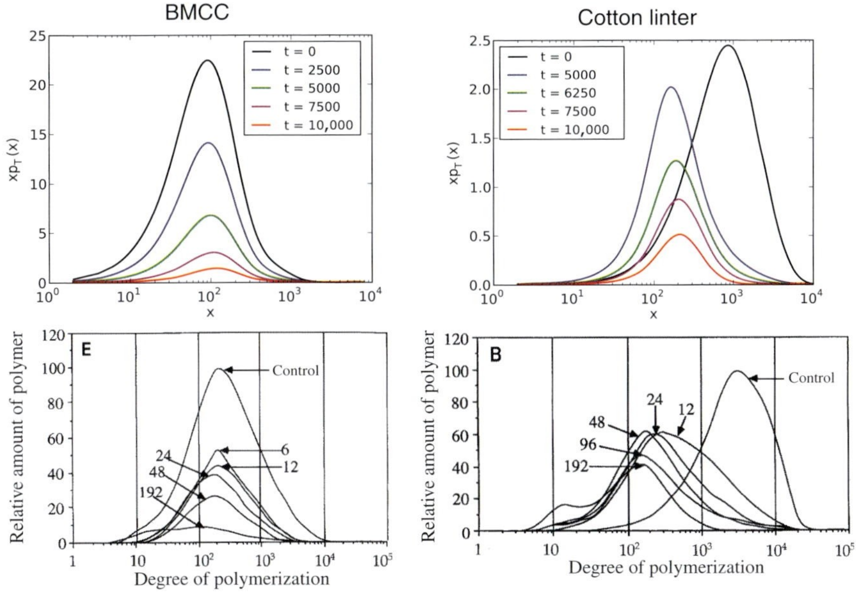

FIGURE 4.3 Qualitative comparison of mechanistic model predictions (top row, from Griggs et al. (2012b)) with experimental results (bottom row, from Srisodsuk et al. (1998)) for the changing DP distribution of cellulose during enzymatic hydrolysis by EG_I and CBH_I. The left column compares results for bacterial microcrystalline cellulose (BMCC) with a relatively low initial DP, and the right column compares results for cotton linter with a relatively high initial DP.

Bioprocessing of Renewable Resources to Commodity Bioproducts, First Edition.
Edited by Virendra S. Bisaria and Akihiko Kondo.

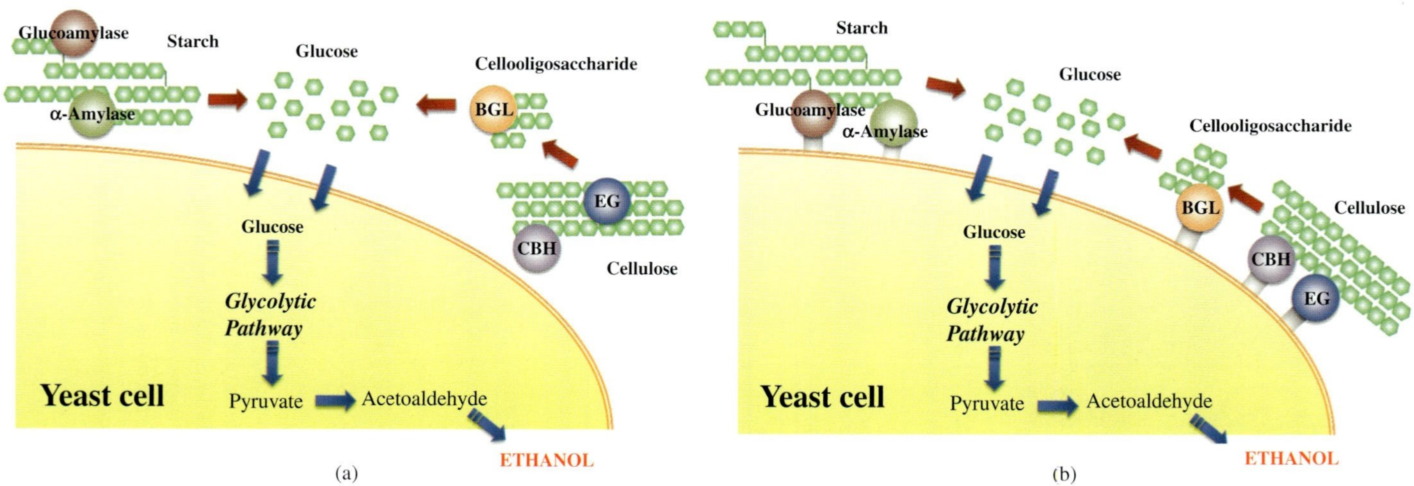

FIGURE 8.2 Direct conversion of starchy and lignocellulosic biomass to ethanol using enzyme-secreting (a) or enzyme surface–displaying (b) yeast cells.

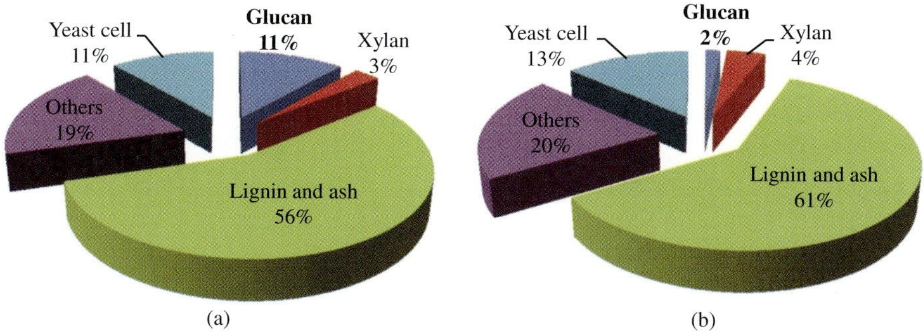

FIGURE 8.3 Composition of residual matter obtained after the fermentation of 200 g-dry-weight/L rice straw, hydrothermally pretreated (Matano et al., 2012), in the presence of 10 FPU/g-biomass cellulase with wild-type (a) and cellulase-displaying *Saccharomyces cerevisiae* strain, NBRC1440/B-EC3 (b).

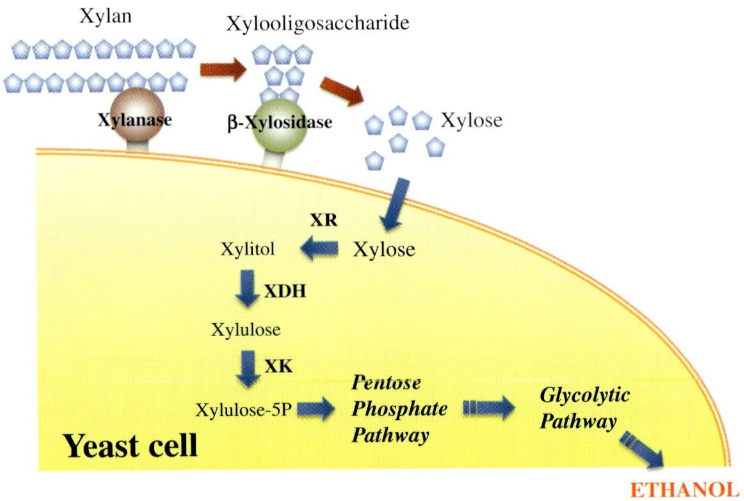

FIGURE 8.5 A recombinant yeast strain displaying hemicellulolytic enzymes such as xylanase and β-xylosidase on the cell surface and expressing xylose-assimilating enzymes such as XR, XDH, and XK.

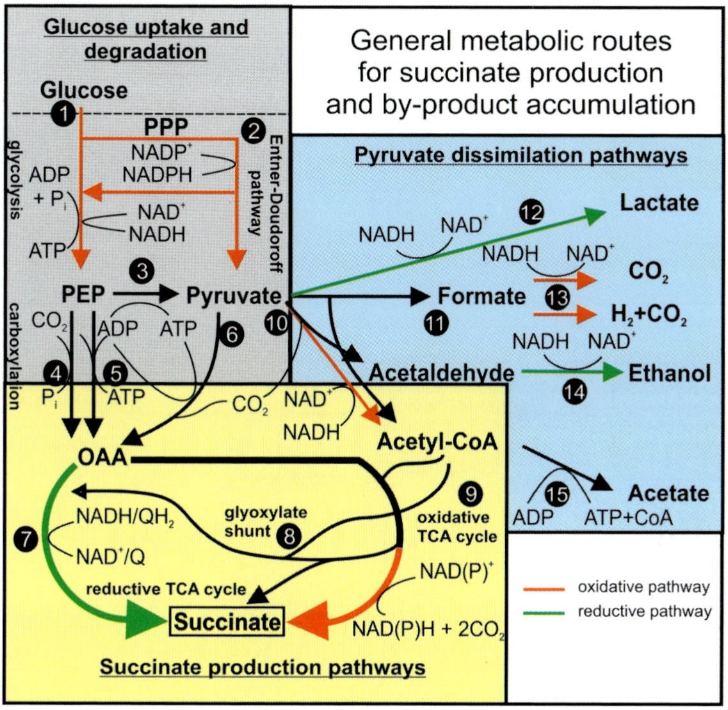

FIGURE 16.2 Metabolic pathways relevant in the context of succinate production from glucose with focus on the type of reaction or reaction sequence—oxidative reactions are indicated by red arrows, reductive reactions by green arrows. The cofactors involved are also shown. The pathways are grouped according to their major purpose in three boxes: grey box—initial substrate degradation pathways, yellow box—succinate-producing pathways, blue box—pyruvate dissimilation pathways. The reactions or reaction sequences are numbered as follows: 1, glucose uptake and glycolysis; 2, oxidative pentose phosphate pathway (PPP) branching with the Entner-Doudoroff pathway; 3, pyruvate kinase; 4, PEP carboxylase; 5, PEP carboxykinase (note that some enzymes used GDP/GTP rather than ADP/ATP); 6, pyruvate carboxylase; 7, reductive TCA cycle including malate dehydrogenase, fumarase, and fumarate reductase or succinate:quinone oxidoreductase; 8, glyoxylate shunt including isocitrate lyase and malate synthase; 9, oxidative TCA cycle including citrate synthase, aconitase, isocitrate dehydrogenase, 2-oxoglutarate dehydrogenase complex, and succinyl-CoA synthetase; 10, pyruvate dehydrogenase complex, pyruvate decarboxylase; 11, pyruvate formate lyase; 12, lactate dehydrogenase; 13, formate dehydrogenase (NAD^+ dependent) or formate hydrogen lyase; 14, alcohol dehydrogenase; 15, phosphotransacetylase and acetate kinase.

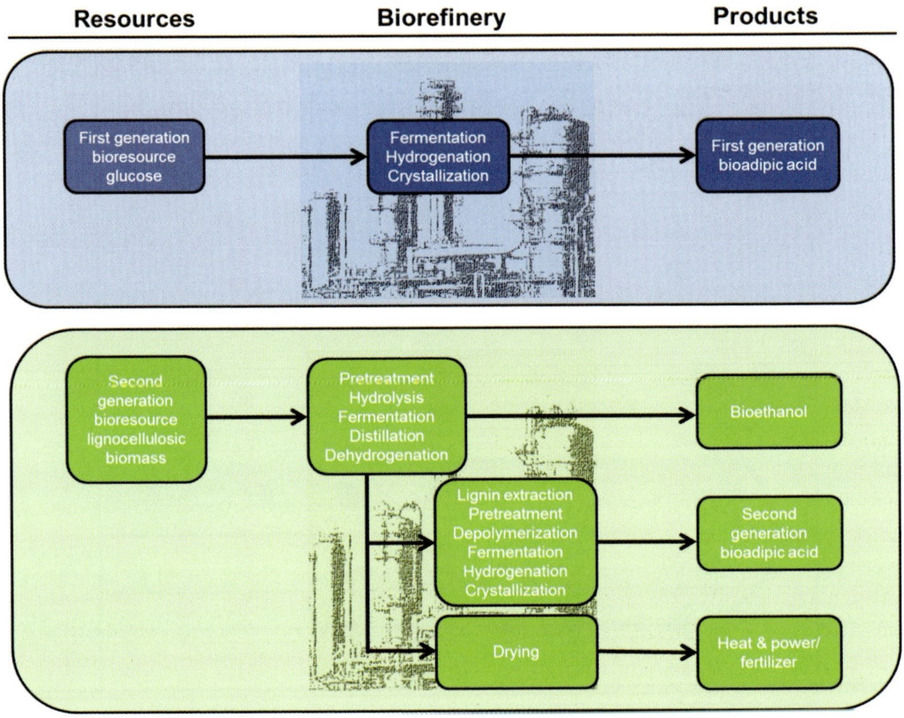

FIGURE 19.3 Biorefinery concepts that integrate the first and second generation bio-adipic acid.

the cost of biomass-based ethanol has decreased drastically. Hong Thuy et al. (2011) reported that the production cost of petroleum-based ethanol was \$1.3/kg, and the cost of biomass-based ethanol was \$0.76/kg in 2009, in the United States. Although large amounts of bioethanol (9 billion gallons per year in 2008) are produced from corn in the United States, exhaustion of food crops will cause many difficulties such as escalation of food prices. Hence, the United States has launched a project where, by 2030, more than 38 billion gallons of biofuel will be produced with less than half produced from corn grain, and large amounts of ethanol are to be produced from lignocellulosic biomass (Kazi et al., 2010).

As described previously, there are many steps in the production of ethanol from lignocellulosic biomass. Gonzalez et al. (2011) reported that the main cost drivers in producing ethanol from lignocellulosic biomass are feedstock, enzyme, tax, fuel (gasoline), depreciation, and labor. Gnansounou and Dauriat (2010) carried out a techno-economic assessment of lignocellulosic ethanol production costs from the standpoint of feedstock. They evaluated the production costs from four kinds of typical feedstock: straw, eucalyptus, poplar, and switchgrass, taking into account feedstock transport costs, feedstock yield, ethanol yield, electricity cost and so on. As a result, the cost from switchgrass (\$0.77/L) was the highest and that from eucalyptus (\$0.56/L) was the lowest. Because ethanol production cost varied drastically even among four typical kinds of feedstock, the selection of feedstock suitable for region and country would appear to be an important consideration to realize cost-effective lignocellulosic ethanol production.

Enzyme usage is the second highest cost driver in lignocellulosic ethanol production. Because lignocellulosic biomass is used for many kinds of biofuels and bio-based chemicals, cellulases are currently the third largest industrial enzyme produced in the world (Wilson, 2009). Most industrial cellulases are produced by cellulolytic fungi such as *T. reesei* and *Humicola insolens*. The Department of Energy (DOE) in the United States has been supporting research by enzyme-producing companies such as Genencor and Novozymes to reduce enzyme production costs for over several decades. As a result, enzyme costs have been drastically cut. Although information is limited and accurate calculation is difficult, enzyme cost is estimated at about 50 cents per gallon of ethanol.

Usage of gasoline, which also greatly influences lignocellulosic ethanol production costs, can be adjusted depending on the pretreatment method. Kazi et al. (2010) evaluated the typical costs of biomass pretreatment methods such as diluted acid, hot water, and ammonia fiber expansion. Their evaluation was carried out taking into account capital investment, installation costs, ethanol yield, electricity export and so on. Diluted acid pretreatment showed the lowest production cost (\$1.36/L of gasoline equivalent [LGE]), and hot water pretreatment showed the highest production cost (\$1.77/LGE). It should be noticed that this is a techno-economic assessment calculated based on data from the literature, and the most cost-effective pretreatment is likely to depend on many factors such as feedstock and pretreatment conditions.

There are many reports concerning the techno-economic assessment of lignocellulosic ethanol production costs. However, industrial lignocellulosic ethanol production is a developing industry and sufficient data for techno-economic assessment is lacking. Most reports have attempted to offset missing data with many valid hypothesizes. More accurate techno-economic assessments are likely to be conducted as the lignocellulosic ethanol production industry matures.

8.5 FUTURE PROSPECTS

Acquisition of biomass hydrolysis activity in yeast through genetic engineering has paved the way to achieving a groundbreaking process for ethanol production from starchy and lignocellulosic biomass by SSF and CBP. In particular, cell surface engineering is a powerful tool to effectively hydrolyze recalcitrant cellulosic materials by displaying various cellulolytic enzymes on the yeast cell surface. By displaying novel enzymes on the cell surface, efficient whole-cell biocatalysts are constructed. The displayed enzymes may be regarded as a kind of self-immobilized enzyme on the cell surface. Although the use of recombinant microbes requires investment costs for their physical separation, CBP has an important benefit in the elimination of the enzyme production process. To further engineer cellulolytic recombinant strains for use in ethanol production, system-wide modifications of intracellular metabolism using advanced engineering tools such as minimal hosts, vectors, genetic controllers, and characterized enzymes are needed, which would improve the potential of not only target productivities but also cell growth and viability during the fermentation. Recent advances in multi-omics techniques promise to offer powerful insight into the effects of environmental perturbation on yeast metabolism, and then to design rational strategies for further engineering of yeast strains through metabolic engineering. Comparative genome analysis using resequenced genome information could identify causal genes and polymorphisms responsible for specific metabolic functions in non-model industrial yeast strains. Transcriptomics and metabolomics promises to become a useful diagnostic tool to illuminate bottlenecks or problems in cellular processes that could be overcome by engineering to improve microbe performance. By employing system-wide modification of intracellular metabolism as well as advanced cell engineering technologies, a super yeast with extreme performance for cost-effective bioethanol production could be designed and developed.

REFERENCES

Chandel AK, Chandrasekhar G, Silva MB, Silvério da Silva S. (2012). The realm of cellulases in biorefinery development. *Crit Rev Biotechnol*, 32, 187–202.

D'Amore T, Celotto G, Russell I, Stewart GG. (1989). Selection and optimization of yeast suitable for ethanol production at 40°C. *Enzyme Microb Technol*, 11, 411–416.

Demain AL. (2000). Small bugs, big business: the economic power of the microbe. *Biotechnol Adv*, 18, 499–514.

den Haan R, Rose SH, Lynd LR, van Zyl WH. (2007). Hydrolysis and fermentation of amorphous cellulose by recombinant *Saccharomyces cerevisiae*. *Metab Eng*, 9, 87–94.

Eksteen JM, ran Rensburg P, Otero RRC, Pretorius IS. (2003). Starch fermentation by recombinant *Saccharomyces cerevisiae* strains expressing the α-amylase and glucoamylase genes from *Lipomyces kononenkoae* and *Saccharomyces fibuligera*. *Biotechnol Bioeng*, 84, 639–646.

Fujita Y, Takahashi S, Ueda M, Tanaka A, Okada H, Morikawa Y, Kawaguchi T, Arai M, Fukuda H, Kondo A. (2002). Direct and efficient production of ethanol from cellulosic material with a yeast strain displaying cellulolytic enzymes. *Appl Environ Microbiol*, 68, 5136–5141.

Fujita Y, Ito J, Ueda M, Fukuda H, Kondo A. (2004). Synergistic saccharification, and direct fermentation to ethanol of amorphous cellulose by use of engineered yeast strain codisplaying three types of cellulolytic enzyme. *Appl Environ Microbiol*, 70, 1207–1212.

Fujitomi K, Sanda T, Hasunuma T, Kondo A. (2012). Deletion of the PHO13 gene in *Saccharomyces cerevisiae* improves ethanol production from lignocellulosic hydrolysate in the presence of acetic and formic acids, and furfural. *Bioresour Technol*, 111, 161–166.

Galazka JM, Tian C, Beeson WT, Martinez B, Glass NL, Cate JHD. (2010). Cellodextrin transport in yeast for improved biofuel production. *Science*, 330, 84–86.

Galbe M, Zacchi G. (2007). Pretreatment of lignocellulosic materials for efficient bioethanol production. *Adv Biochem Eng Biotechnol*, 108, 41–65.

Gnansounou E. (2009). Fuel Ethanol. In: Pandey A, editor. *Handbook of Plant-Based Biofuels*. CRC Press.

Gnansounou E, Dauriat A. (2010). Techno-economic analysis of lignocellulosic ethanol: a review. *Bioresour Technol*, 101, 4980–4991.

Gonzalez R, Treasure T, Phillips R, Jameel H, Saloni D, Abt R, Wright J. (2011). Converting Eucalyptus biomass into ethanol: financial and sensitivity analysis in a co-current dilute acid process. Part II. *Biomass Bioenergy*, 35, 767–772.

Hahn-Hägerdal B, Karhumaa K, Jeppsson M, Gorwa-Grauslund MF. (2007). Metabolic engineering for pentose utilization in *Saccharomyces cerevisiae*. *Adv Biochem Eng Biotechnol*, 108, 147–177.

Hasunuma T, Kondo A. (2012a). Consolidated bioprocessing and simultaneous saccharification and fermentation of lignocellulose to ethanol with thermotolerant yeast strains. *Process Biochem*, 47, 1287–1294.

Hasunuma T, Kondo A. (2012b). Development of yeast cell factories for consolidated bioprocessing of lignocellulose to bioethanol through cell surface engineering. *Biotechnol Adv*, 30, 1207–1218.

Hasunuma T, Sanda T, Yamada R, Yoshimura K, Ishii J, Kondo A. (2011a). Metabolic pathway engineering based on metabolomics confers acetic and formic acid tolerance to a recombinant xylose-fermenting strain of *Saccharomyces cerevisiae*. *Microb Cell Fact*, 10, 2.

Hasunuma T, Sung KM, Sanda T, Yoshimura K, Matsuda F, Kondo A. (2011b). Efficient fermentation of xylose to ethanol at high formic acid concentrations by metabolically engineered *Saccharomyces cerevisiae*. *Appl Microbiol Biotechnol*, 90, 997–1004.

Heer D, Heine D, Sauer U. (2009). Resistance of *Saccharomyces cerevisiae* to high concentrations of furfural is based on NADPH-dependent reduction by at least two oxidoreductase. *Appl Environ Microbiol*, 75, 7631–7638.

Hong Thuy NT, Kikuchi Y, Sugiyama H, Noda M, Hirao M. (2011). Techno-economic and environmental assessment of bioethanol-based chemical process: a case study on ethyl acetate. *Environ Prog Sustain Energy*, 30, 675–684.

Ito J, Kosugi A, Tanaka T, Kuroda K, Shibasaki S, Ogino C, Ueda M, Fukuda H, Doi RH, Kondo A. (2009). Regulation of the display ratio of enzymes on the *Saccharomyces cerevisiae* cell surface by the immunoglobulin G and cellulosomal enzyme binding domains. *Appl Environ Microbiol*, 75, 4149–4154.

Jeffries TW. (1983). Utilization of xylose by bacteria, yeasts, and fungi. *Adv Biochem Eng Biotechnol*, 27, 1–32.

Jeon E, Hyeon JE, Eun LS, Park BS, Kim SW, Lee J, Han SO. (2009). Cellulosic alcoholic fermentation using recombinant *Saccharomyces cerevisiae* engineered for the production of *Clostridium cellulovorans* endoglucanase and *Saccharomycopsis fibuligera* β-glucosidase. *FEMS Microbiol Lett*, 301, 130–136.

Karhumaa K, Sanchez RG, Hahn-Hägerdal B, Gorwa-Grauslund MF. (2007). Comparison of the xylose reductase-xylitol dehydrogenase and the xylose isomerase pathways for xylose fermentation by recombinant *Saccharomyces cerevisiae*. *Microb Cell Fact*, 6, 5.

Katahira S, Fujita Y, Mizukie A, Fukuda H, Kondo A. (2004). Construction of a xylan fermenting yeast strain through codisplay of xylanolytic enzymes on the surface of xylose utilizing *Saccharomyces cerevisiae* cells. *Appl Environ Microbiol*, 70, 5407–5414.

Kazi FK, Fortman JA, Anex RP, Hsu DD, Aden A, Dutta A, Kothandaraman G. (2010). Techno-economic comparison of process technologies for biochemical ethanol production from corn stover. *Fuel*, 89, 20–28.

Kitagawa T, Tokuhiro K, Sugiyama H, Kohda K, Isono N, Hisamitsu M, Takahashi H, Imaeda T. (2010). Construction of β-glucosidase expression system using the multistress-tolerant yeast *Issatchenkia orientalis*. *Appl Microbiol Biotechnol*, 87, 1841–1853.

Kondo A, Ishii J, Hara KY, Hasunuma T, Matsuda F. (2013). Development of microbial cell factories for bio-refinery through synthetic bioengineering, J Biotechnol, 163, 204–216.

Kondo A, Ueda M. (2004). Yeast cell-surface display-application of molecular display. *Appl Microbiol Biotechnol*, 64, 28–40.

Kondo A, Shigechi H, Abe M, Uyama K, Matsumoto T, Takahashi S, Ueda M, Tanaka A, Kishimoto M, Fukuda H. (2002). High level production from starch by a flocculent *Saccharomyces cerevisiae* strain displaying cell surface glucoamylase. *Appl Microbiol Biotechnol*, 58, 291–296.

Kotaka A, Bando H, Kaya M, Kato-Murai M, Kuroda K, Sahara H, Hata Y, Kondo A, Ueda M. (2008). Direct ethanol production from barley β-glucan by sake yeast displaying *Aspergillus oryzae* β-glucosidase and endoglucanase. *J Biosci Bioeng*, 105, 622–627.

Lesage G, Bussey H. (2006). Cell wall assembly in *Saccharomyces cerevisiae*. *Microbiol Mol Biol Rev*, 70, 317–343.

Li BZ, Yuan YJ. (2010). Transcriptome shifts in response to furfural and acetic acid in *Saccharomyces cerevisiae*. *Appl Microbiol Biotechnol*, 86, 1915–1924.

Lilly M, Fierobe HP, van Zyl WH, Volschenk H. (2009). Heterologous expression of a Clostridium minicellulosome in *Saccharomyces cerevisiae*. *FEMS Yeast Res*, 9, 1236–1249.

Liu ZL, Ma M, Song M. (2009). Evolutionarily engineered ethanologenic yeast detoxifies lignocellulosic biomass conversion inhibitors by reprogrammed pathways. *Mol Genet Genomics*, 282, 233–244.

Madhavan A, Tamalampudi S, Srivastava A, Fukuda H, Bisaria VS, Kondo A. (2009a). Alcoholic fermentation of xylose and mixed sugars using recombinant *Saccharomyces cerevisiae* engineered for xylose utilization. *Appl Microbiol Biotechnol*, 82, 1037–1047.

Madhavan A, Tamalampudi S, Ushida K, Kanai D, Katahira S, Srivastava A, Fukuda H, Bisaria VS, Kondo A. (2009b). Xylose isomerase from polycentric fungus *Orpinomyces*: gene sequencing, cloning and expression in *Saccharomyces cerevisiae* for bioconversion of xylose to ethanol. *Appl Microbiol Biotechnol*, 82, 1067–1078.

Madhavan A, Srivastava A, Kondo A, Bisaria VS. (2012). Bioconversion of lignocellulose-derived sugars to ethanol by engineered *Saccharomyces cerevisiae*. *Crit Rev Biotechnol*, 32, 22–48.

Martín C, Marcet M, Almazán O, Jönsson LJ. (2007). Adaptation of a recombinant xylose-utilizing *Saccharomyces cerevisiae* to a sugarcane bagasse hydrolysate with high content of fermentation inhibitors. *Bioresour Technol*, 98, 1767–1773.

Matano Y, Hasunuma T, Kondo A. (2013a). Cell recycle batch fermentation of high-solid lignocellulose using a recombinant cellulase-displaying yeast strain for high yield ethanol production in consolidated bioprocessing. *Bioresour Technol*, 135,403–409.

Matano Y, Hasunuma T, Kondo A. (2012). Display of cellulases on the cell surface of *Saccharomyces cerevisiae* for high yield ethanol production from high-solid lignocellulosic biomass. *Bioresour Technol*, 108, 128–133.

Matano Y, Hasunuma T, Kondo A. (2013b). Simultaneous improvement of saccharification and ethanol production from crystalline cellulose by alleviation of irreversible adsorption of cellulase with a cell surface-engineered yeast strain. *Appl Microbiol Biotechnol*, 97, 2231–2237.

Matsumoto N, Fukunishi O, Miyanaga M, Kakihara K, Nakajima E, Yoshizumi H. (1982). Industrialization of a non-cooking system for alcoholic fermentation from grains. *Agric Biol Chem*, 46, 1549–1558.

Menon V, Rao M. (2012). Trends in bioconversion of lignocellulose: biofuels, platform chemicals & biorefinery concept. *Progr Energ Combust Sci*, 38, 522–550.

Morosoli R, Zalce E, Durand S. (1993). Secretion of a *Cryptococcus albidus* xylanase in *Pichia stipitis* resulting in a xylan fermenting transformant. *Curr Genet*, 24, 94–99.

Murai T, Ueda M, Yamamura Y, Atomi H, Shibasaki Y, Kamasawa N, Osumi M, Amachi T, Tanaka A. (1997). Construction of a starch utilizing yeast by cell surface engineering. *Appl Environ Microbiol*, 63, 1362–1366.

Mussatto SI, Dragone G, Guimarães PM, Silva JP, Carneiro LM, Roberto IC, Vicente A, Domingues L, Teixeira JA. (2010). Technological trends, global market, and challenges of bio-ethanol production. *Biotechnol Adv*, 28, 817–830.

Nair SU, Ramachandran S, Pandey A. (2009). Bioethanol from starchy biomass: Part I Production of starch saccharifying enzymes. In: Pandey A, editor. *Handbook of Plant-Based Biofuels*. CRC Press.

Nakamura N, Yamada R, Katahira S, Tanaka T, Fukuda H, Kondo A. (2008). Effective xylose/cellobiose co-fermentation and ethanol production by xylose-assimilating *S. cerevisiae* via expression of β-glucosidase. *Enzyme Microb Technol*, 43, 233–236.

Nevoigt E. (2008). Progress in metabolic engineering of *Saccharomyces cerevisiae*. *Microbiol Mol Biol Rev*, 72, 379–412.

Olson DG, McBride JE, Shaw AW, Lynd LR. (2011). Recent progress in consolidated bioprocessing. *Curr Opin Biotechnol*, 23, 1–10.

Pérez JA, Ballesteros I, Ballesteros M, Sáez F, Negro MJ, Manzanares P. (2008). Optimizing liquid hot water pretreatment conditions to enhance sugar recovery from wheat straw for fuel-ethanol production. *Fuel*, 87, 3640–3647.

Pimenova NV, Hanley TR. (2004). Effect of corn stover concentration on rheological characteristics. *Appl Biochem Biotechnol*, 113–116, 347–360.

Pittet M, Conzelmann A. (2007). Biosynthesis and function of GPI proteins in the yeast *Saccharomyces cerevisiae. Biochim Biophys Acta*, 1771, 405–420.

Ragauskas AJ, Williams CK, Davison BH, Britovsek G, Cairney J, Eckert CA, Frederick Jr WJ, Hallett JP, Leak DJ, Liotta CL, Mielenz JR, Murphy R, Templer R, Tschaplinski T. (2006). The path forward for biofuels and biomaterials. *Science*, 311, 484–489.

Saha BC. (2003). Hemicellulose bioconversion. *J Ind Microbiol Biotechnol*, 30, 279–291.

Sakamoto T, Hasunuma T, Hori Y, Yamada R, Kondo A. (2011). Direct ethanol production from hemicellulosic materials of rice straw by use of an engineered yeast strain codisplaying three types of hemicellulolytic enzymes on the surface of xylose-utilizing *Saccharomyces cerevisiae* cells. *J Biotechnol*, 158, 203–210.

Sánchez OJ, Cardona CA. (2008). Trends in biotechnological production of fuel ethanol from different feedstocks. *Bioresour Technol*, 99, 5270–5295.

Senthilkumar V, Gunasekaran P. (2009). Bioethanol from biomass: production of ethanol from molasses. In: Pandey A, editor. *Handbook of Plant-Based Biofuels*. CRC Press.

Shigechi H, Uyama K, Fujita T, Matsumoto T, Ueda M, Tanaka A, Fukuda H, Kondo A. (2002). Efficient ethanol production from starch through development of novel flocculent yeast strains displaying glucoamylase and co-displaying or secreting α-amylase. *J Mol Catal B Enzym*, 17, 179–187.

Shigechi H, Fujita Y, Koh J, Ueda M, Fukuda H, Kondo A. (2004a). Energy saving direct ethanol production from low temperature cooked corn starch using a cell surface engineered yeast strain co-displaying glucoamylase and α-amylase. *Biochem Eng J*, 350, 477–484.

Shigechi H, Koh J, Fujita Y, Matsumoto T, Bito Y, Ueda M, Satoh E, Fukuda H, Kondo A. (2004b). Direct ethanol production from raw corn starch via fermentation by use of novel surface engineered yeast strain co-displaying glucoamylase and α-amylase. *Appl Environ Microbiol*, 70, 5037–5040.

Szczodrak J, Targoński Z. (1988). Selection of thermotolerant yeast strains for simultaneous saccharification and fermentation of cellulose. *Biotechnol Bioeng*, 31, 300–303.

Tamalampudi S, Fukuda H, Kondo A. (2009). Bioethanol from starchy biomass: Part II hydrolysis and fermentation. In: Pandey A, editor. *Handbook of Plant-Based Biofuels*. CRC Press.

Tanino T, Hotta A, Ito T, Ishii J, Yamada R, Hasunuma T, Ogino C, Ohmura N, Ohshima T, Kondo A. (2010). Construction of a xylose-metabolizing yeast by genome integration of xylose isomerase gene and investigation of the effect of xylitol on fermentation. *Appl Microbiol Biotechnol*, 88, 1215–1221.

Tsai SL, Oh J, Singh S, Chen R, Chen W. (2009). Functional assembly of minicellulosomes on the *Sacchromyces cerevisiae* cell surface for cellulose hydrolysis and ethanol production. *Appl Environ Microbiol*, 75, 6087–6093.

Tsai SL, Goyal G, Chen W. (2010). Surface display of a functional minicellulosome by intracellular complementation using a synthetic yeast consortium and its application to cellulose hydrolysis and ethanol production. *Appl Environ Microbiol*, 76, 7514–7520.

Tucker MP, Kim KH, Newman MM, Nguyen QA. (2003). Effects of temperature and moisture on dilute-acid steam explosion pretreatment of corn stover and cellulase enzyme digestibility. *Appl Biochem Biothechnol*, 105, 165–177.

Ueda M, Tanaka A. (2000). Cell surface engineering of yeast: construction of arming yeast with biocatalyst. *J Biosci Bioeng*, 90, 125–136.

van Maris AJA, Winkler AA, Kuyper M, de Laat WTAM, van Dijken JP, Pronk JT. (2007). Development of efficient xylose fermentation in *Saccharomyces cerevisiae*: xylose isomerization as a key component. *Adv Biochem Eng Biotechnol*, 108, 179–204.

van Rensburg P, van Zyl WH, Pretorius IS. (1998). Engineering yeast for efficient cellulose degradation. *Yeast*, 14, 67–76.

van Vleet JH, Jeffries TW. (2009). Yeast metabolic engineering for hemicellulosic ethanol production. *Curr Opin Biotechnol*, 20, 300–306.

van Zyl WH, Lynd LR, den Haan R, McBride JE. (2007). Consolidated bioprocessing for bioethanol production using *Saccharomyces cerevisiae*. *Adv Biochem Eng Biotechnol*, 108, 205–235.

Voronovsky AY, Rohuya OV, Abbas CA, Sibirny AA. (2009). Development of strains of the thermotolerant yeast *Hansenula polymorpha* capable of alcoholic fermentation of starch and xylan. *Metab Eng*, 11, 234–242.

Wen F, Sun J, Zhao H. (2010). Yeast surface display of trifunctional minicellulosomes for simultaneous saccharification and fermentation of cellulose to ethanol. *Appl Environ Microbiol*, 76, 1251–1260.

Wilson DB. (2009). Cellulases and biofuels. *Curr Opin Biotechnol*, 20, 295–299.

Wingren A, Galbe M, Zacchi G. (2003). Techno-economic evaluation of producing ethanol from softwood a comparison of SSF and SHF and identification of bottlenecks. *Biotechnol Prog*, 19, 1109–1117.

Yamada R, Tanaka T, Ogino C, Fukuda H, Kondo A. (2010a). Novel strategy for yeast construction using δ-integration and cell fusion to efficiently produce ethanol from raw starch. *Appl Microbiol Biotechnol*, 85, 1491–1498.

Yamada R, Taniguchi N, Tanaka T, Ogino C, Fukuda H, Kondo A. (2010b). Cocktail δ-integraion: a novel method to construct cellulolytic enzyme expression ratio-optimized yeast strain. *Microb Cell Fact*, 9, 32.

Yamada R, Taniguchi N, Tanaka T, Ogino C, Fukuda H, Kondo A. (2011a). Direct ethanol production from cellulosic materials using a diploid strain of *Saccharomyces cerevisiae* with optimized cellulase expression. *Biotechnol Biofuels*, 4, 8.

Yamada R, Yamakawa S, Tanaka T, Ogino C, Fukuda H, Kondo A. (2011b). Direct and efficient ethanol production from high-yielding rice using a *Saccharomyces cerevisiae* strain that express amylases. *Enzyme Microb Technol*, 48, 393–396.

Yamakawa S, Yamada R, Tanaka T, Ogino C, Kondo A. (2012). Repeated fermentation from raw starch using *Saccharomyces cerevisiae* displaying both glucoamylase and α-amylase. *Enzyme Microb Technol*, 50, 343–347.

Yanase S, Hasunuma T, Yamada R, Tanaka T, Ogino C, Fukuda H, Kondo A. (2010a). Direct ethanol production from cellulosic materials at high temperature using the thermotolerant yeast *Kluyveromyces marxianus* displaying cellulolytic enzymes. *Appl Microbiol Biotechnol*, 88, 381–388.

Yanase S, Yamada R, Kaneko S, Noda H, Hasunuma T, Tanaka T, Ogino C, Fukuda H, Kondo A. (2010b). Ethanol production from cellulosic materials using cellulase-expressing yeast. *Biotechnol J*, 5,449–455.

Zhang J, Chu D, Huang J, Yu Z, Dai G, Bao J. (2010). Simultaneous saccharification and ethanol fermentation at high corn stover solids loading in a helical stirring bioreactor. *Biotechnol Bioeng*, 105, 718–728.

Zheng DQ, Wu XC, Tao XL, Wang PM, Li P, Chi XQ, Li YD, Yan QF, Zhao YH. (2011). Screening and construction of *Saccharomyces cerevisiae* strains with improved multi-tolerance and bioethanol fermentation performance. *Bioresour Technol*, 102, 3020–3027.

CHAPTER 9

Fermentative Biobutanol Production: An Old Topic with Remarkable Recent Advances

YI WANG

Department of Food Science and Human Nutrition, University of Illinois at Urbana-Champaign, Urbana, IL, USA
Institute for Genomic Biology, University of Illinois at Urbana-Champaign, Urbana, IL, USA

HOLGER JANSSEN

Department of Food Science and Human Nutrition, University of Illinois at Urbana-Champaign, Urbana, IL, USA

HANS P. BLASCHEK

Department of Food Science and Human Nutrition, University of Illinois at Urbana-Champaign, Urbana, IL, USA
Institute for Genomic Biology, University of Illinois at Urbana-Champaign, Urbana, IL, USA
Center for Advanced Bioenergy Research (CABER), University of Illinois at Urbana-Champaign, Urbana, IL, USA

Bioprocessing of Renewable Resources to Commodity Bioproducts, First Edition.
Edited by Virendra S. Bisaria and Akihiko Kondo.

Abstract

Butanol (stands for *n*-butanol in this chapter, if not otherwise indicated) can be used as an excellent fuel source and a valuable chemical feedstock for various industries. In the past decades, almost all butanol were produced through the petrochemical synthesis route. The high price of the crude oil and various environmental problems caused by consumption of fossil fuels have renewed the interest in biobutanol production through the acetone–butanol–ethanol (ABE) fermentation process from renewable resources. ABE fermentation has a history of over 150 years. With the new attention on this old topic, remarkable advances occurred recently in both academia and industry fields. In this chapter, the basic aspects of ABE fermentation, such as the history, fermentation microorganism, and physiological pathway have been introduced; recent advances in transcriptomic analyses, proteomic studies, and genetic engineering concerning the solventogenic clostridia have been reviewed. Although, from the economic standpoint, the industrial scale production of biobutanol through the traditional ABE fermentation process is not currently viable, it is expected that, with the rapid developments in molecular biology and fermentation engineering fields, this process becomes competitive soon in the near future.

9.1 INTRODUCTION

Butanol (stands for *n*-butanol in this chapter, if not otherwise indicated) can be used as an excellent transportation fuel and a valuable chemical feedstock for various industries. As a fuel, butanol has high volumetric energy content, which is comparable to gasoline; butanol has low vapor pressure, which makes it safer to handle; butanol is less corrosive, which means it can be delivered through existing gasoline supply infrastructure; butanol can be used as a replacement for gasoline at any percentage, and no modification is needed for the existing engine (Dürre, 2007). While as an industry feedstock, butanol can be used in many industries, such as latex surface coating, enamels and lacquers industries, antibiotic, vitamin and pharmaceuticals manufactures, and food and flavor industries. In the past decades, almost all butanol were produced from petrochemical synthesis routes. Recently, the skyrocketing of crude oil price and concerns about environmental problems have renewed the interests in butanol production through biological fermentation processes. The biobutanol production through the traditional acetone–butanol–ethanol (ABE) fermentation using solventogenic clostridia has a history of over 150 years. It was an important industry even until the first part of the twentieth century, before it was diminished by the competition of the petrochemical industry in the early 1960s. The basic fermentation metabolism has been well characterized by many researchers from long time ago (Jones and Woods, 1986). Recently, with the rapid development of technology and availability of innovative study tools, novel advances have been reported on this old topic, including genomic sequencing, transcriptomic analyses, metabolomics studies, metabolic engineering, utilization of cheap substrates, development of novel fermentation processes, and product recovery techniques, etc. Although this traditional fermentative butanol production process is not commercially viable yet from an economic standpoint, with the remarkable advances of the research on this old topic, it is expected that the process could be soon competitive with petrochemical synthesis routes.

9.2 BUTANOL AS A FUEL AND CHEMICAL FEEDSTOCK

As a liquid fuel, butanol has many superior properties over other biofuels, including ethanol. Butanol has higher energy content, 30% more than that of ethanol and is closer to that of gasoline (29.2 MJ/L for butanol versus 21.2 MJ/L for ethanol; and 32.0 MJ/L for gasoline) (Table 9.1); butanol has low vapor pressure, six times less "evaporative" than ethanol and 13.5 times less than gasoline, which makes it safer to handle and especially when used in hot area and hot weather; butanol is less corrosive, and thus it can be delivered through existing gasoline supply infrastructure, while ethanol must be transported via rail, barge, or truck; butanol can be used as a replacement for gasoline gallon for gallon (i.e., 100% or any other percentage), and no modification is needed for the existing engine, while ethanol can only be an additive to gasoline up to about 85% after significant modification made to the engine (Dürre, 2007; Davis and Morton, 2008). In addition, butanol can also be used safely and advantageously up to high blending ratios with the diesel fuel in the diesel engine (Mehta et al., 2010; Rakopoulos et al., 2010); Furthermore, just recently an attempt to make renewable jet fuel from biobutanol has been funded by the US Naval Air Warfare Center Weapons Division (NAWCWD) (Militaryaerospace.com, 2012). On the other hand, butanol is not hygroscopic (thus does not pick up water), which allows being blended with gasoline at the refinery, ahead of storage and distribution, while blending ethanol with gasoline must occur shortly before use. This feature of butanol also prevents contamination of groundwater in case of spills (Dürre, 2007). In this

TABLE 9.1 Physical Properties of Butanol, Comparing to Ethanol and Gasoline

	Butanol	Ethanol	Gasoline
Density (g/mL)	0.81	0.79	0.7–0.8
Boiling point (°C)	118	78	27–221
Energy content (MJ/L)	29.2	21.2	32.0
Water solubility at 25°C, %	9.1	100.0	<0.01

sense, butanol has been considered to be the next generation biofuel after ethanol, and a fuel source much superior over ethanol.

Besides as a promising biofuel, more importantly, butanol could also be used as a valuable feedstock for various industries. The industrial applications of butanol include but are not limited to: (1) solvents: for paints, varnishes, resins, gums, dyes, camphor, vegetable oils, fats, waxes, resins, shellac, rubbers, and alkaloids; (2) plasticizers: to improve how a plastic material processes; (3) coatings: for a variety of applications, such as curable lacquers and cross-linked baking finishes. Now, about half of the butanol is used in latex surface coating, enamels and lacquers industries in the form of butyl acrylate and methacrylate esters (Kirschner, 2006); (4) chemical intermediate or raw material: for producing many other chemicals and plastics, including safety glass, hydraulic fluids, and detergent formulations; (5) textiles: as a swelling agent and manufacturing garments from coated fabric; (6) cosmetics: including eye makeup, foundations, lipsticks, nail care products, personal hygiene products, and shaving products; (7) many others, like flotation agents, cleaners, floor polishes, brake fluid, drugs and antibiotics, hormones, and vitamins (Dow chemical Company, 2002). Currently, the United States' annual consumption of butanol is around 740,000 metric tons, representing a 1.47 billion dollar market at the current price. The projected annual growth rate for butanol in the United States is 2.2% (SRI Consulting, 2009). Global annual consumption of butanol is currently around 2.9 million metric tons, which accounts for a 5.7 billion dollar market, with a predicted annual growth of 4.7% (SRI Consulting, 2009). The butanol market demand is expected to increase dramatically if butanol can be produced economically from low-cost biomass resources.

9.3 HISTORY OF ABE FERMENTATION

For the first time, in 1861, Louis Pasteur described a microorganism, named as *Vibrion butyrique*, and its ability to synthesize butyric acid under oxygen-free conditions (Pasteur, 1861). This is the first documentation of an organism with an oxygen-free metabolism and led to a definition that is today known as "anaerobic growth". Furthermore, Louis Pasteur observed that the same microorganism also produced, beside butyric acid, the solvent butanol. At the end of the nineteenth century, further microorganisms were observed by Albert Fitz and Martinus Beijerinck. Both biologists described the formation of butyric acid and butanol by the microorganisms *Bacillus butylicus* (Fitz, 1876) or *Granulobacter butylicus* and *Granulobacter saccharobutyricus* (Beijerinck, 1893). The research interests in butanol-forming microorganisms continued at the beginning of the twentieth century. At this time, the demand on natural rubber drastically increased and thus, companies were seeking for possibilities to produce synthetic rubber based on chemicals, for example, butanol or isoamyl alcohol. These strategies were the origin of the today's known ABE fermentation and were closely linked to the name of Chaim Weizmann, a Russian Jew who became the first president of Israel in 1949. In 1904, Chaim Weizmann moved from the University of Geneva to the laboratory of the chemist Prof. W.H. Perkin at Manchester University (Beijerinck, 1893). In 1910, supported by the company Strange and Graham Ltd. (England), Chaim Weizmann and coworkers started to examine strategies to isolate butanol- or isoamyl alcohol-forming microorganisms applicable for synthetic rubber production (e.g., butadiene and isoprene). Consequently, several different cultures were isolated and documented, that is, a culture named BY, later renamed as *Clostridium acetobutylicum*. This isolated strain showed outstanding characteristics in utilization of starchy substances and production of solvents, that is, butanol and acetone. The impressive progress and advantages of the whole Weizmann process resulted in a patent application in 1915 (Gabriel, 1928).

The further development of the ABE fermentation was significantly influenced by the outbreak of the First World War in 1914. The requirement of cordite (smokeless gunpowder), mainly produced from acetone, drastically increased. At this time, calcium acetate was used as feedstock for acetone production, but the import of significant amounts of calcium acetate became limited for the British government. Therefore, new strategies were developed, and the British government made efforts to cooperate with different distilleries to produce acetone by microbial fermentation using existing plants, and butanol became a useless by-product stored in vast tanks. Moreover, different feedstocks, that is, potatoes or chestnuts, were tested for the Weizmann process, whereas maize was the most promising substrate. Between 1916 and 1917, the British government decided to transfer the Weizmann process from England to Canada (e.g., Toronto) and started also projects with the United States to profit with available and suitable substrates, that is, in the Midwest Corn Belt in the United States (Ross, 1961). In 1918, after the armistice of the First World War, the interest in acetone production significantly declined and interestingly, butanol got into the focus of the rapidly increasing automobile industry in the United States. For the automobile industry, it turned out that butanol and its ester, butyl acetate, are ideal solvents for the nitrocellulose lacquers. Thus, fermentation process facilities in the British Empire, Canada, or United States were reinstalled, as well as new plants expanded rapidly to provide the automobile industry with butanol (Killeffer, 1927). In 1927, a total of 148 fermentors were operating in several US plants, that is, 96 production fermentors in Peoria with a volume of 50,000 gallon each, producing about 100 tons of solvents per day in empiric batch fermentations (Dürre, 1998; Biebl, 1999). During the Second World War, the interest in acetone synthesis using the Weizmann process rose again, and fermentation capacities in the United States (e.g., in Philadelphia), France (e.g., in Usines de Melle), and United Kingdom expanded again in order to deliver acetone for the manufacture of munitions, partly by commandeering alcohol distilleries. In 1936, C. Weizmann's patent expired and consequently further commercial production plants were established by new founded companies, that is, U.S. Industrial Chemicals Co., Western Condensing Co., or Publiker Industrial Inc. (Jones and Woods, 1986). Furthermore, between 1935 and 1941, several new clostridia strains were isolated utilizing more efficiently different carbon sources (i.e., molasses) and with higher concentrations of solvents.

The ending of the Second World War, in 1945, was the beginning of a drastically decline of the industrial butanol and acetone production by ABE fermentation. Between the years of 1950 and 1960 the solvent production by fermentation almost ceased in the United States and United Kingdom based on two major reasons: (1) industry of petrochemical butanol production increased rapidly and was in competition to ABE fermentation processes and (2) molasses, at this time the major feedstock for ABE fermentation, became very expensive as it was used in significant amounts in cattle feed (Jones and Woods, 1986). Thus, many ABE fermentation plants were closed in the western hemisphere.

The next crucial event influencing the history of the ABE fermentation was the oil crisis and its consequences at the end of the 1970s. A further revival of the industrial scale fermentation process resulted in new efforts to optimize the physiology of the fermentation and new developed downstream processes. Nevertheless, during the 1980s different industrial plants, for example, in South Africa, Brazil, or Russia, were closed in consequence to critical shortage of molasses, maize, or wheat. However, butanol production based on ABE fermentation still possesses a great potential concerning the substantial increase of oil demand in combination with shortage of fossil fuels. Thus, in 2006, DuPont and British Petroleum announced to reinstall industrial plants for ABE fermentation in the United Kingdom and also basic research interests and activities significantly increased and are still in good progress with focus on genetic engineering of suitable clostridia strains to improve carbon utilization and/or product (butanol:acetone) ratios (Lütke-Eversloh and Bahl, 2011).

9.4 PHYSIOLOGY OF CLOSTRIDIAL ABE FERMENTATION

9.4.1 The Clostridial Cell Cycle

C. acetobutylicum is well known for its biphasic ABE fermentation metabolism and consequently rose as a model organism for nonpathogenic clostridia in the past 30 years (Lütke-Eversloh and Bahl, 2011). Thus, the clostridial ABE fermentation is well characterized, whereas the detailed regulation between acid and solvent formation still remains unknown. However, the typical ABE fermentation of clostridial strains can be divided into three major growth phases (Figure 9.1). The first growth phase is highlighted by exponentially growing vegetative cells converting carbohydrates by substrate-level phosphorylation to the acids acetate and butyrate. Thus, the exponential growth phase is also named as "acidogenesis". Consequently, the accumulation of acetate and butyrate in the medium leads to a significantly decrease of the pH (<5). In reaction to a lowering pH, solventogenic clostridia switch their metabolism and cellular morphology. This second growth phase, well known as stationary phase, is characterized by markedly swollen cigar-shaped cells which accumulate a carbohydrate reserve material known as granulose. These cigar-shaped cells are named "clostridial stages" (Jones et al., 1982). Interestingly, for several clostridial strains, for example, *Clostridium pasteurianum*, the intracellular storage compound granulose was determined as α-1,4-linked polyglucan with less than 2% of branched α-1,6-linked chains (Hobson and Nasr, 1951; Whyte and Strasdine, 1972; Darvill et al., 1977), whereas for other strains, for example, *Clostridium saccharobutylicum* (formerly known, *C. acetobutylicum* P262) (Shaheen et al., 2000; Keis et al., 2001), only α-1,4-linked polyglucan units

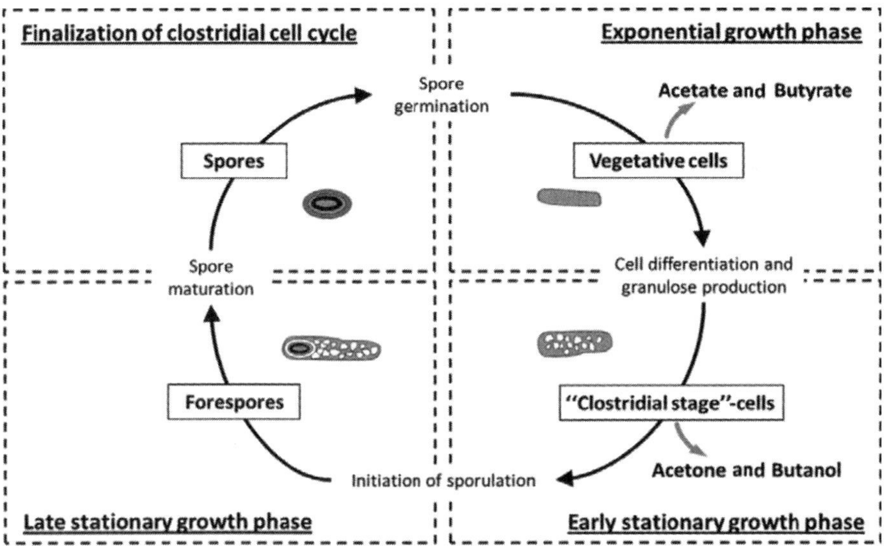

FIGURE 9.1 The cell cycle of *Clostridium acetobutylicum* with its different growth phases, respective cell forms, and major fermentation products.

were determined (Reysenbach et al., 1986). Thus, granulose shows similar characteristics like amylopectin or glycogen. However, during stationary growth phase, the characteristic "clostridial stage" cells also re-assimilate prior produced acids, acetate, and butyrate, and convert them to the neutral solvents, acetone, butanol, and ethanol. This growth phase is well known as "solventogenesis." The characteristic change in product formation is also named "metabolic switch" and is dependent on several environmental factors, that is, change of the pH, the concentration of the carbon source, a threshold of produced acids, acetate, and butyrate, and/or the intracellular NADH/NAD^+ ratio (Jones and Woods, 1986). Concomitant with the "metabolic switch" the microorganism initiates the sporulation process and uses the stored carbohydrate granulose to finalize the energy-consuming steps of spore formation in the third growth phase, at the end of late stationary phase. Furthermore, spores are able to survive bad environmental conditions caused by different stress factors, for example, UV light, heat, drought, or frost. However, the germination of spores under better growth conditions, for example, improved availability of specific nutrition, finalizes the clostridial cell cycle (Figure 9.1).

9.4.2 Physiology and Enzymes of the Central Metabolic Pathway

Solventogenic clostridia, for example, *C. acetobutylicum* and *Clostridium beijerinckii*, are able to consume a variety of sugars. At the beginning of the fermentation process exponential growing cells using the Embden–Meyerhof–Parnas Pathway convert a carbohydrate, for example, glucose, to pyruvate by generation of two molecules ATP and two molecules NADH (Figure 9.2). In general, the enzyme

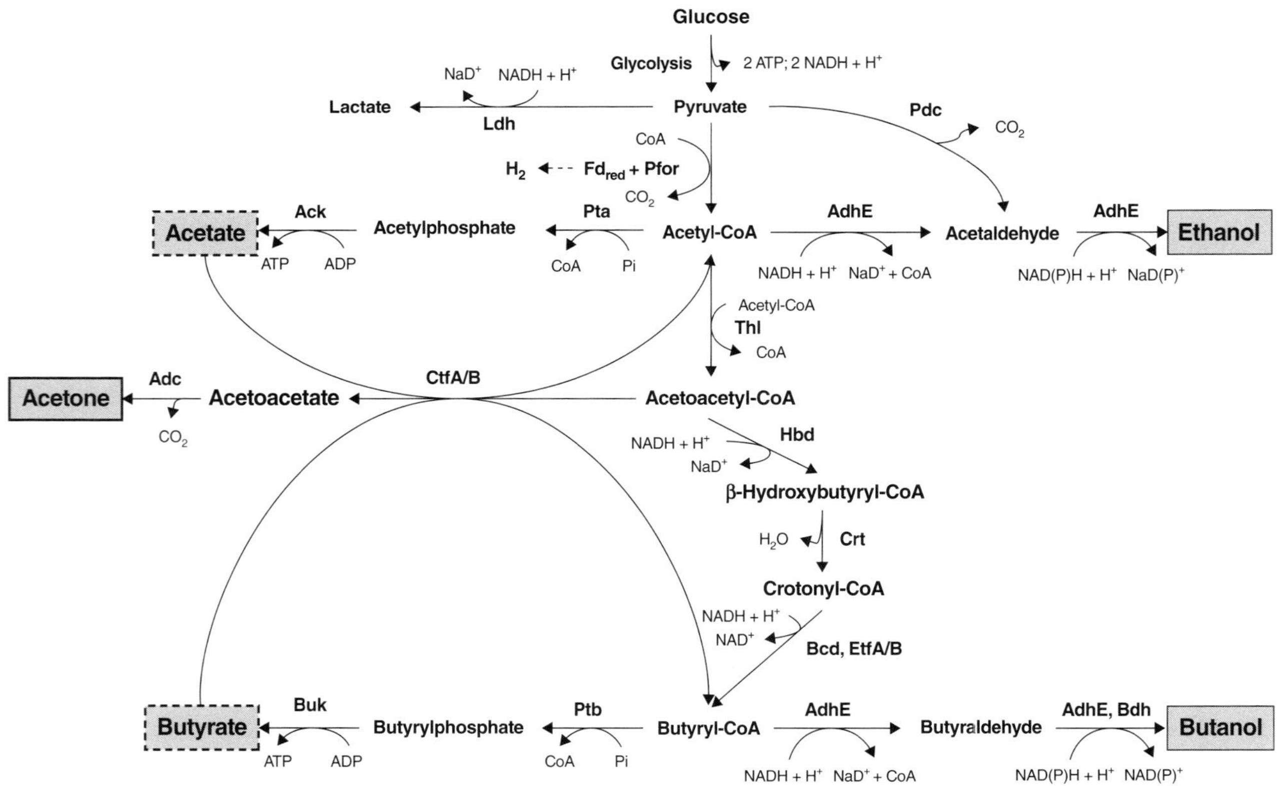

FIGURE 9.2 Physiology of ABE fermentation metabolism of *Clostridium acetobutylicum* with the respective enzymes and products. CoA, coenzyme A; Ldh, lactate dehydrogenase; Pdc, pyruvate decarboxylase; Pfor, pyruvate::ferredoxin oxidoreductase; Fdred, ferredoxin reduced; Thl, thiolase; Hbd, β-hydroxybutyryl-CoA dehydrogenase; Crt, crotonase; Bcd, butyryl-CoA dehydrogenase; Etf, electron transfer flavoprotein; Pta, phosphotransacetylase; Ack, acetate kinase; Ptb, phosphotransbutyrylase; Buk, butyrate kinase; Ctf A/B, acetoacetyl-CoA:acyl-CoA transferase; Adc, acetoacetate decarboxylase; AdhE, aldehyde/alcohol dehydrogenase; Bdh, butanol dehydrogenase.

pyruvate ferredoxin::oxidoreductase (Pfor) utilizes the formed pyruvate using a molecule CoA to release acetyl-CoA and CO_2, whereas the lactate dehydrogenase (Ldh) only shows remarkable influences under certain growth conditions, that is, iron limitation (Bahl and Gottschalk, 1984; Bahl et al., 1986). At this time the organism also produces significant amounts of hydrogen by interplay of Pfor, reduced and oxidized ferredoxins and hydrogenases. Furthermore, during acidogenic growth the central molecule acetyl-CoA is converted to the acids, acetate, and butyrate. Here, the enzymes phosphotransacetylase (Pta) and acetate kinase (Ack) are responsible for the synthesis of acetate (Hartmanis and Gatenbeck, 1984; Boynton et al., 1996). First, Pta converts acetyl-CoA to acetylphosphate, and finally Ack metabolizes acetylphosphate to acetate. On the other hand, for the butyrate production the enzyme thiolase (Thl) combines two molecules of acetyl-CoA to acetoacetyl-CoA. Furthermore, acetoacetyl-CoA is converted in sequential steps to butyryl-CoA by the enzymes β-hydroxybutyryl-CoA dehydrogenase (Hbd), crotonase (Crt), and butyryl-CoA dehydrogenase (Bcd). Here, the enzymes phosphotransbutyrylase (Ptb) and butyrate kinase (Buk) convert butyryl-CoA in a two-step process to butyrate (Hartmanis, 1987; Walter et al., 1993). Ptb produces butyrylphosphate which is used by Buk to finalize butyrate production. Notably, each acid-forming pathway generates one molecule of the important energy molecule ATP.

As mentioned above, the significant production of acids, acetate, and butyrate leads to a drastically decrease of the external pH. Interestingly, solventogenic clostridia are unable to maintain a constant intracellular pH, as described for other bacteria (Booth, 1985; Krulwich et al., 2011). Thus, the internal pH follows the external pH with a difference of $\Delta pH{\sim}1$ (Gottwald and Gottschalk, 1985; Terracciano and Kashket, 1986). Consequently, in response to the changed pH the organism switches its metabolism from acidogenesis to solventogenesis. Therefore, the prior produced acids, acetate, and butyrate are re-assimilated by the enzyme acetoacetyl-CoA:acyl-CoA transferase (CtfA/B). The CtfA/B enzyme converts acetate or butyrate, based on a molecule acetoacetyl-CoA, to their respective CoA-derivate, acetyl- or butyryl-CoA, and to a molecule acetoacetate (Wiesenborn et al., 1989). In a further step, the acetoacetate decarboxylase (Adc) metabolizes acetoacetate to acetone (Petersen and Bennett, 1990). The synthesized CoA molecules, acetyl- and butyryl-CoA, are the precursors for the neutral solvents ethanol and butanol, respectively. In a NAD(P)H-dependent two-step process the bifunctional aldehyde/alcohol dehydrogenase (AdhE) forms acetaldehyde and/or butyraldehyde, and furthermore the respective alcohols ethanol and butanol. Interestingly, solventogenic clostridia possess more than one AdhE (e.g., AdhE1 and AdhE2 in *C. acetobutylicum*) and further additional single alcohol dehydrogenases or butanol dehydrogenases (BdhA and BdhB) (Walter et al., 1992). The activity of different Bdh's and AdhE's, involved in butanol or ethanol production, is dependent on NADH or NADPH as well as on influenced redox potentials (Chen, 1995). Recently, a NAD(P)H pool-influenced mutant of *C. acetobutylicum* (*C. acetobutylicum rex*::int; redox-sensing protein) demonstrated earlier solvent production and in consequence higher final ABE concentrations (Wietzke and Bahl, 2012). In general, the butanol:acetone ratio is defined as 2:1. However, cultures with blocked hydrogen production and consequently with a reduced redox potential, for example, by gassing with carbon-monoxide or addition of methyl-viologen, showed significant increased butanol:acetone ratios by different expressed AdhE's (Fontaine et al., 2002; Hönicke et al., 2012). Nevertheless, further investigations are required to unravel detailed regulation of the different aldehyde/alcohol dehydrogenases in solventogenic clostridia strains under certain growth conditions.

9.5 ABE FERMENTATION PROCESSES, BUTANOL TOXICITY, AND PRODUCT RECOVERY

9.5.1 ABE Fermentation Processes

An ABE batch fermentation usually starts with a heatshock of the spore stocks in order to induce expression of heat shock proteins that enhance butanol tolerance and production and avoid possible culture degeneration (Ezeji et al., 2007c). Various carbon sources and different media have been tested for ABE production. Nevertheless, besides the carbon source, proper buffer, essential minerals, and vitamins usually compose the synthetic medium. An acetone:butanol:ethanol ratio of 3:6:1 with small variations is usually obtained from the fermentation process with solventogenic clostridia. ABE fermentation is a very complex process and is influenced by many factors. A number of studies have been carried out to investigate the effects of different factors on butanol production, including pH, temperature, agitation, and so on (Nishio et al., 1983; Welsh and Veliky, 1984; Soni et al., 1992; Geng and Park, 1993; Salleh et al., 2008). Wang and Blaschek (2011) investigated the individual and interactive effects of pH, initial substrate concentration, and agitation rate on ABE fermentation using a multifactorial experimental design approach and determined the optimum conditions for butanol production when mixed sugars from tropical maize stalk juice were used as the carbon source. Notwithstanding, the effects of various operational parameters on ABE fermentation vary significantly depending on different strains, carbon sources, and media compositions.

Batch fermentation is the most often used operation mode in many industries due to its high efficiency and easy operation. However, fed-batch fermentation is advantageous in cases where there is a potential substrate inhibition at the beginning of fermentation (Ezeji et al., 2004). Fed-batch fermentation is an industrial technique where the reactor is started with a relatively low substrate concentration, and as the substrate is consumed, more concentrated substrate solution is added at a low rate. Through the dilution effect, the toxicity from the end products could be meanwhile alleviated, while the substrate concentration in the reactor is kept below the toxic level. Although it requires continuous and long-term manual work for the operation, continuous fermentation has various advantages over the batch and fed-batch fermentation processes. First, it only needs one inoculum for a long time of operation; in addition, it saves time for sterilization and re-inoculation and thus enhances the overall productivity (Kumar and Gayen, 2011). The cultivation of *C. beijerinckii*

BA101 free cells in degermed corn-based P2 medium for continuous butanol production has been investigated (Ezeji et al., 2007b). Results indicated that long-term continuous cultivation of *C. beijerinckii* BA101 in a degermed corn-based medium was not successful because of the "retrogradation" of the gelatinized degermed corn starch; however, continuous fermentation of saccharified degermed corn with normal and half P2 medium nutrients were successful, and an ABE concentration of up to 14.28 g/L was obtained during 504 hours of operation, after which the experiment was stopped intentionally. Cell immobilization is often employed in continuous fermentation in order to maintain high cell density in the reactor during the fermentation process. In a study for the continuous production of isopropanol and butanol with *C. beijerinckii* DSM 6423 (*C. beijerinckii* NRRL B593), both single- and two-stage continuous fermentation modes were investigated, where wood pulp was employed for holding the cell. While the single-stage continuous fermentation demonstrated much higher solvent productivity (5.52 g/L·h in single stage vs. 0.84 g/L·h in two stage), the two-stage fermentation generated higher solvent yield (0.32 g/g) and final solvent titer (7.51 g/L) (Survase et al., 2011). Li et al. (2011) compared the performance of batch, fed-batch, and continuous ABE fermentation processes under pH-controlled conditions (Li et al., 2011). They reported that while the batch mode provides the highest overall solvent yield, the continuous fermentation worked better in terms of butanol yield and productivity. Interestingly, they also concluded that the fed-batch mode is not suggested for solvent production due to the long time needed for passing from acidogenesis to solventogenesis.

9.5.2 Butanol Toxicity and Butanol-Tolerant Strains

During ABE fermentation, solvents in the fermentation broth become very toxic when the level is beyond 2%. Butanol is the most toxic end product for solventogenic clostridia, and rare butanol-producing strains can tolerate more than 2% butanol (Lin and Blaschek, 1983). Butanol toxicity has been attributed to the chaotropic effects on the cell membrane and the inhibitory effects on sugar uptake, nutrient transport, membrane-bound ATPase activity, disruption of the proton motive force (or intracellular pH), etc. (Bowles and Ellefson, 1985; Ounine et al., 1985). It was reported that 20–30% increase in membrane fluidity was observed when *C. acetobutylicum* was exposed to 1% butanol (Vollherbstschneck et al., 1984). The incompetence of butanol tolerance of the fermentative microorganism thus leads to a very low final solvent titer in the ABE fermentation, which results in a high energy cost for the following solvent recovery process. It was estimated that if the final butanol concentration in the fermentation broth is increased from 1% to 2%, the energy requirement for recovering butanol by distillation would be lowered by 62% (Phillips and Humphrey, 1983). The poor tolerance (and thus the low final solvent titer) in the traditional ABE fermentation process is a major limitation for scaling up the butanol fermentation to an economic viable industrial scale.

Great efforts have been made to improve the butanol tolerance of the solventogenic clostridia stains. A butanol-tolerant derivative (called SA-1) of *C. beijerinckii*

NCIMB 8052 was obtained by serial enrichment with butanol challenge (Lin and Blaschek, 1983). The mutant was superior over other mutants in terms of growth rate, butanol tolerance, butanol production, and butanol/acetone ratio. Later, another butanol-tolerant mutant *C. beijerinckii* SA-2 was obtained through similar procedure by the same group (Baer et al., 1987). The consequences of butanol challenge on cell membrane of SA-2 mutant were investigated; results demonstrated that SA-2 strain developed a mechanism for maintaining a more stable membrane environment by synthesizing an increased amount of saturated fatty acids, which might be closely related to the enhanced butanol tolerance. Recently, in *C. acetobutylicum* DSM 1731, a functionally unknown protein with a hypothetical alcohol-interacting domain was identified, and its disruption led to an increased butanol tolerance; thus it can be considered as a potential target for engineering alcohol tolerance (Jia et al., 2012).

9.5.3 Fermentation Products Recovery

Numerous endeavors have been implemented in fermentation process development in order to alleviate the solvent-induced inhibitory effects and thus improve the fermentation productivity. Usually, the selective removal of fermentation products was integrated simultaneously into the fermentation process in order to maintain a low solvent concentration in the fermentation broth. Many online butanol removal techniques have been reported with various advantages and efficiencies, including liquid–liquid extraction, perstraction, gas stripping, pervaporation, adsorption, reverse osmosis, etc.

For the liquid–liquid extraction technique, a water-insoluble organic extractant is mixed with the fermentation broth. Since butanol is more soluble in the organic extractant than in the fermentation broth, high-concentrated solvents can be obtained from the organic extractant phase, which is immiscible and separated from the fermentation broth. Thus, the solvents can then be recovered by distillation with a relatively low energy cost. Oleyl alcohol, which is relatively nontoxic to the fermentative microorganism, has been proved to be a good extractant for this purpose (Ezeji et al., 2007c). Recently, nonionic surfactants have also been reported to effectively separate butanol from the fermentation broth and thus significantly enhanced the butanol production in an ABE batch fermentation with *C. pasteurianum* (Dhamole et al., 2012).

However, liquid–liquid extraction can form emulsions, and the extractant is usually expensive and sometimes toxic to the cell culture (Dürre, 1998). To solve these problems, the perstraction technique was developed based on the liquid–liquid extraction concept. In the perstraction separation process, the fermentation broth and the extractant are separated by a membrane, where two immiscible phases can exchange butanol. In this process, butanol can diffuse preferentially across the membrane, while other components (substrates, cell culture, and other fermentation products) are retained in the fermentation broth. This strategy effectively avoids the potential problems in the liquid–liquid extraction system, but it needs to point out that the membrane can decrease the butanol extraction rate as a physical barrier (Ezeji et al., 2007c).

Gas stripping is another attractive method for integrated product recovery in the ABE fermentation process. In this system, the gas is directly sparged through the fermentation system, and thus the evaporable solvents can be easily recovered and other essential components for the fermentation and cell culture can be retained in the broth. Inert gas, like nitrogen, could be employed for this purpose, and the mixed gases (carbon dioxide and hydrogen) generated during the fermentation could also be collected and used in order to save the cost for the exogenous nitrogen (Qureshi et al., 2008). Gas stripping is very simple to operate and usually leads to a low chance of clogging or fouling for the fermentation and thus became very attractive recently (Shen et al., 2011; Lee et al., 2012). Actually, till now, the gas stripping process has been proved to be the most effective process for online recovery of butanol (Zheng et al., 2009).

Recently, a novel process with simultaneous ABE fermentation and *in situ* product recovery with vacuum was reported (Mariano et al., 2011). Results indicated that fermentation coupled with *in situ* vacuum recovery led to complete substrate utilization, greater solvent productivity, and improved cell growth.

9.6 METABOLIC ENGINEERING AND "OMICS"—ANALYSES OF SOLVENTOGENIC CLOSTRIDIA

9.6.1 Development and Application of Metabolic Engineering Techniques

In the past 20 years significant genetic engineering technologies were developed to overcome the genetic inaccessibility of solventogenic clostridia species. The major focus of metabolic engineering is to improve butanol tolerance and to increase the final butanol concentration (Ezeji et al., 2010). Further approaches deal with the detailed regulation of the metabolic switch between acidogenesis and solventogenesis, for example, involved regulators, influencing enzyme activities or blocking whole cellular pathways (Papoutsakis, 2008). Here, pioneering genetic work was done by the laboratory of E.T. Papoutsakis by *in vivo* methylation of plasmids for overexpression purposes (Mermelstein et al., 1992; Mermelstein and Papoutsakis, 1993) as well as the development of a gene expression reporter system (Tummala et al., 1999). First techniques for gene deletion were developed based on a plasmid integration technology (Green et al., 1996) followed by gene knockdown techniques using antisense RNA constructs (Tummala et al., 2003a,b). The group of P. Soucaille developed a chromosomal manipulation technique for gene deletion by homologous recombination (Soucaille et al., 2007) and a gene knockout technology for clostridia (TargeTron) based on a mobile group II intron system was developed and published in parallel by Heap et al. (2007, 2010) and Shao et al. (2007). Recently, an anhydrotetracycline-inducible gene expression system for *C. acetobutylicum* was published and discussed as useful tool for further strain engineering approaches (Dong et al., 2012). All these efforts markedly influenced the progress of metabolic engineering of solventogenic clostridia, especially targeting the biphasic fermentative metabolism, and are reflected in several published single as well as first double knockout mutants based on different genetic technologies (Table 9.2).

One major goal of metabolic engineering is to diminish the by-product formation of acetone to generate a homo-butanol fermentative producing strain. Therefore, several groups deleted the respective gene *adc*, whose product converts acetoacetate to acetone (Jiang et al., 2009; Han et al., 2011; Lehmann et al., 2012a). Interestingly, all *adc* knockout mutant strains are still able to produce acetone although in significant different amounts (0.34–8.0 g/L). This phenomenon can be explained by a nonenzymatic decarboxylation of acetoacetate to acetone and CO_2. The first acetone-negative phenotype of *C. acetobutylicum* was published by Lehmann et al. (2012a). Here, the authors generated a acetoacetyl-CoA:acyl-CoA transferase subunit A (*ctfA*::int) mutant and resulted an acetone-negative phenotype concomitant with an accumulation of acetate. In general, the whole enzyme complex CtfA/B is responsible for the reassimilation of the acids acetate and butyrate. Thus, the *ctfA*::int mutant is unable to convert acetate to acetyl-CoA and acetoacetate, the precursor of acetone. However, the *ctfA*::int mutant is still able to reassimilate butyrate with concomitant butanol synthesis, although the *ctfA*::int mutant reaches only the half amount of butanol compared to the wild-type strain. This phenotype suggests that *C. acetobutylicum* is able to convert butyrate to butanol independently of CtfAB. Furthermore, the authors suggest that Buk and Ptb convert butyrate to butanol during their reverse reactions (Lehmann et al., 2012a). Another approach to succeed in decreasing the acetone amount is the enzymatic conversion of acetone to higher-value alcohols. For the first time two research groups documented an engineered isopropanol–butanol–ethanol (IBE)-producing *C. acetobutylicum* strain without further acetone accumulation. Here, *C. acetobutylicum* was engineered by overexpressing a NADPH-dependent secondary alcohol dehydrogenase of *C. beijerinckii* NRRL B 593 to convert acetone in a further reaction to isopropanol (Dai et al., 2012; Lee et al., 2012).

Further progress was achieved by targeting the acid-formation pathways to examine the detailed role of the respective acids acetate and butyrate for solvent production. So far, different mutants were investigated that were affected in the acetate pathway (*pta* or *ack*) (Green et al., 1996; Zhao et al., 2005; Kuit et al., 2012; Lehmann et al., 2012a) or butyrate pathway (*ptb* or *buk*) (Green et al., 1996; Zhao et al., 2005; Lehmann et al., 2012b). Furthermore, first double knockout mutants targeting acetate production and reassimilation (*pta::ctfA*) or acetate and acetone production (*pta::adc*) were published (Lehmann et al., 2012a). An overview is given in Table 9.2. Interestingly, further research elucidated that *C. acetobutylicum* is able to grow independently of the entire butyrate/butanol (C_4) metabolic pathway. Here, the authors were able to disrupt the gene of the β-hydroxybutyryl-CoA dehydrogenase (*hbd*) and switched the solventogenic organism to a significant ethanol producer (Lehmann and Lütke-Eversloh, 2011). However, further efforts have to focus on the genetic control of the metabolic switch between acidogenesis and solventogenesis, especially to

TABLE 9.2 Selected Single or Double KO Mutants of *Clostridium acetobutylicum* (*C. ac.*) or *Clostridium beijerinckii* (*C. bei.*) Targeting the Biphasic Fermentative Metabolism and Documented Fermentation Products Under Certain Batch Growth Conditions

Parental strain	Affected gene[a]	Affected pathway(s)	Acetone (g/L)	Butanol (g/L)	Ethanol (g/L)	Reference
C. ac. WUR	*ack*	Acetyl-phosphate → Acetate	5.7	11.6	1.6	Kuit et al. (2012)
C. ac. ATCC 824	*pta*	Acetyl-CoA → Acetyl-phosphate	3.5	8.7	0.6	Green et al. (1996)
C. ac. ATCC 824	*pta*	Acetyl-CoA → Acetyl-phosphate	1.2	3.3	n.d.	Zhao et al. (2005)
C. ac. ATCC 824	*pta*	Acetyl-CoA → Acetyl-phosphate	2.9	11.8	1.2	Lehmann et al. (2012a)
C. ac. ATCC 824	*buk*	Butyryl-CoA → Butyryl-phosphate	1.9	10.5	0.7	Green et al. (1996)
C. ac. ATCC 824	*buk*	Butyryl-CoA → Butyryl-phosphate	2.9	9.3	n.d.	Zhao et al. (2005)
C. ac. ATCC 824	*ptb*	Butyryl-CoA → Butyryl-phosphate	0.1/4.2[b]	3.4/7.8[b]	0.3/32.4[b]	Lehmann et al. (2012b)
C. ac. ATCC 824	*hbd*	Acetoacetyl-CoA → 3-Hydroxybutyryl-CoA	1.6/2.5[b]	0.0	16.2/33.1[b]	Lehmann and Lütke-Eversloh (2011)
C. ac. ATCC 824	*adc*	Acetoacetate → Acetone	0.5	5.5	0.8	Lehmann et al. (2012a)
C. ac. EA2018	*adc*	Acetoacetate → Acetone	0.34	12.2	3.86	Jiang et al. (2009)
C. bei. NCIMB 8052	*adc*	Acetoacetate → Acetone	8.0	12.0	2.0	Han et al. (2011)
C. ac. ATCC 824	*ctfA*	Acetate → Acetyl-CoA Butyrate → Butyryl-CoA	0.0	7.4	1.0	Lehmann et al. (2012a)
C. ac. ATCC 824	*pta*[1]::*adc*[2]	[1]Acetyl-CoA → Acetyl-phosphate [2]Acetoacetate → Acetone	0.1	3.0	0.4	Lehmann et al. (2012a)
C. ac. ATCC 824	*pta*[1]::*ctfA*[2]	[1]Acetyl-CoA → Acetyl-phosphate [2]Acetate → Acetyl-CoA [2]Butyrate → Butyryl-CoA	0.0	0.7	0.3	Lehmann et al. (2012a)

n.d., no values documented.

[a]*ack*, acetate kinase; *pta*, phosphotransacetylase; *buk*, butyrate kinase; *ptb*, phosphotransbutyrylase; *hbd*, β-hydroxybutyryl-CoA dehydrogenase; *adc*, acetoacetate decarboxylase; *ctfA*, acetoacetyl-CoA:acyl-CoA transferase subunit A.

[b]Second value based on glucose fed-batch fermentation.

unravel interplay between initiation of solventogenesis with different cellular mechanisms of granulose formation, quorum sensing, and/or spore formation (Paredes et al., 2005; Jones et al., 2011; Steiner et al., 2011; Tracy et al., 2011; Steiner et al., 2012).

9.6.2 Butanol Production by Engineered Microbes

As discussed above, the genetic accessibility to solventogenic clostridia is still limited, although recently new techniques were developed (see section 9.6.1), for example, to improve the clostridial butanol production or to unravel the genetic regulation of the metabolic switch (Sillers et al., 2008; Lütke-Eversloh and Bahl, 2011). Thus, several approaches were done and are still under investigation to apply selected genes of the clostridial solventogenic pathway to other model microorganisms for butanol production (Peralta-Yahya and Keasling, 2010; Zhang et al., 2011). The most prominent workhorse for production of several industrial important products (e.g., biofuels) due to well-developed genetic tools is *Escherichia coli* (Clomburg and Gonzalez, 2010). Several attempts were done to introduce solventogenic genes (*crt*, *hbd*, *bcd*, *etfA*, *etfB*, *thil*, and *adhE*) into *E. coli* by different overexpression strategies which resulted in significant different final butanol titers (1.3 mg/L to 30 g/L). An overview of different metabolic engineering approaches, respective butanol titer, and references are listed in Table 9.3. So far, only a few approaches resulted in butanol titer of ~15 g/L or more and thus are comparable to solvent-producing clostridia (Ezeji et al., 2007a; Green, 2011). Nevertheless, further encouraging improvements were undertaken by increasing NADH and acetyl-CoA pools of the cell to enforce the flux and reducing power in direction of NAD(P)H-dependent butanol production (Shen et al., 2011). Therefore, several *E. coli* host strains were also engineered by deleting native NADH-consuming pathways to increase reducing power, that is, native *adhE* for ethanol production, native *ldhA* for lactate production or overexpression of a native pyruvate dehydrogenase complex (*aceEF–lpd*) (Atsumi et al., 2008; Shen and Liao, 2008; Bond-Watts et al., 2011; Garza et al., 2012). Beside butanol production in *E. coli*, several other microorganisms were successfully engineered with also different final butanol titer (Table 9.3), for example, *Saccharomyces cerevisiae* (2.5 mg/L) (Steen et al., 2008), *Lactobacillus brevis* (0.3 g/L) (Berezina et al., 2010), *Bacillus subtilis* (24 mg/L), and *Pseudomonas putida* (0.12 g/L) (Nielsen et al., 2009). Furthermore, a few research groups were successful to introduce butanol synthesis to non-butanol-producing clostridia, for example, *Clostridium tyrobutyricum* using glucose or mannitol as carbon source (10–20.5 g/L) or *Clostridium ljungdahlii* based on the Wood–Ljungdahl Pathway using syngas (CO/H_2) or fructose (0.15 g/L) (Köpke et al., 2010; Yu et al., 2011; Yu et al., 2012). Further promising results by Lan and Liao (2011, 2012) have shown the first autotrophic butanol production by a cyanobacterium *Synechococcus elongates*. Despite the fact that the butanol titer is relatively low (14.5–29.9 mg/L), the group was able to establish butanol production based on light reaction and CO_2. Nevertheless, for the future it will be a challenging but also important and

TABLE 9.3 Summary of Engineered Microbes for *n*-Butanol Production

Parent host strain	Characteristics of engineered strain[a]	Carbon source	Butanol titer	Reference
Escherichia coli JM109	*crt, bcd, etfB, etfA, hbd, adhE, thiL* ↑	Glucose	1.2 g/L	Inui et al. (2008)
E. coli DH10B	*crt, hbd, bcd, etfA, etfB, adhE2, atoB, gapA* ↑	Glycerol	0.58 g/L	Nielsen et al. (2009)
E. coli BW25113	*crt, bcd, etfA, etfB, hbd, atoB, adhE2* ↑ / *adhE*[b], *ldhA, frdBC, fnr, pta* ↓	Glycerol	0.55 g/L	Atsumi et al. (2008)
E. coli BW25113	*kivd, adh2, cimA, leuABCD* ↑ / *ilvI, ilvB* ↓	Glucose	0.4 g/L	Atsumi and Liao (2008)
E. coli BW25113	*kivd, adh2, ilvA, leuABCD, thrABC* ↑ / *metA, tdh, ilvB, ilvI, adhE*[b] ↓	Glucose	0.9 g/L	Shen and Liao (2008)
E. coli BW25113	*fdh, ter, atoB, adhE2, crt, hbd* ↑ / *adhE*[b], *ldhA, frd, pta* ↓	Glucose	15–30 g/L	Shen et al. (2011)
E. coli DH1	*phaA, hbd, crt, ter, adhE2, aceEF-lpd* ↑	Glucose	4.65 g/L	Bond-Watts et al. (2011)
E. coli B	*thiL, adhE2, crt, bcd, etfA, etfB, hbd* ↑ / *adhE*[b], *frdABCD, ldhA, ackA, pflB, mgsA* ↓ / *aceEF-lpd* ↑	Glucose	1.3 mg/L	Garza et al. (2012)
Bacillus subtilis KS438	*crt, bcd, etfA, etfB, hbd, thiL, adhE2* ↑	Glucose or glycerol	24 mg/L	Nielsen et al. (2009)

(continued)

TABLE 9.3 (*Continued*)

Parent host strain	Characteristics of engineered strain[a]	Carbon source	Butanol titer	Reference
Saccharomyces cerevisiae BY4742	*erg10*, *hbd*, *crt*, *ccr*, *adhE2* ↑	Galactose	2.5 mg/L	Steen et al. (2008)
Pseudomonas putida S12	*crt*, *bcd*, *etfA*, *etfB*, *hbd*, *thiL*, *adhE1* ↑	Glycerol	0.12 g/L	Nielsen et al. (2009)
Lactobacillus brevis ATCC 367	*crt*, *bcd*, *etfB*, *etfA*, *hbd* ↑	Glucose	0.3 g/L	Berezina et al. (2010)
Clostridium tyrobutyricum ATCC 25755	*adhE2* ↑ / *ack* ↓	Glucose	10.0 g/L	Yu et al. (2011)
C. tyrobutyricum ATCC 25755	*adhE2* ↑	Mannitol	20.5 g/L	Yu et al. (2012)
C. ljungdahlii DSM 13528	*bcd*, *hbd*, *crt*, *thlA*, *bdhA*, *adhE* ↑	Fructose or CO/H_2	0.15 g/L	Köpke et al. (2010)
Butyribacterium methylotrophicum ATCC 33266	CO-adapted mutant strain	CO	0.33 g/L	Shen et al. (1999)
Synechococcus elongatus PCC 7942	*ter*, *atoB*[b], *adhE2*, *crt*, *hbd* ↑	CO_2	14.5 mg/L	Lan and Liao (2011)
S. elongatus PCC 7942	*ter*, *nphT7*, *bldh*, *yqhD*, *phaJ*, *phaB* ↑	CO_2, light	29.9 mg/L	Lan and Liao (2012)

[a] ↑, Homo- or heterologous gene expression; ↓, gene deletion or disruption of native genes of the host strain. Genes of solventogenic Clostridia (*C. acetobutylicum* or *C. beijerinckii*): *crt*, crotonase; *bcd*, butyryl-CoA dehydrogenase; *etfA/B*, electron transfer flavoprotein subunit A/B; *ack*, acetate kinase, *hbd*, β-hydroxybutyryl-CoA dehydrogenase; *adhE or adhE1 or adhE2*, aldehyde/alcohol dehydrogenases; *thiL* or *thlA*, thiolase. Genes of host strains: *atoB*, acetyl-CoA acetyltransferase; *gapA*, glyceraldehyde 3-phosphate dehydrogenase; *ldhA*, lactate dehydrogenase; *frd*, fumarate reductase; *fnr*, global anaerobic regulator; *pta*, phosphotransacetylase; *aceEF-lpd*, pyruvate dehydrogenase complex; *ackA*, acetate kinase; *pflB*, pyruvate formate lyase; *mgsA*, methylglyoxal synthase; *erg10*, thiolase; *leuABCD*, leucin biosynthesis; *ilvI*, acetohydroxy acid synthase III; *ilvB*, acetohydroxy acid synthase I; *ilvA*, threonine dehydratase; *thrABC*, aspartokinase; *metA*, homoserine O-succinyltransferase; *tdh*, threonine dehydrogenase. Genes of other microorganisms: *kivd*, 2-keto acid decarboxylase from *Lactococcus lactis*; *adh2*, alcohol dehydrogenase of *S. cerevisiae*; *cimA*, (R)-citramalate synthase from *Methanococcus jannaschii*; *fdh*, formate dehydrogenase from *Candida boidinii*; *ter*, trans-enoyl-coenzyme A (CoA) reductase from *Trepenoma denticola*; *phaA*, thiolase from *Ralstonia eutropha*; *ccr*, crotonyl-CoA reductase from *Streptomyces collinus*; *nphT7*, acetoacetyl-CoA synthase from *Streptomyces* sp. strain CL190; *bldh*, CoA-acylating butyraldehyde dehydrogenase from *C. beijerinckii*; *yqhD*, NADPH-dependent alcohol dehydrogenase from *E. coli*; *phaJ*, (R)-specific enoyl-CoA hydratase from *Aeromonas caviae*; *phaB*, acetoacetyl-CoA reductase from *R. eutropha*.

[b] *adhE*, alcohol dehydrogenase from host strain; *atoB*, acetyl-CoA acetyltransferase from *E. coli*.

interesting way to produce butanol directly from CO_2 or industrial waste gases (CO/H_2) to solve upcoming energy and environmental problems.

9.6.3 Global Insights into Solventogenic Metabolism Based on "Transcriptomics" and "Proteomics"

Recent sequencing of the genomes from the solventogenic strains *C. acetobutylicum* (Nolling et al., 2001) and *C. beijerinckii* (Laboratory H.P. Blaschek; JGI project ID 3634512, 2007) led to advances in various high-throughput "omics" technologies. In the past years the most famous technology used was the "transcriptomics" approach, which measures mRNA transcript levels by DNA microarray analyses or by RNA-sequencing technology. First extensive DNA microarray data for *C. acetobutylicum* were published by the laboratory of E.T. Papoutsakis. The group focused on the transcriptional behavior of the general clostridial batch growth (Alsaker and Papoutsakis, 2005; Jones et al., 2008), transcriptional program of sporulation (Tomas et al., 2003a), as well as on the transcriptional response of different butanol stress approaches (Alsaker et al., 2004; Tomas et al., 2004; Alsaker et al., 2010). Thus, they were able to develop first platforms of genes of interests, for example, heat shock genes *groESL*, for further metabolic engineering approaches (Tomas et al., 2003b). Further transcriptomic results were published in correlation with generated mutant strains, for example, for the hyper-butanol-producing strain *C. beijerinckii* BA101 (Shi and Blaschek, 2008). Recently, the same group investigated the transcriptome structure of *C. beijerinckii* NCIMB 8052 using batch cultures to unravel the transcriptional profile of the switch between acidogenesis and solventogenesis by RNA-Seq technology (Wang et al., 2011, 2012). For further insights of the metabolic switch between acid and solvent production, phosphate-limited continuous cultures of *C. acetobutylicum* were applied. Here, research focussed on steady-state growing cells at different pH values and on butanol-adapted or butanol-stressed cells (Janssen et al., 2012; Schwarz et al., 2012). A detailed transcriptional program of the dynamic switch and the onset of solventogenesis was published by Grimmler and coworkers (2011). Here, an interesting feature of solventogenic chemostat cells growing on glucose as sole carbon source was the significant upregulation of genes with function in degradation of alternative carbon sources, for example, amylases, xylanases, or a cellulosome-like operon (CAC0910–CAC0919), although *C. acetobutylicum* is unable to grow on cellulose (Lopez-Contreras et al., 2003). Thus, further transcriptional research focused also on diauxic growth with alternative carbon sources, for example, xylose, to elucidate genes involved in the catabolic repression of *C. acetobutylicum* (Grimmler et al., 2010). Further insights into the catabolic regulations of *C. acetobutylicum* revealed a catabolite control protein A (CcpA)-mutant strain. In general, the CcpA protein is known to mediate carbon catabolite repression (CCR) of the carbon metabolism of the organism (Ren et al., 2010). However, extensive DNA microarray studies revealed that CcpA also plays an important role in solvent production, for example, by positive regulation of the *sol* operon (*adhE1–ctfA–ctfB*), and sporulation of *C. acetobutylicum* (Ren et al., 2012). Moreover, Servinsky et al. (2010) investigated in a global DNA microarray approach the transcriptional regulation of several sugar-specific mechanisms and genes involved in transport of 11 different carbohydrates including pentoses (xylose, arabinose), hexoses (glucose, mannose, galactose, and fructose), disaccharides (sucrose, lactose, maltose and cellobiose), and starch were highlighted.

Nevertheless, transcriptional analyses alone cannot fully unravel the complexities of fundamental microbial biology (Zhang et al., 2010). Therefore, beside comprehensive transcriptome studies different groups also started with quantification of protein abundances, so called "proteomics." Consequently, first proteomes of acidogenic and solventogenic batch culture cells of *C. acetobutylicum* using complex media have been investigated and compared by Terracciano et al. (1988), whereas Sullivan and Bennett (2006) combined proteomic analyses with a *spo0A*-mutant strain. The first comprehensive proteome maps for cytoplasmic and membrane proteins from *C. acetobutylicum* growing under batch culture conditions were published by Mao and coworkers (2010, 2011). Furthermore, Mao et al. (2010, 2011) highlighted significant proteomic differences between a wild-type strain and its more butanol-tolerant mutant strain *C. acetobutylicum* Rh8. Thus, they started an approach for the systematic understanding of the effect of butanol on cellular behavior of solventogenic clostridia. First proteome analyses of phosphate-limited continuous culture cells were published by Pich et al. (1990). Pich and coworkers highlighted the significant induction of specific heat-shock proteins (e.g., DnaK, GroEL) during the onset of solventogenesis, before solvents were detected in the medium. Accumulation of stress proteins is not surprising as during the metabolic shift different stresses arise. Acidic stress is initiated by a decreasing pH and thus by an intracellular increased concentration of undissociated acids. Furthermore, the metabolic switch of chemostat cells, so-called "onset of solventogenesis", is also highlighted by a significant induction of Adc (acetoacetate decarboxylase) (Schaffer et al., 2002). However, steady-state cells of *C. acetobutylicum* did not show significant changes of heat shock proteins or Adc between acidogenesis and solventogenesis, probably due to adaptation processes of exponentially growing cells in a chemostat culture (Janssen et al., 2010).

Processed data sets based on multiple "omics" analyses are valuable for further system biology approaches, that is, genome scale or metabolic network models, for better understanding of the clostridial metabolic switch and its mechanism of butanol production (Lee et al., 2008; Senger and Papoutsakis, 2008a,b; Haus et al., 2011; Milne et al., 2011; McAnulty et al., 2012).

9.7 ECONOMIC OUTLOOK

Economic studies on the production of biobutanol from corn determined that distillation recovery of butanol from the dilute ABE fermentation broth is not economical comparing to the butanol production from the current petrochemical route (Ezeji et al., 2007c). However, the life-cycle analysis on the corn-based butanol production indicated that the use of corn-based butanol as a biofuel can lead to substantial fossil energy savings relative to the use of conventional petroleum gasoline, and it can also avoid a large amount of GHG emission burdens (Wu et al., 2008).

Since substrate normally accounts for most of the total cost associated with the ABE fermentation (Qureshi and Blaschek, 2001), the search and utilization of low-value and abundant substrates would offer tremendous economic benefits. These include the cellulosic residues or by-products and wastes from agriculture production and municipal activities. Second, the development of advanced fermentation process and recovery technology will also make a big difference. The cost of fermentation products recovery is also a large part of the total cost for the industrial fermentation process. The development of novel recovery techniques, especially product recovery-integrated fermentation processes can decrease the cost for the whole process significantly. Actually, the recent development on novel fermentation process has already shown its magnitude in the biofuel industry (Mariano et al., 2008; Mariano et al., 2011; biofuelsdigest.com, 2012). On the other hand, although strain improvement usually has lower direct effects on butanol industry economics, comparing to the cost of substrate and product recovery process, it is not trivial. The improvement of fermentative strains is the most effective way to directly improve the butanol yield and productivity. Once significant improvement has been made to the strains, which is often the bottleneck of the whole fermentation process, experiences from other related industry processes can be referred and introduced for the biobutanol production industry and ultimately the whole economics of the process can be improved.

The market demand is also essential for the economics of butanol fermentation. It is expected that after biobutanol is adapted as a liquid fuel, the demand will astonishingly increase. Nonetheless, butanol is even more important as a chemical feedstock rather than as a fuel (Mascal, 2012). So, development of novel feasible industry routes for butanol-derivative products could significantly increase the market demand for butanol, and thus draw closer to the commercial production of biobutanol using the biological fermentation process. Furthermore, the recovery and utilization of the fermentation by-products (nutritional waste stream, cell biomass, CO_2, and H_2) can also contribute substantially to the economics of biobutanol production. For example, H_2 can be used as a clean energy source, while the nutritional waste stream and cell biomass can be used for some agricultural purposes after proper treatment processes.

Although the commercial production of biobutanol through the traditional ABE fermentation process is not currently or in the near future economically viable, the policy makers should take account the tremendous credit from the reduction of waste and the green house gases due to the renewable production processes. More attention and investments on the basic research are needed, in order to advance the whole biobutanol production pipeline and make it ultimately commercially feasible in the near future.

9.8 CURRENT STATUS AND FUTURE PROSPECTS

In the past few decades, almost all butanol were produced via petrochemical routes. In recent years, increasing crude oil price and concerns about environmental problems have renewed the interests in butanol production through biological fermentation processes. Great efforts have been focused on the academic research on ABE fermentation with solventogenic clostridia. Transcriptional analyses and metabolic engineering have been extensively performed on *C. acetobutylicum* (Green et al., 1996; Alsaker et al., 2004; Alsaker and Papoutsakis, 2005; Shao et al., 2007; Papoutsakis, 2008; Alsaker et al., 2010). A bacterial group II intron-based genetic engineering system (ClosTron or TargeTron) has been recently developed for directly constructing mutation in *Clostridium* species (Heap et al., 2007), and various applications based on this system have been reported (Cooksley et al., 2012; Lehmann et al., 2012a,b). In addition, conversion of low-value feedstocks like agricultural residues and optimization of fermentation process and downstream manipulations have also been studied (Ezeji et al., 2005, 2007a,b; Qureshi et al., 2008; Mariano et al., 2011). In industry, in recent years, a number of big companies started to market butanol and develop the biobutanol processes. Many venture companies have also been established focusing on the commercialization of biobutanol. Several major players include Cathay Industrial Biotech (Shanghai, China), Cobalt Technologies (Mountain view, CA, USA), Green Biologics (Abingdon, England), METabolic EXplorer (Clermont-Ferrand, France), etc. (Table 9.4) (Mascal, 2012). With the blending wall issue for bioethanol industry and the expiration of the blenders tax credit (NYTimes, 2012), many ethanol producers are considering to retrofit and transform their existing ethanol production infrastructures for butanol production (WSJ, 2009; Ethanolproducer.com, 2012). The prospects for biological butanol production look really bright.

However, there are still many challenges for the large-scale industrial production of butanol through clostridia fermentation processes. For example, high cost of the substrates, low yield and final titer, diversion of substrates to other unwanted bioproducts (like acetone and acids), high energy requirement for butanol recovery from dilute streams. Great efforts on overcoming these challenges should be made in order to significantly improve the butanol production process. First, the search and utilization of low-value and abundant substrates that do not compete with food and feed supplies would offer tremendous economic benefits. These include the cellulosic residues or by-products from agricultural and industrial processes, municipal wastes, and other energy biomasses. Effective pretreatment and biomass conversion techniques must be developed to satisfy the efficient fermentation requirement. Second, search or development of new strains and improvement of the existing strains should be performed in order to acquire better degradation capabilities on vast ranges of substrates, higher butanol yield, titer, and productivity. Recently, several new solventogenic clostridia strains have been reported, and the genomes have been sequenced (Bao et al., 2011; Hu et al., 2011; Wu et al., 2012). For strain improvement, transcriptional analysis, metabolic engineering, and system biology tools can play key roles. On one hand, butanol producers such as solventogenic clostridia can be mutated or engineered for desired properties. On the other, butanol production pathways can be introduced into desired industrial microorganisms for butanol production purpose. For example, *E. coli* has been attempted as the host and engineered for this objective (Ranganathan and Maranas, 2010; Shen et al., 2011). Adapted *P. putida* strains that can grow in the presence of up to 6% (vol/vol) butanol has been reported, which could be potential hosts for biobutanol production (Ruhl et al., 2009). Third,

TABLE 9.4 Major Companies Currently Marketing *n*-Butanol

Company	Time founded	Location	Technology	Development status
Butylfuel, LLC	1991	Gahanna, Ohio, USA	Based on a two-stage fermentation that decouples butyric acid production from butanol formation	A pilot plant incorporating this technology in a fibrous bed bioreactor is under development. Green Biologics merged with Butylfuel in January 2012 (Green Biologics press release, 2012)
Cathay Industrial Biotech	1997	Shanghai, China	Continuous clostridia ABE fermentation, with corn starch as the feedstock; optimized process reduced water and energy usage significantly	Currently the largest biobutanol producer in the world, with a production capacity of 21 million gallons of biobutanol per year
METabolic EXplorer	1999	Clermont-Ferrand, France	Using its combined expertise in molecular biology, metabolic engineering, and bioinformatics to design high-performance microorganisms that can transform plant-derived raw materials into an existing bulk chemical, including butanol	With a broad product portfolio, they are developing "an extremely flexible and competitive fermentation process for butanol production," which can utilize sugar, starch, and hemicellulosic feedstocks
Green Biologics	2003	Abington, UK	The technology platform includes a large collection of native and genetically modified clostridia, fermentation process design, production strain to feedstock matching, and separations	Merged with Butylfuel in January 2012; is aiming at retrofit existing ethanol plants and international market expansion including China, Brazil, etc. (Ethanolproducer.com, 2012)

(continued)

TABLE 9.4 *(Continued)*

Company	Time founded	Location	Technology	Development status
Cobalt Technologies	2005	Mountain View, CA, USA	An immobilized cell technology using a proprietary, nongenetically modified clostridia strain with high productivity and ability to convert both C5 and C6 sugars into butanol for operation in a continuous fermentation reactor	Currently operating a 5000 GPY pilot facility since June 2009. Cobalt is now in the process to commission a 470,000 GPY cellulosic biobutanol facility
Tetravitae Bioscience (Eastman Chemical)	2006	Chicago, IL, USA	The platform is based on a stable, nongenetically modified *C. beijerinckii* strain which offers high butanol yields, reduced product inhibition, and the ability to utilize both C5 and C6 sugars	Trying to adapt the technology to process cellulosic feedstocks; acquired by Eastman Chemical Company in November 2011 (Eastman press release, 2011)
Butalco	2007	Canton of Zug, Switzerland	Based on genetically optimized yeasts for production of advanced biofuels like ethanol and butanol by using pentose sugars	Working together with partners to develop integrated production processes (Butalco.com, 2012)

development of novel fermentation processes and improvement of the downstream processing would be significant for improving the butanol production economics. With the advancement of separation technology and fermentation engineering, more efficient and cost-effective downstream processes are expected in the future. With the rapid development in molecule biology and process engineering, significant advances have been and will be made on this old traditional process; it is expected that the fermentative production of biobutanol from renewable resources could become commercially viable in the near future.

REFERENCES

Alsaker KV, Papoutsakis ET. (2005). Transcriptional program of early sporulation and stationary-phase events in *Clostridium acetobutylicum*. *J Bacteriol*, 187, 7103–7118.

Alsaker KV, Spitzer TR, Papoutsakis ET. (2004). Transcriptional analysis of spo0A overexpression in *Clostridium acetobutylicum* and its effect on the cell's response to butanol stress. *J Bacteriol*, 186, 1959–1971.

Alsaker KV, Paredes C, Papoutsakis ET. (2010). Metabolite stress and tolerance in the production of biofuels and chemicals: gene-expression-based systems analysis of butanol, butyrate, and acetate stresses in the anaerobe *Clostridium acetobutylicum*. *Biotechnol Bioeng*, 105, 1131–1147.

Atsumi S, Cann AF, Connor MR, Shen CR, Smith KM, Brynildsen MP, Chou KJY, Hanai T, Liao JC. (2008). Metabolic engineering of *Escherichia coli* for 1-butanol production. *Metab Eng*, 10, 305–311.

Atsumi S, Liao JC. (2008). Metabolic engineering for advanced biofuels production from *Escherichia coli.* Curr Opin Biotechnol, 19, 414–419.

Baer SH, Blaschek HP, Smith TL. (1987). Effect of butanol challenge and temperature on lipid-composition and membrane fluidity of butanol-tolerant *Clostridium acetobutylicum*. *Appl Environ Microbiol*, 53, 2854–2861.

Bahl H, Gottschalk G. (1984). Parameters affecting solvent production by *Clostridium acetobutylicum* in continuous culture. *Biotechnol Bioeng*, 14, 215–223.

Bahl H, Gottwald M, Kuhn A, Rale V, Andersch W, Gottschalk G. (1986). Nutritional factors affecting the ratio of solvents produced by *Clostridium acetobutylicum*. *Appl Environ Microbiol*, 52, 169–172.

Bao GH, Wang RJ, Zhu Y, Dong HJ, Mao SM, Zhang YP, Chen ZG, Li Y, Ma YH. (2011). Complete genome sequence of *Clostridium acetobutylicum* DSM 1731, a solvent-producing strain with multireplicon genome architecture. *J Bacteriol*, 193, 5007–5008.

Beijerinck MW. (1893). Über die butylalkoholgährung und das butylferment. *Verh K Akad Wet*, 10, 1–51.

Berezina OV, Zakharova NV, Brandt A, Yarotsky SV, Schwarz WH, Zverlov VV. (2010). Reconstructing the clostridial n-butanol metabolic pathway in *Lactobacillus brevis*. *Appl Microbiol Biotechnol*, 87, 635–646.

Biebl H. (1999). *Clostridium acetobutylicum*. In: Robinson RK, Batt CA, Patel PD, editors. *Encyclopedia of Food Microbiology*. London: Academic Press, pp. 445–451.

biofuelsdigest.com (2012). Available at: http://www.biofuelsdigest.com/bdigest/2012/01/26/researcher-says-flash-fermentation-discovery-predates-gevo-patent/ (Last accessed: 1/12/2014).

Bond-Watts BB, Bellerose RJ, Chang MCY. (2011). Enzyme mechanism as a kinetic control element for designing synthetic biofuel pathways. *Nat Chem Biol*, 7, 222–227.

Booth IR. (1985). Regulation of cytoplasmic pH in bacteria. *Microbiological Reviews*, 49, 359–378.

Bowles LK, Ellefson WL. (1985). Effects of butanol on *Clostridium acetobutylicum*. *Appl Environ Microbiol*, 50, 1165–1170.

Boynton ZL, Bennett GN, Rudolph FB. (1996). Cloning, sequencing, and expression of genes encoding phosphotransacetylase and acetate kinase from *Clostridium acetobutylicum* ATCC 824. *Appl Environ Microbiol*, 62, 2758–2766.

butalco.com (2012). Available at: http://butalco.com/ (Last accessed: 1/12/2014).

Chen JS. (1995). Alcohol dehydrogenase: multiplicity and relatedness in the solvent-producing clostridia. *Fems Microbiol Rev*, 17, 263–273.

Clomburg JM, Gonzalez R. (2010). Biofuel production in *Escherichia coli*: the role of metabolic engineering and synthetic biology. *Appl Microbiol Biotechnol*, 86, 419–434.

Cooksley CM, Zhang Y, Wang H, Redl S, Winzer K, Minton NP. (2012). Targeted mutagenesis of the *Clostridium acetobutylicum* acetone–butanol–ethanol fermentation pathway. *Metab Eng*, 14, 630–641.

Dai Z, Dong H, Zhu Y, Zhang Y, Li Y, Ma Y. (2012). Introducing a single secondary alcohol dehydrogenase into butanol-tolerant *Clostridium acetobutylicum* Rh8 switches ABE fermentation to high level IBE fermentation. *Biotechnol Biofuels*, 5, 44.

Darvill AG, Hall MA, Fish JP, Morris JG. (1977). The intracellular reserve polysaccharide of *Clostridium pasteurianum*. *Can J Microbiol*, 23, 947–953.

Davis SE, Morton SA. (2008). Investigation of ionic liquids for the separation of butanol and water. *Sep Sci Technol*, 43, 2460–2472.

Dhamole PB, Wang Z, Liu Y, Wang B, Feng H. (2012). Extractive fermentation with non-ionic surfactants to enhance butanol production. *Biomass Bioenerg*, 40, 112–119.

Dong H, Tao W, Zhang Y, Li Y. (2012). Development of an anhydrotetracycline-inducible gene expression system for solvent-producing *Clostridium acetobutylicum*: a useful tool for strain engineering. *Metab Eng*, 14, 59–67.

Dow chemical company. (2002). N-butanol product information. Form no. 327–00014–1001.

Dürre P. (1998). New insights and novel developments in clostridial acetone/butanol/isopropanol fermentation. *Appl Microbiol Biotechnol*, 49, 639–648.

Dürre P. (2007). Biobutanol: an attractive biofuel. *Biotechnol J*, 2, 1525–1534.

ethanolproducer.com (2012). Available at: http://www.ethanolproducer.com/articles/8578/target-for-transformation (Last accessed: 1/12/2014).

Eastman press release (2011). Available at: http://www.eastman.com/company/news_center/2011/pages/eastman_acquires_tetravitae_bioscience_assets.aspx (Last accessed: 1/12/2014).

Ezeji TC, Qureshi N, Blaschek HP. (2004). Acetone butanol ethanol (ABE) production from concentrated substrate: reduction in substrate inhibition by fed-batch technique and product inhibition by gas stripping. *Appl Microbiol Biotechnol*, 63, 653–658.

Ezeji TC, Karcher PM, Qureshi N, Blaschek HP. (2005). Improving performance of a gas stripping-based recovery system to remove butanol from *Clostridium beijerinckii* fermentation. *Bioprocess Biosyst Eng*, 27, 207–214.

Ezeji T, Qureshi N, Blaschek HP. (2007a). Butanol production from agricultural residues: impact of degradation products on *Clostridium beijerinckii* growth and butanol fermentation. *Biotechnol Bioeng*, 97, 1460–1469.

Ezeji T, Qureshi N, Blaschek HP. (2007b). Production of acetone-butanol-ethanol (ABE) in a continuous flow bioreactor using degermed corn and *Clostridium beijerinckii*. *Process Biochem*, 42, 34–39.

Ezeji TC, Qureshi N, Blaschek HP. (2007c). Bioproduction of butanol from biomass: from genes to bioreactors. *Curr Opin Biotechnol*, 18, 220–227.

Ezeji T, Milne C, Price ND, Blaschek HP. (2010). Achievements and perspectives to overcome the poor solvent resistance in acetone and butanol-producing microorganisms. *Appl Microbiol Biotechnol*, 85, 1697–1712.

Fitz A. (1876). Über die gährung des glycerins. *Ber Dtsch Chem Ges*, 9, 1348–1352.

Fontaine L, Meynial-Salles I, Girbal L, Yang XH, Croux C, Soucaille P. (2002). Molecular characterization and transcriptional analysis of *adhE2*, the gene encoding the NADH-dependent aldehyde/alcohol dehydrogenase responsible for butanol production in alcohologenic cultures of *Clostridium acetobutylicum* ATCC 824. *J Bacteriol*, 184, 821–830.

Gabriel CL. (1928). Butanol fermentation process. *Ind Eng Chem*, 20, 1063–1067.

Garza E, Zhao JF, Wang YZ, Wang JH, Iverson A, Manow R, Finan C, Zhou SD. (2012). Engineering a homobutanol fermentation pathway in *Escherichia coli* EG03. *J Ind Microbiol Biotechnol*, 39, 1101–1107.

Geng QH, Park CH. (1993). Controlled-pH batch butanol-acetone fermentation by low acid producing *Clostridium acetobutylicum* B18. *Biotechnol Lett*, 15, 421–426.

Gottwald M, Gottschalk G. (1985). The internal pH of *Clostridium acetobutylicum* and its effect on the shift from acid to solvent formation. *Arch Microbiol*, 143, 42–46.

Green Biologics press release. (2012). Available at: http://www.greenbiologics.com/pr-butylfuel.php (Last accessed: 1/12/2014).

Green EM. (2011). Fermentative production of butanol—the industrial perspective. *Curr Opin Biotechnol*, 22, 337–343.

Green EM, Boynton ZL, Harris LM, Rudolph FB, Papoutsakis ET, Bennett GN. (1996). Genetic manipulation of acid formation pathways by gene inactivation in *Clostridium acetobutylicum* ATCC 824. *Microbiology*, 142, 2079–2086.

Grimmler C, Held C, Liebl W, Ehrenreich A. (2010). Transcriptional analysis of catabolite repression in *Clostridium acetobutylicum* growing on mixtures of D-glucose and D-xylose. *J Biotechnol*, 150, 315–323.

Grimmler C, Janssen H, Krausse D, Fischer RJ, Bahl H, Drre P, Liebl W, Ehrenreich A. (2011). Genome-wide gene expression analysis of the switch between acidogenesis and solventogenesis in continuous cultures of *Clostridium acetobutylicum*. J Mol Microbiol Biotechnol, 20, 1–15.

Han B, Gopalan V, Ezeji TC. (2011). Acetone production in solventogenic *Clostridium* species: new insights from non-enzymatic decarboxylation of acetoacetate. *Appl Microbiol Biotechnol*, 91, 565–576.

Hartmanis MG. (1987). Butyrate kinase from *Clostridium acetobutylicum*. *J Biol Chem*, 262, 617–621.

Hartmanis MGN, Gatenbeck S. (1984). Intermediary metabolism in *Clostridium acetobutylicum*: levels of enzymes involved in the formation of acetate and butyrate. *Appl Environ Microbiol*, 47, 1277–1283.

Haus S, Jabbari S, Millat T, Janssen H, Fischer R-J, Bahl H, King J, Wolkenhauer O. (2011). A systems biology approach to investigate the effect of pH-induced gene regulation on solvent production by *Clostridium acetobutylicum* in continuous culture. *BMC Syst Biol*, 5, 10.

Heap JT, Pennington OJ, Cartman ST, Carter GP, Minton NP. (2007). The ClosTron: a universal gene knock-out system for the genus *Clostridium*. *J Microbiol Methods*, 70, 452–464.

Heap JT, Kuehne SA, Ehsaan M, Cartman ST, Cooksley CM, Scott JC, Minton NP. (2010). The ClosTron: mutagenesis in *Clostridium* refined and streamlined. *J Microbiol Methods*, 80, 49–55.

Hobson PN, Nasr H. (1951). An amylopectin-type polysaccharide synthesised from sucrose by *Clostridium butyricum*. *J Chem Soc*, 407, 1855–1857.

Hönicke D, Janssen H, Grimmler C, Ehrenreich A, Lütke-Eversloh T. (2012). Global transcriptional changes of *Clostridium acetobutylicum* cultures with increased butanol:acetone ratios. *New Biotech*, 29, 485–493.

Hu SY, Zheng HJ, Gu Y, Zhao JB, Zhang WW, Yang YL, Wang SY, Zhao GP, Yang S, Jiang WH. (2011). Comparative genomic and transcriptomic analysis revealed genetic characteristics related to solvent formation and xylose utilization in *Clostridium acetobutylicum* EA 2018. *BMC Genomics*, 12. doi: 10.1186/1471-2164-12-93.

Yukawa H. (2008). Expression of *Clostridium acetobutylicum* butanol synthetic genes in *Escherichia coli*. Appl Microbiol Biotechnol, 77, 1305-16.

Janssen H, Doring C, Ehrenreich A, Voigt B, Hecker M, Bahl H, Fischer RJ. (2010). A proteomic and transcriptional view of acidogenic and solventogenic steady-state cells of *Clostridium acetobutylicum* in a chemostat culture. *Appl Microbiol Biotechnol*, 87, 2209–2226.

Janssen H, Grimmler C, Ehrenreich A, Bahl H, Fischer R-J. (2012). A transcriptional study of acidogenic chemostat cells of *Clostridium acetobutylicum*—solvent stress caused by a transient n-butanol pulse. *J Biotechnol*, 161, 354–365.

Jia KZ, Zhang YP, Li Y. (2012). Identification and characterization of two functionally unknown genes involved in butanol tolerance of *Clostridium acetobutylicum*. *PLoS One*, 7, e38815.

Jiang Y, Xu CM, Dong F, Yang YL, Jiang WH, Yang S. (2009). Disruption of the acetoacetate decarboxylase gene in solvent-producing *Clostridium acetobutylicum* increases the butanol ratio. *Metab Eng*, 11, 284–291.

Jones DT, Woods DR. (1986). Acetone-butanol fermentation revisited. *Microbiological Reviews*, 50, 484–524.

Jones DT, van der Westhuizen A, Long S, Allcock ER, Reid SJ, Woods DR. (1982). Solvent production and morphological changes in *Clostridium acetobutylicum*. *Appl Environ Microbiol*, 43, 1434–1439.

Jones SW, Paredes CJ, Tracy B, Cheng N, Sillers R, Senger RS, Papoutsakis ET. (2008). The transcriptional program underlying the physiology of clostridial sporulation. *Genome Biol*, 9, R114.

Jones SW, Tracy BP, Gaida SM, Papoutsakis ET. (2011). Inactivation of 6^F in *Clostridium acetobutylicum* ATCC 824 blocks sporulation prior to asymmetric division and abolishes

σ^E and σ^G protein expression but does not block solvent formation. *J Bacteriol*, 193, 2429–2440.

Keis S, Sullivan JT, Jones DT. (2001). Physical and genetic map of the *Clostridium saccharobutylicum* (formerly *Clostridium acetobutylicum*) NCP 262 chromosome. *Microbiology*, 147, 1909–1922.

Killeffer DH. (1927). Butanol and acetone from corn-a description of the fermentation process. *Industrial & Engineering Chemistry*, 19, 46–50.

Kirschner M, (2006). n-Butanol. *Chemical Market Reporter*, January 30–February 5, ABI/INFORM Global 42.

Köpke M, Held C, Hujer S, Liesegang H, Wiezer A, Wollherr A, Ehrenreich A, Liebl W, Gottschalk G, Dürre P. (2010). *Clostridium ljungdahlii* represents a microbial production platform based on syngas. *Proc Natl Acad Sci U S A*, 107, 13087–13092.

Krulwich TA, Sachs G, Padan E. (2011). Molecular aspects of bacterial pH sensing and homeostasis. *Nat Rev Microbiol*, 9, 330–343.

Kuit W, Minton N, López-Contreras A, Eggink G. (2012). Disruption of the acetate kinase (*ack*) gene of *Clostridium acetobutylicum* results in delayed acetate production. *Appl Microbiol Biotechnol*, 94, 729–741.

Kumar M, Gayen K. (2011). Developments in biobutanol production: new insights. *Appl Energy*, 88, 1999–2012.

Lan EI, Liao JC. (2011). Metabolic engineering of cyanobacteria for 1-butanol production from carbon dioxide. *Metab Eng*, 13, 353–363.

Lan EI, Liao JC. (2012). ATP drives direct photosynthetic production of 1-butanol in cyanobacteria. *Proc Natl Acad Sci U S A*, 109, 6018–6023.

Lee J, Yun H, Feist AM, Palsson BO, Lee SY. (2008). Genome-scale reconstruction and *in silico* analysis of the *Clostridium acetobutylicum* ATCC 824 metabolic network. *Appl Microbiol Biotechnol*, 80, 849–862.

Lee J, Jang Y-S, Choi SJ, Im JA, Song H, Cho JH, Seung DY, Papoutsakis ET, Bennett GN, Lee SY. (2012). Metabolic engineering of *Clostridium acetobutylicum* ATCC 824 for isopropanol-butanol-ethanol fermentation. *Appl Environ Microbiol*, 78, 1416–1423.

Lehmann D, Lütke-Eversloh T. (2011). Switching *Clostridium acetobutylicum* to an ethanol producer by disruption of the butyrate/butanol fermentative pathway. *Metab Eng*, 13, 464–473.

Lehmann D, Hönicke D, Ehrenreich A, Schmidt M, Weuster-Botz D, Bahl H, Lütke-Eversloh T. (2012a). Modifying the product pattern of *Clostridium acetobutylicum*. *Appl Microbiol Biotechnol*, 94, 743–754.

Lehmann D, Radomski N, Lütke-Eversloh T. (2012b). New insights into the butyric acid metabolism of *Clostridium acetobutylicum*. *Appl Microbiol Biotechnol*, 96, 1325–1339.

Li SY, Srivastava R, Suib SL, Li Y, Parnas RS. (2011). Performance of batch, fed-batch, and continuous A-B-E fermentation with pH-control. *Bioresour Technol*, 102, 4241–4250.

Lin YL, Blaschek HP. (1983). Butanol production by a butanol-tolerant strain of *Clostridium acetobutylicum* in extruded corn broth. *Appl Environ Microbiol*, 45, 966–973.

Lopez-Contreras AM, Martens AA, Szijarto N, Mooibroek H, Claassen PAM, van der Oost J, de Vos WM. (2003). Production by *Clostridium acetobutylicum* ATCC 824 of CelG, a cellulosomal glycoside hydrolase belonging to family 9. *Appl Environ Microbiol*, 69, 869–877.

Lütke-Eversloh T, Bahl H. (2011). Metabolic engineering of *Clostridium acetobutylicum*: recent advances to improve butanol production. *Curr Opin Biotechnol*, 22, 634–647.

Mao S, Luo Y, Zhang T, Li J, Bao G, Zhu Y, Chen Z, Zhang Y, Li Y, Ma Y. (2010). Proteome reference map and comparative proteomic analysis between a wild type *Clostridium acetobutylicum* DSM 1731 and its mutant with enhanced butanol tolerance and butanol yield. *J Proteome Res*, 9, 3046–3061.

Mao SM, Luo YM, Bao GH, Zhang YP, Li Y, Ma YH. (2011). Comparative analysis on the membrane proteome of *Clostridium acetobutylicum* wild type strain and its butanol-tolerant mutant. *Mol Biosyst*, 7, 1660–1677.

Mariano AP, de Angelis DdF, Maugeri Filho F, Atala DIP, Wolf Maciel MR, Maciel Filho R. (2008). An alternative process for butanol production: continuous flash fermentation. *Chemical Product and Process Modeling*, 3, 1934–2659.

Mariano AP, Qureshi N, Maciel R, Ezeji TC. (2011). Bioproduction of butanol in bioreactors: new insights from simultaneous in situ butanol recovery to eliminate product toxicity. *Biotechnol Bioeng*, 108, 1757–1765.

Mascal M. (2012). Chemicals from biobutanol: technologies and markets. *Biofuels Bioprod Biorefining*, 6, 483–493.

McAnulty M, Yen J, Freedman B, Senger R. (2012). Genome-scale modeling using flux ratio constraints to enable metabolic engineering of clostridial metabolism *in silico*. *BMC Syst Biol*, 6, 42.

Mehta RN, Chakraborty M, Mahanta P, Parikh PA. (2010). Evaluation of fuel properties of butanol-biodiesel-diesel blends and their impact on engine performance and emissions. *Ind Eng Chem Res*, 49, 7660–7665.

Mermelstein LD, Papoutsakis ET. (1993). In vivo methylation in *Escherichia coli* by the *Bacillus subtilis* phage φ3T I methyltransferase to protect plasmids from restriction upon transformation of *Clostridium acetobutylicum* ATCC 824. *Appl Environ Microbiol*, 59, 1077–1081.

Mermelstein LD, Welker NE, Bennett GN, Papoutsakis ET. (1992). Expression of cloned homologous fermentative genes in *Clostridium acetobutylicum* ATCC 824. *Bio-Technology*, 10, 190–195.

militaryaerospace.com (2012). Available at: http://www.militaryaerospace.com/articles/2012/03/renewable_bio_n-butanoljetfuelbeingproducedbycobaltandnavalairwa.html (Last accessed: 1/12/2014).

Milne C, Eddy J, Raju R, Ardekani S, Kim P-J, Senger R, Jin Y-S, Blaschek H, Price N. (2011). Metabolic network reconstruction and genome-scale model of butanol-producing strain *Clostridium beijerinckii* NCIMB 8052. *BMC Syst Biol*, 5, 130.

Nielsen DR, Leonard E, Yoon SH, Tseng HC, Yuan C, Prather KLJ. (2009). Engineering alternative butanol production platforms in heterologous bacteria. *Metab Eng*, 11, 262–273.

Nishio N, Biebl H, Meiners M. (1983). Effect of pH on the production of acetone and butanol by *Clostridium acetobutylicum* in a minimum medium. *J Ferment Technol*, 61, 101–104.

Nolling J, Breton G, Omelchenko MV, Makarova KS, Zeng QD, Gibson R, Lee HM, Dubois J, Qiu DY, Hitti J, Wolf YI, Tatusov RL, Sabathe F, Doucette-Stamm L, Soucaille P, Daly MJ, Bennett GN, Koonin EV, Smith DR. (2001). Genome sequence and comparative analysis of the solvent-producing bacterium *Clostridium acetobutylicum*. *J Bacteriol*, 183, 4823–4838.

NYTimes (2012). Available at: http://www.nytimes.com/2012/01/02/business/energy-environment/after-three-decades-federal-tax-credit-for-ethanol-expires.html (Last accessed: 1/12/2014).

Ounine K, Petitdemange H, Raval G, Gay R. (1985). Regulation and butanol inhibition of D-xylose and D-glucose uptake in *Clostridium acetobutylicum*. *Appl Environ Microbiol*, 49, 874–878.

Papoutsakis ET. (2008). Engineering solventogenic clostridia. *Curr Opin Biotechnol*, 19, 420–429.

Paredes CJ, Alsaker KV, Papoutsakis ET. (2005). A comparative genomic view of clostridial sporulation and physiology. *Nat Rev Microbiol*, 3, 969–978.

Pasteur L. (1861). Animacules infusoires vivant sans oxygène libre et déterminant des fermentations. *C R Hebd Séances Acad Sci*, 52, 344–347.

Peralta-Yahya PP, Keasling JD. (2010). Advanced biofuel production in microbes. *Biotechnol J*, 5, 147–162.

Petersen DJ, Bennett GN. (1990). Purification of acetoacetate decarboxylase from *Clostridium acetobutylicum* ATCC 824 and cloning of the acetoacetate decarboxylase gene in *Escherichia coli*. *Appl Environ Microbiol*, 56, 3491–3498.

Phillips JA, Humphrey AE. (1983). An overview of process technology for the production of liquid fuels and chemical feedstocks via fermentation. In: Wise DL, editor. *Organic Chemicals from Biomass*. Menlo Park, CA: Benjamins/Cummings Publishing, pp. 249–304.

Pich A, Narberhaus F, Bahl H. (1990). Induction of heat-shock proteins during initiation of solvent formation in *Clostridium acetobutylicum*. *Appl Microbiol Biotechnol*, 33, 697–704.

Qureshi N, Blaschek HP. (2001). ABE production from corn: a recent economic evaluation. *J Ind Microbiol Biotechnol*, 27, 292–297.

Qureshi N, Sahaa BC, Hector RE, Hughes SR, Cotta MA. (2008). Butanol production from wheat straw by simultaneous saccharification and fermentation using *Clostridium beijerinckii*: Part I-batch fermentation. *Biomass Bioenerg*, 32, 168–175.

Rakopoulos DC, Rakopoulos CD, Giakoumis EG, Dimaratos AM, Kyritsis DC. (2010). Effects of butanol-diesel fuel blends on the performance and emissions of a high-speed DI diesel engine. *Energy Conv Manag*, 51, 1989–1997.

Ranganathan S, Maranas CD. (2010). Microbial 1-butanol production: identification of non-native production routes and *in silico* engineering interventions. *Biotechnol J*, 5, 716–725.

Ren C, Gu Y, Hu SY, Wu Y, Wang P, Yang YL, Yang C, Yang S, Jiang WH. (2010). Identification and inactivation of pleiotropic regulator CcpA to eliminate glucose repression of xylose utilization in *Clostridium acetobutylicum*. *Metab Eng*, 12, 446–454.

Ren C, Gu Y, Wu Y, Zhang W, Yang C, Yang S, Jiang W. (2012). Pleiotropic functions of catabolite control protein CcpA in butanol-producing *Clostridium acetobutylicum*. *BMC Genomics*, 13, 349.

Reysenbach AL, Ravenscroft N, Long S, Jones DT, Woods DR. (1986). Characterization, biosynthesis, and regulation of granulose in *Clostridium acetobutylicum*. *Appl Environ Microbiol*, 52, 185–190.

Ross D. (1961). The acetone-butanol fermentation. *Prog Ind Microbiol*, 3, 73–85.

Ruhl J, Schmid A, Blank LM. (2009). Selected *Pseudomonas putida* strains able to grow in the presence of high butanol concentrations. *Appl Environ Microbiol*, 75, 4653–4656.

Salleh MM, Tsuey LS, Bin Ariff A. (2008). The profile of enzymes relevant to solvent production during direct fermentation of sago starch by *Clostridium saccharobutylicum* P262 utilizing different pH control strategies. *Biotechnol Bioprocess Eng*, 13, 33–39.

Schaffer S, Isci N, Zickner B, Durre P. (2002). Changes in protein synthesis and identification of proteins specifically induced during solventogenesis in *Clostridium acetobutylicum. Electrophoresis*, 23, 110–121.

Schwarz KM, Kuit W, Grimmler C, Ehrenreich A, Kengen SWM. (2012). A transcriptional study of acidogenic chemostat cells of *Clostridium acetobutylicum*—cellular behavior in adaptation to n-butanol. *J Biotechnol*, 161, 366–377.

Senger RS, Papoutsakis ET. (2008a). Genome-scale model for *Clostridium acetobutylicum*: Part I. Metabolic network resolution and analysis. *Biotechnol Bioeng*, 101, 1036–1052.

Senger RS, Papoutsakis ET. (2008b). Genome-scale model for *Clostridium acetobutylicum*: Part II. Development of specific proton flux states and numerically determined sub-systems. *Biotechnol Bioeng*, 101, 1053–1071.

Servinsky MD, Kiel JT, Dupuy NF, Sund CJ. (2010). Transcriptional analysis of differential carbohydrate utilization by *Clostridium acetobutylicum. Microbiology*, 156, 3478–3491.

Shaheen R, Shirley M, Jones DT. (2000). Comparative fermentation studies of industrial strains belonging to four species of solvent-producing clostridia. *J Mol Microbiol Biotechnol*, 2, 115–124.

Shao L, Hu S, Yang Y, Gu Y, Chen J, Jiang W, Yang S. (2007). Targeted gene disruption by use of a group II intron (targetron) vector in *Clostridium acetobutylicum. Cell Res*, 17, 963–965.

Shen CR, Liao JC. (2008). Metabolic engineering of *Escherichia coli* for 1-butanol and 1-propanol production via the keto-acid pathways. *Metab Eng*, 10, 312–320.

Shen CR, Lan EI, Dekishima Y, Baez A, Cho KM, Liao JC. (2011). Driving forces enable high-titer anaerobic 1-butanol synthesis in *Escherichia coli. Appl Environ Microbiol*, 77, 2905–2915.

Shen GJ, Shieh JS, Grethlein AJ, Jain MK, Zeikus JG. (1999). Biochemical basis for carbon monoxide tolerance and butanol production by *Butyribacterium methylotrophicum*. Appl Microbiol Biotechnol, 51, 827–32.

Shi Z, Blaschek HP. (2008). Transcriptional analysis of *Clostridium beijerinckii* NCIMB 8052 and the hyper-butanol-producing mutant BA101 during the shift from acidogenesis to solventogenesis. *Appl Environ Microbiol*, 74, 7709–7714.

Sillers R, Chow A, Tracy B, Papoutsakis ET. (2008). Metabolic engineering of the non-sporulating, non-solventogenic *Clostridium acetobutylicum* strain M5 to produce butanol without acetone demonstrate the robustness of the acid-formation pathways and the importance of the electron balance. *Metab Eng*, 10, 321–332.

Soni BK, Kapp C, Goma G, Soucaille P. (1992). Solvent production from starch: effect of pH on α-amylase and glucoamylase localization and synthesis in synthetic medium. *Appl Microbiol Biotechnol*, 37, 539–543.

Soucaille P, Figge R, Croux C. (2007). Process for chromosomal integration and DNA sequence replacement in clostridia. US Patent 8,168,434 (April 18, 2007).

SRI Consulting (2009). World petrochemicals report, OXO alcohols. Englewood, CA: SRI Consulting.

Steen EJ, Chan R, Prasad N, Myers S, Petzold CJ, Redding A, Ouellet M, Keasling JD. (2008). Metabolic engineering of *Saccharomyces cerevisiae* for the production of n-butanol. *Microb Cell Fact*, 7, 36.

Steiner E, Dago AE, Young DI, Heap JT, Minton NP, Hoch JA, Young M. (2011). Multiple orphan histidine kinases interact directly with spo0A to control the initiation of endospore formation in *Clostridium acetobutylicum*. *Mol Microbiol*, 80, 641–654.

Steiner E, Scott J, Minton NP, Winzer K. (2012). An *agr* quorum sensing system that regulates granulose formation and sporulation in *Clostridium acetobutylicum*. *Appl Environ Microbiol*, 78, 1113–1122.

Sullivan L, Bennett GN. (2006). Proteome analysis and comparison of *Clostridium acetobutylicum* ATCC 824 and spo0A strain variants. *J Ind Microbiol Biotechnol*, 33, 298–308.

Survase SA, Jurgens G, van Heiningen A, Granstrom T. (2011). Continuous production of isopropanol and butanol using *Clostridium beijerinckii* DSM 6423. *Appl Microbiol Biotechnol*, 91, 1305–1313.

Terracciano JS, Kashket ER. (1986). Intracellular conditions required for initiation of solvent production by *Clostridium acetobutylicum*. *Appl Environ Microbiol*, 52, 86–91.

Terracciano JS, Rapaport E, Kashket ER. (1988). Stress phase-associated and growth phase-associated proteins of *Clostridium acetobutylicum*. *Appl Environ Microbiol*, 54, 1989–1995.

Tomas CA, Alsaker KV, Bonarius HPJ, Hendriksen WT, Yang H, Beamish JA, Paredes CJ, Papoutsakis ET. (2003a). DNA array-based transcriptional analysis of asporogenous, nonsolventogenic *Clostridium acetobutylicum* strains SKO1 and M5. *J Bacteriol*, 185, 4539–4547.

Tomas CA, Welker NE, Papoutsakis ET. (2003b). Overexpression of *groESL* in *Clostridium acetobutylicum* results in increased solvent production and tolerance, prolonged metabolism, and changes in the cell's transcriptional program. *Appl Environ Microbiol*, 69, 4951–4965.

Tomas CA, Beamish J, Papoutsakis ET. (2004). Transcriptional analysis of butanol stress and tolerance in *Clostridium acetobutylicum*. *J Bacteriol*, 186, 2006–2018.

Tracy BP, Jones SW, Papoutsakis ET. (2011). Inactivation of σ^E and σ^G in *Clostridium acetobutylicum* illuminates their roles in clostridial-cell-form biogenesis, granulose synthesis, solventogenesis, and spore morphogenesis. *J Bacteriol*, 193, 1414–1426.

Tummala SB, Welker NE, Papoutsakis ET. (1999). Development and characterization of a gene expression reporter system for *Clostridium acetobutylicum* ATCC 824. *Appl Environ Microbiol*, 65, 3793–3799.

Tummala SB, Junne SG, Papoutsakis ET. (2003a). Antisense RNA downregulation of coenzyme a transferase combined with alcohol-aldehyde dehydrogenase overexpression leads to predominantly alcohologenic *Clostridium acetobutylicum* fermentations. *J Bacteriol*, 185, 3644–3653.

Tummala SB, Welker NE, Papoutsakis ET. (2003b). Design of antisense RNA constructs for downregulation of the acetone formation pathway of *Clostridium acetobutylicum*. *J Bacteriol*, 185, 1923–1934.

Vollherbstschneck K, Sands JA, Montenecourt BS. (1984). Effect of butanol on lipid-composition and fluidity of *Clostridium acetobutylicum* ATCC 824. *Appl Environ Microbiol*, 47, 193–194.

Walter KA, Bennett GN, Papoutsakis ET. (1992). Molecular characterization of two *Clostridium acetobutylicum* ATCC 824 butanol dehydrogenase isozyme genes. *J Bacteriol*, 174, 7149–7158.

Walter KA, Nair RV, Cary JW, Bennett GN, Papoutsakis ET. (1993). Sequence and arrangement of two genes of the butyrate-synthesis pathway of *Clostridium acetobutylicum* ATCC 824. *Gene*, 134, 107–111.

Wang Y, Blaschek HP. (2011). Optimization of butanol production from tropical maize stalk juice by fermentation with *Clostridium beijerinckii* NCIMB 8052. *Bioresour Technol*, 102, 9985–9990.

Wang Y, Li X, Mao Y, Blaschek H. (2011). Single-nucleotide resolution analysis of the transcriptome structure of *Clostridium beijerinckii* NCIMB 8052 using RNA-seq. *BMC Genomics*, 12, 479.

Wang Y, Li X, Mao Y, Blaschek H. (2012). Genome-wide dynamic transcriptional profiling in *Clostridium beijerinckii* NCIMB 8052 using single-nucleotide resolution RNA-seq. *BMC Genomics*, 13, 102.

Welsh FW, Veliky IA. (1984). Production of acetone-butanol from acid whey. *Biotechnol Lett*, 6, 61–64.

Whyte JNC, Strasdine GA. (1972). An intracellular α-D-glucan from *Clostridium botulinum*, type E. *Carbohydr Res*, 25, 435–441.

Wiesenborn DP, Rudolph FB, Papoutsakis ET. (1989). Coenzyme A transferase from *Clostridium acetobutylicum* ATCC 824 and its role in the uptake of acids. *Appl Environ Microbiol*, 55, 323–329.

Wietzke M, Bahl H. (2012). The redox-sensing protein Rex, a transcriptional regulator of solventogenesis in *Clostridium acetobutylicum*. *Appl Microbiol Biotechnol*, 96, 749–761.

WSJ (2009). Available at: http://online.wsj.com/article/SB125427984449051595.html (Last accessed: 1/12/2014).

Wu M, Wang M, Liu JH, Huo H. (2008). Assessment of potential life-cycle energy and greenhouse gas emission effects from using corn-based butanol as a transportation fuel. *Biotechnol Prog*, 24, 1204–1214.

Wu Y-R, Li Y, Yang K-L, He J. (2012). Draft genome sequence of butanol-acetone-producing *Clostridium beijerinckii* strain G117. *J Bacteriol*, 194, 5470–5471.

Yu MR, Zhang YL, Tang IC, Yang ST. (2011). Metabolic engineering of *Clostridium tyrobutyricum* for n-butanol production. *Metab Eng*, 13, 373–382.

Yu MR, Du YM, Jiang WY, Chang WL, Yang ST, Tang IC. (2012). Effects of different replicons in conjugative plasmids on transformation efficiency, plasmid stability, gene expression and n-butanol biosynthesis in *Clostridium tyrobutyricum*. *Appl Microbiol Biotechnol*, 93, 881–889.

Zhang WW, Li F, Nie L. (2010). Integrating multiple 'omics' analysis for microbial biology: application and methodologies. *Microbiology*, 156, 287–301.

Zhang F, Rodriguez S, Keasling JD. (2011). Metabolic engineering of microbial pathways for advanced biofuels production. *Curr Opin Biotechnol*, 22, 775–783.

Zhao YS, Tomas CA, Rudolph FB, Papoutsakis ET, Bennett GN. (2005). Intracellular butyryl phosphate and acetyl phosphate concentrations in *Clostridium acetobutylicum* and their implications for solvent formation. *Appl Environ Microbiol*, 71, 530–537.

Zheng Y, Li L, Xian M, Ma Y, Yang J, Xu X, He D. (2009). Problems with the microbial production of butanol. *J Ind Microbiol Biotechnol*, 36, 1127–1138.

CHAPTER 10

Bio-based Butanediols Production: The Contributions of Catalysis, Metabolic Engineering, and Synthetic Biology

XIAO-JUN JI and HE HUANG

State Key Laboratory of Materials-Oriented Chemical Engineering, College of Biotechnology and Pharmaceutical Engineering, Nanjing Tech University, People's Republic of China

Abstract

Butanediols (BDs) are important in the chemical industry with many applications. At present, almost all BD isomers are produced from fossil feedstock. Due to the shortage of the fossil resources, with the technology developed, some BD isomers are approaching to be commercialized for economical bio-based production. In this chapter, two important bio-based BDs' (2,3-BD and 1,4-BD) production are described.

Bioprocessing of Renewable Resources to Commodity Bioproducts, First Edition.
Edited by Virendra S. Bisaria and Akihiko Kondo.

The key technologies involved in their production processes, that is, catalysis, metabolic engineering, and the emerging synthetic biology methods, are highlighted, as they would offer significant opportunities to further decrease the bio-based BDs' production costs, which would accelerate the bio-based BDs' industrialization.

10.1 INTRODUCTION

As crude oil reserves become increasingly scarce, biorefinery systems that integrate biomass conversion processes and equipment to produce fuels, power, and chemicals from annually renewable bio-resources are at the stage of worldwide development (Kamm and Kamm, 2004; Ragauskas et al., 2006; Li et al., 2010b). Many chemicals that could only be produced from the petroleum feedstock in the past can now have the potential to be generated using renewable bioresources (Danner and Braun, 1999; Hatti-Kaul et al., 2007). Production of butanediols (BDs) from biomass is one such example. BDs have four stable isomers, that is, 1,2-BD, 1,3-BD, 2,3-BD, and 1,4-BD. Depending on the configurations of the chiral carbon atoms in the molecule, the BDs can exist as different optical forms except 1,4-BD (Figure 10.1).

BDs are one type of diol with many applications. 1,2-BD can be used to produce polyester resins and plasticizers (Hasegawa and Hayashi, 1986; Anonymous, 1995), and it is a potential feedstock for the industrial production of α-ketobutyric acid, a precursor to many amino acids (Imanari et al., 1992). In contrary to the chemical route from the butadiene, the bio-based 1,2-BD can be obtained as the by-product of the catalytic hydrocracking of starches and sugars such as sorbitol to ethylene glycol and 1,2-propanediol. 1,3-BD is commonly used as a solvent for food-flavouring agents and is a comonomer used in certain polyurethane and polyester resins (Amada et al., 2012). In biology, 1,3-BD can be used as a hypoglycemic agent. Traditionally, 1,3-BD is produced via 3-hydroxybutyraldehyde from the aldol condensation of acetaldehyde that is produced by the Wacker oxidation of ethylene (Burgard et al., 2011). Recently,

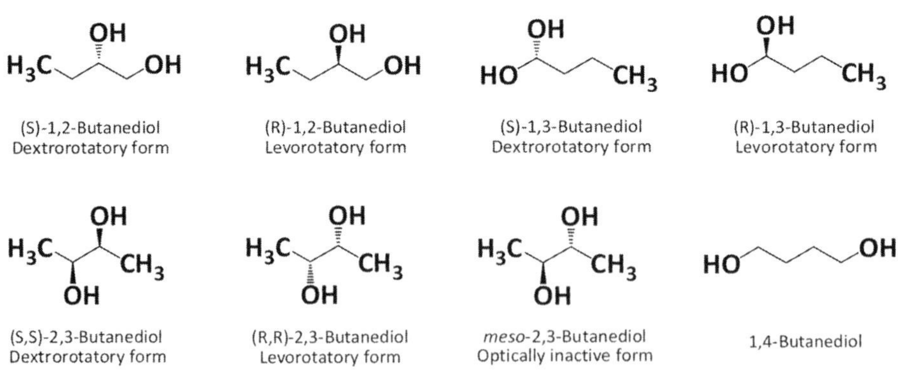

Fig. 10.1

FIGURE 10.1 Molecular structures of different butanediol isomers

it was reported that a Japanese company named Kokyu Alcohol Kogyo Co. Ltd. has developed bio-based 1,3-BD from the sugarcanes with ethanol as the intermediate (Anonymous, 2007).

Among the BD isomers, 2,3-BD and 1,4-BD are two most important commodities. Three stereoisomers of 2,3-BD exist—dextro-[(S, S)-] and levo- [(R, R)-] forms that are optically active, as well as an optically inactive meso form (Figure 10.1). The levo isomer has a low freezing point of −60°C which makes it interesting for commercial use as an antifreeze. Both the levo and dextro 2,3-BD isomers are excellent chiral components for asymmetric synthesis or liquid crystals (Xiao et al., 2010; Liu et al., 2011; Zeng and Sabra, 2011; Shen et al., 2013). Due to the structure of 2,3-BD, there are several interesting chemical reactions, including dehydration, dehydrogenation, ketalization, esterification, and polymerization, which make convenient 2,3-BD derivatives preparation possible (Ji et al., 2011a; Nie et al., 2011; Gubbels et al., 2013). The dehydration of 2,3-BD yields the industrial solvent methyl ethyl ketone (MEK) (Tran and Chambers, 1987; Ji et al., 2012; Zhang et al., 2012). Further dehydration produces 1,3-butadiene that is the building block of synthetic rubber (van Haveren et al., 2008). The dehydrogenation of 2,3-BD could form two high-value compounds, acetoin and diacetyl (Faveri et al., 2003; Ji et al., 2013). Also the high octane rating of 2,3-BD makes it a potential aviation fuel (Celińska and Grajek, 2009; Ji et al., 2012). Additionally, 2,3-BD may serve as a feedstock for the production of numerous valuable chemicals, including printing inks, perfumes, fumigants, moistening and softening agents, explosives, plasticizers, foods, and pharmaceuticals (Garg and Jain, 1995; Syu, 2001; Ji et al., 2011a). The bio-based 2,3-BD production would alleviate the dependence on oil supply for the production of platform chemicals. Therefore, interest in this bioprocess has been increasing recently.

1,4-BD is an important commodity chemical used to manufacture over 2.5 million tons of valuable polymers annually. In addition, 1,4-BD is used industrially as a solvent and in the synthesis of some other chemicals. In the presence of phosphoric acid and high temperature, it dehydrates to the important solvent tetrahydrofuran (Hunter et al., 2006). At about 200°C, in the presence of soluble ruthenium catalysts, 1,4-BD undergoes dehydrogenation to form γ-butyrolactone (Zhao and Hartwig, 2005). The global market for these products is estimated to exceed $4 billion, and they are principally used to make polyesters, polyurethanes, spandex, and biodegradable plastics. 1,4-BD is currently produced exclusively through feedstock derived from oil and natural gas. Recently, some bio-based route for 1,4-BD production has been successfully established (Burk, 2010; Minh et al., 2010; Yim et al., 2011). This would alleviate the dependence on oil supply for the production of platform chemicals.

In this chapter, the state-of-the-art developments and technological achievements of the biological route for the two important diols (2,3-BD and 1,4-BD) production have been summarized. Especially, the key technologies involved in the production of the two bio-based diols, including catalysis, metabolic engineering, and synthetic biology were highlighted. Furthermore, the economic outlook and future prospects on the bio-based BDs' production are discussed.

10.2 BIO-BASED 2,3-BUTANEDIOL

10.2.1 Via Catalytic Hydrogenolysis

The catalysis-mediated method for 2,3-BD production uses the sugar alcohol as substrate. This process has been put into a large scale production by a Chinese company, Changchun Dacheng Group (located in Changchun, Jilin Province, China). The process developed was aimed to produce C2–C4 dihydroxy alcohol and polyol, which generally consisted of about 50–60% 1,2-propanediol, 25% ethylene glycol, and about 25% 1,2- and 2,3-BD as coproducts.

By the independent innovation and the use of biomass as raw material, Changchun Dacheng Group has successfully developed the catalyst technology needed in this project (Liu and Xu, 2006). The catalyst has a high selectivity to the unique hydroxyl structure of biomass molecules and can turn the biomass materials like corn starch into glucose by hydrolysis. The glucose can be converted into sorbitol after catalytic hydrogenation, which subsequently cracks into the mixture of C2–C4 dihydroxy alcohol and polyol. After the heavy ends (including organic salts) being removed, distillated, and refined, 1,2-propanediol, ethylene glycol, and 2,3-BD, with purity of more than 97% and the yield of about 5%, can be obtained (Figure 10.2) (Xu, 2005; Liu et al., 2008, 2009; Ge et al., 2011). The corn-based bio-polyol project with the annual output of 200,000 tons has been put into production by the company, and it is planning to expand the annual production capacity of bio-polyol to 400,000 tons. Also, a corn industrial park with annual production of 1 million tons is under preparation, which can produce 60,000 tons 2,3-BD annually. At the same time, this company is also doing cutting edge research on substituting the substrate with the corn stover, which is the abundant residue that is left behind after corn grain harvest.

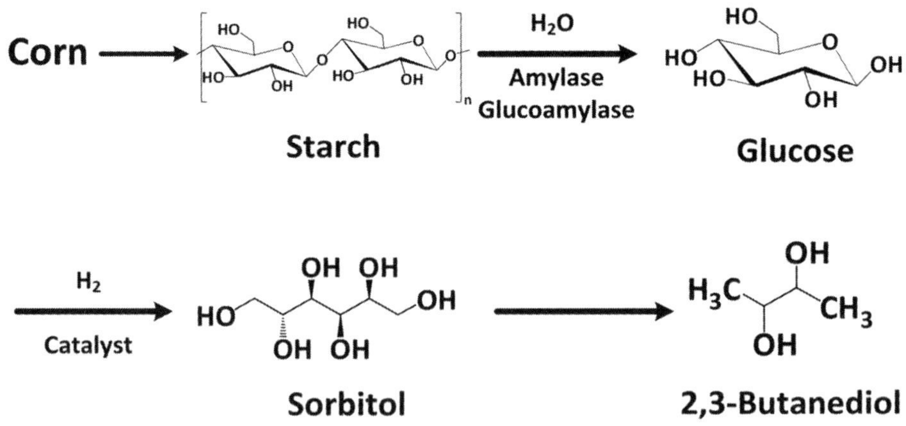

FIGURE 10.2 A new route for the production of 2,3-butanediol (modified from Ge et al., 2011).

10.2.2 Via Sugar Fermentation

10.2.2.1 Microorganisms and Metabolic Pathway A number of microorganisms are able to accumulate 2,3-BD but only a few do so in what might be considered significant quantities. Species which are noted for this ability include those belonging to the genera of Klebsiella, Enterobacter, Paenibacillus, Bacillus, and Serratia, which are considered to be of industrial importance in the production of 2,3-BD (Maddox, 1996).

Until now, *Klebsiella pneumoniae*, *Klebsiella oxytoca*, and *Paenibacillus polymyxa* have been unbeatable in the efficient production of 2,3-BD. *Enterobacter aerogenes* and *Serratia marcescens* are also considered promising microorganisms for this application. As indicated in section 10.1, there are three different 2,3-BD stereoisomers. However, in any given process, the isomer produced is dependent upon the particular microorganism employed. Different microorganisms produce different stereoisomers, but a mixture of two stereoisomers is generally formed (Table 10.1) (Maddox, 1996; Celińska and Grajek, 2009; Shen et al., 2013). The mechanism for this phenomenon has been reviewed previously (Ji et al., 2011a). The product yields and conversion efficiencies of bacteria commonly used for 2,3-BD production are summarized in Table 10.2.

A variety of monosaccharides (hexoses or pentoses) can be fermented to produce 2,3-BD (Syu, 2001). In bacterial metabolism, monosaccharides must first be converted to pyruvate before generation of major products. From glucose, pyruvate is formed in a relatively simple manner via the Embden–Meyerhof pathway (glycolysis). In contrast, the production of pyruvate from pentoses must proceed via a combination of the pentose phosphate and Embden–Meyerhof pathways (Jansen and Tsao, 1983). In addition to 2,3-BD, the pyruvate produced from the monosaccharides is then channeled into a mixture of acetate, lactate, formate, succinate, acetoin, and ethanol, through the mixed acid-2,3-BD fermentation pathway (Ji et al., 2011a).

TABLE 10.1 2,3-Butanediol Stereoisomers Produced by Different Microorganisms

	2,3-Butanediol stereoisomers[a]		
Microorganisms	*L* (+)-	*D* (–)-	*meso*-
Klebsiella pneumoniae	+ (1)	–	+ (9)
Klebsiella oxytoca	+ (1)	–	+ (9)
Enterobacter aerogenes	+ (1)	–	+ (9)
Serratia marcescens	–	–	+
Panebacillus polymyxa	–	+	–
Bacillus subtilis	–	+ (3)	+ (2)
Bacillus amyloliquefaciens	–	+	+

[a] Based on Celińska and Grajek (2009) and Maddox (1996). The relative amounts of the isomers are given in parentheses which are relative to 10.

TABLE 10.2 Microbial 2,3-Butanediol Production Using Different Bacterial Species

	Microbial 2,3-butanediol production[a]						
			Diol concentration (g/L)		Diol productivity	Diol yield	
Microorganism	Substrate	Method	2,3-Butanediol	Acetoin	(g/L • h	(g/g)	Reference
Klebsiella pneumoniae	Glucose	Fed-batch	150	10	4.21	0.43	Ma et al. (2009)
K. pneumoniae	Glucose	Fed-batch	92.4	13.1	2.10	0.49	Qin et al. (2006)
K. pneumoniae	Corn-cob molasses	Fed-batch	78.9	3.6	1.35	0.41	Wang et al. (2010)
K. pneumoniae	Glycerol	Fed-batch	49.2	–	0.18	0.36	Petrov and Petrova (2009)
K. pneumoniae	Glycerol	Fed-batch	70.0	–	0.47	0.39	Petrov and Petrova (2010)
K. pneumoniae	Jerusalem artichoke tuber	Fed-batch; SSF	84.0	7.6	2.29	0.32	Sun et al. (2009)
K. pneumoniae	Jerusalem artichoke stalk and tuber	Fed-batch; SSF	67.4	13.1	1.18	1.81[b]	Li et al. (2010a)
Klebsiella oxytoca	Corn-cob hydrolysate	Fed-batch	35.7	–	0.59	0.50	Cheng et al. (2010)
K. oxytoca	Glucose	Fed-batch	130	1.4	1.64	0.48	Ji et al. (2010)
K. oxytoca	Glucose	Batch	95.5	1.9	1.74	0.49	Ji et al. (2009a)
K. oxytoca	Molasses	Cell recycle	118	2.3	2.40	0.42	Afschar et al. (1991)
K. oxytoca	Corn-cob cellulose	Batch; SSF	25	–	0.36	0.31	Cao et al. (1997)
K. oxytoca	Glucose	Fed-batch	85.5	6.4	3.22	0.50	Qureshi and Cheryan (1989)
K. oxytoca	Acid hydrolysates of Jatropha hulls	Batch	36.31[c]		1.13	0.46	Jiang et al. (2012)

Enterobacter aerogenes	Glucose	Fed-batch	110[c]		5.40	0.49	Zeng et al. (1991)
E.aerogenes	Glucose	Fed-batch	118.1	–	2.19	–	Jung et al. (2012)
Enterobacter cloacae	Cassava powder	Fed-batch; SSF	93.9	5.3	2.11	–	Wang et al. (2012)
Serratia marcescens	Sucrose	Fed-batch	152	0.0	2.67	0.41	Zhang et al. (2010a)
S. marcescens	Sucrose	Fed-batch	139.9	6.7	3.49	0.47	Zhang et al. (2010b)
Paenibacillus polymyxa	Jerusalem artichoke tuber	Batch	36.9	–	0.88	0.50	Gao et al. (2010)
Bacillus subtilis	Glucose	Batch	2.5	0.6	0.33	0.38	Moes et al. (1985)
B. subtilis	Glucose	Batch	6.1	–	0.20	0.34	Biswas et al. (2012)
Bacillus licheniformis	Glucose	Flask without shaking	8.7	–	0.47	0.12	Nilegaonkar et al. (1992)
B. licheniformis	Xylose	Batch	13.8	0.7	2.53	0.48	Wang et al. (2012)
Bacillus amyloliquefaciens	Glucose	Batch	33	–	–	0.33	Alam et al. (1990)
B. amyloliquefaciens	Glucose	Fed-batch	61.4	–	1.71	0.38	Yang et al. (2012)
B. amyloliquefaciens	Glucose	Fed-batch	92.3	–	0.96	0.46	Yang et al. (2011)

SSF: simultaneous saccharification and fermentation.

[a] Diol means 2,3-butanediol plus acetoin.

[b] Means g diol per (3 g stalks + 4 g tubers).

[c] The total production of 2,3-butanediol and acetoin.

10.2.2.2 Metabolic Engineering for Improved 2,3-Butanediol Production Metabolic engineering for efficient microbial 2,3-BD production involves improving both native 2,3-BD producers, that is, homologous hosts, as well as microorganisms with acquired ability to form 2,3-BD via genetic manipulations, that is, heterologous hosts.

Homologous Hosts The conventional 2,3-BD fermentation process is not economically competitive because by-products such as ethanol, acetoin, lactic acid, and acetic acid are simultaneously produced, which inhibit 2,3-BD production (Fond et al., 1985; Zeng et al., 1990), reduce 2,3-BD yield and increase costs of product recovery and purification (Kosaric et al., 1992). To improve the economics of the fermentation process, it is desirable to increase 2,3-BD yield while reducing the production of by-products. Previously, a 2,3-BD hyper-producing *K. oxytoca* ME-XJ-8 was successfully metabolically engineered by altering the mixed acid-2,3-BD fermentation pathway (Ji et al., 2008; 2010). Using the modified proton-suicide method followed by gene insertional inactivation technology, a mutant strain with reduced by-product (ethanol, acetoin, and lactic and acetic acids) release during 2,3-BD fermentation was obtained from the wild-type strain and has proven to be a promising strain for economical 2,3-BD production. By fed-batch culturing of the mutant, a final 2,3-BD titer of up to 130 g/L with a very low level of by-products was obtained, a new record of 2,3-BD fermentation by *K. oxytoca*. Insertional inactivation technology was also successfully applied to reduce by-production of lactic acid for improved 2,3-BD production by *Bacillus licheniformis*, *E. aerogenes,* and *K. oxytoca* (Yang et al., 2007; Jung et al., 2012; Wang et al., 2012b). Zhang et al. (2010a) found that during 2,3-BD fermentation using the strain *S. marcescens*, serrawettin W1, which is a surface-active exolipid, was produced at the same time and thus resulted in foam formation. To avoid excessive addition of antifoam agent and microbial contamination, *S. marcescens* mutants deficient in serrawettin W1 formation were successfully constructed through insertional inactivation of the *swr*W gene coding for serrawettin W1 synthase. Shake-flask batch experiments suggested that disruption of the *swr*W gene led to significant reduction of the foam formation and improved 2,3-BD production at the same time. To enhance the efficiency of 2,3-BD fermentation, besides reducing by-product production, attention should also be paid to the 2,3-BD productivity. A fast glucose-consuming and efficient 2,3-BD-producing strain *K. pneumoniae* SDM was obtained by ion beam mutation (Ma et al., 2009). A record 2,3-BD productivity value of 4.21 g/L·h was obtained in fed-batch culture.

It is normally desirable to use cheaper substrates not requiring pretreatment, such as starch for commercial 2,3-BD production. In line with this, Zheng et al. (2008) cloned the gene *mal*S coding for α-amylase into *K. pneumoniae* and secretory overexpression of α-amylase was carried out in the engineered *K. pneumoniae*. Finally, the one-step production of 2,3-BD from starch was achieved. This strategy simplified the process considerably by enabling starch with minimal pretreatment to be converted to 2,3-BD directly, thereby reducing the production cost significantly.

As good physiological performance of industrial microbes is crucial for successful bioprocesses (Zhang et al., 2009), the *vgb* gene, encoding *Vitreoscilla* (bacterial) hemoglobin (VHb) was introduced into the 2,3-BD producing *E. aerogenes* and *S. marcescens* to improve the physiological functionality of 2,3-BD production in these strains (Wei et al., 1998; Geckil et al., 2004). In both cases, the 2,3-BD production was enhanced. However, the mechanism of VHb for improving 2,3-BD production is still unknown. Three putative reasons have been proposed. The first is that VHb may induce changes in the ratios of $NADH/NAD^+$ and ATP/ADP *in vivo*. Thus, VHb makes the interior of the cell more oxidized, thereby changing the carbon flux within the central carbon metabolic pathways. An alternative interpretation is that the cells which remained alive longer, remained metabolically active longer, which could be the reason for the higher yields. The third opinion is that VHb helped increase oxygen supply to the cells (Celińska and Grajek, 2009).

Heterologous Hosts It is to be noted that chiral synthesis and separation remain costly steps in chemical synthesis and hence using enzymes or microbes to synthesize compounds with high enantiomeric purity represents an alternative and effective approach. Constructing the enantiomerically pure 2,3-BD-producing recombinant organism is one such example (Table 10.3). *Escherichia coli* has frequently been used as a heterologous host engineered for enantiopure 2,3-BD production. Ui et al. (1997) first cloned a gene fragment, encoding genes for three enzymes (ALS, ALDC, and meso-BDH) involved in meso-2,3-BD formation in *K. pneumoniae,* into *E. coli*, and found that the recombinant *E. coli* was able to produce meso-2,3-BD from glucose with no contamination of the (S, S)-form. The highest meso-2,3-BD concentration obtained in recombinant *E. coli* was 17.7 g/L with a starting glucose concentration of 100 g/L. Similarly, a pure meso-2,3-BD biosynthetic pathway was also successfully constructed by other scientists in genetically engineered *E. coli* (Li et al., 2010c; Nielsen et al., 2010; Lee et al., 2012a; Shin et al., 2012). Recently, recombinant *E. coli* strains were engineered to produce (R, R)- and meso-2,3-BD as enantiopure products from glucose by complementing the 2,3-BD biosynthetic pathway of *K. pneumoniae* with secondary alcohol dehydrogenases (sADHs) of differing stereoselectivity. By using engineered whole-cell biocatalysts with strict stereospecificity, which express sADHs, instead of native producers, the enantiomeric selectivity in both (R, R)- and meso-2,3-BD biosynthesis was enhanced and the enantiomeric purity reached over 99% (Yan et al., 2009). Similarly, (R, R)-2,3-BD production was fulfilled in *Clostridium acetobutylicum* by heterologous expression of an acetoin reductase from *Clostridium beijerinckii* (Siemerink et al., 2011). To achieve pure (S, S)-2,3-BD production, Xiao et al. (2010) constructed a whole-cell biocatalyst with an NAD^+ dependent enzyme, ((R, R)-2,3-BD dehydrogenase from *Bacillus subtilis*, EC 1.1.1.4) and an NAD^+ regenerating enzyme, (H_2O-producing NADH oxidase from *Lactobacillus brevis,* EC 1.6.99.3). This biocatalyst displayed significantly high biocatalytic efficiency in terms of production of chiral acetoin and (S, S)-2,3-BD from the commercially available racemic mixture of 2,3-BD. Li et al. (2012) cloned the gene encoding 2,3-BD dehydrogenase from an *Enterobacter cloacae* ssp. dissolvens strain SDM, and expressed it in *E. coli*. Then the whole cells of the recombinant *E. coli* were used to produce (S, S)-2,3-BD from diacetyl. Under optimal conditions, high optical purity (S, S)-2,3-BD (purity >99%) was

TABLE 10.3 Biosynthesis of Different 2,3-Butanediol Stereoisomers with Engineered Heterologous Hosts

Product	Purity (%)	Concentration (g/L)	Yield (g/g)	Substrate	Key enzyme	Heterologous host	Reference
(S, S)-2,3-BD	95	3.70	0.37	Racemic AC	(S, S)-2,3-BDH from *Brevibacterium saccharolyticum*	*Escherichia coli* JM109	Ui et al. (2001)
(S, S)-2,3-BD	92	1.35	0.14	2,3-BD mixture	Glycerol dehydrogenase from *Hansenula polymorpha*	*E. coli* HB101	Yamada-Onodera et al. (2002)
(S, S)-2,3-BD	98	2.20	0.93	Diacetyl	*meso*-2,3-BDH ((*3R*)-AC forming) from *Klebsiella pneumoniae* (S, S)-2,3-BDH from *B. saccharolyticum*	*E. coli* JM109	Ui et al. (2004)
(S, S)-2,3-BD	>99	2.40	0.12	2,3-BD mixture	(R, R)-2,3-BDH from *Bacillus subtilis*; NADH oxidase from *Lactobacillus brevis*	*E. coli* BL21	Xiao et al. (2010)
(S, S)-2,3-BD	>99	26.80	0.67	Diacetyl	(S, S)-2,3-BDH from *Enterobacter cloacae* ssp. dissolvens	*E. coli* BL21(DE3)	Li et al. (2012)
(R, R)-2,3-BD	>99	6.10	0.31	Glucose	ALS, ALDC from *B. subtilis*, sADH from *Thermoanaerobacter brockii*	*E. coli* JCL260	Yan et al. (2009)
(R, R)-2,3-BD	>99	5.80	0.30	Glucose	ALS, ALDC, and sADH from *B. subtilis*	*E. coli* JCL260	Yan et al. (2009)
(R, R)-2,3-BD	>99	5.10	0.29	Glucose	ALS, ALDC from *B. subtilis*, sADH from *Clostridium beijerinckii*	*E. coli* JCL260	Yan et al. (2009)
(R, R)-2,3-BD	>99	9.54	0.33	Glycerol	ALS, ALDC from *B. subtilis*, sADH from *C. beijerinckii*	*E. coli BW25113/ ackA, poxB*	Shen et al. (2012)

(R, R)-2,3-BD	92	1.98	0.034	Glucose	ACR from *C. beijerinckii*	*Clostridium acetobutylicum* ATCC 824	Siemerink et al. (2011)
meso-2,3-BD	98	17.70	0.18	Glucose	ALS, ALDC, and *meso*-2,3-BDH from *K. pneumoniae*	*E. coli* JM109	Ui et al. (1997)
meso-2,3-BD	–	1.12	0.29	Glucose	ALS from *E. coli*, ALDC from *Lactobacillus lactis*, *meso*-BDH from *Saccharomyces cerevisiae*	*E. coli* YYC202 (DE3) *ldhA*$^-$*ilvC*$^-$	Nielsen et al. (2010)
meso-2,3-BD	–	13.00	0.43	Glucose	ALS, ALDC from *B. subtilis*, *meso*-2,3-BDH from *K. pneumoniae*	*E. coli* JM109 *ldhA*$^-$ *pta*$^-$*adhE*$^-$*poxB*$^-$	Li et al. (2010c)
meso-2,3-BD	99	15.70	0.31	Glucose	ALS from *E. coli*, ALDC and *meso*-sADH from *K. pneumonia*	*E. coli* W3110	Lee et al. (2012a)
meso-2,3-BD	99	6.90	0.21	Crude glycerol	ALS from *E. coli*, ALDC and *meso*-sADH from *K. pneumonia*	*E. coli* W3110	Lee et al. (2012a)
meso-2,3-BD	–	5.50	0.80	Cellodextrin	ALS, ALDC from *B. subtilis*, *meso*-2,3-BDH from *K. pneumonia*, cellodextrinase from *Saccharophagus degradans*	*E. coli* MG1655 *ldhA*$^-$*pta*$^-$*ackA*$^-$ *poxB*$^-$	Shin et al. (2012)

AC, acetoin; 2,3-BD, 2,3-butanediol; ACR, acetoin reductase (EC 1.1.1.4); ALS, α-acetolactate synthase (EC 4.1.3.18); ALDC, α-acetolactate decarboxylase (EC 4.1.1.5); 2,3-BDH, 2,3-butanediol dehydrogenase (EC 1.1.1.76); sADH, secondary alcohol dehydrogenase; 2,3-BD mixture, contains (S,S)-, (R,R)-, and meso- 2,3-BD form.

obtained with concentrations of 16.1 g/L and 26.8 g/L in batch and fed-batch cultures, respectively.

10.2.2.3 Alternative Substrates As raw material are a large part of microbial 2,3-BD production cost, a variety of work with biomass shows promise in helping to reduce the costs of fermentation. In past decades, the relatively high cost of conventional starch or sugar substrates has been identified as a major factor affecting the economic viability of 2,3-BD fermentation. Therefore, cheaper alternatives for 2,3-BD production were pursued. There is a good indication that with proper conditions, large quantities of 2,3-BD can be made from the cheaper alternative substrates. Generally, these substrates can be classified as two types: noncellulosic and lignocellulosic.

Noncellulosic Substrates One promising alternative noncellulosic substrate is food industry residue. Perego et al. (2000; 2003) selected food industry wastes because of their high sugar content and compatibility with biological processes. The wastes investigated included starch hydrolysate coming from corn transformation, both raw and decolored molasses from sugar extraction of beets, and whey from cheese manufacture. They found that waste starch hydrolysate was the most promising raw material, ensuring the highest product yield and volumetric productivity, which was nearly twice as estimated for synthetic glucose solutions. This confirmed the existence in the waste material of growth factors that stimulate microbial metabolism; thus, waste starch hydrolysate could potentially be used for 2,3-BD industrial production with no need for additional growth factor supplementation.

Whey permeate, which is a by-product of the dairy industry, has attracted interest as an alternative substrate for 2,3-BD fermentation because of problems with disposal, lactose content, and availability in many countries. The use of whey permeate in 2,3-BD fermentation has been investigated by a number of researchers. However, compared with starch and sugar substrates, whey permeate has proven to be a relatively poor substrate considering overall reactor productivities in batch fermentation (Speckman and Collins, 1982; Barrett et al., 1983; Lee and Maddox, 1984; Champluvier et al., 1989a; Perego et al., 2000). To enhance productivity using whey permeate as substrate, cell immobilization technology was performed and higher 2,3-BD productivity was successfully achieved (Lee and Maddox, 1986; Champluvier et al., 1989b).

Another food industry residue that has been investigated is molasses, the residue left after the crystallization of sucrose from sugarcane juice. The material can be stored in its concentrated form for long time periods and can be diluted to the optimum sugar content prior to use. Normally, very little nutrient supplementation is required for the conversion of molasses. Afschar et al. (1991, 1993) found that *K. oxytoca* could ferment molasses at high concentrations. In one batch experiment, quantities of molasses as high as 280 g/L was converted to 118 g/L of 2,3-BD.

Another potential low-cost noncellulosic substrate is hexose-rich plants such as Jerusalem artichokes and sugarcane. The Jerusalem artichoke (*Helianthus tuberosus*) is a perennial herbaceous plant widely available as non-grain raw material. This plant is of interest due to its high yield, which can reach upto 50 tons of tubers per hectare, and its resistance to low temperatures. The high carbohydrate (inulin and sucrose) content of the tubers, which is up to 85% of the dry weight, makes the Jerusalem artichoke an excellent source of fermentable sugars that could be used as another alternative cheap substrate for 2,3-BD fermentation. The carbohydrates present in the tuber occur mainly in the form of short oligomeric fructans which have an inulinic structure and must be hydrolyzed by acid or inulinase before being fermented.

Sun et al. (2009) successfully developed a process for 2,3-BD production using *K. pneumoniae* from Jerusalem artichoke tubers. Various technologies, including separate hydrolysis and fermentation (SHF) and simultaneous saccharification and fermentation (SSF), were investigated, during which the pretreated Jerusalem artichoke tubers were hydrolyzed to glucose and fructose with exogenous inulinase. The concentration of 2,3-BD reached 81.59 g/L and 91.63 g/L after 40 hours in batch and fed-batch SSF processes, respectively. The developed SSF methods could not only shorten the overall process time, but could also relieve end product inhibition and eliminate the costs of sugar separation. To use the whole Jerusalem artichoke, Li et al. (2010a) further established a 2,3-BD fermentation process using both the Jerusalem artichoke stalk and tuber as feedstock. After adding Jerusalem artichoke tuber to the hydrolysate of the stalk to increase the sugar concentration, fed-batch SSF was then performed, and 80.5 g/L 2,3-BD plus acetoin was obtained in 68 hours. To use the Jerusalem artichoke tuber directly for 2,3-BD production while avoiding the need for supplementation with inulinase, *P. polymyxa,* which is able to secrete inulinase was applied. This could potentially reduce the hydrolysis cost and make the fermentation process more economical. Moreover, the *P. polymyxa* strain can produce the valuable chemical (R, R)-2,3-BD at an optical purity higher than 98% (de Mas et al., 1988). Fages et al. (1986) optimized the important oxygen transfer rate parameter, which affects the kinetics of *P. polymyxa* (R, R)-2,3-BD production from the Jerusalem artichoke. The best results (44 g/L (R, R)-2,3-BD and productivity of 0.79 g/L·h were obtained by utilizing the juice extracted from the tubers without any hydrolysis. Recently, Gao et al. (2010) developed a medium for one-step fermentation of raw inulin extract from Jerusalem artichoke tubers by *P. polymyxa* to produce (R, R)-2,3-BD. Apart from K_2HPO_4 and NH_4Cl, no other nutritional supplementation was necessary. In batch fermentation under optimum conditions, 36.92 g/L of (R, R)-2,3-BD at more than 98% optical purity was achieved. This process featured one-step fermentation of inulin without hydrolysis, which greatly decreased the raw material cost and thus facilitated its practical application.

In Brazil, the 2,3-BD fermentation process using *K. pneumoniae* and locally abundant sugarcane juice, which is rich in carbohydrates, including sucrose, fructose, and glucose, was established using ammonium phosphate as the sole nutritional supplement (Berbert-Molina et al., 2001). With a sugarcane-juice-based medium containing approximately 180 g/L sucrose and 8 g/L $(NH_4)_2HPO_4$, over 70 g/L 2,3-BD plus acetoin were formed. This result was comparable to that achieved with a pure glucose medium containing several nutrients.

The third potential low-cost noncellulosic substrate is biodiesel by-produced glycerol. With the obligatory application of biodiesel fuels in large commercial scale worldwide, enormous quantities of glycerol deposits are likely to be available in the market in the near future. Conversion of glycerol to various high-value-added products thus attracts a lot of interest (Papanikolaou et al., 2008). Glycerol can be used as a carbon source for conversion to 1,3-propanediol, 2,3-BD, and ethanol (Biebl et al., 1998). In fact, 1,3-propanediol is normally the major product of glycerol fermentations. During *K. pneumonia* fermentation with decreasing pH, acetic acid formation by *K. pneumonia* was replaced by formation of 2,3-BD. Biebl et al. (1998) proposed that under microaerobic and low pH conditions, the production of 1,3-propanediol could be reduced and therefore glycerol can be converted to 2,3-BD only. Aiming to direct glycerol conversion predominantly toward production of 2,3-BD, Petrov and Petrova (2009; 2010) investigated the main parameters influencing the fed-batch culture of *K. pneumoniae* and found that pH was the most important factor affecting 2,3-BD production. Spontaneous pH changes and product formation with time were investigated by performing fermentations with noncontrolled pH starting at different levels of initial pH. In the absence of external maintenance, the microorganism attempted to control pH using acetate/2,3-BD alternations of the oxidative pathway of glycerol catabolism, which resulted in pH fluctuations. That is to say, the formation of 2,3-BD is a result of an adaptive mechanism of pH self-control, in response to spontaneous pH drops during glycerol fermentation. Enhancement of 2,3-BD production (by significant increase of glycerol utilization) was thus achieved by developing a new method of "forced pH fluctuations," which was realized by consecutive elevations of pH using a definite Δ pH value, at exact time intervals, allowing multiple variations. Under the defined optimal conditions for maximal glycerol consumption, 70 g/L 2,3-BD was produced, which is the highest recorded amount obtained from glycerol as a sole carbon source so far.

The fourth potential low-cost noncellulosic substrate for 2,3-BD production is carbon-monoxide-containing industrial waste gas. Recently, the researchers in LanzaTech, a New Zealand clean technology company, successfully developed the technology for producing 2,3-BD from steel industry waste gases. They present evidence of 2,3-BD production by nonpathogenic acetogenic Clostridium species using carbon-monoxide-containing industrial waste gases or syngas as the sole source of carbon and energy, and proposed the complete metabolic pathway from carbon monoxide to 2,3-BD through analysis of the *Clostridium* genome (Köpke et al., 2011). The LanzaTech company is aiming to commercialize the production of 2,3-BD using industrial waste gas carbon monoxide as feedstock by the end of 2014 (de Guzman and Chang, 2012). This technology using acetogens as producers, and industrial waste gases as substrates, therefore, has two potentially important commercial implications. Firstly, it decouples the biological production of 2,3-BD from a reliance on farmed or food sugars as the primary carbon and energy sources for product synthesis. Secondly, the acetogens used for the production of 2,3-BD are each risk group 1 (nonpathogenic) organisms. The use of risk group 1 organisms is greatly preferred because of the comparatively lower health and safety risks and associated culture containment and handling costs (Köpke et al., 2011).

Lignocellulosic Substrates Lignocellulose is the most abundant biomass on earth. It has attracted considerable attention as an alternative feedstock for bioprocess and energy resource because of its ready availability and renewable nature. The most attractive use of lignocelluloses is in the production of chemicals using biotechnological means (Ragauskas et al., 2006). The main components of lignocellulose biomass are lignin, cellulose, and hemicellulose. Cellulose is a polymer of glucose, while hemicellulose is a polymer containing mostly pentoses including xylose, arabinose, and ribose (Saha, 2003; Ji et al., 2012). In past decades, the use of low-cost lignocellulosic biomass as an alternative substrate for 2,3-BD production to improve the economics of the process received considerable attention (Rogers et al., 2006). Productivity is not improved when compared to the bioprocess using glucose, but the fact that this feedstock is cheap and abundant makes these optimized fermentations economically attractive.

Corn cob, one kind of lignocellulosic biomass derived from corn processing, is a low cost and widely available agricultural residue. Its use as a substrate for 2,3-BD fermentation has been investigated. In a study, after pretreatment with dilute ammonia and hydrochloric acid, 90% of cellulose in corn cob was hydrolyzed to glucose, which was then used as the substrate for 2,3-BD fermentation (Cao et al., 1997). Lignin and alkaline extractives during the treatment were solubilized and separated from cellulose and hemicellulose. Hemicellulose was then hydrolyzed and separated from cellulose. The SSF using pretreated corn cob was then conducted by adding cellulases and *K. oxytoca* cells. However, this process for converting biomass into 2,3-BD took advantage of the cellulose fraction only, while the hemicellulose fraction was lost. To make biomass conversion economically feasible, it is essential that the hemicellulose fraction also be efficiently converted into 2,3-BD. Cheng et al. (2010) detoxified the acid hydrolysate of corn cob (hemicellulose components-rich stream) by sequentially boiling, over-liming, and adsorbing the hydrolysate onto activated charcoal, and then using the pentose-rich hydrolysate as a substrate for 2,3-BD production by *K. oxytoca*. Under optimum conditions, a maximal 2,3-BD concentration of 35.7 g/L was obtained after 60 hours of fed-batch fermentation, giving a yield of 0.5 g/g reducing sugar and a productivity of 0.59 g/L·h. Corn fiber, another hemicellulose-rich by-product in corn processing, was also successfully used as a substrate for 2,3-BD production by Saha and Bothast (1999). They isolated a 2,3-BD producing *E. cloacae* from decaying wood/corn soil samples, which was found to utilize sugars from acid plus enzyme-saccharified corn fiber and produced 2,3-BD (0.35 g/g available sugars). The strain was also able to produce 2,3-BD from dilute-acid-pretreated corn fiber by SSF (0.34 g/g theoretical sugars). Corn-cob molasses, a waste by-product in xylitol production that contained high concentrations of hemicellulose-derived arabinose and xylose, was recently reused as a substrate for production of 2,3-BD using *K. pneumoniae* (Wang et al., 2010). The maximum 2,3-BD concentration was 78.9 g/L after 61 hours of fed-batch fermentation. This approach, using corn-cob molasses for 2,3-BD production, not only cut down the feedstock cost, but also provided a method to exploit xylitol industry by-products, which could reduce environmental pollution. Recently, the xylose syrup obtained in SSF of alkali/peracetic acid-pretreated sugarcane bagasse was successfully converted to 2,3-BD by *K. pneumoniae* (Zhao et al., 2011). Similarly, the acid hydrolysates of Jatropha hulls (hemicelluloses component) were successfully used for the first time as raw material for the production of 2,3-BD with *K. oxytoca* (Jiang et al., 2012).

2,3-BD fermentations using other sources of lignocellulosic materials such as wood have also been reported (Frazer and McCaskey, 1989; Grover et al., 1990). As early as the 1980s, Yu et al. (1982, 1985c) developed the processes using both acid and enzymatic hydrolyzed wood hemicellulose for 2,3-BD production by *K. pneumoniae*. 2,3-BD yields of 0.4–0.5 g/g were obtained from hydrolyzed wood hemicellulose. The authors proposed that such high yields were partially due to the ability of *K. pneumoniae* to simultaneously ferment uronic acids, such as D-glucuronic and D-galacturonic acid, present in the wood samples. Other compounds in wood (such as furfural and lignin derivatives) were inhibitory to the bacteria if the wood hydrolysate was added at a high concentration (Yu et al., 1982; Frazer and McCaskey, 1991; Nishikawa et al., 1998). A more efficient approach using woody biomass for 2,3-BD production was SSF with the addition of culture filtrates of *Trichoderma harzianum* as a source of hydrolytic enzymes (Yu and Saddler, 1985; Yu et al., 1985b). Using the SSF method developed, the sugars released from both cellulose and hemicellulose could be used for 2,3-BD production; yields of approximately 30% of the theoretical value were obtained, thus eliminating the costs of separating the sugars. To co-utilize both the cellulose and hemicellulose, a sequential coculture approach was also established for the conversion of lignocellulosic substrates to 2,3-BD (Yu et al., 1985a). Firstly, *Clostridium thermocellum* was induced to secrete extracellular endoglucanase (EC 3.2.1.4) and xylanase (EC 3.2.1.8) into the culture medium. *K. pneumoniae* was inoculated as a sequential culture for the rapid utilization of the accumulated sugars and formation of 2,3-BD. This approach was applicable to the use of mixed cellulose and hemicellulose substrates. To further reduce the production costs, co-utilization of both the cellulose and hemicellulose fractions to produce 2,3-BD could be accomplished by simultaneous saccharification and co-fermentation process (Du et al., 2009; Ji et al., 2009b; 2011).

10.3 BIO-BASED 1,4-BUTANEDIOL

10.3.1 Via Catalytic Hydrogenation

In 2004, the US Department of Energy elaborated a list of the top 12 most attractive candidates to focus research efforts in future years (Werpy and Petersen, 2004). Succinic acid was shortlisted as one of such top 12 platform chemicals that could be produced from carbohydrates and was still on a revisited list of top platform chemicals in 2010 (Bozell and Petersen, 2010). Over the last decades, the anaerobic fermentation technology for the production of succinic acid from sugars has been the subject of intensive development (Lin et al., 2012). Recently, the commercial production of bio-based succinic acid has been announced by DSM and Roquette (Heerlen, 2011). This makes succinic acid a new bio-derived building-block chemical that could replace the current petrochemical-derived maleic anhydride platform. Therefore, a two-step 1,4-BD production route based on the bio-based succinic acid was proposed (Figure 10.3). In the reported production route to bio-based 1,4-BD, an aqueous-phase hydrogenation of bio-based succinic acid to 1,4-BD over supported bimetallic catalysts was used (Minh et al., 2010; Chen and Patel, 2012).

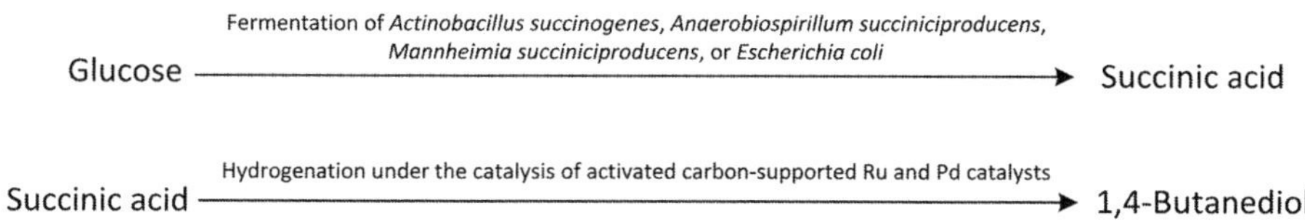

FIGURE 10.3 Two-step production of bio-based 1,4-butanediol

Earlier in 2012, BioAmber Inc., which is a next generation chemicals company located in Minneapolis, Minnesota, announced that it had scaled up its hydrogenation catalyst technology under the license from DuPont and converted multiton quantities of bio-based succinic acid into 1,4-BD. Using bio-based succinic acid made at its French facility as a starting material, BioAmber successfully completed a 1,4-BD production campaign in collaboration with one of the world's leading commercial catalyst suppliers and a toll manufacturer possessing large-scale hydrogenation reactors and distillation columns. Furthermore, it is developing an integrated plant engineering design that combines the production of bio-based succinic acid and 1,4-BD on a single site (Anonymous, 2012a).

10.3.2 Via Sugar Fermentation

1,4-BD is a nonnatural compound not synthesized by any known microbe, and there are no complete biosynthetic pathways that could be used for 1,4-BD production. Therefore, the emerging synthetic biology method, which incorporates the design and construction of new biological components, such as enzymes, genetic circuits, and cells, or the redesign of existing biological systems, will contribute to synthesize this unnatural organic molecule biologically (Keasling, 2008). Recently, the researchers in Genomatica, a San Diego-based company, reported the first direct biocatalytic route to 1,4-BD from renewable carbohydrate feedstock (Yim et al., 2011).

In order to construct a synthetic metabolic pathway for the microbial production of the industrially important 1,4-BD from sugars, an algorithm based on transformation of the functional groups by known specific chemistry was used to design *E. coli* capable of producing 1,4-BD. It predicted more than 10,000 possible routes of four to six steps starting from central metabolites such as acetyl-CoA, α-ketoglutarate, glutamate, and succinyl-CoA (Cho et al., 2010). Two synthetic pathways for 1,4-BD production were selected by evaluating the maximum theoretical 1,4-BD yield, pathway distance, thermodynamic feasibility, and the number of nonnative and new steps (Figure 10.4) (Yim et al., 2011). The first route starts from the tricarboxylic acid (TCA) cycle intermediate succinate that is activated as succinyl-CoA by the native *E. coli* enzyme succinyl-CoA synthetase. After two sequential reduction steps catalyzed by CoA-dependent succinate semialdehyde dehydrogenase and 4-hydroxybutyrate (4HB) dehydrogenase, respectively, the CoA derivative converts to 4HB. Then, 4HB was converted to 1,4-BD through two reduction steps catalyzed by dehydrogenases. The second route for 1,4-BD synthesis branches from *E. coli* central metabolism

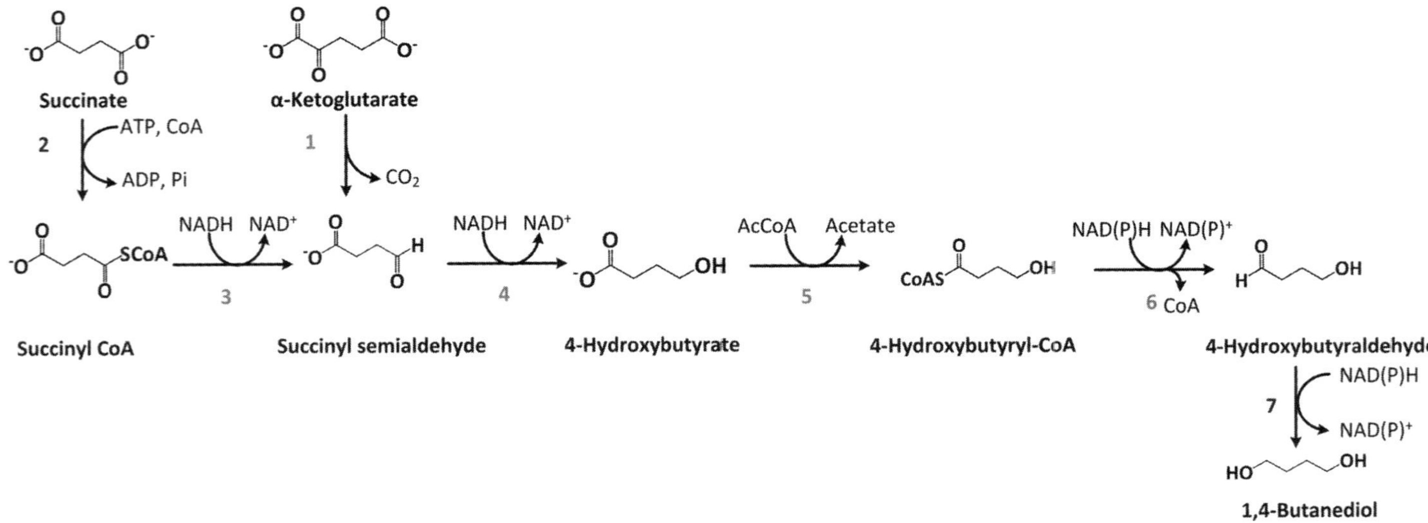

FIGURE 10.4 1,4-Butanediol biosynthetic pathways constructed and engineered into *E. coli* (modified from Yim et al., 2011). Enzymes for each numbered step are as follows: (1) 2-oxoglutarate decarboxylase; (2) succinyl-CoA synthetase; (3) CoA-dependent succinate semialdehyde dehydrogenase; (4) 4-hydroxybutyrate dehydrogenase; (5) 4-hydroxybutyryl-CoA transferase; (6) 4-hydroxybutyryl-CoA reductase; (7) alcohol dehydrogenase. Steps 2 and 7 occur naturally in *E. coli*, whereas others are encoded by heterologous genes introduced.

at the key oxidative TCA-cycle intermediate, α-ketoglutarate. This pathway is thermodynamically more favorable than the succinate route, owing to the irreversible decarboxylation step, and it also consumes one less reducing equivalent. Therefore, the Genomatica researchers chose the second route for 1,4-BD synthesis. They further engineered the *E. coli* host to enhance anaerobic operation of the oxidative TCA cycle, thereby generating reducing power to drive the 1,4-BD pathway guided by a genome-scale metabolic model (Burgard et al., 2009; Burk et al., 2010; Yim et al., 2011; Jiang et al., 2012). The construction of these synthetic pathways in *E. coli* led to the first biological production of 18 g/L of this highly reduced, nonnatural chemical from glucose. In the work, for the first time, synthetic metabolic pathways and correspondingly engineered *E. coli* strains have been reported for the microbial production of the industrially important 1,4-BD from sugars (Zeng and Sabra, 2011; Lee et al., 2012b). This excellent work represents a breakthrough. Commercial production of 1,4-BD from sugar by the engineered *E. coli* is expected to bring about substantial cost advantages compared to current petrochemical processes next to lower energy use and greenhouse gas emissions (Burk, 2010). As indicated by the Genomatica, the developed process consumes 32,000 BTU per pound of 1,4-BD (75 MJ/kg), far less than the acetylene-based process, and does not have any by-products. Notably, Genomatica will begin producing 1,4-BD in a commercial-scale plant in 2013, and 1 year later, larger plants will undertake 1,4-BD production in the United States, Europe, and Asia (Anonymous, 2012b,c).

10.4 ECONOMIC OUTLOOK

In the past years, research and development spending for bio-based BDs has increased substantially, with the most important drivers being price increases for fossil fuels, supply security considerations, and greenhouse gas emission reduction. Therefore, recently, many bio-based BDs production technologies have been developed, of which some are already or nearly to be in industrial production.

The bio-based 2,3-BD and 1,4-BD production described in this chapter would have a bright future. Due to the special structure of 2,3-BD, it could be used as a novel platform chemical to produce many derivatives to enter the chemical industrial chain. Therefore, the bio-based 2,3-BD production would alleviate the dependence on oil supply for the production of platform chemicals. For example, MEK and butadiene, which could be easily derived from 2,3-BD, are two important intermediates in chemical industry. Based on the production cost advantage and continuing improvements of the biological processes, bio-based 2,3-BD is expected to hit the industrial market in the foreseeable future (Zeng and Sabra, 2011). While the developed bio-based 1,4-BD production processes, including catalytic hydrogenating bio-based succinic acid and fermenting sugars using genetically engineered *E. coli* with synthetic 1,4-BD metabolic pathway, will enable low-cost manufacture of 1,4-BD with substantially reduced energy consumption and carbon dioxide emissions relative to its petroleum-based route.

10.5 FUTURE PROSPECTS

BDs are important in chemical industry with many applications. At present, almost all BD isomers are produced from fossil oil, but the petroleum-based system is currently facing global crises, such as climate change and fossil resource depletion. Thus, there is an increasing demand for sustainable production of bio-based BDs. Currently, with the technology developed, some BD isomers, such as 2,3-BD and 1,4-BD, are approaching to be commercialized for economical bio-based production. These bio-based BDs are produced by the catalytic process or the sugar fermentations of microorganisms. It is worth notice that some key technologies are involved in these processes, that is, catalytic hydrogenation and hydrogenolysis of the sugar derivatives, strain improvement by using the metabolic engineering and the emerging synthetic biology methods. In the future, improving these key technologies would offer significant opportunities to further decrease the bio-based BDs' production costs, which would accelerate the industrialization of these important bio-based products.

ACKNOWLEDGMENTS

Funding for our research was provided by the National Science Foundation for Distinguished Young Scholars of China (No. 21225626), the National Basic Research Program of China (No. 2011CBA00807), the National Natural Science Foundation of China (Nos. 21006049, 21376002), the National High Technology Research and Development Program of China (No. 2011AA02A207), the Key Project of Jiangsu Natural Science Foundation (No. BK2011031), the Program for Changjiang Scholars and Innovative Research Team in University (No. IRT 1066), and the Priority Academic Program Development of Jiangsu Higher Education Institutions.

REFERENCES

Afschar AS, Bellgardt KH, Rossell CE, Czok A, Schaller K. (1991). The production of 2,3-butanediol by fermentation of high test molasses. *Appl Microbiol Biotechnol*, 34, 582–585.

Afschar AS, Vaz Rossell CE, Jonas R, Quesada Chanto A, Schaller K. (1993). Microbial production and downstream processing of 2,3-butanediol. *J Biotechnol*, 27, 317–329.

Alam S, Capit F, Weigand WA, Hong J. (1990). Kinetics of 2,3-butanediol fermentation by *Bacillus amyloliquefaciens*: effect of initial substrate concentration and aeration. *J Chem Technol Biotechnol*, 47, 71–84.

Amada Y, Watanabe H, Hirai Y, Kajikawa Y, Nakagawa Y, Tomishige K. (2012). Production of biobutanediols by the hydrogenolysis of erythritol. *ChemSusChem*, 5, 1991–1999.

Anonymous. (2012a). BioAmber produces bio-based 1,4-Butanediol. Available at: http://www.biofpr.com/details/news/1697275/BioAmber_produces_bio-based_14-Butanediol.html. Accessed on Jan 8, 2014.

Anonymous. (2012b). Genomatica takes its 1,4-butanediol to Asia. *Chem Eng News*, 90, 15–16.

Anonymous. (2007). Plant-derived butylene glycol launched on market. *Japan Chem Week*, 48(2418):5

Anonymous. (2012c). Ramping toward commercial roll-out. Available at: www.genomatica.com/products/bdo/

Anonymous. (1995). SIDS Initial Assessment Report, Geneva: United Nations Environment Programme. Available at: http://www.inchem.org/documents/sids/sids/584032.pdf. Accessed on Jan 8, 2014.

Barrett EL, Collins EB, Hall BJ, Matoi SH. (1983). Production of 2,3-butylene glycol from whey by *Klebsiella pneumoniae* and *Enterobacter aerogenes*. *J Dairy Sci*, 66, 2507–2514.

Berbert-Molina MA, Sato S, Silveira MM. (2001). Ammonium phosphate as a sole nutritional supplement for the fermentative production of 2,3-butanediol from sugar cane juice. *Z Naturforsch C*, 56, 787–791.

Biebl H, Zeng AP, Menzel K, Deckwer WD. (1998). Fermentation of glycerol to 1,3-propanediol and 2,3-butanediol by *Klebsiella pneumoniae*. *Appl Microbiol Biotechnol*, 50, 24–29.

Biswas R, Yamaoka M, Nakayama H, Kondo T, Yoshida K, Bisaria VS, Kondo A. (2012). Enhanced production of 2,3-butanediol by engineered *Bacillus subtilis*. *Appl Microbiol Biotechnol*, 94, 651–658.

Bozell JJ, Petersen GR. (2010). Technology development for the production of biobased products from biorefinery carbohydrates-the US Department of Energy's "Top 10" revisited. *Green Chem*, 12, 539–554.

Burgard AP, Burk MJ, Pharkya P. (2011). Methods and organisms for converting synthesis gas or other gaseous carbon sources and methanol to 1,3-butanediol. US Patent 20110129904.

Burgard AP, Van Dien SJ, Burk MJ. (2009). Methods and organisms for the growth coupled production of 1,4-butanediol. US Patent 20090047719A1.

Burk MJ. (2010). Sustainable production of industrial chemicals from sugars. *Int Sugar J*, 112, 30–35.

Burk MJ, Burgard AP, Osterhout RE, Sun J. (2010). Microorganisms for the production of 1,4-butanediol. US Patent 20100112654 A1.

Cao N, Xia Y, Gong CS, Tsao GT. (1997). Production of 2,3-butanediol from pretreated corn cob by *Klebsiella oxytoca* in the presence of fungal cellulose. *Appl Biochem Biotechnol*, 63, 129–139.

Celińska E, Grajek W. (2009). Biotechnological production of 2,3-butanediol–current state and prospects. *Biotechnol Adv*, 27, 715–725.

Champluvier B, Decallonne J, Rouxhet PG. (1989a). Influence of sugar source (lactose, glucose, galactose) on 2,3-butanediol production by *Klebsiella oxytoca* NRRL-B199. *Arch Microbiol*, 152, 411–414.

Champluvier B, Francart B, Rouxhet PG. (1989b). Co-immobilization by adhesion of β-galactosidase in nonviable cells of *Kluyveromyces lactis* with *Klebsiella oxytoca*: Conversion of lactose into 2,3-butanediol. *Biotechnol Bioeng*, 34, 844–853.

Chen GQ, Patel MK. (2012). Plastics derived from biological sources: present and future: a technical and environmental review. *Chem Rev*, 112, 2082–2099.

Cheng KK, Liu Q, Zhang JA, Li JP, Xu JM, Wang GH. (2010). Improved 2,3-butanediol production from corncob acid hydrolysate by fed-batch fermentation using *Klebsiella oxytoca*. *Process Biochem*, 45, 613–616.

Cho A, Yun H, Park J, Lee S, Park S. (2010). Prediction of novel synthetic pathways for the production of desired chemicals. *BMC Syst Biol*, 4, 35.

Danner H, Braun R. (1999). Biotechnology for the production of commodity chemicals from biomass. *Chem Soc Rev*, 28, 395–405.

de Guzman D, Chang J. (2012). Bio-based 2,3-BDO set for 2014 sales. ICIS Chemical Business. Available at: http://www.lanzatech.com/sites/default/files/imce_uploads/lanzatech_story.pdf. Accessed on Aug 13, 2013.

de Mas CD, Jansen NB, Tsao GT. (1988). Production of optically active 2,3-butanediol by *Bacillus polymyxa*. *Biotechnol Bioeng*, 31, 366–377.

Du J, Ji XJ, Huang H, Nie ZK, Ren X, Hu N, Li S. (2009). Simultaneous determination of the substrates of glucose and xylose and the metabolites in the 2,3-butanediol fermentation broth by ion-exclusion liquid chromatography. *Chin J Anal Chem*, 37, 681–684.

Fages J, Mulard D, Rouquet JJ, Wilhelm JL. (1986). 2,3-Butanediol production from *Jerusalem artichoke*, Helianthus tuberosus, by *Bacillus polymyxa* ATCC 12321. Optimization of $k_L a$ profile. *Appl Microbiol Biotechnol*, 25, 197–202.

Faveri DD, Torre P, Molinari F, Perego P, Converti A (2003). Carbon material balances and bioenergetics of 2,3-butanediol bio-oxidation by *Acetobacter hansenii*. *Enzyme Microb Technol*, 33, 708–719.

Fond O, Jansen NB, Tsao GT. (1985). A model of acetic acid and 2,3-butanediol inhibition of the growth and metabolism of *Klebsiella oxytoca*. *Biotechnol Lett*, 7, 727–732.

Frazer FR, McCaskey TA. (1991). Effect of components of acid-hydrolysed hardwood on conversion of D-xylose to 2,3-butanediol by *Klebsiella pneumoniae*. *Enzyme Microb Technol*, 13, 110–115.

Frazer FR, McCaskey TA. (1989). Wood hydrolyzate treatments for improved fermentation of wood sugars to 2,3-butanediol. *Biomass*, 18, 31–42.

Gao J, Xu H, Li QJ, Feng XH, Li S. (2010). Optimization of medium for one-step fermentation of inulin extract from *Jerusalem artichoke* tubers using *Paenibacillus polymyxa* ZJ-9 to produce R,R-2,3-butanediol. *Bioresour Technol*, 101, 7087–7093.

Garg S, Jain A. (1995). Fermentative production of 2,3-butanediol: a review. *Bioresour Technol*, 51, 103–109.

Ge L, Wu X, Chen J, Wu J. (2011). A new method for industrial production of 2,3-butanediol. *J Biomat Nanobiotechnol*, 2, 335–336.

Geckil H, Barak ZE, Chipman DM, Erenler SO, Webster DA, Stark BC. (2004). Enhanced production of acetoin and butanediol in recombinant *Enterobacter aerogenes* carrying *Vitreoscilla* hemoglobin gene. *Bioproc Biosyst Eng*, 6, 325–330.

Grover BP, Garg SK, Verma J. (1990). Production of 2,3-butanediol from wood hydrolysate by *Klebsiella pneumoniae*. *World J Microbiol Biotechnol*, 6, 328–332.

Gubbels E, Jasinska-Walc L, Koning CE. (2013). Synthesis and characterization of novel renewable polyesters based on 2,5-furandicarboxylic acid and 2,3-butanediol. *J Polym Sci Part A: Polym Chem*, 51, 890–898.

Hasegawa R, Hayashi K. (1986). Polyester containing impure 1,2-butanediol. US Patent. 4596886.

Hatti-Kaul R, Tornvall U, Gustafsson L, Borjesson P. (2007). Industrial biotechnology for the production of bio-based chemicals – a cradle-to-grave perspective. *Trends Biotechnol*, 25, 119–124.

Heerlen NL. (2011). DSM and Roquette to open commercial scale bio-based succinic acid plant in 2012. Available at: http://www.dsm.com/en_US/cworld/public/media/pages/

press-releases/36_11_dsm_and_roquette_to_open_commercial_scale_bio_based_succinic_acid_plant_2012.jsp. Accessed on Aug 13, 2013.

Hunter SE, Ehrenberger CE, Savage PE. (2006). Kinetics and mechanism of tetrahydrofuran synthesis via 1,4-butanediol dehydration in high-temperature water. *J Org Chem*, 71, 6229–6239.

Imanari M, Iwane H, Suzuki M, Suzuki N. (1992). Process for preparing α-ketobutyric acid. US Patent 5155263.

Jansen NB, Tsao GT. (1983). Bioconversion of pentoses to 2,3-butanediol by *Klebsiella pneumoniae*. *Adv Biochem Eng Biotechnol*, 27, 85–99.

Ji XJ, Huang H, Du J, Zhu JG, Ren LJ, Hu N, Li S. (2009a). Enhanced 2,3-butanediol production by *Klebsiella oxytoca* using a two-stage agitation speed control strategy. *Bioresour Technol*, 100, 3410–3414.

Ji XJ, Huang H, Du J, Zhu JG, Ren LJ, Li S, Nie ZK. (2009b). Development of an industrial medium for economical 2,3-butanediol production through co-fermentation of glucose and xylose by *Klebsiella oxytoca*. *Bioresour Technol*, 100, 5214–5218.

Ji XJ, Huang H, Li S, Du J, Lian M. (2008). Enhanced 2,3-butanediol production by altering the mixed acid fermentation pathway in *Klebsiella oxytoca*. *Biotechnol Lett*, 30, 731–734.

Ji XJ, Huang H, Nie ZK, Qu L, Xu Q, Tsao GT. (2012). Fuels and chemicals from hemicellulose sugars. *Adv Biochem Eng Biotechnol*, 128, 199–224.

Ji XJ, Huang H, Ouyang PK. (2011a). Microbial 2,3-butanediol production: a state-of-the-art review. *Biotechnol Adv*, 29, 351–364.

Ji XJ, Huang H, Zhu JG, Ren LJ, Nie ZK, Du J, Li S. (2010). Engineering *Klebsiella oxytoca* for efficient 2,3-butanediol production through insertional inactivation of acetaldehyde dehydrogenase gene. *Appl Microbiol Biotechnol*, 85, 1751–1758.

Ji XJ, Nie ZK, Huang H, Ren LJ, Peng C, Ouyang PK. (2011b). Elimination of carbon catabolite repression in *Klebsiella oxytoca* for efficient 2,3-butanediol production from glucose-xylose mixtures. *Appl Microbiol Biotechnol*, 89, 1119–1125.

Ji XJ, Xia ZF, Fu NH, Nie ZK, Shen MQ, Tian QQ, Huang H. (2013). Cofactor engineering through heterologous expression of an NADH oxidase and its impact on metabolic flux redistribution in *Klebsiella pneumoniae*. *Biotechnol Biofuels*, 6, 7.

Jiang LQ, Fang Z, Guo F, Yang LB. (2012). Production of 2,3-butanediol from acid hydrolysates of Jatropha hulls with *Klebsiella oxytoca*. *Bioresour Technol*, 107, 405–410.

Jung MY, Ng CY, Song H, Lee J, Oh MK. (2012). Deletion of lactate dehydrogenase in *Enterobacter aerogenes* to enhance 2,3-butanediol production. *Appl Microbiol Biotechnol*, 95, 461–469.

Kamm B, Kamm M. (2004). Principles of biorefineries. *Appl Microbiol Biotechnol*, 64, 137–145.

Keasling JD. (2008). Synthetic biology for synthetic chemistry. *ACS Chem Biol*, 3, 64–76.

Köpke M, Mihalcea C, Liew F, Tizard JH, Ali MS, Conolly JJ, Al-Sinawi B, Simpson SD. (2011). 2,3-Butanediol production by acetogenic bacteria, an alternative route to chemical synthesis using industrial waste gas. *Appl Environm Microbiol*, 77, 5467–5475.

Kosaric N, Magee RJ, Blaszczyk R. (1992). Redox potential measurement for monitoring glucose and xylose conversion by *K. pneumoniae*. *Chem Biochem Eng Q*, 6, 145–152.

Lee HK, Maddox IS. (1986). Continuous production of 2,3-butanediol from whey permeate using *Klebsiella pneumoniae* immobilized in calcium alginate. *Enzyme Microb Technol*, 8, 409–411.

Lee HK, Maddox IS. (1984). Microbial production of 2,3-butanediol from whey permeate. *Biotechnol Lett*, 6, 815–818.

Lee S, Kim B, Park K, Um Y, Lee J. (2012a). Synthesis of pure meso-2,3-butanediol from crude glycerol using an engineered metabolic pathway in *Escherichia coli*. *Appl Biochem Biotechnol*, 166, 1811–1813.

Lee JW, Na D, Park JM, Lee J, Choi S, Lee SY. (2012b). Systems metabolic engineering of microorganisms for natural and non-natural chemicals. *Nat Chem Biol*, 8, 536–546.

Li D, Dai JY, Xiu ZL. (2010a). A novel strategy for integrated utilization of *Jerusalem artichoke* stalk and tuber for production of 2,3-butanediol by *Klebsiella pneumoniae*. *Bioresour Technol*, 101, 8342–8347.

Li ZJ, Ji XJ, Kan SL, Qiao HQ, Jiang M, Lu DQ, Wang J, Huang H, Jia HH, Ouyang PK, Ying HJ. (2010b). Past, present, and future industrial biotechnology in China. *Adv Biochem Eng Biotechnol*, 122, 1–42.

Li ZJ, Jian J, Wei XX, Shen XW, Chen GQ. (2010c). Microbial production of meso-2,3-butanediol by metabolically engineered *Escherichia coli* under low oxygen condition. *Appl Microbiol Biotechnol*, 87, 2001–2009.

Li L, Wang Y, Zhang L, Ma C, Wang A, Tao F, Xu P. (2012). Biocatalytic production of (2S,3S)-2,3-butanediol from diacetyl using whole cells of engineered *Escherichia coli*. *Bioresour Technol*, 115, 111–116.

Lin CSK, Luque R, Clark JH, Webb C, Du C. (2012). Wheat-based biorefining strategy for fermentative production and chemical transformations of succinic acid. *Biofuels Bioprod Bioref*, 6, 88–104.

Liu J, Qi H, Xu Z. (2009). Separation of diols from a mixture comprising diols and polyols. US Patent 7893305.

Liu J, Qi H, Xu Z. (2008). Systems and methods for continuous multiphase reaction and separation. US Patent 20100084350.

Liu J, Xu J. (2006). A kind of high carbon polyol hydrocracking catalyst, its preparation method and application. China Patent 200610165881.8.

Liu Z, Qin JY, Gao C, Hua DL, Ma CQ, Li LX, Wang Y, Xu P. (2011). Production of (2S,3S)-2,3-butanediol and (3S)-acetoin from glucose using resting cells of *Klebsiella pneumoniae* and *Bacillus subtilis*. *Bioresour Technol*, 102, 10741–10744.

Ma CQ, Wang AL, Qin JY, Li LX, Ai XL, Jiang TY, Tang HZ, Xu P. (2009). Enhanced 2,3-butanediol production by *Klebsiella pneumoniae* SDM. *Appl Microbiol Biotechnol*, 82, 49–57.

Maddox IS. (1996). Microbial production of 2,3-butanediol. In: Rehm HJ, Reed G, editors. *Biotechnology*, Vol. 6. Weinheim: VCH Verlagsgesellschaft mbH, pp. 269–291.

Minh DP, Besson M, Pinel C, Fuertes P, Petitjean C. (2010). Aqueous-phase hydrogenation of biomass-based succinic acid to 1,4-butanediol over supported bimetallic catalysts. *Top Catal*, 53, 1270–1273.

Moes J, Griot M, Keller J, Heinzle E, Dunn IJ, Bourne JR. (1985). A microbial culture with oxygen-sensitive product distribution as a potential tool for characterizing bioreactor oxygen transport. *Biotechnol Bioeng*, 27, 482–489.

Nie ZK, Ji XJ, Huang H, Du J, Li ZY, Qu L, Zhang Q, Ouyang PK. (2011). An effective and simplified fed-batch strategy for improved 2,3-butanediol production by *Klebsiella oxytoca*. *Appl Biochem Biotechnol*, 163, 946–953.

Nielsen DR, Yoon SH, Yuan CJ, Prather KLJ. (2010). Metabolic engineering of acetoin and meso-2,3-butanediol biosynthesis in *E. coli*. *Biotechnol J*, 5, 274–284.

Nilegaonkar SS, Bhosale SB, Kshirsagar DC, Kapadi AH. (1992). Production of 2,3-butanediol from glucose by *Bacillus licheniformis*. *World J Microbiol Biotechnol*, 8, 378–381.

Nishikawa NK, Sutcliffe R, Saddler JN. (1998). The effect of wood-derived inhibitors on 2,3-butanediol production by *Klebsiella pneumoniae*. *Biotechnol Bioeng*, 31, 624–627.

Papanikolaou S, Fakas S, Fick M, Chevalot I, Galiotou-Panayotou M, Komaitis M, Marc I, Aggelis G. (2008). Biotechnological valorisation of raw glycerol discharged after bio-diesel (fatty acid methyl esters) manufacturing process: production of 1, 3-propanediol, citric acid and single cell oil. *Biomass Bioenerg*, 32, 60–71.

Perego P, Converti A, Del Borghi M. (2003). Effects of temperature, inoculum size and starch hydrolysate concentration on butanediol production by *Bacillus licheniformis*. *Bioresour Technol*, 89, 125–131.

Perego P, Converti A, Del Borghi A, Canepa P. (2000). 2,3-Butanediol production by *Enterobacter aerogenes*: selection of the optimal conditions and application to food industry residues. *Bioproc Biosys Eng*, 23, 613–620.

Petrov K, Petrova P. (2010). Enhanced production of 2,3-butanediol from glycerol by forced pH fluctuations. *Appl Microbiol Biotechnol*, 87, 943–949.

Petrov K, Petrova P. (2009). High production of 2,3-butanediol from glycerol by *Klebsiella pneumoniae* G31. *Appl Microbiol Biotechnol*, 84, 659–665.

Qin JY, Xiao ZJ, Ma CQ, Xie NZ, Liu PH, Xu P. (2006). Production of 2,3-butanediol by *Klebsiella pneumoniae* using glucose and ammonium phosphate. *Chin J Chem Eng*, 14, 132–136.

Qureshi N, Cheryan M. (1989). Production of 2,3-butanediol by *Klebsiella oxytoca*. *Appl Microbiol Biotechnol*, 30, 440–443.

Ragauskas AJ, Williams CK, Davison BH, Britovsek G, Cairney J, Eckert CA, Frederick Jr WJ, Hallett JP, Leak DJ, Liotta CL, Mielenz JR, Murphy R, Templer R, Tschaplinski T. (2006). The path forward for biofuels and biomaterials. *Science*, 311, 484–498.

Rogers P, Chen JS, Zidwick MJ. (2006). Organic acid and solvent production. Part III: butanol, acetone and isopropanol; 1,3- and 1,2-propanediol production; and 2,3-butanediol production. *Prokaryotes*, 1, 672–755.

Saha BC. (2003). Hemicellulose bioconversion. *J Ind Microbiol Biotechnol*, 30, 279–291.

Saha BC, Bothast RJ. (1999). Production of 2,3-butanediol by newly isolated *Enterobacter cloacae*. *Appl Microbiol Biotechnol*, 52, 321–326.

Shen MQ, Ji XJ, Nie ZK, Xia ZF, Yang H, Huang H. (2013). Biotechnological production of 2,3-butanediol stereoisomers: synthetic mechanism and realized methods. *Chin J Catal*, 34, 351–360.

Shen X, Lin Y, Jain R, Yuan Q, Yan Y. (2012). Inhibition of acetate accumulation leads to enhanced production of (R, R)-2,3-butanediol from glycerol in *Escherichia coli*. *J Ind Microbiol Biotechnol*, 39, 1725–1729.

Shin HD, Yoon SH, Wu J, Rutter C, Kim SW, Chen RR. (2012). High-yield production of meso-2,3-butanediol from cellodextrin by engineered *E. coli* biocatalysts. *Bioresour Technol*, 118, 367–373.

Siemerink MAJ, Kuit W, Lopez Contreras AM, Eggink G, van der Oost J, Kengen SWM. (2011). D-2,3-Butanediol production due to heterologous expression of an acetoin reductase in *Clostridium acetobutylicum. Appl Environ Microbiol*, 77, 2582–2588.

Speckman RA, Collins EB. (1982). Microbial production of 2,3-butylene glycol from cheese whey. *Appl Environ Microbiol*, 43, 1216–1218.

Sun LH, Wang XD, Dai JY, Xiu ZL. (2009). Microbial production of 2,3-butanediol from *Jerusalem artichoke* tubers by *Klebsiella pneumoniae. Appl Microbiol Biotechnol*, 82, 847–852.

Syu MJ. (2001). Biological production of 2,3-butanediol. *Appl Microbiol Biotechnol*, 55, 10–18.

Tran AV, Chambers RP. (1987). The dehydration of fermentative 2,3-butanediol into methyl ethyl ketone. *Biotechnol Bioeng*, 29, 343–351.

Ui S, Okajirna Y, Mimura A, Kanai H, Kudo T. (1997). Molecular generation of an Escherichia coli strain producing only the meso-isomer of 2,3-butanediol. *J Ferment Bioeng*, 84, 185–189.

Ui S, Takusagawa Y, Ohtsuki T, Mimura A, Ohkuma M, Kudo T. (2001). Stereochemical applications of the expression of the L-2,3-butanediol dehydrogenase gene in *Escherichia coli. Lett Appl Microbiol*, 32, 93–98.

Ui S, Takusagawa Y, Sato T, Ohtsuki T, Mimura A, Ohkuma M, Kudo T. (2004). Production of L-2, 3-butanediol by a new pathway constructed in *Escherichia coli. Lett Appl Microbiol*, 39, 533–537.

van Haveren J, Scott EL, Sanders J. (2008). Bulk chemicals from biomass. *Biofuels, Bioprod Bioref*, 2, 41–57.

Wang AL, Wang Y, Jiang TY, Li LX, Ma CQ, Xu P. (2010). Production of 2,3-butanediol from corncob molasses, a waste by-product in xylitol production. *Appl Microbiol Biotechnol*, 87, 965–970.

Wang AL, Xu YQ, Ma CQ, Gao C, Li LX, Wang Y, Tao F, Xu P. (2012a). Efficient 2,3-butanediol production from cassava powder by a crop-biomass-utilizer, *Enterobacter cloacae* subsp. *dissolvens* SDM. *PLoS ONE*, 7, e40442.

Wang Q, Chen T, Zhao X, Chamu J. (2012b). Metabolic engineering of thermophilic *Bacillus licheniformis* for chiral pure D-2,3-butanediol production. *Biotechnol Bioeng*, 109, 1610–1621.

Wei ML, Webster DA, Stark BC. (1998). Metabolic engineering of *Serratia marcescens* with the bacterial hemoglobin gene: alterations in fermentation pathways. *Biotechnol Bioeng*, 59, 640–646.

Werpy T, Petersen G. (2004). Top value added chemicals from biomass, volume 1: results of screening for potential candidates from sugars and synthesis gas. US Department of Energy. Available at: http://www1.eere.energy.gov/biomass/pdfs/35523.pdf. Last accessed date Jan 8, 2014.

Xiao ZJ, Lv CJ, Gao C, Qin JY, Ma CQ, Liu Z, Liu PH, Li LX, Xu P. (2010). A novel whole-cell biocatalyst with NAD^+ regeneration for production of chiral chemicals. *PLoS ONE*, 5, e8860.

Xu Z. (2005). A Method of producing dihydric alcohol and polyol by sorbital cracking. China Patent 200510008652.0.

Yamada-Onodera K, Yamamoto H, Kawahara N, Tani Y. (2002). Expression of the gene of glycerol dehydrogenase from *Hansenula polymorpha* Dl-1in *Escherichia coli* for the production of chiral compounds. *Acta Biotechnol*, 22, 355–362.

Yan Y, Lee CC, Liao JC. (2009). Enantioselective synthesis of pure (R,R)-2,3-butanediol in *Escherichia coli* with stereospecific secondary alcohol dehydrogenases. *Org Biomol Chem*, 7, 3914–3917.

Yang G, Tian JS, Li JL. (2007). Fermentation of 1,3-propanediol by a lactate deficient mutant of *Klebsiella oxytoca* under microaerobic conditions. *Appl Microbiol Biotechnol*, 73, 1017–1024.

Yang TW, Rao ZM, Zhang X, Lin Q, Xia HF, Xu ZH, Yang ST. (2011). Production of 2,3-butanediol from glucose by GRAS microorganism *Bacillus amyloliquefaciens*. *J Basic Microbiol*, 51, 650–658.

Yang TW, Zhang X, Rao ZM, Gu SH, Xia HF, Xu ZH. (2012). Optimization and scale-up of 2,3-butanediol production by *Bacillus amyloliquefaciens* B10–127. *World J Microbiol Biotechnol*, 28, 1563–1574.

Yim H, Haselbeck R, Niu W, Pujol-Baxley C, Burgard A, Boldt J, Khandurina J, Trawick JD, Osterhout RE, Stephen R, Estadilla J, Teisan S, Schreyer HB, Andrae S, Yang TH, Lee SY, Burk MJ, Van Dien S. (2011). Metabolic engineering of *Escherichia coli* for direct production of 1,4-butanediol. *Nat Chem Biol*, 7, 445–452.

Yu EKC, Chan MKH, Saddler JN. (1985a). Butanediol production from lignocellulosic substrates by *Klebsiella pneumoniae* grown in sequential co-culture with *Clostridium thermocellum*. *Appl Microbiol Biotechnol*, 22, 399–404.

Yu EKC, Deschatelets L, Louis-Seize G, Saddler JN. (1985b). Butanediol production from cellulose and hemicellulose by *Klebsiella pneumoniae* grown in sequential coculture with *Trichoderma harzianum*. *Appl Environ Microbiol*, 50, 924–929.

Yu EKC, Deschatelets L, Tan LUL, Saddler JN. (1985c). A simple method for xylanase preparation used for the hydrolysis and fermentation of hemicellulose to butanediol. *Biotechnol Lett*, 7, 425–430.

Yu EKC, Levitin N, Saddler JN. (1982). Production of 2,3-butanediol by *Klebsiella pneumoniae* grown on acid hydrolyzed wood hemicellulose. *Biotechnol Lett*, 4, 741–746.

Yu EKC, Saddler JN. (1985). Biomass conversion to butanediol by simultaneous saccharification and fermentation. *Trends Biotechnol*, 3, 100–104.

Zeng AP, Biebl H, Deckwer WD. (1990). Effect of pH and acetic acid on growth and 2,3-butanediol production of *Enterobacter aerogenes* in continuous culture. *Appl Microbiol Biotechnol*, 33, 485–489.

Zeng AP, Biebl H, Deckwer WD. (1991). Production of 2,3-butanediol in a membrane bioreactor with cell recycle. *Appl Microbiol Biotechnol*, 34, 463–468.

Zeng AP, Sabra W. (2011). Microbial production of diols as platform chemicals: recent progresses. *Curr Opin Biotechnol*, 22, 749–757.

Zhang LY, Sun JA, Hao YL, Zhu JW, Chu J, Wei DZ, Shen YL. (2010a). Microbial production of 2,3-butanediol by a surfactant (serrawettin)-deficient mutant of *Serratia marcescens* H30. *J Ind Microbiol Biotechnol*, 37, 857–862.

Zhang LY, Yang YL, Sun JA, Shen YL, Wei DZ, Zhu JW, Chu J. (2010b). Microbial production of 2,3-butanediol by a mutagenized strain of *Serratia marcescens* H30. *Bioresour Technol*, 101, 1961–1967.

Zhang WG, Yu DH, Ji XJ, Huang H. (2012). Efficient dehydration of bio-based 2,3-butanediol to butanone over boric acid modified HZSM-5 zeolites. *Green Chem*, 14, 3441–3450.

Zhang YP, Zhu Y, Zhu Y, Li Y. (2009). The importance of engineering physiological functionality into microbes. *Trends Biotechnol*, 27, 664–672.

Zhao J, Hartwig JF. (2005). Acceptorless, neat, ruthenium-catalyzed dehydrogenative cyclization of diols to lactones. *Organometallics*, 24, 2441–2446.

Zhao X, Song Y, Liu D. (2011). Enzymatic hydrolysis and simultaneous saccharification and fermentation of alkali/peracetic acid-pretreated sugarcane bagasse for ethanol and 2,3-butanediol production. *Enzyme Microb Technol*, 49, 413–419.

Zheng Y, Zhang HY, Zhao L, Wei LJ, Ma XY, Wei DZ. (2008). One-step production of 2,3-butanediol from starch by secretory over-expression of amylase in *Klebsiella pneumoniae*. *J Chem Technol Biotechnol*, 83, 1409–1412.

CHAPTER 11

1,3-Propanediol

YAQIN SUN, CHENGWEI MA, HONGXIN FU, YING MU, and ZHILONG XIU

School of Life Science and Biotechnology, Dalian University of Technology, Dalian, PR China

Bioprocessing of Renewable Resources to Commodity Bioproducts, First Edition.
Edited by Virendra S. Bisaria and Akihiko Kondo.

Abstract

1,3-Propanediol, a promising bulk chemicals, has attracted worldwide attention due to its enormous application in polymers, cosmetics, foods, lubricants, medicines, and as an intermediate for the synthesis of heterocyclic compounds. It can be produced from renewable resources using microorganisms focusing mainly on the ecologically friendly process, industrial safety, and sustainable development. In this review various strategies employed for microbial production of 1,3-propanediol are introduced and compared. The overall investigation includes strains, genetic and metabolic engineering, substrate, metabolic regulation, fermentation, process optimization, downstream processing, and economic analysis. The future prospect and challenges of 1,3-propanediol from biotechnology is discussed for its industrial production.

11.1 INTRODUCTION

Nowadays, due to the depletion of fossil fuels, environmental problems and the sustainable development, the chemical industries are seriously exploring bio-based or renewable feedstocks for producing bulk chemicals, such as 1,3-propanediol (1,3-PD), 2,3-butanediol (2,3-BD), lactic acid, succinic acid, etc. Among all these, 1,3-PD is the promising bulk chemical which has attracted worldwide attention because of its enormous applications in polyesters, particularly as a monomer for synthesis of polytrimethylene terephthalate (PTT), polyethers, polyurethanes, and heterocyclic compounds such as indole and quinolines (Barbirato et al., 1995; Saxena et al., 2009; Kaur et al., 2012a). PTT is a biodegradable polyester and has great potential application in textile, carpets, and upholstery manufacturing (Witt et al., 1994). It has superior stretching and stretch recovery characteristics compared to other polyesters (Kurian, 2005).

Previously, PTT was produced based on 1,3-PD by chemical synthesis. The traditional chemical routes of commercial synthesis for 1,3-PD production are from acrolein by DuPont and from ethylene oxide by Shell. The route of DuPont is that acrolein is converted to 3-hydroxypropionaldehyde (3-HPA) by hydration, followed to produce 1,3-PD by hydrogenation (Lawrence and Sullivan, 1972). Shell followed the method of hydroformylation of ethylene oxide to 3-hydroxypropanal. This is subsequently extracted and hydrogenated for the production of 1,3-PD (Sullivan, 1993).

Presently, bioconversion of 1,3-PD is mainly based on two kinds of fermentation routes. One is commercially introduced into the market owing to the efforts of DuPont and Genencor in developing a recombinant *Escherichia coli* strain from glucose (Nakamura and Whited, 2003). Another is to produce 1,3-PD from glycerol by natural microorganisms. The latter one was first described in 1881 but little attention was paid to its biosynthesis for over a century due to practical and economical feasibility of biotechnology in comparison with traditional petrochemical routes (Biebl et al., 1999; Zeng and Biebl, 2002). An increasing interest in microbial production of 1,3-PD from glycerol has been developing since late 1980s because of increased biodiesel production in which glycerol is generated as a by-product in 10% by weight (Zeng and Sabra, 2011). Biotechnological innovations and improvements have been developed continuously for the microbial production of 1,3-PD.

In this review, various operating and metabolic engineering strategies for microbial production of 1,3-PD from glycerol are reviewed and compared. Besides glycerol, other cheap carbon sources which can be converted to 1,3-PD by natural and genetically engineered strains have also been introduced. It also highlights the significance of mathematical-model-based approach for simulation and optimization to describe metabolic process and improve 1,3-PD production. The present state of strategies for the downstream processing and purification of 1,3-PD have also been described according to their advantages and drawbacks. Finally, the economic analysis of this bioprocess are developed in comparison with chemical routes and the future prospect of 1,3-PD is discussed for its industrial production.

11.2 BIOCONVERSION OF GLUCOSE INTO 1,3-PROPANEDIOL

For biological production on large scale, it is always desirable to use cheaper and more abundant substrates such as glucose or starch. Keeping this in view, conversion of glucose into 1,3-PD has been well investigated. However, no natural microorganisms have been found that can directly ferment sugars to 1,3-PD so far. Nowadays, two kinds of approaches have been developed in order to convert glucose to 1,3-PD in a high level. The first one is referred to as a single-step fermentation by genetically engineered strains, which are recombined by introducing genes for both glycerol and 1,3-PD production (Tong and Cameron, 1992; Nakamura and Whited, 2003; Tang et al., 2009). Unfortunately, this strategy is not satisfactory according to reported results that 1,3-PD concentration was lower than 1.2 g/L by metabolically engineered *Saccharomyces cerevisiae* or *Klebsiella pneumoniae* (Zheng et al., 2008a; Ma et al., 2010b). DuPont and Genencor developed a glucose-based process using a recombinant *E. coli* carrying the genes for the production of glycerol from *S. cerevisiae* and the genes for the production of 1,3-PD from *K. pneumoniae*. The final concentration of 1,3-PD reached 135 g/L which was the highest reported concentration for 1,3-PD bioconversion (Nakamura and Whited, 2003). However, this process was realized on the addition of vitamin B_{12} which imposed on the cost of the bioconversion process.

The second approach is referred to as a mixed culture or two-step fermentation involving two microorganisms that one can convert sugars to glycerol and another can convert glycerol to 1,3-PD (Hartlep et al., 2002; Mendes et al., 2011). High concentration of substrate (glucose and glycerol) inhibited cell growth and 1,3-PD formation in mixed cultures. Meanwhile, the cultivation conditions for the mixed cultures were not easily controlled adapting to the growth of different microorganisms (Cameron et al., 1998; Biebl et al., 1999). A two-step fermentation of 1,3-PD was realized by the yeast *Pichia farinosa* or a recombinant *E. coli* strain and *K. pneumoniae* (Hartlep et al., 2002). However, the culture broth from *E. coli* converting glucose to glycerol in the first step contained toxic metabolites that strongly impair the 1,3-PD formation and growth of *K. pneumoniae* converting glycerol to 1,3-PD in the second step. The overall 1,3-PD yield from glucose in the two-stage process with *E. coli* and *K. pneumoniae* reached only 0.17 g/g.

Besides glucose, other carbon sources such as sugarcane molasses have also been used for the two-step 1,3-PD fermentation process (Mendes et al., 2011). Mendes et al. (2011) studied the production of 1,3-PD from glucose and molasses in a two-step process using two recombinant microorganisms of *S. cerevisiae* strain HC42 and *Clostridium acetobutylicum* DG1. A final 1,3-PD concentration of 25.5 g/L, productivity of 0.16 g/L·h and yields based on glycerol and glucose of 0.56 g/g and 0.24 g/g, respectively, at initial glucose concentration of 103 g/L were obtained, which were the highest values reported for a two-step process.

11.3 BIOCONVERSION OF GLYCEROL INTO 1,3-PROPANEDIOL

As mentioned above, glycerol can be obtained by glucose fermentation. However, an enormous amount of glycerol is released as by-product from biodiesel production and fat saponification process (Cheng et al., 2006; Mu et al., 2006). In addition, glycerol is also a minor by-product in ethanol manufacture (Zeng and Biebl, 2002). Crude glycerol, which has been shown to be an excellent substrate for 1,3-PD production, is becoming an abundant and cheap source for the biological production of 1,3-PD (Petitdemange et al., 1995; Papanikolaou et al., 2000; Mu et al., 2006). The utilization of crude glycerol to produce 1,3-PD provides an opportunity which is capable to capitalize on the surplus of waste glycerol.

Glycerol metabolism is a dismutation process occurring through coupled oxidative and reductive pathways as shown in Figure 11.1. In the reductive pathway, glycerol is dehydrated to 3-HPA by glycerol dehydratase (GDHt), and which then is further converted to 1,3-PD by 1,3-PD oxydoreductase (PDOR) under the oxidation of reducing powers afforded by the oxidative pathway. In a coupled oxidative pathway, glycerol is catalyzed by glycerol dehydrogenase (GDH) leading to the formation of dihydroxyacetone (DHA) which is coupled with reducing equivalents (NADH) generation. This intermediate is further phosphorylated by two DHA kinases (ATP-dependent DHAK I and PEP-dependent DHAK II), and the other by-products (succinic acid, lactic acid, 2,3-BD, acetate, ethanol, butyrate, butanol, and so on) are then formed and bioenergy and reducing powers are generated.

11.3.1 Strains

Quite a large number of native microorganisms are capable of converting glycerol into 1,3-PD, including *Klebsiella* (*K. pneumoniae* and *K. oxytoca*), *Clostridia* (*C. butyricum*, *C. pasteurianum*, *C. bifermentans*, *C. saccharobutylicum*, and *C. diolis*), *Enterobacter* (*E. agglomerans*), *Citrobacter* (*C. freundii*), and *Lactobacilli* (*L. brevis*, *L. buchneri*, and *L. diolivorans*). Among these microorganisms, *K. pneumoniae* and *C. butyricum* have been most intensively studied due to their substrate tolerance, high yield, and productivity (Zeng and Biebl, 2002; Xiu et al., 2007; Celinska, 2012). *Clostridium butyricum* is a typically anaerobic microorganism whereas

K. pneumoniae grows rapidly on glycerol under aerobic, microaerobic, and anaerobic conditions. 1,3-PD production using different bacteria species is compared in Table 11.1.

11.3.2 Fermentation

Although the bioconversion of 1,3-PD is an environmentally friendly route, it suffers from some drawbacks including the cost of substrate and other medium components, low conversion yield, inhibition by substrate and product, formation and toxicity of by-products (such as ethanol, 2,3-BD, organic acids, etc.) (Zeng et al., 1994, 1995; Zeng and Deckwer, 1995; Chen et al., 2003a,b). These disadvantages should be overcome to make biological production of 1,3-PD more economically feasible and attractive contrasting to the chemical route. Different strategies which have been employed for enhancing the microbial production of 1,3-PD including selection of substrate, co-fermentation or coproduction, metabolic regulation, optimization of operation modes, and so on are described in detail as follows.

11.3.2.1 Substrate and Co-substrate According to Deckwer (1995), substrate costs as much as 50% of the entire production cost of 1,3-PD. Low-cost raw material, such as industrial crude glycerol, directly from biodiesel production units without purification, have been utilized to produce 1,3-PD due to its economic and abundant advantages (Deckwer, 1995; Papanikolaou et al., 2000, 2004; Mu et al., 2006; Jun et al., 2010). Comparing to pure glycerol, 1,3-PD production was more effective when utilizing raw glycerol. More than 80 g/L of 1,3-PD was produced, which is the highest 1,3-PD concentration reported so far for *K. pneumoniae* using raw glycerol (Jun et al., 2010).

Likewise, to improve 1,3-PD production by an economic and efficient approach, cheap carbon sources and co-substrates such as glucose (Tong and Cameron, 1992; Abbad-Andaloussi et al., 1998; Xiu et al., 2007; Sun et al., 2010), sucrose (Yang et al., 2007), maltose (Ragout et al., 1996), xylose (Tong and Cameron, 1992; Jin et al., 2011a), hemicellulosic hydrolysates (Jin et al., 2011b) were used instead of the fraction of glycerol, resulting in more biomass and reducing equivalents. The pathways of glycerol and glucose co-metabolism are also as shown in Figure 11.1. Abbad-Andaloussi et al. (1998) studied the effect of glucose on glycerol metabolism by *C. butyricum* DSM5431. The results indicated that glucose catabolism was used to produce bioenergy whereas glycerol was used mainly in the utilization of reducing equivalents. According to stoichiometric analysis of glycerol-glucose co-fermentation in *K. pneumoniae*, 100% of glycerol could be transformed into 1,3-PD under anaerobic conditions with a ratio of 0.32 mol glucose per mol glycerol if glycerol was neither oxidized nor used to form biomass (Xiu et al., 2007). In comparison with glycerol used as a sole substrate, the batch co-fermentation of glucose and glycerol showed that the yield of 1,3-PD on glycerol and biomass concentration increased. However, glycerol consumption and 1,3-PD accumulation become slow in co-fermentation. Based on these research, the oscillation behavior of glucose and glycerol co-fermentation by *K. pneumoniae* DSM2026 under microaerobic

TABLE 11.1 Typical Results of Microbial Production of 1,3-Propanediol Using Different Bacteria Species

Strains	Substrate	Fermentation modes	$C_{1,3\text{-PD}}$ (g/L)	$Q_{1,3\text{-PD}}$ (g/L·h)	$Y_{1,3\text{-PD/Gly}}$ (mol/mol)	References
Klebsiella pneumoniae						
DSM2026	PG	Fed-batch in 1 m^3	74.6		0.61	Liu et al. (2007)
	PG	Fed-batch in 20 m^3	63.9		0.50	
DSM2026	PG	Fed-batch	61.9	2.00	0.48	Mu et al. (2006)
	CG	Fed-batch	53.0	1.70	0.47	
DSM2026	PG	Continuous	35.2–48.5	4.9–8.8	0.61	Menzel et al. (1997)
CGMCC 1.6366	PG	Fed-batch in 5 m^3	66.1		0.52	Zheng et al. (2009)
		Fed-batch in 50 m^3	63.3		0.50	
LX3	PG	Fed-batch	56.68	1.73	0.47	Xue et al. (2010)
		Fed-batch with organic acid addition	71.38	2.24	0.59	
ME-308	PG	Fed-batch with pH fluctuation	70	0.97	0.70	Ji et al. (2009)
ME-303	PG, corn straw	Fed-batch	71.58	1.93	0.65	Jin et al. 2011b
DSM4799	CG	Fed-batch	80.2	1.15	0.54	Jun et al. (2010)
Klebsiella oxytoca						
M5al	PG + sucrose	Fed-batch	83.56	1.61	0.62	Yang et al. (2007)
Clostridium butyricum						
AKR 102a	PG	Fed-batch in 1 L	93.7	3.3	0.63	Wilkens et al. (2012)
	CG	Fed-batch in 1 L	76.2	2.3	0.62	
	CG	Fed-batch in 200 L	61.5	2.1	0.64	
DSM 5431	PG	Batch	56.0	2.2	0.62	Biebl et al. (1992)
	PG	Fed-batch	70.3	1.5	0.68	Abbad-Andaloussi et al. (1996)
VPI 1718	CG	Fed-batch	67.9	0.78	0.67	Chatzifragkou et al. (2011)
CNCM1211	PG	Batch	67.0		0.63	Himmi et al. (1999)
F 2b	CG	Continuous, two-stage	46.0	3.4	0.53	Papanikolaou et al. (2000)

Clostridium pasteurianum						
LMG 3285	PG	Batch	3.02		0.4	Kubiak et al. (2012)
	PG	Continuous	10		0.34	
DSM 525	PG	Batch	5.5	1.19		Moon et al. (2011)
Clostridium bifermentans						
CB 29-80	PG	Isolation	9.78–16.78		0.24–0.41	Myszka et al. (2012)
Clostridium saccharobutylicum						
NRRL B-643	PG	Batch	6.7		0.36	Gungormusler et al. (2010), Kubiak et al., 2012
Clostridium diolis						
DSM15410	PG	Fed-batch	78.5	2.82	0.64	Otte et al. (2009)
DSM15410	PG	Repeated-batch	67.8	1.04		Kaur et al. (2012b)
Enterobacter agglomerans						
CNCM1210	PG	Batch	18.0		0.29	Barbirato et al. (1996)
Citrobacter freundii						
DSM30040	PG	Continuous, two-stage	41.5	1.38	0.62	Boenigk et al. (1993)
Lactobacillus diolivorans						
DSM14421	PG	Batch	41.7	0.85	0.79	Pfluegl et al. (2012)
	$m_{Glc/PG} = 0.1$ mol/mol	Fed-batch	68.3	0.48	0.73	
	$m_{Glc/PG} = 0.1$ mol/mol, vitamin B12	Fed-batch	84.5	0.45	0.76	

PG, pure glycerol; CG, crude glycerol; Glc, glucose.

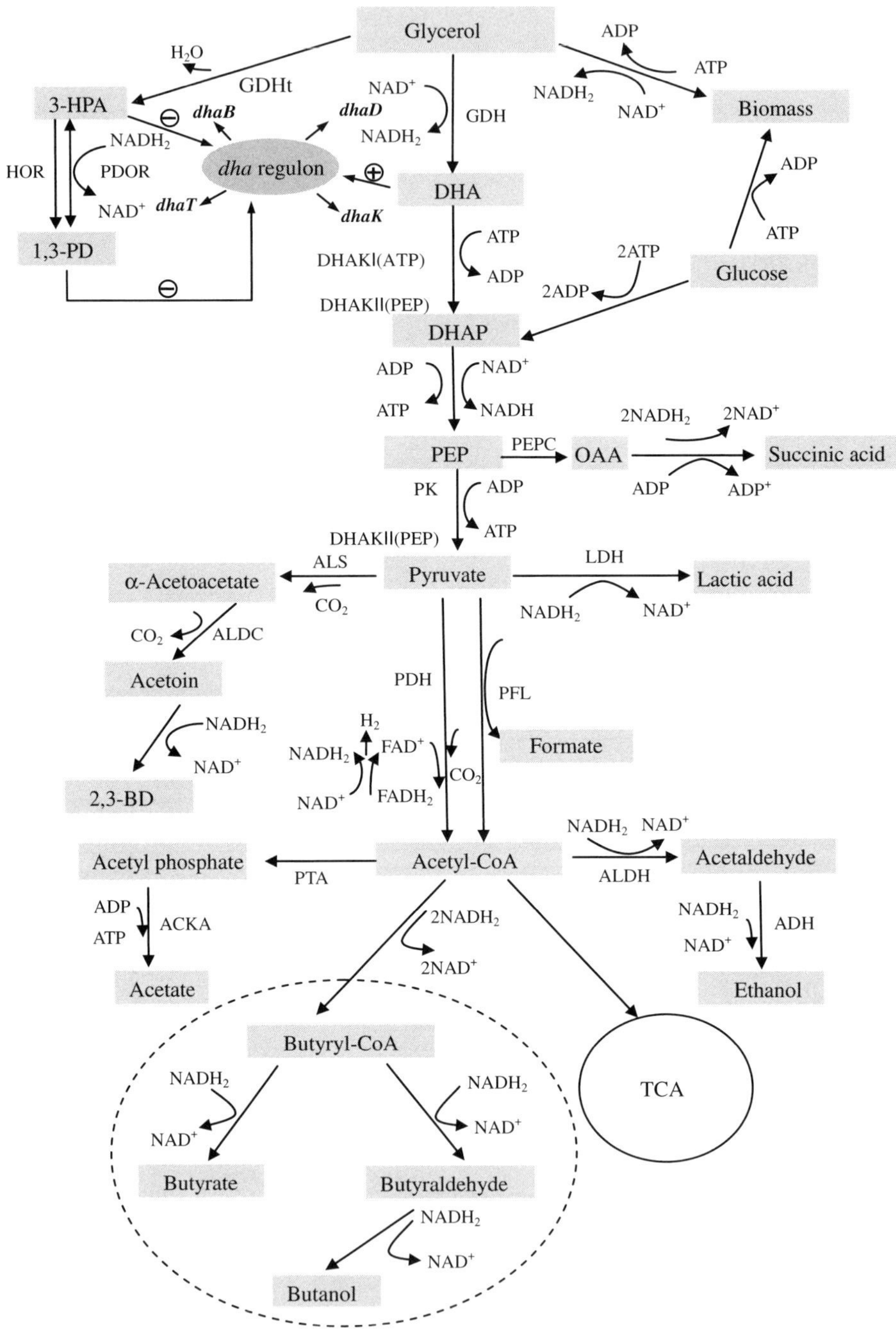

FIGURE 11.1 Metabolic pathway of glycerol conversion by *Klebsiella* spp. and *Clostridium* spp. The dotted lines show the pathway by Clostridium spp.

conditions was investigated (Sun et al., 2010). Comparing with a steady state at 80 g/L of glycerol as single carbon source, oscillations occurred at 80 g/L of glycerol and 35 g/L glucose as co-substrate. The accumulation of 3-HPA, which is toxic to cell growth, might have led to oscillation behavior.

When glycerol and xylose co-fermentation was carried out under fed-batch conditions in the case of *K. pneumoniae*, the final 1,3-PD concentration and conversion from glycerol were increased significantly by 9.1% and 20%, respectively. The NADH was increased by 1–3 mg/g cell dry weight (CDW) compared with that from glycerol alone. Moreover, 2,3-BD was also doubly produced as a major by-product (Jin et al., 2011a). Glycerol and xylose or glucose co-fermentation was also investigated by the engineered strain, in which the *dha* regulon of *K. pneumoniae* was expressed in *E. coli* (Tong and Cameron, 1992). The yield of 1,3-PD from glycerol was improved from 0.46 mol/mol with glycerol alone to 0.63 mol/mol with glucose as co-substrate and 0.55 mol/mol with xylose as co-substrate.

Using hemicellulosic hydrolysate as co-substrate, the final 1,3-PD production and yield on glycerol increased by 17.8% and 25.0% than that from glycerol alone in fed-batch fermentations (Jin et al., 2011b). Moreover, the effects of primary degradation products such as individual sugars (xylose, glucose, mannose, arabinose, and galactose) and major inhibitors (furfural, acetate, and formate) on *K. pneumoniae* growth and 1,3-PD production were also investigated in this study. Xylose and mannose could efficiently promote the 1,3-PD production and cell growth. Low concentration of furfural (0.28 g/L) and sodium acetate (1.46 g/L) were not inhibitory to cell growth, rather they had stimulatory effects on the growth and 1,3-PD biosynthesis, especially acetate.

11.3.2.2 Metabolic Regulation The strategy of metabolic regulation including the aeration, pH regulation, organic acids addition, and so on, was also applied to enhance the yield, productivity, and concentration of 1,3-PD. The influencing factors were studied and optimized by a number of research groups (Cheng et al., 2004; Lin et al., 2005; Liu et al., 2007; Zheng et al., 2008c, 2010; Ji et al., 2009; Xue et al., 2010; Petrov and Stoyanov, 2012).

The biosynthesis of 1,3-PD by *K. pneumoniae* could be accomplished under anaerobic, aerobic, and microaerobic conditions. According to stoichiometric analysis of glycerol bioconversion to 1,3-PD by *K. pneumoniae*, the maximum theoretical yield of 1,3-PD to glycerol could reach to 0.85 mol/mol under microaerobic conditions rather than 0.72 mol/mol under anaerobic conditions (Chen et al., 2003a). It was reported that 1,3-PD productivity enhanced from 1.62 to 2.94 g/L·h by changing the fermentation from anaerobic to microaerobic. The other reported studies showed that oxygen as an exogenous electron acceptor, could enhance 1,3-PD production (Chen et al., 2003b; Cheng et al., 2004). Moreover, the metabolic flux of 1,3-PD production by *K. pneumoniae* was analyzed and optimized in continuous cultures under microaerobic conditions with different glycerol concentrations (Wang et al., 2011). The results indicated that more carbon flux was shifted to the oxidative pathway with increasing glycerol concentrations under microaerobic conditions. A maximum 1,3-PD concentration of 20.22 g/L was obtained under high glycerol concentration of 69.92 g/L with air sparging at 0.04 vvm. Microaerobic fermentation for 1,3-PD production from crude glycerol was also applied in scale of 1 m^3 using *K. pneumoniae* (Liu et al., 2007). The final 1,3-PD concentration, molar yield, and productivity were 72 g/L, 0.57 mol/mol, and 2.1 g/L·h, respectively, which approached to those of 75 g/L, 0.61 mol/mol, and 2.2 g/L·h under anaerobic conditions. The experimental results achieved in 1 m^3 stirred bioreactor demonstrate that microaerobic fermentation is suitable for the production of 1,3-PD from glycerol by *K. pneumoniae*. In view of less investment and lower operation costs, microaerobic fermentation will be more attractive for the industrial production of 1,3-PD.

Some research of metabolic regulation had focused on the addition of organic acids, such as fumarate, citrate, succinate, etc. Fumarate addition was beneficial for 1,3-PD production by *K. pneumoniae*, but the 1,3-PD productivity still remained low (Lin et al., 2005). A mixture of organic acids including citrate, fumarate, and succinate was added into 1,3-PD fermentation by *K. pneumoniae* under fed-batch culture, which led to significant increase in cell growth and 1,3-PD production, and sharp decrease in the formation of by-products, especially lactic acid and ethanol (Xue et al., 2010).

Most research supported the pH maintenance at 7.0 and some suggested pH maintenance at two different levels (6.3 and 7.0) to decrease the by-products formation (Ji et al., 2009). The forced pH fluctuation applied for 1,3-PD production was described (Petrov and Stoyanov, 2012). The method was realized by consecutive raisings of pH ranging from 1.0 to 2.0 between 2 and 4 hours during fed-batch fermentation. The maximal 1,3-PD concentration, productivity, and molar yield on glycerol increased by 10%, 22%, and 29% compared with constant pH under fed-batch conditions, respectively. In addition, the formation of by-products was significantly decreased, especially that of lactic and acetic acid.

11.3.2.3 Operation Modes Different operation modes including batch, fed-batch, continuous and multistage cultivation, cell recycling, and immobilization have been implemented and optimized to maximize 1,3-PD production from glycerol. These operating strategies are illustrated in detail as follows.

Although a batch cultivation appears as a simple strategy for production, it brings the problem of low 1,3-PD concentration at the end of fermentation due to substrate limitation and product inhibition. The final concentration of 1,3-PD under batch cultures was generally lower than 50.0 g/L by *C. butyricum* and *K. pneumoniae* (Homann et al., 1990; Barbirato et al., 1998; Himmi et al., 1999). However, the yield on glycerol was high and could reach to 0.69 mol/mol using raw glycerol by *C. butyricum* CNCM1211 (Barbirato et al., 1998).

Comparing with batch cultures, more glycerol was used and more biomass was obtained under fed-batch fermentations, which led to high 1,3-PD concentration and productivity. Fed-batch fermentations coupled with other strategies such as co-substrate, pH control, ammonium feeding, different aeration strategies, and medium composition have been well developed by researchers (Reimann and Biebl, 1996; Cheng et al., 2004; Ji et al., 2009; Dong et al., 2010; Jin et al., 2011a). Generally, over 60 g/L of 1,3-PD could be achieved under fed-batch cultures from crude or pure glycerol by *C. butyricum* and *K. pneumoniae*. Some typical results of fed-batch fermentations are summarized in Table 11.1. Among the previous reports, a fed-batch strategy combining a low base-driven glycerol addition with constant on-line glycerol measurement was developed for a high substrate and product tolerant strain. A final 1,3-PD concentration of 87 g/L with an overall 1,3-PD productivity of 1.9 g/L h was achieved using refined glycerol in comparison with 80.1 g/L and 1.8 g/L·h using raw glycerol (Hirschmann et al., 2005). Moreover, fed-batch cultures with cell immobilization in a fixed-bed reactor were performed to enhance 1,3-PD production. The productivity of 1,3-PD increased with the cycle number, for example, 1.06 g/L·h versus 1.61 g/L·h at the first and fourth cycles, respectively, in a repeated fed-batch fermentation due to successful cell immobilization. During 46 cycles of fed-batch fermentation taking place over 1460 hours, a stable and reproducible 1,3-PD production performance was observed with both pure and raw glycerol (Jun et al., 2010).

Furthermore, fed-batch fermentation for 1,3-PD production had been successfully scaled up to 1 m^3 and 5 m^3 for *K. pneumoniae* (Cheng et al., 2007; Liu et al., 2007). The final 1,3-PD concentration, molar yield, and productivity of 75 g/L, 0.61 mol/mol, and 2.2 g/L·h were achieved on 1 m^3 scale under anaerobic condition, respectively. Through three-stage liquid inoculum development and a pulse addition of $(NH_4)_2SO_4$ and yeast extract at 30 hours of fermentation, the best results in a 5 m^3 fermentation were achieved leading to 58.8 g/L 1,3-PD with a yield of 0.53 mol/mol and productivity of 0.92 g/L·h. Although fed-batch fermentation is beneficial for obtaining high 1,3-PD concentration and productivity, it is not a good choice for industrial production on large scale primarily due to lack of suitable online, real time sensors for the measurement of key process variables (e.g., glycerol, 1,3-PD, biomass) to achieve the optimum concentration (Kaur et al., 2012b).

In contrast to batch and fed-batch fermentations, continuous fermentations could eliminate product inhibition (e.g., 1,3-PD, ethanol, organic acids). Menzel et al. (1997) investigated the effects of initial glycerol concentration and dilution rate on 1.3-PD production by *K. pneumoniae* DSM2026 under continuous cultures. The results indicated that 1,3-PD production was a function of initial glycerol concentration and dilution rate. Using the volumetric productivity of 1,3-PD as an optimization target, Xiu et al. (2004) investigated the optimal conditions of continuous glycerol fermentations by *K. pneumoniae*. For continuous fermentations, the optimal dilution rate and initial glycerol concentration in feeding medium were 0.29/h and 67.25 g/L, respectively, with the corresponding productivity of 8.67 g/L·h. Although a continuous culture afforded a significant advantage of high productivity, low product concentration (<50 g/L) was usually obtained mainly due to high substrate feeding. Furthermore, high dilution rates (>0.35/h) burden downstream process pressure and increase the 1,3-PD production cost (Zeng et al., 1993; Menzel et al., 1997; Papanikolaou et al., 2000, 2004; González-Pajuelo et al., 2004, 2005a; Chatzifragkou et al., 2011). To ensure high cell concentration in the bioreactor under continuous culture which may be beneficial for obtaining high product concentration, cell recycling was studied for 1,3-PD production by Reimann et al. (1998). It was the first attempt for improved 1,3-PD bioconversion with cell recycling using hollow fiber modules.

Comparing with the continuous cultivation without cell recycling, the cell density increased 4- to 5-folds for initial glycerol of 32 g/L and 3.4- to 4-folds for initial glycerol of 56 g/L with a retention ratio of 5. The maximum 1,3-PD concentration of 26.6 g/L was obtained with a glycerol concentration of 56 g/L and a retention ratio of 5.

Different from the two-step fermentation process of glucose to 1,3-PD, the aim of multistage fermentation of glycerol to 1,3-PD focuses on improving 1,3-PD concentration, productivity, and yield. A two-stage fermentation under continuous cultures was carried out by *C. freundii* (Boenigk et al., 1993). Biomass was produced in the first fermentor under glycerol limitation conditions and a reduction in the dilution rate was carried out in the second fermentor, in order to further increase the production of 1,3-PD. The obtained maximum 1,3-PD concentration was 41.42 g/L while the highest productivity was 1.38 g/L·h. A higher productivity of 3.4 g/L·h was observed using isolated *C. butyricum* from industrial glycerol under two-stage continuous fermentation (Papanikolaou et al., 2000). This cultivation strategy utilized a high dilution rate in the first fermentor in order to increase the 1,3-PD productivity and a lower dilution rate in the second stage to obtain an increased product concentration. In both fermentations, the dilution rate was chosen to be lower in the second stage than in the first. It has also been reported that in these cases biomass concentration was always less in the second fermentor than in the first one, and this was attributed to bacterial autolysis in the second stage of the culture due to either microbial propagation under permanently unfavorable conditions (increased contact time between the cells and the metabolic products) or excretion of autolytic enzymes (Papanikolaou et al., 2000). Moreover, to eliminate the toxicity of 3-HPA and maintain the expected high productivity of 1,3-PD, a two-stage fed-batch strategy was investigated using *K. pneumoniae* (Zheng et al., 2008b). The batch culture stage was conducted at initial glycerol concentration and stirring rate of 40 g/L and 250 rpm, respectively, and the feeding culture stage was performed at stirring rate of 300 rpm. A final 1,3-PD concentration of 74.07 g/L was obtained with the yield and productivity of 0.62 mol/mol and 3.09 g/L·h. A multistage fed-continuous culture was evaluated by *K. pneumoniae* under microaerobic and anaerobic conditions (Wang, 2010). In the first stage, a high biomass was obtained under microaerobic condition (0.04 vvm air). In the second stage, glycerol feeding concentration was chosen to be low (40 g/L) at a high dilution rate (0.2/h) or high (70–110 g/L) at a low dilution rate (0.1/h) in order to prevent limitation or inhibition effect of substrate on cell growth. In the third stage, the residual glycerol was consumed to low level (1–13 g/L) and further the concentration and yield of 1,3-PD enhanced under anaerobic condition (0.04 vvm N_2). The final 1,3-PD concentration of 36.7–46.2 g/L was obtained with the productivity of 4.0–8.1 g/L·h.

11.3.2.4 By-products During fed-batch fermentation of 1,3-PD by *K. pneumoniae*, sequential synthesis of products are exhibited, including the target product (1,3-PD) and some valuable by-products, such as 2,3-BD, succinate, lactate, ethanol, hydrogen, and 3-HPA. The dominate flux distribution showed that the metabolic flux was shifted from acetate formation to 1,3-PD formation in early exponential growth phase, and then to lactate synthesis in the late exponential growth phase in fed-batch fermentation (Xu et al., 2009).

2,3-BD is normally less than 20 g/L in fed-batch fermentation of 1,3-PD. High concentration of 2,3-BD (60 g/L) was obtained using sucrose as co-substrate in fermentation of 1,3-PD by a lactate-deficient mutant of *K. oxytoca* under microaerobic conditions, accompanying with high concentration of 1,3-PD (83 g/L) (Yang et al., 2007).

Lactic acid is often accumulated under anaerobic conditions, and its concentration is usually below 10 g/L. However, the concentrations of 1,3-PD and lactic acid were up to 62.6 g/L and 33.4 g/L, respectively, using a newly isolated *K. pneumoniae* by plasma (Song et al., 2012).

Hydrogen and 1,3-PD coproduction from glycerol was investigated in heat-treated mixed cultures under anaerobic conditions (Selembo et al., 2009). The molar yields of 1,3-PD and hydrogen were 0.69 mol/mol and 0.28 mol/mol using pure glycerol, respectively. A similar result was obtained using raw glycerol (70% glycerol content) with yields of hydrogen and 1,3-PD of 0.31 mol/mol and 0.59 mol/mol, respectively.

The intermediate of glycerol metabolism, 3-HPA, is likely to be accumulated under excessive substrate concentration and microaerobic conditions. Due to its toxicity to cell growth, it is desired to be transformed to low or nontoxic compounds, for example, 3-hydroxypropionic acid (3-HP). In theory, simultaneous production of 1,3-PD and 3-HP might balance the production and consumption of the essential cofactor $NADH/NAD^+$. Nevertheless, the overall product yield for both 1,3-PD and 3-HP were less than 0.61 mol/mol (Ashok et al., 2011; Huang et al., 2012). Additionally, a significant quantity of by-products was produced, including organic acids and alcohols. Furthermore, the coproduction of 1,3-PD and 3-HP was studied using resting cells of a recombinant *K. pneumoniae* J2B strain (Kumar et al., 2012). The concentration of 15.92 g/L 1,3-PD and 11.30 g/L 3-HP was obtained in 12 hours using 1.0 g CDW/L of recombinant strain under anaerobic fed-batch conditions. The use of resting cells improved substantially the carbon conversion for product formation, for example, 0.71 mol/mol of 1,3-PD and 3-HP on glycerol.

11.3.3 Bioprocess Optimization and Control

One of the approaches to optimize the microbial production of 1,3-PD from glycerol is process optimization based on kinetic models of glycerol metabolism. The fermentation of glycerol by *K. pneumoniae* is a complex bioprocess, since microbial growth is subjected to multiple inhibitions of substrate and products, for example, glycerol, 1,3-PD, ethanol, and acetate (Zeng et al., 1994). An excess kinetic model for substrate consumption and product formation was established in previous studies (Zeng et al., 1994, 1995; Zeng and Deckwer, 1995). Later, the model was improved to describe substrate consumption and products formation in a large range of feed glycerol concentrations (Xiu et al., 2000). The improved model can not only describe continuous and batch glycerol fermentation by *K. pneumoniae,* but also predict the occurrence of multiplicity in this bioprocess. Subsequently, batch and continuous glycerol fermentations by *K. pneumoniae* were optimized by considering the growth kinetics of multiple inhibitions and the metabolic overflow of substrate consumption and product formation (Xiu et al., 2004). For batch cultures, the optimal initial glycerol concentration was found to be 88.3 g/L, which led to the highest 1,3-PD volumetric productivity of 4.0 g/L·h. For continuous fermentations, the optimal dilution rate and initial glycerol concentration in feed were 0.29/h and 67.3 g/L, respectively. The corresponding productivity was 8.7 g/L·h that was more than twice the productivity of an optimal batch culture. Furthermore, optimization was carried out for a two-stage continuous process, in which the first stage was operated at the optimal conditions and the second one was used to consume the residual glycerol. The production of 1,3-PD by *C. butyricum* from raw glycerol was successfully simulated by other analytical models, for example, Contois-type model, without considering inhibitions due to strong tolerance of strains against high concentrations of substrate, 1,3-PD or butyric acid (Papanikolaou and Aggelis, 2003; Papanikolaou et al., 2004). In these "black box" kinetic models, the intermediate and intracellular substances or enzymes of glycerol metabolism were not taken into consideration. A novel mathematical model was proposed to describe the continuous and batch fermentations of glycerol by *K. pneumoniae*, in which the enzyme-catalytic kinetics on the reductive pathway, the transport of glycerol and diffusion of 1,3-PD across cell membrane, and the inhibition by 3-HPA of GDHt and PDOR were all taken into consideration (Sun et al., 2008). The analysis results indicated that the inhibition by 3-HPA of PDOR was stronger than that of GDHt. In addition, a fourteen-dimensional nonlinear dynamic system for gene regulatory dynamics of glycerol bioconversion to 1,3-PD was presented to describe the continuous culture and multiplicity analysis, in which two regulated negative-feedback mechanisms of repression and enzyme inhibition were investigated (Sun et al., 2012). The model describing the expression of gene–mRNA–enzyme product was established according to the repression of the *dha* regulon by 3-HPA. The results of simulation and analysis indicated that 3-HPA accumulation would repress the expression of the *dha* regulon at the transcriptional level. These kinetic models give us new insights into the regulation of glycerol metabolism in *K. pneumoniae* and explain some of the experimental observations.

11.4 METABOLIC ENGINEERING

11.4.1 Stoichiometric Analysis/MFA

A stoichiometric analysis considering the complete set of products including gases and cell mass may provide first evidence for the suggested metabolic pathways. Therefore, stoichiometric analysis of 1.3-PD production, considering energy and reducing equivalent and product balances, has been developed under different operation modes (Zeng et al., 1993; Chen et al., 2003a; Xiu et al., 2007). First, according to reducing equivalent balance and product formation, the theoretical maximum molar yield of 1,3-PD from glycerol was considered to be 0.72 mol/mol under anaerobic conditions with no formation of hydrogen and ethanol (Zeng et al., 1993). Subsequently, stoichiometric analysis results indicated that microaerobic condition was benefit for the cell growth and the product formation (Chen et al., 2003a). The theoretical yield of 1,3-PD from glycerol under microaerobic conditions depended not only on the molar fraction of reducing equivalent oxidized completely by molecular oxygen in tricarboxylic acid (TCA) cycle, but also on the molar fraction of TCA cycle from acetyl-CoA metabolism. The maximum theoretical yield of 1,3-PD from glycerol could reach to 0.85 mol/mol under microaerobic condition if all acetyl-CoA entered into TCA cycle with no formation of hydrogen, ethanol, acetate, and no consumption of oxygen. However, the practical yield of 1,3-PD was usually much lower compared with the theoretical maximum (Menzel et al., 1997; Jun et al., 2010), because some part of glycerol was taken up to produce biomass and by-products. To enhance the yield of 1,3-PD from glycerol and decrease the cost of production, a cheap carbon source, such as glucose was used as an H-donor substrate instead of glycerol to provide both reducing equivalents for 1,3-PD formation and ATP for biomass. Therefore, stoichiometric analysis of glycerol–glucose co-fermentation in *K. pneumonia* was developed according to energy, reducing equivalent, and product balances (Xiu et al., 2007). The theoretical analysis revealed that the yield of 1,3-PD from glycerol under microaerobic conditions depended not only on the ratio of glucose to glycerol initially added, but also on the molar fraction of reducing equivalent oxidized completely by molecular oxygen in TCA cycle and the molar fraction of TCA cycle from acetyl-CoA metabolism. The maximum ratio of 0.32 mol glucose per mol glycerol was needed to convert glycerol completely to 1,3-PD under anaerobic conditions if glycerol neither enters oxidation pathways nor forms biomass. The ratio can be reduced under microaerobic conditions. These results of stoichiometric analysis are helpful for experimental design and control.

The measured fluxes obtained under steady-state conditions were used to estimate intracellular fluxes and identify the robustness of branch points of the anaerobic glycerol metabolism in *K. pneumoniae* for the production of 1,3-PD by metabolic flux analysis (MFA) (Zhang et al., 2008). The flux distribution revealed that the branch points of glycerol and DHAP were rigid to the environmental conditions. However, the pyruvate and acetyl-CoA metabolisms gave cells the flexibility to regulate the energy and intermediate fluxes under various environmental conditions. Later on, a model that utilizes existing knowledge of oxygen and redox sensing/regulatory system to assist elementary flux modes (EFMs) was developed to predict the metabolic potential of *K. pneumoniae* for the production of 1,3-PD under anaerobic and aerobic conditions (Zhang and Xiu, 2009). It was found that the theoretical optimal 1,3-PD yield could reach to 0.84 mol/mol if the pentose phosphate pathway (PPP), and transhydrogenase had a high flux under anaerobic condition. However, PPP had little influence on the theoretical 1,3-PD yield, and the flux through TCA cycle was high under aerobic conditions. Although flux balance analysis (FBA) is an effective tool in MFA, the accurate prediction depends on the reasonable objective function. Gong et al. (2009) proposed two nonlinear bilevel programming models on anaerobic glycerol metabolism in *K. pneumoniae* for 1,3-PD production. Numerical results revealed some interesting conclusions, for example, biomass production was the main force to drive *K. pneumoniae* metabolism, and the objective functions, which were obtained in terms of several different groups of flux distributions, were similar.

A limitation of the classical method for MFA is the requirement for isotopic steady state. To extend the scope of flux determination from stationary to nonstationary systems, Antoniewicz et al. (2007) presented a novel modeling strategy that combines key ideas from isotopomer spectral analysis (ISA) and stationary MFA. The novel modeling strategy was illustrated with a nonstationary system that closely resembles industrial production conditions, that is, fed-batch fermentation of *E. coli* that produces 1,3-PD. As such, for the first time, detailed time profiles of *in vivo* fluxes were established. It was found that intracellular fluxes changed significantly during the fed-batch fermentation. The intracellular flux associated with 1,3-PD pathway increased by 10%. Concurrently, it was observed that there was a decrease in the split ratio between glycolysis and PPP from 70/30 to 50/50 as a function of time. The TCA cycle flux, on the other hand, remained constant throughout the fermentation. Another new method for dynamic metabolic flux analysis (DMFA) was proposed for systems that are not at metabolic steady states (Leighty and Antoniewicz, 2011). The advantages of the DMFA method are (1) time series of metabolite concentration data can be applied directly for estimating dynamic fluxes, making data smoothing, and estimation of average extracellular rates unnecessary; (2) flux estimation is achieved without integration of ODEs, or iterations; (3) characteristic metabolic phases in the fermentation data are identified automatically by the algorithm, rather than selected manually/arbitrarily. In the case of a commercial glucose-limited fed-batch process for 1,3-PD production, the DMFA method accurately captured the dynamic behavior of the fed-batch fermentation and identified characteristic metabolic phases.

11.4.2 Pathway Engineering

11.4.2.1 Utilization of Cofactor I/II In the *K. pneumoniae* reduction pathway for 1,3-PD synthesis, glycerol is first dehydrated to 3-HPA and then reduced to 1,3-PD with NADH consumption (Figure 12.1). Rapid conversion of 3-HPA to 1,3-PD is one of the ways to improve the yield of 1,3-PD from glycerol and to avoid 3-HPA accumulation, which depends on enzyme activity of the reaction and the amount of reducing equivalents available from the oxidative pathway of glycerol. In the work by Ma et al. (2010a), relaxation of the coenzyme specificity of 1,3-PD oxidoreductase for both NADH and NADPH was accomplished by rational protein design. First, Asp41 was illustrated to be the key residue responsible for the coenzyme specificity by computational alanine-scanning mutagenesis of the active site residues. Then, the relaxation was successfully realized by site-directed mutagenesis.

In the study by Vaidyanathan et al. (2011), *Lactobacillus reuteri* has been metabolically engineered for improving 1,3-PD production by the expression of an NADP-dependent *E. coli* alcohol dehydrogenase, *yqhD*, that is known to efficiently convert the precursor 3-HPA to 1,3-PD. The engineered strain exhibited significantly altered formation rates for the product and other metabolites during the fermentation. An increase in the 1,3-PD specific productivity of 34% and molar yield by 13% was achieved in the clone, relative to the native strain. Zhuge et al. (2010) expressed the *yqhD* gene from *E. coli* and the *dhaT* gene (encoding 1,3-PD oxidoreductase) from

K. pneumoniae individually in *K. pneumoniae* using the *tac* promoter expression plasmid. Expression of *dhaT* reduced formation of by-products (ethanol and lactic acid) and increased molar yield of 1,3-PD slightly, while expression of *yqhD* did not enhance molar yield of 1,3-PD, but increased ethanol concentration in broth because NADPH participation in transforming 3-HPA to 1,3-PD allowed more cellular NADH to be used to produce ethanol. Co-expression of both genes therefore decreased by-products and increased the molar yield of 1,3-PD by 11.8%, by catalyzing 3-HPA conversion to 1,3-PD using two cofactors (NADH and NADPH).

Zhang et al. (2009) reported the effect of introducing a heterogeneous NADH regeneration system into *K. oxytoca* on cell growth and glycerol metabolism. Expression of *fdh* gene (encoding formate dehydrogenase) from *Candida boidinii* in *K. oxytoca* resulted in higher intracellular concentrations of both NADH and NAD^+ during the metaphase fermentation, with the ratio of NADH to NAD^+ unaltered and cell growth unaffected, interestingly different from that in engineered *E. coli*, *Lactococcus lactis*, and others. MFA revealed that the fluxes to 1,3-PD, ethanol, and lactate were all increased, suggesting both the oxidative and reductive metabolisms of glycerol were enhanced. An aldehyde dehydrogenase ALDHk homologous to *E. coli* ALDH but with NAD^+-dependent propionaldehyde dehydrogenase activity was identified in *K. pneumoniae* (Luo et al., 2011). Overexpression of ALDHk in the glycerol oxidative pathway-deficient mutant AK/pVOTHk resulted in a 3.6-fold increase (0.57–2.07 g/L) in the production of 3-HP and a 1.1-fold enhancement (8.43–9.65 g/L) of 1,3-PD synthesis, when glycerol was provided as the carbon source, compared to the levels synthesized by the control strain (AK/pVOT).

11.4.2.2 *Elimination of By-products Formation* Glycerol metabolism is a typical biological oxidoreductive reaction. 1,3-PD is the final product of the reductive branch, while acetate, succinate, lactate, 2,3-butanediol, and ethanol are produced in the oxidative branch (Figure 11.1). The first step in the formation of ethanol from acetyl-CoA is catalyzed by aldehyde dehydrogenase (ALDH), an enzyme that competes with 1,3-PD oxidoreductase for the cofactor NADH. In order to improve the production of 1,3-PD by genetic engineering, an inactivation mutation of the *aldA* gene encoding ALDH in *K. pneumoniae* YMU2 was generated by insertion of a tetracycline resistance marker (Zhang et al., 2006). The final titer, the productivity of 1,3-PD and the yield of 1,3-PD relative to glycerol of the mutant strain reached 70.50 g/L, 1.07 g/L·h and 0.70 mol/mol, respectively, which were much higher than those of the parent strain.

Mutants lacking D-lactate pathway were constructed by knocking out the *ldhA* gene encoding D-lactate dehydrogenase (LDH) of K. pneumoniae HR526 (Xu et al., 2009). In experiments using pure glycerol as feedstock, the 1,3-PD concentrations, conversion, and productivity increased from 95.39 g/L, 0.48 and 1.98 g/L·h to 102.06 g/L, 0.52 mol/mol and 2.13 g/L·h, respectively. The diol (1,3-PD and 2,3-butanediol) conversion increased from 0.55 mol/mol to a maximum of 0.65 mol/mol. Lactate would not accumulate until 1,3-PD exceeded 84 g/L, and the final lactate concentration decreased dramatically from more than 40 g/L to less than 3 g/L. In another work (Yang et al., 2007), four lactate-deficient mutants were obtained by knocking out the *ldhA* gene in *K. oxytoca* M5al. The results showed that higher 1,3-PD concentration, productivity, and conversion from glycerol to 1,3-PD could be gained than those of the wild-type strain and no lactic acid was produced under both anaerobic and microaerobic conditions. In fed-batch fermentation with sucrose as co-substrate under microaerobic conditions, the 1,3-PD concentration, conversion, and productivity were improved significantly to 83.56 g/L, 0.62 mol/mol, and 1.61 g/L·h, respectively. Furthermore, 60.11 g/L 2,3-butanediol was also formed as a major by-product in the broth.

2,3-butanediol, which has similar properties of high boiling point and water solubility with 1,3-PD, not only contests the carbon flow and NADH with 1,3-PD but also serves as an obstacle for obtaining high purity 1,3-PD in downstream processes. A 2,3-butanediol pathway-deficient mutant of *K. oxytoca* ZG36 was constructed by knocking out the *budA* gene (encoding α-acetolactate decarboxylase) of the wild-type strain M5al (Zhang et al., 2012). The indexes of the 1,3-PD titer, the conversion from glycerol to 1,3-PD, and the productivity per CDW increased by 42%, 62%, and 46%, respectively, compared with the M5al, and the yield of the by-products (e.g., lactate, succinate, acetate, and ethanol) also increased obviously.

In the study by Horng et al. (2010), *dhaD* and *dhaK*, the genes for GDH and DHA kinase, respectively, were inactivated by homologous recombination in *K. pneumoniae*. Analysis of the cell-free supernatant revealed elimination of lactate and 2,3-butanediol, as well as ethanol accumulation in the *dhaD*/*dhaK* double mutant, compared with the wild-type strain. Furthermore, 1,3-PD productivity was increased in the strain expressing GDHt and 1,3-PD dehydrogenase regulated by the arabinose PBAD promoter. The oxidative branch of glycerol metabolism was inactivated by constructing two mutant strains in the work presented by Seo et al. (2009). In one of the mutant strains, the structural genes encoding GDH and DHA kinase were deleted from the chromosomal DNA, whereas in the second mutant strain *dhaR*, which is a putative transcription factor that activates gene expression, was deleted from the chromosomal DNA. The production yield was higher in the recombinant strain (0.57 mol/mol) than the wild strain (0.47 mol/mol). Oh et al. (2012) also generated a genetically engineered *K. pneumoniae* strain (AK-VOT) to eliminate by-product formation during the production of 1,3-PD from glycerol. As expected, by-product formation was completely absent (except for acetate). However, 1,3-PD productivity was severely reduced owing to a delay in cell growth attributable to a low rate of glycerol consumption. This problem was solved by establishing a two-stage process separating cell growth from 1,3-PD production.

11.4.2.3 Construction of 1,3-Propanediol Formation Pathway A genetic means of increasing the enzyme activities of 1,3-PD formation pathway through the overexpression of 1,3-PD operon was performed in *K. pneumoniae* S6 (Ma et al., 2012). The recombinant strain S6-PD showed 27- and 15-fold increase in activities of GDHt and 1,3-PDOR, respectively, with respect to the wild-type strain. However, it failed to improve the 1,3-PD yield due to the inadequacy of cofactor NADH. Therefore, an NADH regeneration system was constructed by heterologous expression of NAD^+-dependent formate dehydrogenase gene (*fdh1*) from *C. boidinii* and introduced into S6-PD in order to improve the glycerol utilization and 1,3-PD formation. The 1,3-PD, ethanol, and lactate for S6-FPD showed an increase of 18.3%, 30.0%, and 19.6% than the control strain S6-PD, respectively. In an earlier study by Ma et al. (2010a), *dha* operon from *K. pneumoniae* was cloned and expressed in *E. coli* and *S. cerevisiae* W303-1A, respectively. The recombinant *E. coli* harboring the *dha* operon could produce 8–9 g/L 1,3-PD from glycerol while 1,3-PD in recombinant strain W303-1A-*dha* could not be detected.

To obtain a "vitamin B_{12}-free" biological process, González-Pajuelo et al. (2005b) developed a metabolic engineering strategy in *C. acetobutylicum*. The 1,3-PD pathway from *C. butyricum* was introduced on a plasmid in several mutants of *C. acetobutylicum* altered in product formation. In a fed-batch culture, the DG1(pSPD5) strain was able to produce 1,3-PD at a higher concentration (83.90 g/L) and productivity than the natural producer *C. butyricum* VPI 3266. Furthermore, this strain was also successfully used for very long term continuous production of 1,3-PD at high volumetric productivity (3 g/L·h) and titer (59.89 g/L). Dabrowski et al. (2012) constructed a recombinant *E. coli* strain producing 1,3-PD from glycerol by introducing genes of the *dha* operon from *C. butyricum* 2CR371.5. The *E. coli* strain produced 1,3-PD of 3.7 g/L with the yield of 0.3 g/g glycerol consumed. A novel polycistronic operon under the control of the temperature-sensitive lambda phage P(L)P(R) promoter regulated by the cIts857 repressor was engineered and expressed in *E. coli* K-12 ER2925 (Tang et al., 2009). The genes for the production of 1,3-PD in *C. butyricum*, *dhaB1* and *dhaB2*, which encode the vitamin B_{12}-independent GDHt DhaB1 and its activating factor, DhaB2, were tandemly arrayed with the *E. coli yqhD* gene. 1,3-PD from glycerol by this recombinant *E. coli* strain was studied in a two-stage fermentation process. During the two-stage process, the overall 1,3-PD concentration and productivity reached 104.4 g/L and 2.61 g/L·h, respectively, and the conversion of glycerol to 1,3-PD reached 90.2% (g/g).

In order to engineer an *E. coli* strain that can produce 1,3-PD directly from glucose, Liang et al. (2011) constructed a stress-induced metabolic pathway by the expression of *gpd1* and *gpp2* genes from *S. cerevisiae* and *dha* operon from *K. pneumoniae*, respectively. Batch cultivation of the recombinant *E. coli* showed that 12.1 g/L 1,3-PD was accumulated in the culture without using any inducer. According to the report by Nakamura and Whited (2003), the carbon pathway of the engineered *E. coli* utilizes glycerol 3-phosphate dehydrogenase and glycerol 3-phosphate phosphatase genes, obtained from *S. cerevisiae*, to provide glycerol. GDHt (*dhaB1*, *dhaB2*, *dhaB3*) and its reactivating factors (*dhaBX*, *orfX*), obtained from *K. pneumoniae*, enable the conversion of glycerol to 3-HPA. A previously uncharacterized oxidoreductase endogenous to *E. coli* (*yqhD*) catalyzes 3-HPA into 1,3-PD. In addition, two fundamental changes imposed on the *E. coli*-based strain are the elimination of D-glucose transport by the phosphotransferase system (PTS) and downregulation of glyceraldehyde 3-phosphate dehydrogenase. The end result is a metabolically engineered organism that provides 1,3-PD at a rate of 3.5 g/L·h, a titer of 135 g/L, and a weight yield of 51% in D-glucose fed-batch fermentations.

11.5 DOWN-PROCESSING OF 1,3-PROPANEDIOL

Recovery and purification of 1,3-PD from complex and dilute fermentation broth meet severe challenges, which is mainly attributed to its low volatility, great affinity for water, and the presence of residual glycerol, macromolecules (e.g., proteins, nucleic acids, polysaccharides), salts, and some by-products (e.g., 2,3-BD, ethanol, acetate, lactate, succinate) (Xiu and Zeng, 2008). This becomes the bottleneck of the commercialization of 1,3-PD from renewable resources.

Considerable progress has been made in the separation of 1,3-PD, including evaporation, distillation, membrane filtration, pervaporation, molecular sieves, chromatography, liquid–liquid extraction, reactive extraction, and salting-out extraction (SOE). However, all these methods have some limitations.

The evaporation and distillation techniques are widely used for separation of liquid mixtures, therefore they have been attempted for the recovery of 1,3-PD. Sanz et al. (2001) determined the necessary vapor–liquid equilibrium (VLE) data of water + 1,3-PD + glycerol for computer simulation of the 1,3-PD distillation process. The patents applied by Ames (2002) disclosed the purification processes for biologically produced 1,3-PD and emphasized that an alkaline environment could reduce impurity (esters, pigments) formation during the distillation process. However, the necessity for a large amount of heat energy led to a high cost of the target product purified in this way. In addition, desalination and deproteinization are required before distillation which makes the process complicated and nonprofitable. Otherwise, the denatured macromolecules and crystallized salts would result in low efficiency of evaporation/distillation and low yield of the target product. Flocculation, charcoal adsorption, and electrodialysis can be used for deproteinization and desalination before evaporation (Roturier et al., 2002; Gong et al., 2004; Wu et al., 2011). The macromolecular impurities as well as salts have been precipitated by adding alcohol into the concentrated broth after evaporation due to alcohol precipitation and dilution crystallization (Gao et al., 2007).

Adkesson et al. (2005) disclosed a method for purifying 1,3-PD from the fermentation broth generated by a bioengineered *E. coli*. The broth was subjected successively to microfiltration, ultrafiltration, and nanofiltration, removing cellular biomass, macromolecules with molecule weights greater than about 5000 Daltons, and small molecules with molecule weights about 200–400 Daltons, respectively. Subsequent purification steps include a mixed ion exchange and a series of distillations. Finally, product with high purity (99.99%) was obtained.

Pervaporation has advantages over evaporation and distillation due to its low energy demand and inexpensive equipment (Li et al., 2001a,b,c). A series of zeolite membranes were prepared and used to separate 1,3-PD from binary, ternary, quaternary (1,3-PD/glycerol/glucose/water) systems, and a cell-free fermentation broth. The separation of 1,3-PD is attributed to adsorption and diffusion. A thorough verification revealed that the 1,3-PD/glycerol selectivity was mainly controlled by preferential adsorption, and the 1,3-PD/glucose selectivity was mainly controlled by differences in diffusion rates. The method entailed a low flux and selectivity as well as indispensable membrane regeneration step when a cell-free fermentation broth was treated.

Electrodialysis membrane has been employed for desalination before evaporation (Gong et al., 2004; Wu et al., 2011). About 90% of the organic acid salts were successfully removed by electrodialysis. However, the loss of 1,3-PD due to diffusion, membrane fouling, high investment, and operational cost make this process undesirable.

Molecular sieves techniques, especially adsorption on hydrophobic zeolites such as silicalite or de-aluminous NaY zeolites, or even active carbon, were considered for 1,3-PD purification (Corbin and Norton, 2003). However, the capacity is quite low. Furthermore, Corbin and Norton (2003) disclosed an in-suit separation method which connecting the fermentor to a column filled with zeolite extrudate through a cross-flow filtration unit. Here, 1,3-PD in the permeate fermentation broth was cyclically adsorbed and then desorbed by ethanol and water. An indispensible need for dewatering and a high chance of contamination restricted its industrial application. In the patent applied by Roturier et al. (2002), the deproteinized and desalinated clarified solution was obtained by passing the fermentation broth through a charcoal column, a strong cationic resin, and then a weakly basic anionic resin of the acrylic type. This solution was subjected to a strongly acidic cation exchange resin of the polystyrene sulfonic acid type crosslinked with divinylbenzene, the elution being carried out using water. During this process, the sample was diluted 45 times by water, leading to a high energy demand for the dewatering afterward. Another strong cation exchange resin of polystyrene sulfonate in the Na form was contrived to separate 1,3-PD from the impurities (Hilaly and Binder, 2002). This process was conducted using a column containing 100 mL of the said resin. The original feed solution was diluted using water by ten times. The experiments resulted in a product with purity of 87% and the recovery was 95.7%. Ninety-eight percent of overall ionic species were removed by exposing 1,3-PD broth to a strong acidic cation exchange resin followed by a weak basic anion exchange resin (Adkesson et al., 2005). Residual salts and color after evaporation were significantly removed when fed to a mixture of strong acid cation exchange resin and strong base anion resin. Ion exchange resin must be regenerated more frequently due to a large amount of anionic and cationic molecules in fermentative broths. This method was, however, difficult to obtain 1,3-PD with high purity because of the lack of ionic properties in the target compound.

A chromatography column packed with silica resin was studied to separate 1,3-PD from a top phase containing 1,3-PD and 1,2-PD after extraction using ethyl acetate (Cho et al., 2006). The optimal elution condition comprised of 98% ethyl acetate and 2% methanol. This process gave a yield of 82% and a purity of 98%. Gradient chromatography loaded with silica gel was applied to glycerol-based fermentation broth after filtration, charcoal treatment, and vacuum distillation, using a mixture of chloroform and methanol as mobile phase. The overall yield of 1,3-PD was 75.45% with 98% of purity (Anand et al., 2011). Process chromatography was tested for removing 1,3-PD *in situ* with the aim to avert feedback inhibition on cell growth and product formation (Wilkins and Lowe, 2004). High cost for scaling up made these methods unattractive.

Compared to distillation, solvent extraction is considered as an energy-efficient and large-scale separation method. Liquid–liquid extraction with organic solvents can be directly applied to the recovery of the target product from dilute solutions, if a suitable solvent can be found. Malinowski (1999) evaluated the liquid–liquid extraction for the separation of 1,3-PD from dilute aqueous solutions. Experimental results showed the distribution of 1,3-PD into extraction solvents, which appeared to be not good enough for developing a simple and efficient extraction process. An attempt to separate 1,3-PD from a dilute solution by normal physical or complex extraction was also not successful due to its strong hydrophility (Xiang et al., 2001). Although many solvent extractants were listed in a patent (Baniel et al., 2006), including alcohols, ketones, esters, acids, ethers, or vegetable oils with log *P* value from 0.8 to 7.7, the hydrophilic 1,3-PD in dilute broths is not apt to enter into hydrophobic solvents, except for adding a large amount of solvent into a concentrated broth. Similarly, a hydrophobic solvent, ethyl acetate, was also used in phase separation of 1,3-PD from a mixture containing 1,3-PD, 1,2-PD, glycerol, and glucose (Cho et al., 2006). Ethanol was attempted as a cosolvent with ethyl acetate for improving the extraction efficiency of biologically derived 1,3-PD. However, the distribution coefficient of 1,3-PD only increased from 0.14 to 0.20 with actual fermentation broth (Boonsongsawat et al., 2010). Conventional liquid–liquid extraction process requires the handling of large quantities of solvents and, in particular, 1,3-PD does not distribute into organic solvents sufficiently to make simple liquid–liquid extraction efficient.

A possible avenue toward easier recovery of 1,3-PD was proposed by Broekhuis et al. (1994), where formaldehyde and acetaldehyde were reacted with 1,3-PD to form its dioxolane derivative and then recovering it by means of extraction using aromatic hydrocarbons. A reactive extraction process combining reversible reaction of 1,3-PD with acetaldehyde to 2-methyl-1,3-dioxane (2-MD) and a simultaneous extraction of 2-MD by organic solvent such as o-xylene, toluene, or ethylbenzene was studied (Malinowski, 2000; Xiang et al., 2001; Fang and Zhou, 2006). 1,3-PD was finally obtained by hydrolyzing 2-MD. It was reported that the yield of 2-MD was 91–92%, the overall conversion of 1,3-PD was 98–99.1%, and the recovery of dioxane into the organic extractant was 75–77%. However, the impurities in real fermentation broths are apt to cause inactivation of the catalyst for reaction between 1,3-PD and acetaldehyde, for example, a strongly acidic cation exchange resin, therefore, electrodialysis was necessary. Solid heterogeneous catalysts, for example, ZrO_2–MoO_3, might be improved but could only be used five cycles for the acetalization of 1,3-PD from dilute solutions (Wu et al., 2012). Furthermore, many substances, such as ethanol, 2,3-BD, glycerol as well as soluble proteins, in the broth can react with aldehyde, thus reducing the specificity of this method (Hao et al., 2005). Hao et al. (2006) later proposed a similar but improved reactive extraction using propionaldehyde, butyraldehyde, and isobutyraldehyde as both reactant and extractant. Proteins and cell debris must be removed and ethanol is best removed due to its hydrophilic character before the reactive extraction. The acetals produced were hydrolyzed in a reactive distillation column using a strongly acidic cation exchange resin as catalyst. Because of the additional need to regenerate 1,3-PD from its dioxolane derivative, the complexity and the cost of the chemicals used, and the toxicity of extractant (*o*-xylene, toluene, or ethylbenzene) make the extraction and purification process prohibitive.

A novel method has been recently developed and applied for recovery of compounds bearing multiple hydrophilic groups efficiently from aqueous solutions (Li et al., 2009, 2011). The use of short-chain alcohols or hydrophilic organic solvents and salts to form SOE (formerly called aqueous two-phase extraction) systems for separation and purification of 1,3-PD. Compared with the traditional solvent extractions and reactive extractions, hydrophilic organic solvents, and salts are green in terms of carcinogenic and toxic effects. Our experiments showed that the SOE could be used to extract 1,3-PD from fermentative broth (Li et al., 2009, 2011) as shown in Table 11.2. An SOE composed of 35% (v/v) methanol and saturated phosphate was found to give a highest partition coefficient (38.3) and recovery of 1,3-PD (98.1%) in single-step extraction with simultaneous extraction of 2,3-BD and removal of cells and proteins with high efficiency. In addition, methanol not only extracted 1,3-PD effectively from the broth, but also allowed recovery of phosphate (94.7%) by dilution crystallization through adjusting the pH to 4.5. Aydogan et al. (2010) studied the SOE system of dipotassium hydrogen phosphate/ethanol, where the optimal partition coefficient (20.28) and yield (97.20%) of 1,3-PD were obtained by using the response surface method. Another SOE system consisted of hydrophobic ionic liquids and a mixture of K_2HPO_4 and KH_2PO_4 were developed and investigated for the extraction of 1,3-PD, the distribution coefficients were between 1.5 and 22.5, which can be attributed to the polarity or hydrogen-bound accepting strength of the anion or cation (Mueller and Gorak, 2012). However, the possibility of subsequent product separation and the influence of other components have not been considered. Wu and Wang (2012) evaluated the salting-out effect using pentanol as solvent on recovery of 1,3-PD from fermentation broth. One- and two-salt systems were tested, the optimal distribution coefficient and recovery were 3.72% and 72% using sodium phosphate (0.4 g/mL), and 47.2% and 92.5% using both sodium phosphate (0.4 g/mL) and sodium sulfate (0.2 g/mL), respectively at 60°C. Stripping 1,3-PD into water from the organic phase at a low temperature was executed, and the stripping percentage was 91.7%. Salt can be separated using a solubility difference. The application of this approach for the recovery of 1,3-PD appeared noneconomic due to the requirement of large amounts of energy for dewatering and controlling operating temperature. It is worth mentioning that the novel SOE systems with short-chain alcohols or hydrophilic organic solvents deserve more attention in the future owing to its low cost, simple operation, and environment-friendly character.

11.6 INTEGRATED PROCESSES

11.6.1 Biodiesel and 1,3-Propanediol

Biodiesel (fatty acid methyl esters), which is derived from triacylglycerols by transesterification with short-chain alcohols, has attracted considerable attention during the

TABLE 11.2 Salting-Out Extraction of 1,3-Propanediol

				R, %[a]							
SOES composition	Treatment of fermentation broths	Y, %[a]	K[a]	Cells	Proteins	Formic acid	Lactic acid	Acetic acid	Succinic acid	Fumaric acid	References
35% Methanol (v/v)/saturated phosphate (pH10.7)	None	98.1	38.3	99.85	92.4	90.1	76.7	87	88.6	81	Li et al. (2011)
24% Methanol /14% K_2HPO_4 (w/w)	Filtered	98.2	7.5	–	–	–	–	–	–	–	Li et al. (2011)
46% Ethanol (v/v)/saturated $(NH_4)_2SO_4$	None	93.7	4.8	–	–	–	–	–	–	–	Li et al. (2009)
46% Ethanol/17% $(NH_4)_2SO_4$ (w/w)	None	–	–	99	83	–	–	–	–	–	Li et al. (2009)
35.39% Ethanol/28.40% K_2HPO_4 (w/w)	None	97.2	20.3	96.3	95.4	–	60	42	–	–	Aydogan et al. (2010)
Pentanol/0.4 g/mL Na_3PO_4 + 0.2 g/mL Na_2SO_4 (pentanol/fermentation broth = 1/1, v/v)	Centrifuged	92.5	47.2	–	–	–	–	–	–	–	Wu and Wang (2012)

[a]Y is recovery yield of 1,3-propanediol; K is partition coefficient of 1,3-propanediol between the top and bottom phase; R is removal ratio of cells, proteins, or organic acids from the fermentation broth.

past decade as a renewable, biodegradable, and nontoxic fuel (Ma and Hanna, 1999; Andrade and Vasconcelos, 2003). Glycerol is the main by-product during biodiesel production. It represents 10% (w/w) of the esters and has to be used to increase the profitability of the overall process. Glycerol is easily adsorbed on the surface of catalyst, resulting in serious negative effect on the transesterification (Dossat et al., 1999; Xu et al., 2003). To avoid the glycerol inhibition on rate and extent of the conversion, dialysis, and adsorption were used to remove glycerol by Bélafi-Bakó et al. (2002) and Stevenson et al. (1994), respectively.

Glycerol is a commodity chemical used in cosmetics, liquid soaps, foods, pharmaceuticals, lubricants, antifreeze solutions, tobacco, and so on (Chotani et al., 2000). It is surplusing on the world market due to the obligatory application of biodiesel on a large commercial scale. In the last years, crude glycerol has been treated as typical industrial waste water in various Western Union countries. Therefore, various ways of (crude) glycerol valorization by fermentation technology attract significant attention, for example, transformation of glycerol into citric acid (Papanikolaou et al., 2008), hydrogen and ethanol (Ito et al., 2005), microbial lipid (Papanikolaou et al., 2008; Scott et al., 2011), or 1,3-PD (Mu et al., 2006; Papanikolaou et al., 2008; Chatzifragkou et al., 2011; Wilkens et al., 2012; Song et al., 2012).

Mu *et al.* (2008) examined a combined process of lipase-catalyzed biodiesel production and 1,3-PD production, in which glycerol was passed through a membrane from lipase-catalyzed reactor to fermentor for 1,3-PD production. This technology has the ability to continuously remove glycerol from transesterification reaction mixture and convert it to 1,3-PD by *K. pneumoniae* directly. So it avoids glycerol inhibition on lipase, reduces the production cost, and raises productivity for biodiesel and 1,3-PD. Partial methanol penetrated into fermentation broth with glycerol, which could be the drawback of this process, but the effects of methanol on cell growth and products formation were insignificant. The molar yield of 1,3-PD from glycerol of 0.47 mol/mol and the volumetric productivity of 1.7 g/L·h were obtained.

11.6.2 Glycerol and 1,3-Propanediol

For large-scale biological production of 1,3-PD, it is desirable to use cheaper and more abundant substrates such as starch or glucose. However, no natural microorganisms have been so far found to be able to directly convert glucose into 1,3-PD.

To produce 1,3-PD directly from glucose, one strategy is to use a genetic approach, by which the genes directing conversion of glucose to glycerol as well as glycerol to 1,3-PD can be combined and expressed in one organism (Cameron et al., 1998; Emptage et al., 2003). Other strategy is to use one microorganism to convert sugars to glycerol and another to convert glycerol to 1,3-PD. This can be achieved through a one-step process using mixed cultures, but microbial repression of glucose on 1,3-PD production appears to be unfavorable. Genetic modification of microorganisms is needed to suppress the repression of 1,3-PD formation(Biebl et al., 1999; Cameron et al., 1998). In a patent application, Haynie and Wagner (1997) described the conversion of glucose to 1,3-PD in mixed or linked fermentation by yeast and *K. pneumoniae*, *C. freundii*, or recombinant *E. coli*. Moderate yields and a maximum 1,3-PD concentration of 5 g/L were reported.

Two-step fermentation processes, that is, fermentation of glucose by a yeast strain to produce glycerol in the first step and fermentation of glycerol by *K. pneumoniae* to 1,3-PD in the second step, can be carried out or can be combined into one process by a recombinant *E. coli* (Hartlep et al., 2002; Huang et al., 2002; Cheng et al., 2006; Mendes et al., 2011). According to Hartlep et al. (2002), the overall 1,3-PD yield from glucose in the two-step process was 0.17 g/g. Huang et al. (2002) described a two-step fermentation process by Saccharomyces LORRE Y8 and *K. pneumoniae*. The final concentration of 1,3-PD was 39.4 g/L, and the yield of 1,3-PD to glycerol was 0.54 mol/mol with a productivity of 1.1 g/L·h. After desalination treatment of glycerol broth containing 152 g/L from *Candida krusei* culture by electrodialysis, a fed-batch fermentation was developed by *K. pneumonia* to obtain 53 g/L of 1,3-PD with a yield of 0.41 g/g and a productivity of 0.94 g/L·h (Cheng et al., 2006).

11.6.3 1,3-Propanediol and Biogas

Clostridia butyricum can convert glycerol into 1,3-PD, thereby generating acetate, formate, and butyrate as by-products. The removal of organic acids from the system, especially the most inhibitive acetic acid, would benefit this biotransformation process. A novel mixed culture comprising *C. butyricum* and a methane-producing bacterium, *Methanosarcina mazei*, was proposed to relieve the inhibition and to utilize the by-products for biogas production (Freidmann et al., 2009; Bizukojc et al., 2010). The organic acids, particularly acetic and formic acids produced by *C. butyricum* would be utilized efficiently by *M. mazei* to produce methane. This would improve the overall biotransformation process in the favor of 1,3-PD production because of relieving the inhibition of these organic acids on the formation of 1,3-PD. In this mixed culture process, the toxic by-products from glycerol bioconversion are converted to methane by methanogenic organisms, simultaneously in the same bioreactor or in a subsequent stage. This concept was successfully demonstrated in laboratory scale and is being scaled up in pilot plant. Using crude glycerol (80% glycerol) as a carbon source and inocula adapted from a local wastewater treatment plant, 1,3-PD can be produced as the main product at concentration as high as 60 g/L in a not-yet-optimized semi-batch culture. A high yield of 1,3-PD is 0.6 mol/mol, which is close to the theoretical maximal yield of anaerobic glycerol conversion, has been achieved (Sabra et al., 2010). In order to examine the efficiency of such a mixed culture, metabolic modeling of the culture system was performed (Bizukojc et al., 2010). Methanol was considered as a co-substrate, since it may exist in glycerol solution from biodiesel production. The calculations revealed that if methanol is present, the methane production can increase by 130%. *M. mazei* can scavenge over 70% of the acetate secreted by *C. butyricum*.

11.7 ECONOMIC OUTLOOK

Like other fermentative products, such as ethanol, butanol, lactic acid, and succinic acid, the cost of 1,3-PD production depends also on the costs of substrate, fermentation, and separation, in which the substrate cost is the main part, even up to more than 50% (Deckwer, 1995; Hermann and Patel, 2007). The two biological routes are expected to be competitive with the petrochemical route to produce 1,3-PD in the view of economics. Comparing with sugar-based 1,3-PD production, the glycerol-based 1,3-PD production cost decrease greatly as the use of crude glycerol generated as "waste" from the biodiesel plants. The techno-economic sensitivity on bioconversion of glycerol into 1,3-PD showed that the cost of glycerol was the key factor influencing the profit of 1,3-PD bioproduction. Through applying the integrated production of biodiesel and 1,3-PD, the production cost of 1,3-PD could be decreased while the retained profit was increased by 31% (Mu et al., 2009). Moreover, the economic analysis indicated that the increase in the productivity would not simultaneously bring more profits for continuous fermentations. On the contrary, the increase in the yield of 1,3-PD from glycerol would bring growth to the profit, but this growth became less pronounced once the yield was higher than 0.68 mol/mol. With the availability of cheap and abundant crude glycerol, the cost of downstream processing would make a very high portion in the total production cost, mounting up to about 50–70% (Xiu and Zeng, 2008). The recovery of 1,3-PD from complex fermentation broth represents a true bottleneck in the development of a commercially viable bioprocess. It is worthwhile to develop a simple and economically viable purification process from fermentation broth. Furthermore, energy integration for fermentation and downstream processing must be emphasized, which could further enhance the efficiency of microbial production of 1,3-PD on large scale.

11.8 FUTURE PROSPECTS

The microbial production of 1,3-PD is an ecologically friendly, economically feasible process, and sustainable development technology while minimizing the dependence on fossil resources. A successful bioconversion of glucose into 1,3-PD has been realized by DuPont on industrial scale. Glycerol-based fermentation has been also explored on pilot scale, but some biotechnological bottlenecks are still needed to break. Comparing with the sugar-based route, the researchers should make their many efforts on glycerol-based strain improvement in light of system biology. Although some important improvements in metabolic engineering of bioconversion of glycerol to 1,3-PD have been obtained in the past decade, the understanding of metabolic regulation is still not deep enough to achieve target product as much as that from the sugar-based route. For the two bioroutes, the utilization of cheap materials should be taken into consideration, for example, crude glycerol from biodiesel production and hydrolysates of lignocelluloses. These cheap substrates would bring some new problems for bioprocess and biosystem engineering, including strain, fermentation, and separation. It would be essential to develop integrated bioprocesses, and simple and effective separation methods. The above challenges are to be confronted for industrial production of 1,3-PD in the future. Undoubtedly, microbial production of 1,3-PD would be a good example for bio-based chemicals by bioroute.

ACKNOWLEDGMENTS

This work was financially supported by "863" projects (2012AA021202-3) of the Ministry of Science and Technology of the People's Republic of China.

A LIST OF ABBREVIATIONS

Acetyl-CoA	acetyl coenzyme A
ACKA	acetatekinase
ADH	alcohol dehydrogenase
aldA	gene encoding aldehyde dehydrogenase;
ALDC	acetolactate decarboxylase
ALDH	acetaldehyde dehydrogenase
ALS	acetolactatesynthase
2,3-BD	2,3-butanediol
budA	gene encoding α-acetolactate decarboxylase
CDW	cell dry weight
DHA	dihydroxyaceton
dhaB1	gene encoding the vitamin B_{12}-independent glycerol dehydratase
dhaB2	gene encoding activating factor of the vitamin B_{12}-independent glycerol dehydratase
dhaD	gene encoding glycerol dehydrogenase
dhaK	gene encoding dihydroxyacetone kinase
DHAK I	ATP-dependent dihydroxyacetone kinases
DHAK II	PEP-dependent dihydroxyacetone kinases
DHAP	dihydroxyacetonephosphate
dhaR	the putative transcription factor
dhaT	gene encoding 1,3-propanediol oxidoreductase
DMFA	dynamic metabolic flux analysis
EFMs	elementary flux modes
FBA	flux balance analysis
fdh	gene encoding formate dehydrogenase
GDH	glycerol dehydrogenase
GDHt	glycerol dehydratase
gpd1	genen encoding glycerol-3-P dehydrogenase
gpp2	gene encoding glycerol-3-P phosphatase
3-HP	3-hydroxypropionic acid
3-HPA	3-hydroxypropionaldehyde
HOR	hypothetical oxidoreductase
ISA	isotopomer spectral analysis
LDH	lactate dehydrogenase
ldhA	gene encoding D-lactate dehydrogenase
MFA	metabolic flux analysis
OAA	oxaloacetic acid
1,3-PD	1,3-propanediol

PDH	pyruvate dehydrogenase
PDOR	1,3-propanediol oxidoreductase
PEP	phosphoenolpyruvate
PEPC	phosphoenolpyruvate carboxylase
PFL	pyruvate format lyase
PK	pyruvate kinase
PPP	pentose phosphate pathway
PTA	phosphotransacetylase
PTS	phosphotransferase system
PTT	polytrimethylene terephthalate
SOE	salting-out extraction
TCA	tricarboxylic acid cycle
VLE	vapor liquid equilibrium
yqhD	gene-encoding alcohol dehydrogenase

REFERENCES

Abbad-Andaloussi S, Durr C, Raval G, Petitdemange H. (1996). Carbon and electron flow in *Clostridium butyricum* in chemostate culture on glycerol and glucose. *J Appl Microbiol*, 142, 1149–1158.

Abbad-Andaloussi S, Amine J, Gerard P, Petitdemange H. (1998). Effect of glucose on glycerol metabolism by *Clostridium butyricum* DSM 5431. *J Appl Microbiol*, 84, 515–522.

Adkesson DM, Alsop AW, Ames TT, Chu LA, Disney JM, Dravis BC, Fitzgibbon P, Gaddy JM, Gallagher FG, Lehnhardt WF, Lievense JC, Luyben ML, Seapan M, Trotter RE, Wenndt GM, Yu EK. (2005). Purification of biologically-produced 1,3-propanediol. US Patent 7919658.

Ames TT. (2002). Process for isolation of 1,3-propanediol from fermentation broth. US Patent 6361983.

Anand P, Saxena RK, Marwah RG. (2011). A novel downstream process for 1,3-propanediol from glycerol-based fermentation. *Appl Microbiol Biotechnol*, 90, 1267–1276.

Andrade JC, Vasconcelos I. (2003). Continuous cultures of Clostridium acetobutylicum: culture stability and low-grade glycerol utilisation. *Biotechnol Lett*, 25, 121–125.

Antoniewicz MR, Kraynie DF, Laffend LA, Gonzalez-Lergier J, Kelleher JK, Stephanopoulos G. (2007). Metabolic flux analysis in a nonstationary system: fed-batch fermentation of a high yielding strain of E. coli producing 1,3-propanediol. *Metab Eng*, 9, 277–292.

Ashok S, Raj SM, Rathnasingh C, Park S. (2011). Development of recombinant *Klebsiella pneumoniae* Delta dhaT strain for the co-production of 3-hydroxypropionic acid and 1,3-propanediol from glycerol. *Appl Microbiol Biotechnol*, 90, 1253–1265.

Aydogan O, Bayraktar E, Mehmetoglu U, Kaeding T, Zeng AP. (2010). Selection and optimization of an aqueous two-phase system for the recovery of 1,3-propandiol from fermentation broth. *Eng Life Sci*, 10, 121–129.

Baniel AM, Jansen RP, Vitner A, Baiada A. (2006). Process for producing 1, 3-propanediol. US Patent 7056439.

Barbirato F, Camarasca-Claret C, Bories A, Grivet JP. (1995). Description of the glycerol fermentation by a new 1,3-propanediol producing microorganism. *Appl Microbiol Biotechnol*, 43, 786–793.

Barbirato F, Grivet JP, Soucaille P, Bories A. (1996). 3-Hydroxypropionaldehyde, an inhibitory metabolite of glycerol fermentation to 1,3-propanediol by enterobacterial species. *Appl Environ Microbiol*, 62, 1448–1451.

Barbirato F, Himmi EH, Conte T, Bories A. (1998). 1,3-propanediol production by fermentation: an interesting way to valorize glycerin from the ester and ethanol industries. *Ind Crop Prod*, 7, 281–289.

Bélafi-Bakó K, Kovács F, Gubicza L, Hancsók J. (2002). Enzymatic biodiesel production from sunflower oil by *Candida antarctica* lipase in a solvent-free system. *Biocatal Biotransform*, 20, 437–439.

Biebl H, Marten S, Hippe H, Deckwer WD. (1992). Glycerol conversion to 1,3-propanediol by newly isolated *Clostrida*. *Appl Microbiol Biotechnol*, 36, 592–597.

Biebl H, Menzel K, Zeng AP, Deckwer WD. (1999). Microbial production of 1,3-propanediol. *Appl Microbiol Biotechnol*, 52, 289–297.

Bizukojc M, Dietz D, Sun J, Zeng AP. (2010). Metabolic modelling of syntrophic-like growth of a 1,3-propanediol producer, *Clostridium butyricum*, and a methanogenic archeon, *Methanosarcina mazei*, under anaerobic conditions. *Bioproc Biosyst Eng*, 33, 507–523.

Boenigk R, Bowien S, Gottschalk G. (1993). Fermentation of glycerol to 1,3-propanediol in continuous cultures of *Citrobacter freundii*. *Appl Microbiol Biotechnol*, 38, 453–457.

Boonsongsawat T, Shotipruk A, Tantayakom V, Prasitchoke P, Chandavasu C, Boonnoun P, Muangnapoh C. (2010). Solvent extraction of biologically derived 1,3-propanediol with ethyl acetate and ethanol cosolvent. *Sep Sci Technol*, 45, 541–547.

Broekhuis RR, Lynn S, King CJ. (1994). Recovery of propylene glycol from dilute aqueous solutions via reversible reaction with aldehydes. *Ind Eng Chem Res*, 33, 3230–3237.

Cameron DC, Altaras NE, Hoffman ML, Shaw AJ. (1998). Metabolic engineering of propanediol pathways. *Biotechnol Prog*, 14, 116–125.

Celinska E. (2012). *Klebsiella spp* as a 1, 3-propanediol producer—the metabolic engineering approach. *Crit Rev Biotechnol*, 32, 274–288.

Chatzifragkou A, Papanikolaou S, Dietz D, Doulgeraki AI, Nychas GJE, Zeng AP. (2011). Production of 1,3-propanediol by *Clostridium butyricum* growing on biodiesel-derived crude glycerol through a non-sterilized fermentation process. *Appl Microbiol Biotechnol*, 91, 101–112.

Chen X, Xiu ZL, Wang JF, Zhang DJ, Xu P. (2003a). Stoichiometric analysis and experimental investigation of glycerol bioconversion to 1,3-propanediol by *Klebsiella pneumoniae* under microaerobic conditions. *Enzyme Microbiol Technol*, 33, 386–394.

Chen X, Zhang DJ, Qi WT, Gao SJ, Xiu ZL, Xu P. (2003b). Microbial fed-batch production of 1,3-propanediol by *Klebsiella pneumoniae* under micro-aerobic conditions. *Appl Microbiol Biotechnol*, 63, 143–146.

Cheng KK, Liu DH, Sun Y, Liu WB. (2004). 1,3-Propanediol production by *Klebsiella pneumoniae* under different aeration strategies. *Biotechnol Lett*, 26, 911–915.

Cheng KK, Zhang JA, Liu DH, Sun Y, Yang MD, Xu JM. (2006). Production of 1,3-propanediol by *Klebsiella pneumoniae* from glycerol broth. *Biotechnol Lett*, 28, 1817–1821.

Cheng KK, Zhang JA, Liu DH, Sun Y, Liu HJ, Yang MD, Xu JM. (2007). Pilot-scale production of 1,3-propanediol using *Klebsiella pneumoniae*. *Process Biochem*, 42, 740–744.

Cho MH, Joen SI, Pyo SH, Mun S, Kim JH. (2006). A novel separation and purification process for 1,3-propanediol. *Process Biochem*, 41, 739–744.

Chotani G, Dodge T, Hsu A, Kumar M, LaDuca R, Trimbur D, Weyler W, Sanford K. (2000). The commercial production of chemicals using pathway engineering. *Biochim Biophys Acta*, 1543, 434–455.

Corbin DR, Norton T. (2003). Process to separate 1,3-propanediol or glycerol, or a mixture there of from a biological mixture. US Patent 6603048.

Dabrowski S, Pietrewicz-Kubicz D, Zablotna E, Dlugolecka A. (2012). 1,3-propanediol production by *Escherichia coli* expressing genes of dha operon from *Clostridium butyricum* 2CR371.5. *Acta Biochim Pol*, 59, 357–361.

Deckwer WD. (1995). Microbial conversion of glycerol into 1,3-propanediol. *FEMS Microbiol Rev*, 16, 143–149.

Dong XY, Xiu ZL, Li S, Hou YM, Zhang DJ, Ren CS. (2010). Dielectric barrier discharge plasma as a novel approach for improving 1,3-propanediol production in *Klebsiella pneumoniae*. *Biotechnol Lett*, 32, 1245–1250.

Dossat V, Combes D, Marty A. (1999). Continuous enzymatic transesterification of high oleic sunflower oil in a packed bed reactor: influence of the glycerol production. *Enzyme Microbiol Technol*, 25, 194–200.

Emptage M, Haynie SL, Laffend LA, Pucci JP, Whited G. (2003). Process for the biological production of 1,3-propanediol with high titer. US Patent, 6514733.

Fang YJ, Zhou P. (2006). Study on reactive extraction kinetics of 1,3-propanediol in dilute aqueous solutions. *Sep Sci Technol*, 41, 329–340.

Freidmann H, Zeng AP. (2009). Method and device for microbial production of a certain product and methane. WO 2009047275.

Gao SJ, Zhang DJ, Sun YQ, Xiu ZL. (2007). Separation of 1,3-propanediol from glycerol-based fermentations of *Klebsiella pneumoniae* by alcohol precipitation and dilution crystallization. *Front Chem Eng China*, 1, 202–207.

Gong Y, Tang Y, Wang XL, Yu LX, Liu DH. (2004). The possibility of the desalination of actual 1,3-propanediol fermentation broth by electrodialysis. *Desalination*, 16, 169–178.

Gong ZH, Liu CY, Feng EM, Zhang QR. (2009). Computational method for inferring objective function of glycerol metabolism in *Klebsiella pneumoniae*. *Comput Biol Chem*, 33, 1–6.

González-Pajuelo M, Andrade J, Vasconcelos I. (2004). Production of 1,3-propanediol by *Clostridium butyricum* VPI 3266 using a synthetic medium and raw glycerol. *J Ind Microbiol Biotechnol*, 31, 442–446.

González-Pajuelo M, Andrade JC, Vasconcelos I. (2005a). Production of 1,3-Propanediol by *Clostridium butyricum* VPI 3266 in continuous cultures with high yield and productivity. *J Ind Microbiol Biotechnol*, 32, 391–396.

González-Pajuelo M, Meynial-Salles I, Mendes F, Andrade JC, Vasconcelos I, Soucaille P. (2005b). Metabolic engineering of *Clostridium acetobutylicum* for the industrial production of 1,3-propanediol from glycerol. *Metab Eng*, 7, 329–336.

Gungormusler M, Gonen C, Ozdemir G, Azbar N. (2010). 1,3-Propanediol production potential of Clostridium saccharobutylicum NRRL B-643. *New Biotechnol*, 27, 782–788.

Hao J, Liu HJ, Liu DH. (2005). Novel route of reactive extraction to recover 1,3-propanediol from a dilute aqueous solution. *Ind Eng Chem Res*, 44, 4380–4385.

Hao J, Xu F, Liu HJ, Liu DH. (2006). Downstream processing of 1,3-propanediol fermentation broth. *Chem Technol Biotechnol*, 81, 102–108.

Hartlep M, Hussmann W, Prayitno N, Meynial-Salles I, Zeng AP. (2002). Study of two-stage processes for the microbial production of 1,3-propanediol from glucose. *Appl Microbiol Biotechnol*, 60, 60–66.

Haynie SL, Wagner LW. (1997). Process for making 1,3-propanediol from carbohydrates mixed microbial culture. US Patent, 5599689.

Hermann BG, Patel MK. (2007). Today's and tomorrow's bio-based bulk chemicals from white biotechnology. *Appl Biochem Biotechnol*, 136, 361–388.

Hilaly AK, Binder TP. (2002). Method of recovering 1,3-propanediol from fermentation broth. US Patent, 6479716.

Himmi EH, Bories A, Barbirato F. (1999). Nutrient requirements for glycerol conversion to 1,3-propanediol by *Clostridium butyricum*. *Bioresource Technol*, 67, 123–128.

Hirschmann S, Baganz K, Koschik I, Vorlop KD. (2005). Development of an integrated bioconversion process for the production of 1,3-propanediol from raw glycerol waters. *Landbauforsch Volk*, 55, 261–267.

Homann T, Tag C, Biebl H, Deckwer WD, Schink B. (1990). Fermentation of glycerol to 1,3-propanediol by *Klebsiella* and *Citrobacter* strains. *Appl Microbiol Biotechnol*, 33, 121–126.

Horng YT, Chang KC, Chou TC, Yu CJ, Chien CC, Wei YH, Soo PC. (2010). Inactivation of dhaD and dhaK abolishes by-product accumulation during 1,3-propanediol production in *Klebsiella pneumoniae*. *J Ind Microb Biotechnol*, 37, 707–716.

Huang H, Gong CS, Tsao GT. (2002). Production of 1,3-propanediol by *Klebsiella pneumoniae*. *Appl Biochem Biotechnol*, 98–100, 687–698.

Huang YN, Li ZM, Shimizu K, Ye Q. (2012). Simultaneous production of 3-hydroxypropionic acid and 1,3-propanediol from glycerol by a recombinant strain of *Klebsiella pneumoniae*. *Bioresource Technol*, 103, 351–359.

Ito T, Nakashimada Y, Senba K, Matsui T, Nishio N. (2005). Hydrogen and ethanol production from glycerol-containing wastes discharged after biodiesel manufacturing process. *J Biosci Bioeng*, 100, 260–265.

Ji XJ, Huang H, Zhu JG, Hu N, Li S. (2009). Efficient 1,3-propanediol production by fed-batch culture of *Klebsiella pneumoniae*: the role of pH fluctuation. *Appl Biochem Biotechnol*, 159, 605–613.

Jin P, Lu SG, Huang H, Luo F, Li S. (2011a). Enhanced reducing equivalent generation for 1,3-propanediol production through cofermentation of glycerol and xylose by *Klebsiella pneumoniae*. *Appl Biochem Biotechnol*, 165, 1532–1542.

Jin P, Li S, Lu SG, Zhu JG, Huang H. (2011b). Improved 1,3-propanediol production with hemicellulosic hydrolysates (corn straw) as cosubstrate: impact of degradation products on *Klebsiella pneumoniae* growth and 1,3-propanediol fermentation. *Bioresource Technol*, 102, 1815–1821.

Jun SA, Moon C, Kang CH, Kong SW, Sang BI, Um Y. (2010). Microbial fed-batch production of 1,3-propanediol using raw glycerol with suspended and immobilized *Klebsiella pneumoniae*. *Appl Biochem Biotechnol*, 161, 491–501.

Kaur G, Srivastava AK, Chand S. (2012a). Advances in biotechnological production of 1,3-propanediol. *Biochem Eng J*, 64, 106–118.

Kaur G, Srivastava AK, Chand S. (2012b). Simple strategy of repeated batch cultivation for enhanced production of 1,3-propanediol using *Clostridium diolis*. *Appl Biochem Biotechnol*, 167, 1061–1068.

Kubiak P, Leja K, Myszka K, Celinska E, Spychala M, Szymanowska-Powalowska D, Czaczyk K, Grajek W. (2012). Physiological predisposition of various *Clostridium species* to synthetize 1,3-propanediol from glycerol. *Process Biochem*, 47, 1308–1319.

Kumar V, Sankaranarayanan M, Jae KE, Durgapal M, Ashok S, Ko Y, Sarkar R, Park S. (2012). Co-production of 3-hydroxypropionic acid and 1,3-propanediol from glycerol using resting cells of recombinant *Klebsiella pneumoniae* J2B strain overexpressing aldehyde dehydrogenase. *Appl Microbiol Biotechnol*, 96, 373–383.

Kurian JV. (2005). A new polymer platform for the future-Sorona (R) from corn derived 1,3-propanediol. *J Polym Environ*, 13, 159–167.

Lawrence FR, Sullivan RH. (1972). Process for making a dioxane. US Patent, 3687981.

Leighty RW, Antoniewicz MR. (2011). Dynamic metabolic flux analysis (DMFA): a framework for determining fluxes at metabolic non-steady state. *Metab Eng*, 13, 745–755.

Li SG, Tuan VA, Falconer JL. (2001a). Separation of 1,3-propanediol from aqueous solutions using pervaporation through an X-type zeolite membrane. *Ind Eng Chem Res*, 40, 1952–1959.

Li SG, Tuan VA, Falconer JL. (2001b). Effects of zeolite membrane structure on the separation of 1,3-propanediol from glycerol and glucose by pervaporation. *Chem Mater*, 13, 1865–1873.

Li SG, Tuan VA, Falconer JL, Noble RD. (2001c). Separation of 1,3-propanediol from glycerol and glucose using a ZSM-5 zeolite membrane. *J Membrane Sci*, 191, 53–59.

Li ZG, Jiang B, Zhang DJ, Xiu ZL. (2009). Aqueous two-phase extraction of 1, 3-propanediol from glycerol-based fermentation broths. *Sep Purif Technol*, 44, 112–117.

Li ZG, Teng H, Xiu ZL. (2011). Extraction of 1,3-propanediol from glycerol-based fermentation broths with methanol/phosphate aqueous two-phase system. *Process Biochem*, 46, 586–591.

Liang QF, Zhang HJ, Li SN, Qi QS. (2011). Construction of stress-induced metabolic pathway from glucose to 1,3-propanediol in *Escherichia coli*. *Appl Microbiol Biotechnol*, 89, 57–62.

Lin RH, Liu HJ, Hao J, Cheng K, Liu DH. (2005). Enhancement of 1,3-propanediol production by *Klebsiella pneumoniae* with fumarate addition. *Biotechnol Lett*, 27, 1755–1759.

Liu HJ, Zhang DJ, Xu YH, Mu Y, Sun YQ, Xiu ZL. (2007). Microbial production of 1,3-propanediol from glycerol by *Klebsiella pneumoniae* under micro-aerobic conditions up to a pilot scale. *Biotechnol Lett*, 29, 1281–1285.

Luo LH, Seo JW, Oh BR, Seo PS, Heo SY, Hong WK, Kim DH, Chul HK. (2011). Stimulation of reductive glycerol metabolism by overexpression of an aldehyde dehydrogenase in a recombinant *Klebsiella pneumoniae* strain defective in the oxidative pathway. *J Ind Microbiol Biotechnol*, 38, 991–999.

Ma FR, Hanna MA. (1999). Biodiesel production: a review. *Bioresource Technol*, 70, 1–15.

Ma CW, Zhang L, Dai JY, Xiu ZL. (2010a). Relaxing the coenzyme specificity of 1,3-propanediol oxidoreductase from *Klebsiella pneumoniae* by rational design. *J Biotechnol*, 146, 173–178.

Ma Z, Rao ZM, Xu LY, Liao XR, Fang HY, Zhuge B, Zhuge J. (2010b). Expression of *dha* operon required for 1,3-propanediol formation in *Escherichia coli* and *Saccharomyces cerevisiae*. *Curr Microbiol*, 60, 191–198.

Ma Z, Shentu XP, Bian YY, Yu X. (2012). Effects of NADH availability on the *Klebsiella pneumoniae* strain with 1,3-propanediol operon over-expression. *J Basic Microbiol*. doi:10.1002/jobm.201100580

Malinowski JJ. (1999). Evaluation of liquid extraction potentials for downstream separation of 1,3-propanediol. *Biotechnol Tech*, 13, 127–130.

Malinowski JJ. (2000). Reactive extraction for downstream separation of 1,3-propanediol. *Biotechnol Progr*, 16, 76–79.

Mendes FS, Gonzalez-Pajuelo M, Cordier H, Francois JM, Vasconcelos I. (2011). 1,3-Propanediol production in a two-step process fermentation from renewable feedstock. *Appl Microbiol Biotechnol*, 92, 519–527.

Menzel K, Zeng AP, Deckwer WD. (1997). High concentration and productivity of 1,3-propanediol from continuous fermentation of glycerol by *Klebsiella pneumoniae*. *Enzyme Microbiol Technol*, 20, 82–86.

Moon C, Lee CH, Sang BI, Um Y. (2011). Optimization of medium compositions favoring butanol and 1,3-propanediol production from glycerol by Clostridium pasteurianum. *Bioresource Technol*, 102, 10561–10568.

Mu XJ, Teng H, Sun YQ, Xiu ZL. (2009). Techno-economic analysis on bioconversion of glycerol into 1,3-propanediol. *Chinese J Process Eng*, 9, 947–952.

Mu Y, Teng H, Zhang DJ, Wang W, Xiu ZL. (2006). Microbial production of 1,3-propanediol by *Klebsiella pneumoniae* using crude glycerol from biodiesel preparations. *Biotechnol Lett*, 28, 1755–1759.

Mu Y, Xiu ZL, Zhang DJ. (2008). A combined bioprocess of biodiesel production by lipase with microbial production of 1,3-propanediol by *Klebsiella pneumoniae*. *Biochem Eng J*, 40, 537–541.

Mueller A, Gorak A. (2012). Extraction of 1,3-propanediol from aqueous solutions using different ionic liquid-based aqueous two-phase systems. *Sep Purif Technol*, 97, 130–136.

Myszka K, Leja K, Olejnik-Schmidt AK, Czaczyk K. (2012). Isolation process of industrially useful *Clostridium bifermentans* from natural samples. *J Biosci Bioeng*, 113, 631–633.

Nakamura CE, Whited GM. (2003). Metabolic engineering for the microbial production of 1,3-propanediol. *Curr Opin Biotechnol*, 14, 454–459.

Oh BR, Seo JW, Heo SY, Hong WK, Luo LH, Son JH, Park DH, Kim CH. (2012). Fermentation strategies for 1,3-propanediol production from glycerol using a genetically engineered *Klebsiella pneumoniae* strain to eliminate by-product formation. *Bioproc Biosyst Eng*, 35, 159–165.

Otte B, Grunwaldt E, Mahmoud O, Jennewein S. (2009). Genome shuffling in *Clostridium diolis* DSM 15410 for improved 1,3-propanediol production. *Appl Environ Microbiol*, 75, 7610–7616.

Papanikolaou S, Aggelis G. (2003). Modelling aspects of the biotechnological valorization of raw glycerol: production of citric acid *Yarrowia lipolytica* and 1,3-propanediol by *Clostridium butyricum*. *J. Chem. Technol. Biotechnol*, 78, 542–547.

Papanikolaou S, Ruiz-Sanchez P, Pariset B, Blanchard F, Fick M. (2000). High production of 1,3-propanediol from industrial glycerol by a newly isolated *Clostridium butyricum* strain. *J Biotechnol*, 77, 191–208.

Papanikolaou S, Fick M, Aggelis G. (2004). The effect of raw glycerol concentration on the production of 1,3-propanediol by *Clostridium butyricum*. *J Chem Technol Biotechnol*, 79, 1189–1196.

Papanikolaou S, Fakas S, Fick M, Chevalot I, Galiotou-Panayotou M, Komaitis M, Marc I, Aggelis G. (2008). Biotechnological valorisation of raw glycerol discharged after bio-diesel (fatty acid methyl esters) manufacturing process: production of 1,3-propanediol, citric acid and single cell oil. *Biomass Bioenerg*, 32, 60–71.

Petitdemange E, Dürr C, Abbad-Andaloussi S, Raval G. (1995). Fermentation of raw glycerol to 1,3-propanediol by new strains of *Clostridium butyricum*. *Ind Microbiol Biotechnol*, 15, 498–502.

Petrov K, Stoyanov A. (2012). Accelerated production of 1,3-propanediol from glycerol by *Klebsiella pneumoniae* using the method of forced pH fluctuations. *Bioproc Biosyst Eng*, 35, 317–321.

Pfluegl S, Marx H, Mattanovich D, Sauer M. (2012). 1,3-Propanediol production from glycerol with *Lactobacillus diolivorans*. *Bioresource Technol*, 119, 133–140.

Ragout A, Sineriz F, Diekmann H, deValdez GF. (1996). Shifts in the fermentation balance of *Lactobacillus reuteri* in the presence of glycerol. *Biotechnol Lett*, 18, 1105–1108.

Reimann A, Biebl H. (1996). Production of 1,3-propanediol by *Clostridium butyricum* DSM 5431 and product tolerant mutants in fed-batch culture: feeding strategy for glycerol and ammonium. *Biotechnol Lett*, 18, 827–832.

Reimann A, Biebl H, Deckwer WD. (1998). Production of 1,3-propanediol by *Clostridium butyricum* in continuous culture with cell recycling. *Appl Microbiol Biotechnol*, 49, 359–363.

Roturier JM, Fouache C, Berghmans E. (2002). Process for the purification of 1,3-propanediol from a fermentation medium. US Patent, 6428992.

Sabra W, Dietz D, Tjahjasari D, Zeng AP. (2010). Biosystems analysis and engineering of microbial consortia for industrial biotechnology. *Eng Life Sci*, 10, 407–421.

Sanz MT, Blanco B, Beltran S, Beltran S, Cabezas JL, Coca J. (2001). Vapor liquid equilibria of binary and ternary systems with water, 1,3-propanediol, and glycerol. *J Chem Eng Data*, 46, 635–639.

Saxena RK, Anand P, Saran S, Isar J. (2009). Microbial production of 1,3-propanediol: recent developments and emerging opportunities. *Biotechnol Adv*, 27, 895–913.

Scott SD, Armenta RE, Berryman KT, Norman AW. (2011). Use of raw glycerol to produce oil rich in polyunsaturated fatty acids by a thraustochytrid. *Enzyme Microbiol Technol*, 48, 267–272.

Selembo PA, Perez JM, Lloyd WA, Logan BE. (2009). Enhanced hydrogen and 1,3-propanediol production from glycerol by fermentation using mixed cultures. *Biotechnol Bioeng*, 104, 1098–1106.

Seo MY, Seo JW, Heo SY, Baek JO, Rairakhwada D, Oh BR, Seo PS, Choi MH, Kim CH. (2009). Elimination of by-product formation during production of 1,3-propanediol in *Klebsiella pneumoniae* by inactivation of glycerol oxidative pathway. *Appl Microbiol Biotechnol*, 84, 527–534.

Song ZY, Sun YQ, Wei BC, Xiu ZL. (2012). Two-step salting-out extraction of 1,3-propanediol and lactic acid from fermentation broths of biodiesel-derived crude glycerol by *Klebsiella pneumoniae*. *Eng Life Sci* (Accepted).

Stevenson DE, Stanley RA, Furneaux RH. (1994). Near-quantitative production of fatty-acid alkyl esters by lipase-catalyzed alcoholysis of fats and oils with adsorption of glycerol by silica-gel. *Enzyme Microbiol Technol*, 16, 478–484.

Sullivan CJ. (1993). *Propanediols Ullmann's Encyclopedia of Industrial Chemistry*. vol. A22, Weinheim: VCH, pp. 163–171.

Sun YQ, Qi WT, Teng H, Xiu ZL, Zeng AP. (2008). Mathematical modeling of glycerol fermentation by *Klebsiella pneumoniae*: concerning enzyme-catalytic reductive pathway and transport of glycerol and 1,3-propanediol across cell membrane. *Biochem Eng J*, 38, 22–32.

Sun LH, Song ZY, Sun YQ, Xiu ZL. (2010). Dynamic behavior of glycerol-glucose co-fermentation for 1,3-propanediol production by *Klebsiella pneumoniae* DSM 2026 under micro-aerobic conditions. *World J Microbiol Biotechnol*, 26, 1401–1407.

Sun YQ, Ye JX, Mu XJ, Teng H, Feng EM, Zeng AP, Xiu ZL. (2012). Mathematical simulation and analysis of *Dha* regulon for glycerol metabolism in *Klebsiella pneumoniae*. *Chinese J Chem Eng*, 20, 958–970.

Tang XM, Tan YS, Zhu H, Zhao K, Shen W. (2009). Microbial conversion of glycerol to 1,3-propanediol by an engineered strain of *Escherichia coli*. *Appl Environ Microbiol*, 75, 1628–1634.

Tong IT, Cameron DC. (1992). Enhancement of 1,3-propanediol production by cofermentation in *Escherichia coli* expressing *Klebsiella pneumoniae* dha regulon genes. *Appl Biochem Biotechnol*, 34–35, 149–159.

Vaidyanathan H, Kandasamy V, Gopal Ramakrishnan G, Ramachandran K, Jayaraman G, Ramalingam S. (2011). Glycerol conversion to 1, 3-propanediol is enhanced by the expression of a heterologous alcohol dehydrogenase gene in *Lactobacillus reuteri*. *AMB Express*, 1, 37–44.

Wang YH. 2010. Metabolic flux analysis of 1,3-propanediol fermentation by *Klebsiella pneumonia* under microaeroic conditions. Doctoral Dissertation of Dalian University of Technology.

Wang YH, Teng H, Xiu ZL. (2011). Effect of aeration strategy on the metabolic flux of *Klebsiella pneumoniae* producing 1,3-propanediol in continuous cultures at different glycerol concentrations. *J Ind Microbiol Biotechnol*, 38, 705–715.

Wilkens E, Ringel AK, Hortig D, Willke T, Vorlop KD. (2012). High-level production of 1,3-propanediol from crude glycerol by *Clostridium butyricum* AKR102a. *Appl Microbiol Biotechnol*, 93, 1057–1063.

Wilkins AE, Lowe DJ. (2004). Product removal process for use in a biofermentation system. US Patent 7166460.

Witt U, Mueller R-J, Augusta J, Widdecke H, Deckwer W-D. (1994). Synthesis, properties and biodegradability of polyesters based on 1,3-propanediol. *Macromol Chem Phys*, 195, 793–802.

Wu HS, Wang YJ. (2012). Salting-out effect on recovery of 1,3-propanediol from fermentation broth. *Ind Eng Chem Res*, 51, 10930–10935.

Wu M, Li CL, Zhang J, Miao CC, Zheng YP, Sun YM. (2012). ZrO_2-MoO_3 for the acetalization of 1,3-propanediol from dilute solutions. *Ind Eng Chem Res*, 51, 6304–6309.

Wu RC, Xu YZ, Song YQ, Luo JA, Liu D. (2011). A novel strategy for salts recovery from 1,3-propanediol fermentation broth by bipolar membrane electrodialysis. *Sep Purif Technol*, 83, 9–14.

Xiang BT, Chen SF, Liu D. (2001). Extraction of 1,3-propanediol from in dilute fermentation broth. *J Tsinghua Univ* (Science and Technology), 41, 53–55.

Xiu ZL, Zeng AP. (2008). Present state and perspective of downstream processing of biologically produced 1,3-propanediol and 2,3-butanediol. *Appl Microbiol Biotechnol*, 78, 917–926.

Xiu ZL, Zeng AP, An LJ. (2000). Mathematical modeling of kinetics and research on multiplicity of glycerol bioconversion to 1,3-propanediol. *J Dalian Univ Technol*, 40, 428–433.

Xiu ZL, Song BH, Wang ZT, Sun LH, Feng EM, Zeng AP. (2004). Optimization of biodissimilation of glycerol to 1,3-propanediol by *Klebsiella pneumoniae* in one- and two-stage continuous anaerobic cultures. *Biochem Eng J*, 19, 189–197.

Xiu ZL, Chen X, Sun YQ, Zhang DJ. (2007). Stoichiometric analysis and experimental investigation of glycerol-glucose cofermentation in *Klebsiella pneumoniae* under microaerobic conditions. *Biochem Eng J*, 33, 42–52.

Xu YY, Du W, Liu DH, Zeng J. (2003). A novel enzymatic route for biodiesel production from renewable oils in a solvent-free medium. *Biotechnol Lett*, 25, 1239–1241.

Xu YZ, Guo NN, Zheng ZM, Ou XJ, Liu HJ, Liu DH. (2009). Metabolism in 1,3-propanediol fed-batch fermentation by a D-lactate deficient mutant of *Klebsiella pneumoniae*. *Biotechnol Bioeng*, 104, 965–972.

Xue XD, Li W, Li ZM, Xia YL, Ye Q. (2010). Enhanced 1,3-propanediol production by supply of organic acids and repeated fed-batch culture. *J Ind Microbiol Biotechnol*, 37, 681–687.

Yang G, Tian JS, Li JL. (2007). Fermentation of 1,3-propanediol by a lactate deficient mutant of *Klebsiella oxytoca* under microaerobic conditions. *Appl Microbiol Biotechnol*, 73, 1017–1024.

Zeng AP. (1995). A kinetic model for product formation of microbial and mammalian cells. *Biotechnol Bioeng*, 46, 314–324.

Zeng AP, Biebl H. (2002). Bulk chemicals from biotechnology: the case of 1,3-propanediol production and the new trends. *Adv Biochem Eng Biotechnol*, 74, 239–259.

Zeng AP, Deckwer WD. (1995). A kinetic model for substrate and energy consumption of microbial growth under substrate-sufficient conditions. *Biotechnol Progr*, 11, 71–79.

Zeng AP, Sabra W. (2011). Microbial production of diols as platform chemicals: Recent progresses. *Curr Opin Biotechnol*, 22, 749–757.

Zeng AP, Biebl H, Schlieker H, Deckwer WD. (1993). Pathway analysis of glycerol fermentation by *Klebsiella pneumoniae*: regulation of reducing equivalent balance and product formation. *Enzyme Microbiol Technol*, 15, 770–779.

Zeng AP, Rose A, Biebl H, Tag C, Guenzel B, Deckwer WD. (1994). Multiple product inhibition and growth modeling of *Clostridium butyricum* and *Klebsiella pneumoniae* in glycerol fermentation. *Biotechnol Bioeng*, 44, 902–911.

Zhang QR, Xiu ZL. (2009). Metabolic pathway analysis of glycerol metabolism in *Klebsiella pneumoniae* incorporating oxygen regulatory system. *Biotechnol Progr*, 25, 103–115.

Zhang YP, Li Y, Du CY, Liu M, Cao ZA. (2006). Inactivation of aldehyde dehydrogenase: a key factor for engineering 1,3-propanediol production by *Klebsiella pneumoniae*. *Metab Eng*, 8, 578–586.

Zhang QR, Teng H, Sun YQ, Xiu ZL, Zeng AP. (2008). Metabolic flux and robustness analysis of glycerol metabolism in *Klebsiella pneumoniae*. *Bioproc Biosyst Eng*, 31, 127–135.

Zhang YP, Huang ZH, Du CY, Li Y, Cao ZA. (2009). Introduction of an NADH regeneration system into *Klebsiella oxytoca* leads to an enhanced oxidative and reductive metabolism of glycerol. *Metab Eng*, 11, 101–106.

Zhang G, Yang G, Wang X, Guo QJ, Li Y, Li JL. (2012). Influence of blocking of 2,3-butanediol pathway on glycerol metabolism for 1,3-propanediol production by *Klebsiella oxytoca*. *Appl Biochem Biotechnol*, 168, 116–128.

Zheng Y, Zhao L, Zhang JG, Zhang HY, Ma XY, Wei DZ. (2008a). Production of glycerol from glucose by coexpressing glycerol-3-phosphate dehydrogenase and glycerol-3-phosphatase in *Klebsiella pneumoniae*. *J Biosci Bioeng*, 105, 508–512.

Zheng ZM, Cheng KK, Hu QL, Liu HJ, Guo NN, Liu DH. (2008b). Effect of culture conditions on 3-hydroxypropionaldehyde detoxification in 1,3-propanediol fermentation by *Klebsiella pneumoniae*. *Biochem Eng J*, 39, 305–310.

Zheng ZM, Hu QI, Hao J, Xu F, Guo NN, Sun Y, Liu DH. (2008c). Statistical optimization of culture conditions for 1,3-propanediol by *Klebsiella pneumoniae* AC 15 via central composite design. *Bioresource Technol*, 99, 1052–1056.

Zheng ZM, Guo NN, Hao J, Cheng KK, Sun Y, Liu DH. (2009). Scale-up of micro-aerobic 1,3-propanediol production with Klebsiella pneumonia CGMCC 1.6366. *Process Biochem*, 44, 944–948.

Zheng ZM, Xu YZ, Wang TP, Dong CQ, Yang YP, Liu DH. (2010). Ammonium and phosphate limitation in 1,3-propanediol production by *Klebsiella pneumoniae*. *Biotechnol Lett*, 32, 289–294.

Zhuge B, Zhang C, Fang HY, Zhuge J, Permaul K. (2010). Expression of 1,3-propanediol oxidoreductase and its isoenzyme in *Klebsiella pneumoniae* for bioconversion of glycerol into 1,3-propanediol. *Appl Microbiol Biotechnol*, 87, 2177–2184.

CHAPTER 12

Isobutanol

BERNHARD J. EIKMANNS

Institute of Microbiology and Biotechnology, University of Ulm, Ulm, Germany

BASTIAN BLOMBACH

Institute of Biochemical Engineering, University of Stuttgart, Stuttgart, Germany

Abstract

The future society and petrochemical-based industry are faced to energy and resource limitation and environmental problems due to the steadily decreasing availability of fossil fuels. Besides hydro-, wind-, and solar power, the biotechnological production of fuels and chemicals from renewable resources is regarded as key technology to

Bioprocessing of Renewable Resources to Commodity Bioproducts, First Edition.
Edited by Virendra S. Bisaria and Akihiko Kondo.

satisfy the world's demand for sustainable energy and goods. Microbial fermentation of renewable feedstocks found its way in white biotechnology, replacing more and more traditional crude oil-based processes for the production of chemicals and fuels. Due to the high economic and environmental impact, recent research activities strongly focused on the development of processes for the fermentative production of higher alcohols. Among them, the group of branched-chain alcohols including 2-methyl-1-butanol, 3-methyl-1-butanol, 2-methyl-1-propanol (isobutanol), and 1-butanol has attracted much attention since these alcohols possess chemical features close to gasoline. Especially isobutanol was shown to be a feasible candidate, which can be produced with maximal yield from renewable biomass. The production of these nonnatural alcohols was made accessible by implementation of the so-called Ehrlich pathway into relevant host organisms and was further improved by sophisticated metabolic engineering and process optimization. The Ehrlich pathway, consisting of 2-ketoacid decarboxylase and alcohol dehydrogenase reactions, converts the branched-chain ketoacids 2-ketomethylvalerate, 2-ketoisocaproate, 2-ketoisovalerate, and 2-ketovalerate (i.e., the direct precursors of L-isoleucine, L-leucine, L-valine, and norvaline, respectively) into the respective alcohols. So far, this strategy was successfully applied to transform a number of organisms such as *Escherichia coli*, *Corynebacterium glutamicum*, *Saccharomyces cerevisiae*, *Ralstonia eutropha*, or *Synechococcus elongatus* into (efficient) producers of higher alcohols.

12.1 INTRODUCTION

The depletion of earth's fossil energy resources, accompanied by the strong impact of their use on the environment, particularly in the form of higher CO_2 emissions, raises the demand for sustainable, safe, and efficient substitution of hitherto crude oil-derived fuels, chemicals, and chemical building blocks from renewable resources. Besides chemical manufacturing of renewable feedstocks to valuable compounds, biotechnological processes afford more and more opportunities to produce fuels, building blocks, and solvents in a cost-effective way from biomass (Bozell and Petersen, 2010).

Bioethanol is currently the major biofuel with an annual worldwide production of about 2.9×10^{10} gallons (Renewable Fuels Association, 2011). Established industrial ethanol fermentation processes for the first-generation feedstocks (e.g., sucrose and starch) are state of the art and will soon be completely available for feedstocks of the second generation (cellulose, hemicellulose, lignocellulose). However, ethanol is not considered as optimal substitute for gasoline, since it possess a lower vapor pressure, a higher octane number, and only 65% of the energy content of gasoline. Ethanol is hygroscopic and therefore not completely compatible with existing engines and pipelines (Table 12.1; Li et al., 2010). In contrast, higher alcohols such as 2-methyl-1-butanol, 3-methyl-1-butanol, 2-methyl-1-propanol (isobutanol), and 1-butanol possess properties which are much more comparable to gasoline and therefore are recognized as superior biofuels compared to ethanol. Among these alcohols,

TABLE 12.1 Chemical Properties of Ethanol, Isobutanol, and Gasoline

Fuel	Ethanol	Isobutanol	Alkanes (gasoline)
Volumetric energy density (MJ/L)	20.8	29.0	32.0
Vapor pressure (psi)	1.1	0.17	0.1–30
Average octane number	116	110	90
Hygroscopicity	High	Low	Low
Fits current infrastructure?	No	Yes	Yes

Source: Adapted from Li et al. (2010).

isobutanol has gained much attention since Atsumi et al. (2008) for the first time engineered *Escherichia coli* by implementation of the so-called Ehrlich pathway (Ehrlich, 1907, see section 12.2) for the efficient production of this branched-chain alcohol. In contrast to ethanol, isobutanol has chemical properties close to gasoline (Table 12.1) and therefore fits criteria for a relevant next-generation biofuel. Furthermore, isobutanol can be enzymatically or chemocatalytically dehydrated to isobutene, which nowadays is exclusively produced by petroleum refining in a scale of more than 10 million metric tons per year (van Leeuwen et al., 2012). Isobutene represents a key precursor for a variety of chemicals, for example, as fuel additive, for butyl rubbers or isoprene, and numerous other products (van Leeuwen et al., 2012). Taken together, isobutanol represents a highly attractive commodity chemical, and its biotechnological production would represent an essential step toward a sustainable bio-economy. However, the cost-effective bio-based production of this alcohol is a most relevant goal for the future and has to meet economic and environmental requirements. Therefore, the microbial production systems have to perform excellent with regard to yield, productivity, and substrate flexibility. This chapter summarizes the current knowledge and achievements of metabolic engineering strategies and process optimization for the microbial production of isobutanol.

12.2 THE ACCESS CODE FOR THE MICROBIAL PRODUCTION OF BRANCHED-CHAIN ALCOHOLS: 2-KETOACID DECARBOXYLASE AND AN ALCOHOL DEHYDROGENASE

Already in the beginning of the nineteenth century, Felix Ehrlich identified higher alcohols, also known as fusel oils, in alcoholic fermentations with yeast and showed that these alcohols are formed from amino acids without generating free ammonia (Ehrlich, 1907). Further studies identified the reactions of the Ehrlich pathway: (i) deamination of an amino acid by a transamination reaction, (ii) decarboxylation of the 2-ketoacid to the corresponding aldehyde, and (iii) the reduction of the aldehyde to the respective alcohol (Figure 12.1; Sentheshanmuganathan and Elsden, 1958; Sentheshanmuganathan, 1960, reviewed in Hazelwood et al., 2008).

In 2008, Atsumi et al. made use of the Ehrlich pathway by implementation of the last two reactions of this pathway into *E. coli* and marked the initial step toward the

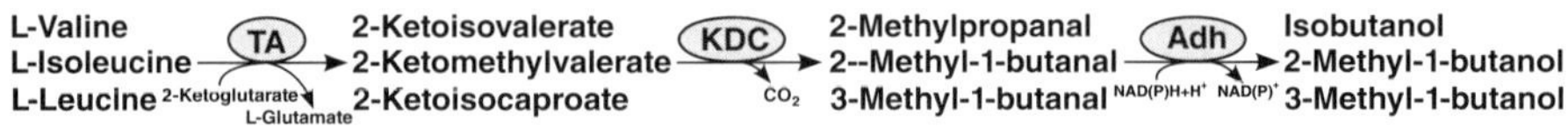

FIGURE 12.1 The Ehrlich pathway exemplified for the conversion of the branched-chain amino acids L-valine, L-isoleucine, and L-leucine to the corresponding alcohols isobutanol, 2-methyl-1-butanol, and 3-methyl-1-butanol. Adh, alcohol dehydrogenase; KDC, 2-ketoacid decarboxylase; TA, transaminase.

directed biotechnological production of branched-chain alcohols (Figure 12.2). The authors expressed different combinations of genes coding for 2-ketoacid decarboxylases (KDC; EC 4.1.1.72) and alcohol dehydrogenase (Adh) enzymes (EC 1.1.1.1). One of the tested combinations led to the formation of significant amounts of 2-methyl-1-butanol, 3-methyl-1-butanol, isobutanol, and 1-butanol from glucose, with isobutanol as the most abundant. The relevant combination consisted of the KDC from *Lactococcus lactis* (encoded by the *kivd* gene) and an Adh (encoded by *adh2*) from *Saccharomyces cerevisiae* (Atsumi et al., 2008). A further study identified NADPH-dependent broad-substrate range aldehyde reductase YqhD from *E. coli* rather than NADH-dependent Adh2 to be responsible for the conversion of the aldehyde to the alcohol (Atsumi et al., 2009a; Jarboe, 2011).

The combined expression of *kivd* and an Adh gene was later on the most common feature of metabolic engineering strategies to transform relevant microorganisms into higher alcohol producers (Cann and Liao, 2008; Connor and Liao, 2008; Atsumi et al., 2009b; Smith et al., 2010, Blombach et al., 2011; Li et al., 2011, 2012a, b; Lu et al., 2012; Brat et al., 2012; Higashide et al., 2011). The key element is the KDC from *L. lactis*. This enzyme is unique and until now, no other KDC enzyme of this enzyme class was identified to be useful for the production of higher alcohols. Only the acetohydroxyacid synthase (AHAS) from *Bacillus subtilis*, encoded in this organism

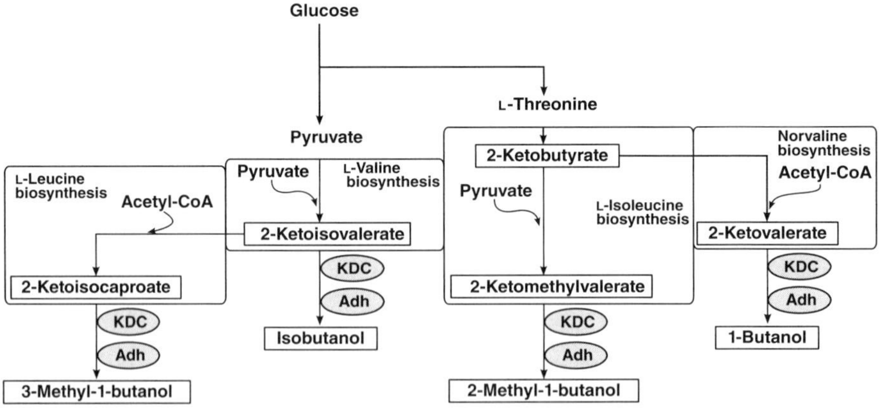

FIGURE 12.2 Formation of the higher alcohols 3-methyl-1-butanol, isobutanol, 2-methyl-1-butanol, and 1-butanol derived from intermediates of the branched-chain amino acid pathways. Adh, alcohol dehydrogenase; KDC, 2-ketoacid decarboxylase.

by the *alsS* gene and usually catalyzing the conversion of two pyruvate molecules to acetolactate, was shown to possess 2-ketoisovalerate decarboxylase side activity (Atsumi et al., 2009c). In contrast, Adh enzymes with a broad (aldehyde) substrate range are widely distributed and different Adhs, for example, from *E. coli*, *L. lactis,* or *Corynebacterium glutamicum*, have been proven to be useful for higher alcohol production (Atsumi et al., 2009a; Smith et al., 2010; Blombach et al., 2011; Liu et al., 2012). Table 12.2 gives an overview of the biochemical properties of relevant enzymes with KDC or Adh activity used for the production of higher alcohols.

Besides implementation of the Ehrlich pathway, a lot of attention has been paid on the optimization of precursor availability, the carbon flux to the desired alcohol, and on process development in dependence of the genetic background of the respective organism. However, microbial isobutanol production is currently the most promising process, since metabolic engineering approaches resulted in strains capable to produce this alcohol with significant higher yield and titers compared to the other branched-chain alcohols 2-methyl-1-butanol, 3-methyl-1-butanol, and 1-butanol (Table 12.3; Atsumi et al., 2008,Atsumi and Liao, 2008; Cann and Liao, 2008; Shen and Liao, 2008; Connor et al., 2010; Bastian et al., 2011). Therefore, the following part of this chapter concentrates on metabolic engineering strategies to develop relevant and attractive microbial isobutanol producer strains and fermentative processes.

12.3 METABOLIC ENGINEERING STRATEGIES FOR DIRECTED PRODUCTION OF ISOBUTANOL

12.3.1 Isobutanol Production with *Escherichia coli*

The initial step toward the biotechnological production of isobutanol was done by Atsumi et al. (2008) with *E. coli*. Implementation of *kivd* from *L. lactis* and *adh2* from *S. cerevisiae*, and additional overexpression of the *E. coli* L-valine biosynthetic pathway genes *ilvIHCD*, encoding AHAS, acetohydroxyacid isomeroreductase (AHAIR), and dihydroxyacid dehydratase (DHAD) resulted in an improved yield due to an increased drain-off of pyruvate toward 2-ketoisovalerate (Figure 12.3). To avoid by-product formation and to further increase pyruvate availability, the genes *adhE*, *ldhA*, *frdAB*, *pta*, and *fnr* encoding the bi-functional AdhE, D-lactate dehydrogenase (LDH), fumarate reductase (FRD), phosphotransacetylase (PTA), and the fumarate and nitrate reductase regulator FNR, respectively, were stepwise deleted (Figure 12.3). The resulting strain produced isobutanol with a yield of about 0.21 g of isobutanol per gram of glucose and was further improved by overexpression of *alsS* encoding AHAS from *B. subtilis* instead of *ilvIH*. The latter improvement in combination with the deletion of the *pflB* gene, encoding pyruvate formate lyase (PFL), resulted in *E. coli* JCL260/pSA55/pSA69 and led to the production of about 300 mM (22 g/L) with a yield of about 0.86 mol/mol (0.35 g/g) in minimal medium with yeast extract and glucose under microaerobic conditions (Atsumi et al., 2008; Table 12.3).

Applying anaerobic conditions, the maximal theoretical yield can be calculated to be 0.41 g isobutanol per gram of glucose (1 mole/mole; Li et al., 2010).

TABLE 12.2 Overview of the Biochemical Properties of Relevant Enzymes with 2-Ketoacid Decarboxylase or Alcohol Dehydrogenase Activity

Enzyme (organism)	Substrate	Cofactor	K_M (mM)	k_{cat}/s	k_{cat}/K_M	Reference
KDC (*L. lactis*)	2-Ketoisovalerate	TPP	2.2 ± 0.9	38.3 ± 9.8	17	Zhang et al. (2008)
AlsS (*B. subtilis*)	2-Ketoisovalerate	TPP	300 ± 35	8.9 ± 1.2	0.03	Atsumi et al. (2009c)
YqhD (*E. coli*)	Isobutyraldehyde	NADPH	1.8 ± 1.3	1 ± 0.1	0.7	Atsumi et al. (2009a)
Adh2 (*S. cerevisiae*)	Isobutyraldehyde	NADH	385 ± 31	0.9 ± 0.1	2.2×10^{-3}	Atsumi et al. (2009a)
AdhA (*L. lactis*)	Isobutyraldehyde	NADH	9.1 ± 2.9	6.6 ± 0.2	0.8	Atsumi et al. (2009a)
AdhA (*C. glutamicum*)	Isobutyraldehyde	NADH	–	–	–	Smith et al. (2010)

Adh, alcohol dehydrogenase; AlsS, acetohydroxyacid synthase; KDC, 2-ketoacid decarboxylase; YqhD, broad-range alcohol dehydrogenase; TPP, thiamine pyrophosphate.

TABLE 12.3 Relevant Yields and Titers of Isobutanol, 2-Methyl-1-Butanol, 3-Methyl-1-Butanol, 1-Butanol Derived of the Metabolic Pathways of Branched-Chain Amino Acids

Strain	Process	Product	$Y_{P/S}$ (mole/mole), (g/g)	Titer (g/L)	Reference
E. coli JCL260/pSA55/pSA69	Flask	Isobutanol	0.86, 0.35	22	Atsumi et al. (2008a)
E. coli 1993 (pGVferm6)	Flask	Isobutanol	1.03, 0.42	13.4	Bastian et al. (2011)
E. coli CRS22	Flask	2-Methyl-1-butanol	0.35, 0.17	1.3	Cann and Liao (2008)
E. coli AL2	Flask	3-Methyl-1-butanol	0.22, 0.11	9.5	Connor et al. (2010)
E. coli CRS-BuOH 23	Flask	1-Butanol	~0.09, 0.04	0.8	Shen and Liao (2008)

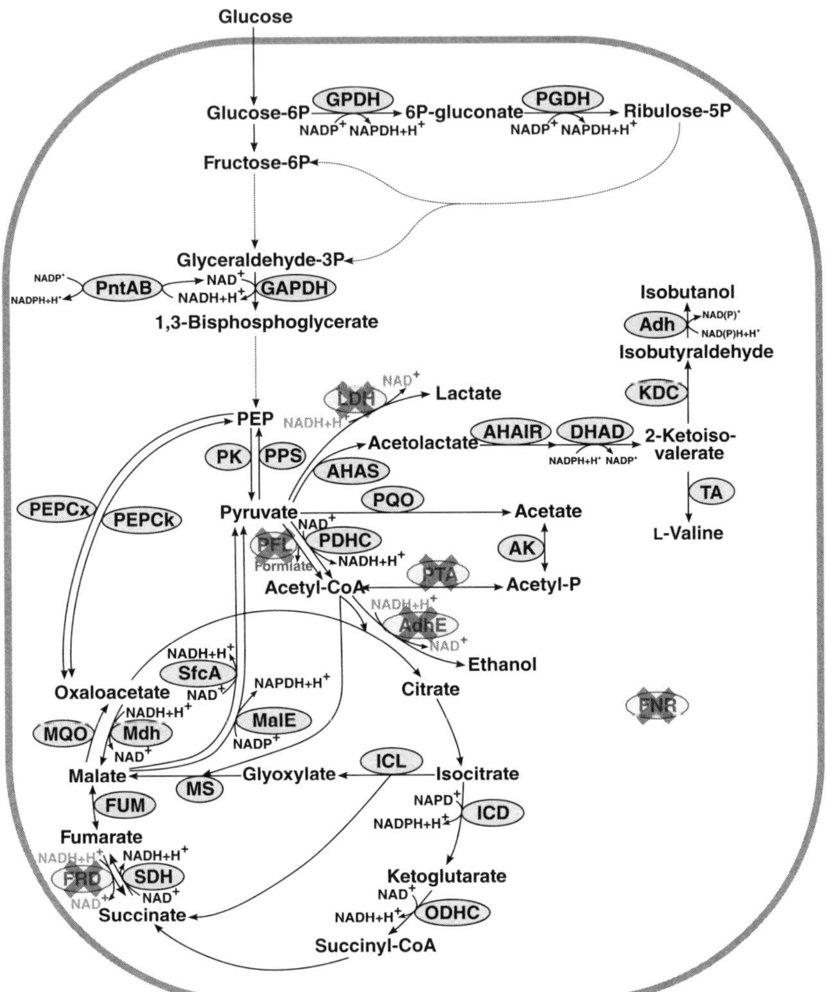

FIGURE 12.3 The enzymes of the central metabolism with the biosynthetic pathway of L-valine in *E. coli* and the syntethic pathway from 2-ketoisovalerate to isobutanol. Adh, alcohol dehydrogenase; AdhE, alcohol dehydrogenase E; AHAIR, acetohydroxyacid isomeroreductase; AHAS, acetohydroxyacid synthase; AK, acetate kinase; DHAD, dihydroxyacid dehydratase; FNR, fumarate and nitrate reductase regulator; FUM, fumarase; FRD, fumarate reductase; GAPDH, glyceraldehyde-3P dehydrogenase; GPDH, G6P dehydrogenase; ICD, isocitrate dehydrogenase; ICL, isocitrate lyase; KDC, 2-ketoacid decarboxylase from *L. lactis*; LDH, D-lactate dehydrogenase; MalE, NADPH-dependent malic enzyme; Mdh, malate dehydrogenase; MQO, malate:quinone oxidoreductase; MS, malate synthase; ODHC, oxoglutarate dehydrogenase complex; PDHC, pyruvate dehydrogenase complex; PEP, phosphoenolpyruvate; PEPCk, PEP carboxykinase; PEPCx, PEP carboxylase; PFL, pyruvate formate lyase; PGDH, 6P-gluconate dehydrogenase; PK, pyruvate kinase; PntAB, membrane-bound transhydrogenase; PPS, PEP synthetase; PTA, phosphotransacetylase; PQO, pyruvate:quinone oxidoreductase; SDH, succinate dehydrogenase; SfcA, NADH-dependent malic enzyme; TA, transaminase B.

However, maintaining a balanced redox state is crucial for an efficient production process under oxygen limitation. AHAIR enzymes are predominantly NADPH dependent, whereas different types of Adhs accept either NADH+H^+ or NADPH+H^+. Thus, the formation of one mole isobutanol from pyruvate requires one mole NADH+H^+ and one mole NADPH+H^+ or 2 moles NADPH+H^+, respectively. Regarding that most bacteria generate 2 moles NADH+H^+ per mole of glucose in the course of glycolysis, the conversion of NADH+H^+ to NADPH+H^+ or the construction of a completely NADH-dependent isobutanol-forming pathway is essential for efficient production (Bastian et al., 2011). Both strategies were compared in an approach to optimize cofactor utilization for isobutanol production with *E. coli* (Bastian et al., 2011). Protein engineering of the NADPH-dependent AHAIR identified a mutated variant (IlvC^{A71S,R76D,S78D,Q110V}) that showed a strong preference for NADH+H^+ over NADPH+H^+. Furthermore, the catalytic efficiency and the affinity toward isobutyraldehyde of the NADH-dependent AdhA from *L. lactis* was significantly improved by random mutagenesis and recombination of useful mutations. Overexpression of the engineered genes for AHAIR and AdhA as well as for AHAS (from *B. subtilis*), DHAD, and KDC (from *L. lactis*) in *E. coli* with inactivated LdhA, AdhE, FRD, PFL, AHAIR, and PTA resulted in *E. coli* 1993 (pGVferm6), producing isobutanol with the maximal theoretical yield of 1.03 mole isobutanol per mole of glucose (0.42 g/g) under anaerobic conditions (Table 12.3). Also overexpression of the membrane-bound transhydrogenase PntAB in the same strain background with a completely NADPH+H^+-dependent pathway, overexpressing native *ilvC* (encoding AHAIR), and *yqhd* (encoding Adh), resulted in the maximal theoretical yield, however, with reduced productivity compared to *E. coli* 1993 (pGVferm6) (Bastian et al., 2011).

All these results demonstrate that *E. coli* is a very promising host for biotechnological production of isobutanol. However, industrial relevant processes also have to result in high final titers, which do not match to the known cytotoxicity of higher alcohols on microorganisms (Knoshaug and Zhang, 2009; Ezeji et al., 2010). As outlined further in section 12.4, the maximal tolerance of *E. coli* against isobutanol is lower compared to that of some Gram-positive bacteria, such as *C. glutamicum* or *B. subtilis* (Brynildsen and Liao, 2009; Smith et al., 2010; Li et al., 2011) or to that of *S. cerevisiae*. Consequently, several metabolic engineering approaches focused on these alternative hosts.

12.3.2 Isobutanol Production with *Corynebacterium glutamicum*

Corynebacterium glutamicum is a Gram-positive, facultative anaerobic organism that grows on a variety of sugars, organic acids, and alcohols as single or combined carbon and energy sources (Liebl, 2006; Nishimura et al., 2007; Takeno et al., 2007). The organism is generally regarded as safe (GRAS status) and is traditionally employed for large-scale production of amino acids such as L-glutamate (>2 million tons/year) and L-lysine (>1.4 million tons/year) but also a few thousand tons of L-valine and L-isoleucine per year (Takors et al., 2007; Ajinomoto Co., 2010, 2011). Since isobutanol derives from 2-ketoisovalerate, the precursor of L-valine, efforts have been made to transform *C. glutamicum* into an efficient producer by the use of the current knowledge on amino acid production.

Smith et al. (2010) overexpressed the genes coding for AHAS from *B. subtilis* and the native AHAIR and DHAD as well as KDC from *L. lactis* and found that this approach led to the formation of significant amounts of higher branched-chain alcohols, with isobutanol as the most dominant one. Further analysis identified the native AdhA to be responsible for the conversion of the isobutyraldehyde (2-methylpropanal) to isobutanol. The overexpression of the native *adhA* gene as well as deletion of genes coding for pyruvate carboxylase and L-LDH resulted in the formation of about 4.9 g isobutanol/L with a yield of about 23% of the theoretical maximum. The attempt to increase NADPH+H^+ availability by inactivation of the glucose-6P-isomerase gene *pgi* to redirect the carbon flux into the NADPH+H^+-generating pentose phosphate pathway failed, probably due to an imbalance in the redox state of the cell (Smith et al., 2010).

In a further approach, Blombach et al. (2011) engineered *C. glutamicum* for the efficient production of isobutanol by the use of an existing 2-ketoisovalerate production strain, which was deduced from previously constructed L-valine producers (Krause et al., 2010, Blombach et al., 2007, 2008). This *C. glutamicum* strain carries deletions of the *aceE*, *pqo*, and *ilvE* genes, coding for the E1-subunit of the pyruvate dehydrogenase complex (PDHC), the pyruvate:quinone oxidoreductase (PQO), and transaminase B (TA), respectively (Figure 12.4). To channel the carbon from pyruvate more efficiently toward 2-ketoisovalerate, the native *ilvBNCD* (genes encoding AHAS, AHAIR, and DHAD, respectively) were overexpressed from plasmid (Krause et al., 2010). The additional expression of *kivd* and *adh2* did not result in the formation of isobutanol but in the secretion of L-lactate and succinate under oxygen deprivation conditions. Inactivation of the L-LDH abolished L-lactate formation and resulted in the production of 26 mM (1.9 g/L) isobutanol with a yield of 0.22 mole per mole of glucose. This strain was further improved by deletion of the *mdh* gene encoding malate dehydrogenase (Mdh) and heterologous expression of the *pntAB* genes coding for the membrane-bound transhydrogenase from *E. coli*. Both modifications led to reduced formation of succinate, increased glucose consumption probably due to a more balanced redox state of the cell, and to a significantly higher yield of 0.6 mole per mole of glucose and titers of about 42 mM (3.1 g/L). Blombach et al. (2011) finally overexpressed the native *adhA* gene instead of *adh2* from *S. cerevisiae* and the resulting strain, *C. glutamicum* Δ*aceE* Δ*pqo* Δ*ilvE* Δ*ldhA* Δ*mdh* (pJC4*ilvBNCD-pntAB*) (pBB1*kivd-adhA*) (designated as *C. glutamicum* Iso7), produced about 82 mM (6.1 g/L) isobutanol with a yield of about 0.77 mole per mole of glucose in shaken bottles under oxygen deprivation conditions (Table 12.4). Interestingly, the authors found strong indications that aside from the transhydrogenase PntAB, a transhydrogenase-like route consisting of pyruvate carboxylase/PEP carboxylase, Mdh, and malic enzyme significantly contributes to the NADPH+H^+ supply (Figure 12.4; Blombach et al., 2011; Blombach and Eikmanns, 2011). These results show that *C. glutamicum* aside from *E. coli* is a very attractive host for industrial isobutanol production.

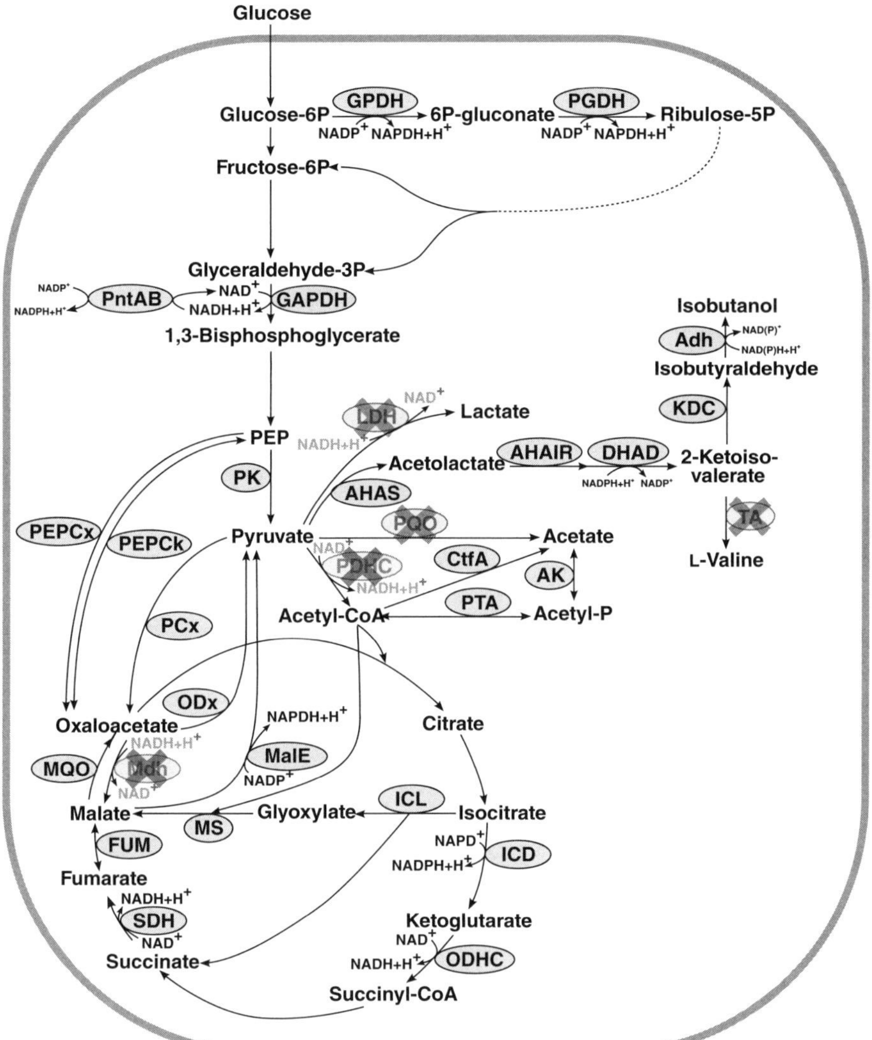

FIGURE 12.4 The enzymes of the central metabolism with the biosynthetic pathway of L-valine in *Corynebacterium glutamicum* and the syntethic pathway from ketoisovalerate to isobutanol. For most abbreviations see Figure 12.3; *additional abbreviations*: LDH, L-lactate dehydrogenase; PCx, pyruvate carboxylase; PntAB, membrane-bound transhydrogenase from *E. coli*.

12.3.3 Isobutanol Production with *Bacillus subtilis*

In *B. subtilis*, the combined (over)expression of the genes encoding native AHAS, AHAIR, DHAD, and KDC from *L. lactis* and Adh2 from *S. cerevisiae* resulted in significant isobutanol formation. However, the overall yield was only about 0.2 mole isobutanol per mole of glucose, and the best producing strain *B. subtilis* UL03

TABLE 12.4 Relevant Characteristics of Selected Isobutanol-Producing Strains of *E. coli*, *C. glutamicum*, *B. subtilis*, *R. euthropha*, *S. cerevisiae*, *S. elongates*, and *C. cellulolyticum*

Strain	Process	Substrate	$Y_{P/S}$ (mole/mole), (g/g)	Titer (g/L)	Reference
E. coli JCL260/pSA55/pSA69	Flask	Glc	0.86, 0.35	22	Atsumi et al. (2008a)
	Fed-batch	Glc	0.68, 0.28	50.8[a]	Baez et al. (2011)
E. coli 1993 (pGVferm6)	Flask	Glc	1.03, 0.42	13.4	Bastian et al. (2011)
C. glutamicum Iso 7	Bottle	Glc	0.77, 0.32	6.1	Blombach et al. (2011)
	Fed-batch	Glc, Ace	0.48, 0.20	13.0	
B. subtilis UL05	Flask	Glc, Ace	0.44, 0.18	2.3	Li et al. (2012b)
	Fed-batch	Glc, Ace	0.47, 0.19	5.5	
R. eutropha 2425/pJL26	Flask	Fructose	n.g.	0.3	Lu et al. (2012)
	Flask	Fructose	n.g.	14[b]	
S. cerevisiae Isoy8	Flask	Glc	0.04, 0.02	0.6	Brat et al. (2012)
S. elongatus SA579	Bottle	CO_2	–	0.45	Atsumi et al. (2009b)
C. cellulolyticum H10 pWH320	Flask	Cellulose	–	0.66	Higashide et al. (2012)

Glc, Glucose; Ace, acetate; n.g., not given in the publication.

[a]Titer was obtained with *in situ* product removal by gas stripping (see text, section 12.5).

[b]Given titer represents the sum of isobutanol and 3-methyl-1-butanol produced during a semicontinous cultivation of *R. eutropha* 2425/pJL26.

secreted significant amounts of the by-products acetate, lactate, and ethanol under microaerobic conditions (Li et al., 2011). Elementary mode analysis identified LDH and the PDHC as potential targets to improve isobutanol production with *B. subtilis* UL03 (Li et al., 2012b). Experimental verification indeed revealed that the newly constructed LDH- and PDHC-deficient strain *B. subtilis* UL05 showed an elevated performance and produced about 31 mM (2.3 g/L) isobutanol with a yield of about 0.44 mole per mole of glucose under microaerobic conditions (Table 12.4; Li et al., 2012b). These results are promising and indicate that *B. subtilis* might also be a suitable host for sustainable isobutanol production.

12.3.4 Isobutanol Production with *Clostridium cellulolyticum*

Biofuels of the second generation require the use of nonfood substrates. Therefore, the production of chemicals and fuels from cellulosic feedstocks will be a most relevant goal for a future sustainable bioeconomy. The direct production from cellulose, hemi-, or lignocellulose is attractive since this consolidated process is supposed to reduce the overall production costs compared to fermentation processes which are dependent on the degradation of these polymers into the monomers (Higashide et al., 2011). This strategy was realized with *C. cellulolyticum* which is naturally able to consume cellobiose and cellulose. In a metabolic engineering approach the plasmid-encoded expression of *alsS* from *B. subtilis*, *ilvC*, *ilvD*, and *yqhd* from *E. coli*, and *kivd* and *adhA* from *L. lactis* resulted in the formation of up to 5 mM (0.36 g isobutanol/L) on cellobiose and 9 mM (0.66 g isobutanol/L) on cellulose (Table 12.4; Higashide et al., 2011).

12.3.5 Isobutanol Production with *Ralstonia eutropha*

Ralstonia eutropha is well known as a natural producer of polyhydroxybutyrate (PHB) as intracellular carbon storage. In the absence of organic substrates, this facultative lithoautotrophic bacterium possesses the ability to use H_2 and CO_2 as sole carbon and energy source (Cramm, 2009). This property and the known metabolic engineering tools for *R. eutropha* make this bacterium an interesting candidate for the production of fuels and chemicals.

Lu et al. (2012) successfully redirected the carbon flux in *R. eutropha* from PHB toward isobutanol and 3-methyl-1-butanol. PHB synthesis was shut down by the deletion of the *phaCAB* genes encoding the polymer biosynthesis enzymes ß-ketothiolase (PhaA), acetoacetyl-CoA reductase (PhaB), and PHB synthase (PhaC). Since under aerobic conditions native Adhs of *R. eutropha* are not active, mutated *R. eutropha* strains possessing constitutively expressed *adh* genes were screened for their ability to produce branched-chain alcohols. For this purpose, the *phaCAB* genes were deleted, and the resulting mutants were transformed with an overexpression plasmid carrying the genes encoding native AHAS, AHAIR, and DHAD as well as KDC from *L. lactis*. Among the identified strains, *R. eutropha* 2401/pJL26 showed the best performance and produced up to 2.7 mM isobutanol (0.2 g/L) and 0.35 mM 3-methyl-1-butanol (0.03 g/L) from fructose as carbon source. The stepwise deletion of *ilvE*, *bkdAB*, and *aceE*, encoding L-valine-specific transaminase, branched-chain ketoacid dehydrogenase complex, and PDHC, respectively, further improved alcohol production due to improved precursor availability. In semicontinuous flask cultivations the final strain *R. eutropha* 2425/pJL26 produced 14 g of total accumulated alcohols (isobutanol and 3-methyl-1-butanol) per liter within 50 days from fructose (Table 12.4; Lu et al., 2012).

Li et al. (2012a) recently developed an integrated electromicrobial process to convert CO_2 to higher alcohols with engineered *R. eutropha*. The idea was to reduce CO_2 electrochemically to formate which then is converted to higher alcohols. Formate is highly soluble in water and serves carbon (CO_2) and NADH for the microbial alcohol production system. In order to implement this system, *R. eutropha* was engineered for alcohol production by inactivation of the PHB biosynthesis pathway (see above), genomic integration of *alsS* from *B. subtilis*, *ilvC* and *ilvD* from *E. coli*, and plasmid-bound overexpression of *kivd* from *L. lactis* and *yqhD* from *E. coli*. The resulting strain *R. eutropha* LH74D was produced in a pH-coupled formic acid feeding fermenter about 1.4 g fuels/L. The strain finally was employed for an integrated process, in which simultaneously CO_2 is reduced to formate which is immediately converted to isobutanol and 3-methyl-1-butanol, cumulatively reaching about 0.14 g/L. These results demonstrate that *R. eutropha* might be a relevant candidate to produce biofuels from CO_2- and H_2-rich gas streams or, as described above, in a combined electromicrobial approach.

12.3.6 Isobutanol Production with *Synechococcus elongatus*

Recycling of CO_2 into fuels and chemicals using photosynthesis is also a promising approach. Atsumi et al. (2009b) engineered the cyanobacterium *S. elongatus* PCC7942 to produce isobutyraldehyde (isobutanal) and isobutanol from CO_2 and H_2O. Genomic integration of *alsS* from *B. subtilis*, *ilvC* and *ilvD* from *E. coli*, and *kivd* from *L. lactis* enabled *S. elongatus* to produce 10 mM isobutyraldehyde (723 mg/L) within 12 days. Isobutanol was not formed, indicating that *S. elongatus* does not possess an endogenous Adh able to reduce the aldehyde under the conditions employed. Additional integration of the *rbcLS* genes coding for the ribulose-bisphosphate carboxylase from *S. elongatus* PCC6301 resulted in the formation of about 15 mM isobutyraldehyde (1.1 g/L) within 8 days with *S. elongatus* SA665, showing that CO_2 fixation is one of the bottlenecks for isobutyraldehyde and isobutanol production (Atsumi et al., 2009b). To transform the isobutyraldehyde-forming strain *S. elongatus* SA665 into an isobutanol producer, the *yqhD* gene from *E. coli* was finally integrated into the genome, which allowed the production of 6 mM isobutanol (0.45 g/L) from CO_2 (Table 12.4). However, the relatively low tolerance of *S. elongatus* for isobutanol (about 0.75 g/L) and the relatively low final titer strongly limit the application on industrial scale (Atsumi et al., 2009b). However, the efficient and cost-effective biotechnological production of chemicals or fuels from solar energy, CO_2, and H_2O undoubtedly represents a future key technology.

12.3.7 Isobutanol Production with *Saccharomyces cerevisiae*

Since *S. cerevisiae* is the workhorse in industrial ethanol fermentations and has the natural capacity to secrete fusel alcohols as by-products (reviewed in Hazelwood et al., 2008), this organism is supposed to be an attractive host for the production of higher alcohols. Furthermore, *S. cerevisiae* tolerates up to 20% ethanol and also shows a comparatively high tolerance against butanol (Knoshaug and Zhang, 2009; Chen et al., 2011).

So far, the common approach to engineer *S. cerevisiae* for isobutanol production was to overexpress the native genes of the L-valine biosynthetic pathway (*Ilv2*, *Ilv3*, *Ilv5*) or of genes coding for truncated enzyme versions lacking the N-terminal mitochondrial targeting sequence and therefore re-located from the mitochondrial matrix to the cytosol (Chen et al., 2011; Brat et al., 2012; Lee et al., 2012; Matsuda et al., 2012; Kondo et al., 2012). These strains were further improved by overexpression of the native 2-ketoisovalerate decarboxylase gene *Aro10* or *kivd* from *L. lactis*, the native Adh genes *Adh2* or *Adh6* as well as the *Bat2* gene, encoding yeast cytoplasmic branched-chain amino acid transferase (Chen et al., 2011; Brat et al., 2012; Lee et al., 2012; Kondo et al., 2012). Improving pyruvate availability by inactivation of a major pyruvate decarboxylase gene *Pdc1* was also beneficial for isobutanol production with *S. cerevisiae* (Kondo et al., 2012). However, the currently best producing strain *S. cerevisiae* Isoy8 only produces about 8 mM isobutanol (0.6 g/L) with a yield of nearly 15 mg per gram of glucose which has to be significantly improved in future studies.

12.4 OVERCOMING ISOBUTANOL CYTOTOXICITY

As outlined above, several microorganisms have been engineered to produce high yields of isobutanol (up to the theoretical maximum of 1 mole per mole of glucose). However, the toxicity of this alcohol obviously limits volumetric productivity and the final product titers and thus, increasing the isobutanol tolerance of relevant microorganisms and adaptation to exogenous isobutanol is a most relevant goal to achieve industrially relevant production processes (Zhang et al., 2009).

Several studies analyzed the tolerance against higher alcohols (including biofuels) in industrially relevant microorganisms, and also the effects of higher alcohols on bacterial cells and/or the mechanisms of alcohol toxicity were investigated (e.g., Fischer et al., 2008; Brynildsen and Liao, 2009; Rühl et al., 2009; Knoshaug and Zhang, 2009; Nielsen et al., 2009; Atsumi et al., 2010; Smith et al., 2010; Chen et al., 2011; reviewed in Nicolaou et al., 2010; Ezeji et al., 2010, and Minty et al., 2011). However, due to the numerous mechanisms of toxicity and to the versatile cellular responses to alcohol stress, microbial tolerance against higher concentrations of alcohols is a very complex and multigenic property and therefore, knowledge-based and rational engineering for improving the alcohol tolerance of industrial microorganisms is still in its infancy.

Isobutanol has been shown to severely impair the growth of *E. coli* cells at concentrations between 5 and 10 g/L; complete growth arrest and loss of viability were observed at 10–15 g/L (Brynildsen and Liao, 2009; Smith et al., 2010; Minty et al., 2011). A direct comparison of isobutanol sensitivity with *E. coli* and *C. glutamicum* revealed that the latter organism throughout showed about a twofold higher viability on exposure to different concentrations of isobutanol (Smith et al., 2010). Also, Blombach et al. (2011) reported that *C. glutamicum* showed only a relatively weak reduction of the growth rate with 20 g isobutanol/L (0.28/h vs. 0.40/h), substantiating that this organism is more tolerant and thus might be a favorable production host. However, the observed effects of exogenous isobutanol on growth alone do not necessarily be relevant for isobutanol production since (i) the intracellular isobutanol concentrations in producer strains may be quite different from the extracellular concentration and (ii) at least for *E. coli* it has been shown that isobutanol production can continue for a while after growth has ceased in an isobutanol production experiment (Atsumi et al., 2008).

So far, only with *E. coli* efforts have been made to analyze and map the cellular response to isobutanol stress in detail and to improve producer strains in a directed way (Brynildsen and Liao, 2009; Atsumi et al., 2010; Minty et al., 2011). Brynildsen and Liao (2009) analyzed the transcriptional response of *E. coli* to isobutanol stress under aerobic conditions and found a different transcriptional response compared to that to ethanol. The authors postulated an isobutanol response network with quinone malfunction as key element altering the respiratory machinery and leading to complex reductive stress response. To overcome isobutanol toxicity, Atsumi et al. (2010) isolated an isobutanol-tolerant *E. coli* strain by a serial transfer method. Whole genome sequencing in combination with reconstruction identified five mutations (*arcA*, *gatY*, *tnaA*, *yhbJ*, and *marCRAB*) to be responsible for increased isobutanol tolerance. Furthermore, enhanced membrane synthesis by improved availability of glucosamine-6-phosphate had a positive effect on isobutanol tolerance. However, the final strain obtained was more tolerant to isobutanol, but showed much lower isobutanol formation, when compared to the parental strain *E. coli* JCL260/pSA55/pSA69 (Atsumi et al., 2010). More recently, Minty et al. (2011) used experimental evolution combined with genome re-sequencing to identify the genotypic adaptations of *E. coli* lineages with increased isobutanol tolerance. Thereby, the authors identified a set of mutations (*marC*, *hfq*, *mdh*, *acrAB*, *gatYZABCD*, *rph*) common in several isobutanol tolerant lineages, and they speculated that *rpoS* and post-transcriptional regulators such as the RNA-binding protein Hfq are promising targets to improve isobutanol production with *E. coli*. However, these results demonstrate again the complexity of the cellular response to isobutanol stress and indicate that improving the isobutanol tolerance does not *per se* result in an improvement of the whole production process. Based on these findings it can be again speculated that *E. coli* might not be the optimal host for high-titer isobutanol production. Therefore, approaches to improve the tolerance of organisms which naturally show a higher tolerance against solvents (e.g., *C. glutamicum*, *B. subtilis*, or *S. cerevisiae*) are more promising. *Pseudomonas putida* has recently been shown to grow in the presence of very high concentrations (6%, v/v) of butanol (Rühl et al., 2009), making this organism an attractive target for metabolic engineering approaches. A different approach to overcome the isobutanol tolerance problem from the bioprocess perspective is the *in situ* product removal as exemplified by the gas stripping method applied by Baez et al. (2011) and described in the following paragraph.

12.5 PROCESS DEVELOPMENT FOR THE PRODUCTION OF ISOBUTANOL

All of the above-mentioned examples for fermentative isobutanol production were performed in simple flasks or bottles. However, the development of feasible production processes is challenging since not only common upscale problems arise, but also the cytotoxicity of higher alcohols is a striking problem for the transfer into the industrial scale.

For the most promising bacterial production strains, fed-batch processes were developed differing in the conditions applied. For *E. coli* JCL260/pSA55/pSA69 strictly aerobic conditions were used, with *in situ* product removal by gas stripping with air and subsequent sampling of isobutanol by condensing (Baez et al., 2011). Such an approach was already successfully applied for the production of up to 2.38 M isopropanol (143 g/L) with an engineered *E. coli* strain (Inokuma et al., 2010). Also for isobutanol production with *E. coli* a stripping approach was useful and resulted in titers of more than 675 mM (50 g/L) with a yield of 0.68 mole isobutanol per mole of glucose within 72 hours, at a cultivation temperature of 30°C (Table 12.4). However, in spite of aerobic conditions during the whole process, cells stopped growing after 10 hours, reaching a maximum cell density of 6.7 g/L, probably due to isobutanol toxicity. Increasing the temperature from 30°C to 37°C, in order to increase the vapor pressure for a more efficient stripping process, failed. At 37°C, *E. coli* JCL260/pSA55/pSA69 showed a drastically reduced final isobutanol titer and yield (Baez et al., 2011). Thus, the isobutanol-induced growth arrest obviously limits the overall productivity for an industrial scale application.

For the best *C. glutamicum* producer strain Iso7, it was observed that the performance under aerobic conditions was much worse than under oxygen deprivation conditions in sealed bottles (Blombach and Eikmanns, 2011). To transfer the process conditions from the bottle to a bioreactor, a two-phase fermentation was developed. *C. glutamicum* Iso7 was cultivated in the first phase under aerobic conditions. In this phase the cells grew with glucose and acetate to high cell densities and produced no isobutanol. After complete consumption of the acetate, the cells stopped growing and the production phase was started by switching off aeration. The residual oxygen in the culture was rapidly consumed, and *C. glutamicum* Iso7 produced up to about 180 mM isobutanol (13 g/L) with a yield of 0.48 mole per mole of glucose (0.20 g/g) (Figure 12.5; Table 12.4). The reason for the production stop at about 180 mM isobutanol remains unclear but might also be attributed to isobutanol toxicity for the cells (Blombach et al., 2011). The negative effect might be overcome by developing an integrated stripping process with nitrogen, which has been successfully applied for 1-butanol production with Clostridia (Ezeji et al., 2005). Furthermore, a significantly

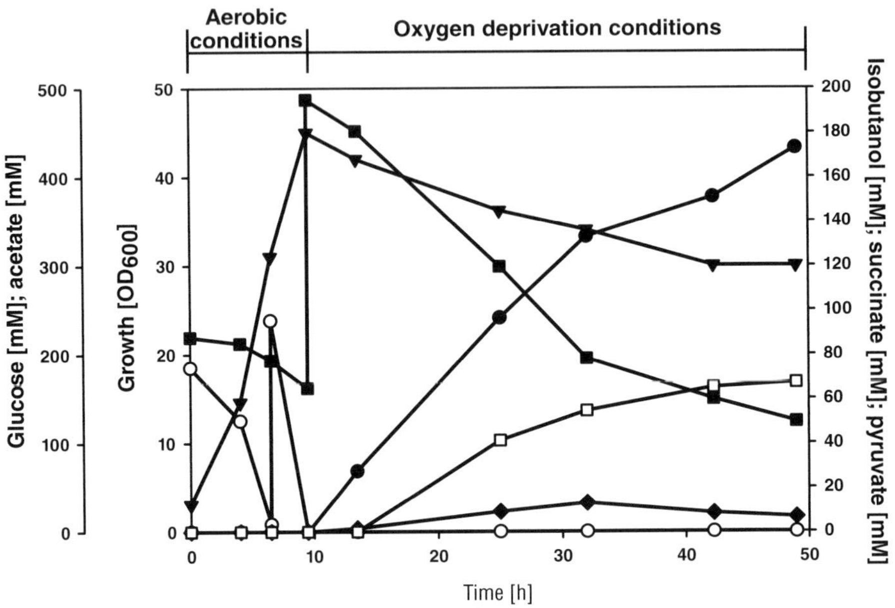

FIGURE 12.5 Isobutanol accumulation during a representative fed-batch fermentation of *Corynebacterium glutamicum* Iso7 [*C. glutamicum* Δ*aceE* Δ*pqo* Δ*ilvE* Δ*ldhA* Δ*mdh* (pJC4*ilvBNCD-pntAB*) (pBB1*kivd-adhA*)] on CGXII medium initially containing 4% (w/v) glucose, 1% (w/v) acetate, 0.5% (w/v) yeast extract, and 2 mM L-valine, L-isoleucine, and L-leucine, respectively. After 9.5 hours, the aeration was switched off and the stirring speed was reduced to 300 rpm. ▼, OD_{600}; ■, glucose; ○, acetate; □, succinate; ♦, pyruvate; ●, isobutanol.

reduced yield was observed in the fed-batch fermentations with *C. glutamicum* Iso7 when compared to the cultivations in shaken bottles (Table 12.4; Blombach et al., 2011). However, the lower yield of *C. glutamicum* Iso7 was constant in the whole production phase, and therefore, this effect can hardly be explained by isobutanol toxicity but indicates that the physiological state of the cell during the transition from aerobic to oxygen-deprived conditions may have an influence on the overall production behavior (Blombach et al., 2011). Recently, Martínez et al. (2010) investigated the role of the transition from aerobic to anaerobic conditions in a succinate production process with *E. coli* and showed that introducing a microaerobic phase at the end of the aerobic growth phase led to an adjustment of the enzymatic machinery and to improved succinate production under anaerobic conditions. In accordance, it might be possible to improve the *C. glutamicum* production process by introducing microaerobic conditions at the end of the aerobic growth phase. However, the physiological changes of *C. glutamicum* during the transition from aerobic to anaerobic conditions have so far not been investigated. A deep insight in the metabolic adaptation of the cells to such alternating culture conditions will certainly help to further optimize isobutanol production with *C. glutamicum* by novel metabolic engineering approaches and by adapting the fermentation process conditions.

In analogy to the process with *C. glutamicum* Iso7, a two-phase fermentation process was developed with PDHC-deficient *B. subtilis* UL05 yielding 74 mM isobutanol (5.5 g/L). In this case, the yield obtained [0.47 mole per mole (0.19 g/g); Li et al., 2012b] was about the same (or even slightly higher) when compared to that observed in flask batch fermentations (Table 12.4).

As outlined above (section 12.4), all microorganisms so far engineered and employed for isobutanol production suffer from a relatively low tolerance to this alcohol (in general <2%). Therefore, *in situ* product removal by gas stripping, pervaporation, or extraction with, for example, oleyl-alcohol during the fermentation has been established to keep the alcohol concentrations in the cultures at levels below cytotoxic concentrations for the respective microorganism (Inokuma et al., 2010; Connor et al., 2010; Ezeji et al., 2010; Baez et al., 2011). In the case that the produced isobutanol should be processed to isobutene (a widely used commodity chemical with a wide range of applications), another possible and elegant approach to bypass the problem with isobutanol toxicity might be the *in situ* enzymatic dehydration of isobutanol to isobutene by oleate hydrolases (EC 4.2.1.53), as recently described by van Leeuwen et al. (2012). Since isobutene is gaseous it can easily be recovered from the fermenter. Furthermore, it has a low aqueous solubility which minimizes product toxicity to the microorganisms (van Leeuwen et al., 2012). The same holds true for isobutyraldehyde which shows a comparatively high volatility and, therefore, can efficiently be stripped out of the medium and subsequently condensed (Rodriguez and Atsumi, 2012). *E. coli* was recently engineered for the production of this aldehyde by use of the isobutanol-producing strain *E. coli* JCL260 (see section 12.3.1), overexpressing *alsS*, *ilvC*, *ilvD*, and *kivd*. Interestingly it was not sufficient to delete the native Adh gene *yqhd* to achieve a relevant yield. The additional deletion of *adhP*, *eutG*, *yjgB*, *betA*, *adhP*, and *fucO*, all encoding Adhs, was essential to develop an efficient isobutyraldehyde producer (Rodriguez and Atsumi, 2012). With the final strain, *E. coli* AL626, up to 35 g isobutyraldehyde/L within 5 days were produced in a long-term experiment with *in situ* product removal by gas stripping (Rodriguez and Atsumi, 2012). As isobutene, isobutyraldehyde is a commodity chemical with a manifold application range and can, in contrast to isobutanol, be more easily removed from the culture broth (van Leeuwen et al., 2012; Rodriguez and Atsumi, 2012).

12.6 ECONOMIC OUTLOOK

The current and future society doubtless strongly relies on mobility, whose hunger for energy is mainly satisfied by combustion of fossil fuels at the present time. The requirements for future biofuels, replacing gasoline, diesel, and jet fuels, are to be cheap, available in large quantities, and to possess the adequate energy density. Therefore, production processes have to perform excellent with regard to yield and productivity. As stated in section 12.1 the attributes of isobutanol are superior over ethanol and close to that of gasoline. Therefore, isobutanol can be regarded as "an ideal biofuel" (Savage, 2011). Metabolic engineering strategies are readily available to develop producer strains, and the proof of concept has been shown for several biofuel production approaches, however, mainly in lab-scale so far. The most striking challenge for the future biofuel economy using higher alcohols instead of ethanol is the use of non-food carbon sources (see section 12.7), the development of cost-effective product recovery, and the scale-up of economically relevant processes to produce biofuels at "a rate we currently dig out oil from the underground" (Savage, 2011). Taking current biofuel production numbers into account, which amount to about 3% of the overall worldwide transport fuel consumption (International Energy Agency, 2011) with a negligible contribution of higher alcohols, it becomes obvious that it is still a long way to a "green" advanced fuel economy.

12.7 FUTURE PROSPECTS

The implementation of the last two steps of the Ehrlich pathway, that is, a KDC in combination with an Adh reaction, has proven to be a generally applicable metabolic engineering approach to transform relevant host, such as *E. coli*, *B. subtilis*, *C. glutamicum*, *R. eutropha*, or *S. elongatus* into (efficient) isobutanol producers. However, to reach relevant isobutanol yields, productivities, and titers, overcoming the isobutanol inhibitory effects is an essential prerequisite in the upscaling process, aside from optimization of the process conditions. Isobutanol toxicity might be overcome by integrated product removal and/or the use of highly tolerant strains. In addition, environmental friendly and sustainable bioprocesses require the use of second-generation feedstocks, such as cellulose, ligno- and hemicellulose, CO_2 and H_2O, synthesis gas, or microbial protein (hydrolysates), which do not compete for feed or food. Therefore, either expanding the substrate spectrum (e.g., to pentoses) of relevant hosts or the use of organisms, which are able to efficiently utilize these alternative substrates, is a prerequisite for the bio-based production of higher alcohols. As outlined above, several hosts and engineering strategies are at the beginning of their development, which allow the production of isobutanol from second-generation feedstocks, for example, from cellobiose and cellulose with *Clostridium* (Higashide et al., 2011), from CO_2, H_2O, and solar energy with *Synechococcus* (Atsumi et al., 2009b), or from CO_2 by electromicrobial conversion with *Ralstonia* (Li et al., 2012a).

Synthesis gas, a mixture of mainly CO, CO_2, and H_2, has been used in chemical industry as feedstock and can be generated by gasification of coal and oil but also from biomass, municipal waste, or by recycling of used plastics (Köpke et al., 2010). Isobutanol production from synthesis gas has so far not been reported. However, Köpke et al. (2010) engineered *Clostridium ljungdahlii*, which is naturally able to use synthesis gas as carbon and energy source, for the production of 1-butanol by implementation of the CoA-dependent 1-butanol synthesis pathway from *Clostridium acetobutylicum*. The final titer of about 0.5 mM 1-butanol was rather low; however, this approach demonstrated the feasibility to produce fuels and chemicals from synthesis gas.

Just recently, Huo et al. (2011, 2012) engineered *E. coli* for the protein-based higher alcohol production. This was achieved by directed evolution using chemical mutagenesis to improve the utilization of amino acids, by inactivation of the ammonia assimilation (Δ*glnA*, Δ*gdhA*), introduction of three transamination and deamination cycles by expressing the L-leucine dehydrogenase genes *leuDH* from *Thermoactinomyces intermedius*, *ilvE*, *avtA*, *ilvA*, and *sdaB* from *E. coli*, deletion of the quorum-sensing gene *lsrA*, and overexpression of *alsS* from *B. subtilis*, *ilvC*, *ilvD*, and *yqhd* from *E. coli*, and *kivd* and *adhA* from *L. lactis*. The final strain produced up to 4 g higher alcohols per liter (C4 and C5; 50% isobutanol) with 56% of the theoretical yield on yeast extract and also relevant titers of alcohols on algal and bacterial proteins as feedstock (Huo et al., 2011).

Taken together, the above-described examples substantiate the fast and innovative progress toward an efficient and sustainable bio-based microbial production of higher alcohols with the most prominent example isobutanol. However, these approaches also hint to the future challenges for biotechnological production processes: optimization of downstream processing and *in situ* product removal, development of highly tolerant strains or engineering natural highly tolerant strains and especially, development of processes based on second-generation feedstocks, and complete and efficient consumption of the monomers of complex substrates.

NOMENCLATURE

The support of the Fachagentur Nachwachsende Rohstoffe (FNR) of the Bundesministerium für Ernährung, Landwirtschaft und Verbraucherschutz (FNR grant 220-095-08A; BioProChemBB project, ERA-IB program), and the Deutsche Forschungsgemeinschaft (DFG, grant TA 241/5-1) is gratefully acknowledged.

ABBREVIATIONS

Adh	alcohol dehydrogenase (encoded by *adh2*, *adh6*, *yqhD*, *adhA*, *adhE*, *adhP*, *eutG*, *yjgB*, *betA*, *fucO*)
AHAIR	acetohydroxyacid isomeroreductase (encoded by *ilvC*, *Ilv5*)
AHAS	acetohydroxyacid synthase (encoded by *ilvBN*, *ilvIH*, *alsS*, *Ilv2*)
AK	acetate kinase
BAT	branched-chain amino acid transferase (encoded by *Bat2*)
avtA	transamination gene
BCKDC	branched-chain ketoacid dehydrogenase complex (encoded by *bkdAB*)
DFG	Deutsche Forschungsgemeinschaft
DHAD	dihydroxyacid dehydratase (encoded by *ilvD*, *Ilv3*)
FNR	fumarate and nitrate reductase regulator (encoded by *fnr*)
FUM	fumarase
FRD	fumarate reductase (encoded by *frdAB*)
GAPDH	glyceraldehyde-3P dehydrogenase

glnA	glutamine synthetase
gdhA	glutamate dehydrogenase
GPDH	G6P dehydrogenase
ICD	isocitrate dehydrogenase
ICL	isocitrate lyase
ilvA	transamination gene
KDC	2-ketoacid decarboxylase (encoded by *kivD*, *Aro10*)
LDH	lactate dehydrogenase (encoded by *ldhA*)
LLDH	L-leucine dehydrogenase (encoded by *leuDH*)
lsrA	quorum-sensing gene
MalE	NADPH-dependent malic enzyme
Mdh	malate dehydrogenase (encoded by *mdh*)
mM	milli-molar
MQO	malate:quinone oxidoreductase
MS	malate synthase
NAD^+	nicotinamide adenine dinucleotide oxidized
$NADH+H^+$	nicotinamide adenine dinucleotide reduced
$NADP^+$	nicotinamide adenine dinucleotide phosphate oxidized
$NADPH+H^+$	nicotinamide adenine dinucleotide phosphate reduced
ODHC	oxoglutarate dehydogenase complex
PDC	pyruvate decarboxylase (encoded by *Pdc1*)
PDHC	pyruvate dehydrogenase complex (encoded by *aceE*, *aceF*, *lpd*)
PEP	phosphoenolpyruvate
PEPCk	PEP carboxykinase
PEPCx	PEP carboxylase
PCx	pyruvate carboxylase
PFL	pyruvate formate lyase (encoded by *pflAB*)
PGDH	6P-gluconate dehydrogenase
PhaA	β-ketothiolase (encoded by *phaA*)
PhaB	acetoacetyl-CoA reductase (encoded by *phaB*)
PhaC	PHB synthase (encoded by *phaC*)
PHB	polyhydroxybutyrate
PK	pyruvate kinase
PntAB	membrane-bound transhydrogenase
PPS	PEP synthetase
PTA	phosphotransacetylase (encoded by *pta*)
PQO	pyruvate:quinone oxidoreductase (encoded by *poxB*, *pqo*)
RBC	ribulose-bisphosphate carboxylase (encoded by *rbcLS*)
sdaB	transamination gene
SDH	succinate dehydrogenase
SfcA	NADH-dependent malic enzyme
TA	transaminase B (encoded by *ilvE*).

REFERENCES

Ajinomoto Co. (2010). Food Products Business. Available at: http://www.ajinomoto.com/ir/pdf/Food-Oct2010.pdf; last accessed date January 10th, 2014.

Ajinomoto Co. (2011). Feed-Use Amino Acid Business. Available at: http://www.ajinomoto.com/ir/pdf/Feed-useAA-Oct2011.pdf; last accessed date January 10th, 2014.

Atsumi S, Liao JC. (2008). Directed evolution of *Methanococcus jannaschii* citramalate synthase for biosynthesis of 1-propanol and 1-butanol by *Escherichia coli*. *Appl Environ Microbiol*, 74, 7802–7808.

Atsumi S, Hanai T, Liao JC. (2008). Non-fermentative pathways for synthesis of branched-chain higher alcohols as biofuels. *Nature*, 451, 86–89.

Atsumi S, Wu TY, Eckl EM, Hawkins SD, Buelter T, Liao JC. (2009a). Engineering the isobutanol biosynthetic pathway in *Escherichia coli* by comparison of three aldehyde reductase/alcohol dehydrogenase genes. *Appl Microbiol Biotechnol*, 85, 651–657.

Atsumi S, Higashide W, Liao JC. (2009b). Direct photosynthetic recycling of carbon dioxide to isobutyraldehyde. *Nat Biotechnol*, 27, 1177–1182.

Atsumi S, Li Z, Liao JC. (2009c). Acetolactate synthase from *Bacillus subtilis* serves as a 2-ketoisovalerate decarboxylase for isobutanol biosynthesis in *Escherichia coli*. *Appl Environ Microbiol*, 75, 6306–6311.

Atsumi S, Wu T-Y, Machado IMP, Huang W-C, Chen P-Y, Pellegrini M, Liao JC. (2010). Evolution, genomic analysis, and reconstruction of isobutanol tolerance in *Escherichia coli*. *Mol Syst Biol*, 6, 449.

Baez A, Cho K-M, Liao JC. (2011). High-flux isobutanol production using engineered *Escherichia coli*: a bioreactor study with *in situ* product removal. *Appl Microbiol Biotechnol*, 90, 1681–1690.

Bastian S, Liu X, Meyerowitz JT, Snow CD, Chen MMY, Arnold FH. (2011). Engineered ketol-acid reductoisomerase and alcohol dehydrogenase enable anaerobic 2-methylpropan-1-ol production at theoretical yield in *Escherichia coli*. *Metab Eng*, 13, 345–352.

Blombach B, Eikmanns BJ. (2011). Current knowledge on isobutanol production with *Escherichia coli*, *Bacillus subtilis* and *Corynebacterium glutamicum*. *Bioeng Bugs*, 2, 346–350.

Blombach B, Schreiner ME, Holatko J, Bartek T, Oldiges M, Eikmanns BJ. (2007). L-Valine production with pyruvate dehydrogenase complex-deficient *Corynebacterium glutamicum*. *Appl Environ Microbiol*, 73, 2079–2084.

Blombach B, Schreiner ME, Bartek T, Oldiges M, Eikmanns BJ. (2008). *Corynebacterium glutamicum* tailored for high-yield L-valine production. *Appl Microbiol Biotechnol*, 79, 471–479.

Blombach B, Riester T, Wieschalka S, Ziert C, Youn JW, Wendisch VF, Eikmanns BJ. (2011). *Corynebacterium glutamicum* tailored for efficient isobutanol production. *Appl Environ Microbiol*, 77, 3300–3310.

Bozell JJ, Petersen GR. (2010). Technology development for the production of biobased products from biorefinery carbohydrates—the US Department of Energy's "Top 10" revisited. *Green Chem*, 12, 539–554.

Brat D, Weber C, Lorenzen W, Bode HB, Boles E. (2012). Cytosolic re-localization and optimization of valine synthesis and catabolism enables increased isobutanol production with the yeast *Saccharomyces cerevisiae*. *Biotechnol Biofuels*, 5, 65.

Brynildsen MP, Liao JC. (2009). An integrated network approach identifies the isobutanol response network of *Escherichia coli*. *Mol Syst Biol*, 5, 277.

Cann AF, Liao JC. (2008). Production of 2-methyl-1-butanol in engineered *Escherichia coli*. *Appl Microbiol Biotechnol*, 81, 89–98.

Chen X, Nielsen KF, Borodina I, Kielland-Brandt MC, Karhumaa K. (2011). Increased isobutanol production in *Saccharomyces cerevisiae* by overexpression of genes in valine metabolism. *Biotechnol Biofuels*, 4, 21.

Connor MR, Liao JC. (2008). Engineering of an *Escherichia coli* strain for the production of 3-methyl-1-butanol. *Appl Environ Microbiol*, 74, 5769–5775.

Connor MR, Cann AF, Liao JC. (2010). 3-Methyl-1-butanol production in *Escherichia coli*: random mutagenesis and two-phase fermentation. *Appl Microbiol Biotechnol*, 86, 1155–1164.

Cramm R. (2009). Genomic view of energy metabolism in *Ralstonia eutropha* H16. *J Mol Microbiol Biotechnol*, 16, 38–52.

Ehrlich F. (1907). Über die Bedingungen der Fuselölbildung und über ihren Zusammenhang mit dem Eiweißaufbau der Hefe. *Ber Dtsch Chem Ges*, 40, 1027–1047.

Ezeji TC, Karcher PM, Qureshi N, Blaschek HP. (2005). Improving performance of a gas stripping-based recovery system to remove butanol from *Clostridium beijerinckii* fermentation. *Bioprocess Biosyst Eng*, 27, 207–214.

Ezeji TC, Milne C, Price ND, Blaschek HP. (2010). Achievements and perspectives to overcome the poor solvent resistance in acetone and butanol-producing microorganisms. *Appl Microbiol Biotechnol*, 85, 1697–1712.

Fischer CR, Klein-Marcuschamer D, Stephanopoulos G. (2008). Selection and optimization of microbial hosts for biofuels production. *Metab Eng*, 10, 295–304.

Hazelwood LA, Daran J-M, van Maris AJA, Pronk JT, Dickinson JR. (2008). The Ehrlich pathway for fusel alcohol production: a century of research on *Saccharomyces cerevisiae* metabolism. *Appl Environ Microbiol*, 74, 2259–2266.

Higashide W, Li Y, Yang Y, Liao JC. (2011). Metabolic engineering of *Clostridium cellulolyticum* for production of isobutanol from cellulose. *Appl Environ Microbiol*, 77, 2727–2733.

Huo Y-X, Cho KM, Rivera JGL, Monte E, Shen CR, Yan Y, Liao JC. (2011). Conversion of proteins into biofuels by engineering nitrogen flux. *Nat Biotechnol*, 29, 346–351.

Huo Y-X, Wernick DG, Liao JC. (2012). Towards nitrogen neutral biofuel production. *Curr Opin Biotechnol*, 23, 406–413.

Inokuma K, Liao JC, Okamoto M, Hanai T. (2010). Improvement of isopropanol production by metabolically engineered *Escherichia coli* using gas stripping. *J Biosci Bioeng*, 110, 696–701.

International Energy Agency. (2011). Available at: http://www.iea.org/topics/biofuels; last accessed date January 10th, 2014.

Jarboe LR. (2011). YqhD: a broad-substrate range aldehyde reductase with various applications in production of biorenewable fuels and chemicals. *Appl Microbiol Biotechnol*, 89, 249–257.

Knoshaug EP, Zhang M. (2009). Butanol tolerance in a selection of microorganisms. *Appl Biochem Biotechnol*, 153, 13–20.

Kondo T, Tezuka H, Ishii J, Matsuda F, Ogino C, Kondo A. (2012). Genetic engineering to enhance the Ehrlich pathway and alter carbon flux for increased isobutanol production from glucose by *Saccharomyces cerevisiae*. *J Biotechnol*, 159, 32–37.

Köpke M, Held C, Hujer S, Liesegang H, Wiezer A, Wollherr A, Ehrenreich A, Liebl W, Gottschalk G, Dürre P. (2010). *Clostridium ljungdahlii* represents a microbial production platform based on syngas. *Proc Natl Acad Sci U S A*, 107, 13087–13092.

Krause F, Blombach B, Eikmanns BJ. (2010). Metabolic engineering of *Corynebacterium glutamicum* for 2-ketoisovalerate production. *Appl Environ Microbiol*, 76, 8053–8061.

Lee W-H, Seo S-O, Bae Y-H, Nan H, Jin Y-S, Seo J-H. (2012). Isobutanol production in engineered *Saccharomyces cerevisiae* by overexpression of 2-ketoisovalerate decarboxylase and valine biosynthetic enzymes. *Bioprocess Biosyst Eng*, 35, 1467–1475.

Li H, Cann AF, Liao JC. (2010). Biofuels: biomolecular engineering fundamentals and advances. *Ann Rev Chem Biomol Eng*, 1, 19–36.

Li S, Wen J, Jia X. (2011). Engineering *Bacillus subtilis* for isobutanol production by heterologous Ehrlich pathway construction and the biosynthetic 2-ketoisovalerate precursor pathway overexpression. *Appl Microbiol Biotechnol*, 91, 577–589.

Li H, Opgenorth PH, Wernick DG, Rogers S, Wu T-Y, Higashide W, Malati P, Huo Y-X, Cho KM, Liao JC. (2012a). Integrated electromicrobial conversion of CO_2 to higher alcohols. *Science*, 335, 1596.

Li S, Huang D, Li Y, Wen J, Jia X. (2012b). Rational improvement of the engineered isobutanol-producing *Bacillus subtilis* by elementary mode analysis. *Microb Cell Fact*, 11, 101.

Liebl W. (2006). The genus *Corynebacterium*-nonmedical. In: Dworkin, M., Falkow, S., Rosenberg E, Schleifer KH, Stackebrandt E. editors. *The Prokaryotes*, 3rd ed. vol. 3, New York: Springer, pp. 796–818.

Liu X, Bastian S, Snowa CS, Brustada EM, Saleskia TE, Xub J-H, Meinhold P, Arnold FH. (2012). Structure-guided engineering of *Lactococcus lactis* alcohol dehydrogenase LlAdhA for improved conversion of isobutyraldehyde to isobutanol. *J Biotechnol*, 164, 188–194.

Lu J, Brigham CJ, Gai CS, Sinskey AJ. (2012). Studies on the production of branched-chain alcohols in engineered *Ralstonia eutropha*. *Appl Microbiol Biotechnol*, 96, 283–297.

Martínez I, Bennett GN, San K-Y. (2010). Metabolic impact of the level of aeration during cell growth on anaerobic succinate production by an engineered *Escherichia coli* strain. *Metab Eng*, 12, 499–509.

Matsuda F, Kondo T, Ida K, Tezuka H, Ishi J, Kondo A. (2012). Construction of an artificial pathway for isobutanol biosynthesis in the cytosol of *Saccharomyces cerevisiae*. *Biosci Biotechnol Biochem*, 76, 2139–2141.

Minty JJ, Lesnefsky AA, Lin F, Chen Y, Zaroff TA, Veloso AB, Bie B, McConnell CA, Ward RJ, Schwartz DR, Rouillard J-M, Gao Y, Gulari E, Lin XN. (2011). Evolution combined with genomic study eludicates genetic bases of isobutanol tolerance in *Escherichia coli*. *Microb Cell Fact*, 10, 18.

Nicolaou SA, Gaida SM, Papoutsakis ET. (2010). A comparative view of metabolite and substrate stress and tolerance in microbial bioprocessing: from biofuels and chemicals, to biocatalysis and remediation. *Metab Eng*, 12, 307–331.

Nielsen DR, Leonard E, Yoon SH, Tseng HC, Yuan C, Prather KLJ. (2009). Engineering alternative butanol production platforms in heterologous bacteria. *Metab Eng*, 11, 262–273.

Nishimura T, Vertès AA, Shinoda Y, Inui M, Yukawa H. (2007). Anaerobic growth of *Corynebacterium glutamicum* using nitrate as a terminal electron acceptor. *Appl Microbiol Biotechnol*, 75, 889–897.

Renewable Fuels Association. (2011). World Fuel Ethanol Production 2011. Available at: http://www.ethanolrfa.org/pages/World-Fuel-Ethanol-Production; last accessed date January 10th, 2014.

Rodriguez GM, Atsumi S. (2012). Isobutyraldehyde production from *Escherichia coli* by removing aldehyde reductase activity. *Microb Cell Fact*, 11, 90.

Rühl J, Schmid A, Blank LM. (2009). Selected *Pseudomonas putida* strains able to grow in the presence of high butanol concentrations. *Appl Environ Microbiol*, 75, 4653–4656.

Savage N. (2011). The ideal biofuel. *Nature*, 474, 9–11.

Sentheshanmuganathan S. (1960). The mechanism of the formation of higher alcohols from amino acids by *Saccharomyces cerevisiae*. *Biochem J*, 74, 568–576.

Sentheshanmuganathan S, Elsden SR. (1958). The mechanism of the formation of tyrosol by *Saccharomyces cerevisiae*. *Biochem J*, 69, 210–218.

Shen CR, Liao JC. (2008). Metabolic engineering of *Escherichia coli* for 1-butanol and 1-propanol production via the keto-acid pathways. *Metab Eng*, 10, 312–320.

Smith K, Cho K, Liao JC. (2010). Engineering *Corynebacterium glutamicum* for isobutanol production. *Appl Microbiol Biotechnol*, 87, 1045–1055.

Takeno S, Ohnishi J, Komatsu T, Masaki T, Sen K, Ikeda M. (2007). Anaerobic growth and potential for amino acid production by nitrate respiration in *Corynebacterium glutamicum*. *Appl Microbiol Biotechnol*, 75, 1173–1182.

Takors R, Bathe B, Rieping M, Hans S, Kelle R, Huthmacher K. (2007). Systems biology for industrial strains and fermentation processes–example: amino acids. *J Biotechnol*, 129, 181–190.

van Leeuwen BNM, van der Wulp AM, Duijnstee I, van Maris AJA, Straathof AJJ. (2012). Fermentative production of isobutene. *Appl Microbiol Biotechnol*, 93, 1377–1387.

Zhang K, Sawaya MR, Eisenberg DS, Liao JC. (2008) Expanding metabolism for biosynthesis of nonnatural alcohols. *Proc Natl Acad Sci USA*, 105, 20653–20658.

Zhang Y, Zhu Y, Zhu Y, Li Y. (2009). The importance of engineering physiological functionality into microbes. *Trends Biotechnol*, 27, 664–672.

CHAPTER 13

Lactic Acid

KENJI OKANO

Department of Biotechnology, Graduate School of Engineering, Osaka University, Osaka, Japan

TSUTOMU TANAKA AND AKIHIKO KONDO

Department of Chemical Science and Engineering, Graduate School of Engineering, Kobe University, Kobe, Japan

Abstract

Lactic acid (LA) is an important and versatile chemical that can be produced from renewable resources such as biomass. LA is used in the food, pharmaceutical, and polymers industries and is produced by microorganism fermentation; however, most microorganisms cannot directly utilize biomass such as starchy materials and cellulose. Here, we summarize LA and LA-based polymers production using several kinds of genetically modified microorganisms, such as lactic acid bacteria, *Escherichia coli*, *Corynebacterium glutamicum*, and yeasts. Using gene manipulation and metabolic

Bioprocessing of Renewable Resources to Commodity Bioproducts, First Edition.
Edited by Virendra S. Bisaria and Akihiko Kondo.

engineering, the yield and optical purity of LA produced from biomass have been significantly improved, as well as LA-based polymers. The drawbacks as well as improvements of LA production by fermentation are discussed.

13.1 HISTORY OF LACTIC ACID

Lactic acid (LA), 2-hydroxypropionic acid, is the simplest hydroxyl acid. It has a long history, and it has existed from at least four to five thousand years ago for preservation of human foodstuffs by fermentation (Davidson et al., 1995). LA was first discovered in sour milk by Scheele in 1780, who initially considered it a milk component. In 1789, Lavoisier named this milk component *acide lactique*, which became the possible origin of the current terminology for LA. And later, Pasteur discovered that LA was not a milk component but a fermentation metabolite generated by certain microorganisms (Wee et al., 2006). In 1839, Fremy demonstrated fermentative production of LA from various kinds of carbohydrates such as sucrose, lactose, mannitol, starch, and dextrin. The first commercial production of LA started in the United States by a microbial process in 1881 (Vijayakumar et al., 2008).

13.2 APPLICATIONS OF LACTIC ACID

LA is an important and versatile chemical with many potential applications. There are four major categories for the current uses and applications of LA: food, cosmetic, pharmaceutical, and chemical applications. The potential applications of LA are summarized by Wee et al. (2006) and illustrated in Figure 13.1. Emerging applications of LA are the use as a feedstock monomer for chemical conversions. LA has both a carboxylic acid function and a hydroxyl group, making it reactive in a variety of chemical conversions to useful products (Figure 13.2). By decarboxylation, hydrogenation, dehydration, condensation, and self-esterification, LA is converted to acetaldehyde, propylene glycol, acrylic acid, 2,3-petanedione, and lactide, respectively. Lactide is then converted to poly lactic acid (PLA), which can be used as bio-based plastics, by ring-opening polymerization (Varadarajan and Miller, 1999). Now, production of PLA has become promising application of LA. With the shortage and/or unstable supply of crude oil, its astonishing price variation, and the concerns of its environmental impact and the greenhouse gases released by petroleum-based plastic and chemical industries, bio-based plastics including PLA have gained significant attention (Jem et al., 2010).

13.3 POLY LACTIC ACID

PLA is a sustainable alternative to petrochemical-derived products, since LA can be produced on a mass scale by the microbial fermentation of agricultural products/by-products mainly the carbohydrate-rich substances (Nampoothiri et al., 2010). In

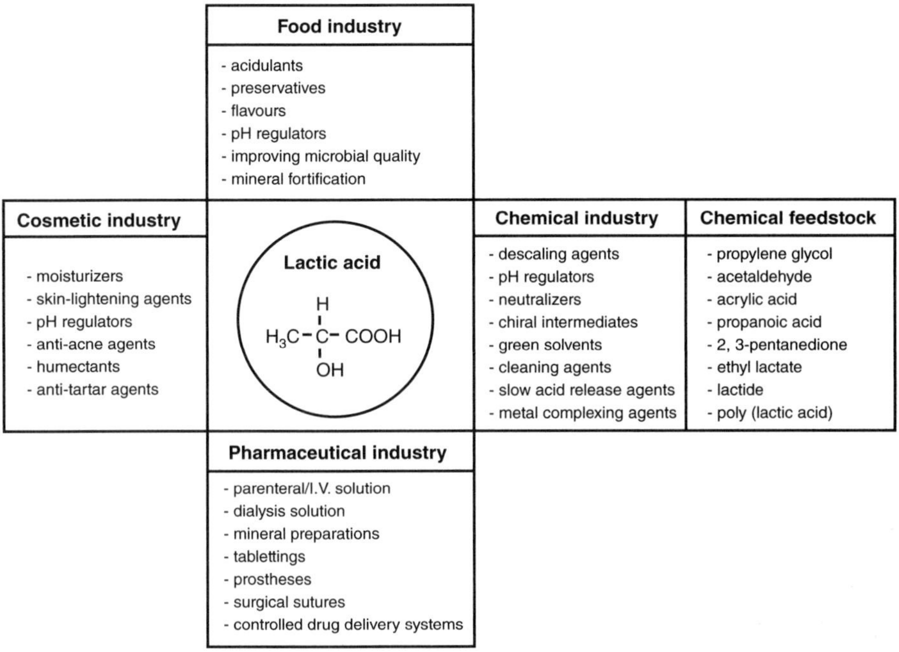

FIGURE 13.1 The potential applications of lactic acid.

comparison to other biopolymers, the production of PLA has numerous advantages including: (1) fixation of significant quantities of carbon dioxide via agricultural production; (2) significant energy savings; (3) the ability to recycle back to LA by hydrolysis or alcoholysis; (4) reduction of landfill volumes; (5) improvement of agricultural economy; and (6) the all-important ability to tailor physical properties through material modifications (Jamshidian et al., 2010). According to two bulk PLA producers, NatureWorks and PURAC, PLA or PLA components have the substitution potential to partially replace low-density polyethylene, high-density polyethylene, polypropylene, polyamide, and polyethylene terephthalate, as well as seeing possibilities to substitute for poly methyl methacrylate and polyurethane (Shen et al., 2009). From the point of view of biological LA production, selective production of optically pure D- or L-LA is an important challenge to improve the physical properties of PLA as optical purity of LA is crucial to the physical properties of PLA. In addition, stereocomplex PLA, which is composed of both poly D- and L-LA, is known to have high thermostability. Stereocomplex-type polymers show a high melting point (~230°C), that is, approximately 50°C higher compared with the respective single polymers (Ikada et al., 1987). Therefore, demand for both D- and L-LA has been occurred, and LA was produced by microbial fermentation generally employing LA bacteria (LAB) not by chemical synthesis occurring mixture of both D- and L-LA (Wee et al., 2006).

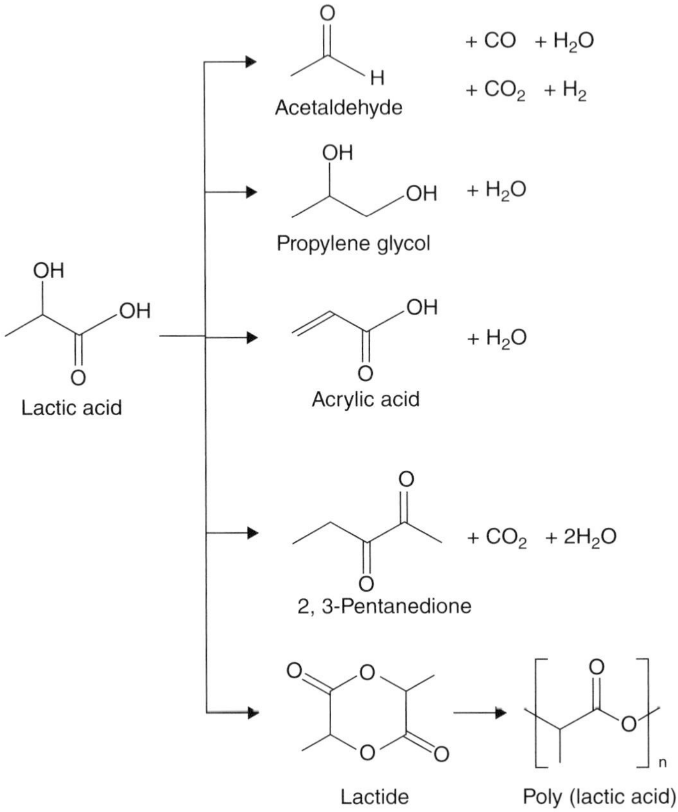

FIGURE 13.2 Chemical conversions of lactic acid to useful products.

13.4 CONVENTIONAL LACTIC ACID PRODUCTION

LAB can be divided into two groups based on the fermentation products of glucose (Hofvendahl and Hahn-Hägerdal, 2000). The homofermentative LAB fermented glucose via the Embden–Meyerhof pathway, and 1 mole of glucose is theoretically converted to 2 mole of LA (1:1 in gram order). On the other hand, heterofermentative LAB converted glucose to equimolar amounts of LA, carbon dioxide, and ethanol or acetate via the phosphoketolase pathway (PK pathway) (Figure 13.3). Thus strains that can be used in the commercial production of LA are almost limited in those belonged to the genus *Lactobacillus* (*Lb.*), *Lactococcus* (*Lc.*), and *Enterococcus* (*Ent.*). In addition, optical purity of LA has to be taken into account as it is crucial to the physical properties of PLA. Kylä-Nikkilä et al. (2000) reported *Lactobacillus helveticus* exclusively produced LA from lactose, but the produced LA was the mixture of both D- and L-LA (Table 13.1). Such observation was also reported in LA fermentation by *Lactobacillus plantarum* NCIMB 8826 (Okano et al., 2009c). On the other hand, several species have been reported to exclusively produce L-LA. One

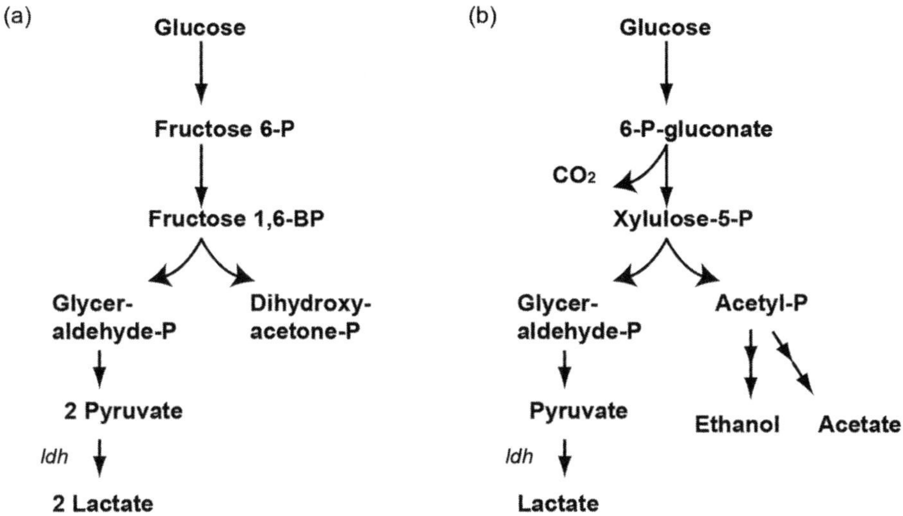

FIGURE 13.3 Schematic illustration of metabolic pathway of glucose in lactic acid bacteria. Homofermentative pathway (a), heterofermentative pathway (b).

of these are *Lactobacillus rhamnosus* CASL (Wang et al., 2010), *Lactococcus lactis* subsp. *lactis* ATCC 19435 (Åkerberg et al., 1998), and *Enterococcus faecalis* (Wee et al., 2004), and they produced L-LA with high optical purity (98%, 98%, and 97.4%, respectively) (Table 13.1). Different to L-LA-producing LAB, there are fewer strains known to produce optically pure D-LA. Calabia and Tokiwa (2007) reported D-LA production from sugarcane molasses by using *Lactobacillus delbrueckii* subsp. *lactis* JCM 1148, and 107 g/L of D-LA was produced from 119 g/L sugars with relatively high optical purity of 94.4%.

Although most investigations of LA production were carried out with LAB, filamentous fungi such as *Rhizopus* have also been utilized for LA production. Fungal fermentation has some advantages in that *Rhizopus oryzae* requires only a simple medium and can produce L-LA from starchy materials with the aid of its own amylolytic enzyme activity, but it also requires vigorous aeration (Yin et al., 1997; Table 13.1).

Detailed information concerning conventional LA fermentation was excellently reviewed by Hofvendahl and Hahn-Hägerdal (2000) and Wee et al. (2006).

13.5 LACTIC ACID PRODUCTION FROM RENEWABLE RESOURCES

There have been many attempts to produce LA from cheap raw materials such as starchy and cellulosic materials, whey, and molasses. Among these, starchy and lignocellulosic materials are currently receiving a great deal of attention, because they are cheap, abundant, and renewable (Wee et al., 2006). One bottleneck for LA production from starchy and lignocellulosic materials is the cost of pretreatment of raw materials.

TABLE 13.1 Various Parameters in LA Fermentation in Conventional LA Production

Strain	Substrate	Initial sugar (g/L)	Sugar consumed (g/L)	LA produced (g/L)	D, L Type	Optical purity (%)[a]	Yield (–)[b]	Productivity (g/L·h)[c]	Reference
Lb. plantarum NCIMB 8826	Glucose	100	96.3	86.0	D/L	12.6	0.89	4.51	Okano et al. (2009c)
Lactobacillus herveticus CNRZ 32	Lactose	81	–	77.0	D/L	15.0	–	–	Kylä-Nikkilä et al. (2000)
Lb. rhamnosus CASL	Glucose	150	–	129.5	L	98–99	0.86	–	Wang et al. (2010)
Lc. lactis subsp. *lactis* ATCC 19435	Whole wheat flour hydrolysate	79	–	70	L	98.0	–	4.0	Åkerberg et al. (1998)
Ent. faecalis RKY 1	Molasses	102	100.7	95.7	L	97.4	0.95	4.0	Wee et al. (2004)
Lb. delbrueckii subsp. *lactis* JCM 1148	Sugarcane molasses	119	117.8	107.0	D	94.3	0.90	–	Calabia and Tokiwa (2007)
R. oryzae NRRL 395	Liquefied corn starch	120	116.0	95.1	L	–	0.82	–	Yin et al. (1997)

[a] |D-LA–L-LA|/|D-LA + L-LA|×100.

[b] Grams of LA produced per gram of total sugar consumed.

[c] Maximum volumetric productivity.

Currently, starchy materials are saccharificated by physicochemical and enzymatic treatment using amylolytic enzymes before LA fermentation because most LAB cannot utilize starchy materials directly (Narita et al., 2004). As pretreatments make the whole process cost-inefficient, direct fermentation of starchy materials would reduce the cost of LA production. Utilization of lignocellulosic materials is more challenging. Although lignocellulose is inedible and represents one of the most abundant and inexpensive biomass sources in the world, its utilization is more difficult than starchy materials. This is because lignocellulose contains cellulose as its main component, which is a persistent polymer (Okano et al., 2009d), and its secondary abundant component, hemicellulose, contains considerable amounts of pentose sugars such as xylose and arabinose, which are unavailable to most microorganisms (Tanaka et al., 2002).

The other bottlenecks for cost-efficient LA production are the separation and purification processes of LA after fermentation. A low pH has inhibitory effects on cell growth and LA production during LA fermentation, and most *Lactobacillus* sp. cannot grow and produce LA below pH 4, although the pKa of LA is 3.78 (Adachi et al., 1998). Therefore, neutralizing agents such as $CaCO_3$, NaOH, or NH_4OH must be added to keep the pH at a constant value. This requires processing for the regeneration of precipitated lactates (Porro et al., 1995). Moreover, the use of complex media such as yeast extract and corn steep liquor hampers not only separation but also purification of LA.

As described above, the ideal LA production can be achieved by direct LA fermentation of starchy or lignocellulosic materials in simple medium and nonneutralizing condition. However, satisfying such requirements is very difficult through the traditional use of LAB. Therefore, improving LAB via gene modification has become an essential and interesting research area. In addition, movements to use other microorganisms such as *Escherichia coli*, *Corynebacterium glutamicum*, and yeast for LA production via gene modification have also arisen (Zhou et al., 2003a; Okino et al., 2005; Ishida et al., 2005).

In this section, we mainly focused on LA production using genetically modified (GM) microorganisms and discussed recent achievements and limits that concern the production of enantiomeric pure LA from renewable resources.

13.5.1 Lactic Acid Bacteria

As described above, LAB has been traditionally used for LA production and is still the predominant candidate for industrial LA production as a raw material for PLA. The salient features of LAB are its high yield and high productivity of LA production (Litchfield, 1996). Interestingly, in some environments, many wild amylolytic LAB (ALAB) have been isolated (Giraud et al., 1994; Guyot et al., 2000; Narita et al., 2004). Using such ALAB, many researchers have attempted direct LA production from starchy materials. As shown in Table 13.2, although direct LA production from starch has been achieved, LA productivity is not practical according to most reports. *Lactobacillus manihotivorans* LMG 18010 can produce L-LA from starch with a high optical purity of 98.0%, but its yield is significantly low at 0.67 gram per gram of consumed sugar) (Guyot et al., 2000), and *Lb. plantarum* A6 can produce LA with a

TABLE 13.2 Various Parameters in LA Fermentation from Oligo or Polymer Substrate Using Wild or Recombinant LAB

Strain	Substrate	Initial sugar (g/L)	Sugar consumed (g/L)	LA produced (g/L)	D, L type	Optical purity (%)[a]	Yield (–)[b]	Productivity (g/L·h)[c]	Reference
Wild type									
Lactobacillus Manihotivorans LMG 18010	Soluble starch	17.5	–	12.6	L	98.0	0.67	0.50	Guyot et al. (2000)
Lb. plantarum A6	Soluble starch	17.1	–	15.2	D/L	14	0.84	2.10	Guyot et al. (2000)
Str. bovis 148	Raw starch	20	16.7	14.7	L	95.6	0.88	–	Narita et al. (2004)
Recombinant									
Lc. lactis IL 1403/pCUSαA	Soluble starch	20	17.4	15.6	L	99.2	0.89	1.57	Okano et al. (2007)
Lb. plantarum Δ*ldhL1*	Glucose	100	96.7	86.6	D	99.7	0.89	4.54	Okano et al. (2009c)
Lb. plantarum Δ*ldhL1*/pCUSαA	Raw starch	100	86.2	73.2	D	99.6	0.85	3.86	Okano et al. (2009c)
Lb. plantarum Δ*ldhL1*/pCU-CelA	Cellohexaose	2	–	1.27	D	99.5	–	–	Okano et al. (2009d)
Lb. plantarum Δ*ldhL1*/pCU-CelA	β-Glucan[d]	2	–	1.47	D	99.7	–	–	Okano et al. (2009d)

[a] |D-LA–L-LA|/|D-LA + L-LA|×100.
[b] Grams of LA produced per gram of total sugar consumed.
[c] Maximum volumetric productivity.
[d] β-Glucosidase from *A. aculeatus* produced by recombinant *A. oryzae* was added.

high yield of 0.84, but it produces a mixture of L- and D-LA that is not suitable for PLA synthesis (Guyot et al., 2000). In one of the few successful studies, *Streptococcus* (*Str.*) *bovis* 148 was found to produce L-LA from raw corn starch with a high yield of 0.88 and a relatively high optical purity of 95.6% (Narita et al., 2004). However, screening for useful ALAB is both time consuming and difficult.

To overcome such problems, recent studies have applied recombinant strategies. Okano et al. (2007) attempted expression of α-amylase from *Str. bovis* 148 (AmyA) (Satoh et al., 1993) that efficiently degraded raw starch with the aid of a C-terminal starch-binding domain (Matsui et al., 2007) in *Lc. lactis* IL 1403 that produces L-LA with high yield and high optical purity (Hofvendahl and Hahn-Hägerdal, 1997). The resulting recombinant *Lc. lactis* strain IL 1403/pCUSαA successfully secreted AmyA and directly produced L-LA from soluble starch with a high yield of 0.89 and an optical purity of 99.2% (Table 13.2). This successful result provided a new methodology to achieve efficient and direct LA production from starch through the combination of an efficient LA producer and known useful amylolytic enzymes.

D-LA as well as L-LA can be produced from starch based on recombinant strategies. Since there has been no report about D-LA-producing ALAB, and the fact that gene manipulation of D-LA-producing LAB is known to be difficult (Serror et al., 2002), the construction of D-LA-producing LAB from gene manipulated D,L-LA-producing strains has been attempted. Okano et al. (2009c) disrupted the L-lactate dehydrogenase (LDH) gene of *Lb. plantarum* NCIMB 8826. Since the resulting strain *Lb. plantarum* Δ*ldhL1* possesses only D-LDH, *Lb. plantarum* Δ*ldhL1* produces exclusively D-LA from glucose and its optical purity is significantly high (99.7%) (Table 13.2). Moreover, after introduction of an AmyA-expressing plasmid into *Lb. plantarum* Δ*ldhL1*, the resulting strain *Lb. plantarum* Δ*ldhL1*/pCUSαA produced 73.2 g/L of D-LA from raw corn starch with a high yield of 0.85 and an optical purity of 99.6% (Table 13.2). Thus, direct enantiomeric pure LA production from starch has become feasible.

Direct fermentation from cellulose is an important challenge. Although simultaneous saccharification and fermentation of cellulose has been attempted (Yáñez et al., 2003; Yoon, 1997), there has been no report demonstrating direct LA fermentation from not only cellulose but also from cellooligosaccharides consisting of more than four glucose units (Adsul et al., 2007). By introducing an expression plasmid for endoglucanase (EG) from *Clostridium thermocellum* (CelA) in *Lb. plantarum* Δ*ldhL1*, direct D-LA production from cellopentaose and cellohexaose was achieved (Okano et al., 2009d). The resulting strain *Lb. plantarum* Δ*ldhL1*/pCU-CelA produced 1.27 g/L of D-LA from a medium-containing 2 g/L of cellopentaose and cellohexaose (Table 13.2). Moreover, by addition of β-glucosidase (BGL) from *Aspergillus aculeatus* (BGL1) produced by recombinant *Aspergillus oryzae*, *Lb. plantarum* Δ*ldhL1*/pCU-CelA produced 1.47 g/L of D-LA from β-glucan (Table 13.2).

For utilization of hemicellulose, utilization of pentose sugars such as xylose and arabinose is a major problem. Some LAB such as *Lactobacillus pentosus* (Bustos et al., 2005), *Lactobacillus brevis* (Chaillou et al., 1998), *Lb. plantarum* (Helanto et al., 2007), and *Leuconostoc* (*Leuc.*) *lactis* (Ohara et al., 2006) are known to ferment either or both arabinose and xylose. The metabolic pathway of pentose is

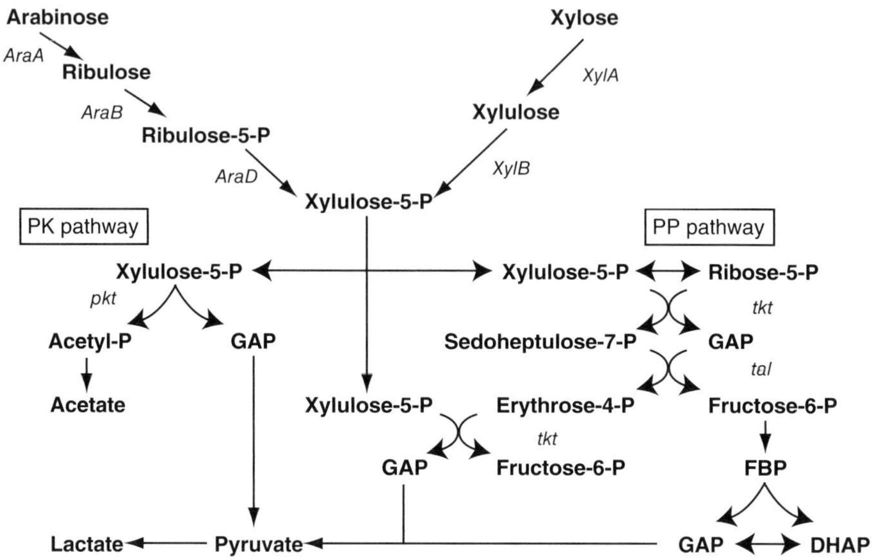

FIGURE 13.4 Schematic illustration of metabolic pathway of pentose in lactic acid bacteria.

shown in Figure 13.4. In these microorganisms, arabinose and xylose are converted to xylulose-5-phosphate (X5P) through several enzymatic reactions (Helanto et al., 2007; Tanaka et al., 2002). This common intermediate, X5P, is converted to equimolar amounts of LA and acetic acid (AA) (3:2 in gram order) via the phosphoketolase pathway (PK pathway) (Tanaka et al., 2002). This low yield of LA is an undesirable feature. Recently, Tanaka et al. (2002) have shown that *Lc. lactis* IO-1 has another pathway for X5P assimilation, the pentose phosphate pathway (PP pathway), in addition to the PK pathway. Since 5 mole of LA is produced from 3 mole of X5P through this pathway, *Lc. lactis* IO-1 is capable of producing L-LA at high yield of 0.68 over 0.60 that is the theoretical yield via the PK pathway (Table 13.3). Abdel-Rahman et al. (2011) found that *Enterococcus mundtii* QU 25 also had both pathways and preferentially utilized the PP pathway compared to the PK pathway. Using this strain, efficient L-LA production was achieved with a high yield of 0.90, and an optical purity of >99.9%. These interesting findings indicated the possibility of homogenous LA fermentation from pentose sugars by metabolic engineering.

Shinkawa et al. (2011) carried out addition of xylose assimilation ability to *Lc. lactis* IL 1403 by introducing a plasmid carrying the *xylRAB* gene from *Lc. lactis* IO-1 (pXylRAB) and the resulting strain was shown to utilize xylose via both the PK and PP pathways. Using this strain, redirection of the PK pathway to the PP pathway was examined. The endogenous phosphoketolase gene (*ptk*) that encodes phosphoketolase was replaced with an endogenous transketolase gene (*tkt*). In the resulting strain *Lc. lactis* IL 1403 *ptk*::*tkt*/pXylRAB, AA production was almost abolished, and this strain produced 50.1 g/L of L-LA from 60 g/L of xylose with a high yield of 0.95, and the optical purity of L-LA was 99.6% while the produced AA was only 0.3 g/L (Table 13.3).

TABLE 13.3 Various Parameters in LA Fermentation from Pentose Sugars Using Wild or Recombinant LAB

Strain	Substrate	Initial sugar (g/L)	LA produced (g/L)	AA produced (g/L)	D, L type	Optical purity (%)[a]	Yield (–)[b]	Productivity (g/L·h)[c]	Reference
Wild type									
Lc. lactis IO-1	Xylose	70.3	33.3	11.7	L	95.6	0.68	–	Tanaka et al. (2002)
Ent. mundtii QU 25	Xylose	50.1	44.1	–	L	>99.9	0.90	3.84	Abdel-Rahman et al. (2011)
Recombinant									
Lc. lactis IL 1403 *ptk::tkt* /pXylRAB	Xylose	60	50.1	0.3	L	99.6	0.95		Shinkawa et al. (2011)
Lb. plantarum Δ*ldhL1-xpk1::tkt*	Arabinose	50	38.6	0.4	D	99.9	0.82	3.78	Okano et al. (2009a)
Lb. plantarum Δ*ldhL1-xpk1::tkt*-Δ*xpk2*/pCU-PXylAB	Xylose	50	41.2	1.0	D	99.2	0.89	1.60	Okano et al. (2009b)
Lb. plantarum ldhL1::PxylAB-xpk1::tkt-xpk2::PxylAB	Xylose, arabinose, glucose	80	61.2	0.1	D	99.5	0.80	4.90	Yoshida et al. (2011)

[a] |D-LA–L-LA|/|D-LA + L-LA|×100.
[b] Grams of LA produced per gram of total sugar consumed.
[c] Maximum volumetric productivity.

Such strategy was also applied to D-LA production from pentose sugars. In *Lb. plantarum* Δ*ldhL1*, which originally assimilate arabinose via the PK pathway and exclusively produce D-LA (Okano et al., 2009c), redirection of the PK pathway to the PP pathway was examined. The endogenous phosphoketolase 1 gene (*xpk1*) that encodes phosphoketolase was replaced with a heterologous *tkt* from *Lc. lactis* IL 1403 (Okano et al., 2009a). In the resulting strain *Lb. plantarum* Δ*ldhL1-xpk1*::*tkt*, AA production was almost abolished, and this strain produced 38.6 g/L of D-LA from 50 g/L of arabinose with a high yield of 0.82, and the optical purity of D-LA was 99.9%, while the produced AA was only 0.4 g/L (Table 13.3). Homogenous D-LA production from xylose was also achieved by disruption of the phosphoketolase 2 gene (*xpk2*) that is presumed to be expressed in the presence of xylose and with the introduction of a plasmid for the expression of a xylose isomerase gene (*xylA*) and xylulose kinase gene (*xylB*) from *Lb. pentosus* NRIC 1069 in *Lb. plantarum* Δ*ldhL1-xpk1*::*tkt* (Okano et al., 2009a). Using the resultant strain *Lb. plantarum* Δ*ldhL1-xpk1*::*tkt*-Δ*xpk2*/pCU-PXylAB, 41.2 g/L of D-LA was produced from xylose with a high yield of 0.89 and an optical purity of 99.2%, while the produced AA was only 1.0 g/L (Table 13.3).

In an effort to test industrial applications, Yoshida et al. (2011) constructed *Lb. plantarum* Δ*ldhL1-xpk1*::*tkt*-Δ*xpk2* derivative having two copies of the *xylA* and *xylB* expression cassette on the disrupted *ldhL1* and *xpk2* loci. Since this *ldhL1*::*PxylAB-xpk1*::*tkt-xpk2*::*PxylAB* strain possesses *xylA* and *xylB* in the chromosome, it does not require cost-inefficient selective culture with antibiotics to retain the plasmid. This strain showed almost the same level of xylose consumption ability and D-LA productivity as that of Δ*ldhL1-xpk1*::*tkt*-Δ*xpk2*/pCU-PXylAB strain. Interestingly, *ldhL1*::*PxylAB-xpk1*::*tkt-xpk2*::*PxylAB* strain could simultaneously utilize xylose (25 g/L), arabinose (5 g/L), and glucose (50 g/L), and 61.2 g/L of D-LA was produced with a high yield of 0.80 and an optical purity of 99.5% (Table 13.3). As industrial effluents such as lignocellulose hydrolysates are composed of a mixture of these sugars, co-utilization of several sugars is favorable characteristic for economically feasible processes.

As described above, direct production of optically pure LA from a huge variety of mono-, oligo-, or polysaccharides may therefore be possible with a high yield. Thus, cost-efficient LA production is feasible from the view of LA fermentation. However, LAB requires complex nutrients due to their limited ability to synthesize B vitamins and amino acids (Hofvendahl and Hahn-Hägerdal, 2000), and this results in cost-inefficient LA purification. From both a view of metabolic engineering and purification, solutions to this problem should be studied.

13.5.2 *Escherichia coli*

Compared to LAB, *E. coli* does not require a complex medium and can grow in a simple mineral salt medium. In addition, *E. coli* can naturally utilize both hexose and pentose sugars (Zhou et al., 2003a). The main problem for LA production in *E. coli* is improving LA yield, since *E. coli* produces a mixture of organic acids (D-LA, AA, succinic acid (SA), and formic acid (FA)) and ethanol to accommodate the reducing equivalents generated during glycolysis (Figure 13.5) (Zhou et al., 2003a).

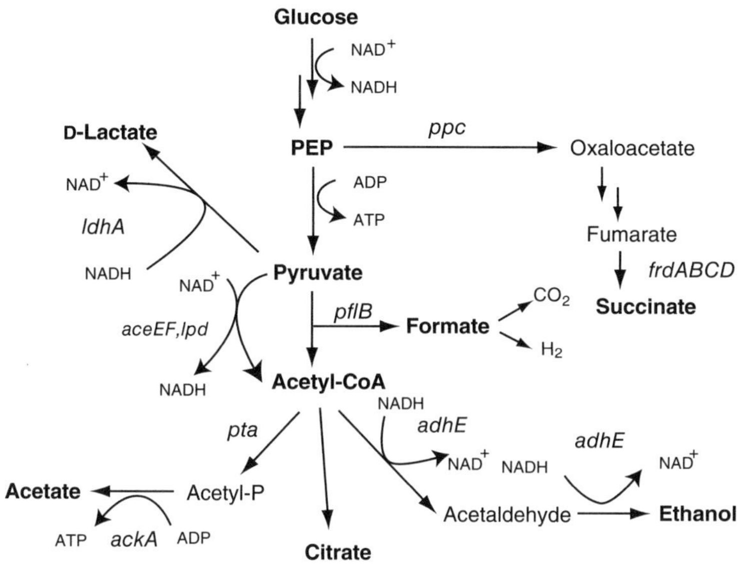

FIGURE 13.5 Synthesis of different fermentation products in native *E. coli*.

Successful D-LA production using recombinant *E. coli* was first reported by Chang et al. (1999). A mutation only in the phosphotransacetylase gene (*pta*) significantly improved LA yield. Mutant strain JP201 produced approximately sixfold more D-LA than parental strain RR1. Interestingly, not only production of AA but also production of FA and ethanol was almost blocked. This phenomenon is considered to be due to reduced levels of pyruvic acid (PA)–FA lyase (PFL) in the *pta* mutant (Chang et al., 1999). As a result, production of AA, FA, and ethanol was suppressed, and accumulating PA provoked activation of LDH (Tarmy and Kaplan, 1968). In fed-batch fermentation from glucose, this strain produced 60 g/L of D-LA (Table 13.4) with a yield of 0.8 while SA was simultaneously produced (~9 g/L). To prevent accumulation of SA, a mutation in the gene for phosphoenolpyruvate carboxylase (*ppc*), the branch point leading to SA synthesis, was introduced into JP201. Using the resultant *pta ppc* double mutant, JP203, D-LA production was carried out. As expected, SA production was almost inhibited, and JP203 produced 62.2 g/L of D-LA with a high yield of 0.9 (Table 13.4). Although successful D-LA production was achieved, the *ppc* mutant has an auxotrophic requirement for tricarboxylic acid pathway intermediates or amino acids. This is one of the drawbacks when using *E. coli* as a host for LA production.

On the other hand, Zhou et al. (2003a) suggested another strategy for homogenous LA fermentation that has no auxotrophic requirement. They succeeded in inhibiting FA, AA, and ethanol production by a mutation in the PFL gene (*pflB*) of *E. coli* W3110. In addition, SA production was inhibited by a mutation in fumarate reductase (*frdBC*), not *ppc*, while the resultant strain SZ40 remained prototrophic. Although SZ40 produced 51.8 g/L of D-LA with an extremely high yield of 0.99 and an optical

TABLE 13.4 Various Parameters in LA Fermentation Using Recombinant *E. coli*

Strain	Substrate	LA produced (g/L)	Fermentation time (h)	D, L type	Optical purity (%)[a]	Yield (–)[b]	Reference
E. coli							
JP201 (*pta*)	Glucose	60.0	60[c]	D	–	0.80	Chang et al. (1999)
JP203 (*pta pcc*)	Glucose	62.2	60[c]	D	–	0.90	Chang et al. (1999)
SZ40 (*pflB frdBC*)	Glucose	51.8	192	D	>99	0.99	Zhou et al. (2003a)
SZ62 (*pflB frdBC adhE ackA*)	Glucose	48.6	168	D	–	0.98	Zhou et al. (2003a)
SZ79 (*pflB frdBC adhE ackA ldhA::ldhL*)	Glucose	43.1	408	L	–	0.91	Zhou et al. (2003b)
SZ85 (mutant of SZ79 with increased Expression of *ldhL*)	Glucose	45.5	120	L	99.5	0.95	Zhou et al. (2003b)
SZ85 (mutant of SZ79 with increased Expression of *ldhL*)	Xylose	40.0	312	L	–	0.93	Zhou et al. (2003b)

[a]|D-LA–L-LA|/|D-LA + L-LA|×100.

[b]Grams of LA produced per gram of total sugar consumed.

purity of >99% in M9 mineral salts medium, measurable amounts of AA and ethanol were also produced (0.26 and 0.32 g/L, respectively) (Table 13.4). In addition, the growth of SZ40 was significantly hampered by the mutation, and fermentation time reached 192 hours. To eliminate the production of AA and ethanol, a further mutation was introduced into SZ40. Mutations in the alcohol dehydrogenase gene (*adhE*) and acetate kinase (*ackA*) led to inhibition of ethanol and AA production, and the resultant strain SZ63 produced 48.6 g/L of D-LA with a high yield of 0.98 (Table 13.4). Surprisingly, SZ63 had partially improved growth inhibition, and fermentation time was reduced to 168 hours (Table 13.4). This improvement was assumed to be due to increasing pools of acetyl phosphate and acetyl coenzyme A. Thus a prototrophic, homogenous, fermentative *E. coli* was constructed. Another favorable feature is that SZ63 contained no antibiotic resistance genes or plasmids.

L-LA production was also achieved using derivatives of SZ63 (Zhou et al., 2003b). A part of the D-LDH gene of *E. coli* (*ldhA*) was replaced with the L-LDH gene of *Pediococcus acidilactici* (*ldhL*). Although the resultant strain SZ79 produced 43.1 g/L of L-LA with a high yield of 0.91, SZ79 showed poor growth and fermentation time reached 408 hours (Table 13.4). This is considered to be due to weak expression of the *ldhL* gene. Mutants of SZ79 that showed improved growth were readily isolated and one mutant, SZ85, exhibited a 30-fold increase in L-LDH activity in comparison to SZ79. Several mutations in the upstream, coding, and terminator regions of *ldhL* were confirmed in SZ85. Using SZ85, more rapid L-LA fermentation than that of SZ79 was achieved. SZ85 produced 45.5 g/L of L-LA with a high yield of 0.95 and an optical purity of 99.5% in 120 hours of fermentation in M9 mineral salts medium (Table 13.4). Moreover, L-LA fermentation from xylose in M9 mineral salts medium was also achieved, and 40.0 g/L of L-LA was produced with yield of 0.93 in 312 hours of fermentation (Table 13.4).

In addition to metabolic engineering of LA-producing pathway, PLA synthesis in microbial cell has been attempted in *E. coli*. PLA is a representative bio-based polyester, which is chemically synthesized by ring-opening polymerization of a cyclic diester (lactide) of LA, produced by microbial fermentation. By introducing variations in molecular weight and crystallinity, PLA is turned into highly valuable materials for biomedical, food, and general purpose applications, as described in numerous patents. Thus, PLA combines inexpensive large-scale fermentation with chemical processing capacity to produce a value-added polymer product. The paradigm shift from the chemo-process to the bio-process for PLA production is thus preferable to overcome this problem.

The first biosynthetic system for LA-based polyesters was created in recombinant *E. coli* (Taguchi et al., 2008). A coenzyme A (CoA) transferase gene and the engineered polyhydroxyalkanoate synthase gene were introduced into *E. coli*, and the one-step biosynthesis of the LA-incorporated copolyester, P(6 mol% LA-co-94 mol% 3HB), with a number-average molecular weight of 1.9×10^5, was successfully demonstrated. To date, the *E. coli* platform has been used to produce various LA-based polymers incorporating 3-hydroxybutyrate (3HB), 3-hydroxyvalerate, 3-hydroxyhexanoate, and 2-hydroxy acids(Shozui et al., 2009; Yamada et al., 2009; Han et al., 2011; Matsumoto et al., 2011).

Taken together, *E. coli* can produce both D-LA and L-LA with high optical purity and extremely high yields even in mineral salts medium. In addition, the assimilation capacity of pentose sugars is an attractive feature. However, the low productivity of LA and low acidic tolerance (normally cultivated around pH 7.0) should be improved.

13.5.3 *Corynebacterium glutamicum*

Corynebacterium glutamicum is an aerobic Gram-positive bacterium that has been widely used for the industrial production of amino acids such as L-glutamate and L-lysine (Hermann, 2003; Leuchtenberger et al., 2005). Under oxygen deprivation, cell growth of this bacterium is arrested, while it retains the capability to produce mix-organic acids such as L-LA, SA, and AA from glucose in mineral salts medium. Using this phenomenon, Okino et al. (2005) suggested a novel system for organic acids production containing LA that based on the use of a reactor filled with a high-density cells derived from an aerobic culture, leading to a bioprocess with high volumetric productivity. Using the *C. glutamicum* R strain (Yukawa et al., 2007), L-LA production was achieved with high volumetric productivity of 42.9 g/L·h at a cell concentration of 60 g-dry-cell/L although significant SA was simultaneously produced (11.7 g/L·h) (Okino et al., 2005).

D-LA was also produced using the same system. By the expression of the D-LDH encoding gene from *Lb. delbrueckii* in the *C. glutamicum* Δ*ldhA* strain, which is an internal L-LDH null mutant, D-LA production from glucose was achieved (Okino et al., 2008). In fed-batch fermentation, the *C. glutamicum* mutant produced 120 g/L of D-LA with a high optical purity of >99.9% at a cell concentration of 60 g-dry-cell/L at 30 hours.

Although LA production from hexose sugars such as glucose and sucrose is possible, utilization of pentose sugars such as xylose and arabinose is not possible. To enable *C. glutamicum* to utilize xylose, Kawaguchi et al. (2006) examined the expression of *xylA* from *E. coli* that encodes xylose isomerase and *xylB* from *E. coli* that encodes xylulokinase in the *C. glutamicum* R strain using a multicopy plasmid under the control of the constitutive promoter *trc*. Only the expression of *xylA* (CRX1 strain) enables xylose assimilation, as *C. glutamicum* R has a putative endogenous xylulokinase gene (Kawaguchi et al., 2006). The fact that the xylulokinase gene-null mutant, CRX3, hardly grows in xylose-containing medium supports the existence of xylulokinase in *C. glutamicum* R (Kawaguchi et al., 2006). Such a xylulokinase gene is also observed in the genome of *C. glutamicum* ATCC 13032, although its function has not been investigated (Kalinowski et al., 2003). Both the expression of *xylA* and *xylB* (CRX2) further improved the growth rate (Kawaguchi et al., 2006). Using CRX2, L-LA production from xylose was achieved (29 mmol/L·h) with a yield of 0.53 accompanied with SA production (productivity of 17 mmol/L·h and yield of 0.25). Moreover, L-LA production from mixed sugars of glucose and xylose was also achieved, while repression of xylose metabolism by glucose was found.

Arabinose utilization has also been examined. The expression of *E. coli* genes *araA*, *araB*, and *araD* encoding arabinose isomerase, ribulokinase, ribulose-5-phosphate 4-epimerase, respectively, was carried out in the *C. glutamicum* R strain (Kawaguchi et al., 2008). In the resultant strain CRA1, arabinose was successfully consumed (3.4 mmol/h·g-dry-cell) and L-LA was produced while SA and AA were simultaneously produced. Moreover, L-LA production from mixed sugars of glucose and arabinose was also achieved, although the arabinose consumption rate was significantly lower (0.06 g/h·g-dry-cell) than that of glucose (0.76 g/h·g-dry-cell).

In real lignocellulose hydrolysate, oligosaccharides such as cellobiose are also found in addition to monosaccharides of hexose and pentose sugars (Katahira et al., 2006), and they should be simultaneously utilized. Sasaki et al. (2008) constructed a recombinant *C. glutamicum* strain that simultaneously consumed glucose, xylose, and cellobiose. Chromosomal integration of a *xylA–xylB* gene cluster from *E. coli* under the control of the *trc* promoter was first carried out in the *C. glutamicum* R strain. The X5 strain that possesses five copies of the *xylA–xylB* genes in nonessential regions for cell growth in the genome showed rapid xylose consumption (41.2 mmol/L·h) compared to CRX2 (19.3 mmol/L·h). In order to apply cellobiose utilization to strain X5, $bglF^{317A}$–*bglA* genes, which encode the phosphoenolpyruvate phosphotransferase system BGL-specific enzyme IIBCA component and phospho-β-glucosidase, respectively, controlled by the constitutive *tac* promoter, were integrated into the chromosome of strain X5. The $bglF^{317A}$–*bglA* genes were obtained from the *C. glutamicum* R-CEL strain which is a cellobiose-adaptive mutant (Kotrba et al., 2001), and a single mutation of V317A of *bglF* enabled cellobiose assimilation (Kotrba et al., 2003). The resultant strain X5C1 strain could simultaneously consume glucose, xylose, and cellobiose and produced LA, SA, and AA (Sasaki et al., 2008). Surprisingly, xylose was consumed at a constant rate different from that of fermentation using the CRX2 strain (Kawaguchi et al., 2006).

In considering practical applications of LA-based polyester, especially for food-grade and biomedical demands, the use of endotoxin-free Gram-positive bacteria is preferable. Endotoxin-free production system for LA-based polyester using *C. glutamicum* was developed through enzymatic reactions including (1) generation of D-LA-CoA by D-LDH and propionyl-CoA transferase, (2) 3HB-CoA generation catalyzed by β-ketothiolase and NADPH-dependent acetoacetyl-CoA reductase, and (3) copolymerization of LA-CoA and 3HBCoA catalyzed by LA-polymerizing enzyme (Song et al., 2012). For the synthesis of P(LA-co-3HB), these three steps were needed to be functional.

Thus, production of optically pure LA from a variety of sugars has become possible using mineral salts-based media with high volumetric productivity and extremely high optical purity. In addition, utilization of polysaccharides may be feasible. Direct lysine production from starch using GM *C. glutamicum*, which secretes amylase or displays it on the cell surface, has already been reported (Tateno et al., 2007a,b). By applying such technology to this GM *C. glutamicum* system, direct LA production from polysaccharides may be possible. Similar to *E. coli*, the acid tolerance of *C. glutamicum* is extremely low as its LA fermentation operates around a pH of 7.0. The low yield of LA accompanied with SA and AA production also should be improved.

13.5.4 Yeasts

Compared to bacterial species, yeasts such as *Saccharomyces* (*Sac.*) *cerevisiae*, which produce ethanol in anaerobic cultivation, are robust and more tolerant to low pH (Skory, 2003). This may enable construction of a nonneutralizing fermentation process and eliminate the regeneration of precipitated lactates. In ethanol fermentation, PA is converted into ethanol by two enzymatic reactions by pyruvate decarboxylase (PDC) and alcohol dehydrogenase (ADH). Recent research has shown that a transgenic yeast expressing heterologous L-LDH can produce LA from PA. Figure 13.6 shows the LA production strategy using this yeast. Such metabolically engineered yeasts were first reported by Dequin and Barre (1994) and Porro et al. (1995). They examined the expression of the *Lactobacillus casei* or bovine LDH genes using a multicopy plasmid and recombinant *Sac. cerevisae* strains and succeeded in producing 10–20 g/L of LA (Table 13.5), although a significant amount of ethanol was simultaneously produced. A similar result was obtained using wine yeast as a host (Dequin et al., 1999; Table 13.5). In such transgenic yeasts, improvement of LA yield by inhibiting ethanol production is an important goal, and several metabolic engineering approaches have been examined.

There are three structural *PDC* genes, *PDC1*, *PDC5*, and *PDC6*, in the *Sac. cerevisiae* genome (Hohmann, 1991), and several mutants lacking single- or multi-*PDC*s have been reported. Adachi et al. (1998) constructed a *pdc1* mutant and expressed the bovine LDH gene under the *ADH1* promoter. However, no major effect was observed as the ethanol yield only slightly decreased (from 0.35 to 0.26) and the LA yield slightly increased (from 0.16 to 0.20) (Table 13.5). This is due to incomplete inactivation of PDC activity as the *PDC1* deletion leads to a great increase in *PDC5* promoter-driven mRNA expression, while *PDC5* seems to be not, or only poorly, expressed in wild-type cells (Hohmann and Cederberg, 1990). In *PDC1* mutants, approaches that improve LDH activity have been examined. Ishida et al. (2005) used the *Sac. cerevisiae* OC-2T strain (Saitoh et al., 1996) as a host, which is a diploid and homothallic yeast, and two copies of the bovine LDH gene were integrated into the *PDC1* locus. As a result, *LDH* was expressed under the control of the native *PDC1* promoter, and *PDC1* was completely inactive. The resulting strain YIBO-7A showed approximately fivefold higher LDH activity than that of the OC-2T strain expressing LDH using a multicopy plasmid and produced 50.6 g/L of LA with a high yield of 0.65 (Table 13.5). LA production was further improved by increasing the copy number

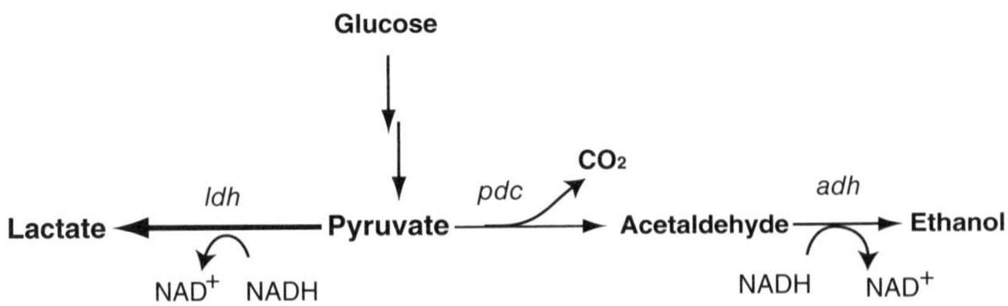

FIGURE 13.6 Schematic illustration of lactic acid production strategy in yeast.

TABLE 13.5 Various Parameters in LA Fermentation from Glucose Using Recombinant Yeast

Strain	Vector	Promoter	Source of LDH	pH control	LA produced (g/L)	Yield (–)[a]	Reference
Batch							
Sac. cerevisiae	YEp	*ADH1*	*L. casei*	–	12.0	0.24	Dequin and Barre (1994)
Sac. cerevisiae (wine yeast)	YEp	*ADH1*	*L. casei*	–	8.6	0.04	Dequin et al. (1999)
Sac. cerevisiae	YEp	*ADH1*	Bovine	+	7.9	0.16	Adachi et al. (1998)
Sac. cerevisiae (*pdc1*)	YEp	*ADH1*	Bovine	+	–	0.20	Adachi et al. (1998)
Sac. cerevisiae (wine yeast, *pdc1*)	YEp	*PDC1*	Bovine (2 copies)	–	16.1	0.20	Ishida et al. (2005)
Sac. cerevisiae (wine yeast, *pdc1*)	YIp	*PDC1*	Bovine (2 copies)	–	50.6	0.65	Ishida et al. (2005)
Sac. cerevisiae (wine yeast, *pdc1*)	YIp	*PDC1*	Bovine (6 copies)	+	68.0	–	Saitoh et al. (2005)
Sac. cerevisiae (wine yeast, *pdc1*)	YIp	*PDC1*	*Leuc. mesenteroides* (2 copies)	+	61.5	0.61	Ishida et al. (2006b)
Sac. cerevisiae (wine yeast, *pdc1 pdc5*)	YIp	*PDC1*	Bovine (2 copies)	+	82.3	0.82	Ishida et al. (2006a)
Sac. cerevisiae (wine yeast, *pdc1 adh1*)	YIp	*PDC1 ADH1*	Bovine (2 copies) Bovine (2 copies)	+	71.8	0.74	Tokuhiro et al. (2009)
Sac. cerevisiae (*pdc1*)	YIp	*PDC1*	Bovine	–	38.0	–	Suzuki et al. (2013)
Sac. cerevisiae (*pdc1 dse2 scw11 eaf3 sed1*)	YIp	*PDC1*	Bovine	–	48.0	–	Suzuki et al. (2013)
Fed-batch							
Sac. cerevisiae	YEp	*ADH1*	Bovine	–	20.0	–	Porro et al. (1995)
K. lactis (*klpdc1 klpda1*)	YRp	*KlPDC1*	Bovine	+	60.0	0.85	Bianchi et al. (2001)

[a]Grams of LA produced per gram of total sugar consumed.

of *LDH*. Saitoh et al. (2005) integrated four more copies of *LDH* into the genome of YIBO-7A, and the resulting strain possessing six copies of *LDH* produced 68.0 g/L of LA from 100 g/L of glucose (Table 13.5), which is 1.28 times higher than that produced by YIBO-7A. Moreover, when LA fermentation was carried out in a sugar juice-based medium containing 200 g/L of glucose, 122 g/L of L-LA was produced with an extremely high optical purity of >99.9%. D-LA can also be produced using the same strategy. D-LDH of *Leuc. mesenteroides* was integrated instead of bovine LDH into the OC-2T strain (Ishida et al., 2006b). The resultant strain produced 61.5 g/L of D-LA with a yield of 0.61 and an extremely high optical purity of 99.9%.

Double inactivation of *pdc1* and *pdc5* (Hohmann and Cederberg, 1990; Ishida et al., 2006a) or triple inactivation of *pdc1*, *pdc5*, and *pdc6* (Hohmann, 1991) has also been reported. These resultant mutants had strongly impaired growth on glucose medium. In fact, the *pdc1 pdc5* double mutant of *Sac. cerevisiae* with two copies of the bovine LDH gene in the *PDC1* locus produced 82.3 g/L of LA with a high yield of 0.82, and ethanol production was repressed to only 2.8 g/L (Ishida et al., 2006a; Table 13.5). However, it took a long incubation time of 192 hours. Thus, double inactivation of *pdc1* and *pdc5* led to serious growth inability. To achieve both a high yield of LA and productivity, the use of a *pdc1* and *adh1* mutant has been suggested. Among the five ADHs, the *ADH1* gene product is the major enzyme responsible for the conversion of acetaldehyde to ethanol (Leskovac et al., 2002). While a single *adh1* mutant expressing the *R. oryzae* LDH gene has been reported (Skory, 2003), the *adh1* mutation led to poor growth and a decrease in LA yield (from 0.44 to approximately 0.20). This was attributed to the accumulation of acetaldehyde to toxic level as a result of the *ADH1* disruption. Tokuhiro et al. (2009) hypothesized that by decreasing ADH activity in the *pdc1* mutant, acetaldehyde accumulation would be lowered because the metabolic flux from pyruvate to acetaldehyde is reduced by the *PDC1* disruption; therefore a *pdc1 adh1* double mutant that possesses four bovine *LDH* genes was constructed. The resultant strain produced 71.8 g/L of LA with a high yield of 0.74 in 63 hours (Table 13.5), which is quite rapid compared to fermentation using the *pdc1 pdc5* double mutant (192 hours), although the yield of LA was slightly lower than that of the *pdc1 pdc5* double mutant (Table 13.5).

As an effort to improve tolerant to low pH, a genome-wide screening for strains showing hyper-resistance to 6% LA (pH 2.6) was performed using the gene deletion collection of *Sac. cerevisiae*. Suzuki et al. (2013) identified 94 genes whose disruption led to resistance to 6% LA in rich medium from 4828 genes. Among them, *DSE2*, *SCW11*, *EAF3*, and *SED1* disruption gave strong resistance to 6% LA compared with the parental strain BY4742. They further found that multiple combinations of *DSE2*, *SCW11*, *EAF3*, and *SED1* disruption led to enhanced resistance to LA. Combined with *PDC1* disruption by integrating the bovine LDH gene, LA fermentation under nonneutralizing condition was carried out. While *LDH*-harboring wild-type strain (SH6764) produced 38.0 g/L of L-LA, *LDH*-harboring *dse2 scw11 eaf3 sed1* quadruple mutant (SH6779) produced 48.0 g/L of L-LA after 72 hours of fermentation (Table 13.5).

The other approach for LA production involves the use of a Crabtree-negative yeast such as *Kluyveromyces lactis*. Bianchi et al. (2001) used *K. lactis* strains lacking the

KlPDC1 gene, which is a single *PDC* gene expressing PDC activity, and transformed them with the plasmid having the bovine *LDH* gene under the control of the promoter of *KlPDC1*. Moreover, the pyruvate dehydrogenase E1α subunit gene (*KlPDA1*) was deleted. Using the resultant strain, 60.0 g/L of L-LA was produced with a high yield of 0.85 in fed-batch fermentation (Table 13.5). However, similar to the *Sac. cerevisiae pdc1 pdc5* double mutant, it required a long fermentation time (500 hours).

In addition to LA fermentation from glucose using GM yeasts, the use of a variety of sugars is an important goal. Cellobiose is one of the main components of oligosaccharides obtained from cellulose degradation and a potent inhibitor of CBHs (Tokuhiro et al., 2008), which are key enzymes for degradation of crystalline cellulose. However, the yeast *Sac. cerevisiae* cannot assimilate cellobiose. Thus, rapid assimilation of cellobiose is of primary importance for developing a cellulose-utilizing yeast. Tokuhiro et al. (2008) constructed *Sac. cerevisiae* that possesses eight bovine LDH genes in its genome. Also, this yeast expresses BGL from *A. aculeatus* (BGL1) (Kawaguchi et al., 1996) fused with N-terminal anchor domain of alpha-agglutinin in the genome, and BGL1 is stably expressed and displayed on the cell surface. The recombinant successfully produced LA from 95 g/L of cellobiose with a yield of 0.70. Moreover, productivity of LA from cellobiose (2.8 g/L·h) is comparable to that from glucose (3.0 g/L·h).

In yeast LA fermentation, the improvement of LA yield that is not accompanied by growth inability and increased LA production rate is an essential goal. Further improvement is expected.

13.6 ECONOMIC OUTLOOK

PLA was first synthesized over 150 years ago. However, no immediate application was found due to economic and technological reasons. Its usefulness in medical applications became apparent in 1960s. Efforts to develop PLA as a commodity plastic were first made in the late 1980s and early 1990s by DuPont, Coors Brewing, and Cargill (Shen et al., 2009). Currently, bioplastics globally make up nearly 300,000 metric tons of the plastics market, and NatureWorks LLC, which is the leading company in PLA technology and markets, has the 150,000 metric tons per year of PLA plant. Even though it sounds a lot, it only accounts for less than 1% of the 181 million metric tons of synthetic plastics the world produces each year. Nevertheless, the bioplastic market is growing by 20–30% each year, but not sufficient to meet everyone's need (Nampoothiri et al., 2010). This low production volume is due for the most part to high manufacturing costs. On an industrial scale, the manufacturing cost of the LA monomer is targeted to be less than US$0.8/kg, as the selling price of PLA must decrease by roughly half of its present price of US$2.2/kg to compete with fossil-fuel-based plastics (Wee et al., 2006). Thus, the majority of the cost of manufacturing PLA is occupied by LA monomer production costs. The techniques described above for direct production of optically pure LA from biomass will become important techniques to cut the cost for LA monomer production.

13.7 FUTURE PROSPECTS

In this chapter, the development of GM microorganisms for enantiomeric pure LA production from renewable resources was highlighted. As described in LA fermentation using LAB, direct production of optically pure LA from starch is on the verge of possible commercialization. Moreover, recent achievement demonstrated a strong possibility of direct LA production from β-glucan and other cellulosic compounds by co-expression of EG and other cellulolytic enzymes such as BGL and cellobiohydrolase (Okano et al., 2009d). Steady progress has also been achieved in homogenous LA production from pentose sugars in many microorganisms such as LAB, *E. coli*, and *C. glutamicum*. There are a few LAB, such as *Leuc. lactis* SHO-47 and SHO-54, that can assimilate xylooligosaccharides such as xylobiose to xylohexaose (Ohara et al., 2006). By introducing such characteristics into a homogenous LA producer for pentose, homogenous LA production from larger components of hemicellulose could be possible.

To further consolidate the processes for LA-based polymer production, introduction of the biosynthetic system for LA-based polyesters observed in *E. coli* and *C. glutamicum* will be a breakthrough. This may bring in not only simplification of LA-based polymer production process but also elimination of need for neutralizing agent by inhibiting accumulation of LA in fermentation medium.

PLA and LA-based polymers are undoubtedly at the forefront of bio-based polymer innovation, and full-scale production of PLA is expected to contribute to the construction of a bio-refinery society. Integration of knowledge described in this chapter is expected.

NOMENCLATURE

AA	acetic acid
ADH	alcohol dehydrogenase
ALAB	amylolytic lactic acid bacteria
BGL	β-glucosidase
CoA	coenzyme A
EG	endoglucanase
FA	formic acid
3HB	3-hydroxybutyrate
LA	lactic acid
LAB	lactic acid bacteria
LDH	lactate dehydrogenase
PA	pyruvic acid
PDC	pyruvate decarboxylase
PFL	pyruvate–formate lyase
PK pathway	phosphoketolase pathway
PP pathway	pentose phosphate pathway
PLA	poly lactic acid

SA succinic acid
X5P xylulose-5-phosphate

REFERENCES

Abdel-Rahman MA, Tashiro Y, Zendo T, Hanada K, Shibata K, Sonomoto K. (2011). Efficient homofermentative L-(+)-lactic acid production from xylose by a novel lactic acid bacterium, *Enterococcus mundtii* QU 25. *Appl Environ Microbiol*, 77, 1892–1895.

Adachi E, Torigoe M, Sugiyama M, Nikawa J, Shimizu K. (1998). Modification of metabolic pathways of *Saccharomyces cerevisiae* by the expression of lactate dehydrogenase and deletion of pyruvate decarboxylase genes for the lactic acid fermentation at low pH value. *J Ferment Bioeng*, 86, 284–289.

Adsul M, Khire J, Bastawde K, Gokhale D. (2007). Production of lactic acid from cellobiose and cellotriose by *Lactobacillus delbrueckii* mutant Uc-3. *Appl Environ Microbiol*, 73, 5055–5057.

Åkerberg C, Hofvendahl K, Zacchi G, Hahn-Hägerdal B. (1998). Modeling the influence of pH, temperature, glucose and lactic acid concentration on the kinetics of lactic acid production by *Lactococcus lactis* ssp. *lactis* ATCC 19435 in whole-wheat flour. *Appl Microbiol Biotechnol*, 49, 682–690.

Bianchi MM, Brambilla L, Protani F, Liu CL, Lievense J, Porro D. (2001). Efficient homolactic fermentation by *Kluyveromyces lactis* strains defective in pyruvate utilization and transformed with the heterologous *LDH* gene. *Appl Envion Microbiol*, 67, 5621–5625.

Bustos G, Moldes AB, Cruz JM, Domínguez JM. (2005). Influence of the metabolism pathway on lactic acid production from hemicellulosic trimming vine shoots hydrolyzates using *Lactobacillus pentosus*. *Biotechnol Prog*, 21, 793–798.

Calabia BP, Tokiwa Y. (2007). Production of D-lactic acid from sugarcane molasses, sugarcane juice and sugar beet juice by *Lactobacillus delbrueckii*. *Biotechnol Lett*, 29, 1329–1332.

Chaillou S, Bor YC, Batt CA, Postma PW, Pouwels PH. (1998). Molecular cloning and functional expression in *Lactobacillus plantarum* 80 of *xylT*, encoding the D-xylose-H^+ symporter of *Lactobacillus brevis*. *Appl Environ Microbiol*, 64, 4720–4728.

Chang DE, Jung HC, Rhee JS, Pan JG. (1999). Homofermentative production of D- or L-lactate in metabolically engineered *Escherichia coli* RR1. *Appl Environ Microbiol*, 65, 1384–1389.

Davidson BE, Llanos RM, Cancilla MR, Redman NC, Hillier AJ. (1995). Current research on the genetics of lactic acid production in lactic acid bacteria. *Int Dairy J*, 5, 763–784.

Dequin S, Barre P. (1994). Mixed lactic acid-alcoholic fermentation by *Saccharomyces cerevisiae* expressing the *Lactobacillus casei* L(+)-LDH. *Biotechnology*, 12, 173–177.

Dequin S, Baptista E, Barre P. (1999). Acidification of grape musts by *Saccharomyces cerevisiae* wine yeast strains genetically engineered to produce lactic acid. *Am J Enol Vitic*, 50, 45–50.

Giraud E, Champailler A, Raimbault M. (1994). Degradation of raw starch by a wild amylolytic strain of *Lactobacillus plantarum*. *Appl Environ Microbiol*, 60, 4319–4323.

Guyot JP, Calderon M, Morlon-Guyot J. (2000). Effect of pH control on lactic acid fermentation of starch by *Lactobacillus manihotivorans* LMG 18010^T. *J Appl Microbiol*, 88, 176–182.

Han X, Satoh Y, Satoh T, Matsumoto K, Kakuchi T, Taguchi S, Dairi T, Munekata M, Tajima K. (2011) Chemo-enzymatic synthesis of polyhydroxyalkanoate (PHA) incorporating 2-hydroxybutyrate by wild-type class I PHA synthase from *Ralstonia eutropha*. *Appl Microbiol Biotechnol*, 92, 509–517.

Helanto M, Kiviharju K, Leisola M, Nyyssölä A. (2007). Metabolic engineering of *Lactobacillus plantarum* for production of L-ribulose. *Appl Environ Microbiol*, 73, 7083–7091.

Hermann T. (2003). Industrial production of amino acids by coryneform bacteria. *J Biotechnol*, 104, 155–172.

Hofvendahl K, Hahn-Hägerdal B. (1997). L-lactic acid production from whole wheat flour hydrolysate using strains of *Lactobacilli* and *Lactococci*. *Enzyme Microb Technol*, 20, 301–307.

Hofvendahl K, Hahn-Hägerdal B. (2000). Factors affecting the fermentative lactic acid production from renewable resources. *Enzyme Microb Technol*, 26, 87–107.

Hohmann S. (1991). Characterization of *PDC6*, a third structural gene for pyruvate decarboxylase in *Saccharomyces cerevisiae*. *J Bacteriol*, 173, 7963–7969.

Hohmann S, Cederberg H. (1990). Autoregulation may control the expression of yeast pyruvate decarboxylase structural genes *PDC1* and *PDC5*. *Eur J Biochem*, 188, 615–621.

Ikada Y, Jamshidi K, Tsuji H, Hyon SH. (1987). Stereocomplex formation between enantiomeric poly (lactides). *Macromolecules*, 20, 904–906.

Ishida N, Saitoh S, Tokuhiro K, Nagamori E, Matsuyama T, Kitamoto K, Takahashi H. (2005). Efficient production of L-lactic acid by metabolically engineered *Sacchromyces cerevisiae* with a genome-integrated L-lactate dehydrogenase gene. *Appl Environ Microbiol*, 71, 1964–1970.

Ishida N, Saitoh S, Onishi T, Tokuhiro K, Nagamori E, Kitamoto K, Takahashi H. (2006a). The effect of pyruvate decarboxylase gene knockout in *Sacchromyces cerevisiae* on L-lactic acid production. *Biosci Biotechnol Biochem*, 70, 1148–1153.

Ishida N, Suzuki T, Tokuhiro K, Nagamori E, Onishi T, Saitoh S, Kitamoto K, Takahashi H. (2006b). D-lactic acid production by metabolically engineered *Sacchromyces cerevisiae*. *J Biosci Bioeng*, 101, 172–177.

Jamshidian M, Tehrany EA, Imran M, Jacquot M, Desobry S. (2010). Poly-lactic acid: production, applications, nanocomposites, and release studies. *Compr Rev Food Sci Food Saf*, 9, 552–571.

Jem KJ, van der Pol JF, de Vos S. (2010). Microbial lactic acid, its polymer poly(lactic acid), and their industrial applications. *Microbiol Monogr*, 14, 323–346.

Kalinowski J, Bathe B, Bartels D, Bischoff N, Bott M, Burkovski A, Dusch N, Eggeling L, Eikmanns BJ, Gaigalat L, et al. (2003). The complete *Corynebacterium glutamicum* ATCC 13032 genome sequence and its impact on the production of L-aspartate-derived amino acids and vitamins. *J Biotechnol*, 104, 5–25.

Katahira S, Mizuike A, Fukuda H, Kondo A. (2006). Ethanol fermentation from lignocellulosic hydrolysate by a recombinant xylose- and cellooligosaccharide-assimilating yeast strain. *Appl Microbiol Biotechnol*, 72, 1136–1143.

Kawaguchi T, Enoki T, Tsurumaki S, Sumitani J, Ueda M, Ooi T, Arai M. (1996). Cloning and sequencing of the cDNA encoding b-glucosidase 1 from *Aspergillus aculeatus*. *Gene*, 173, 287–288.

Kawaguchi H, Vertès AA, Okino S, Inui M, Yukawa H. (2006). Engineering of a xylose metabolic pathway in *Corynebacterium glutamicum*. *Appl Environ Microbiol*, 72, 3418–3428.

Kawaguchi H, Sasaki M, Vertès AA, Inui M, Yukawa H. (2008). Engineering of an L-arabinose metabolic pathway in *Corynebacterium glutamicum*. *Appl Microbiol Biotechnol*, 77, 1053–1062.

Kotrba P, Inui M, Yukawa H. (2001). The *ptsI* gene encoding enzyme I of the phosphotransferase system of *Corynebacterium glutamicum*. *Biochem Biophys Res Commun*, 289, 1307–1313.

Kotrba P, Inui M, Yukawa H. (2003). A single V317A or V317M substitution in Enzyme II of a newly identified b-glucoside phosphotransferase and utilization system of *Corynebacterium glutamicum* R extends its specificity towards cellobiose. *Microbiology*, 149, 1569–1580.

Kylä-Nikkilä K, Hujanen M, Leisola M, Palva A. (2000). Metabolic engineering of *Lactobacillus helveticus* CNRZ32 for production of pure L-(+)-lactic acid. *Appl Environ Microbiol*, 66, 3835–3841.

Leskovac V, Trivić S, Pericin D. (2002). The three zinc-containing alcohol dehydrogenases from baker's yeast, *Saccharomyces cerevisiae*. *FEMS Yeast Res*, 2, 481–494.

Leuchtenberger W, Huthmacher K, Drauz K. (2005). Biotechnological production of amino acids and derivatives: current status and prospects. *Appl Microbiol Biotechnol*, 69, 1–8.

Litchfield JH. (1996). Microbiological production of lactic acid. *Adv Appl Microbiol*, 42, 45–95.

Matsui Y, Okada S, Uchimura T, Kondo A, Satoh E. (2007). Determination and analysis of the starch binding domain of *Streptococcus bovis* 148 raw-starch-hydrolyzing a-amylase. *J Appl Glycosci*, 54, 217–222.

Matsumoto K, Ishiyama A, Sakai K, Shiba T, Taguchi S (2011) Biosynthesis of glycolate-based polyesters containing medium chain-length 3-hydroxyalkanoates in recombinant *Escherichia coli* expressing engineered polyhydroxyalkanoate synthase. *J Biotechnol*, 156, 214–217.

Nampoothiri KM, Nair NR, John RP. (2010). An overview of the recent developments in polylactide (PLA) research. *Bioresour Technol*, 101, 8493–8501.

Narita J, Nakahara S, Fukuda H, Kondo A. (2004). Efficient production of L-(+)-lactic acid from raw starch by *Streptococcus bovis* 148. *J Biosci Bioeng*, 97, 423–425.

Ohara H, Owaki M, Sonomoto K. (2006). Xylooligosaccharide fermentation with *Leuconostoc lactis*. *J Biosci Bioeng*, 101, 415–420.

Okano K, Kimura S, Narita J, Fukuda H, Kondo A. (2007). Improvement in lactic acid production from starch using a-amylase-secreting *Lactococcus lactis* cells adapted to maltose or starch. *Appl Microbiol Biotechnol*, 75, 1007–1013.

Okano K, Yoshida S, Tanaka T, Ogino C, Fukuda H, Kondo A. (2009a). Homo D-lactic acid fermentation from arabinose by redirection of phosphoketolase pathway to pentose phosphate pathway in L-lactate dehydrogenase gene-deficient *Lactobacillus plantarum*. *Appl Environ Microbiol*, 75, 5175–5178.

Okano K, Yoshida S, Tanaka T, Ogino C, Fukuda H, Kondo A. (2009b). Homo D-lactic acid fermentation from xylose by introduction of xylose assimilation genes and redirection of the phosphoketolase pathway to the pentose phosphate pathway in L-lactate dehydrogenase gene-deficient *Lactobacillus plantarum*. *Appl Environ Microbiol*, 75, 7858–7861.

Okano K, Zhang Q, Shinkawa S, Yoshida S, Tanaka T, Fukuda H, Kondo A. (2009c). Efficient production of optically pure D-lactic acid from raw corn starch by using genetically modified L-lactate dehydrogenase gene-deficient and a-amylase-secreting *Lactobacillus plantarum* strain. *Appl Environ Microbiol*, 75, 462–467.

Okano K, Zhang Q, Yoshida S, Tanaka T, Ogino C, Fukuda H, Kondo A. (2009d). D-Lactic acid production from cellooligosaccharides and b-glucan using L-LDH gene-deficient and endoglucanase-secreting *Lactobacillus plantarum*. *Appl Microbiol Biotechnol*, 85, 643–650.

Okino S, Inui M, Yukawa H. (2005). Production of organic acids by *Corynebacterium glutamicum* under oxygen deprivation. *Appl Microbiol Biotechnol*, 68, 475–480.

Okino S, Suda M, Fujikura K, Inui M, Yukawa H. (2008). Production of D-lactic acid by *Corynebacterium glutamicum* under oxygen deprivation. *Appl Microbiol Biotechnol*, 78, 449–454.

Porro D, Brambilla L, Ranzi BM, Martegani E, Alberghina L. (1995). Development of metabolically engineered *Saccharomyces cerevisiae* cells for the production of lactic acid. *Biotechnol Prog*, 11, 294–298.

Saitoh S, Mieno Y, Nagashima T, Kumagai C, Kitamoto K. (1996). Breeding of a new type of baker's yeast by d-integration for overproduction of glucoamylase using a homothallic yeast. *J Ferment Bioeng*, 81, 98–103.

Saitoh S, Ishida N, Onishi T, Tokuhiro K, Nagamori E, Kitamoto K, Takahashi H. (2005). Genetically engineered wine yeast produces a high concentration of L-lactic acid of extremely high optical purity. *Appl Environ Microbiol*, 71, 2789–2792.

Sasaki M, Jojima T, Inui M, Yukawa H. (2008). Simultaneous utilization of D-cellobiose, D-glucose, and D-xylose by recombinant *Corynebacterium glutamicum* under oxygen-deprived conditions. *Appl Microbiol Biotechnol*, 81, 691–699.

Satoh E, Niimura Y, Uchimura T, Kozaki M, Komagata K. (1993). Molecular cloning and expression of two a-amylase genes from *Streptococcus bovis* 148 in *Escherichia coli*. *Appl Environ Microbiol*, 59, 3669–3673.

Serror P, Sasaki T, Ehrlich SD, Maguin E. (2002). Electrotransformation of *Lactobacillus delbrueckii* subsp. *bulgaricus* and *L. delbrueckii* subsp. *lactis* with various plasmids. *Appl Environ Microbiol*, 68, 46–52.

Shen L, Haufe J, Patel MK. (2009). Product overview and market projection of emerging bio-based plastics. PRO-BIP Report.

Shinkawa S, Okano K, Yoshida S, Tanaka T, Ogino C, Fukuda H, Kondo A. (2011). Improved homo L-lactic acid fermentation from xylose by abolishment of the phosphoketolase pathway and enhancement of pentose phosphate pathway in genetically modified xylose-assimilating *Lactococcus lactis*. *Appl Microbiol Biotechnol*, 91, 1537–1544.

Shozui F, Matsumoto K, Nakai T, Yamada M, Taguchi S. (2009). Biosynthesis of novel terpolymers poly(lactate-co-3-hydroxybutyrateco-3-hydroxyvalerate)s in lactate-overproducing mutant *Escherichia coli* JW0885 by feeding propionate as a precursor of 3-hydroxyvalerate. *Appl Microbiol Biotechnol*, 85, 949–954.

Skory CD. (2003). Lactic acid production by *Saccharomyces cerevisiae* expressing a *Rhizopus oryzae* lactate dehydrogenase gene. *J Ind Microbiol Biotechnol*, 30, 22–27.

Song Y, Matsumoto K, Yamada M, Gohda A, Brigham CJ, Sinskey AJ, Taguchi S. (2012) Engineered *Corynebacterium glutamicum* as an endotoxin-free platform strain for lactate-based polyester production. *Appl Microbiol Biotechnol*, 93, 1917–1925.

Suzuki T, Sakamoto T, Sugiyama M, Ishida N, Kambe H, Obata S, Kaneko Y, Takahashi H, Harashima S. (2013). Disruption of multiple genes whose deletion causes lactic-acid resistance improves lactic-acid resistance and productivity in *Saccharomyces cerevisiae*. *J Biosci Bioeng*, 115(5), 467–474.

Taguchi S, Yamada Y, Matsumoto K, Tajima K, Satoh Y, Munekata M, Ohno K, Kohda K, Shimamura T, Kambe H, Obata S. (2008). A microbial factory for lactate-based polyesters using a lactate-polymerizing enzyme. *Proc Natl Acad Sci U S A*, 105, 17323–17327.

Tanaka K, Komiyama A, Sonomoto K, Ishizaki A, Hall SJ, Stanbury PF. (2002). Two different pathways for D-xylose metabolism and the effect of xylose concentration on the yield coefficient of L-lactate in mixed-acid fermentation by the lactic acid bacterium *Lactococcus lactis* IO-1. *Appl Microbiol Biotechnol*, 60, 160–167.

Tarmy EM, Kaplan NO. (1968). Kinetics of *Escherichia coli* B D-lactate dehydrogenase and evidence for pyruvate controlled change in conformation. *J Biol Chem*, 243, 2587–2596.

Tateno T, Fukuda H, Kondo A. (2007a). Production of L-lysine from starch by *Corynebacterium glutamicum* displaying a-amylase on its cell surface. *Appl Microbiol Biotechnol*, 74, 1213–1220.

Tateno T, Fukuda H, Kondo A. (2007b). Direct production of L-lysine from raw corn starch by *Corynebacterium glutamicum* secreting *Streptococcus bovis* a-amylase using *cspB* promoter and signal sequence. *Appl Microbiol Biotechnol*, 77, 533–541.

Tokuhiro K, Ishida N, Kondo A, Takahashi H (2008) Lactic fermentation of cellobiose by a yeast strain displaying b-glucosidase on the cell surface. *Appl Microbiol Biotechnol*, 79, 481–488.

Tokuhiro K, Ishida N, Nagamori E, Saitoh S, Onishi T, Kondo A, Takahashi H. (2009). Double mutation of the *PDC1* and *ADH1* genes improves lactate production in the yeast *Saccharomyces cerevisiae* expressing the bovine lactate dehydrogenase gene. *Appl Microbiol Biotechnol*, 82, 883–890.

Varadarajan S, Miller DJ. (1999). Catalytic upgrading of fermentation-derived organic acids. *Biotechnol Prog*, 15, 845–854.

Vijayakumar J, Aravindan R, Viruthagiri T. (2008). Recent trends in the production, purification and application of lactic acid. *Chem Biochem Eng Q*, 22, 245–264.

Wang L, Zhao B, Liu B, Yang C, Yu B, Li Q, Ma C, Xu P. (2010). Efficient production of L-lactic acid from cassava powder by *Lactobacillus rhamnosus*. *Bioresour Technol*, 101, 7895–7901.

Wee YJ, Kim JN, Yun JS, Ryu HW. (2004). Utilization of sugar molasses for economical L(+)-lactic acid production by batch fermentation of *Enterococcus faecalis*. *Enzyme Microb Technol*, 35, 568–573.

Wee YJ, Kim JN, Ryu HW. (2006). Biotechnological production of lactic acid and its recent applications. *Food Technol Biotechnol*, 44, 163–172.

Yamada M, Matsumoto K, Nakai T, Taguchi S (2009) Microbial production of lactate-enriched poly[(R)-lactate-co-(R)-3-hydroxybutyrate] with novel thermal properties. *Biomacromolecules*, 10, 677–681.

Yáñez R, Moldes AB, Alonso JL, Parajó JC. (2003). Production of D(-)-lactic acid from cellulose by simultaneous saccharification and fermentation using *Lactobacillus coryniformis* subsp. *torquens*. *Biotechnol Lett*, 25, 1161–1164.

Yin P, Nishina N, Kosakai Y, Yahiro K, Park YS, Okabe M. (1997). Enhanced production of L(+)-lactic acid from corn starch in a culture of *Rhizopus oryzae* using an air-lift bioreactor. *J Ferment Bioeng*, 84, 249–253.

Yoon HH. (1997). Simultaneous saccharification and fermentation of cellulose for lactic acid production. *Biotechnol Bioprocess Eng*, 2, 101–104.

Yoshida S, Okano K, Tanaka T, Ogino C, Kondo A. (2011). Homo-D-lactic acid production from mixed sugars using xylose-assimilating operon-integrated *Lactobacillus plantarum*. *Appl Microbiol Biotechnol*, 92, 67–76.

Yukawa H, Omumasaba CA, Nonaka H, Kos P, Okai N, Suzuki N, Suda M, Tsuge Y, Watanabe J, Ikeda Y, Vertès AA, Inui M. (2007). Comparative analysis of the *Corynebacterium glutamicum* group and complete genome sequence of strain R. *Microbiology*, 153, 1042–1058.

Zhou S, Causey TB, Hasona A, Shanmugam KT, Ingram LO. (2003a). Production of optically pure D-lactic acid in mineral salts medium by metabolically engineered *Escherichia coli* W3110. *Appl Environ Microbiol*, 69, 399–407.

Zhou S, Shanmugam KT, Ingram LO. (2003b). Functional replacement of the *Escherichia coli* D-(-)-lactate dehydrogenase gene (*ldhA*) with the L-(+)-lactate dehydrogenase gene (*ldhL*) from *Pediococcus acidilactici*. *Appl Environ Microbiol*, 69, 2237–2244.

CHAPTER 14

Microbial Production of 3-Hydroxypropionic Acid From Renewable Sources: A Green Approach as an Alternative to Conventional Chemistry

VINOD KUMAR, SOMASUNDAR ASHOK, and SUNGHOON PARK

Department of Chemical and Biomolecular Engineering, Pusan National University, Jangjeon-dong, Geumjeong-gu, Busan, Republic of Korea

Bioprocessing of Renewable Resources to Commodity Bioproducts, First Edition.
Edited by Virendra S. Bisaria and Akihiko Kondo.

Abstract

3-Hydroxypropionic acid (3-HP) is a valuable platform chemical that can be produced from glucose or glycerol by biological processes. This chapter provides an overview and the current status of microbial 3-HP production. It also describes various constraints of microbial 3-HP production and possible solutions. Finally, economic outlook and future prospects of biological 3-HP production are discussed.

14.1 INTRODUCTION

In the last decade, microbial production of fuels and chemicals from renewable resources has drawn a great deal of attention due to the serious concerns pertaining to global warming and rapid increases in oil price (Celińska and Grajek, 2009). Many chemicals produced by chemical processes in the past are now produced or being investigated for possible production from renewable resources by biological methods (Danner and Braun, 1999; Hatti-Kaul et al., 2007). 3-Hydroxypropionic acid (3-HP), a bifunctional molecule containing carboxyl and hydroxyl groups, is one such chemical. 3-HP is a platform chemical that secures the third position in a list of the top 12 platform chemicals that can be derived from biomass according to the United States Department of Energy (Werpy and Petersen, 2004; Kumar et al., 2013b). 3-HP can be used for the production of a wide range of commercially important chemicals such as acrylic acid, 1,3-propanediol (PDO), methyl acrylate, acrylamide, malonic acid, propiolactone, and acrylonitrile. 3-HP is also used as a cross-linking agent for polymer coatings and metal lubricants, and as an antistatic agent for textiles (CEP, 2003; Gokarn et al., 2007). In addition, 3-HP can be polymerized to biodegradable polymers, poly (3-hydroxypropionate) and poly (3-hydroxypropionate-co-3-hydroxybutyrate) (Zhou et al., 2011; Meng et al., 2012). The polymer [poly(3-hydroxypropionate)] can also be used in the manufacture of surgical and drug-releasing materials due to its high biocompatibility and biodegradability. Despite many useful properties, the application of 3-HP has been impeded due to its high production cost (Suthers and Cameron, 2005; Jiang et al., 2009; Pina et al., 2009). Realizing the importance of 3-HP, Cargill (http://www.cargill.com) and the United States Department of Energy have invested over US$6 million to investigate production of 3-HP from renewable resources.

This chapter describes the recent progress in biological production of 3-HP using different microorganisms and substrates. Biochemical pathways responsible for 3-HP production are examined in detail, and the state-of-the-art of the development of recombinant strains and fermentation processes is presented. In addition, major hurdles associated with currently available microbial 3-HP production technologies and the strategies to overcome these hurdles are discussed. Finally, future prospects of microbial 3-HP production are presented.

14.2 NATURAL MICROBIAL PRODUCTION OF 3-HP

Several microorganisms have the ability to synthesize 3-HP naturally (Table 14.1). The best-understood process is its production as an intermediate in the 3-hydroxypropionate and 3-hydroxypropionate/4-hydroxybutyrate cycle, autotrophic pathways for carbon dioxide fixation (Strauss et al., 1992; Strauss and Fuchs, 1993; Ishii et al., 2004). The pathway includes carboxylation of acetyl-CoA into malonyl-CoA at the expense of one ATP, which is catalyzed by biotin-dependent acetyl-CoA carboxylase. The malonyl-CoA is reduced into 3-HP via malonate semialdehyde using two NADPH molecules. 3-HP is further reductively converted to propionyl-CoA. *Chloroflexus aurantiacus*, a thermophilic photosynthetic bacterium, possesses this cycle and secretes 3-HP during phototrophic growth. The maximum amount of 3-HP secreted was reported to be 350 μM when cells were cultivated in the presence of propionate (Holo, 1989). Some other microorganisms such as *Acidianus brierleyi*, *Metallosphaera sedula*, *Acidianus ambivalens*, *Sulfolobus* sp. strain VE6, and *Sulfolobus metallicus* (Hügler et al., 2003) also have this cycle. The autotrophic 3-HP production should be advantageous since the cost of the carbon source often accounts for the majority of the costs associated with manufacturing bulk organic acids (Jiang et al., 2009). However, as a metabolic intermediate, 3-HP is not accumulated in high concentrations from this cycle.

3-HP can also be produced via degradative pathways. For example, *Byssochlamys* sp. can produce 3-HP (4.8% v/v) from the degradation of acrylic acid (Takamizawa et al., 1993). *Alcaligenes faecalis* also produces 3-HP from dimethylsulfoniopropionate (DMSP). In this process, DMSP is degraded to dimethylsulfide and acrylic acid, after which the acrylic acid is converted to 3-HP (Ansede et al., 1999). Dave et al. (1996) observed transient accumulation of 3-HP during acrylic acid degradation by a fungal strain of *Geotrichum* sp. and *Trichoderma* sp. Anderson et al. (2008) also reported 3-HP production from degradation of pyrimidine base uracil. The uracil catabolism (URC) pathway produces uridine monophosphate and urea as intermediates and 3-HP, ammonia, and carbon dioxide as the final products.

Some microorganisms produce 3-HP as an end product. Various strains of *Lactobacillus* species (*L.* strain 208-A, *L. reuteri*, and *L. collinoides*) can transform glycerol into 3-HP as an end-product. The strain *L. collinoides* 17 accumulated 31.3 mM 3-HP and 37.1 mM PDO from glycerol. In these strains, glycerol is first converted to 3-hydroxypropionaldehyde (3-HPA) by glycerol dehydratase, after which 3-HPA is disproportionated into 3-HP and PDO (Soblov and Smiley, 1960; Talarico et al., 1988; Garai-Ibabe et al., 2008). Some endophytic fungi excrete 3-HP as a nematicide. Schwarz et al. (2004) extracted 3-HP from the culture filtrate of several endophytic fungi isolated from above-ground plant organs. Biochemical analysis revealed that 3-HP was produced by *Phomopsis phaseoli* and four strains of *Melanconium betulinum*.

Although 3-HP can be naturally produced from various microorganisms, these microorganisms cannot be used directly for 3-HP production on a large scale since the level of production is very low and/or the substrate (e.g., acrylate) is too expensive.

TABLE 14.1 3-HP Production by Natural Microorganisms

Organism	Carbon source	Immediate precursor of 3-HP	References
3-HP as an intermediate of 3-hydroxypropionate or 3-hydroxypropionate/4-hydroxybutyrate cycle			
Chloroflexus aurantiacus	Carbon dioxide	Malonate semialdehyde	Ishii et al. (2004)
Metallosphaera sedula	Carbon dioxide	Malonate semialdehyde	Berg et al. (2007)
3-HP as an end-product of 3-HPA-mediated oxidative pathway			
Lactobacillus strain 208-A	Glycerol	3-HPA	Soblov and Smiley (1960)
L. reuteri	Glycerol	3-HPA	Talarico et al. (1988)
L. collinoides	Glycerol	3-HPA	Garai-Ibabe et al. (2008)
L. reuteri	Glycerol	3-HPA	Krauter et al. (2012)
3-HP as an end-product of degradative pathway			
Byssochlamys sp.	Acrylic acid	Acrylic acid	Takamizawa et al. (1993)
Geotrichum sp. and *Trichoderma* sp.	Acrylic acid	Acrylic acid	Dave et al. (1996)
Alcaligenes faecalis	Dimethylsulfoniopropionate	Acrylic acid	Ansede et al. (1999)
Saccharomyces kluyveri	Uracil	Malonate semialdehyde	Anderson et al. (2008)
3-HP production as a nematicide			
Endophytic fungi	Not known	Not known	Schwarz et al. (2004)

14.3 PRODUCTION OF 3-HP FROM GLUCOSE BY RECOMBINANT MICROORGANISMS

The US-based agricultural products company, Cargill, initiated research on biological 3-HP production with the purpose of commercial production (Cargill, 2012). As a result of their studies, they proposed and patented seven metabolic pathways for 3-HP production from glucose (Gokarn et al., 2001; Liao et al., 2005, 2007; Marx et al., 2007). These pathways are mediated by either pyruvate or phosphoenolpyruvate (PEP), both of which are common intermediates of sugar metabolism (Figure 14.1). The maximum theoretical yield (mol/mol) of 3-HP from glucose, based on pathway stoichiometry and/or comparison of energy content between substrate (glucose) and product (3-HP), is 2.0 (Dugar and Stephanopoulos, 2011). To be practically feasible, the pathway leading to 3-HP production should be redox-balanced and generate net ATP for use in cell growth, maintenance, product excretion, and so on. All biosynthetic pathways shown in Figure 14.1 are redox-balanced; however, many do not generate ATP and some are even ATP-consuming.

The ATP yield and thermodynamic feasibilities (predicted on the basis of group contribution method) of these pathways are shown in Table 14.2. The ATP yield for a pathway can vary depending upon the enzyme(s) involved in the pathway. For example, the carboxylation of PEP into oxaloacetate produces one ATP if mediated by PEP carboxykinase; however, the same reaction does not generate ATP when catalyzed by PEP carboxylase. Henry et al. (2010) compared and evaluated Cargill's seven pathways along with several new pathways constructed by BNICE (Biochemical Network Integrated Computational Explorer). Based on several criteria including pathway length, thermodynamic feasibility, maximum achievable yield of 3-HP from glucose, and maximum achievable activity (intracellular concentration), they identified four pathways which were promising, one from Cargill and three from BNICE (Henry et al., 2010; Figure 14.2). In addition, they claimed that, among these four pathways, the one developed by Cargill which occurred via α- and β-alanines (pyruvate→α-alanine→β-alanine→3-oxopropanoate→3-HP) was most promising. In this pathway, the energy consuming step was bypassed by the use of alanine-2,3-aminomutase, an enzyme that catalyzes the transfer of an amino group from the α- to β-carbon (Jiang et al., 2009). This allowed the highest possible yield (2.0 mol 3-HP from 1.0 mol glucose) (Willke and Vorlop, 2004; Henry et al., 2010). This pathway was tested by Liao et al. (2007b) in a recombinant *Escherichia coli* overexpressing a vitamin B_{12}-dependent lysine 2,3-aminomutase, an isozyme of alanine-2,3-aminomutase. However, detailed information pertaining to titer, yield, and productivity of 3-HP have not been revealed.

Three other pathways suggested to be promising by BNICE (Figure 14.2) also deserve some attention. The pathway via lactate (pyruvate→lactate→3-HP) is attractive since it involves the fewest number of reactions. However, the conversion of lactate into 3-HP is thermodynamically unfavorable and can only proceed in the forward direction when the lactate concentration is very high and the 3-HP is very low (Herrmann et al., 2005; Jiang et al., 2009; Henry et al., 2010). This condition can be

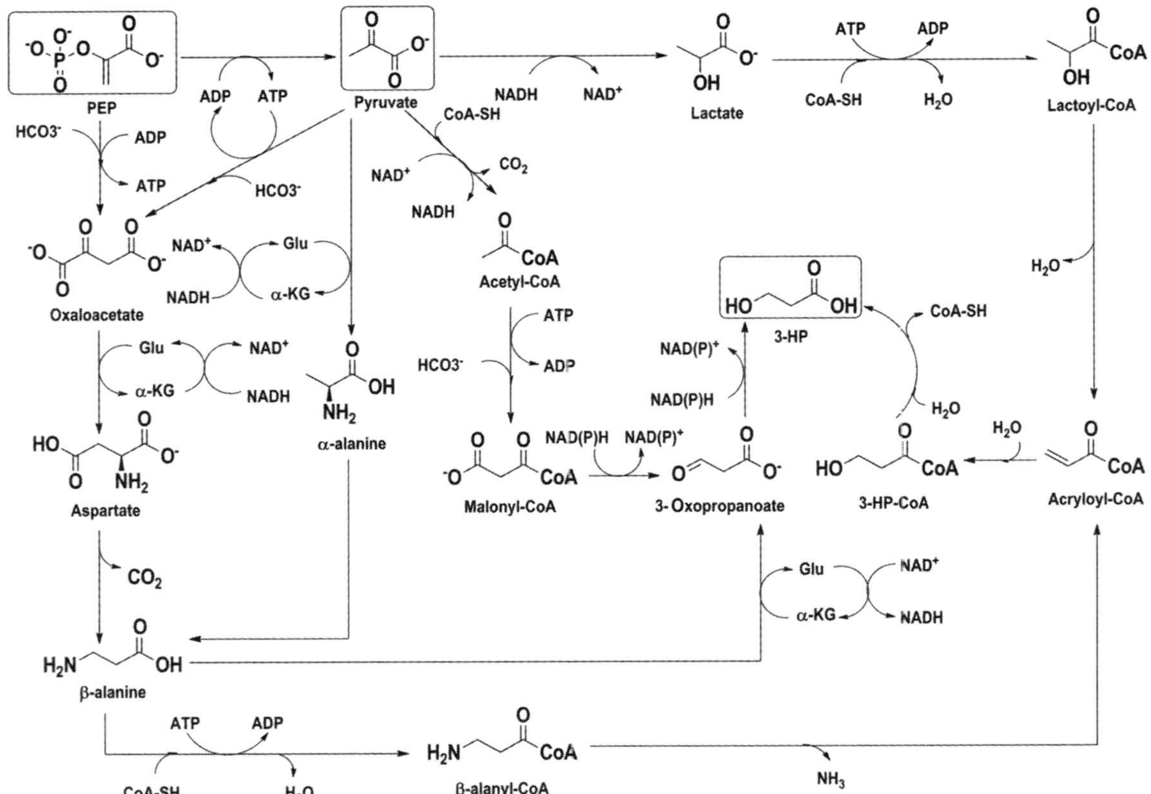

FIGURE 14.1 Biochemical pathways for 3-HP production. Glu, glutamate; α-KG, α-ketoglutarate. (Jiang et al., 2009; Henry et al., 2010).

TABLE 14.2 The ATP Yield and Thermodynamic Feasibility for Various Metabolic Pathways Starting with Glucose for the Production of 3-HP

Metabolic pathway	Net ATP yield (mol/mol 3-HP)	Thermodynamic feasibility
Pyruvate→lactate→lactoyl-CoA→acryloyl-CoA→3-HP-CoA→3-HP	1/0	Unfavorable
Pyruvate→acetyl-CoA→malonyl-CoA→3-oxopropanoate→3-HP	0	Favorable
Pyruvate/PEP→OAA→aspartate→β-alanine→3-oxopropanoate→3-HP	1/0	Favorable
Pyruvate/PEP→OAA→aspartate→β-alanine→β-alanyl-CoA→acryloyl-CoA→3-HP-CoA→3-HP	0/-1	Unfavorable
Pyruvate→α-alanine→β-alanine→3-oxopropanoate→3-HP	1	Favorable
Pyruvate→α-alanine→β-alanine→β-alanyl-CoA→acryloyl-CoA→3-HP-CoA→3-HP	1/0	Unfavorable

Source: Adapted from Jiang et al. (2009) and Henry et al. (2010).

met when lactate is accumulated at high concentrations and 3-HP is quickly excreted as soon as it is produced. The other two pathways suggested by BNICE occur via α- and β-alanines (pyruvate→α-alanine→β-alanine→propenoate→3-HP). These pathways are similar to the one proposed by Cargill in that all intermediary compounds are the same except for 3-oxopropanoate, which is replaced by propenoate. The aminating and deaminating reactions in these two pathways are also somewhat different.

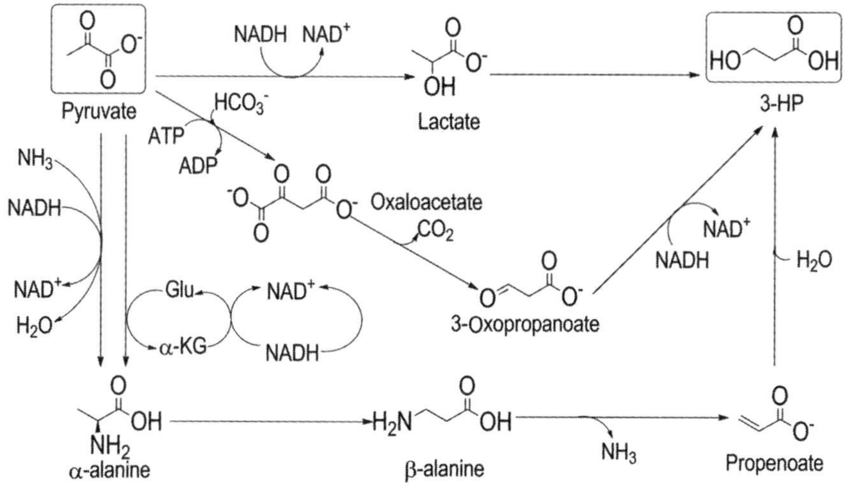

FIGURE 14.2 Novel pathways for 3-HP biosynthesis designed by Biochemical Network Integrated Computational Explorer (BNICE) (Henry et al., 2010).

Specifically, glutamate and α-ketoglutarate along with transaminases are involved in Cargill's pathway, while ammonia along with aminase and deaminase are involved in the BNICE pathway (Henry et al., 2010). However, the pathways suggested by BNICE have not been tested experimentally and their practical applications have not been demonstrated. Beside these three pathways, one more pathway constructed by BNICE is worthy of discussion. The pathway consists of three steps only and 3-HP production in this pathway takes place via oxaloacetate and 3-oxopropanoate (pyruvate→oxaloacetate→3-oxopropanoate→3-HP). This pathway is thermodynamically feasible, but requires the consumption of one ATP; thus, the yield is as low as 1.22 mol 3-HP/mol glucose (Henry et al., 2010).

Interestingly, the only pathway that has been seriously studied experimentally for the production of 3-HP from glucose is the one that occurs via malonyl-CoA. This pathway was not thought to be promising by Henry et al. (Figure 14.1). However, Rathnasingh et al. (2012) developed a recombinant *E. coli* strain overexpressing *mcr*, *pntAB*, and *accADBC* encoding malonyl-CoA reductase, nicotinamide nucleotide transhydrogenase, and acetyl-CoA carboxylase and biotinilase, respectively. The *mcr* gene was derived from *C. aurantiacus*, while the *pntAB* and *accADBC* genes were from *E. coli* K-12. In this pathway, acetyl-CoA is carboxylated into malonyl-CoA (by acetyl-CoA carboxylase and biotinilase), which is then reduced into 3-HP by malonyl-CoA reductase using two molecules of NADPH as electron donors. The conversion of NADH into NADPH was also facilitated by nicotinamide nucleotide transhydrogenase (*pntAB*). The recombinant strain yielded 2.16 mM 3-HP from glucose in 24 hours. It should be noted that malonyl-CoA is a well-known intermediate for fatty acid synthesis. To improve 3-HP production via the malonyl-CoA reductase pathway, Lynch et al. (2011) attempted to increase carbon flux from malonyl-CoA toward 3-HP synthesis by reducing the expression level of enoyl-ACP reductase (*fabI*) (key enzyme of fatty biosynthesis). They could produce 5.8 g/L 3-HP in shake flask cultures and 20.7 g/L 3-HP in bioreactor cultures at 38.5 hours. Further modification of the recombinant *E. coli* by deleting competing pathways leading to by-products formation (lactate, ethanol, acetate, and methylglyoxal) resulted in a drastic improvement in 3-HP production to 49 g/L in 69 hours, which is the maximum titer reported to date using glucose as the substrate (Lynch et al., 2011).

14.4 PRODUCTION OF 3-HP FROM GLYCEROL BY RECOMBINANT MICROORGANISMS

Currently, crude glycerol is generated in excess as an inevitable by-product from biodiesel industries (da Silva et al., 2009). Crude glycerol is a very inexpensive (<US$200 per ton) and good carbon source from which many biochemicals and biofuels can be produced (Dharmadi et al., 2006). Accordingly, the use of glycerol as a carbon source can add profitability and sustainability to the biodiesel industry (da Silva et al., 2009; Jiang et al., 2009; Saxena et al., 2009). The production of 3-HP from glycerol has been extensively studied in recent years. The maximum theoretical yield (mol/mol) of 3-HP from glycerol is 1.0.

14.4.1 Glycerol Metabolism for the Production of 3-HP and Cell Growth

Glycerol is more reduced than traditional carbohydrates such as glucose, xylose, and so on. Due to its reduced nature, conversion of glycerol to pyruvate generates twice the amount of reducing equivalents than conversion of glucose or xylose. Therefore, the microorganisms that can grow on glycerol under anaerobic conditions, including *Citrobacter*, *Clostridium*, *Enterobacter*, *Klebsiella*, and *Lactobacillus* species, require disposal of the two extra hydrogen atoms, which necessitates the formation of a by-product such as PDO to act as an electron sink. The production of PDO from glycerol proceeds via an aldehyde, 3-HPA, which can also be converted to 3-HP by proper aldehyde dehydrogenase(s) as observed in some strains of *Klebsiella* and *Lactobacillus* (Johnson and Lin, 1987; Nakamura and Whited, 2003; Dharmadi et al., 2006; Ashok et al., 2011). If these strains are genetically modified to overexpress a proper aldehyde dehydrogenase and block the enzymes leading to the production of PDO, more 3-HP can be produced.

Biosynthetic pathways for the production of 3-HP from glycerol can be classified as either CoA-dependent or CoA-independent depending on the mechanism for the conversion of 3-HPA to 3-HP. Both the pathways generate NADH during the conversion of glycerol to 3-HP and considered "energy surplus" pathways (Dugar and Stephanopoulos, 2011). In both pathways, glycerol is first dehydrated to 3-HPA by either glycerol dehydratase encoded by *dhaB* or diol dehydratase encoded by *pduCDE*, both of which are vitamin B_{12}-dependent (Bobik et al., 1999; Ashok et al., 2011; Luo et al., 2012). Once 3-HPA is produced, it is further converted to 3-HP in a CoA-dependent or CoA-independent manner. In the CoA-dependent pathway, 3-HPA is oxidized to 3-HP-CoA followed by phosphorylation to 3-HP–phosphate and subsequent dephosphorylation to 3-HP (Luo et al., 2011a, 2012). The CoA-dependent pathway is closely related to the degradation pathway of 1,2-propanediol in *Salmonella enterica*. In this pathway, 1,2-propanediol is converted to propionaldehyde followed by disproportionation to propanol and propionic acid. The disproportionation occurs through a sequence of reactions catalyzed by CoA-dependent propionaldehyde dehydrogenase (PduP), phosphotransacylase (PduL), propionate kinase (PduW), and propanol dehydrogenase (PduQ) (Figure 14.3). The conversion of propionaldehyde to propionyl-CoA requires NAD^+ as a cofactor, whereas that of propionyl phosphate to propanol requires NADH. When propionate is produced instead of propanol, ATP is generated. All of the genes necessary for consumption of 1,2-propanediol by *S. enteric* are found in the propanediol utilization (*pdu*) locus as a single cluster (Leal et al., 2003; Liu et al., 2007; Xue et al., 2008). Unlike the CoA-dependent pathway, the CoA-independent pathway directly oxidizes 3-HPA to 3-HP by NAD^+-dependent aldehyde dehydrogenase (Figure 14.3) and does not produce ATP.

When glycerol is used as a carbon and energy source, it should be oxidized to generate PEP and pyruvate via an oxidative pathway (Figure 14.4). In many microorganisms that can grow on glycerol under anaerobic conditions, the oxidative pathway always appears along with another reductive pathway leading to the production of

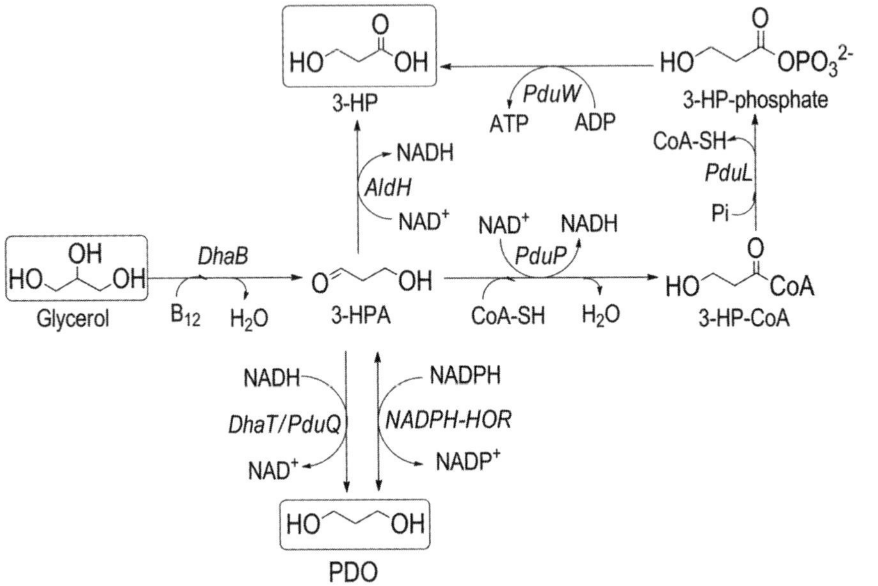

FIGURE 14.3 Biochemical reactions showing production of 3-HP from glycerol by CoA-dependent and CoA-independent pathways (Xue et al., 2008; Ashok et al., 2011; Luo et al., 2011a).

PDO. The reductive pathway from which 3-HP is also derived is highly dependent on the oxidative pathway for the generation of ATP and maintenance of redox balance. Under fermentative conditions, the oxidative metabolism starts with the dehydrogenation of glycerol to dihydroxyacetone (DHA) by the NAD^+-dependent enzyme glycerol dehydrogenase. DHA is then phosphorylated by the PEP/ATP-dependent dihydroxyacetone kinase to dihydroxyacetone phosphate (DHAP), which is subsequently funneled to glycolysis (Figure 14.4). In few microorganisms and the species belonging to the *Klebsiella* genus also have an alternative oxidative pathway that normally operates under respiratory conditions. This branch contains glycerol kinase(s), which catalyze the reaction of glycerol to *sn*-glycerol-3-phosphate and glycerol-3-phosphate dehydrogenases for the conversion of *sn*-glycerol-3-phosphate to DHAP. Further conversion of DHAP results in the formation of various organic acids and reduced metabolites such as acetic acid, lactic acid, succinic acid, ethanol, 2,3-butanediol, and formic acid (Ashok et al., 2011).

14.4.2 Synthesis of 3-HP from Glycerol Through the CoA-Dependent Pathway

The production of 3-HP by various *Lactobacillus* species (*L.* strain 208-A, *L. reuteri*, and *L. collinoides*), *S. enteric*, and *Klebsiella pneumoniae* relies on a CoA-dependent pathway. Yasuda et al. (2007) showed that resting cells of the mutant strain of *L. reuteri* lacking glycerol dehydrogenase resulted in the production of 7.2 g/L of PDO

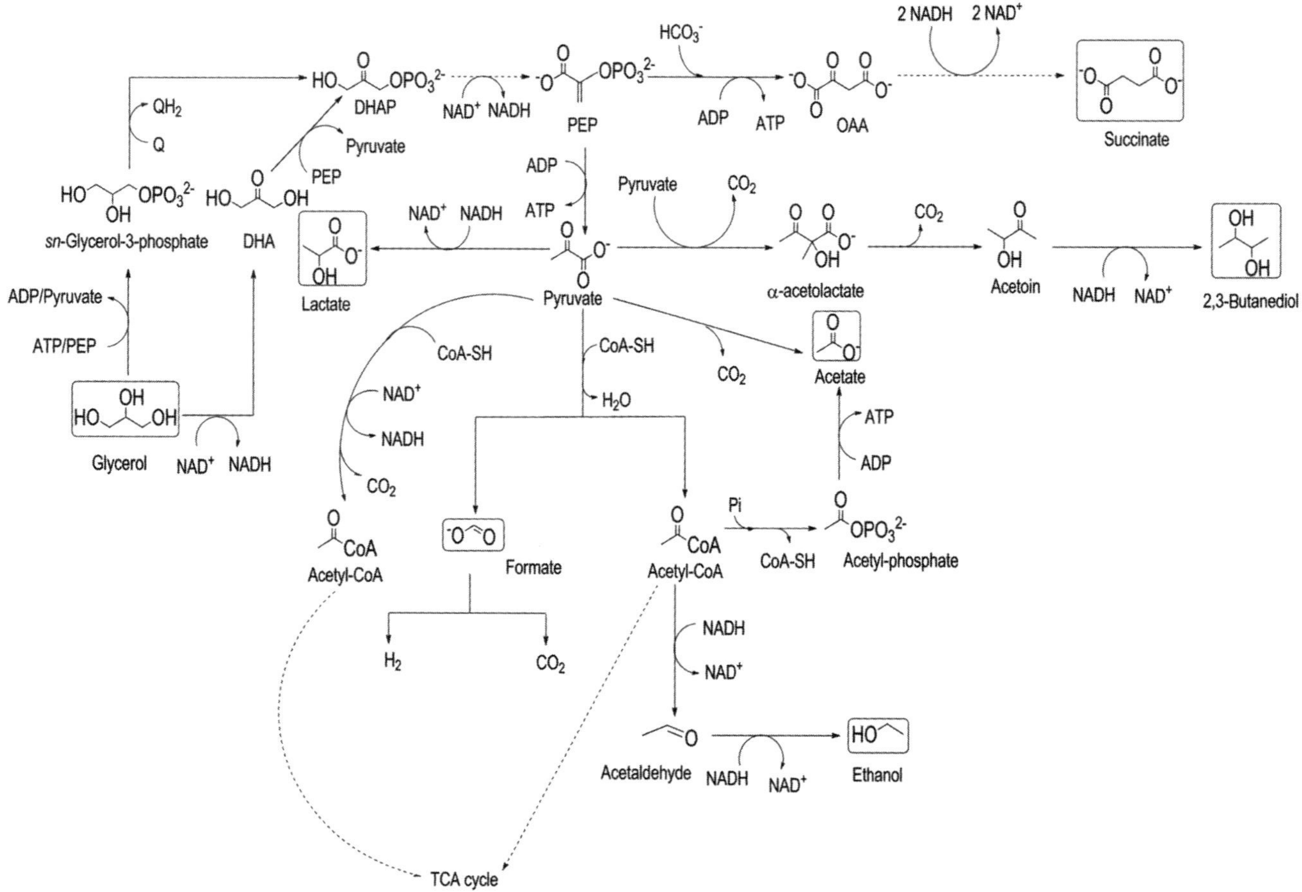

FIGURE 14.4 Metabolic pathway depicting oxidative metabolism of glycerol (Ashok et al., 2011; Kumar et al., 2012).

and 8.5 g/L of 3-HP. Luo et al. (2011a) cloned and characterized the CoA-dependent propionaldehyde dehydrogenase (PduP) of *L. reuteri*. The PduP enzyme showed broad substrate specificity for aliphatic aldehydes including 3-HPA and could use both NAD^+ and $NADP^+$ as a cofactor. Among the substrates tested, the highest specific activity was noticed with propionaldehyde. Specifically, the K_m and V_{max} values were 1.18 mM and 0.35 U/mg, respectively, at pH 7.8 and 37°C. *Klebsiella pnuemoniae* can also naturally produce 3-HP from glycerol and has been shown to possess the *pdu* operon (Luo et al., 2012). To verify the role of PduP, a mutant *K. pneumoniae* lacking aldehyde dehydrogenase (PuuC) was developed (Luo et al., 2011b). The mutant strain could produce 3-HP to the same extent as the wild type, indicating that the metabolic pathway for 3-HP synthesis was unaltered. Furthermore, overexpression of *pduP* enhanced 3-HP production, confirming that PduP is involved in 3-HP production in *K. pneumoniae*. In their next study, Luo et al. (2012) constructed a *pduP* deletion mutant strain of *K. pneumoniae*. The growth achieved by both the wild-type and mutant strains was almost the same until 15 hours, but the cell viability of the mutant strain declined remarkably after 15 hours. The mutant strain produced about 1.0 g/L 3-HP in 15 hours without any further production, while the wild-type strain continued to synthesize about 2.0 g/L 3-HP in 25 hours. The low 3-HP titer and cell death in the case of the mutant strain was attributed to the accumulation of cytotoxic intermediate metabolite 3-HPA (~6 mM). Upon homologous overexpression of *pduP* in the mutant strain, the level of 3-HPA was reduced and normal glycerol metabolism and cell growth were restored. These studies clearly suggest that, in natural strains, *K. pneumoniae* utilizes the CoA-dependent pathway to produce 3-HP from glycerol and PduP plays a vital role in 3-HP production. However, in other studies, *K. pneumoniae* was shown to efficiently produce 3-HP from glycerol via the CoA-independent pathway when a proper ALDH such as PuuC was overexpressed (see section 14.4.3). At present, 3-HP production by the CoA-dependent pathway is in the preliminary phase and far from commercialization. More extensive studies are needed to understand the detailed catalytic mechanism underlying 3-HP synthesis through PduP.

14.4.3 Synthesis of 3-HP From Glycerol Through the CoA-Independent Pathway

The CoA-independent production of 3-HP from glycerol can be accomplished via a two-step reaction catalyzed by glycerol dehydratase (GDHt) and aldehyde dehydrogenase (ALDH). Two types of strains, a nonnatural (*E. coli*) and a natural producer (*K. pneumonia*) of the target product have been investigated for CoA-independent 3-HP production. *Klebsiella pneumoniae* has all of the genes for GDHt (e.g., DhaB) and ALDH, which are required for the conversion of glycerol to 3-HP. In addition, *K. pneumoniae* can naturally produce coenzyme B_{12}, which is an essential cofactor for DhaB. However, the level and/or mode of expression of the relevant genes for the CoA-independent pathway, especially ALDH, in wild-type *K. pneumoniae* are not appropriate for the optimal production of 3-HP. Conversely, *E. coli* cannot produce coenzyme B_{12} and does not have the genes for GDHt. Additionally, the expression level of 3-HPA-specific ALDH is low in wild-type *E. coli*. However, *E. coli* has well-developed genetic tool boxes and is easy to culture.

14.4.3.1 Escherichia coli Suthers and Cameron (2005) reported the construction of a recombinant *E. coli* by heterologously overexpressing the *dhaB* of *K. pneumoniae* and an ALDH of *Saccharomyces cerevisiae*. The developed organism could produce 3-HP from glycerol at 0.1 g/L. Raj et al. (2008) screened a highly active ALDH enzyme (AldH) for the conversion of 3-HPA to 3-HP from *E. coli* K-12 MG1655. They developed the recombinant *E. coli* SH254 by introducing AldH (pCDF/*aldH*) along with DhaB (pBAD/*dhaB*) of *K. pneumoniae* DSM 2026 into *E. coli* BL21 DE3.When cultivated in M9 minimal medium supplemented with vitamin B_{12}, strain SH254 could produce 3-HP at 0.58 g/L in flask culture and 2.5 g/L in a bioreactor experiment, respectively. However, the specific activities of the two enzymes in strain SH254 declined with time and the 3-HP production rate decreased in the later period. Furthermore, the imbalance between the activities of two enzymes (DhaB and AldH) caused intracellular accumulation of toxic 3-HPA. To address these problems, they developed a new recombinant strain, *E. coli* SH-BGA1, by cloning *aldH* in a high-copy plasmid and *dhaB* in a low-copy plasmid (Raj et al., 2009). They also introduced glycerol dehydratase reactivase (GdrAB), which is known to improve the stability of DhaB by replacing damaged coenzyme B_{12} molecules with new intact molecules during the dehydration reaction of glycerol (Mori et al., 1997). Furthermore, they briefly optimized fermentation conditions such as pH, Isopropyl β-D-1-thiogalactopyranoside (IPTG) concentration, liquid-to-flask volume ratio, and substrate concentration. As a result, they could produce 4.4 g/L 3-HP at the flask level in 48 hours and 31 g/L 3-HP in a glycerol fed-batch bioreactor culture in 72 hours with a molar yield of 0.35 on glycerol. To further improve 3-HP production, they constructed another recombinant strain, *E. coli* SH-BGK1, by cloning α-ketoglutaric semialdehyde dehydrogenase (KGSADH) of *Azospirillum brasilense* in place of AldH of the recombinant *E. coli* SH-BGA1 (Rathnasingh et al., 2009). In its purified form, KGSADH (16.6 ± 1.43 U/mg protein) was less active than AldH (38.10 ± 0.76 U/mg protein) with 3-HPA as substrate. However, the AldH activity in crude cell extract of SH-BGK1 (1.94 ± 0.15 U/mg protein) was higher than that of SH-BGA1 (1.56 ± 0.08 U/mg protein). The higher ALDH activity in SH-BGK1 was expected to reduce the accumulation of toxic intermediate 3-HPA during the conversion of glycerol to 3-HP. In a glycerol fed-batch bioreactor experiment, strain SH-BGK1 produced 38.7 g/L 3-HP in 72 hours, which is the highest titer reported to date using glycerol as substrate. Recently, Kwak et al. (2012) developed a recombinant *E. coli* by expressing DhaB and DhaR (reactivation factor of glycerol dehydratase) from *Lactobacillus brevis* and AldH from *E. coli*. The fed-batch cultivation of this recombinant strain with a two-step feeding strategy produced 14.3 g/L 3-HP under aerobic conditions (dissolved oxygen >10%). Acetate (4.2 g/L) and PDO (3.9 g/L) were the main by-products. Initially the cells were grown on glucose and once the culture optical density (OD) reached 100, the cells were fed with a mixture of glucose and glycerol.

14.4.3.2 Klebsiella pneumoniae *Klebsiella pneumoniae* has a *de novo* biosynthetic pathway for coenzyme B_{12} and well-developed metabolic pathways for glycerol assimilation (Ashok et al., 2011; Celińska, 2011). To divert 3-HPA to 3-HP in *K. pneumoniae*, oxidoreductases converting 3-HPA to PDO should be disrupted and an ALDH catalyzing the reaction of 3-HPA to 3-HP should be overexpressed. However, the conversion of 3-HPA to 3-HP is always accompanied with NADH generation (Jo et al., 2008); therefore, continuous regeneration of NAD^+ is critically important for uninterrupted 3-HP production. The regeneration of NAD^+ from NADH can be accomplished most efficiently via the electron transport chain (ETC) under fully aerobic conditions (Richardson, 2000). However, there are several concerns about adopting aerobic conditions for 3-HP production with *K. pneumoniae*. First, in the presence of oxygen, the production of coenzyme B_{12} is significantly suppressed (Keuth and Bisping, 1994; Ye et al., 1996). In addition, the DhaB enzyme is inactivated by oxygen (Xu et al., 2009), and the entire reductive pathway for glycerol metabolism is downregulated. Therefore, the use of *K. pneumoniae* as a host strain for 3-HP production is challenging despite the important advantage that the strain can naturally synthesize vitamin B_{12}.

In one study, Ashok et al. (2013a) developed recombinant *K. pneumoniae* by (i) overexpressing γglutamyl-γ-aminobutyraldehyde dehydrogenase (PuuC) of *K. pneumoniae* and (ii) disrupting two major oxidoreductases, DhaT and YqhD. PuuC, similar to AldH and KGSADH, was reported to have a high enzymatic activity for the conversion of 3-HPA to 3-HP, with NAD^+ as a cofactor (Raj et al., 2010). The recombinant *K. pneumoniae ΔdhaT ΔyqhD* (PuuC) could produce ~3.6 g/L 3-HP under microaerobic conditions in 12 hours of flask culture. However, when aeration was either increased or decreased, 3-HP production was seriously decreased due to insufficient production of coenzyme B_{12} (under high aeration) or reduced regeneration of NAD^+ (under low aeration). These findings demonstrate that maintaining proper aeration is critical to successful production of 3-HP with *K. pneumoniae*. In glycerol fed-batch bioreactor cultivation under proper aeration conditions, the strain could produce 3-HP at >28 g/L in 48 hours with a glycerol carbon yield >40%. In another study, Ashok et al. (2013b) attempted 3-HP production under anaerobic conditions in the presence of nitrate as an alternative electron acceptor to regenerate NAD^+. The addition of nitrate facilitated NAD^+ regeneration and reduced the intracellular NADH level. However, the presence of this external electron acceptor decreased DhaB activity and 3-HP production. High carbon flux through the anaerobic respiratory pathway initiated by glycerol kinase (GlpK) was also observed in the presence of nitrate (Figure 14.4). To reduce carbon flux through the oxidative pathway and improve 3-HP production, *glpK* and *dhaT* were eliminated. The resulting strain showed 3-HP production of 22 g/L in 42 hours in an anaerobic bioreactor experiment in the presence of nitrate. Good cell growth and improved intracellular NAD^+ levels were also observed.

14.4.4 Coproduction of 3-HP and PDO From Glycerol

In *K. pneumoniae*, 3-HP and PDO are derived from the same intermediate, 3-HPA. The fate of 3-HPA depends on the activity of relevant enzymes (oxidoreductases and 3-HPA-specific ALDH) and the availability of redox cofactors (NAD^+ and NADH). If a single product, either 3-HP or PDO, should be produced at high yield, one of the other types of enzymes should be disrupted and one kind of cofactor should be regenerated efficiently. When compared to gene disruptions, cofactor regeneration is much more challenging. In *K. pneumoniae*, cofactor regeneration is performed by oxidative metabolism of glycerol and, to this end, the carbon flow through oxidative pathway should become very high. One approach to deal with this redox balance problem is to produce both 3-HP and PDO together. If NADH or NAD^+ can be recycled within the reductive pathway by producing 3-HP and PDO together, the carbon flow through the oxidative pathway is reduced and a higher carbon yield of coproduction can be achieved. Furthermore, since anaerobic or microaerobic conditions can be employed for this coproduction, the problems associated with the expression of genes for vitamin B_{12} (*cob*) and glycerol assimilation (*dha*) in *K. pneumoniae* can be substantially alleviated. Several groups have adopted this approach and reported successful results with *K. pneumoniae*.

Zhu et al. (2009) developed a recombinant *K. pneumoniae* by introducing an ALDH from *E. coli*. The recombinant strain was grown under three different aeration conditions and the highest amount of 3-HP (2.8 g/L) and PDO (9.8 g/L) was achieved under microaerobic conditions. Ashok et al. (2011) examined coproduction using recombinant *K. pneumoniae* DSM 2026 overexpressing PuuC and lacking DhaT. The recombinant strain showed greatly improved DhaB activity; thus, the carbon flux through DhaB-mediated pathway was high. The removal of DhaT did not eliminate PDO production, but instead resulted in balanced production of 3-HP and PDO. These findings suggested that NADPH-dependent YqhD was responsible for PDO production in this recombinant strain. In one flask experiment conducted under microaerobic conditions with an initial glycerol concentration of 100 mM, equimolar 3-HP and PDO (40 mM each) were obtained in 24 hours. The overall yield of 3-HP and PDO was >80% and the production of by-products such as ethanol, acetate, and lactate was negligible. The fed-batch bioreactor cultivation of the same recombinant strain produced both compounds at a similar level (3-HP, 16.0 g/L; PDO, 16.8 g/L) within 24 hours. However, the cumulative yield of 3-HP and PDO on glycerol (mol/mol) was as low as 51% due to production of a significant amount of by-product acids and alcohols. In a coproduction study, Huang et al. (2012) investigated the effects of various AldH on the coproduction of 3-HP and PDO in *K. pneumoniae* and found that γ-glutamyl-γ-aminobutyraldehydedehydrogenase of *E. coli* was most efficient. Bioreactor cultivation of the recombinant *K. pneumoniae* overexpressing this enzyme yielded 3-HP and PDO at 24.4 and 49.3 g/L, respectively, in 24 hours under anaerobic conditions. The cumulative molar yield of the two metabolites was 0.61 (0.18 on 3-HP and 0.43 on PDO). In their next study, the recombinant strain was cultivated under anaerobic, microaerobic, and aerobic conditions. They found that the concentrations and yields (mol/mol) of 3-HP improved continuously with increase in aeration rate under microaerobic conditions (Huang et al., 2013). The maximum titer of 3-HP (48.9 g/L) was achieved at an aeration rate of 1.5 vvm along with 25.3 g/L PDO and 24.9 g/L lactic acid in 28 hours. The molar yields of 3-HP, PDO, and lactate were 0.41, 0.25, and 0.22, respectively. The concentration (48.9 g/L) and volumetric productivity (1.75 g/L·h) of 3-HP achieved in this investigation were the highest reported to date.

Although Ashok et al. (2011) and Huang et al. (2012) reported successful coproduction of 3-HP and PDO from glycerol by growing cells of *K. pneumoniae*, a significant amount of organic acids and alcohols were produced as by-products. To avoid the formation of by-products and improve the yield, Kumar et al. (2012b) employed resting cells of recombinant *K. pneumoniae* J2B strain overexpressing *KGSADH* for the coproduction of 3-HP and PDO. They grew cells under microaerobic conditions to maximize the whole-cell 3-HP production. The cells were then harvested and used for 3-HP production under nongrowing conditions. They found that fed-batch bioconversion in a 1.5-L bioreactor with 1.0 g-CDW/L under anaerobic conditions resulted in 11.3 g/L 3-HP and 15.2 g/L PDO in 12 hours with a cumulative yield of 0.71 (mol/mol). In addition to 3-HP and PDO, significant amount of lactic acid was also accumulated. In the next study, they constructed another recombinant strain of *K. pneumoniae* J2B by deleting lactate dehydrogenase (*ldhA*) and overexpressing *KGSADH*. The lactic acid production was completely abolished in this new recombinant. In addition, the final titer of 3-HP and PDO improved to 22.7 and 23.5 g/L, respectively, within 60 hours and the cumulative product yield was enhanced to 0.77 (Kumar et al., 2013).

14.5 MAJOR CHALLENGES FOR MICROBIAL PRODUCTION OF 3-HP

Bioconversion offers a safer and more viable alternative with the opportunity to produce a variety of chemicals under milder conditions. However, there are currently only a very few examples of bio-based processes outcompeting petro-based process such as yeast-based ethanol, yeast- or lactic acid bacteria-based lactic acid and *E. coli*-based PDO (Patel et al., 2006). For these bio-based chemicals to be economically feasible, a high product concentration (above 100 g/L), productivity (over 2 g/L·h) and substrate-to-product yield (above 50%) are required. For the production of 3-HP, several biological routes and recombinant strains have already been developed and examined. However, as summarized in Table 14.3, the results with most recombinant strains are still not satisfactory for commercial production of 3-HP. There are several barriers to microbiological production of 3-HP that differ depending on the kind of substrates and/or pathways employed. The following section covers these challenges and offers some possible solutions while focusing on processes that use glycerol as substrate.

14.5.1 Toxicity and Tolerance

In general, the toxic effects of organic acids are related to their ability to diffuse across the cell membrane. Inside the cell, these compounds dissociate and disrupt pH as well as the anion pool in the cytoplasm. The acidification of cytoplasm can ruin the integrity of purine bases and result in denaturing of essential enzymes, both of which seriously impair cell viability. The export of organic acids requires high energy, especially under low extracellular pH. van Maris et al. (2004) reported that under low pH, the metabolic energy requirement for product export may equal or exceed the metabolic energy yielded from product formation, causing the metabolic energy for growth and other essential cellular functions to become deficient. Once cells cannot maintain their intracellular pH within a physiologically acceptable range, serious

TABLE 14.3 3-HP Production by Different *Klebsiella pneumoniae* and *Escherichia coli* strains

Recombinant strain	Carbon source	Titer[a] (g/L)	$Y_{P/S}$ (mol/mol)	Q_P (g/L·h)	Aeration condition	Reference
Batch cultivation						
E. coli BL21_*dhaB_aldH*	Glycerol	0.6	0.48	0.02	Aerobic	Raj et al. (2008)
K. pneumoniae ME-308_*aldA*	Glycerol	2.8 (9.8)	0.10	0.12	Microaerobic	Zhu et al. (2009)
K. pneumoniae AK_*pduP*	Glycerol	1.4 (8.4)	0.07	0.06	Aerobic	Luo et al. (2011b)
E. coli BL21_*mcr_acc_pntAB*	Glucose	0.2	0.02	0.01	Aerobic	Rathnasingh et al. (2012)
Fed-batch cultivation						
E. coli BL21_*dhaB_aldH*	Glycerol	31.0	0.35	0.43	Aerobic	Raj et al. (2009)
E. coli SH254_*dhaB_KGSADH*	Glycerol	38.7	0.36	0.54	Aerobic	Rathnasingh et al. (2009)
E. coli BX3_0240	Glucose	49.0	0.46	0.71	Aerobic	Lynch et al. (2011)
E. coli BL21_*dhaB_dhaR_aldH*	Glucose and glycerol	14.3 (3.9)	0.15	0.26	Aerobic	Kwak et al. (2012)
K. pneumoniae DSM 2026 Δ*dhaT_puuC*	Glycerol	16.0 (16.3)	0.23	0.67	Aerobic	Ashok et al. (2011)

(continued)

TABLE 14.3 (*Continued*)

Recombinant strain	Carbon source	Titer[a] (g/L)	$Y_{P/S}$ (mol/mol)	Q_P (g/L·h)	Aeration condition	Reference
K. pneumoniae AK_*aldHk*	Glycerol	6.8 (22.7)	–	0.23	Aerobic	Luo et al. (2011a)
K. pneumoniae Cu	Glycerol	1.9 (7.8)	0.10	0.08	Aerobic	Luo et al. (2012)
K. pneumoniae WM3*pUC18kan_aldHec*	Glycerol	24.4 (49.3)	0.18	1.02	Anaerobic	Huang et al. (2012)
K. pneumoniae WM3 *pUC18kan_aldHec*	Glycerol	48.9 (25.3)	0.41	1.75	Microaerobic	Huang et al. (2013)
K. pneumoniae J2B Δ*dhaT_KGSADH*	Glycerol	16.3 (5.8)	0.40	0.30	Aerobic	Ko et al. (2012)
K. pneumoniae Δ*dhaT*Δ*yqhD_dhaB_puuC*	Glycerol	28.1 (3.3)	0.40	0.58	Aerobic	Ashok et al. (2013a)
K. pneumoniae Δ*glpK*Δ*dhaT_puuC*	Glycerol	22.0 (5.9)	0.30	0.46	Anaerobic	Ashok et al. (2013b)
K. pneumoniae J2B_*KGSADH*	Glycerol	11.3 (15.9)	0.27	0.94	Anaerobic	Kumar et al. (2012a)
K. pneumoniae J2BΔ*ldhA_KGSADH*	Glycerol	22.7 (23.5)	0.35	0.38	Aerobic	Kumar et al. (2012b)

PDO, 1,3-propanediol; 3-HP, 3-hydroxypropionic acid.

[a]The values shown in the parentheses are the PDO concentrations obtained along with 3-HP in coproduction studies.

damage to cellular metabolism and finally cell death occur (Holyoak et al., 1996; Brul and Coote, 1999; Halm et al., 2004). This situation is further aggravated when the metabolic pathway leading to product formation has no net ATP production. However, the mechanisms of most of the transport processes and their energy requirements are still not clear. Transport phenomena such as substrate uptake and product export are important factors that must be investigated extensively to improve most acid production processes. The problems associated with acid toxicity can be partly solved by adapting the microorganisms to low pH and conducting fermentation at pH values lower than the pKa of the acid. This would also circumvent the use of large amounts of the acid titrant, thereby lowering the overall production cost. However, the final 3-HP concentration may be lower (Sauer et al., 2008).

Similar to other acids, 3-HP causes pH-based growth inhibition. In addition, anion-specific interference with cellular metabolism has been observed. Warnecke et al. (2010) reported that 3-HP toxicity was related to inhibition of the chorismate and threonine super-pathways. Specifically, they tested the 3-HP tolerance of 10 different clones homologously expressing the enzyme(s) involved in chorismate and threonine super-pathways and observed a significant increase in 3-HP tolerance when measured by the minimum inhibitory concentration (MIC). Furthermore, they found that 3-HP tolerance was improved when some metabolites in the chorismate or threonine super-pathways were added to the culture medium. Taken together, the results of their study suggest that alleviation of the inhibition of either of these super-pathways can mitigate the toxicity of 3-HP.

3-HPA is a toxic intermediary compound produced during glycerol-based 3-HP production and its accumulation should be avoided. The accumulation of 3-HPA at 15–30 mM often completely stops 3-HP production. Although several hypotheses have been proposed for its bactericidal action, the exact mechanism is unclear (Barbirato et al., 1996; Rasch, 2002). 3-HPA accumulation is caused by an imbalance between the activities of DhaB and AldH and/or PDOR. In addition, PDO buildup can contribute to the accumulation of 3-HPA since PDOR can catalyze the reverse reaction (PDO→3-HPA) under high PDO concentrations. During 3-HP production from glycerol, 3-HPA accumulation can be avoided by balancing the rates of its production and consumption. This can be accomplished by either reducing DhaB activity or enhancing the activity of AldH (Celińska, 2011). The former strategy would reduce the overall rate of production; therefore, the latter is considered to be more desirable.

14.5.2 Redox Balance and By-products Formation

Regeneration of NAD^+ is a challenging issue in glycerol-based 3-HP production. Redox balance affects a broad range of genes, cellular functions, and metabolite profiles. Continuous regeneration of NAD^+ is necessary for high 3-HP production, and the easiest way to accomplish this without wasting carbons is to conduct the cultivation under fully aerobic conditions. However, the maintenance of a high oxygen concentration suppresses the expression of the *dha* operon (Celińska, 2011) and synthesis of vitamin B_{12}. If oxygen levels are reduced to microaerobic/anaerobic conditions, a large amount of by-products (lactic acid, acetic acid, and ethanol) are accumulated and 3-HP yield is decreased. It is challenging to determine and maintain optimal oxygen levels that allow continuous regeneration of NAD^+ while not interfering with the synthesis of vitamin B_{12} or proper carbon flux toward 3-HP production. Anaerobic 3-HP production in the presence of nitrate is one strategy to deal with this problem. The potential of this approach was demonstrated by the high 3-HP final titer, ~22 g/L, in one recombinant *K. pneumoniae* (Ashok et al., 2013). In this case, it is essential to maintain high nitrogenase activity to actively regenerate NAD^+ and prevent accumulation of the toxic nitrite. The NADH accumulation can also be alleviated by introducing NADH oxidase, which converts NADH into NAD^+ using molecular oxygen (Auzat et al., 1999; Hummel and Riebel, 2003). This method will potentially be useful when the rate of NADH oxidation by NADH dehydrogenase(s) in the ETC is limited by a low ATP synthesis rate due to low ATP synthetase activity. NADH oxidation can also be limited when intracellular ATP levels are high due to the rapid production of ATP by oxidative phosphorylation. Overexpression of transhydrogenase and interconversion of NADH and NADPH will also help to some extent, although these methods have never been tested.

14.5.3 Vitamin B_{12} Supply

Glycerol dehydratase (DhaB) is a vitamin B_{12}-dependent enzyme that catalyzes the free radical–mediated conversion of glycerol into 3-HPA. Coenzyme B_{12} is often inactivated during this reaction and should therefore be supplied continuously (Toraya, 2002). In many microorganisms, including *K. pneumoniae*, coenzyme B_{12} is not synthesized well under aerobic conditions, which are needed for active regeneration of NAD^+. *Pseudomonas denitrificans* is known to produce vitamin B_{12} under aerobic conditions. The use of *P. denitrificans* as a host can be an effective strategy for solving the problems associated with aerobic vitamin B_{12} synthesis. However, glycerol metabolism has not been reported in *P. denitrificans*, and this organism has not been studied as a recombinant host and no genetic tool box is currently available for this strain. The other option is introduction of the complete vitamin B_{12} biosynthetic pathway of *P. denitrificans* in *E. coli*. However, this will also be challenging since more than 20 genes of the vitamin B_{12} biosynthetic pathway should be properly expressed in *E. coli*.

14.6 ECONOMIC OUTLOOK

It is hard to predict the economical outlook for 3-HP as a bulk chemical because its market has not been established yet. However, the potential of biological 3-HP production can be evaluated based on the market of acrylic acid. Acrylic acid, to which 3-HP can be easily converted, is a high value and high volume chemical used in the production of broad range of materials such as plastic, fiber, coatings, paints, and super-adsorbent diapers. At present, around 4.5 million ton of acrylic acid per annum (worth US$11 billion) is produced and its growth rate is 4% every year

(Cargill, 2012). The market for super-adsorbents used in personal care (produced from acrylic acid) alone is more than one million ton per year. Currently, acrylic acid is derived by catalytic oxidation of propylene and its market price is approximately \$2.6/kg (varies according to the oil price).

Chemical dehydration of 3-HP to acrylic acid is simple, but results in loss of one water molecule (20% weight loss). Considering the cost of chemical conversion of 3-HP to acrylic acid and the weight loss, it is suggested that the price of 3-HP should not be more than 60% of that of acrylic acid, that is, \$1.5/kg. With the current bioprocess technology, production cost of acrylic acid using biologically synthesized 3-HP has been reported to be approximately \$1.1/kg from dextrose and \$0.84/kg from sucrose, respectively (OPXBio, 2012). The exact performance data for fermentation process used for the cost estimation are not known, but they are close to; final titer of 75 g/L, molar yield of 50%, and productivity of 1.5 g/L·h. If cheap crude glycerol (2–3 times less expensive than glucose) is used as carbon source, the production cost of 3-HP can be further reduced. Therefore, it is believed that biological 3-HP production will find its way to full commercialization in a few years.

14.7 FUTURE PROSPECTS

Over the last 10 years, biological 3-HP production has shown significant progress, and its commercial production is expected to be realized in the near future. However, many important issues that require further investigations still remain. For example, 3-HP tolerance in 3-HP-producing strains should be further improved to enhance its titer in culture broth. With glycerol as substrate, synthesis of vitamin B_{12}, regeneration of NAD^+, and 3-HPA accumulation should be carefully studied individually or in an integrated manner. The use of *P. denitrificans* as a recombinant host or anaerobic NAD^+ regeneration via nitrate reduction will be studied in greater detail to reconcile the requirement of the opposite conditions for vitamin B_{12} synthesis (anaerobic) and NAD^+ regeneration (aerobic). The careful examination and manipulation of by-products formation while fine-tuning internal redox balance will help improve the yield of 3-HP production from the carbon source used. Finally, an efficient and economical downstream process for separation and purification of 3-HP from the culture broth will be important. To simplify the downstream processes, it is desirable to conduct cultivation in medium with low pH (Jiang et al., 2009). This may be possible through development of yeast or *Lactobacillus* as recombinant hosts.

ACKNOWLEDGMENT

This study was supported financially by Korean Ministry of Education, Science and Technology through Advanced Biomass R&D Center (ABC) (Grant no. 2010-0029799), KAIST, Korea. We are grateful to the Brain Korea 21 program and the 2012 Post-Doc Development Program (Dr Vinod Kumar), Pusan National University for their financial assistance.

LIST OF ABBREVIATIONS

AccADBC	Acetyl-CoA carboxylase
ALDH	Aldehyde dehydrogenase
AldH	A specific aldehyde dehydrogenase from *E. coli*
BNICE	Biochemical Network Integrated Computational Explorer
DHA	Dihydroxyacetone
DHAP	Dihydroxyacetone phosphate
DhaB	Glycerol dehydratase from *K. pneumoniae*
DhaT	1,3-Propanediol oxidoreductase from *K. pneumoniae*
GlpK	Glycerol kinase
Glu	Glutamate
3-HPA	3-Hydroxypropionaldehyde
3-HP	3-Hydroxypropionic acid
α-KG	α-Ketoglutarate
KGSADH	α-Ketoglutaric semialdehyde dehydrogenase
LdhA	Lactate dehydrogenase
Mcr	Malonyl-CoA reductase
OAA	Oxaloacetate
PDO	1,3-Propanediol
PduL	Phosphotransacylase
PduP	Propionaldehyde dehydrogenase
PduQ	Propanol dchydrogcnase
PduW	Propionate kinase
PEP	Phosphoenolpyruvate
PntAB	Nicotinamide nucleotide transhydrogenase
PuuC	NAD^+-dependent gamma-glutamyl-gamma-aminobutyraldehyde dehydrogenase
Q_P	Volumetric productivity (g/L·h)
$Y_{P/S}$	Product yield (mol product/mol substrate)
YqhD	NADPH-dependent hypothetical oxidoreductase

REFERENCES

Andersen G, Björnberg O, Polakova S, Pynyaha Y, Rasmussen A, Møller K, Hofer A, Moritz T, Sandrinl MPB, Merico AM, Compagno C, Åkerlund HE, Gojkovic Z, Piškur J. (2008). A second pathway to degrade pyrimidine nucleic acid precursors in eukaryotes. *J Mol Biol*, 380(4), 656–666.

Ansede JH, Pellechia PJ, Yoch DC. (1999). Metabolism of acrylate to β-hydroxypropionate and its role in dimethylsulfoniopropionate lyase induction by a salt marsh sediment bacterium, *Alcaligenes faecalis* M3A. *Appl Environ Microbiol*, 65, 5075–5081.

Ashok S, Raj SM, Rathnasingh C, Park S. (2011). Development of recombinant *Klebsiella pneumoniae ΔdhaT* strain for the co-production of 3-hydroxypropionic acid and 1,3-propanediol from glycerol. *Appl Microbiol Biotechnol*, 90, 1253–1265.

Ashok S, Sankaranarayanan M, Ko Y, Jae K, Ainala SK, Kumar V, Park S. (2013a). Production of 3-hydroxypropionic acid from glycerol by recombinant *Klebsiella pneumoniae* *ΔdhaTΔyqhD* which can produce vitamin B_{12} naturally. *Biotechnol Bioeng*, 110, 511–524.

Ashok S, Raj SM, Ko Y, Sankaranarayanan M, Zhou S, Kumar V, Park S. (2013b). Effect of nitrate on glycerol metabolism and anaerobic production of 3-hydroxypropionic acid by recombinant *Klebsiella pneumoniae* overexpressing *puuC*. *Metab Eng*, 15, 10–24.

Auzat I, Regaud SC, Bras GL, Santos DD, Ogunniyi AD, Thomas IL, Garel JR, Paton JC, Trombe MC. (1999). The NADH oxidase of *Streptococcus pneumoniae*: its involvement in competence and virulence. *Mol Microbiol*, 34, 1018–1028.

Barbirato F, Soucaille P, Bories A. (1996). Physiologic mechanisms involved in accumulation of 3-hydroxypropionaldehyde during fermentation of glycerol by *Enterobacter agglomerans*. *Appl Environ Microbiol*, 62, 4405–4409.

Berg IA, Kockelkorn D, Buckel W, Fuchs G. (2007). A 3-hydroxypropionate/4-hydroxybutyrate autotrophic carbon dioxide assimilation pathway in Archaea. *Science*, 318, 1782–1786.

Bobik TA, Havemann GD, Busch RJ, Williams DS, Aldrich HC. (1999). The propanediol utilization (*pdu*) operon of *salmonella enteric* serovar typhimurium LT2 includes genes necessary for formation of polyhedral organelles involved in coenzyme B_{12}-dependent 1,2-propanediol degradation. *J Bacteriol*, 181, 5967–5975.

Brul S, Coote P. (1999). Preservative agents in food: mode of action and microbial resistance mechanisms. *Int J Food Microbiol*, 50, 1–17.

Cargill. (2012). Cargill and Novozymes have signed an agreement to develop technologies to produce acrylic acid from renewable raw materials. Available at: http://www.basf.com/group/corporate/en/function/conversions:/publish/content/news-and-media-relations/news-releases/downloads/2012/P363_NZ_BASF_Cargill_press-release_e.pdf.

Celińska E. (2011). *Klebsiella spp* as a 1, 3-propanediol producer—the metabolic engineering approach. *Crit Rev Biotechnol*, 32(3), 274–288.

Celińska E, Grajek W. (2009). Biotechnological production of 2,3-butanediol—current state and prospects. *Biotechnol Adv*, 27, 715–725.

CEP. (2003). *Engineering 1,3-Carbon Molecules*. New York: AIChE. Available at: www.cepmagazine.org.

Danner H, Braun R. (1999). Biotechnology for the production of commodity chemicals from biomass. *Chem Soc Rev*, 28, 395–405.

da Silva GP, Mack M, Contiero J. (2009). Glycerol: a promising and abundant carbon source for industrial microbiology. *Biotechnol Adv*, 27, 30–39.

Dave H, Ramakrishna C, Desai JD. (1996). Degradation of acrylic acid by fungi from petrochemical activated sludge. *Biotechnol Lett*, 18, 963–964.

Dharmadi Y, Murarka A, Gonzalez R. (2006). Anaerobic fermentation of glycerol by *Escherichia coli*: a new platform for metabolic engineering. *Biotechnol Bioeng*, 94(5), 821–829.

Dugar D, Stephanopoulos G. (2011). Relative potential of biosynthetic pathways for biofuels and bio-based products. *Nat Biotechnol*, 29(12), 1074–1078.

Garai-Ibabe G, Ibarburu I, Berregi A, Claisse O, Lonvaud-Funel A, Irastorza A, Dueñas MT. (2008). Glycerol metabolism and bitterness producing lactic acid bacteria in cidermaking. *Int J Food Microbiol*, 121, 253–261.

Gokarn RR, Selifonova OV, Jessen HJ, Steven JG, Selmer T, Buckel W. (2001). 3-hydroxypropionic acid and other organic compounds. Patent Application PCT/US2001/043607.

Gokarn RR, Selifonova OV, Jessen HJ, Gort SJ, Selmer T, Buckel W. (2007). 3-hydroxypropionic acid and other organic compounds. US Patent 7,186,541 B2.

Halm M, Hornbæk T, Arneborg N, Sefa-Dedeh S, Jespersen L. (2004). Lactic acid tolerance determined by measurement of intracellular pH of single cells of *Candida krusei* and *Saccharomyces cerevisiae* isolated from fermented maize dough. *Int J Food Microbiol*, 94, 97–103.

Hatti-Kaul R, Tornvall U, Gustafsson L, Borjesson P. (2007). Industrial biotechnology for the production of bio-based chemicals—a cradle-to-grave perspective. *Trends Biotechnol*, 25, 119–124.

Henry CS, Broadbelt LJ, Hatzimanikatis V. (2010). Discovery and analysis of novel metabolic pathways for the biosynthesis of industrial chemicals: 3-hydroxypropanoate. *Biotechnol Bioeng*, 106, 462–473.

Herrmann G, Selmer T, Jessen HJ, Gokarn RR, Selifonova O, Gort SJ, Buckel W. (2005). Two beta-alanyl-CoA:ammonia lyases in *Clostridium propionicum*. *FEBS J*, 272, 813–821.

Holo H. (1989). *Chloroflexus aurantiacus* secretes 3-hydroxypropionate, a possible intermediate in the assimilation of CO_2 and acetate. *Arch Microbiol*, 151, 252–256.

Holyoak CD, Stratford M, McMullin Z, Cole MB, Crimmins K, Broen AJP, Coote P. (1996). Activity of the plasma membrane H^+-ATPase and optimal glycolytic flux are required for rapid adaptation and growth in the presence of the weak acid preservative sorbic acid. *Appl Environ Microbiol*, 62, 3158–3164.

Huang Y, Li Z, Shimizu K, Ye Q. (2012). Simultaneous production of 3-hydroxypropionic acid and 1,3-propanediol from glycerol by a recombinant strain of *Klebsiella pneumoniae*. *Bioresour Technol*, 103, 351–359.

Huang Y, Li Z, Shimizu K, Ye Q. (2013). Co-production of 3-hydroxypropionic acid and 1,3-propanediol by *Klebseilla pneumoniae* expressing *aldH* under microaerobic conditions. *Bioresour Technol*, 128, 505–512.

Hügler M, Huber H, Stetter KO, Fuchs G. (2003). Autotrophic CO_2 fixation pathways in archaea (Crenarchaeota). *Arch Microbiol*, 179, 160–173.

Hummel W, Riebel B. (2003). Isolation and biochemical characterization of a new NADH oxidase from *Lactobacillus brevis*. *Biotechnol Lett*, 25, 51–54.

Ishii M, Chuakrut S, Arai H, Igarashi Y. (2004). Occurrence, biochemistry and possible biotechnological application of the 3-hydroxypropionate cycle. *Appl Microbiol Biotechnol*, 64, 605–610.

Jiang X, Meng X, Xian M. (2009). Biosynthetic pathways for 3-hydroxypropionic acid production. *Appl Microbiol Biotechnol*, 82, 995–1003.

Jo JE, Raj SM, Rathnasingh C, Selvakumar E, Jung WC, Park SH. (2008). Cloning, expression, and characterization of an aldehyde dehydrogenase from *Escherichia coli* K-12 that utilizes 3-hydroxypropionaldehyde as a substrate. *Appl Microbiol Biotechnol*, 81, 51–60.

Johnson EA, Lin ECC. (1987). *Klebsiella pneumoniae* 1, 3-propanediol: NAD^+ oxidoreductase. *J Bacteriol*, 169, 2050–2054.

Keuth S, Bisping B. (1994). Vitamin B_{12} production by *Citrobacter freundii* or *Klebsiella pneumoniae* during tempeh fermentation and proof of enterotoxin absence by PCR. *Appl Environ Microbiol*, 60(5), 1495–1499.

Krauter H, Willke T, Vorlop K-D. (2012) Production of high amounts of 3-hydroxypropionaldehyde from glycerol by *Lactobacillus reuteri* with strongly increased biocatalyst lifetime and productivity. *N Biotechnol*, 29(2), 211–217.

Kumar V, Mugesh S, Jae K, Durgapal M, Ashok S, Ko Y, Sarkar R, Park S. (2012). Co-production of 3-hydroxypropionic acid and 1,3-propanediol from glycerol using resting cells of *Klebsiella pneumoniae* J2B strain with overexpression of KGSADH. *Appl Microbiol Biotechnol*, 96, 373–383.

Kumar V, Mugesh S, Durgapal M, Zhou S, Ko Y, Ashok S, Sarkar R, Park S. (2013a). Simultaneous production of 3-hydroxypropionic acid and 1,3-propanediol from glycerol using resting cell system of D-lactic acid deficient mutant of *Klebsiella pneumoniae* with overexpression of KGSADH. *Bioresour Technol*, 135, 555–563.

Kumar V, Ashok S, Park S. (2013b). Recent advances in biological production of 3-hydroxypropionic acid. *Biotechnol Adv*, 31, 945–961.doi:10.1016/j.biotechadv.2013.02.008

Ko Y, Ashok S, Zhou S, Kumar V, Park S. (2012) Aldehyde dehydrogenase activity is important to the production of 3-hydroxypropionic acid from glycerol by recombinant *Klebsiella pneumoniae. Process Biochem*, 47, 1135–1143.

Kwak S, Park Y-C, Seo J-H. (2012). Biosynthesis of 3-hydroxypropionic acid from glycerol in recombinant *Escherichia coli* expressing *Lactobacillus brevis dhaB* and *dhaR* gene clusters and *E. coli* K-12 *aldH. Bioresour Technol*, 135, 432–439.

Leal NA, Havemann GD, Bobik TA. (2003). PduP is a coenzyme-A-acylating propionaldehyde dehydrogenase associated with the polyhedral bodies involved in B_{12}-dependent 1, 2-propanediol degradation by a *Salmonella enterica* serovar Typhimurium LT12. *Arch Microbiol*, 180, 353–361.

Liao HH, Gokarn RR, Gort SJ, Jessen HJ, Selifonova OV. (2005). Alanine 2,3-aminomutase. Patent Application EP1575881.

Liao HH, Gokarn RR, Gort SJ, Jessen HJ, Selifonova OV. (2007a). Production of 3-hydropropionic acid using beta-alanine/pyruvate aminotransferase. Patent Application 20070107080.

Liao HH, Gokarn RR, Gort SJ, Jessen HJ, Selifonova OV. (2007b). Alanine 2,3-Aminomutase. US Patent 7,309,597 B2.

Liu YN, Leal A, Sampson EM, Johnson CL, Havemann GD, Bobik TA. (2007). PduL is an evolutionarily distinct phosphotransacylase involved in B_{12}-dependent 1,2-propanediol degradation by *Salmonella enterica* serovar Typhimurium LT2. *J Bacteriol*, 189, 1589–1596.

Luo LH, Seo JW, Baek JO, Oh BR, Heo SY, Hong WK, Kim DH, Kim CH. (2011a). Identification and characterization of the propanediol utilization protein PduP of *Lactobacillus reuteri* for 3-hydroxypropionic acid production from glycerol. *Appl Microbiol Biotechnol*, 89, 697–703.

Luo LH, Seo JW, Oh BR, Seo PS, Heo SY, Hong WK, Kim DH, Kim CH. (2011b). Stimulation of reductive glycerol metabolism by overexpression of an aldehyde dehydrogenase in a recombinant *Klebsiella pneumoniae* strain defective in the oxidative pathway. *J Ind Microbiol Biotechnol*, 38, 991–999.

Luo LH, Kim CH, Heo SY, Oh BR, Hong WK, Kim S, Kim DH, Seo JW. (2012). Production of 3-hydroxypropionic acid through propionaldehyde dehydrogenase PduP mediated biosynthetic pathway in *Klebsiella pneumoniae. Bioresour Technol*, 103, 1–6.

Lynch MD, Gill RT, Warnecke LT. (2011). Method for producing 3-hydroxypropionic acid and other products. PCT Patent 038,364 A1.

Marx A, Wendisch VF, Rittmann D, Buchholz S. (2007). Microbiological production of 3-hydroxypropionic acid. Patent Application WO/2007/042494.

Meng D-C, Shi Z-Y, Wu L-P, Zhou Q, Wu Q, Chen J-C, Chen G-Q. (2012). Production and characterization of poly(3-hydroxypropionate-co- 4-hydroxybutyrate) with fully controllable structures by recombinant *Escherichia coli* containing an engineered pathway. *Metab Eng*, 14, 317–324.

Mori K, Tobimatsu T, Hara T, Toraya T. (1997). Characterization, sequencing, and expression of the genes encoding a reactivating factor for glycerol-inactivated adenosylcobalamin-dependent diol dehydratase. *J Biol Chem*, 272, 32034–32041.

Nakamura CE, Whited GM. (2003). Metabolic engineering for the microbial production of 1,3-propanediol. *Curr Opin Biotechnol*, 14, 454–459.

OPXBio. (2012). Bio-acrylic acid on the way. Available at: http://greenchemicalsblog.com/2012/09/01/5060/.

Patel M, Crank M, Dornburg V, Hermann B, Roes L, Hüsing B, Overbeek L, Terragni F, Recchia E. (2006). Medium and long-term opportunities and risks of the biotechnological production of bulk chemicals from renewable resources—the potential of white biotechnology. European Commission. Available at: http://www.chem.uu.ul/brew/.

Pina CD, Falletta E, Rossi M. (2011). A green approach to chemical building blocks. The case of 3-hydroxypropanoic acid. *Green Chem*, 13, 1624–1632.

Raj SM, Rathnasingh C, Jo JE, Park S. (2008). Production of 3-hydroxypropionic acid from glycerol by a novel recombinant *Escherichia coli* BL21 strain. *Process Biochem*, 43, 1440–1446.

Raj SM, Rathnasingh C, Jung WC, Park SH. (2009). Effect of process parameters on 3-hydroxypropionic acid production from glycerol using a recombinant *Escherichia coli*. *Appl Microbiol Biotechnol*, 84, 649–657.

Raj SM, Rathnasingh C, Jung WC, Selvakumar E, Park S. (2010). A Novel NAD^+-dependent aldehyde dehydrogenase encoded by the *puuC* gene of *Klebsiella pneumoniae* DSM 2026 that utilizes 3-hydroxypropionaldehyde as a substrate. *Biotechnol Bioprocess Eng*, 15(1), 131–138.

Rasch M. (2002). The influence of temperature, salt and pH on the inhibitory effect of reuterin on *Escherichia coli*. *Int J Food Microbiol*, 72, 225–231.

Rathnasingh C, Raj SM, Jo JE, Park SH. (2009). Development and evaluation of efficient recombinant *Escherichia coli* strains for the production of 3-hydroxypropionic acid from glycerol. *Biotechnol Bioeng*, 104, 729–739.

Rathnasingh C, Raj SM, Lee Y, Catherine C, Ashok S, Park S. (2012). Production of 3-hydroxypropionic acid via malonyl-CoA pathway using recombinant *Escherichia coli* strains. *J Biotechnol*, 157(4), 633–640.

Richardson DJ. (2000). Bacterial respiration: a flexible process for a changing environment. *Microbiology*, 146, 551–571.

Sauer M, Porro D, Mattanovich D, Branduardi P. (2008). Microbial production of organic acids: expanding the markets. *Trends Biotechnol*, 26, 100–108.

Saxena RK, Anand P, Saran S, Isar J. (2009). Microbial production of 1,3-propanediol: recent developments and emerging opportunities. *Biotechnol Adv*, 27, 895–913.

Schwarz M, Köpcke B, Weber R, Sterner O, Anke H. (2004). 3-Hydroxypropionic acid as a nematicidal principle in endophytic fungi. *Phytochemistry*, 65, 2239–2245.

Sobolov M, Smiley FA. (1960). Metabolism of glycerol by an acrolein-forming *Lactobacillus*. *J Bacteriol*, 79, 261–266.

Strauss G, Fuchs G. (1993). Enzymes of a novel autotrophic CO_2 fixation pathway in the phototrophic bacterium *Chloroflexus aurantiacus*, the 3-hydroxypropionate cycle. *Eur J Biochem*, 215, 633–643.

Strauss G, Eisenreich W, Bacher A, Fuchs G. (1992). ^{13}C-NMR study of autotrophic CO_2 fixation pathways in the sulfur-reducing Archaebacterium *Thermoproteus neutrophilus* and in the phototrophic Eubacterium *Chloroflexus aurantiacus*. *Eur J Biochem*, 205, 853–866.

Suthers PF, Cameron DC. (2005). Production of 3-hydroxypropionic acid in recombinant organisms. US Patent 6,852,517.

Takamizawa K, Horitsu H, Ichikawa T, Kawai K, Suzuki T. (1993). β-hydroxypropionic acid production by *Byssochlamys* sp. grown on acrylic acid. *Appl Microbiol Biotechnol*, 40, 196–200.

Talarico TD, Casas IA, Chung TC, Dobrogosz WJ. (1988). Production and isolation of reuterin, a growth inhibitor produced by *Lactobacillus reuteri*. *Antimicrob Agents Chemother*, 32, 1854–1858.

Toraya T. (2002). Enzymatic radical catalysis: coenzyme B_{12}-dependent diol dehydratase. *Chem Rec*, 2, 352–366.

van Maris AJA, Konings WN, Van Dijken JP, Pronk JT. (2004). Microbial export of lactic and 3-hydroxypropanoic acid: implications for industrial fermentation processes. *Metab Eng*, 6(4), 245–255.

Warnecke TE, Lynch MD, Karimpour-Fard A, Lipscomb ML, Handke P, Mills T, Ramey CJ, Hoang T, Gill RT. (2010). Rapid dissection of a complex phenotype through genomic-scale mapping of fitness altering genes. *Metab Eng*, 12, 241–250.

Werpy T, Petersen G. (2004). Top value added chemicals from biomass, vol 1: results of screening for potential candidates from sugars and synthesis gas. US Department of Energy. Available at: http://www.osti.gov/bridge.

Willke T, Vorlop K-D. (2004). Industrial bioconversion of renewable resources as an alternative to conventional chemistry. *Appl Microbiol Biotechnol*, 66, 131–142.

Xu XL, Zhang GL, Wang LW, Ma BB, Li C. (2009). Quantitative analysis on inactivation and reactivation of recombinant glycerol dehydratase from *Klebsiella pneumoniae* XJPD-Li. *J Mol Catal B Enzym*, 56(2–3), 108–114.

Xue J, Murrieta CM, Rule DC, Miller WK. (2008). Exogenous or L-rhamnose-derived 1,2-propanediol is metabolized via a pduD-dependent pathway in *Listeria innocua*. *Appl Environ Microbiol*, 74, 7073–7079.

Yasuda S, Mukoyama M, Horikawa H, Toraya T, Morita H. (2007). Process for producting 1,3-propanediol and or 3-hydroxypropionic acid. US Patent 20070148749.

Ye K, Shijo M, Jin S, Shimizu K. (1996). Efficient production of vitamin B_{12} from propionic acid bacteria under periodic variation of dissolved oxygen concentration. *J Ferment Bioeng*, 82, 484–491.

Zhou Q, Shi Z-Y, Meng DC, Wu Q, Chen J-C, Chen G-Q. (2011). Production of 3-hydroxypropionate homopolymer and poly(3-hydroxypropionate-co-4-hydroxybutyrate) copolymer by recombinant *Escherichia coli*. *Metab Eng*, 13, 777–785.

Zhu J-G, Ji X-J, Huang H, Du J, Li S, Ding Y-Y Ding. (2009). Production of 3-hydroxypropionic acid by recombinant *Klebsiella pneumoniae* based on aeration and ORP controlled strategy. *Korean J Chem Eng*, 26, 1679–1685.

CHAPTER 15

Fumaric Acid Biosynthesis and Accumulation

ISRAEL GOLDBERG and J. STEFAN ROKEM

Department of Microbiology and Molecular Genetics, The Institute of Medical Research Israel-Canada, The Hebrew University—Hadassah Medical School, Jerusalem, Israel

---

Bioprocessing of Renewable Resources to Commodity Bioproducts, First Edition.
Edited by Virendra S. Bisaria and Akihiko Kondo.

Abstract

With the rapidly increased production cost of fossil-(petroleum)-based chemicals, it has been suggested that the biological production of fumaric acid from renewable resources can replace current petrochemically based maleic anhydride/acid process. This chapter reviews the accumulated knowledge and recent research, as well as a critical analysis of future perspectives to develop economically competitive bio-based fumaric acid processes to replace the petrochemical route.

15.1 INTRODUCTION

In 1817, the dry distillation of malic acid by Braconnet and independently by Vauquelin led to the discovery of two isomeric acids, that is, maleic acid (the *cis*-form) and fumaric acid (the *trans*-form) (Meek, 1975). Fumaric acid is a naturally occurring acid; that is, it is found in plants that belong to the genus *Fumaria* (e.g., *Fumaria officinalis*), a common European herb and a climbing annual plant, hence its name. Fumaric acid (2-butenedioic acid; *trans*-1,2-ethylenedicarboxylic acid, $HO_2CCH{=}CHCO_2H$) is a symmetric, unsaturated dicarboxylic acid.

15.1.1 Uses

Since 1946 fumaric acid is widely used in the food industry. As a food additive, fumaric acid is used as an acidity regulator (compound E297). It is nontoxic (GRAS—generally recognized as safe) and is the least expensive of the food grade acids. Fumaric acid is a weak acid which forms salts and diesters; the salts, like sodium and calcium fumarate, and esters, like dimethyl fumarate, are known as fumarates. This acid undergoes additions across the double bond. Its solubility in water is low, and, therefore, in order to increase its application in various foods a cold water-soluble (CWS) fumaric acid, which contains a wetting agent (e.g., dioctyl sodium sulfosuccinate) is used. Fumaric acid is a valuable intermediate for preparing edible products such as L-malic acid and L-aspartic acid (a component of aspartame) (Goldberg and Rokem, 2009). Fumaric acid can be used as a supplement in cattle feed causing a reduction in methane emission (Roa Engel et al., 2008). The dicarboxylic nature of fumaric acid, as well as the unsaturation of this molecule, makes it an important intermediate for polymer (biodegradable plastic) production. Other industrial uses of fumaric acid are in jet printing inks, plastics surface coating, and paper sizing, and as an intermediate in the preparation of nontoxic, greater hardness, unsaturated polyester, and alkyd resins. Fumaric acid is used by the pharmaceutical industry to produce alexipharmic sodium dimercaptosuccinate and ferrous fumarate, as an optical bleaching agent, in formulations for alternative medicine or as fumaric acid esters (such as ethyl hydrogen fumarate, monoethyl fumarate, and dimethyl fumarate to treat psoriasis patients who are unable to produce fumaric acid in the skin when exposed to light) and multiple sclerosis (Goldberg et al., 1991; Moharregh-Khiabani, 2009; Fox and Gold, 2011). Recently, it was found that fumaric acid is one of the antibacterial components in *Aloe vera* L. (He et al., 2011). Its abilities to be converted into pharmaceutical products and act as starting material for polymerization and esterification reactions have led the US Department of Energy to designate fumaric acid among the top 12 biomass building-block chemicals with potential to significantly enhance the economy (Werpy and Petersen 2004). Food and beverages accounted for 33% of world consumption of fumaric acid in 2009, followed by rosin paper sizes (20%), unsaturated polyester resins (18.6%), and alkyd resins (12.3%) (Chemical Economics Handbook (CEH) report Fumaric Acid, April 2010 http://www.ihs.com/products/chemical/planning/ceh/fumaric-acid.aspx, retrieved October 30, 2012).

15.1.2 Production

Fumaric acid is produced as a colorless, crystalline powder with a fruit-like taste (a fruit acid). It was first prepared from succinic acid (Volhard, 1892). The traditional chemical synthesis involved oxidation of furfural obtained from the processing of maize, using chlorate in the presence of a vanadium-based catalyst (Milas, 1943). In the early 1940s, fumaric acid was made by fermentation on a commercial scale by Chas. Pfizer & Co., (about 4000 tons/year) using a strain of the fungus *Rhizopus* (probably *Rhizopus arrhizus*, later named *Rhizopus oryzae*[1]) (Rhodes et al., 1959, 1962; Goldberg et al., 1991).[2] The biological production of fumaric acid was stopped when the chemical synthesis via maleic acid became economically more attractive (Miall, 1978; Goldberg et al., 1991). Since fumaric acid is a symmetrical molecule having no isomers the biological process offers no specific advantage over the chemical process. Thus, fumaric acid is presently produced through the isomerization of maleic acid (or maleic anhydride), obtained from a catalytic vapor phase oxidation of benzene or C4 hydrocarbons (such as *n*-butane or *n*-butane–*n*-butene mixtures) (Roa Engel et al., 2008). Part of the production pathway can be replaced with enzymatic conversion of maleic acid to fumaric acid using whole cells of *Pseudomonas alcaligenes* sp. or thermostable maleate isomerase from various *Bacillus* sp. This

[1]Throughout the text in this chapter we will use the name *Rhizopus oryzae*.

[2]From a historic perspective, fumaric acid production by *Rhizopus oryzae*, as well as the production of citric and gluconic acids by *Aspergillus niger*, has great importance as the first true examples of industrial scale biotechnological processes (De Jongh and Nielsen, 2007). But even more important is the fact that the successful submerged penicillin production is attributed to the transfer of know-how from the submerged fumaric acid production process using a filamentous fungus. The later process was used as a model for scale-up and fermentation techniques for the penicillin production process (De Jongh and Nielsen, 2007). This historic fact is documented in the letter that was sent on June 6, 1941, by Mr. Gordon Cragwall of Chas. Pfizer & Co., then a small manufacturer of chemicals, to Dr. Martin Henry Dawson, who was the first to employ penicillin parenterally with a therapeutic response as an immediate goal. In this letter Cragwall wrote: "As you know, we probably produce more mold than any other firm in the country ... In addition to citric acid we are also producing gluconic and fumaric acids by fermentative processes ... We are equipped to produce molds in large quantities and maintain pure cultures of a wide variety of organisms. Among those with which we have worked we might mention a large number of strains of penicillia, aspergilla, mucor, rhyzopi, and others" (Hobby, 1985).

enzymatic conversion occurs without the formation of the unfavorable by-products obtained in the chemical process. Thus, a maximum conversion yield (95%) of maleic acid to fumaric acid was achieved (Roa Engel et al., 2008). Recently, a novel technology was described for a new route to synthesize maleic acid from renewable furfural. Using dioxygen as oxidant, the simple copper salts can catalyze oxidation of furfural to maleic acid in aqueous solution (Shi et al., 2011).

There is some discrepancy in annual volumes of fumaric acid produced, whereas Roa Engel et al., (2008) report on world production of 90,000 ton in 2007, a report by a Chinese consulting company, CCM International Limited, states a volume of 105,000 tons/year for the PRC for 2006 (http://www.slideshare.net/ccminternational/production-and-market-of-fumaric-acid-in-china, retrieved May 6, 2012). Between 2006 and 2009, world capacity for fumaric acid grew at an average annual rate of 1.2%, outpacing world consumption, which declined at an average annual rate of 1.1% during the same period (CEH report Fumaric Acid, April 2010). According to Myriant, an integrated biorefining company, the global market for fumaric acid is $1.7 billion (http://www.icis.com/blogs/green-chemicals/2011/05/myriant-files-for-ipo.html, retrieved June 14, 2012).

During recent years, the continuous and the dramatic increase of the petroleum prices, with a forecast for even higher prices in the future, have reintroduced the interest for the biological manufacture (fermentation) of bulk chemicals, such as fumaric acid. The demand and price of the product will determine if the biological process can compete with the chemical production. It is of importance to understand not only what is the potential for production by fermentation, but also the changes that occur in raw material and process conditions for the chemical route. It should be noted that much of such information is not available in the public domain, so rough estimates, based on the available price for raw material and product will be used for this analysis. For bulk chemicals with price of >$1/kg the price is affected mainly by the cost of raw material (Willke and Vorlop, 2004). The fermentation is considered as a sustainable process, which involves net carbon dioxide fixation, may now even be an economically viable process for acid production (Roa Engel et al., 2008; Zhang and Yang, 2011; Zhang et al., 2013; Xu et al., 2012; Ferreira et al., 2013). This chapter reviews the accumulated knowledge and recent research, as well as the developments of technologies to improve the bio-based fumaric acid process and scrutinizes the possibility of the biological process to be economically competitive to the petrochemical route.

15.2 MICROBIAL SYNTHESIS OF FUMARIC ACID

15.2.1 Producer Organisms

There are several organisms that can accumulate fumaric acid when given the proper conditions. *Rhizopus nigricans* was the first fungus described to produce fumaric acid by Ehrlich (1911). The main organism used in industry and most extensively described in the literature is *R. oryzae* earlier termed *R. arrhizus*. A survey of *Mucorales* identified in addition to *Rhizopus* also *Mucor*, *Cunninghamella*, and *Circinella* species as able to accumulate fumaric acid.

Rhizopus oryzae is a group of fungi with considerable variability described for different isolated species. This has led to an ongoing discussion, with frequent suggestions of new species names and with recurrent changes in names. A revision of the genus reported in 1984 used strains of the Dutch CBS culture collection (Schipper 1984). The criteria used were morphological and species delimitation was derived from mating experiments. Abe et al. (2007) suggested to include the organic acid production profile as a criterion for the taxonomy of the genus. Two distinct groups were established based on the analysis of the internal transcribed spacer of rDNA and among others the genes for lactate dehydrogenase B, actin and translation elongation factor-1 α. The two acid patterns (fumaric/malic and lactic) correlated with these sequences, and with the presence of 1 or 2 lactate dehydrogenases (*ldhA* and *ldhB*, see section 15.4). It was suggested to reclassify the fumaric–malic acid producers to *Rhizopus delemar* and keep *R. oryzae* for lactic acid producers. The proposed classification is rarely used in the literature. The synonymy of *R. arrhizus* and *R. oryzae* is well established, and even though the former was described earlier, the preferred name today is *R. oryzae*, which is used in this chapter.

Different strains of *R. oryzae* have been used for research of fumaric acid production. The research in the public domain is on a laboratory scale using either shake flasks or laboratory-sized fermentors. The majority of studies have been performed on two different strains, namely ATCC 10260 (NRRL 1526) and ATCC 20344 (NRRL 6400). It is not clear which of these strains is the favored strain on an industrial scale. It is most probable that mutants with better characteristics will be developed for use in an industrial process. In one study mutants were isolated after treatment with nitrogen ion implantation. Single-layer spore suspensions on plates were implanted by N^+ beams with a dose of 2×10^{15} ions/cm at 10 keV energy. The isolated mutants had enhanced glucoamylase activity. This allowed for the use of starches as direct carbon source for fumaric acid accumulation with simultaneous saccharification and fermentation resulting in 44.1 g/L fumaric acid from an initial total sugar concentration of corn starch of 100 g/L (Deng et al., 2012). Using femtosecond laser irradiation mutants were isolated that produced fumaric acid to a concentration of 49.4 g/L with a yield of 0.56 g fumaric acid per gram glucose (Yu et al., 2012). This should be compared to the final concentrations reported when fatty acid where added either as Tweens (40 and 80) or corn, soybean, or cottonseed oil with a maximal fumaric acid concentration of 60 g/L with a productivity of 0.63 g/L·h in shake flasks (Goldberg and Stieglitz, 1985). In a laboratory size fermentor, 107 g/L of fumaric acid with the highest literature value for productivity of 2 g/L·h were obtained by Ng et al. (1986), as compared to values of 49.4 g/L and 0.59 g/L·h, respectively, for the mutant recently described (Yu et al., 2012).

Roa Engel et al. (2008) reviewed data for fumaric acid accumulation by different strains of *Rhizopus*. The most efficient process was with a rotary biofilm contractor studied by Cao et al. (1996) with volumetric productivities of 4.25 g/L·h compared with 2 g/L·h for a stirred bioreactor (Ng et al., 1986). The fungus is grown to form a biofilm by self-immobilization on plastic discs, the formed acid is absorbed on a resin to avoid feedback inhibition, and the organic acid recovery is integrated with the fermentation. The scaleup of such a system would involve novel technological solutions for an integrated process (Fu et al., 2010).

15.2.2 Carbon Sources

The expense of the feedstock for production, especially the carbon source, has a decisive influence on the price of low value–high volume compounds such as fumaric acid. The main production of fumaric acid is by synthetic means based on petroleum-derived maleic anhydride as the raw material. The price of maleic anhydride is two and a half times more expensive than the main fermentation process raw material, glucose (prices in the beginning of 2012). *Rhizopus oryzae* is capable of utilizing simple sugars such as glucose; however, sucrose is poorly metabolized by *R. oryzae* and xylose as carbon source gives low fumaric acid productivities. There are reports that the fungus can also utilize a range of low cost substrates that are less refined than the simple sugars, such as starches, molasses, corn mash, corn steep liquor, potato flour, rice bran, and cassava bagasse.

15.2.3 Solid-State Fermentations

Processes with solid-state fermentations have been investigated using among others corn distillers grain with solubles, where heat or acid hydrolysis treatment of the raw material was required to obtain high concentrations of fumaric acid (West, 2008). The solid-state fermentation is slower than the submerged fermentation, but the capital cost of the former is considerably lower. Cheap substrates such as acid hydrolysate of dairy manure resulted in low yield of 0.15 g/g with a final concentration of 4.9 g/L of fumaric acid (Liao et al., 2008). The use of hydrolysates of lignocellulosic materials have the inherent problem of diauxie (sequential utilization) resulting from the blend of mainly two sugar monomers (glucose and xylose) in hydrolysates. The productivity of fumaric acid formation is so low that the time required for its accumulation puts the economics of the process in question. To circumvent sequential utilization, xylose was used for growth of *R. oryzae* and glucose for biosynthesis of the fumaric acid (Xu et al., 2010). The initial dilute acid hydrolysis of corn straw released mainly xylose, the main component of hemicelluloses, followed by enzymatic digestion to obtain the glucose from the cellulosic fraction of the lignocellulose. The increase in xylose concentration in hydroylsates had a negative effect on the subsequent fumaric acid accumulation stage, probably due to a higher concentration of inhibitors present in these hydrolysates. The maximal productivity obtained was 0.37 g/L·h using mild acid hydrolysates for growth and adding pure glucose for the second stage of fumaric acid accumulation with a final concentration of 30 g/L of fumaric acid (Xu et al., 2010). When using starchy raw material (potato, cassava, and corn) it would be advantageous to have a simultaneous saccharification and fermentation process (SSF), to save on extra enzyme costs. Isolation of strains with glucose insensitive glucoamylase was achieved by isolating glucose analog resistant strains and yields of 0.49 g/g were obtained for SSF of corn starch without any addition of extra enzymes (Deng et al.,

2012). A potentially cheaper substrate is the utilization of cellulose in an effective manner (Xu et al., 2010). Use of the fast-growing evergreen tree *Eucalyptus globulus* as raw material, with limited lignin content and mainly xylose in the hemicellulose fraction, the fumaric acid yield decreased from 0.71 g/g using synthetic media in a fed batch mode to 0.35 g/g when hydrolysates of *E. globulus* were used (Rodriguez-Lopez et al., 2012). With ion-exchange-treated hydrolysates and addition of 15% of the carbon source as glucose it was possible to increase the yield to 0.44 g/g. These results are representative of similar problems encountered in other fermentations of low value/high volume fermentation products, where hydrolysates of lignocellulose used directly as fermentation substrates contain inhibitors that reduce the yields of the desired products (Liu and Blaschek, 2010).

15.2.4 Submerged Fermentation Conditions

Certain fermentation conditions for fumaric acid are required to obtain high productivity. Among the parameters reported in the literature are pH, morphology, and neutralizing agent. The production media are nitrogen, but not carbon, limited (Goldberg et al., 2006).

15.2.4.1 Effect of pH Fermentation close to neutral pH during the production phase of the fermentation is ideal for high productivity, yield, and final concentration of fumarate. The disadvantages are the waste salts produced. High base concentrations are required to maintain the pH close to neutral. To solubilize the accumulated salt, sulfuric acid is used and the free fumaric acid is obtained with the formation of stoichiometric amounts of gypsum ($CaSO_4$). The possibility to perform the fermentation at lower pH would have a great impact on the economics of a large-scale process for fumaric acid fermentations. At lower pH larger fractions of the undissociated form of the acid would be formed with low solubility enabling separation by crystallization without the formation of waste salts (Roa Engel et al., 2011). In shake flasks, pH had strong effect on fumaric acid accumulation. When pH was kept at 5.0, 30 g/L of fumaric acid accumulated compared to 9.4 g/L at pH 3.0. Incubation at lower pH resulted in high accumulation of byproducts like glycerol and ethanol (Roa Engel et al., 2011).

15.2.4.2 Morphology of Fungus The suggested morphology for fumaric acid accumulation using *R. oryzae* is the formation of small pellets (<1 mm in diameter), found to be problematic to achieve by inoculation of spores directly into fermentor vessels (Roa Engel et al., 2011). It is known that fungal morphology is influenced by many factors such as medium composition, inoculums size, pH, medium shear, additives (polymers, surfactants, and chelators), culture temperature, and medium viscosity (Papagianni, 2004). For *R. oryzae* in shake flasks pH, shaking frequency, and working volume were the important factors that determined the pellet formation and size (Roa Engel et al., 2011). There is no consensus in the literature on the optimal morphology for fumaric acid formation and neither the exact conditions that will lead to the desired pellet formation. Decrease in pH and higher agitation in large shake flasks was beneficial for obtaining smaller pellet diameter (Roa Engel et al., 2011), whereas $CaCO_3$ addition affected the pellet diameter but not pH in another study (Liao et al., 2008). In a soybean peptone medium with inorganic ions growing *R. oryzae* at 30°C at pH 5, a pellet size of 0.55 mm resulted in the highest fumaric acid accumulation (39.6 g/L in shake flasks, 50 mL in 250 mL flask, and 35.4 g/L in a stirred tank of 7 L both at 200 rpm (Zhou et al., 2011). Transfer of the seed culture in several steps at pH between 2.5 and 4 at 0.5 incremental steps resulted in good pellet formation (up to 1.2 mm in diameter; Fu et al., 2009); however, there is no agreement on how to achieve uniform pellets and different strains and environmental conditions have been used. The challenge is to determine optimal morphology and a way to obtain that morphology for fumaric acid accumulation conditions on an industrial scale.

15.2.4.3 Neutralizing Agent Attempts have been made to use other neutralizing agents than $CaCO_3$, which has been shown to be optimal, with negative results in general (Zhou et al., 2002). $CaCO_3$ has several disadvantages, like the low solubility of Ca-fumarate which contributes to the increase in viscosity of the fermentation broth. Sodium salts such as Na_2CO_3 decrease the downstream costs, and thorough cost analysis (ISO 14044, 2006) of the total process is required to determine which neutralizing agent is most economical to use at the industrial level. It should be noted that the solubility of Na-fumarate is 22% w/v, high compared to Ca-fumarate, and its use inhibits fumarate formation (see section 15.5).

15.2.5 Transport of Fumaric Acid

Transport of fumaric acid outside of the fungal cell is poorly understood. It is important to understand how and at what conditions fumaric acid is transported out of the cell. The ambient pH is one of the main factors found to influence the excretion of organic acids in filamentous fungi and there are different hypothesis on the mechanisms involved (overflow metabolism hypothesis, charge balance hypothesis, and aggressive acidification hypothesis all detailed in Vrabl et al., 2012).

15.2.6 Production Processes

The biological production of fumaric acid using *R. oryzae* was carried out on an industrial scale by Pfizer (see section 15.1.2). With the rapid development of the petrochemical industry, the fermentation route to fumaric acid became uneconomical and until today, to the best of our knowledge, all fumaric acid is produced through catalytic oxidation of benzene or butane to maleic anhydride hydrolyzed to maleic acid followed by isomerization to fumaric acid. There is no information available in the public domain to describe current efforts to develop an industrial biological process, but there are several research groups that are reporting on attempts to increase fumaric acid accumulation by isolation of high producing mutants as well as using cheaper substrates and also decrease cost and waste material formation of the downstream processes (Huang et al., 2010; Kang et al., 2010). In no case are the newly isolated mutants as effective in production of fumaric acid as earlier reports using wild-type strains for study of fumaric acid accumulation (Rhodes et al., 1959, 1962; Ng et al., 1986). When corn, at the beginning of this century was still at a surplus (before its use as raw material for biofuel, ethanol) research was funded for its use as a raw material for fumaric acid production (http://www.reeis.usda.gov/web/crisprojectpages/192861.html, retrieved June 14, 2012)

15.3 A PLAUSIBLE BIOCHEMICAL MECHANISM FOR FUMARIC ACID BIOSYNTHESIS AND ACCUMULATION IN *RHIZOPUS*

Fumaric acid is an intermediate in the tricarboxylic acid (TCA) cycle used by cells to produce energy and building blocks (Krebs, 1970). It is formed by the oxidation of succinic acid by the enzyme succinate dehydrogenase. Fumaric acid is further converted to L-malic acid by fumarase. Fumaric acid does not accumulate under normal growth conditions of cells and microorganisms. It may accumulate in mutant eukaryotic cells and in mammals under certain pathological conditions. Uniquely, this acid is synthesized and accumulated in the medium-to-high molar yields (i.e., about 150% of glucose) and concentrations (i.e., more than 100 g/L) by the filamentous fungus *R. oryzae*, under specific stress conditions. This unusual phenomenon appears to be a general situation where, following a specific stimulus, certain filamentous fungi accumulate intermediates of the TCA cycle as end products (e.g., citric and L-malic acids) (Goldberg et al., 1991, 2006; Magnuson and Lasure, 2004). Although fumaric acid was commercially produced on a large scale (see above, 15.1.2), the mechanism(s) by which it is produced are not fully understood and have been the subject of relatively few publications.

According to the present suggested mechanism, fumaric acid accumulation by *R. oryzae* occurs mainly by a cytosolic reductive pathway converting pyruvic to fumaric acid by the successive activities of pyruvate carboxylase, malate dehydrogenase, and fumarase (Osmani and Scrutton, 1985; Kenealy et al., 1986). In the following discussion we will try to answer some of the questions that arise from this general mechanism.

15.3.1 How Can the High Molar Yield of Fumaric Acid be Explained?

Although Felix Ehrlich was the first to show in 1911 that fumaric acid is produced by *R. nigricans*, it took 28 years until Foster and Waksman (1939) identified an overproducer *Rhizopus* strain, which was (probably) used shortly after in the industry for the production of fumaric acid on a large scale (see above, 15.2.1).

Fumaric acid (molar yield of approximately 100%; moles of acid produced per moles of glucose utilized × 100), L-malic acid (15 mol%), and succinic acid (5 mol%) were the major acids formed during the fermentation. C_4 acid (fumaric, L-malic, and succinic acid) molar yields of 120–145%, with a maximal fumaric acid concentration of 107 g/L, were obtained after 4–5 days (Goldberg and Stieglitz, 1985, 1986; Kenealy et al., 1986; Ng et al., 1986). These high yields confirmed earlier data obtained by Rhodes et al. (1959, 1962).

How can we explain the unusually high molar yield of fumaric acid obtained by a *Rhizopus* strain?

Back in 1949, it was demonstrated by Foster et al., with *R. nigricans* that ethanol and carbon dioxide may serve as substrates for fumaric acid production. Experiments with [^{14}C]ethanol indicated that the yield of fumaric acid from ethanol was so high that the TCA cycle was excluded as the sole mechanism in acid formation by *Rhizopus*. These results were taken as presumptive evidence for the operation of a C-2 plus C-2 condensation in fumaric acid synthesis (Foster et al., 1949), similar to the later-discovered sequence of reactions involved in the glyoxylate cycle in bacteria (Kenealy et al., 1986). However, this pathway cannot account for a molar yield exceeding 100%. More recent data appear to rule out the glyoxylate bypass as a significant pathway in the bulk accumulation of fumaric acid by *R. nigricans* in media of high sugar concentration (Wegener and Romano, 1964; Romano et al., 1967).

According to Rhodes et al. (1959) the accumulation of fumaric acid maybe attributed to low activity or absence of fumarase. However, a high *in vitro* activity of fumarase was measured during the fumaric acid production stage (Osmani and Scrutton, 1985; Kenealy et al., 1986).

Alternatively, a C-3 plus C-1 carbon dioxide fixation mechanism was suggested (Romano et al., 1967; Overman and Romano, 1969). This mechanism involves the operation of pyruvate carboxylase yielding oxaloacetic acid, so that C_4 acids can be withdrawn for biosynthesis during the growth phase (Osmani and Scrutton, 1985). Operation of this reaction in conjunction with reactions of the TCA cycle would yield only 1 mole of fumaric acid per mole of glucose utilized if the oxidative reactions of the pathway were used. However, if the reductive reactions of the TCA cycle (see below, 15.3.2) were used, 2 moles of fumaric acid per mole of glucose would be formed.

Direct proof that this reductive TCA pathway participates in fumaric acid production came from ^{13}C NMR experiments using [1-^{13}C] and [U-^{13}C] glucose as a carbon source for *R. oryzae* (Kenealy et al., 1986). The unexpectedly high fumaric acid molar yields can thus be explained in terms of pyruvate carboxylation as the initial reaction of a reductive TCA pathway (Figure 15.1).

Recently, Yu et al. (2012) obtained a stable high-yielding *R. oryzae* FM19 mutant by laser irradiation. The mutant produced fumaric acid approximately 1.56-fold higher than that of the wild type. Metabolic profiling showed that among the different changes in metabolic pathways, compared with the parental strain, the citrate cycle and the reductive pyruvate carboxylation pathways under aerobic conditions were simultaneously elevated, thus providing more ATP, reducing power (NADH), and TCA cycle intermediates, which is necessary to maintain cell metabolism and product synthesis (Kenealy et al., 1986).

15.3.2 Where in the Cell is the Localization of the Reductive Reactions of the TCA Cycle?

Evidence for the carbon dioxide fixation mechanism for fumaric acid production came from two sources. Osmani and Scrutton (1985) discovered that *R. oryzae*

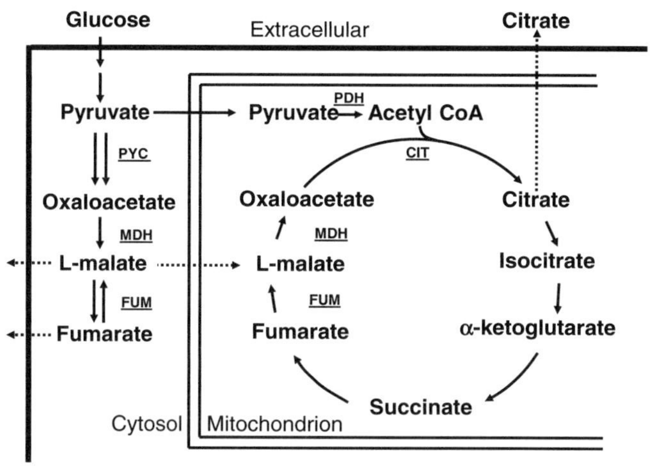

FIGURE 15.1 The reductive reactions of the tricarboxylic acid (TCA) cycle. CIT, citrate synthase; FUM, fumarase; MDH, malate dehydrogenase; PDH, pyruvate dehydrogenase; PYC, pyruvate carboxylase.

harbors cytosolic pyruvate carboxylase, malate dehydrogenase, and fumarase. On the basis of the availability of these enzymes in the cytosol, these authors suggested the existence of an extramitochondrial pathway leading from pyruvic to oxaloacetic acid which is converted to L-malic and then to fumaric acid.

Electrophoretic studies of fumarase were carried out in *R. oryzae*. The analyses revealed two fumarase isoenzymes, one localized solely in the cytosol and the other found both in the cytosol and in the mitochondrial fraction. The activity of the cytosolic isoenzyme of fumarase was higher during the acid production stage than during growth. Addition of cycloheximide-inhibited fumaric acid production and decreased the activity of the cytosolic isoenzyme of fumarase. These results suggested that *de novo* synthesis is required for the increase in the activity of the cytosolic isoenzyme and that such an increase in activity is essential for fumaric acid accumulation (Peleg et al., 1989b).

But once the notion of a cytosolic reductive TCA pathway is accepted, the question is why the experimentally obtained yields of C_4 organic acids (120–145%, see above, 15.3.1) are lower than the theoretical maximal yield of 200% calculated for this pathway. A plausible explanation for this is that the fungus must divert some of the pyruvic acid carbon to mitochondria and to the oxidative TCA cycle to obtain energy required for maintenance requirements and limited cellular growth (Figure 15.1).

15.3.3 What is the Role of Cytosolic Fumarase in Fumaric Acid Accumulation in *Rhizopus* Strain?

The operation of the cytosolic reductive TCA pathway for fumaric acid production in *R. oryzae* raises the question as to the unique role of the cytosolic fumarase in this fungus.

It was known that the fumarase enzymes in eukaryotic organisms such as *Saccharomyces cerevisiae*, rat, and human are encoded in each organism by a single nuclear gene, yet the enzymes are found both in the cytosol and mitochondria. It was found that in *S. cerevisiae* (which accumulates up to 2 g/L of L-malic acid but not fumaric acid) fumarase translation initiates only from the 5′ proximal AUG codon (Sass et al., 2001). Thus, mitochondrial and cytosolic isoforms of yeast fumarase are derivatives of a single translation product and have identical amino acids and kinetic properties (low K_m value for fumaric acid). Whereas these findings can explain L-malic acid accumulation in yeast, an intriguing question that remains to be answered is the mechanism by which large amounts of fumaric acid are produced by the eukaryotic fungus *R. oryzae*.

To shed more light on this question, the yeast fumarase gene FUM1 was used as a probe to clone a homologous fumarase gene from *R. oryzae*. Accordingly, the alignment of the amino acid sequence of this *R. oryzae* fumarase with other eukaryotic (class II) fumarases indicated a high degree of sequence similarity (75–80%). The gene designated *fumR* was shown to produce a single transcript, and as predicted from the DNA sequence, a protein with an apparent molecular mass of 50 kDa was detected, employing *S. cerevisiae*–fumarase antiserum. However, this protein did not appear to be responsible for the approximately sixfold increase in fumarase activity upon transfer of *Rhizopus* cells into acid production medium. This is because the level of the protein as detected by western blotting changed only slightly upon this shift (Peleg et al., 1989b; Friedberg et al., 1995). Furthermore; antiserum against yeast fumarase only partially neutralized the fumarase activity in *R. oryzae* cell extracts, suggesting the existence of an additional nonhomologous enzyme (E. Battat, O. Pines, and I. Goldberg, unpublished data cited in Goldberg et al., 2006).

The *fumR* gene (100% identity with the above gene) was cloned first and overexpressed in *R. oryzae*. All transformants showed significant increase in fumarase activity during both growth and fumaric acid production stages. However, fumarase overexpression in *R. oryzae* yielded more malic acid, instead of fumaric acid, in the fermentation because the overexpressed fumarase catalyzed the hydration of fumaric acid to malic acid (Zhang and Yang, 2012).

Additional suggestive evidence for the existence of a second distinct fumarase in this fungus comes from the analysis of fumarase activity in cell lysates. Fumarase in lysates of *R. oryzae* from medium B (growth medium) has a lower K_m value for fumaric acid (0.78 mM) than for L-malic acid (2.9 mM), similar to fumarase from lysates of *S. cerevisiae* (Pines et al., 1996). Fumarase activities (with L-malic acid as the substrate) in both these lysates were not inhibited by fumaric acid. In sharp contrast, fumarase activity measured in extracts prepared from *R. oryzae* cells incubated in medium C (production medium), but not with *S. cerevisiae*, was completely inhibited by 2 mM fumaric acid (E. Battat and I. Goldberg, unpublished data cited in Goldberg et al., 2006).

It should be mentioned that with another *Rhizopus* strain, Ding et al. (2011) showed in cell extracts, in accordance to previous findings that lowering the urea concentrations in the medium from 2.0 to 0.1 g/L caused an increase of 300% in the cytosolic fumarase activity, accompanied with an increase in fumaric acid production.

However, the activity of fumarase (measured with fumaric acid as the substrate) was not inhibited by fumaric acid.

Based on these results, Goldberg et al. (2006) suggested two possibilities to explain these findings. The first is that in *R. oryzae* the cytosolic fumarase is kinetically different from the mitochondrial isoenzyme, due to distinct posttranslational modifications, or to specific conditions in the two compartments. The second possibility is that *R. oryzae* harbors two genes encoding two different fumarases, one in mitochondria, which catalyzes the conversion of fumaric to L-malic acid, and a cytosolic enzyme, which catalyzes the conversion of L-malic to fumaric acid. Upon transfer into medium C, a fumarase with unique characteristics is induced. L-malic acid's conversion to fumaric acid is enhanced by the induced fumarase and when the concentration of fumaric acid in the cell exceeds 2 mM, the reverse reaction to L-malic acid is fully inhibited. Thus, this property of the unique fumarase, whose existence was then only hypothesized, can ensure that fumaric acid is accumulated.

This question was further studied by Song et al. (2011) who isolated the *fumR* gene (GenBank accession number GU013473) by reverse transcription (RT)-(PCR) during the period of maximal fumaric acid accumulation by *R. oryzae*. The *fumR* gene was then expressed in recombinant *Escherichia coli* BL21(DE3) and resulted in the accumulation of a fusion protein. SDS-PAGE analysis confirmed the recombinant fumarase to have an apparent subunit molecular mass of 50 kDa. The specific activity of FUMR in crude extracts was 4.13 U/mg and was 63.52 U/mg for the purified enzyme, using L-malate (50 mM) as a substrate. The K_m values of the purified FUMR protein for L-malic acid and fumaric acid were 0.46 and 3.07 mM, respectively. The activity of FUMR catalyzing hydration of fumaric to L-malic acid was completely inhibited by 2 mM fumaric acid.

These elegant results, which were similar in part to those obtained in crude extracts of *R. oryzae*, provided a direct proof for the unique enzymatic properties of FUMR. The lower K_m value of FUMR for L-malic acid suggested its preference for conversion of L-malic acid to fumaric acid. Coupled with the observed inhibition of the reverse reaction from fumaric to L-malic acid by concentrations of fumaric acid exceeding 2 mM, would ensure the accumulation of fumaric acid in *R. oryzae*. However, as Gajewski et al. (1985) and Meussen (2012) pointed out, the $\Delta G0'$ of the conversion of L-(+)-malic acid to fumaric acid is 3.6 kJ/mol, indicating that at equilibrium, the L-(+)-malic acid concentration is higher than the fumaric acid concentration. Therefore, a dicarboxylic acid transporter with a high selectivity for fumaric acid may also be required for fumaric acid production in *R. oryzae.*

Thus, the cytosolic FUMR may be an important step for the accumulation of high concentrations of fumaric acid in *R. oryzae*. This conclusion can be better confirmed if fumaric acid production will increase as a result of the overexpression of FUMR in *R. oryzae* (Song et al., 2011; Zhang and Yang, 2012).

In order to further study the possible mechanisms that can explain fumarase distribution in *R. oryzae*, the deduced amino acid sequences of the two fumarase genes mentioned above were compared (GenBank accession number X78576—Friedberg et al., 1995, and GU013473—Song et al., 2011). The results confirmed that the later *fumR* had a deletion of 15-amino acid sequence in the N-terminal region. It was assumed that the difference in the amino acid sequences may cause the different characterization and distribution of the fumarase isozymes in *R. oryzae* (Song et al., 2011).

As mentioned above, *R. oryzae* is one of the few fumaric acid producer organisms described in the literature which can accumulate more than 100 g/L of fumaric acid as the main product. Thus, it is not surprising to note that the cytosolic fumarase in *R. oryzae* appears to have unique characteristics that differentiate it from previously reported cytosolic fumarases from other eukaryotic organisms. For example, the cytosolic fumarase in *S. cerevisiae* exhibited a 17-fold higher affinity for fumaric acid than for L-malic acid, while the K_m for L-malic acid was very high. This indicated that the cytosolic fumarase, like the mitochondrial enzyme, catalyzed the conversion of fumaric to L-malic acid and not the reverse reaction, which explains the fact that this yeast accumulates small amounts of L-malic acid but not fumaric acid (Peleg et al., 1990; Pines et al., 1996). A similar conclusion was drawn for the mechanism responsible for L-malic acid accumulation (fumaric acid comprises only up 2% of L-malic acid) in *Aspergillus flavus* (Peleg et al., 1988; Battat et al., 1991). While these findings strengthen the important role of the cytosolic fumarase in fumaric acid production by *R. oryzae*, more research is needed to ascertain if indeed the difference in 15-amino acids is responsible for the role of this enzyme in this unique fungus.

15.4 TOWARD ENGINEERING *RHIZOPUS* FOR FUMARIC ACID PRODUCTION

Metabolic engineering and systems biology include tools that are being used to enhance the formation of a desired metabolite. Incurring changes of the cellular metabolic flux toward the desired product with molecular biology tools may improve productivity. Systems biology tools such as transcriptomics, proteomics, metabolomics, and fluxomics would enable a better understanding of the effects on changes incurred, where different constructs are compared using the omics tools mentioned above. The pathways in charge for fumaric acid accumulation are known (see section 15.3) and the unique biology of certain *R. oryzae* strains makes them excellent candidates for industrial use, even though other routes are suggested (use of *Aspergillus niger* as platform organism—see below, 15.6.2).

Three different transformation systems for *R. oryzae* have been described: one using the T-DNA of *Agrobacterium tumefaciens*, able to complement auxotrophic mutants of *R. oryzae* (Michielse et al., 2004). However, it was found that a gene conversion, without the introduction of any additional DNA gave restoration of the auxotroph. It was not possible to have a gene of interest expressed and this is not a transformation method that will allow for heterologous gene expression. In the same report the "conventional" $CaCl_2$ PEG transformation method was used with both circular and linear plasmids. The transformants had unstable phenotypes and no vector DNA integrated to the chromosomal DNA. Use of the microprojectile bombardment system, biolistic methods, developed for plant cells, enabled transformation of the sequenced strain, *R. oryzae* 99-880 (Skory and Ibrahim, 2007). The transformed clones were around three times slower in using the C-source compared with non-transformants. The majority of the transformants accumulated lactic acid, not present at all for the wild type or controls, in which the fumaric acid concentration was higher than in the transformants. The introduction of an *ldhA* gene (lactic dehydrogenase present in *R. oryzae* strains accumulating lactic acid) changed the acid profile and more lactic acid accumulated. With the use of an RNA interference method it was shown that by silencing the two genes coding for lactic dehydrogenases (*ldhA* and *ldhB*) it was possible to increase the concentration of ethanol on account of the diminished accumulation of lactic acid (Gheinani et al., 2011). The gene-silencing technique was found to be transient and is therefore not appropriate for industrial applications. No attempts have been reported on experiments with use of molecular biology techniques to improve the formation of fumaric acid in *R. oryzae*, even though the genome of one strain has been sequenced (Ma et al., 2009). Transfer of the pathway into another host is suggested below (see section 15.6.2). To obtain yields that are competitive, the relative levels of the enzymes of the pathway need to be coordinated. The use of transcription factor(s) that sense a key intermediate and dynamically regulate the gene expression, described for biodiesel production, is a novel and useful method to diminish unwanted side effects, such as metabolic imbalance, bottlenecks, or toxicity of intermediates or end products to the host (Zhang et al., 2012a). This may allow for the desired metabolic pathway to be regulated dynamically in response to the physiological state of the cell. No attempts for heterologous expression of the metabolic pathway for fumaric acid production are reported in the literature. It seems there are still obstacles for the effective use of the modern tools mentioned above. In the literature attempts are described to use other hosts, such as *E. coli*, to accumulate large quantities of organic acids, mainly succinic and lactic acid, using anaerobic conditions (Cao et al., 2011; see section 15.6.2).

A systems approach was used and intracellular metabolites were determined for a mutant, obtained by random mutagenesis with femtosecond laser irradiation, and compared with the wild-type strain of *R. oryzae* "wild 1.22" (derived from ATCC 20344; Yu et al., 2012; see also section 15.3.1). The mutant was propagated for 10 generations, and kept its capability for fumaric acid accumulation. The ethanol concentration in the fermentation of the mutant was lower than the wild type, indicating that the mutant had impaired capability to make ethanol and probably channeled the pyruvate toward fumaric acid. The metabolic profiling of the two strains showed that amino acids, fatty acids, and TCA cycle intermediates changed, comparing the two strains. No detailed analysis of the different profiles obtained were performed, and it is not clear if the random mutagenesis method used, will result in multiple mutations, of which some may not be related to fumaric acid formation. It should be noted that the mutant accumulated a final concentration of 49.6 g/L with a productivity of 0.29 g/L·h, concentrations and productivities much lower than from earlier reports on fumaric acid accumulation (Ng et al., 1986; Cao et al., 1996, see also 15.2.1). It would be of great interest to have metabolic profiles of defined mutants, to get a clearer picture of the changes that are present in metabolites for an overproducer, as compared to wild type. To improve production of a specific acid means to intensify metabolic flux through other dicarboxylic acids and other rate-limiting steps. For redox neutral C4 di-acids accumulation, it would result in a maximum theoretical yield of 2 moles C4 di-acids per mole of glucose. Metabolic profiling to improve fumaric acid production was reported (Wang et al., 2013) with a 14% increase in fumaric acid accumulation to 56.5 g/L.

The first report of actual metabolic engineering was just published (Zhang et al., 2012b). The endogenous gene for pyruvate carboxylase (*pyc*) of the strain used in the study (i.e., *R. oryzae* M 16, a transformant of the sequenced strain 99880 (NRRL 43880) and phosphoenol pyruvate carboxylase (*pepC*) from *E. coli* DH5α, were overexpressed to increase formation of oxaloacetate. *R. oryzae* transformed with plasmids containing the *pyc* gene had 56–83% increased enzyme activity compared to wild type and *pepC* transformants also resulted in activity of its gene product, of an enzyme not present naturally in *R. oryzae*. However, growth kinetics and morphology was different comparing wild type and transformants. There was much less growth especially for the *pyc* transformant measured in seed medium after 24 hours. The wild type formed mycelial clumps, whereas pellets formed for both types of transformants, large compact pellets for the *pyc* transformants and smaller for the *pepC* transformants. The ultimate test is whether the strains with higher oxaloacetate formation capability also gave higher yields and rates of fumaric acid formation. The wild type accumulates 20 g/L of fumaric acid in 100 hours with little by-product formation (malic acid and ethanol), whereas the *pyc* transformant produced more malic acid (3 g/L) and less that 1 g/L of fumaric acid and the *pepC* transformant showed an increase of 26% in fumaric acid accumulation. The negligible performance of the *pyc* transformants is probably due to the oxygen limitation caused by their growth mode as large compact pellets (Zhan et al., 2012b). It should be noted that the amount of fumaric acid accumulating for the best transformants is 25 g/L in shake flasks as compared to 107 g/L in small fermentors using a different "wild-type" strain (NRRL 1526, Ng et al., 1986).

15.5 ECONOMIC OUTLOOK

To be competitive in price, the production of bulk chemicals by biological means requires among others, the use of available, cheap, and metabolically acceptable raw material. These should allow for high productivity and final concentration of the desired metabolite. For simple and straightforward downstream processing, negligible formation of side products, and minimal use of other chemicals are important.

The economic analysis of fumaric acid production is based on literature values (Gangl et al., 1990). Their analysis is the basis for looking at the prospects for production of fumaric acid by biological means with more current information. A serious breakdown of different costs was made to compare the chemical and different biological routes and showed that the chemical route resulted in 50% lower price for the fumaric acid. For biological production the return differential of various alternatives was compared, based on both experimental and idealized values. The higher yields and productivities obtained for fermentations using $CaCO_3$ (Ng et al., 1986) were used for the Na_2CO_3 "case" where no new solids are formed and with the possibility for reuse of the cells. The use of Na_2CO_3 is economically favorable, with less use of chemicals, less waste produced, and much simpler downstream processing. However, even with a theoretical optimal formation by biological means, using neutralizing agent Na_2CO_3, and repeated reuse of cells (five times) the biological route could not compete in price with the chemical route. In earlier attempts to produce soluble sodium or potassium fumarates, inhibition occurred when the concentration of fumarates reached 3.5–4% (Rhodes et al., 1962), which is similar to results from shake flask experiments where both growth and sporulation was inhibited when the sodium fumarate concentration reached 4% (w/v) and its accumulation was completely inhibited at concentrations above 10% (Rhodes et al., 1962). In the study of Gangl et al. (1990) using Na_2CO_3 as neutralizing agent, 9.9% sodium fumarate was accumulating in the broth with no signs of inhibition.

The principal operating costs associated with fermentation processes of low value compounds are the costs of the fermentation substrate, which for fumaric acid amounts to about 50–55% of total production costs. The small differential that normally exists between substrate and the product prices is one of the major intrinsic limitations of commercial fumaric acid fermentation. The break-even price for production is dependent on the price that can be obtained for the product, compared to the price paid for the substrate. The current prices commanded by purpose-grown crops in most developed countries make it difficult for the fermentation route to compete economically. The use of lower value agricultural, industrial, and domestic wastes could help improve the economics but the variable nature of these types of materials makes them much less attractive as fermentation substrates. For the idealized case described by Gangl et al. (1990), substrate cost was 75% of the total cost, mainly due to reduction in equipment cost and also taking into account the higher productivity for formation of fumaric acid.

The following factors are suggested to be taken into account to enable a fair comparison of the chemical and biological manufacturing alternatives today:

(a) It is hard to get reliable estimates of the actual volume of the market for fumaric acid today. For other bulk commodities economics of scale and novel markets led to establishment of large-scale facilities in the fermentation industry. A recent example is the setting up of a production plant by NatureWorks LLC (formerly Cargill Dow LLC) in 2002, with a capacity of almost three times the production volume (~180,000 tons/year; about 140,000 tons/year of polylactic acid (PLA) plastic) of the then existing lactic acid world market (~60,000 tons/year). This was possible since NatureWorks product, NatureWorks™ PLA, is a novel product and does not compete with other uses of lactic acid as well as the large scale of the factory (Vink et al., 2004). In 1995, the price of PLA produced at a volume of ~4500 tons/year sold for \$6.6/kg and with the new factory the price came down to \$2.9/kg in 2002 and \$1.9/kg in 2007 (Schut, 2007). Change to feedstock such as corn stover (lignocellulose) was forecast to lower the price to \$0.77/kg (Schut, 2007). However, it should also be noted that the new plant accounts for only 0.1% of the total market volume of plastics produced every year (Fahey, 2001).

(b) The price of raw material is crucial when comparing the chemical and biological processes. The immediate raw material for the chemical process is maleic anhydride, obtained from petrochemicals. For a long time benzene was the immediate raw material, and still is in the People's Republic of China. The cheaper butane is less toxic and has replaced benzene in the developed world (Roa Engel et al., 2008). The selling price of fumaric acid on the market (all chemically produced) is quoted as $1100 per ton for non-food grade fumaric acid (http://www.quamnet.com/media/IPO/LTN20120321012.pdf, retrieved April 30, 2012).

(c) For equipment costs there are no real estimates after those described by Gangl et al. (1990). It is possible that not only the raw material but also the actual process has undergone drastic changes since 1990. Looking at the bulk price of maleic anhydride, the direct precursor of fumaric acid, its price has been quite stable for the last 22 years (1990 and 2012—US$1500 per ton); however, the price of oil was $21.42 per barrel (January 1990) http://www.ioga.com/Special/crudeoil_Hist.htm, retrieved June 14, 2012, and ~$96 per barrel (June 2012) http://www.oil-price.net/, retrieved June 14, 2012. In 2012 the price of maleic anhydride is ~$1.5/kg similar to the price quoted by Roa Engel et al. (2008). There is a discrepancy here, since the price of oil today is almost double what was considered the break-even point comparing the two processes (Gangl et al., 1990). One explanation might be that benzene which was the main raw material for maleic anhydride has been changed to butane, which is cheaper and/or process changes leading to a lower price.

(d) Economics of scale is a very decisive factor (see (a)). The scale related to in the paper of Gangl et al. (1990) was for a factory with an annual production of 2300 tons. For comparison production of bulk lactic acid (used for PLA production) the annual production volume is about 180,000 tons, and a smaller scale would have made the operation unprofitable (Vink et al., 2004). The question of what volumes will saturate the market (decline in price) arises. The use of fumaric acid for polymer manufacture requires a lower purity than for food use and can therefore be made cheaper.

(e) It has been proposed that only for products with a selling price of over $2/kg, biotechnology is an economical option (Willke and Vorlop, 2004).

(f) In the analysis of Gangl et al. (1990) the price for a barrel of oil that would allow the biological route to be competitive was $61, whereas in April 2012 it is $104, and to the best of our knowledge all fumaric acid is still produced by chemical methods.

To enable the biological route, the producing organism should be able to accumulate higher final concentrations of fumaric acid, at a lower pH, to simplify the downstream processing, together with finding new niches for fumaric acid use, from cheap raw material such as lignocellulose and waste.

15.6 FUTURE PERSPECTIVES

15.6.1 Biorefinery

Due to the rapidly increased production cost of fossil (petroleum)-based chemicals, fermentatively produced fumaric acid from renewable resources could replace current petrochemically based maleic acid as unsaturated dibasic acid mainly in the polyester resin industry but also in medicine and food industries in the nearby future. However, this can be achieved only if the bio-based production process for fumaric acid would be economically competitive with the current fossil-based process. This change requires the improvement of the large, past-operating fermentation processes for acid production in many aspects, such as production of free acid at a low pH, product yield and productivity, cheap and renewable raw material, and problems related to cell morphology and mass transfer.

This book promotes the concept of the biorefinery, which is a facility that integrates biomass conversion processes and equipment to produce fuels, power, and chemicals from biomass. The biorefinery concept is analogous to today's petroleum refineries, which produce multiple fuels and products from petroleum (Sauer et al., 2008). The classical corn wet milling is an example for the basis of the biorefinery concept whereby one raw material is processed into various products.

To realize the concept of biorefinery, renewable resources need to be converted efficiently by biological means to several commodity chemicals. From the standpoint of the microbiologist who is responsible for the operation of this biorefinery, it is desirable to have only one microorganism in the biorefinery which is versatile enough to produce the various products. This situation is preferable to the (impossible) situation where different microorganisms are utilized simultaneously at the same production plant.

15.6.2 Platform Microorganisms

Hence, the concept of using a "platform microorganism" in the biorefinery, in a similar way to using one raw material, has an enormous advantage. According to De Jongh and Nielsen (2007) a platform organism is "an industrial organism that is extensively used to produce several commodity products. The main advantages of these organisms are that they are well understood and have been used by industry for an extended period of time." Previously, we have suggested that by the use of the right "platform fungus" different organic acids could be produced efficiently (Goldberg et al., 2006).

Few platform organisms were suggested for fumaric acid production in addition to *Rhizopus* strains. *E. coli* has been chosen as a "workhouse" for the production of fumaric acid and many valuable chemicals including other organic acids (e.g., succinic and L-lactic acid). This organism was chosen "due to its clear genetic background, convenient to be genetically modified and good growth properties with simple nutrient requirements" (Cao et al., 2011; Yu et al., 2011). Myriant Company, a development-stage industrial biotechnology company, announced that fumaric acid, as well as succinic, lactic, and adipic acids, can be produced based on its proprietary platform that involves modified (non-GMO) *E. coli* strains (http://www1.eere.energy.gov/biomass/pdfs/ibr_arra_myriant.pdf, retrieved June 14, 2012).

Recently, the yeast *S. cerevisiae* was suggested as a suitable organism for fumaric acid production. This yeast is able to grow at relatively low pH values, although it does not naturally produce fumaric acid. Additional advantages of this organism are that its genome has been fully sequenced and that it is considered as a GRAS organism (Jamalzadeh et al., 2012). In a recent review on a systems-level approach for metabolic engineering of yeast, the authors concluded that retrofitting existing microorganisms for the generation of natural or synthetic products is complex, and that the insertion of heterologous pathways in a microorganism does not imply per se a high-level production of the desired bioproduct (Kim et al., 2012). To this point it is interesting, and maybe encouraging, noting the patent of NatureWorks LLC, USA, on genetically modified yeast species producing lactic acid via fermentation using exogenous lactate dehydrogenase gene. This company is the world's largest producer of L-lactic acid (see section 15.5) and it might be using one of these strains in their production process (Vink et al., 2004; Suominen et al., 2008).

The production of high concentration of the free acid at low pH values would allow for a profound improvement on the past operating fumaric acid production processes. We therefore suggest considering the alternative use of *Aspergillus niger* for the production of this acid, as well as for other organic acids and different chemicals. It should be noted that in addition to the intrinsic advantages of *Aspergillus* over *Rhizopus* (Tables 15.1 and 15.2) (see the reviews of Ghosh and Ray, 2011, and Meussen et al., 2012, on the potentiality of *R. oryzae* to produce a wide spectrum of metabolites, including fumaric and other organic acids, and the reviews by Goldberg et al., 2006; De Jongh and Nielsen, 2007; Vongsangnak and Nielsen, 2010; Andersen

TABLE 15.1 Advantages and Disadvantages of *Rhizopus* (Zygomycota) for the Industrial Process for Organic Acid Production

Advantages of *Rhizopus*
Produces fumaric and L(+) lactic acid at high concentrations
Utilizes various sugar (e.g., pentoses and starch)
Generally recognized as safe (GRAS) in foods
Disadvantages of *Rhizopus*
Produces more than one metabolic end product in significant yields
Sensitive to pH below 4.5—continuous neutralization is required for maximal yield
Product purification is complex and costly
Tendency to filamentous growth in the acid production stage
Molecular biology techniques in development

TABLE 15.2 Advantages and Disadvantages of *Aspergillus* (Ascomycota) for the Industrial Process for Organic Acid Production

Advantages of *Aspergillus*
The largest commercial production of acids (citric acid ~1.5×10^6 tons/year)
Phenomenal productivity (gluconic acid 30 g/L·h)
Phenomenal acid concentration (citric acid ~200 g/L)
Produces free acid and excrete it to the medium
Grow and produce acid at low pH (citric acid <pH 1.0)
Growth and acid production are not inhibited below pH of 4.5
Production of one product according to conditions (citric acid below pH of 3.2)
Simplified product purification
Produce a variety of acids in different media
Genome sequence is known
Molecular biology techniques are well implemented
Grows in small pellets
Generally recognized as safe (GRAS) in foods
Disadvantages of *Aspergillus*
Does not produce organic acids, except for citric acid, at high concentrations

et al., 2011; and Knuf and Nielsen, 2012 on the advantages of *A. niger*), the biochemical pathways for acid biosynthesis (e.g., fumaric and L-malic acids) and the localization of key enzymes of these pathways are similar in both fungi (Osmani and Scrutton, 1983; Peleg et al., 1989a; Bercovitz et al., 1990; Goldberg et al., 1991, 2006).

Understanding in detail the mechanism for fumaric acid biosynthesis and accumulation is required to pave the way to improve acid production in *R. oryzae*, or in other organisms, for future novel uses. With the rapid development of synthetic biology it will be possible to design the desired platform organism that will be able to produce high concentrations of different free organic acids and other products such as ethanol from cheap and renewable carbon sources on a very large scale (see, e.g., the 2011 research project of Leo de Graaff, Wageningen University, The Netherlands, which aim to use system biology to engineer the *A. niger* metabolic pathways that are involved in the cost-efficient production of organic acids, in particular itaconic acid and fumaric acid—http://www.systemsbiology.nl/FSB/, retrieved June 14, 2012). And although "retrofitting microorganisms for the generation of nonnatural or synthetic products is extremely complex," it seems to be a possibility in the near future.

ACKNOWLEDGMENT

J.S.R. was supported by a grant from The Leo & Boris Ritoff Fund for studies in applied microbiology.

REFERENCES

Abe A, Oda Y, Asano K, Sone T. (2007). *Rhizopus delemar* is the proper name for *Rhizopus oryzae* fumaric-malic acid producers. *Mycologia*, 99, 714–722.

Andersen MR, Salazar MP, Schaap PJ, van de Vondervoort PJ, Culley D, THykaer J, Frisvad JC, Nielsen KF, Albang R, Albermann K, et al. (2011). Comparative genomics of citric-acid-producing *Aspergillus niger* ATCC 1015 versus enzyme-producing CBS 513.88. *Genome Res*, 21, 885–897.

Battat E, Peleg Y, Bercovitz A, Rokem JS, Goldberg I. (1991). Optimization of L-malic acid production by *Aspergillus flavus* in a stirred fermentor. *Biotechnol Bioeng*, 37, 1108–1116.

Bercovitz A, Peleg Y, Battat E, Rokem JS, Goldberg I. (1990). Localization of pyruvate carboxylase in organic acid producing *Aspergillus* strains. *Appl Environ Microbiol*, 56, 1594–1597.

Cao NJ, Du JX, Gong CS, Tsao GT. (1996). Simultaneous production and recovery of fumaric acid from immobilized *Rhizopus oryzae* with a rotary biofilm contactor and adsorption column. *Appl Env Microbiol*, 62, 2926–2931.

Cao Y, Cao Y, Lin X. (2011). Metabolically engineered *Escherichia coli* for biotechnological production of four-carbon 1,4-dicarboxylic acids. *J Ind Microbiol Biotechnol*, 38, 649–656.

De Jongh WA, Nielsen J. (2007). Filamentous fungi as cell factories for metabolite production. In: Dijksterhuis J, editor. *Food Mycology: A Multifaceted Approach to Functional Food.* Boca Raton, FL: CRC Press, 163–182.

Deng Y, Li S, Xu Q, Gao M, Huang H. (2012). Production of fumaric acid by simultaneous saccharification and fermentation of starchy materials with 2-deoxyglucose-resistant mutant strains of *Rhizopus oryzae*. *Bioresour Technol*, 107, 363–367.

Ding Y, Li S, Dou C, Tu Y, Haung H. (2011). Production of fumaric acid by *Rhizopus oryzae*: role of carbon-nitrogen ration. *Appl Biochem Biotechnol*, 164, 1461–1467.

Ehrlich F. (1911). Formation of fumaric acid by means of molds. *Ber Dtsch Chem Ges*, 44, 3737–3742.

Fahey J. (2001). Shucking petroleum. Forbes Magazine, November 26, 2001.

Ferreira JA, Lennartson PR, Edebo L, Taherzadeh MJ. (2013). Zygomycetes-based biorefinery: present status and future prospects. *Bioresour Technol*, 135, 523–532.

Friedberg D, Peleg Y, Monsonego A, Maissi S, Battat E, Rokem JS, Goldberg I. (1995). The *fumR* gene encoding fumarase in the filamentous fungus *Rhizopus oryzae*: cloning, structure and expression. *Gene*, 163, 139–144.

Foster JW, Waksman SA. (1939). The production of fumaric acid by molds belonging to the genus *Rhizopus*. *J Am Chem Soc*, 61, 127–35.

Foster JW, Carson SF, Anthony DS, Davis JB, Jefferson WE, Long MW. (1949). Aerobic formation of fumaric acid in the mold *Rhizopus nigricans*: synthesis by direct 2C2 condensations. *Proc Natl Acad Sci USA*, 35, 663–671.

Fox RJ, Gold R. (2011). Dimethyl fumarate to treat multiple sclerosis. In: Cohen JA Rudick RA, editors. *Multiple Sclerosis Therapeutics*, 4th ed. Cambridge: Cambridge University Press, pp. 387–392.

Fu Y, Xu Q, Li S, Huang H, Chen Y. (2009). A novel multi-stage preculture strategy of *Rhizopus oryzae* ME-F12 for fumaric acid production in a stirred-tank reactor. *World J Microbiol Biotechnol*, 25,1871–1876.

Fu YQ, Li S, Chen Y, Xu Q, Huang H, Sheng XY. (2010). Enhancement of fumaric acid production by *Rhizopus oryzae* using a two-stage dissolved oxygen control strategy. *Appl Biochem Biotechnol*, 162, 1031–1038.

Gajewski E, Goldberg RN, Steckler DK. (1985). Thermodynamics of the conversion of fumarate to L-(−)-malate. *Biophys Chem*, 22,187–195.

Gangl IC, Weigand WA, Keller FA. (1990). Economic comparison of calcium fumarate and sodium fumarate production by *Rhizopus arrhizus*. *Appl Biochem Biotechnol*, 24–25, 663–677.

Gheinani AH, Jahromi NH, Feuk-Lagerstedt E, Taherzadeh MJ. (2011). RNA silencing of lactate dehydrogenase gene in *Rhizopus oryzae*. *J RNAi Gene Silencing*, 7, 443–448.

Ghosh B, Ray RR. (2011). Current commercial perspective of *Rhizopus oryzae*: a review. *J Appl Sci*, 11, 2470–2486.

Goldberg I, Rokem JS. (2009). Organic and fatty acid production, microbial. In: Schaechter M, editor. *Encyclopedia of Microbiology*. Oxford: Elsevier, pp. 421–442.

Goldberg I, Stieglitz B. (1985). Improved rate of fumaric acid production by Tweens and vegetable oils in *Rhizopus arrhizus*. *Biotech Bioeng*, 27, 1067–1069.

Goldberg I, Stieglitz B. (1986). Fermentation process for production of carboxylic acids. US Patent 4,564,594.

Goldberg I, Peleg Y, Rokem JS. (1991). Citric, fumaric and malic acids. In: Goldberg I Williams R, editors, *Biotechnology and Food Ingredients*. New York: Van Nostrand Reinhold, pp. 349–374.

Goldberg I, Rokem JS, Pines O. (2006). Organic acids: old metabolites, new themes. *J Chem Technol Biotechnol*, 81, 1601–1611.

He CL, Fu BD, Shen HQ, Jiang XL, Wei XB. (2011). Fumaric acid, an antibacterial component of Aloe vera L. *African J Biotechnol*, 10, 2973–2977.

Hobby GL. (1985). *Penicillin. Meeting the Challenge*. New Haven, CT: Yale University Press.

Huang L, Wei O, Zang R, Xu Z, Cen P. (2010). High throughput screening of high yield colonies of *Rhizopus oryzae* for enhanced production of fumaric acid. *Ann Microbiol*, 60, 287–292.

ISO 14044: Environmental Management—Life Cycle Assessment—Requirement and Guidelines. (2006). Available at: http://www.iso.org/iso/iso_catalogue/catalogue_tc/catalogue_detail.htm?csnumber=38498 [Accessed April 5, 2012].

Jamalzadeh E, Verheijen PJT, Heijnen JJ, van Gulik WM. (2012). pH-dependent uptake of fumaric acid in *Saccharomyces cerevisiae* under anaerobic conditions. *Appl Environ Microbiol*, 78, 705–716.

Kang SW, Lee H, Kim D, Kim S, Chun GT, Lee J, Kim SW, Park C. (2010). Strain development and medium optimization for fumaric acid production. *Biotech Bioproc Eng*, 15, 761–769.

Kenealy W, Zaady E, dePreez JC, Stieglitz B, Goldberg I. (1986). Biochemical aspects of fumaric acid accumulation by *Rhizopusarrhizus*. *Appl Environ Microbiol*, 52, 128–133.

Kim IK, Rolda A, Siewers V, Nielsen J. (2012). A systems-level approach for metabolic engineering of yeast cell factories. *FEMS Yeast Res*, 12, 228–248.

Knuf C, Nielsen J. (2012). *Aspergilli*: Systems biology and industrial applications. *Biotech J*, 7, 1147–1155.

Krebs HA. (1970). The history of the tricarboxylic acid cycle. *Perspect Biol Med*, 14, 154–170.

Liao W, Liu Y, Frear C, Chen SL. (2008). Co-production of fumaric acid and chitin from a nitrogen rich lignocellulosic material dairy manure using a pelletized filamentous fungi *Rhizopus oryzae* ATTC 20344. *Bioresour Technol*, 96, 2026–2032.

Liu ZL, Blaschek HP. (2010). Lignocellulosic biomass conversion to ethanol by *Saccharomyces*. In: Vertes A, Qureshi N, Yukawa H, Blaschek HP, editors. *Biomass to Biofuels: Strategies for Global Industries*. West Sussex, UK: John Wiley, pp. 17–36.

Ma LJ, Ibrahim AS, Skory C, Grabherr MG, Burger G, Butler, G, Butler M, Elias M, Idnurm A, Lang BF, et al. (2009). Genomic analysis of the basal lineage fungus *Rhizopus oryzae*. *PLoS Genet*, 5(7), e1000549.

Magnuson J, Lasure LL. (2004). Organic acid production by filamentous fungi. In: Lang J, Lang L, editors. *Advances in Fungal Biotechnology for Industry, Agriculture, and Medicine*. New York: Kluwer Academic/Plenum Publishers, pp. 307–340.

Meek JS. (1975). The determination of a mechanism of isomerization of maleic acid to fumaric acid. *J Chem Edu*, 52, 541–543.

Meussen BJ, de Graaf LH, Sanders JPM. Weusthuis RA. (2012). Metabolic engineering of *Rhizopus oryzae* for the production of platform chemicals. *Appl Microbiol Biotechnol*, 94, 875–886.

Miall LM. (1978). Primary products of metabolism: organic acids. In: Rose AH, editor. *Economic Microbiology*. New York: Academic Press. 2, 47–119.

Michielse CB, Salim K, Ragas O, Ram AFJ, Kudla B, Jarry B, Punt PJ, van den Hondel CAMJJ. (2004). Development of a system for integrative and stable transformation of the *Zygomycete Rhizopus oryzae* by *Agrobacterium*-mediated DNA transfer. *Mol Genet Genomics*, 271, 499–510.

Milas NA. (1943). Fumaric acid. *Org Synth*, 2, 302.

Moharregh-Khiabani D, Linker RA, Gold R, Stangel M. (2009). Fumaric acid and its esters: an emerging treatment for multiple sclerosis. *Curr Neuropharmacol*, 7, 60–64.

Ng TK, Hesser RJ, Stieglitz B, Griffiths BS, Ling LB. (1986). Production of tetrahydrofuran/1,4butanediol by a combined biological and chemical process. *Biotech Bioeng Symp*, 17, 344–363.

Osmani SA, Scrutton MC. (1983). The subcellular localization of pyruvate carboxylase and of some other enzymes in *Aspergillus nidulans*. *Eur J Biochem*, 133, 551–560.

Osmani SA, Scrutton MC. (1985). The subcellular localization and regulatory properties of pyruvate carboxylase from *Rhizopus arrhizus*. *Eur J Biochem*, 147, 119–128.

Overman SA, Romano AH. (1969). Pyruvate carboxylase of *Rhizopus nigricans* and its role in fumaric acid production. *Biochem Biophys Res Commun*, 37, 457–470.

Papagianni, M. (2004). Fungal morphology and metabolite production in submerged mycelial processes. *Biotechnol Adv*, 22, 189–259.

Peleg Y, Stieglitz B, Goldberg I. (1988). Malic acid accumulation by *Aspergillus flavus*. I. Biochemical aspects of acid biosynthesis. *Appl Microbiol Biotechnol*, 28, 69–75.

Peleg Y, Barak A, Scrutton MC, Goldberg I. (1989a). Malic acid accumulation by *Aspergillus flavus*.III. 13C-NMR and isoenzyme analysis. *Appl Microbiol Biotechnol*, 32, 334–339.

Peleg Y, Battat E, Scrutton MC, Goldberg I. (1989b). Isoenzyme pattern and subcellular localization of enzymes involved in fumaric acid accumulation by *Rhizopus oryzae*. *Appl Microbiol Biotechnol*, 32, 334–339.

Peleg Y, Rokem JS, Goldberg I, Pines O. (1990). Inducible overexpression of the FUM1 gene in *Saccharomyces cerevisiae*: localization of fumarase and efficient fumaric acid bioconversion to L-malic acid. *Appl Environ Microbiol*, 56, 2777–2783.

Pines O, Even Ram S, Elnathan N, Battat E, Aharonov O, Gibson D, Goldberg I. (1996). The cytosolic pathway of L-malic acid synthesis in *Saccharomyces cerevisiae:* the role of fumarase. *Appl Microbiol Biotechnol*, 46, 393–399.

Rhodes RA, Moyer AJ, Smith ML, Kelley SE. (1959). Production of fumaric acid by *Rhizopus arrhizus*. *Appl Microbiol*, 7, 74–80.

Rhodes RA, Lagoda AA, Misenheimer TJ, Smith ML, Anderson RF, Jackson RW. (1962). Production of fumaric acid in 20-l fermentors. *Appl Microbiol*, 10, 9–15.

Roa Engel CA, Straathof AJJ, Zijlmans TW, Van Gulik WM, van der Wielen LAM. (2008). Fumaric acid production by fermentation. *Appl Microbiol Biotechnol*, 78, 379–389.

Roa Engel CA, van Gulik WM, Marang L, van der Wielen LAM, Straathof AJJ. (2011). Development of a low pH fermentation strategy for fumaric acid production by *Rhizopus oryzae*. *Enzyme Microb Technol*, 48, 39–47.

Rodriguez-Lopez J, Sanchez AJ, Gomez DM, Romani A, Parajo JC. (2012). Fermentative production of fumaric acid from Eucalyptus globulus wood hydrolysate. *J Chem Technol Biotechnol*, 87, 1036–1040.

Romano AH, Bright MM, Scott WE. (1967). Mechanism of fumaric acid accumulation in *Rhizopus nigricans*. *J Bacteriol*, 93, 600–604.

Sass E, Blachinsky E, Karniely S, Pines O. (2001). Mitochondrial and cytosolic isoforms of yeast fumarase are derivatives of a single translation product and have identical amino termini. *J Biol Chem*, 276, 46111–46117.

Sauer M, Porro D, Mattanovich D, Branduardi P. (2008). Microbial production of organic acids: expanding the markets. *Trends Biotechnol*, 26, 100–108.

Schipper MAA. (1984). A revision of the genus *Rhizopus* I. The *Rhizopus* stolonifer-group and *Rhizopus oryzae*. *Stud Mycol*, 25, 1–19.

Schuking FJ. (2001). Petroleum. Forbes Magazine 26 November, 2001.

Schut JH. (2007). Extruding biopolymers: packaging reaps cost benefit of going www.ptonline.com/articles/extruding-biopolymers-packaging-reaps-cost-benefit-of-going-green retrieved October 30, 2012. *Plastics Technol*, Feb 1.

Shi S, Guo H, Yin G. (2011). Synthesis of maleic acid from renewable resources: catalytic oxidation of furfural in liquid media with dioxygen. *Cat Commun*, 12, 731–733.

Skory CD, Ibrahim AS. (2007). Native and modified lactate dehydrogenase expression in a fumaric acid producing isolate *Rhizopus oryzae* 99-880. *Curr Genet*, 52, 23–33.

Song P, Li S, Ding Y, Xu Q, Huang H. (2011). Expression and characterization of fumarase (FUMR) from *Rhizopus oryzae*. *Fungal Biol*, 115, 49–53.

Suominen P, Aristidou A, Pentllila M, Ilmen M, Ruohonen L, Koivuranta K, Robert-Perez K. (2008). Genetically modified yeast of the species *Issatchenkia orientalis* and closely related species and fermentation processes using same. European Patent Application Publication EP1885842 A2.

Vink ETH, Ra'bago KR, Glassner DA, Springs B, O'Connor RP, Kolstad J, Gruber PR. (2004). The sustainability of NatureWorks™polylactide polymers and Ingeo™polylactide fibers: an update of the future. *Macromol Biosci*, 4, 551–564.

Volhard J. (1892). Darstellung von Maleïnsäureanhydrid. *Liebigs Ann Chem*, 268, 255–256.

Vongsangnak W, Nielsen J. (2010). Bioinformatics and system biology of *Aspergillus niger*. In: Machida M, Gomi K, editors. *Aspergillus—Molecular Biology and Genetics*. Norfolk, UK: Caister Academic Press, pp. 61–84.

Vrabl P, Fuchs V, Pichler B, Schinagl CW, Burgstaller W. (2012). Organic acid excretion in *Penicillium ochrochloron* increases with ambient pH. *Front Microbiol*, 3, 1–10.

Wang G, Huang D, Qi H, Wen J, Jia X, Chen Y. (2013). Rational medium optimization based on comparative metabolic profiling analysis to improve fumaric acid production. *Bioresour Technol*, 137, 1–8.

Wegener WS, Romano AH. (1964). Control of isocitratase in *Rhizopus nigricans*. *J Bacteriol*, 87, 156–161.

Werpy T, Petersen G. (2004). Top value added chemicals from biomass, vol. I. Results of screening for potential candidates from sugars and synthesis gas. US Department of Energy, Washington, DC. http://dx.doi.org/10.2172/15008859. Retrieved October 30, 2012.

West TP. (2008). Fumaric acid production by *Rhizopus oryzae* on corn distillers' grains with solubles. *Res J Microbiol*, 3, 35–40.

Willke T, Vorlop KD. (2004). Industrial bioconversion of renewable resources as an alternative to conventional chemistry. *Appl Microbiol Biotechnol*, 66, 131–142.

Xu Q, Li S, Fu Y, Tai C, Huang H. (2010). Two stage utilization of corn straw by *Rhizopus oryzae* for fumaric acid production. *Bioresour Technol*, 101, 6262–6264.

Xu Q, Li S, Huang H, Wen J. (2012). Key technologies for the industrial production of fumaric acid by fermentation. *Biotech Adv*, 30, 1685–1696.

Yu C, Cao Y, Zou H, Xian M. (2011). Metabolic engineering of *Escherichia coli* for biotechnological production of high-value organic acids and alcohols. *Appl Microbiol Biotechnol*, 89, 573–583.

Yu S, Huang D, Wen J, Li S, Chen Y, Jia X. (2012). Metabolic profiling of a *Rhizopus oryzae* fumaric acid production mutant generated by femtosecond laser irraditation. *Biores Technol*, 114, 610–615.

Zhang K, Yang S. (2011). Production of Fumaric Acid in Fermentation by *Rhizopus Oryzae*: Process Design and Economic Analysis. Abstract 626j in the 11 AlChE Annual Meeting, Minneapolis, MN, USA.

Zhang B, Yang ST. (2012). Metabolic engineering of *Rhizopus oryzae*: Effects of overexpressing *fumR* gene on cell growth and fumaric acid biosynthesis from glucose. *Proc Biochem*, 47, 2159–2165.

Zhang F, Carothers JM, Keasling JD. (2012a). Design of a dynamic sensor-regulator system for production of chemicals and fuels derived from fatty acids. *Nat Biotechnol*, 30, 354–359.

Zhang B, Skory CD, Yang ST. (2012b). Metabolic engineering of *Rhizopus oryzae*: Effects of overexpressing *pyc* and *pepc* genes on fumaric acid biosynthesis from glucose. *Metabol Engng*, 14, 512–520.

Zhang K, Zhang B, Yang ST. (2013). Production of Citric, Itaconic, Fumaric, and Malic Acids in Filamentous Fungal Fermentations, in Bioprocessing Technologies in Biorefinery for Sustainable Production of Fuels, Chemicals, and Polymers (eds S.-T. Yang, H. A. El-Enshasy and N. Thongchul), John Wiley & Sons, Inc., Hoboken, NJ, USA. 375–398.

Zhou Y, Du J, Tsao GT. (2002). Comparison of fumaric acid production by *Rhizopus oryzae* using different neutralizing agents. *Bioproc Biosyst Eng*, 25, 179–181.

Zhou Z, DU G, Hua Z, Zhou J, Chen J. (2011). Optimization of fumaric acid production by *Rhizopus delemar* based on the morphology formation. *Biores Technol*, 102, 9345–9349.

CHAPTER 16

Succinic Acid

BORIS LITSANOV*, MELANIE BROCKER, MARCO OLDIGES, and MICHAEL BOTT

IBG-1: Biotechnology, Institute of Bio- and Geosciences, Forschungszentrum Jülich, Jülich, Germany

*current: Institute of Microbiology, Eidgenössische Technische Hochschule Zürich, Zürich, Switzerland

Abstract

Succinic acid has been identified as an important platform chemical that can be produced from renewable carbon sources. The molecules that can be derived from

Bioprocessing of Renewable Resources to Commodity Bioproducts, First Edition.
Edited by Virendra S. Bisaria and Akihiko Kondo.

succinic acid are commodity chemicals such as 1,4-butanediol, 2-pyrrolidinone, and tetrahydrofuran. Even larger markets are offered by the use of succinic acid and its derivatives as precursors for polymers, such as poly(butylenesuccinate). In this review, we summarize the application areas of succinic acid, the microorganisms that are capable of efficient succinate production either naturally or after genetic modifications, the downstream processing, and the companies involved in succinate manufacturing.

16.1 SUCCINATE AS AN IMPORTANT PLATFORM CHEMICAL FOR A SUSTAINABLE BIO-BASED CHEMISTRY

Nowadays most of the building blocks used in the chemical industry are still derived from petroleum and natural gas. However, the continuously increasing crude oil and gas prices accompanied by depleting reserves and the raised public demand for "green" products increase the interest for biotechnological production of precursors for the chemical industry (Hermann and Patel, 2007). One such bio-based building block is succinate, which has been evaluated to be one of the top 12 value-added chemicals from biomass by the US Department of Energy (Werpy and Peterson, 2004). It is a C_4 dicarboxylate which is nowadays produced in a quantity of 20,000–40,000 tons per year mainly from the petroleum derivative maleic anhydride (Cukalovic and Stevens, 2008; http://www.icis.com/Articles/2012/01/30/9527521/chemical-industry-awaits-for-bio-succinic-acid-potential.html). A rather small quantity originates from biotechnological processes and is used exclusively in the food industry (Zeikus et al., 1999). This situation is about to change in the next couple of years because effective biotechnological processes for succinate production in bulk amounts from renewable carbon sources such as carbohydrates or glycerol have been developed (McKinlay et al., 2007) and now enter the phase of industrial application. A number of chemical companies such as BioAmber Inc., Myriant Technologies LLC, Reverdia (DSM & Roquette joint venture), Succinity GmbH (BASF & Purac joint venture) announced their ongoing plans for construction of production facilities and for the start of bio-based succinic acid production between 2012 and 2015 with an additional total output of more than 50,000 tons per year. However, the market is expected to possess much higher growth potential as bio-based succinate or its free acid could serve as an alternative feedstock for a number of different industrial applications.

The first main application field for bio-based succinic acid is the synthesis of three important bulk chemicals, namely 1,4-butanediol (BDO), γ-butyrolactone (GBL), and tetrahydrofuran (THF), with an estimated total production volume in the year 2007 of more than 500,000 tons per year (McKinlay et al., 2007). The worldwide production of GBL was estimated at that time to 250,000 tons per year, and it is used primarily as solvent and as precursor for the production of pharmaceuticals, herbicides, and rubber additives. THF is also a widely used solvent and represents the starting material in numerous chemical syntheses. However, BDO is the most important chemical in this group. It has a constantly growing market with global production estimated to be about 2 million tons per year nowadays (http://www.icis.com/Articles/2012/01/30/9527521/chemical-industry-awaits-for-bio-succinic-acid-potential.html). It is a raw material for the synthesis of THF and GBL or can be used for the production of the high-performance resins polybutylene terephthalate (PBT) and polybutylene succinate (PBS). The production of PBS requires not only BDO but also succinic acid, and this marks a second major application area for bio-based succinic acid—the production of bio-based polymers (Bechthold et al., 2008). PBS and particularly its modified derivatives mPBS have very similar properties to polypropylene and polyvinylchloride and can be used with the existing processing equipment. These features make PBS an interesting alternative to petroleum-based plastics. The application as polymer precursor can tremendously increase the succinic acid market size, with estimates of up to 25×10^6 tonnes per year (Bozell and Petersen, 2010).

Some other important product classes derived from succinic acid are succinate esters and salts, sulfosuccinates and sulfosuccinamates, or pyrrolidones. Succinate esters and salts find very diverse applications, for example, as additives or active ingredients in pharmaceutics, pH modifiers, flavor enhancer, antibacterial agents in the food industry, as additives in fuels or deicers. The pyrrolidones represent another important group of succinic acid derivatives which are used as solvents, starting materials, or additives for the pharmaceutical, cosmetic, and food industry. They are produced directly from succinic acid or from its derivative GBL. Additionally, sulfosuccinates and sulfosuccinamates represent valuable examples of succinate-based surfactants which round up the possible succinate product portfolio (Cukalovic and Stevens, 2008). All these application areas point out the large impact of a cost-effective biotechnological succinate production process for the establishment of a sustainable chemical industry (Figure 16.1).

The biotechnological production of succinate could utilize not only established resources such as different sugar-containing hydrolysates or molasses (Agarwal et al., 2006; Liu et al., 2008; Zheng et al., 2009; Borges and Pereira, 2011), but it could also recycle waste organic materials, for example, whey from cheese and milk industry (Lee et al., 2003a; Wan et al., 2008), cereals from wheat flour production (Dorado et al., 2009), or glycerol from biofuel synthesis (Murarka et al., 2008). Especially glycerol represents a very interesting substrate as it accumulates in high amounts [10% (w/w)] in the course of biodiesel production (da Silva et al., 2009). As a consequence of the constantly increasing rates of biofuel production, glycerol, previously considered to be a valuable side product of the process, nowadays becomes an excess by-product. Its disposal costs represent an important problem which may compromise the cost-effectiveness of biofuel production. The economic viability of the process could be increased by recycling glycerol as substrate for the production of higher value chemicals such as succinate. The coupling between biofuel and bio-succinate production in the concept of a biorefinery will enhance the sustainability of both processes (Yazdani and Gonzalez, 2007).

16.2 MICROORGANISMS FOR BIO-SUCCINATE PRODUCTION—PHYSIOLOGY, METABOLIC ROUTES, AND STRAIN DEVELOPMENT

Succinate is a metabolite occurring in most living organisms, as it is an intermediate of the citric acid or tricarboxylic acid (TCA) cycle. This pathway is used to oxidize

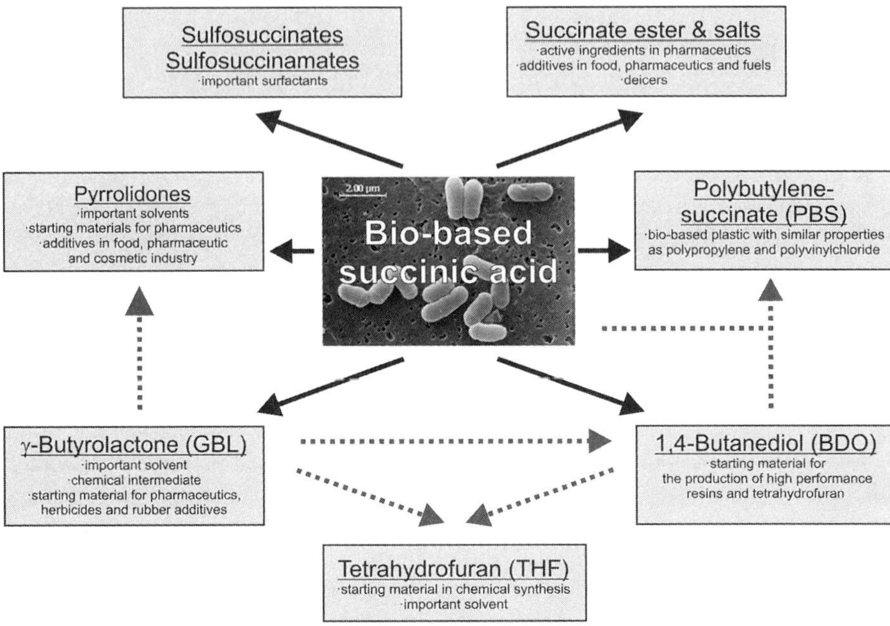

FIGURE 16.1 Classes or groups of chemical compounds which could be derived from bio-based succinic acid. The probable or current applications are given below the name of the compound or its group. Interconversion reactions between GBL, THF, and BDO are displayed with grey dotted arrows. The possibility to convert GBL to pyrrolidones and BDO and succinic acid to PBS are also indicated by dotted arrows.

carbon sources to carbon dioxide, thereby generating reducing equivalents in the form of NADH and reduced quinones. Their reoxidation by the aerobic respiratory chain is used to generate proton-motive force and drive ATP synthesis by F_1F_0-ATP synthase. At the same time, the TCA cycle also provides important precursors for biosynthesis, in particular 2-oxoglutarate and oxaloacetate (Bott, 2007). Under anaerobic fermentative conditions, succinate is often found as an end product of energy metabolism, where it is formed in a reductive pathway from oxaloacetate, allowing reoxidation of NADH by malate dehydrogenase and fumarate reductase. Reduction of fumarate to succinate can also occur separately in bacteria capable of fumarate respiration. Additionally, succinate is also an intermediate of the glyoxylate shunt, a metabolic pathway with anaplerotic function enabling organisms to grow on acetate, ethanol, and fatty acids (Ensign, 2006).

In view of the involvement of succinate in diverse metabolic routes, it is not surprising many different combinations of pathways for succinate accumulation are found in natural producers or were genetically engineered to construct production strains. Figure 16.2 presents a general map for the metabolic pathways leading to production of succinate with emphasis on the used cofactors and the type of reactions involved. Glucose conversion (grey box), carboxylation reactions as a link between C_3 and C_4 intermediates, and succinate production routes (yellow box) are displayed,

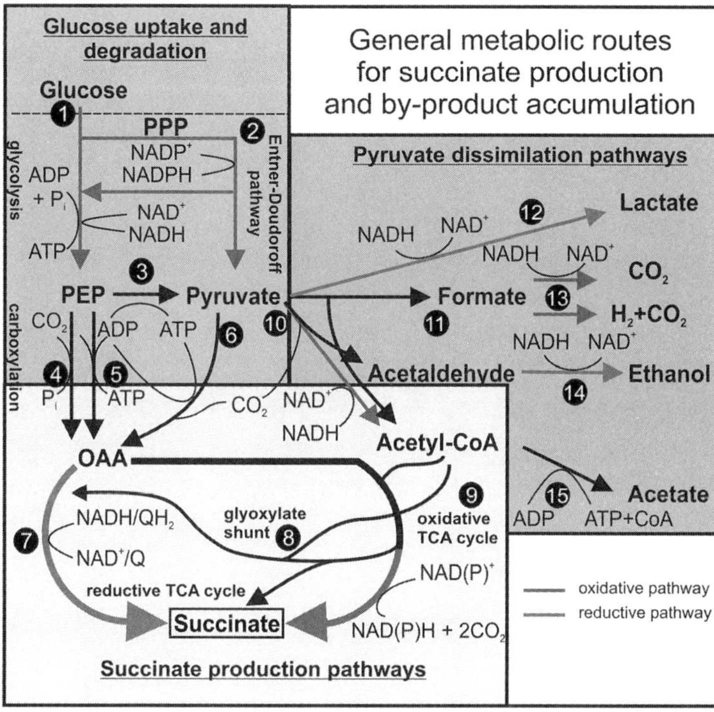

FIGURE 16.2 Metabolic pathways relevant in the context of succinate production from glucose with focus on the type of reaction or reaction sequence—oxidative reactions are indicated by red arrows, reductive reactions by green arrows. The cofactors involved are also shown. The pathways are grouped according to their major purpose in three boxes: grey box—initial substrate degradation pathways, yellow box—succinate-producing pathways, blue box—pyruvate dissimilation pathways. The reactions or reaction sequences are numbered as follows: 1, glucose uptake and glycolysis; 2, oxidative pentose phosphate pathway (PPP) branching with the Entner-Doudoroff pathway; 3, pyruvate kinase; 4, PEP carboxylase; 5, PEP carboxykinase (note that some enzymes used GDP/GTP rather than ADP/ATP); 6, pyruvate carboxylase; 7, reductive TCA cycle including malate dehydrogenase, fumarase, and fumarate reductase or succinate:quinone oxidoreductase; 8, glyoxylate shunt including isocitrate lyase and malate synthase; 9, oxidative TCA cycle including citrate synthase, aconitase, isocitrate dehydrogenase, 2-oxoglutarate dehydrogenase complex, and succinyl-CoA synthetase; 10, pyruvate dehydrogenase complex, pyruvate decarboxylase; 11, pyruvate formate lyase; 12, lactate dehydrogenase; 13, formate dehydrogenase (NAD^+ dependent) or formate hydrogen lyase; 14, alcohol dehydrogenase; 15, phosphotransacetylase and acetate kinase. (*For a color version, see the color plate section.*)

together with the major competing reactions leading to by-product accumulation or connected to cofactor regeneration or energy production (blue box). A detailed description of the specific pathways and features relevant for the different succinate-producing microorganisms is given in the following chapters and the properties of some prominent succinate producers are summarized in Table 16.1.

TABLE 16.1 Selection of Superior Succinate Producer Organisms, the Corresponding Production Conditions and Important Production Parameters of the Resulting Succinate Production Processes Utilizing Two Important Model Substrates for Implementation in the Concept of Biorefinery—Glucose and Glycerol

Succinate producers	Process description	Succinate yield [mol/mol]	Total succinate produced [mM]	Volumetric productivity [mM/h]	Total by-product conc. [mM] (mM carbon)	By-product yield [mol C by-product/mol C glucose]	Reference
Glucose as substrate							
Anaerobiospirillum succiniciproducens ATCC 53488	Anaerobic batch process in complex medium on 80 l-scale	1.37	180	9	109 (205)	0.26	Glassner and Datta (1992)
A. succiniciproducens ATCC 53488	Anaerobic continuous process in complex medium with *in situ* product recovery	1.35	703	88	73 (146)	0.05	Meynial-Salles et al. (2008)
Actinobacillus succinogenes ATCC 55618 FZ 53	Anaerobic batch process in complex medium	1.26	897	12	368 (773)	0.18	Guettler et al. (1996)
Basfia succiniciproducens	Anaerobic batch process in complex medium	0.91	49	13	85 (132)	0.42	Scholten and Dägele (2008)
Corynebacterium glutamicum BOL-3/pAN6-*gap*	Aerobic growth in minimal medium, then anaerobic fed-batch production in saline solution	1.67 (1.36[a])	1134	21 (16[a])	107 (417)	0.10	Litsanov et al. (2012a)
C. glutamicum R Δ*ldhA*-pCRA717	Aerobic growth in complex medium, then anaerobic fed-batch production in minimal medium	1.40 (0.85[a])	1240	27 (21[a])	270 (540)	0.10	Okino et al. (2008)

C. glutamicum BL-1 (pAN6-*pyc*P458S*ppc*)	Aerobic batch process in minimal medium with nitrogen limitation	0.45	90	0.8	58 (195)	0.16	Litsanov et al. (2012b)
Escherichia coli SBS550MG (pHL314)	Aerobic growth in complex medium, then anaerobic fed-batch production in complex medium	1.60 (1.60[a])	350	4 (3[a])	≈50 (50)	0.04	Sanchez et al. (2005b)
E. coli KJ134	Anaerobic batch process in minimal medium	1.53	606	6	72 (192)	0.08	Jantama et al. (2008a)
E. coli HL27659k (pKK313)	Aerobic fed-batch process in complex medium	0.94	494	9	119 (307)	0.09	Lin et al. (2005a)
Mannheimia succiniciproducens LPK7	Anaerobic fed-batch process in complex medium	1.16	444	15	242 (800)	0.35	Lee et al. (2006)
Saccharomyces cerevisiae Δ*sdh1* Δ*sdh2*Δ*idh1*Δ*idp1*	Aerobic batch process in minimal medium	0.11	31	0.2	2 (7)	<0.01	Raab et al. (2010)
S. cerevisiae	Microaerobic process at low pH (final pH 3)	nda	≈790	≈11	185 (536)	nda	See text for web reference

(*continued*)

TABLE 16.1 ***(Continued)***

Succinate producers	Process description	Succinate yield [mol/mol]	Total succinate produced [mM]	Volumetric productivity [mM/h]	Total by-product conc. [mM] (mM carbon)	By-product yield [mol C by-product/mol C glucose]	Reference
Glycerol as substrate							
A. succiniciproducens (ATCC 53488)	Anaerobic fed-batch (feed yeast extract and glycerol) process	1.23	161	1.3	10 (20)	0.05	Lee et al. (2001)
B. succiniciproducens	Anaerobic batch process in complex medium	0.94	54	6	13 (18)	0.12	Scholten and Dägele (2008)
C. glutamicum BL-1 (pVWEx1-*glpFKD*)	Aerobic batch process in minimal medium	0.21	79	4	35 (70)	0.06	Litsanov et al. (2013)
E. coli Δ*adhE* Δ*pta* Δ*poxB* Δ*ldhA* Δ*ppc* (pZS-pyc)	Microaerobic process in minimal medium	0.54	119	2	≈16 (32)	0.06	Blankschien et al. (2010)
E. coli XZ721	Anaerobic batch process in minimal medium	0.80	102	0.7	18 (42)	0.11	Zhang et al. (2010)
Yarrowia lipolytica Y-3314	Aerobic process in unbuffered complex medium at low pH (final pH 3.2)	0.38	144	0.86	nda	nda	Yuzbashev et al. (2010)

nda, no data available.

[a]For two-stage processes, approximate values calculated for the entire process are displayed in parentheses.

16.2.1 *Anaerobiospirillum succiniciproducens*

Anaerobiospirillum succiniciproducens is a strictly anaerobic, Gram-negative spiral-shaped γ-proteobacterium belonging to the family *Succinicivibrionaceae*. Its ability to produce high quantities of succinate was already described in 1981 (Rifkin and Opdyke, 1981). However, this publication mainly discussed the potential of *A. succiniciproducens* for both invasiveness and pathogenicity, at least in compromised hosts, and the observation that succinate is the major product of the anaerobic metabolism of this species remained only a side result. The first study on succinate production with *A. succiniciproducens* was published 10 years later by the Zeikus lab (Samuelov et al., 1991). Growth of *A. succiniciproducens* and the final succinate yield were reported to be strongly dependent on the availability of carbon dioxide and the pH of the culture medium. Based on several enzymatic tests a succinate production pathway was proposed. According to it, *A. succiniciproducens* metabolizes glucose via the Embden–Meyerhof–Parnas pathway (EMP) to phosphoenolpyruvate (PEP), which (i) can be carboxylated by phosphoenolpyruvate carboxykinase to oxaloacetate which then is reduced in the reductive branch of the TCA cycle to succinate, or (ii) can be converted to lactate or ethanol, or (iii) can be used for energy generation by conversion to acetate. The former two routes are associated with reoxidation of NADH formed in the EMP pathway.

The CO_2/HCO_3^- availability and the pH value are the key parameters which influence the fate of PEP at this branching point by influencing the specific activities of the involved enzymes. An increased availability of CO_2/HCO_3^- and a lowered pH were found to be beneficial for succinic acid production and led to a reduced by-product accumulation (Samuelov et al., 1991). The positive effect of a lowered pH might be related to the fact that PEPCk uses CO_2 rather than HCO_3^- as substrate (Cooper et al., 1968). Besides carbon dioxide the addition of H_2 gas as potential additional electron donor was also shown to increase succinate production and to decrease the activity of pyruvate:ferredoxin oxidoreductase, thereby increasing the succinate/acetate ratio (Yoo and Zeikus, 1996; Lee et al., 1999). The first patents on succinic acid production and purification with *A. succiniciproducens* were granted in 1992 (Datta, 1992; Datta et al., 1992; Glassner and Datta, 1992). According to these publications, this species was capable of accumulating 426 mM (50 g/L) succinate from glucose as a carbon source at a pH of 5.9 with a volumetric productivity of 17.8 mM/h reaching a yield of 1.37 mol/mol.

These promising results stimulated further studies on industrial relevant substrates for succinate production with *A. succiniciproducens*. Whey (Samuelov et al., 1999; Lee et al., 2000) and wood hydrolysate together with corn steep liquor as nitrogen base (Lee et al., 2003b) were identified as suitable complex carbon sources delivering yields of 0.90 g succinate per gram of lactose and 0.88 g succinate per gram of glucose, respectively. In addition, *A. succiniciproducens* is able to utilize glycerol for succinate production (Lee et al., 2001), which is nowadays an interesting alternative carbon source as it can be integrated into the concept of a biorefinery. With respect to succinate yield and productivity, the best continuous process with glycerol as carbon substrate resulted in 1.07 mole of succinate per mole of glycerol with a volumetric productivity of 17.8 mM/h at a dilution rate of 0.14/h (Lee et al., 2001). Unfortunately, this process requires yeast extract as a complex medium supplement which is a disadvantage for possible industrial applications as it complicates product purification. Efforts to improve succinate production with *A. succiniciproducens* were also made by optimization of the fermentation process applying continuous culture conditions (Lee et al., 2010b). The best succinate production conditions from glucose reported for *A. succiniciproducens* used a continuous culture with *in situ* product recovery performed in an integrated membrane bioreactor electrodialysis system (Meynial-Salles et al., 2008). Under these conditions *A. succiniciproducens* was able to accumulate 703 mM succinate with high volumetric and specific productivity (88 mM/h and 9 mmol/g DCW/h, respectively) and a succinate yield of 1.35 mole per mole of glucose.

In summary, *A. succiniciproducens* possesses several interesting features as a natural succinate producer: (i) acetate is the only major by-product; (ii) process conditions, in particular the pH and the supply of CO_2/HCO_3^- and H_2, have a high impact on succinate production; (iii) specific productivity is very high; (iv) a variety of industrially relevant carbon sources can be utilized, allowing a flexible feedstock concept; and (v) extensive process optimization know-how is available. However, there are also several challenges which might hinder an industrial application at this time: (i) the species is classified as an S2 organism, requiring higher safety standards; (ii) lack of tools for genetic manipulation still restricts the knowledge of the metabolism of *A. succiniciproducens* and limits the possibilities to manipulate it.

16.2.2 Family *Pasteurellaceae*

Three important natural succinate producers, *Actinobacillus succinogenes*, *Mannheimia succiniciproducens*, and *Basfia succiniciproducens*, all members of the *Pasteurellaceae* family within the γ-proteobacteria, were isolated from rumen contents (Guettler et al., 1999; Lee et al., 2002; Kuhnert et al., 2010). All three species are capnophilic (thrive in the presence of high concentrations of carbon dioxide), facultative anaerobic bacteria which share a similar fermentative metabolism resulting in the production of succinate as major product accompanied by pyruvate dissimilation to lactate, acetate, formate, and in some cases to ethanol as minor products.

16.2.2.1 Actinobacillus succinogenes *Actinobacillus succinogenes* was isolated from a bovine rumen in Michigan, USA, in the course of a search for natural succinic acid producers (Guettler et al., 1999). It naturally produces succinate, formate, acetate, and ethanol as fermentation end products. Already in 1996 a patent was filled by the Michigan Biotechnology Institute (MBI) describing fluoroacetate-resistant mutants of *A. succinogenes* which accumulated decreased levels of acetate and formate as by-products (Guettler et al., 1996). To this day most of the isolates described in this patent still belong to the best natural succinate producers identified so far. For example, the isolate *A. succinogenes* 130Z accumulated 105 g/L succinate from 133 g glucose during 78 hours. Since then succinate production with *A. succinogenes* has been systematically studied and MBI is seeking commercial partnerships for its bio-based succinic acid platform (http://www.mbi.org/mbi-technologies.html).

To investigate the pathways for conversion of glucose to succinate in *A. succinogenes*, enzyme assays (van der Werf et al., 1997) and metabolic flux analysis (McKinlay and Vieille, 2008) were performed, resulting in a framework of central metabolism which in the meantime was further improved by the availability of the genome sequence (McKinlay et al., 2010). Glucose is taken up by a PEP-dependent phosphotransferase system as well as by a glucose permease. In the latter case, a hexokinase is responsible for phosphorylation. Glucose 6-phosphate is metabolized almost exclusively through the EMP, although a complete pentose phosphate pathway (PPP) is encoded in the genome. Carbon flux analysis confirmed the low activity of the PPP and revealed that probably most of the NADPH is formed by a transhydrogenase or through the combined action of malate dehydrogenase and malic enzyme (McKinlay and Vieille, 2008). PEP formed in the EMP is carboxylated to oxaloacetate by PEP carboxykinase, the sole C_3-carboxylating enzyme in *A. succinogenes*, which is also important for energy conservation. Oxaloacetate is then converted to succinate via malate dehydrogenase, fumarase, and fumarate reductase in the reductive TCA cycle. Pyruvate decarboxylation by the pyruvate dehydrogenase complex and by pyruvate–formate lyase coupled to formate dehydrogenase yields additional redox equivalents for reduction of oxaloacetate to succinate. Conversion of acetyl-CoA to acetate and ethanol serves for ATP synthesis and for maintaining the redox balance. Besides PEP carboxykinase, C_4 and C_3 metabolism are linked via malic enzyme and oxaloacetate decarboxylase, which allows high metabolic flexibility in the flux distribution between the C_4 (succinate producing, reductive reactions) and C_3 (NADH and energy producing oxidative reactions) pathways. An incomplete oxidative branch of the TCA cycle lacking citrate synthase and isocitrate dehydrogenase explains the glutamate auxotrophy of *A. succinogenes*.

The genome of *A. succinogenes* encodes for a variety of sugar transporters, providing the molecular basis for a broad substrate spectrum, an important characteristic of *A. succinogenes* as succinate producer. It was shown that glucose, fructose, xylose, arabinose (van der Werf et al., 1997), mannose, and sucrose (Guettler et al., 1999) are efficient substrates for cultivation of *A. succinogenes* and therefore could be used for succinate production. The efficient accumulation of succinate with *A. succinogenes* from a variety of complex carbon and nitrogen sources such as sugarcane bagasse hemicellulose hydrolysate (Borges and Pereira, 2011), corn fiber hydrolysate (Chen et al., 2010, 2011), corncob hydrolysate (Yu et al., 2010), cheese whey (Wan et al., 2008), wheat milling by-products (Dorado et al., 2009), cane molasses (Liu et al., 2008), and spent brewer's yeast cells (Jiang et al., 2010) has been demonstrated. Until now only few reports dealing with metabolic improvements of succinate production by *A. succinogenes* have been published. They describe breeding studies which successfully increased the tolerance of *A. succinogenes* toward high ammonia concentrations (up to 354 mM), which are known to be inhibitory for growth (Ye et al., 2010), or reduced acetate production by selection of fluoroacetate-resistant mutants (Guettler et al., 1996). Genetic tools such as expression plasmids and shuttle vectors were also developed, giving the opportunity for rational strain improvement (Kim et al., 2004b; Jang et al., 2007). Another step toward the industrial application of *A. succinogenes* as succinate producer was the development of a defined minimal medium, which however includes L-glutamate, L-methionine, and L-cysteine, reflecting the auxotrophy of *A. succinogenes* for these amino acids (McKinlay et al., 2005).

16.2.2.2 Mannheimia succiniciproducens *Mannheimia succiniciproducens* is not only taxonomically closely related to *A. succinogenes*, but it also shares many metabolic and succinate-producing traits with it. *Mannheimia succiniciproducens* was isolated from the rumen of a Korean cow in 2002 and since then it has been systematically characterized and improved to establish it as one of the most promising succinate producers (Lee et al., 2002, 2008). Two years after isolation, the genome of *M. succiniciproducens* was already sequenced (Hong et al., 2004). The analysis of the data resulted in the first metabolic network assembled for a capnophilic bacterium. The genome annotation revealed that *M. succiniciproducens* possesses the EMP, the PPP, and the Entner–Doudoroff pathway (EDP). Three enzymes involved in interconversion of C_3 and C_4 metabolites are present, which are PEP carboxykinase, PEP carboxylase, and malic enzyme. Despite of the redundancy in function, it was proven in later deletion experiments that PEP carboxykinase is the major carboxylating enzyme in *M. succiniciproducens*, and the flux through this reaction correlates with the flux from oxaloacetate to succinate via malate dehydrogenase, fumarase, and fumarate reductase (Lee et al., 2006).

Besides the similarities, several major metabolic differences exist between *M. succiniciproducens* and *A. succinogenes* (McKinlay et al., 2010). First of all, *M. succiniciproducens* possesses the oxidative branch of the TCA cycle including citrate synthase and isocitrate dehydrogenase and only succinyl-CoA synthetase was found to be absent. This allows *M. succiniciproducens* to synthesize the amino acids of the glutamate family, in contrary to *A. succinogenes*. However, all other auxotrophies of *A. succinogenes* including the one for L-cysteine and L-methionine were also found in *M. succiniciproducens*. A second important difference concerns the pyruvate dissimilation routes. *Mannheimia succiniciproducens* like *A. succinogenes* converts pyruvate to acetyl-CoA either via the pyruvate dehydrogenase complex or via pyruvate formate lyase. Acetyl-CoA is converted to acetate via phosphotransacetylase and acetate kinase, coupled to ATP synthesis. But to restore the redox balance, *M. succiniciproducens* uses a different strategy than *A. succinogenes*. Although homologues to all putative alcohol dehydrogenase genes from *A. succinogenes* are present in the genome of *M. succiniciproducens*, this species does not produce ethanol. Instead, *M. succiniciproducens* forms lactate via lactate dehydrogenase. The differences in metabolism are relevant for the establishment of chemically defined media for *M. succiniciproducens* (Song et al., 2008) and for metabolic engineering strategies to improve succinate production (Lee et al., 2006).

In order to allow metabolic engineering, genetic tools for introduction of gene deletions in *M. succiniciproducens* were developed based on conditional host-dependent replicative plasmids and counterselection techniques. This genetic toolbox was later expanded by a markerless gene disruption system based on Cre-*lox* recombination (Kim et al., 2008). Relying on some of these tools and the available genome information, a rational strategy for strain development aiming at reduction of the by-product accumulation resulted in the construction of strain *M. succiniciproducens* LPK7 (Lee et al., 2006). In this strain, three major pyruvate dissimilation routes were disturbed by deletion of the genes encoding for lactate dehydrogenase, pyruvate formate lyase, phosphotransacetylase, and acetate kinase. Thereby the production of the common by-products acetate, lactate, and formate was minimized (<10 mM/125 mM glucose) and the succinate yield was increased by 40%. However, pyruvate and malate accumulated extracellularly at high concentrations during fermentation of the LPK7 strain. A proteome comparison between the LPK7 strain and the wild type during succinate fermentation helped to find out that increased supplementation of pantothenate and L-cysteine as precursors for CoA biosynthesis could to some extent reduce the pyruvate accumulation by the LPK7 strain (Lee and Lee, 2010). The accumulation of malate pointed at problems in maintaining the redox balance in the LPK7 strain, which still need to be resolved to achieve the maximal predicted yield of 1.71 mole of succinate per mole of glucose based on a metabolic flux analysis (Hong et al., 2004). Fed-batch fermentation of the LPK7 strain with glucose resulted in the accumulation of 444 mM succinate with a yield of 1.16 mole per mole of glucose and a specific productivity of 20 mmol/gDCW/h, showing the potential for fast, high-yield succinate production. An additional advantage for the industrial application of *M. succiniciproducens* is its ability to utilize many important carbon sources such as glucose, lactose, mannitol, arabitol, fructose, xylose (Lee et al., 2002), and sucrose (Lee et al., 2010a). Consequently, a variety of abundant low-cost complex feed stocks such as starch, cellulose, and hemicellulose hydrolysate (Kim et al., 2004a), or whey (Lee et al., 2003a) may serve as suitable substrates for the industrial production of succinate.

***16.2.2.3* Basfia succiniciproducens** Like the other succinate producers from the same family, *B. succiniciproducens* was isolated from the rumen of a cow by the chemical company BASF. The publicly available knowledge on the physiology and metabolism of this species is scarce and includes only three reports on the identification as a novel species, on the substrate spectrum, and on continuous succinate fermentation from crude glycerol (Scholten and Dägele, 2008; Scholten et al., 2009; Kuhnert et al., 2010). *Basfia succiniciproducens* utilizes many industrially relevant sugars such as sucrose, glucose, fructose, xylose, arabinose, galactose, and mannose. This substrate spectrum probably allows the utilization of cellulose and hemicellulose hydrolysates. A special feature among the described succinate producers from the *Pasteurellaceae* family is the ability to ferment glycerol. *Basfia succiniciproducens* utilizes either pure or crude glycerol in rich medium and accumulates almost exclusively succinate as fermentative product with a yield of 0.93 mole per mole of substrate (Scholten and Dägele, 2008), a value close to the maximal theoretical yield of 1 mole of succinate per mole of glycerol in the absence of growth (Dharmadi et al., 2006). As by-products only traces of acetate (5 mM) and formate (6 mM) were detected. In serum bottles, a high volumetric productivity of 10 mM/h was achieved. In a continuous fermentation, however, the succinate yield was 0.78 mole per mole of substrate, the volumetric productivity only 0.8 mM/h, and the specific productivity 3.17 mmol/g DCW/h (Scholten et al., 2009). In a recent press release from BASF, the foundation of a joint venture company with CSM named Succinity GmbH was announced, which will produce and market succinic acid produced with *B. succiniciproducens* (http://www.basf.com/group/pressemitteilungen/P-12-444).

In summary, all three members of the *Pasteurellaceae* family described above, *A. succinogenes*, *M. succiniciproducens*, and *B. succiniciproducens*, were selected from ruminant content because of their ability to produce succinate and where shown to be suitable organisms for integration into a biorefinery concept for a number of reasons: (i) the ability to convert a wide variety of industrially relevant substrates with high productivity to high succinate titers, (ii) the possibility to improve the production capabilities by metabolic engineering due to available genome sequences, genetic tools, and continuously increasing knowledge about the metabolism and physiology, (iii) wide know-how on fermentation and production processes, and (iv) the nonpathogenicity of the three species. The topic of pathogenicity was for a long time a counterargument for the application of succinate producers from the *Pasteurellaceae* family, as a number of pathogens such as *Haemophilus influenzae*, *Pasteurella multocida*, and *Actinobacillus pleuropneumoniae* belong to the same family and the succinate producers might share common pathogenic features with them. This assumption was proven to be false as the genome sequences of *A. succinogenes* and *M. succiniciproducens* revealed that these species lack a great number of the virulence factors detected in the most important pathogens of this family, and there are no reports of pathogenicity for these bacteria till now (McKinlay et al., 2010). In addition, a comparison between the cytotoxicity of *Mannheimia haemolytica* and *B. succiniciproducens* proved the safety of the latter industrial succinate producer (Kuhnert et al., 2010).

16.2.3 *Escherichia coli*

Since decades *Escherichia coli*, a γ-proteobacterium of the family *Enterobacteriaceae*, serves as the most important model organism for studying bacterial physiology and metabolism. The extensive and detailed knowledge available for this species formed the basis for a multitude of metabolic engineering projects for improving the production of endogenous products such as ethanol (Jarboe et al., 2010) or for construction of strains producing metabolites not naturally occurring in this organism, such as lycopene synthesis (Harada and Misawa, 2012). The organism is used as host for a variety of biotechnological production processes, such as amino acids (Wada and Takagi, 2006; Sprenger, 2007; Dong et al., 2011) or heterologous proteins (Jana and Deb, 2005).

Escherichia coli is a facultative anaerobe which can grow in the absence of oxygen either by various types of anaerobic respiration or by mixed acid fermentation. In the latter type of catabolism, succinate is formed as an end product besides acetate, lactate, ethanol, and formate. In the past years, a multitude of studies was published on the engineering of *E. coli* metabolism for increased succinate production (Wendisch et al., 2006; Thakker et al., 2012). The first efforts to increase the succinate titre aimed at the inhibition of lactate and formate synthesis by inactivation of the genes encoding lactate dehydrogenase and pyruvate formate lyase (Stols et al., 1997; Donnelly et al., 1998). However, this strategy led to poor anaerobic growth and succinate production of the resulting strain NZN111 on glucose, caused by redox imbalances. The pyruvate formed during glucose uptake by the phosphotransferase system could not be used efficiently for NADH oxidation anymore. The possibility to reoxidize NADH via the reductive branch of the TCA cycle was ineffective as *E. coli* only possesses PEP carboxlyase but not pyruvate carboxylase and in contrast to pyruvate the PEP level is low due to the activity of the phosphotransferase system. Therefore, PEP carboxylation to oxaloacetate was insufficient and consequently also the activity of the reductive branch of the TCA cycle (Hong and Lee, 2002; Vemuri et al., 2002). This was confirmed by studies on the overexpression of the native malic enzyme or of a heterologous NAD^+-dependent malic enzyme, which improved the carboxylation of pyruvate to oxaloacetate and NAD^+ regeneration (Stols et al., 1997; Hong and Lee, 2001). Another proof for the high pyruvate level and the redox imbalance in the NZN111 mutant was the isolation of the spontaneous mutant AFP111 with a defect in the glucose-specific phosphotransferase system. This mutant had simultaneously activated an alternative glucose uptake system and used ATP instead of PEP for glucose phosphorylation. Strain AFP111 enabled simultaneous anaerobic growth and succinate production without accumulation of high lactate and formate levels (Donnelly et al., 1998; Chatterjee et al., 2001).

The studies described above pointed out that efficient carboxylation of PEP and pyruvate is key for increased succinate accumulation. Therefore, different heterologous C_3 carboxylation enzymes such as PEP carboxylase from *Sorghum vulgare*, pyruvate carboxylase from *Lactococcus lactis* (Lin et al., 2005d), and PEP carboxykinase from *A. succinogenes* (Kim et al., 2004c) were expressed in *E. coli* to support the native PEP carboxylase and led to significantly increased succinate levels. However, the exclusive use of the reductive TCA cycle for succinate production from glucose limits the possible yield to 1 mol/mol because of the redox balance. Somewhat higher yields are possible due to the anaerobic activity of the pyruvate dehydrogenase complex or formate dehydrogenase (in strains not devoid of pyruvate formate lyase), which supply additional NADH (Sanchez et al., 2005a). Nevertheless, significantly higher yields require an alternative succinate production pathway with a lower NADH demand. This idea was realized and led to the construction of a strain capable of accumulating the maximum succinate yield from glucose of 1.6 mol/mol after activation of the glyoxylate shunt as a second succinate production route by elimination of the repressor gene *iclR*. This reduced the total NADH requirements to 1.25 mole per mole of succinate. The combination of this effect with the deletions of *adhE* (alcohol dehydrogenase), *ldhA* (NAD^+-dependent L-lactate dehydrogenase), *pta–ack* (phosphotransacetylase and acetate kinase), and plasmid-based overexpression of the *pyc* gene (pyruvate carboxylase) from *L. lactis* in *E. coli* strain SBS550MG/pHL314 resulted in an efficient succinate producer accumulating 350 mM succinate with a yield of 1.6 mole per mole of glucose (Sanchez et al., 2005b). In a different study, an anaerobic succinate producer was created without the use of heterologous genes via inactivation of the fermentative pathways competing with succinate formation and was used for evolution and selection of strains with improved anaerobic growth in glucose minimal medium. The resulting superior producer KJ122 accumulated 606 mM succinate with a yield of 1.5 mol/mol (Jantama et al., 2008b). Both strategies showed the potential of metabolic engineering methods for construction of efficient succinate producer strains using a microorganism with an initially low level of succinate formation.

Under aerobic conditions, the requirements for creation of an *E. coli* succinate producer are different of course. Succinate is an intermediate of the oxidative TCA cycle, but inactivation of succinate dehydrogenase (SDH) by deleting the corresponding genes can turn it into a terminal product of oxidative metabolism. Glucose is catabolized via the EMP and the resulting PEP/pyruvate can enter the oxidative TCA cycle as acetyl-CoA formed by the pyruvate dehydrogenase complex and as oxaloacetate formed by PEP carboxylation. After the condensation of acetyl-CoA and oxaloacetate to citrate it is oxidized via *cis*-aconitate, isocitrate, and 2-oxoglutarate to succinate. This route involves two oxidative decarboxylation reactions, leading to the loss of carbon as carbon dioxide, which is a major drawback of the aerobic route and reduces the maximal succinate yield to 1 mole per mole of glucose. However, a highly efficient aerobic producer might possess advantageous features, such as higher productivity than anaerobic ones. To explore this idea, a study for establishing an aerobic succinate production pathway in *E. coli* was performed (Lin et al., 2005a). The inactivation of SDH led to a minor accumulation of succinate and instead a high level of acetate was produced. The combination of further metabolic engineering steps such as the elimination of acetate production routes, the activation of the glyoxylate shunt, and increased C_3 carboxylation via expression of a *Sorghum* sp. PEP carboxylase in a *ptsG*-negative background allowed to obtain a succinate yield of 1 mole per mole of glucose in a complex medium (Lin et al., 2005c). The resulting strain reached a succinate titre of 494 mM in a fed-batch experiment, however, its specific productivity of 0.7 mmol/g DCW/h (Lin et al., 2005b) was not superior compared to known anaerobic producers.

Some of the available *E. coli* succinate producers were tested or genetically modified for the utilization of a variety of different industrially relevant carbon sources such as sucrose, fructose, and xylose (Andersson et al., 2007; Wang et al., 2011). Noteworthy are the studies for constructing microaerobic or anaerobic succinate producers for the utilization of glycerol (da Silva et al., 2009; Blankschien et al., 2010; Zhang et al., 2010). Due to its higher reduction state, glycerol theoretically allows succinate formation in an 1:1 ratio via the reductive branch of the TCA cycle in the presence of carbon dioxide or bicarbonate. However, redox imbalances caused by the fact that glycerol is also used for biomass synthesis led to a very slow fermentative growth. The best strains reached 80% of the theoretical succinate yield from glycerol but with a very low specific productivity (Zhang et al., 2010).

In summary, metabolic engineering of *E. coli* led to a wide spectrum of succinate producer strains, several of which approached the theoretical limits of the succinate yield either under aerobic or anaerobic conditions from different carbon sources such as glucose or glycerol. An important feature of these successful studies was the implementation of the glyoxylate shunt as succinate production route as a second key step besides the enhancement of the C_3 carboxylation reactions.

16.2.4 *Corynebacterium glutamicum*

Corynebacterium glutamicum has a more than 50 years long history as an amino acid-producing bacterium. It is an established industrial bacterium and currently used for the production of more than 2.2 million tons of the flavor enhancer L-glutamate and about 1.5 million tons of the feed additive L-lysine per year. With the aim to improve amino acid production, the physiology, the metabolism, and the regulatory processes of this species were intensively studied, in particular after the genome sequence became available in 2003 (Ikeda and Nakagawa, 2003; Kalinowski et al., 2003). In parallel, a versatile toolbox was developed for genetic engineering of this organism, which most recently was extended by the use of single-cell metabolite sensors as a novel tool for strain development (Kirchner and Tauch, 2003; Binder et al., 2012; Mustafi et al., 2012). Thus, a large body of knowledge is now available for this species (Eggeling and Bott, 2005; Burkovski, 2008; Yukawa and Inui, 2013).

Due to its proven suitability for large-scale L-amino acid production, the capabilities of this organism for the synthesis of other commercially interesting metabolites, such as diamines, ethanol, isobutanol, or D-amino acids, was and is intensively explored (Inui et al., 2004b; Mimitsuka et al., 2007; Smith et al., 2010; Blombach et al., 2011; Stäbler et al., 2011; Becker and Wittmann, 2012). In this context, also the synthesis of organic acids with *C. glutamicum* was studied (Wieschalka et al., 2013). The first hints showing the potential for organic acid production were made in the early 1990s by Dominguez and coworkers (1993). They reported that during batch fermentation under oxygen-limited conditions *C. glutamicum* forms lactate, succinate, and acetate. Further investigations on the metabolism of *C. glutamicum* strain R showed that *C. glutamicum* does not grow under anaerobic conditions without the addition of NO_3^- as electron acceptor (Nishimura et al., 2007; Takeno et al., 2007), but that these resting cells are metabolically active and convert glucose to lactate as major product accompanied by succinate and acetate (Inui et al., 2004a). By addition of HCO_3^- the C_3 carboxylation flux could be improved and thereby the ratio of the organic acids could be changed in favor of succinate. Further improvements were achieved by deleting the *ldhA* gene for the NAD^+-dependent L-lactate dehydrogenase and by overexpression of the native *pyc* gene for pyruvate carboxylase in order to improve C_3 carboxylation. In contrast to *E. coli*, *C. glutamicum* possesses a plethora of enzymes at the PEP–pyruvate–oxaloacetate node including PEPCx, PCx, PEPCk, oxaloacetate decarboxylase, and malic enzyme (Sauer and Eikmanns, 2005). The resulting strain *C. glutamicum* R Δ*ldhA*-pCRA717 accumulated 1240 mM succinate with a yield of 1.4 mole per mole of glucose and high volumetric productivity of 27 mM/h under anoxic conditions in minimal medium, but simultaneously produced also high acetate concentrations (270 mM) as by-product (Okino et al., 2008). Acetate formation from pyruvate contributed to the formation of additional reducing equivalents and thus to the observed succinate yield of above 1 mol/mol (Yasuda et al., 2007).

Recently, a series of strains for anaerobic succinate production from glucose was constructed based on the type strain ATCC13032 (Litsanov et al., 2012a). In this study, acetate formation was minimized by eliminating all known acetate-forming pathways, which include phosphotransacetylase and acetate kinase, pyruvate:menaquinone oxidoreductase, and acetyl-CoA:CoA transferase. All of the corresponding genes *pta-ackA*, *pqo*, and *coaT* were successively deleted in addition to the *ldhA* gene. The *pqo* gene was replaced by a second copy of the *pyc* gene under the control of the strong promoter of the elongation factor TU (P_{tuf}). The encoded PCx contains a point mutation (P458S) that was found to be beneficial for lysine production. In order to maximize the succinate yield, formate was used as an additional donor of reducing equivalents and carbon dioxide. Although *C. glutamicum* possesses an endogenous formate dehydrogenase (Witthoff et al., 2012), this enzyme is apparently not active under anoxic conditions. Therefore, a heterologous NAD^+-dependent formate dehydrogenase from *Mycobacterium vaccae* was introduced. The corresponding *fdh* gene was integrated chromosomally in the Δ*pta-ackA* locus, again under the control of P_{tuf}. Finally, the *gapA* gene encoding glyceraldehyde-3-phosphate dehydrogenase was overexpressed using plasmid pAN6-gap to enable an optimal flux through glycolysis under the pressure of an increased NADH pool caused by formate oxidation. In fed-batch fermentations, the resulting *C. glutamicum* strain BOL-3/pAN6-*gap* accumulated 1134 mM succinate in 53 hours with a succinate yield of 1.67 mole per mole of glucose and a high volumetric productivity of 21 mM/h (Figure 16.3); 772 mM formate was oxidized to CO_2. Analysis of the carbon balance revealed that part of the resulting CO_2 was also incorporated into succinate. In total, 107 mM by-products (corresponding to 417 mM C) were formed (35 mM α-ketoglutarate, 33 mM malate, 20 mM acetate, 13 mM fumarate, and 6 mM pyruvate), which represented only 10% of the total glucose carbon consumed (4074 mM C). These results showed that reduction of by-product formation coupled with additional provision of reducing equivalents are a successful strategy for improving anaerobic succinate formation from glucose.

Besides anaerobic succinate producers of *C. glutamicum*, also aerobic ones were constructed (Litsanov et al., 2012b). By deletion of the *sdhCAB* genes for SDH, succinate became an end product of oxidative metabolism. However, growth of the Δ*sdhCAB* strain in glucose minimal medium resulted in the accumulation of acetate as major product. Therefore, the genes for all known acetate-producing pathways were deleted as described above for the anaerobic producer. Furthermore, C_3 carboxylation was improved by overexpression of the *ppc* gene encoding for PEPCx. Overexpression of the *pyc*^{P458S} gene for pyruvate carboxylase as second carboxylation enzyme or overexpression of the genes *aceA* and *aceB* coding for the glyoxylate shunt enzymes isocitrate lyase and malate synthase had only minor effects on succinate production. In shake flask cultures, the strain BL-1 (Δ*sdhCAB*Δ*cat*, Δ*pqo*, Δ*pta-ackA*) with plasmid pAN6-pyc^{P458S}ppc formed succinate with a yield of 0.36 mole per mole of glucose and a specific productivity of 1.6 mmol/g DCW/h, which is the highest one reported for aerobic bacterial succinate producers. A further increase of the succinate yield to 0.45 mole per mole of glucose was achieved by decoupling of growth and production via limitation of the nitrogen source. These results confirmed

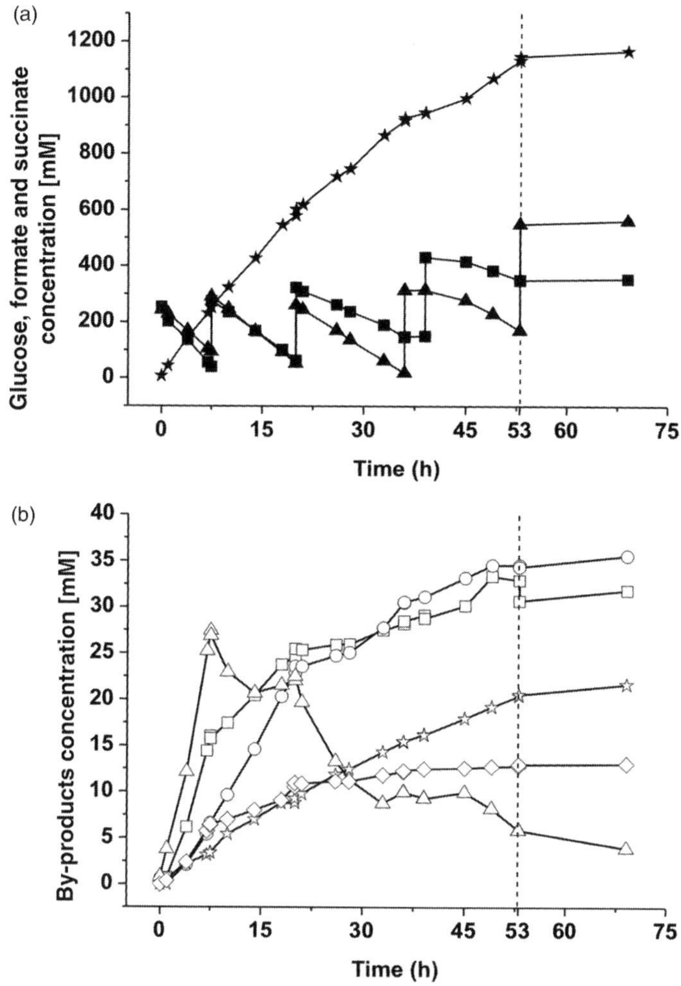

FIGURE 16.3 Anaerobic fed-batch fermentation with *Corynebacterium glutamicum* strain BOL-3/pAN6-*gap* showing (a) succinate production (black stars) during co-utilization of glucose (black squares), and formate (black triangles) and (b) formation of the by-products pyruvate (open triangles), acetate (open stars), α-ketoglutarate (open circles), malate (open squares), and fumarate (open diamonds). The dashed line shown after 53 hours of incubation marks the end of anaerobic succinate production activity. The experiment was performed in a Multifors bioreactor system, and the cells pregrown aerobically were resuspended in 450 mL of a solution containing 0.9% (w/v) NaCl, 220 mM glucose, 250 mM $NaHCO_3$, and 220 mM sodium formate. The cell suspension had an initial OD_{600} of about 50. After 7.5, 20, and 39 hours, 222 mM glucose and 250 mM $NaHCO_3$ were added. After 7.5 and 20 hours, 200 mM sodium formate was added and after 36 and 53 hours 290 mM sodium formate. The pH was kept at pH 6.9 by automated addition of 4 M KOH. After feeding with $NaHCO_3$, the pH increased to 7.2–7.4 and returned to pH 6.9 by the acid production of the cells. Two independent fermentations were performed, both showing comparable results with respect to product yield (5% difference) and titre (10% difference). Taken from reference Litsanov et al. (2012a). Copyright © American Society for Microbiology, Applied and Environmental Microbiology 78: p. 3325–3337, doi:10.1128/AEM.07790-11.

that aerobic succinate production with *C. glutamicum* from glucose is possible, but not as efficient as anaerobic production.

The aerobic succinate producer of *C. glutamicum* offered the opportunity to explore the utilization of glycerol as an alternative substrate. Glycerol cannot be utilized by wild-type *C. glutamicum*, but transformation with plasmid pVWEx1-*glpFKD* carrying the glycerol utilization genes from *E. coli* was previously shown to allow growth on glycerol as sole carbon source (Rittmann et al., 2008). Introduction of this plasmid into the *C. glutamicum* aerobic succinate producer BL-1 allowed succinate formation from glycerol (Litsanov et al., 2013). The strain was able to accumulate 79 mM succinate during aerobic growth, reaching 42% of the theoretical possible yield with a high volumetric productivity (3.59 mM/h). These results showed that not only glucose but also glycerol can be used as feedstock for succinate production with *C. glutamicum*.

Corynebacterium glutamicum is naturally able to grow on a wide variety of carbon sources such as different sugars, sugar acids, sugar alcohols, organic acids, or ethanol and additionally was engineered for the utilization of many industrially relevant substrates such as starch, the hemicellulosic sugars xylose and arabinose, cellobiose, and the whey sugars lactose and galactose in minimal medium (Barrett et al., 2004; Kawaguchi et al., 2006, 2008; Sasaki et al., 2008, 2009; Blombach and Seibold, 2010; Meiswinkel et al., 2013; Uhde et al., 2013). Therefore, a number of different feedstocks are in principle available for succinate production with *C. glutamicum* in the concept of a biorefinery.

In summary, superior anaerobic succinate producers of *C. glutamicum* are available based on metabolically active, but nongrowing cells. Formate can be used as an additional donor of reducing equivalents and CO_2, allowing improved succinate yields. *Corynebacterium glutamicum* was also successfully engineered for aerobic succinate production from glucose or glycerol. Regarding succinate transport, the C_4 dicarboxylate uptake systems DccT and DctA were identified (Youn et al., 2008, 2009) and more recently also the *sucE* gene encoding a succinate exporter (Fukui et al., 2011; Huhn et al., 2011).

16.2.5 Yeast-Based Producers

Yeasts are established hosts for biotechnological production processes for a long time. The availability of a sophisticated toolbox for genetic engineering and a long-term process engineering experience combined with an extraordinary high acid- and osmotolerance make yeasts particularly interesting for industrial large-scale production of acids such as succinate. The metabolic routes for succinate production from glucose are similar in yeasts to those in bacterial producers (Raab and Lang, 2011). On the one hand, succinate can be produced anaerobically via the reductive arm of the TCA cycle, which requires CO_2 fixation and has a limited ATP yield. In *Saccharomyces cerevisiae*, the establishment of such a production mode is quite challenging as some of the TCA cycle enzymes (i) have low affinities for their substrates in the reductive direction, such as fumarase (Pines et al., 1996) or (ii) are subject to glucose repression, as described for an isoform of malate dehydrogenase (Hung et al., 2004). Therefore, excessive metabolic engineering is required to overcome these limitations in the construction of efficient anaerobic succinate producers. On the other hand, an aerobic succinate production route via the oxidative arm of the TCA cycle can be used for succinate production, which releases CO_2 and thus reduces the carbon yield, but leads to a higher energy yield. The formed ATP can be used to maintain an intracellular neutral pH and to actively export succinate into an unbuffered medium. This allows succinate production at low external pH, a rather unique process feature possible with acid-tolerant yeast hosts which facilitates product purification (Yuzbashev et al., 2011). Moreover, the glyoxylate shunt represents an additional succinate production route under both aerobic and anaerobic conditions. A recent theoretical comparison of these two production pathways revealed possible advantages and drawbacks and suggested that a combination of both metabolic routes under microaerobic conditions results in optimal succinate production reaching yields close to the theoretical maximum of 1.71 mole per mole of glucose (Raab and Lang, 2011). A preceding study showed that the application of an aerobic production route via the glyoxylate shunt using a mutant of *S. cerevisiae* lacking SDH and isocitrate dehydrogenase led to the accumulation of only marginal succinate concentrations (31 mM) and a low yield of 0.11 mole per mole of glucose (Raab et al., 2010). In the meantime Reverdia, a newly founded joint venture company of the Dutch biotech company DSM and the French starch producer Roquette Freres reported on the successful construction of a *S. cerevisiae* producer strain which accumulates about 100 g/L succinic acid at an external pH of 3 under microaerobic conditions. This strain was generated by rational metabolic engineering followed by classical strain development steps and produces succinate via a combination of the reductive and oxidative arms of the TCA cycle and the glyoxylate shunt. The succinate production process based on this strain seems to be very effective as the first commercial large-scale production plant is expected to be launched at the end of 2012 (http://www.bio4bio.life.ku.dk/BiorefineryConference2012/~/media/Bio4bio/ABBA/jansen_Presentation%20Bio4Bio%20Conference%20Copenhagen.ashx).

The idea that yeast-based succinate production at low pH could be very promising was also proven by a study with the obligately aerobic yeast *Yarrowia lipolytica*. A succinate dehydrogenase deletion mutant selected on glycerol was able to accumulate 144 mM succinate in non-buffered medium lowering the pH to 3.5 with a yield 0.38 mole per mole of glycerol (Yuzbashev et al., 2010). Although the selected strain lost its ability to utilize glucose during mutagenesis, it might be of interest for application in a biorefinery. In summary the yeast-based succinate producers have the advantage to accumulate succinic acid at low pH. This feature can facilitate downstream processing (see below), but it probably also reduces the production rate and diminishes the total product yield as some part of the substrate is oxidized for the additional energy needed to maintain pH homeostasis and to export the product (Yuzbashev et al., 2011).

16.3 NEUTRAL VERSUS ACIDIC CONDITIONS FOR PRODUCT FORMATION

Several microbial hosts were shown to be capable of efficient succinate production, as presented in the previous sections and summarized in Table 16.1. The metabolic capability of the hosts is intimately linked to the environmental conditions, at which product formation takes place. Two main sets of conditions can be distinguished. Most of the microbial hosts described above, in particular the bacterial ones, are cultivated under neutral or slightly acidic conditions, since they are not tolerant to low pH conditions. The pH is controlled during the fermentation process by addition of base, thereby titrating the two acidic protons of the succinic acid carboxylate groups ($pK_{a1} = 4.16$, $pK_{a2} = 5.61$). Hence, stoichiometric base addition is required during the production phase (Yuzbashev et al., 2011). The final product of fermentations at neutral pH is the succinate dianion and equivalent amounts of base cations.

Yeast-based production hosts tolerate low pH conditions, allowing production of succinic acid and circumventing any base addition for pH control during the fermentation process (Nicolaou et al., 2010). The acidified cultivation medium enables direct crystallization of succinic acid after separation of the cells. However, acidic conditions also pose serious problems for cells producing weak acids (Yuzbashev et al., 2011). At a low extracellular pH of, for example, 3, uncharged succinic acid predominates and may be able to penetrate the cytoplasmic membrane by diffusion. In the cytoplasm, the pH is close to neutral and succinic acid dissociates into the succinate dianion and two protons, causing acidification. In order to maintain a neutral intracellular pH and avoid toxic levels of succinate, protons and succinate have to be exported into the medium again by plasma membrane H^+-ATPases and exporters such as the ABC transporter Pdr12, respectively. In the medium, succinic acid is formed again and can reenter the cell. Thus, a futile cycle of passive influx of succinic acid and active export of succinate and protons may exist, increasing the demand for maintenance energy and reducing product selectivity. However, the fact that Reverdia has launched a *S. cerevisiae*-based succinate production process operating at pH 3 under microaerobic conditions indicates that the above-described problem apparently can be managed, although no data available on parameters like yield or specific productivity. It appears likely that reductions in these parameters have to be accepted as reward for a low pH process.

Another issue associated with the pH is carbon dioxide fixation. As outlined above, probably all succinate production pathways are critically dependent on the net fixation of carbon dioxide by C_3 carboxylation reactions. Enzymes responsible for this conversion are PEP carboxylase, PEPCK, pyruvate carboxylase, and malic enzyme. Studies on the active species of "CO_2" used by these enzymes indicated that PEP carboxylase and pyruvate carboxylase use bicarbonate as substrate, whereas PEP carboxykinase and malic enzyme use CO_2 (for references see Bott and Thauer, 1989). Under acidic conditions, the equilibrium of the reaction $CO_2(aq) + H_2O \leftrightarrows HCO_3^- + H^+$ ($pK_a = 6.33$ at 30°C) is strongly shifted to carbon dioxide, which may positively or negatively impact carbon dioxide fixation, depending on the carboxylating enzymes present.

16.4 DOWNSTREAM PROCESSING

Besides efficient producer strains and optimal fermentation conditions, a suitable purification strategy is a critical issue in the establishment of an economically viable succinate or succinic acid production process, as the downstream processing costs can amount to more than 60% of the total production costs (Bechthold et al., 2008). The first step in downstream processing is the separation of biomass from the product or the product-enriched solution by centrifugation or microfiltration. In a subsequent ultrafiltration step, cell debris, proteins, and other polymers are removed from the cell-free fermentation broth. The separated biomass can be recycled within the process and reused for product formation (Gerigk et al., 2002; Maass et al., 2002). Such an approach could also be beneficial for succinic acid production, in particular for anoxic production processes in which there is no need to avoid oxygen limitation during the recycling process of the biomass. Future activities might also investigate downstream processing steps that are directly applicable to the fermentation suspension without biomass separation.

For the recovery of succinic acid, a variety of methods are available, such as precipitation, electrodialysis, liquid–liquid extraction, or sorption and ion exchange which have recently been reviewed (Kurzrock and Weuster-Botz, 2010; Cheng et al., 2012). Classically, organic acids such as lactic acid or citric acid are isolated from aqueous fermentation broth by precipitation with calcium hydroxide or oxide followed by acidification with concentrated sulfuric acid, which yields equimolar amounts of $CaSO_4$. After filtration, the free acid is purified further, concentrated, and isolated by crystallization steps (Vickroy, 1985; Berovic and Legisa, 2007; Verhoff, 2012). Such processes have also been reported for succinic acid (Datta, 1992; Datta et al., 1992). Besides calcium hydroxide or oxide, also precipitation by ammonia has been described (Berglund et al., 1999; Yedur et al., 2001). Since large amounts of chemicals are required in these processes which lead to low-value by-products like gypsum and/or wastewater with high salt loads, precipitation may not be a viable approach for large-scale production of succinic acid (Hermann and Patel, 2007). Electrodialysis, an established method for organic acid recovery, has also been applied for succinic acid (Glassner and Datta, 1992) but has a number of disadvantages such as low purification yields, high energy consumption, high material costs for the membranes, and low selectivity for succinic acid (McKinlay et al., 2007). Ion exchange techniques were reported to show insufficient selectivity and overall yield and are only regarded as additional purification steps (Pai et al., 2002). Although reactive extraction setups can provide superior performance compared to ion exchange processes, due to higher mass transfer rates (Rüffer et al., 2004), the necessity to use highly cell toxic trialkyl- or tetraalkylamines as carrier molecules limits their application. More recently, an alternative concept for downstream processing was developed which does not aim at the isolation of succinic acid. In this process, a strong anion exchange resin is used not only for succinate recovery by sorption, but also as a catalyst for the conversion of succinate into diethyl succinate via *O*-alkylation by gaseous chloroethane (Lopez-Garzon et al., 2012).

For purification of succinic acid from acidic fermentation broth (pH 3) as obtained by yeast-based fermentation processes, relatively short routes for downstream processing are possible. The protonated form of succinic acid shows much lower solubility in aqueous systems compared to its anionic form. After separation of the cells, the broth is evaporated leading to crystallization of succinic acid, which is then solubilized, purified by ion exchange and crystallized again (Figure 16.4). Since the

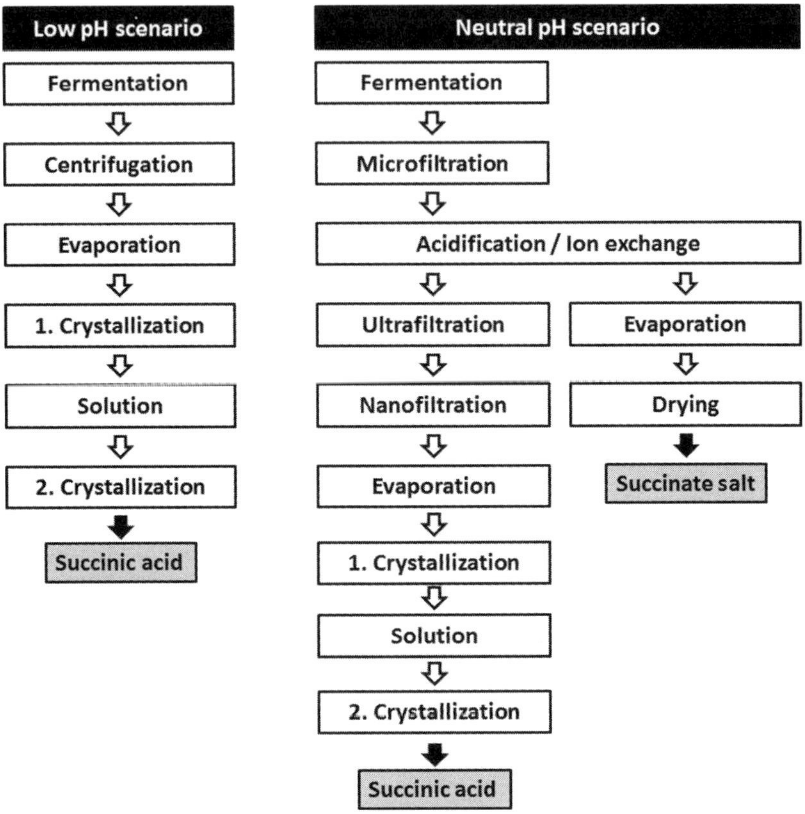

FIGURE 16.4 Display of involved downstream processing steps for purification of succinic acid produced under low and neutral pH scenario.

product is already produced in its protonated form no further by-product streams are formed in this scenario.

16.5 COMPANIES INVOLVED IN BIO-SUCCINIC ACID MANUFACTURING

Introduction of bio-based succinic acid as a significant step toward a sustainable chemical industry depends on two important economic parameters. The first is summarized under the term of the production costs. Major constituents along the production process are the costs and the availability of the feedstock with 10–15%, the fermentation process itself contributing to 20–25% of the costs, and the purification of the final product, which may utilize between 60% and 70% of the total costs. All these parameters can be improved to reduce the succinic acid price. For example, processes and/or producer strains capable of utilizing different kinds of feedstocks can be established (flexible feedstock concept). Increases in the efficiency of the fermentation, such as higher yield and higher productivity, will lower the fermentation costs by reducing the demand of carbon source and the production time. The efforts to decrease the downstream processing costs are directed either toward fermentations at low pH or toward *in situ* product recovery. As described in detail in the previous sections, these areas are subject to intensive research.

The second economic parameter which has high impact on the bio-based succinic acid market is the availability of succinic acid and the demand for succinic acid-based products. Currently the market is evolving very rapidly. For several years bio-based succinic acid was on the future portfolio of only a few companies. The plans for industrial production were still very vague as most of the companies were in the phase of research and development. Meanwhile production organisms and processes have been developed and tested in several pilot plants and in the last 2 years a number of press releases mapped several of the major players (Table 16.2) in the surprisingly fast-growing market of bio-based succinic acid.

16.5.1 Bioamber Inc.

BioAmber Inc. was established in 2008 as a joint venture between ARD (Agro-industrie Recherches et Développements) and DNP Green Technology exclusively for the production of bio-based succinic acid and its derivatives (http://www.bio-amber.com/bioamber/en/home). The initial process utilized genetically modified (GM) *E. coli* strains as production host, but a new generation nonbacterial production host is under development. A pilot plant for bio-based succinate production with a capacity of 2000 tons started already in 2010 in Pomacle, France. Meanwhile another production plant with a total maximal capacity of 34,000 tons succinic acid and 23,000 tons of 1,4-BDO is under construction in Sarnia, Ontario, Canada. Several additional facilities in Thailand and Brasil are planned as a result of a joint venture agreement with Mitsui & Co., a Japanese trading company which will distribute the bio-succinic acid throughout Asia. BioAmber Inc. focuses not only on the production of bio-based succinic acid, but also on its derivatives and consequently searches for partners or customers for such products. Examples are the collaboration with Faurecia and Mitsubishi Chemical Corp. on the production of 100% bio-based automotive plastics (http://biomassmagazine.com/articles/8284/bioamber-faurecia-and-mitsubishi-collaborate-on-bioplastics), the partnership with Lanxess to develop bio-based plasticizers (http://biomassmagazine.com/articles/7395/bioamber-lanxess-partner-to-develop-biobased-plasticizers), or the cooperation with DuPont and Evonik on the catalysis research area to develop sustainable production of 1,4-BDO, GBL, and THF from bio-succinic acid (http://www.bio-amber.com/bioamber/en/products).

16.5.2 Myriant Technologies LLC

A second company involved in bio-based succinic acid production is Myriant Technologies LLC, which was established in 2009 to unify the bio-based chemical business

TABLE 16.2 Overview on the Major Companies Running or Planning to Start a Bio-Succinic Acid Production

Companies	Joint venture partners	Organism	Feedstock	Plant location	Plant capacity planned	Integrated products	Expected start of production	Reference
BioAmber Inc.		Genetically modified (GM) *Escherichia coli* strains	Raw materials from agriculture or forestry	1. Pomacle; France 2. Sarnia, Canada (2013)[a]	1. nda 2. 17,000–34,000 tons of succinic acid, and a 23,000 tons BDO per year	With Evonic: BDO, THF, and GBL with Mitsubishi Chemical: PBS	Since 2010	http://www.bio-amber.com/bioamber/en/products/succinic acid
Myriant Technologies LCC		Modified (non-GM) *E. coli* strains	Sorghum grain and sorghum grits, lignocellulosic sugar hydrolysate, derived from corn stover, sugarcane bagasse, and wood chips	Lake Providence, Louisiana, The United States (2013)[a]	13,600 tons per year	Myrifilm® solvent with Davy Process Technology BDO	First quarter of 2013	http://www.chemicals-technology.com/projects/myriant-plant/

Reverdia	DSM and Roquette	*Saccharomyces cerevisiae*	Starch derivatives; in the longer term: cellulosic biomass	Cassano Spinola, Italy	10,000 tons per year	nda	In 2012	http://www.dsm.com/en_US/cworld/public/media/pages/press-releases/36_11_dsm_and_roquette_to_open_commercial_scale_bio_based_succinic_acid_plant_2012.jsp
Succinity GmbH	BASF and CSM	*Basfia succiniciproducens*	nda	Near Barcelona, Spain (2013)[a]	10,000 tons per year	nda	Late 2013	http://www.basf.com/group/corporate/en/function/conversions:/publish/content/news-and-media-relations/news-releases/downloads/2012/P444_BASF_Purac_2012_Oct4_e.pdf

nda, no data available.
[a]Plants under construction with estimated completion date in parenthesis.

and intellectual portfolio of BioEnergy International LCC (http://www.myriant.com). Myriant is located in the USA and has proved its proprietary fermentation technology in the production of D-lactic acid. However, it also presented technology for the sustainable production of bio-based succinic acid and was granted up to US$50 million from the US Department of Energy (US DOE) to construct a production plant at Lake Providence, Louisiana. The production there is expected to start in the first quarter of 2013 with a capacity of 13,600 tons and options for increase. Like BioAmber, Myriant searches for opportunities to develop and sell bio-succinic acid-derived products, such as the solvent Myrifilm®. In this context, Myriant joined forces with Davy Process Technology to establish a sustainable butanediol production. To open up markets in Asia, Myriant signed partnerships with Sojitz and PTT Chemical Group. PTT Chemical Group, Thailand's largest petrochemical producer, invested US$60 million in the development of the succinic acid production technology (http://www.myriant.com/platform/partners.cfm).

16.5.3 Reverdia

Reverdia, a joint venture between DSM and Roquette, represents another important company which has the ambition to become a global supplier of bio-based succinic acid under the trade mark Biosuccinium™ (http://www.reverdia.com). The production process utilizes GM strains of *S. cerevisiae,* which allows production at low pH and thereby easier purification of the final product. After the successful start of a pilot plant in 2010 in Lestrem, France, Reverdia is now constructing its first production plant in Cassano Spinola, Italy, with a capacity of approximately 10,000 tons per year. Ongoing plans deal with the construction of several other plants until 2020.

16.5.4 Succinity GmbH

Another company involved in industrial bio-based succinic acid production is Succinity GmbH founded in October 2012 (http://www.basf.com/group/pressrelease/P-12-444). This company is a joint venture between BASF and Purac, a subsidiary of CSM. It is aiming to supply the market with succinic acid produced with *B. succiniciproducens*. The first large-scale fermentation plant with a total capacity of 10,000 tons is expected to start production in late 2013 at the Purac site near Barcelona, Spain.

16.6 FUTURE PROSPECTS AND ECONOMIC OUTLOOK

Bio-based succinic acid is a very promising compound which can be used in many different application fields and has the potential to substitute several petroleum-based bulk chemicals. Research within the past years led to the development of a number of different production strains, fermentation strategies, and downstream processes suitable for industrial needs. Based on these results, several companies started bio-succinic acid production at industrial scale. The short overview of these companies given above, which is certainly not complete, and the press releases of the last few years show that the market is very dynamic and fast growing (http://www.icis.com/Articles/2012/01/30/9527521/chemical-industry-awaits-for-bio-succinic-acid-potential.html). Therefore, bio-succinic acid has the potential to become an important component of a sustainable petroleum-independent chemistry in the near future.

The production costs (plus profit) for bio-succinic acid have been estimated to be in the range 1.1–1.6 € per kilogram of succinic acid (dependent on the sugar price) for an annual production of 100,000 tons without telling details about the recovery (Hermann and Patel, 2007). The decision on the best succinate production process with respect to economical and environmental parameters needs an evaluation of all steps involved. Product formation under acidic as well as neutral pH conditions suffers from the challenges described above and can be regarded as a trade-off between production performance and compatibility with subsequent downstream processing procedures to obtain a product of sufficient quality. The current situation indicates that different production processes for succinic acid, succinate salts, or succinate derivatives using different production hosts and different pH values for fermentation are economically viable. A view on industrial succinate production processes running in 5–10 years will give a more conclusive answer on which processes are the most favorable ones.

REFERENCES

Agarwal L, Isar J, Meghwanshi GK, Saxena RK. (2006). A cost effective fermentative production of succinic acid from cane molasses and corn steep liquor by *Escherichia coli*. *J Appl Microbiol*, 100, 1348–1354.

Andersson C, Hodge D, Berglund KA, Rova U. (2007). Effect of different carbon sources on the production of succinic acid using metabolically engineered *Escherichia coli*. *Biotechnol Prog*, 23, 381–388.

Barrett E, Stanton C, Zelder O, Fitzgerald G, Ross RP. (2004). Heterologous expression of lactose- and galactose-utilizing pathways from lactic acid bacteria in *Corynebacterium glutamicum* for production of lysine in whey. *Appl Environ Microbiol*, 70, 2861–2866.

Bechthold I, Bretz K, Kabasci S, Kopitzky R, Springer A. (2008). Succinic acid: a new platform chemical for biobased polymers from renewable resources. *Chem Eng Technol*, 31, 647–654.

Becker J, Wittmann C. (2012). Bio-based production of chemicals, materials and fuels—*Corynebacterium glutamicum* as versatile cell factory. *Curr Opin Biotechnol*, 23, 631–640.

Berglund KA, Yedur S, Dunuwila DD. (1999). Succinic acid production and purification. US Patent 5,958,744.

Berovic M, Legisa M. (2007). Citric acid production. *Biotechnol Annu Rev*, 13, 303–343.

Binder S, Schendzielorz G, Stäbler N, Krumbach K, Hoffmann K, Bott M, Eggeling L. (2012). A high-throughput approach to identify genomic variants of bacterial metabolite producers at the single-cell level. *Genome Biol*, 13.

Blankschien MD, Clomburg JM, Gonzalez R. (2010). Metabolic engineering of *Escherichia coli* for the production of succinate from glycerol. *Metab Eng*, 12, 409–419.

Blombach B, Seibold GM. (2010). Carbohydrate metabolism in *Corynebacterium glutamicum* and applications for the metabolic engineering of L-lysine production strains. *Appl Microbiol Biotechnol*, 86, 1313–1322.

Blombach B, Riester T, Wieschalka S, Ziert C, Youn JW, Wendisch VF, Eikmanns BJ. (2011). *Corynebacterium glutamicum* tailored for efficient isobutanol production. *Appl Environ Microbiol*, 77, 3300–3310.

Borges ER, Pereira N Jr. (2011). Succinic acid production from sugarcane bagasse hemicellulose hydrolysate by *Actinobacillus succinogenes*. *J Ind Microbiol Biotechnol*, 38, 1001–1011.

Bott M. (2007). Offering surprises: TCA cycle regulation in *Corynebacterium glutamicum*. *Trends Microbiol*, 15, 417–425.

Bott M, Thauer RK. (1989). The active species of CO_2 formed by carbon monoxide dehydrogenase from *Peptostreptococcus productus*. *Z Naturforsch C*, 44, 392–396.

Bozell JJ, Petersen GR. (2010). Technology development for the production of biobased products from biorefinery carbohydrates—the US Department of Energy's "Top 10" revisited. *Green Chem*, 12, 539–554.

Burkovski A, editor. (2008). *Corynebacteria: Genomics and Molecular Biology*. Norfolk: Caister Academic Press.

Chatterjee R, Millard CS, Champion K, Clark DP, Donnelly MI. (2001). Mutation of the *ptsG* gene results in increased production of succinate in fermentation of glucose by *Escherichia coli*. *Appl Environ Microbiol*, 67, 148–154.

Chen K, Jiang M, Wei P, Yao J, Wu H. (2010). Succinic acid production from acid hydrolysate of corn fiber by *Actinobacillus succinogenes*. *Appl Biochem Biotechnol*, 160, 477–485.

Chen KQ, Li J, Ma JF, Jiang M, Wei P, Liu ZM, Ying HJ. (2011). Succinic acid production by *Actinobacillus succinogenes* using hydrolysates of spent yeast cells and corn fiber. *Bioresource Technol*, 102, 1704–1708.

Cheng KK, Zhao XB, Zeng J, Wu RC, Xu YZ, Liu DH, Zhang JA. (2012). Downstream processing of biotechnological produced succinic acid. *Appl Microbiol Biotechnol*, 95, 841–850.

Cooper TG, Tchen TT, Wood HG, Benedict CR. (1968). The carboxylation of phosphoenolpyruvate and pyruvate. I. The active species of "CO_2" utilized by phosphoenolpyruvate carboxykinase, carboxytransphosphorylase, and pyruvate carboxylase. *J Biol Chem*, 243, 3857–3863.

Cukalovic A, Stevens CV. (2008). Feasibility of production methods for succinic acid derivatives: a marriage of renewable resources and chemical technology. *Biofuels Bioprod Bioref*, 2, 505–529.

da Silva GP, Mack M, Contiero J. (2009). Glycerol: a promising and abundant carbon source for industrial microbiology. *Biotechnol Adv*, 27, 30–39.

Datta R. (1992). Process for the purification of succinic acid by anaerobic fermentation. US Patent 5,143,833.

Datta R, Glassner DA, Jain MK, Vicky Roy JR. (1992). Fermentation and purification process for succinic acid. US Patent 5,168,055.

Dharmadi Y, Murarka A, Gonzalez R. (2006). Anaerobic fermentation of glycerol by *Escherichia coli*: a new platform for metabolic engineering. *Biotechnol Bioeng*, 94, 821–829.

Dominguez H, Nezondet C, Lindley ND, Cocaign M. (1993). Modified carbon flux during oxygen limited growth of *Corynebacterium glutamicum* and the consequences for amino acid overproduction. *Biotechnol Lett*, 15, 449–454.

Dong X, Quinn PJ, Wang X. (2011). Metabolic engineering of *Escherichia coli* and *Corynebacterium glutamicum* for the production of L-threonine. *Biotechnol Adv*, 29, 11–23.

Donnelly MI, Millard CS, Clark DP, Chen MJ, Rathke JW. (1998). A novel fermentation pathway in an *Escherichia coli* mutant producing succinic acid, acetic acid, and ethanol. *Appl Biochem Biotechnol*, 70–72, 187–198.

Dorado MP, Lin SK, Koutinas A, Du C, Wang R, Webb C. (2009). Cereal-based biorefinery development: utilisation of wheat milling by-products for the production of succinic acid. *J Biotechnol*, 143, 51–59.

Eggeling L, Bott M, editors. (2005). *Handbook of Corynebacterium glutamicum*. Boca Raton, FL: CRC Press.

Ensign SA. (2006). Revisiting the glyoxylate cycle: alternate pathways for microbial acetate assimilation. *Mol Microbiol*, 61, 274–276.

Fukui K, Koseki C, Yamamoto Y, Nakamura J, Sasahara A, Yuji R, Hashiguchi K, Usuda Y, Matsui K, Kojima H, Abe K. (2011). Identification of succinate exporter in *Corynebacterium glutamicum* and its physiological roles under anaerobic conditions. *J Biotechnol*, 154, 25–34.

Gerigk MR, Maass D, Kreutzer A, Sprenger G, Bongaerts J, Wubbolts M, Takors R. (2002). Enhanced pilot-scale fed-batch L-phenylalanine production with recombinant *Escherichia coli* by fully integrated reactive extraction. *Bioprocess Biosyst Eng*, 25, 43–52.

Glassner DA, Datta R. (1992). Process for the production and purification of succinic acid. US Patent 5,143,834.

Guettler MV, Jain MK, Rumler D. (1996). Method for making succinic acid, bacterial variants for use in the process, and methods for obtaining variants. US Patent 5,573,931.

Guettler MV, Rumler D, Jain MK. (1999). *Actinobacillus succinogenes sp.* nov., a novel succinic-acid-producing strain from the bovine rumen. *Int J Syst Bacteriol*, 49, 207–216.

Harada H, Misawa N. (2012). Novel approach in the biosynthesis of functional carotenoids in *Escherichia coli*. *Methods Mol Biol*, 892, 133–141.

Hermann BG, Patel M. (2007). Today's and tomorrow's bio-based bulk chemicals from white biotechnology: a techno-economic analysis. *Appl Biochem Biotechnol*, 136, 361–388.

Hong SH, Lee SY. (2001). Metabolic flux analysis for succinic acid production by recombinant *Escherichia coli* with amplified malic enzyme activity. *Biotechnol Bioeng*, 74, 89–95.

Hong SH, Lee SY. (2002). Importance of redox balance on the production of succinic acid by metabolically engineered *Escherichia coli*. *Appl Microbiol Biotechnol*, 58, 286–290.

Hong SH, Kim JS, Lee SY, In YH, Choi SS, Rih JK, Kim CH, Jeong H, Hur CG, Kim JJ. (2004). The genome sequence of the capnophilic rumen bacterium *Mannheimia succiniciproducens*. *Nat Biotechnol*, 22, 1275–1281.

Huhn S, Jolkver E, Krämer R, Marin K. (2011). Identification of the membrane protein SucE and its role in succinate transport in *Corynebacterium glutamicum*. *Appl Microbiol Biotechnol*, 89, 327–335.

Hung GC, Brown CR, Wolfe AB, Liu J, Chiang HL. (2004). Degradation of the gluconeogenic enzymes fructose-1,6-bisphosphatase and malate dehydrogenase is mediated by distinct proteolytic pathways and signaling events. *J Biol Chem*, 279, 49138–49150.

Ikeda M, Nakagawa S. (2003). The *Corynebacterium glutamicum* genome: features and impacts on biotechnological processes. *Appl Microbiol Biotechnol*, 62, 99–109.

Inui M, Murakami S, Okino S, Kawaguchi H, Vertès A, Yukawa H. (2004a). Metabolic analysis of *Corynebacterium glutamicum* during lactate and succinate productions under oxygen deprivation conditions. *J Mol Microbiol Biotechnol*, 7, 182–196.

Inui M, Kawaguchi H, Murakami S, Vertes AA, Yukawa H. (2004b). Metabolic engineering of *Corynebacterium glutamicum* for fuel ethanol production under oxygen-deprivation conditions. *J Mol Microbiol Biotechnol*, 8, 243–254.

Jana S, Deb JK. (2005). Strategies for efficient production of heterologous proteins in *Escherichia coli*. *Appl Microbiol Biotechnol*, 67, 289–298.

Jang YS, Jung YR, Lee SY, Kim JM, Lee JW, Oh DB, Kang HA, Kwon O, Jang SH, Song H, Lee SJ, Kang KY. (2007). Construction and characterization of shuttle vectors for succinic acid-producing rumen bacteria. *Appl Environ Microbiol*, 73, 5411–5420.

Jantama K, Haupt MJ, Svoronos SA, Zhang X, Moore JC, Shanmugam KT, Ingram LO. (2008a). Combining metabolic engineering and metabolic evolution to develop nonrecombinant strains of *Escherichia coli* C that produce succinate and malate. *Biotechnol Bioeng*, 99, 1140–1153.

Jantama K, Zhang X, Moore JC, Shanmugam KT, Svoronos SA, Ingram LO. (2008b). Eliminating side products and increasing succinate yields in engineered strains of *Escherichia coli* C. *Biotechnol Bioeng*, 101, 881–893.

Jarboe LR, Zhang X, Wang X, Moore JC, Shanmugam KT, Ingram LO. (2010). Metabolic engineering for production of biorenewable fuels and chemicals: contributions of synthetic biology. *J Biomed Biotechnol*, 2010, 761042.

Jiang M, Chen K, Liu Z, Wei P, Ying H, Chang H. (2010). Succinic acid production by *Actinobacillus succinogenes* using spent brewer's yeast hydrolysate as a nitrogen source. *Appl Biochem Biotechnol*, 160, 244–254.

Kalinowski J, Bathe B, Bartels D, Bischoff N, Bott M, Burkovski A, Dusch N, Eggeling L, Eikmanns BJ, Gaigalat L, et al. (2003). The complete *Corynebacterium glutamicum* ATCC 13032 genome sequence and its impact on the production of L-aspartate-derived amino acids and vitamins. *J Biotechnol*, 104, 5–25.

Kawaguchi H, Vertes AA, Okino S, Inui M, Yukawa H. (2006). Engineering of a xylose metabolic pathway in *Corynebacterium glutamicum*. *Appl Environ Microbiol*, 72, 3418–3428.

Kawaguchi H, Sasaki M, Vertes AA, Inui M, Yukawa H. (2008). Engineering of an L-arabinose metabolic pathway in *Corynebacterium glutamicum*. *Appl Microbiol Biotechnol*, 77, 1053–1062.

Kim DY, Yim SC, Lee PC, Lee WG, Lee SY, Chang HN. (2004a). Batch and continuous fermentation of succinic acid from wood hydrolysate by *Mannheimia succiniciproducens* MBEL55E. *Enzyme Microb Technol*, 35, 648–653.

Kim P, Laivenieks M, McKinlay J, Vieille C, Gregory Zeikus J. (2004b). Construction of a shuttle vector for the overexpression of recombinant proteins in *Actinobacillus succinogenes*. *Plasmid*, 51, 108–115.

Kim P, Laivenieks M, Vieille C, Zeikus JG. (2004c). Effect of overexpression of *Actinobacillus succinogenes* phosphoenolpyruvate carboxykinase on succinate production in *Escherichia coli*. *Appl Environ Microbiol*, 70, 1238–1241.

Kim JM, Lee KH, Lee SY. (2008). Development of a markerless gene knock-out system for *Mannheimia succiniciproducens* using a temperature-sensitive plasmid. *FEMS Microbiol Lett*, 278, 78–85.

Kirchner O, Tauch A. (2003). Tools for genetic engineering in the amino acid-producing bacterium *Corynebacterium glutamicum*. *J Biotechnol*, 104, 287–299.

Kuhnert P, Scholten E, Haefner S, Mayor D, Frey J. (2010). *Basfia succiniciproducens* gen. nov., sp. nov., a new member of the family *Pasteurellaceae* isolated from bovine rumen. *Int J Syst Evol Microbiol*, 60, 44–50.

Kurzrock T, Weuster-Botz D. (2010). Recovery of succinic acid from fermentation broth. *Biotechnol Lett*, 32, 331–339.

Lee JW, Lee SY. (2010). Proteome-based physiological analysis of the metabolically engineered succinic acid producer *Mannheimia succiniciproducens* LPK7. *Bioprocess Biosyst Eng*, 33, 97–107.

Lee PC, Lee WG, Kwon S, Lee SY, Chang HN. (1999). Succinic acid production by *Anaerobiospirillum succiniciproducens*: effects of the H_2/CO_2 supply and glucose concentration. *Enzyme Microb Technol*, 24, 549–554.

Lee PC, Lee WG, Kwon S, Lee SY, Chang HN. (2000). Batch and continuous cultivation of *Anaerobiospirillum succiniciproducens* for the production of succinic acid from whey. *Appl Microbiol Biotechnol*, 54, 23–27.

Lee PC, Lee WG, Lee SY, Chang HN. (2001). Succinic acid production with reduced by-product formation in the fermentation of *Anaerobiospirillum succiniciproducens* using glycerol as a carbon source. *Biotechnol Bioeng*, 72, 41–48.

Lee PC, Lee SY, Hong SH, Chang HN. (2002). Isolation and characterization of a new succinic acid-producing bacterium, *Mannheimia succiniciproducens* MBEL55E, from bovine rumen. *Appl Microbiol Biotechnol*, 58, 663–668.

Lee PC, Lee SY, Hong SH, Chang HN. (2003a). Batch and continuous cultures of *Mannheimia succiniciproducens* MBEL55E for the production of succinic acid from whey and corn steep liquor. *Bioprocess Biosyst Eng*, 26, 63–67.

Lee PC, Lee SY, Hong SH, Chang HN, Park SC. (2003b). Biological conversion of wood hydrolysate to succinic acid by *Anaerobiospirillum succiniciproducens*. *Biotechnol Lett*, 25, 111–114.

Lee SJ, Song H, Lee SY. (2006). Genome-based metabolic engineering of *Mannheimia succiniciproducens* for succinic acid production. *Appl Environ Microbiol*, 72, 1939–1948.

Lee SY, Kim JM, Song H, Lee JW, Kim TY, Jang YS. (2008). From genome sequence to integrated bioprocess for succinic acid production by *Mannheimia succiniciproducens*. *Appl Microbiol Biotechnol*, 79, 11–22.

Lee JW, Choi S, Kim JM, Lee SY. (2010a). *Mannheimia succiniciproducens* phosphotransferase system for sucrose utilization. *Appl Environ Microbiol*, 76, 1699–1703.

Lee PC, Lee SY, Chang HN. (2010b). Kinetic study on succinic acid and acetic acid formation during continuous cultures of *Anaerobiospirillum succiniciproducens* grown on glycerol. *Bioprocess Biosyst Eng*, 33, 465–471.

Lin H, Bennett G, San K. (2005a). Genetic reconstruction of the aerobic central metabolism in *Escherichia coli* for the absolute aerobic production of succinate. *Biotechnol Bioeng*, 89, 148–156.

Lin H, Bennett GN, San KY. (2005b). Fed-batch culture of a metabolically engineered *Escherichia coli* strain designed for high-level succinate production and yield under aerobic conditions. *Biotechnol Bioeng*, 90, 775–779.

Lin H, Bennett GN, San KY. (2005c). Metabolic engineering of aerobic succinate production systems in *Escherichia coli* to improve process productivity and achieve the maximum theoretical succinate yield. *Metab Eng*, 7, 116–127.

Lin H, San KY, Bennett GN. (2005d). Effect of *Sorghum vulgare* phosphoenolpyruvate carboxylase and *Lactococcus lactis* pyruvate carboxylase coexpression on succinate production in mutant strains of *Escherichia coli*. *Appl Microbiol Biotechnol*, 67, 515–523.

Litsanov B, Brocker M, Bott M. (2012a). Toward homosuccinate fermentation: metabolic engineering of *Corynebacterium glutamicum* for anaerobic production of succinate from glucose and formate. *Appl Environ Microbiol*, 78, 3325–3337.

Litsanov B, Kabus A, Brocker M, Bott M. (2012b). Efficient aerobic succinate production from glucose in minimal medium with *Corynebacterium glutamicum*. *Microb Biotechnol*, 5, 116–128.

Litsanov B, Brocker M, Bott M. (2013). Glycerol as a substrate for aerobic succinate production in minimal medium with *Corynebacterium glutamicum*. *Microb Biotechnol*, 6, 189–195.

Liu YP, Zheng P, Sun ZH, Ni Y, Dong JJ, Zhu LL. (2008). Economical succinic acid production from cane molasses by *Actinobacillus succinogenes*. *Bioresource Technol*, 99, 1736–1742.

Lopez-Garzon CS, Ottens M, van der Wielen LAM, Straathof AJJ. (2012). Direct downstream catalysis: from succinate to its diethyl ester without intermediate acidification. *Chem Eng J*, 200, 637–644.

Maass D, Gerigk MR, Kreutzer A, Weuster-Botz D, Wubbolts M, Takors R. (2002). Integrated L-phenylalanine separation in an *E. coli* fed-batch process: from laboratory to pilot scale. *Bioprocess Biosyst Eng*, 25, 85–96.

McKinlay JB, Vieille C. (2008). ^{13}C-metabolic flux analysis of *Actinobacillus succinogenes* fermentative metabolism at different $NaHCO_3$ and H_2 concentrations. *Metab Eng*, 10, 55–68.

McKinlay JB, Zeikus JG, Vieille C. (2005). Insights into *Actinobacillus succinogenes* fermentative metabolism in a chemically defined growth medium. *Appl Environ Microbiol*, 71, 6651–6656.

McKinlay JB, Vieille C, Zeikus JG. (2007). Prospects for a bio-based succinate industry. *Appl Microbiol Biotechnol*, 76, 727–740.

McKinlay JB, Laivenieks M, Schindler BD, McKinlay AA, Siddaramappa S, Challacombe JF, Lowry SR, Clum A, Lapidus AL, Burkhart KB, Harkins V, Vieille C. (2010). A genomic perspective on the potential of *Actinobacillus succinogenes* for industrial succinate production. *BMC Genomics*, 11, 680.

Meiswinkel TM, Gopinath V, Lindner SN, Nampoothiri KM, Wendisch VF. (2013). Accelerated pentose utilization by *Corynebacterium glutamicum* for accelerated production of lysine, glutamate, ornithine and putrescine. *Microb Biotechnol*, 6, 131–140.

Meynial-Salles I, Dorotyn S, Soucaille P. (2008). A new process for the continuous production of succinic acid from glucose at high yield, titer, and productivity. *Biotechnol Bioeng*, 99, 129–135.

Mimitsuka T, Sawai H, Hatsu M, Yamada K. (2007). Metabolic engineering of *Corynebacterium glutamicum* for cadaverine fermentation. *Biosci Biotechnol Biochem*, 71, 2130–2135.

Murarka A, Dharmadi Y, Yazdani SS, Gonzalez R. (2008). Fermentative utilization of glycerol by *Escherichia coli* and its implications for the production of fuels and chemicals. *Appl Environ Microbiol*, 74, 1124–1235.

Mustafi N, Grünberger A, Kohlheyer D, Bott M, Frunzke J. (2012). The development and application of a single-cell biosensor for the detection of L-methionine and branched-chain amino acids. *Metab Eng*, 14, 449–457.

Nicolaou SA, Gaida SM, Papoutsakis ET. (2010). A comparative view of metabolite and substrate stress and tolerance in microbial bioprocessing: from biofuels and chemicals, to biocatalysis and bioremediation. *Metab Eng*, 12, 307–331.

Nishimura T, Vertes AA, Shinoda Y, Inui M, Yukawa H. (2007). Anaerobic growth of *Corynebacterium glutamicum* using nitrate as a terminal electron acceptor. *Appl Microbiol Biotechnol*, 75, 889–897.

Okino S, Noburyu R, Suda M, Jojima T, Inui M, Yukawa H. (2008). An efficient succinic acid production process in a metabolically engineered *Corynebacterium glutamicum* strain. *Appl Microbiol Biotechnol*, 81, 459–464.

Pai RA, Doherty MF, Malone MF. (2002). Design of reactive extraction systems for bioproduct recovery. *AIChE Journal*, 48, 514–526.

Pines O, Even-Ram S, Elnathan N, Battat E, Aharonov O, Gibson D, Goldberg I. (1996). The cytosolic pathway of L-malic acid synthesis in *Saccharomyces cerevisiae*: the role of fumarase. *Appl Microbiol Biotechnol*, 46, 393–399.

Raab AM, Lang C. (2011). Oxidative versus reductive succinic acid production in the yeast *Saccharomyces cerevisiae*. *Bioeng Bugs*, 2, 120–123.

Raab AM, Gebhardt G, Bolotina N, Weuster-Botz D, Lang C. (2010). Metabolic engineering of *Saccharomyces cerevisiae* for the biotechnological production of succinic acid. *Metab Eng*, 12, 518–525.

Rifkin GD, Opdyke JE. (1981). *Anaerobiospirillum succiniciproducens* septicemia. *J Clin Microbiol*, 13, 811–813.

Rittmann D, Lindner SN, Wendisch VF. (2008). Engineering of a glycerol utilization pathway for amino acid production by *Corynebacterium glutamicum*. *Appl Environ Microbiol*, 74, 6216–6222.

Rüffer N, Heidersdorf U, Kretzers I, Sprenger GA, Raeven L, Takors R. (2004). Fully integrated L-phenylalanine separation and concentration using reactive-extraction with liquid-liquid centrifuges in a fed-batch process with *E. coli*. *Bioprocess Biosyst Eng*, 26, 239–248.

Samuelov NS, Lamed R, Lowe S, Zeikus JG. (1991). Influence of CO_2-HCO_3^- levels and pH on growth, succinate production, and enzyme activities of *Anaerobiospirillum succiniciproducens*. *Appl Environ Microbiol*, 57, 3013–3019.

Samuelov NS, Datta R, Jain MK, Zeikus JG. (1999). Whey fermentation by *Anaerobiospirillum succiniciproducens* for production of a succinate-based animal feed additive. *Appl Environ Microbiol*, 65, 2260–2263.

Sanchez AM, Bennett GN, San KY. (2005a). Efficient succinic acid production from glucose through overexpression of pyruvate carboxylase in an *Escherichia coli* alcohol dehydrogenase and lactate dehydrogenase mutant. *Biotechnol Prog*, 21, 358–365.

Sanchez AM, Bennett GN, San KY. (2005b). Novel pathway engineering design of the anaerobic central metabolic pathway in *Escherichia coli* to increase succinate yield and productivity. *Metab Eng*, 7, 229–239.

Sasaki M, Jojima T, Inui M, Yukawa H. (2008). Simultaneous utilization of D-cellobiose, D-glucose, and D-xylose by recombinant *Corynebacterium glutamicum* under oxygen-deprived conditions. *Appl Microbiol Biotechnol*, 81, 691–699.

Sasaki M, Jojima T, Kawaguchi H, Inui M, Yukawa H. (2009). Engineering of pentose transport in *Corynebacterium glutamicum* to improve simultaneous utilization of mixed sugars. *Appl Microbiol Biotechnol*, 85, 105–115.

Sauer U, Eikmanns BJ. (2005). The PEP-pyruvate-oxaloacetate node as the switch point for carbon flux distribution in bacteria. *FEMS Microbiol Rev*, 29, 765–794.

Scholten E, Dägele D. (2008). Succinic acid production by a newly isolated bacterium. *Biotechnol Lett*, 30, 2143–2146.

Scholten E, Renz T, Thomas J. (2009). Continuous cultivation approach for fermentative succinic acid production from crude glycerol by *Basfia succiniciproducens* DD1. *Biotechnol Lett*, 31, 1947–1951.

Smith KM, Cho KM, Liao JC. (2010). Engineering *Corynebacterium glutamicum* for isobutanol production. *Appl Microbiol Biotechnol*, 87, 1045–1055.

Song H, Kim TY, Choi BK, Choi SJ, Nielsen LK, Chang HN, Lee SY. (2008). Development of chemically defined medium for *Mannheimia succiniciproducens* based on its genome sequence. *Appl Microbiol Biotechnol*, 79(2), 263–272.

Sprenger GA. (2007). From scratch to value: engineering *Escherichia coli* wild type cells to the production of L-phenylalanine and other fine chemicals derived from chorismate. *Appl Microbiol Biotechnol*, 75, 739–749.

Stäbler N, Oikawa T, Bott M, Eggeling L. (2011). *Corynebacterium glutamicum* as a host for synthesis and export of D-amino acids. *J Bacteriol*, 193, 1702–1709.

Stols L, Kulkarni G, Harris BG, Donnelly MI. (1997). Expression of *Ascaris suum* malic enzyme in a mutant *Escherichia coli* allows production of succinic acid from glucose. *Appl Biochem Biotechnol*, 63–65, 153–158.

Takeno S, Ohnishi J, Komatsu T, Masaki T, Sen K, Ikeda M. (2007). Anaerobic growth and potential for amino acid production by nitrate respiration in *Corynebacterium glutamicum*. *Appl Microbiol Biotechnol*, 75, 1173–1182.

Thakker C, Martinez I, San KY, Bennett GN. (2012). Succinate production in *Escherichia coli*. *Biotechnol J*, 7, 213–224.

Uhde A, Youn JW, Maeda T, Clermont L, Matano C, Kramer R, Wendisch VF, Seibold GM, Marin K. (2013). Glucosamine as carbon source for amino acid-producing *Corynebacterium glutamicum*. *Appl Microbiol Biotechnol*, 97, 1679–1687.

van der Werf MJ, Guettler MV, Jain MK, Zeikus JG. (1997). Environmental and physiological factors affecting the succinate product ratio during carbohydrate fermentation by *Actinobacillus sp.* 130Z. *Arch Microbiol*, 167, 332–342.

Vemuri GN, Eiteman MA, Altman E. (2002). Effects of growth mode and pyruvate carboxylase on succinic acid production by metabolically engineered strains of *Escherichia coli*. *Appl Environ Microbiol*, 68, 1715–1727.

Verhoff FH. (2012). Citric acid. In: *Ullmann's Encyclopedia of Industrial Chemistry*. Weinheim: Wiley-VCH Verlag GmbH & Co. KGaA, pp. 197–202.

Vickroy TB. (1985). Lactic acid. In: Blanch HW, Drew S, Wang DIC, editors. *Comprehensive Biotechnology*. New York: Pergamon Press, pp. 761–776.

Wada M, Takagi H. (2006). Metabolic pathways and biotechnological production of L-cysteine. *Appl Microbiol Biotechnol*, 73, 48–54.

Wan C, Li Y, Shahbazi A, Xiu S. (2008). Succinic acid production from cheese whey using *Actinobacillus succinogenes* 130Z. *Appl Biochem Biotechnol*, 145, 111–119.

Wang J, Zhu J, Bennett GN, San KY. (2011). Succinate production from different carbon sources under anaerobic conditions by metabolic engineered *Escherichia coli* strains. *Metab Eng*, 13, 328–335.

Wendisch VF, Bott M, Eikmanns BJ. (2006). Metabolic engineering of *Escherichia coli* and *Corynebacterium glutamicum* for biotechnological production of organic acids and amino acids. *Curr Opin Microbiol*, 9, 268–274.

Werpy T, Peterson G. (2004). *Top Value Added Chemicals from Biomass*. Oak Ridge: US Department of Energy.

Wieschalka S, Blombach B, Bott M, Eikmanns BJ. (2013). Bio-based production of organic acids with *Corynebacterium glutamicum*. *Microb Biotechnol*, 6, 87–102.

Witthoff S, Eggeling L, Bott M, Polen T. (2012). *Corynebacterium glutamicum* harbours a molybdenum cofactor-dependent formate dehydrogenase which alleviates growth inhibition in the presence of formate. *Microbiology*, 158, 2428–2439.

Yasuda K, Jojima T, Suda M, Okino S, Inui M, Yukawa H. (2007). Analyses of the acetate-producing pathways in *Corynebacterium glutamicum* under oxygen-deprived conditions. *Appl Microbiol Biotechnol*, 77, 853–860.

Yazdani SS, Gonzalez R. (2007). Anaerobic fermentation of glycerol: a path to economic viability for the biofuels industry. *Curr Opin Biotechnol*, 18, 213–219.

Ye GZ, Jiang M, Li J, Chen KQ, Xi YL, Liu SW, Wei P, Ouyang PK. (2010). Isolation of NH_4^+-tolerant mutants of *Actinobacillus succinogenes* for succinic acid production by continuous selection. *J Microbiol Biotechnol*, 20, 1219–1925.

Yedur S, Berglund KA, Dunuwila DD. (2001). Succinic acid production and purification. US Patent 6,265,190.

Yoo JY, Zeikus JG. (1996). Modulation of phosphoenolpyruvate metabolism of *Anaerobiospirillum succiniciproducens* ATCC 29305. *J Microbiol Biotechnol*, 6, 43–49.

Youn JW, Jolkver E, Krämer R, Marin K, Wendisch VF. (2008). Identification and characterization of the dicarboxylate uptake system DccT in *Corynebacterium glutamicum*. *J Bacteriol*, 190, 6458–6466.

Youn JW, Jolkver E, Krämer R, Marin K, Wendisch VF. (2009). Characterization of the dicarboxylate transporter DctA in *Corynebacterium glutamicum*. *J Bacteriol*, 191, 5480–5488.

Yu J, Li Z, Ye Q, Yang Y, Chen S. (2010). Development of succinic acid production from corncob hydrolysate by *Actinobacillus succinogenes*. *J Ind Microbiol Biotechnol*, 37, 1033–1040.

Yukawa H, Inui M, editors. (2013). *Corynebacterium glutamicum: Biology and Biotechnology*. Heidelberg: Springer.

Yuzbashev TV, Yuzbasheva EY, Sobolevskaya TI, Laptev IA, Vybornaya TV, Larina AS, Matsui K, Fukui K, Sineoky SP. (2010). Production of succinic acid at low pH by a recombinant strain of the aerobic yeast *Yarrowia lipolytica*. *Biotechnol Bioeng*, 107, 673–682.

Yuzbashev TV, Yuzbasheva EY, Laptev IA, Sobolevskaya TI, Vybornaya TV, Larina AS, Gvilava IT, Antonova SV, Sineoky SP. (2011). Is it possible to produce succinic acid at a low pH? *Bioeng Bugs*, 2, 115–119.

Zeikus JG, Jain MK, Elankovan P. (1999). Biotechnology of succinic acid production and markets for derived industrial products. *Appl Microbiol Biotechnol*, 51, 545–552.

Zhang X, Shanmugam KT, Ingram LO. (2010). Fermentation of glycerol to succinate by metabolically engineered strains of *Escherichia coli*. *Appl Environ Microbiol*, 76, 2397–2401.

Zheng P, Dong JJ, Sun ZH, Ni Y, Fang L. (2009). Fermentative production of succinic acid from straw hydrolysate by *Actinobacillus succinogenes*. *Bioresource Technol*, 100, 2425–2429.

CHAPTER 17

Glutamic Acid

TAKASHI HIRASAWA

Department of Bioengineering, Tokyo Institute of Technology

HIROSHI SHIMIZU

Department of Bioinformatic Engineering, Graduate School of Information Science and Technology, Osaka University

Bioprocessing of Renewable Resources to Commodity Bioproducts, First Edition.
Edited by Virendra S. Bisaria and Akihiko Kondo.

Abstract

The acidic amino acid glutamic acid is considered as one of the most important building blocks for chemical synthesis. Glutamic acid is currently produced from inexpensive sugars via the glutamic-acid-producing microorganism *Corynebacterium glutamicum*. The molecular mechanism and metabolic engineering of glutamic acid production via *C. glutamicum* have been reported. Moreover, other chemicals from glutamic acid such as γ-aminobutyric acid, poly(γ-glutamic acid), and 5-aminolevulinic acid have also been produced by microorganisms. This chapter reviews the molecular mechanism and metabolic engineering of glutamic acid production by *C. glutamicum* and the production of chemicals from glutamic acid by various microorganisms. The potential use of glutamic acid as a building block for producing chemicals is also briefly discussed.

17.1 INTRODUCTION

Glutamic acid (2-aminopentanedioic acid) is an amino acid with two carboxylic acid groups (Figure 17.1). Because it has a dissociation constant of 4.1, glutamic acid is negatively charged at neutral pH. Glutamic acid is also used as a building block for producing chemicals. The US Department of Energy (DOE) screened potential candidates of top value-added chemicals produced from biomass resources and identified glutamic acid as being one of the top 12 screened chemicals (Werpy and Petersen, 2004). Glutamic acid is also one of the constituents of waste streams from biofuel produced by sugar cane and wheat. Thus, the utilization of glutamic acid as a by-product in the biofuel production process should be considered for producing useful chemicals. In addition, the sodium salt of glutamic acid (monosodium glutamic acid (MSG)) is a known flavor enhancer isolated from seaweed (*konbu* in Japanese).

D- and L-glutamic acid can be chemically synthesized from acrylonitrile but the separation of these isomers is required for further utilization. Another method to obtain glutamic acid is via the hydrolysis of proteins, such as wheat gluten. In 1956, the microorganism *Corynebacterium glutamicum* was discovered to produce glutamic acid (Kinoshita et al., 1957; Udaka, 1960). Currently, MSG has been industrially produced from sugar cane molasses and ammonia using *C. glutamicum* and other microorganisms. The production of MSG is one of the most major industrial glutamic acid production processes. *C. glutamicum* produces and secretes only one isomer, that is, L-glutamic acid. Since its discovery, the analysis of the molecular mechanisms and

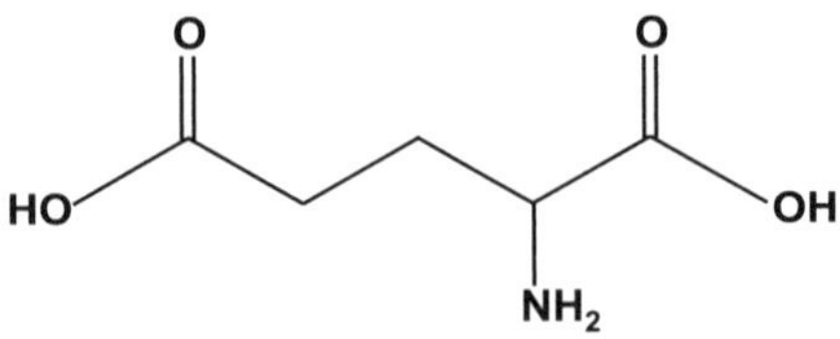

FIGURE 17.1 Structure of L-glutamic acid.

metabolic engineering approaches for glutamic acid production by *C. glutamicum* has been studied.

This chapter reviews the molecular mechanism and metabolic engineering of glutamic acid production by *C. glutamicum* and potential use of glutamic acid as a building block for producing several other chemicals.

17.2 GLUTAMIC ACID PRODUCTION BY *CORYNEBACTERIUM GLUTAMICUM*

In 1908, Dr. Kikunae Ikeda identified MSG in seaweed as a compound with the taste of Umami. Subsequently, MSG was produced on an industrial scale by extracting from wheat hydrolysates. In 1956, Drs. Shukuo Kinoshita and Shigezo Udaka in Kyowa Hakko Kogyo Co., Ltd. (currently Kyowa Hakko Bio Co., Ltd.) isolated the microorganism *Micrococcus glutamicus* (later renamed as *Corynebacterium glutamicum*), which produces a high amount of glutamic acid (Kinoshita et al., 1957; Udaka, 1960). The costs of glutamic acid production from inexpensive sugar cane molasses and ammonia have subsequently decreased. Recently, glutamic acid production by *C. glutamicum* from renewable resources has been performed.

17.2.1 Glutamic Acid Production by *Corynebacterium Glutamicum* and Its Molecular Mechanism

A fermentation process has been developed for the direct production of glutamic acid from sugar and ammonia in *C. glutamicum*. At present, *C. glutamicum* is also a widely used host microorganism for other amino acids, such as lysine, arginine, threonine, and valine. *C. glutamicum* is a high G+C Gram-positive, rod-shaped, and facultative anaerobic bacterium. In addition, the genome DNA of this microorganism has been sequenced (Ikeda and Nakagawa, 2003; Kalinowski et al., 2003; Yukawa et al., 2007). *C. glutamicum* is regarded as a useful host species for producing chemicals, such as organic acids (Okino et al., 2005; Wendisch et al., 2006), alcohols (Blombach et al., 2011), secreted proteins (Kikuchi et al., 2003), and amino acids other than glutamic acid (Nakayama et al., 1961; Sano and Shiio, 1970; Shiio and Nakamori, 1970; Tsuchida et al., 1975; Ohnishi et al., 2002; Blombach et al., 2008; Becker et al., 2011; Hasegawa et al., 2012, 2013). This section reviews the molecular mechanism of glutamic acid production by *C. glutamicum*.

17.2.1.1 Triggers for Glutamic Acid Overproduction in Corynebacterium Glutamicum Glutamic acid is derived from 2-oxogluratrate, a metabolite in the TCA cycle, and produced in a one-step reaction catalyzed by glutamate dehydrogenase (GDH) in the presence of high concentrations of ammonia (Figure 17.2). *C. glutamicum* is a biotin (vitamin B7)-auxotrophic microorganism. A wild-type strains of *C. glutamicum* cannot secrete glutamic acid when excess biotin is in the culture medium. However, at limiting biotin concentrations, *C. glutamicum* produces

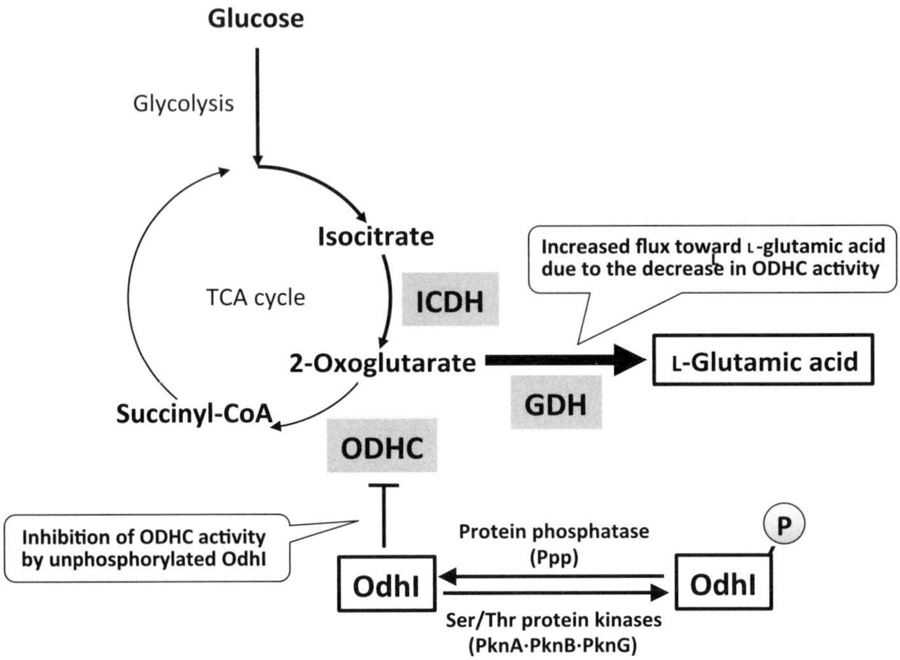

FIGURE 17.2 Metabolic reactions around 2-oxoglutarate node in the TCA cycle. Regulation of 2-oxoglutarate dehydrogenase complex (ODHC) activity via phosphorylation status of OdhI protein is also shown. ICDH, isocitrate dehydrogenase; GDH, glutamate dehydrogenase.

significant amounts of glutamic acid (Shiio et al., 1962). Because significant amounts of biotin are found in the raw material required for glutamic acid production (e.g., molasses), other strategies were required to induce glutamic acid overproduction in the presence of excess biotin. Treatments with fatty acid ester surfactants (e.g., Tween 40) and β-lactam antibiotics (e.g., penicillin) have been reported to induce glutamic acid production by *C. glutamicum* (Nara et al., 1964; Takinami et al., 1965). These treatments are thought to affect the cell surface structure of *C. glutamicum*. Biotin is a cofactor of the enzyme required for the biosynthesis of fatty acids, which are components of cytoplasmic membrane. Tween 40 and penicillin affect the synthesis of fatty acids and peptidoglycans, which are components of the cytoplasmic membrane and cell wall, respectively. Therefore, the glutamic acid produced by *C. glutamicum* has been proposed to passively leak through the cell surface. However, based on differences between intracellular and extracellular glutamic acid levels, this leak model of glutamic acid cannot explain the significant amounts (more than 60–80 g/L) of glutamic acid produced.

17.2.1.2 Metabolic Change and Its Molecular Mechanism During Glutamic Acid Production by Corynebacterium Glutamicum In 1971, a change in the enzyme activity of the 2-oxoglutarate dehydrogenase complex (ODHC) was observed during fermentation of glutamic acid by *C. glutamicum* (Shigu and Terui, 1971). The ODHC is located at branch points between the TCA cycle and glutamic acid biosynthesis catalyzed by GDH. Kawahara et al., later observed that ODHC activity decreased during glutamic acid overproduction induced by biotin limitation as well as the addition of Tween 40 and penicillin (Kawahara et al., 1997). The decrease in ODHC activity induced the accumulation of 2-oxoglutarate, the substrate for GDH. As a result, the metabolic flux toward glutamic acid production increased. In 2006, Niebisch et al. (2006) discovered the protein OdhI, which is involved in decreasing ODHC activity (Figure 17.2). The OdhI protein directly interacted with a catalytic subunit of ODHC, OdhA (E1o), inhibited ODHC activity, and regulated interactions between OdhI and OdhA via phosphorylation. Furthermore, the non-phosphorylated OdhI interacted with OdhA, whereas phosphorylated OdhI could not (Krawczyk et al., 2010). Kinases (PknA, PknB, and PknG) and a phosphatase (Ppp) for phosphorylating and dephosphorylating OdhI were also identified (Niebisch et al., 2006; Schultz et al., 2007, 2009). Intracellular levels of non-phosphorylated OdhI were also enhanced by induction treatments (e.g., biotin limitation as well as the addition of Tween 40 and penicillin) for glutamic acid overproduction by *C. glutamicum* (Boulahya et al., 2010; Kim et al., 2011).

Recently, proteins consisting of AceE, AceF, Lpd, and OdhA form a supercomplex in *C. glutamicum*. The AceE, AceF, and Lpd proteins are the E1p, E2p, and E3p subunits of pyruvate dehydrogenase complex (PDHC), respectively (Hoffelder et al., 2010). In most organisms, ODHC consists of E1o, E2o, and E3o, and E3o is Lpd. However, in the case of *C. glutamicum* and other related species *Mycobacterium tuberculosis*, the gene encoding the E2o protein in the ODHC is not found in the genome. For ODHC activity, the E2 protein in ODHC may potentially be shared with the AceF protein in the supercomplex. During glutamic acid production, PDHC activity decreased as well as ODHC activity (Hasegawa et al., 2008). Additional studies on the relationship between the PDHC and ODHC supercomplexes, OdhI, and ODHC activity in glutamic acid production need to be performed.

Glutamic acid overproduction by *C. glutamicum* is triggered by biotin limitation as well as the addition of Tween 40 and penicillin. These treatments affect the integrity of the *C. glutamicum* cell surface. Therefore, some reports have proposed a model of the passive leakage of glutamic acid through the membrane and investigated the relationship between changes in the cell surface and glutamic acid production (Hoischen and Krämer, 1990; Hirasawa et al., 2000, 2001; Hashimoto et al., 2006). Recently, a mechanosensitive channel protein NCgl1221 was reported to contribute to glutamic acid production in *C. glutamicum*. The NCgl1221 gene was identified in the analysis of the *C. glutamicum odhA* disruptant (Nakamura et al., 2007). The NCgl1221 disruptant could not secrete glutamic acid outside of cells. Mechanosensitive channel proteins can sense changes in membrane tension. The NCgl1221 protein senses changes in membrane tension by its conformational change. Conformational changes in the NCgl1221 protein appear to induce glutamic acid overproduction by *C. glutamicum*, which is induced by biotin limitation as well as the addition of Tween 40 and penicillin, and the NCgl1221 protein functions as an exporter of glutamic acid. Hashimoto et al., reported that the NCgl1221 protein functions as a mechanosensitive channel through which glutamic acid can leak across the cell membrane by passive diffusion (Hashimoto et al., 2010, 2012).

17.2.2 Metabolic Engineering of Glutamic Acid Production by *Corynebacterium Glutamicum*

To improve bioproduction using cells, both genetic modification of metabolic pathways and process operation strategies are important. Metabolic engineering can be defined as a methodology for improvement strategies of production of target product(s) and cellular characteristics by genetic engineering. Here, the metabolic engineering approach of glutamic acid production by *C. glutamicum* is reviewed.

17.2.2.1 Metabolic Flux Redistribution by the Induction of Glutamic Acid Production in Corynebacterium Glutamicum Decrease in ODHC activity occurs during glutamic acid production by *C. glutamicum* and can be induced by limiting the amount of biotin as well as the addition of Tween 40 and penicillin. Thus, the metabolic fluxes toward 2-oxoglutarate are redistributed to succinyl-CoA or glutamic acid biosynthesis by triggering treatments for glutamic acid production.

We analyzed the impact of enhanced enzyme activity around the 2-oxoglutarate node (i.e., isocitrate dehydrogenase (ICDH) and GDH; Figure 17.2) and decreased ODHC activity due to the biotin limitation on the glutamic acid production by *C. glutamicum* by metabolic flux redistribution analysis (Shimizu et al., 2003). There was a small impact of changes in ICDH and GDH activities on metabolic flux redistribution toward glutamic acid production, but the change in ODHC activity had a significant impact on glutamic acid production and metabolic flux redistribution, indicating that the ODHC is an important enzyme in the overproduction of glutamic acid in *C. glutamicum*. Moreover, we also analyzed the metabolic flux redistribution during Tween 40-triggered glutamic acid production by *C. glutamicum* and demonstrated that the metabolic flux is redistributed toward glutamic acid production by the addition of Tween 40 (Shirai et al., 2005).

17.2.2.2 The Importance of Anaplerotic Reactions in Glutamic Acid Production by Corynebacterium Glutamicum In glutamic acid production, a supply of oxaloacetate is required for the TCA cycle to maintain 2-oxoglutarate biosynthesis. Anaplerotic reactions produce oxaloacetate from intermediate metabolites in glycolysis because the intermediate metabolites in the TCA cycle are used for anabolic reactions, including glutamic acid biosynthesis. *C. glutamicum* carries two anaplerotic reactions, each of which is catalyzed by phosphoenolpyruvate carboxylase (PPC) and pyruvate carboxylase (PC). PPC encoded by the *ppc* gene can produce oxaloacetate from phosphoenolpyruvate and CO_2, and PC encoded by *pyc* can also produce oxaloacetate from pyruvate and CO_2 (O'Regan et al., 1989; Peters-Wendisch et al., 1998).

Our group analyzed the roles of PPC and PC in inducing glutamic acid production by biotin limitation. PC requires biotin as a cofactor for enzyme activity (Sato et al.,

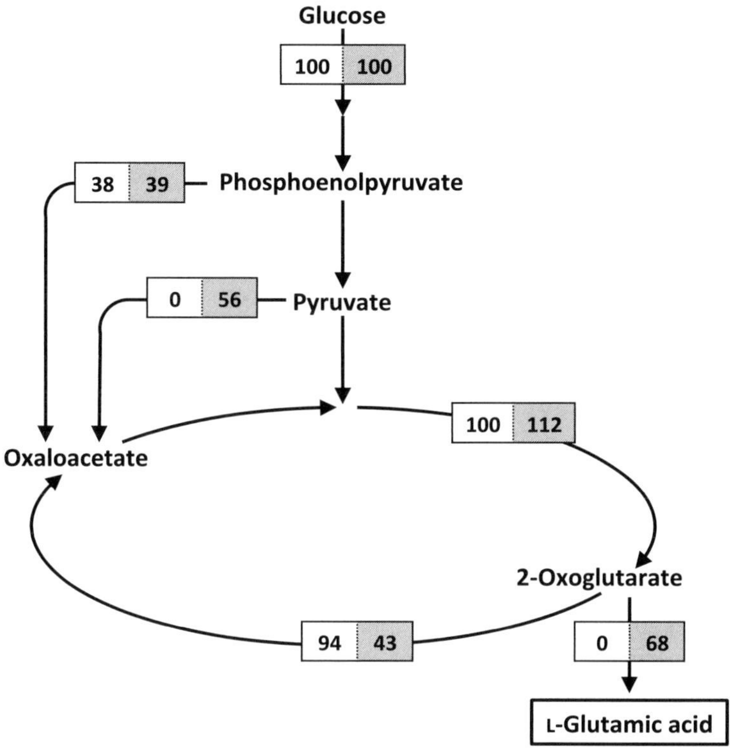

FIGURE 17.3 The involvement of anaplerotic reactions in Tween 40-triggered glutamic acid production by *Corynebacterium glutamicum*. ^{13}C-metabolic flux analysis of *C. glutamicum* during Tween 40-induced glutamic acid overproduction was performed and metabolic flux distributions on anaplerotic reactions and the 2-oxoglutarate branch are shown. The left and right values in each box represent the fluxes in the growth and glutamic acid production phases. The flux for glucose uptake is normalized to 100.

2008). Therefore, under limiting biotin concentrations, PPC was proposed to supply oxaloacetate in glutamic acid production. Indeed, the *ppc* knockout in the *C. glutamicum* strain was not able to produce glutamic acid at limiting biotin concentrations, whereas the *pyc* knockout strain produced glutamic acid. ^{13}C tracer experiments showed that the flux of anaplerotic reactions in the *pyc* knockout mutant was lower than that in the wild-type *C. glutamicum* at limiting biotin concentrations, suggesting that PPC is important for glutamic acid production under biotin limitation conditions. We also analyzed the contribution of anaplerotic reactions to the overproduction of glutamic acid triggered by Tween 40 using precise metabolic flux analysis based on ^{13}C metabolic flux analysis, in which the metabolic flux distribution is estimated from the ^{13}C enrichment in proteinogenic amino acids (Shirai et al., 2007) (Figure 17.3). PPC was observed to be active during cell growth, whereas PC became active by the addition of Tween 40, inducing glutamic acid production.

17.2.2.3 Metabolic Engineering Approaches for the Improvement of Glutamic Acid Production by Corynebacterium Glutamicum The enhancement of glutamic acid production by *C. glutamicum* can be achieved by metabolic engineering approaches. The following are engineering targets for enhancing glutamic acid production: ODHC, regulation of ODHC activity, and anaplerotic reactions. For example, *odhA* disruption without triggers or decreasing OdhA levels by expressing *odhA* antisense RNA with triggering treatments can increase glutamic acid production by *C. glutamicum* (Asakura et al., 2007; Kim et al., 2009). Our research group showed that the overexpression of the *odhI* gene encoding the protein that inhibits ODHC activity increases glutamic acid production by *C. glutamicum* without triggering treatments (Kim et al., 2010). As for anaplerotic reactions, glutamic acid production triggered by Tween 60, which has the same effect as Tween 40, was enhanced by the overexpression of *pyc* and decreased by the *pyc* disruptant (Peters-Wendisch et al., 2001).

17.2.2.4 Glutamic Acid Production by Corynebacterium Glutamicum from Renewable Resources Recent studies on glutamic acid production by *C. glutamicum* using renewable resources demonstrated that metabolic engineering could be applied to *C. glutamicum* to utilize biomass resources and waste by-products produced during industrial chemical production.

C. glutamicum cannot consume pentose sugars, such as xylose and arabinose, which are obtained by hydrolyzing lignocellulosic biomass. Therefore, the construction of recombinant strains of *C. glutamicum* to convert pentose sugars to other chemicals was attempted. A *C. glutamicum* recombinant strain expressing *Escherichia coli xylA* and/or *xylB* genes encoding xylose isomerase and xylulokinase, respectively, was able to grow on xylose (Kawaguchi et al., 2006). Moreover, a *C. glutamicum* strain expressing the *E. coli araBAD* genes was able to grow on arabinose as the sole carbon source (Kawaguchi et al., 2008). Schneider et al. (2011) successfully produced glutamic acid from arabinose in a strain harboring *E. coli araBAD* genes by adding ethambutol, which is an inhibitor of cell wall biosynthesis in *Mycobacteria* and can induce glutamic acid production by *C. glutamicum* (Radmacher et al., 2005). Gopinath et al. (2011) constructed a strain expressing *E. coli xylA* and *araBAD* genes, which could grow on both xylose and arabinose. Moreover, this strain produced glutamic acid upon the addition of ethambutol from a mixture of xylose, arabinose, rice straw, and wheat bran hydrolysates.

Glycerol is one of the major by-products in biodiesel production, and utilizing glycerol is an important objective in metabolic engineering for chemicals production. Because *C. glutamicum* does not grow well on glycerol as the sole carbon source, a *C. glutamicum* recombinant strain that could consume glycerol as a carbon source needed to be constructed. Notably, *E. coli* possesses the glycerol assimilation pathway, and the introduction of the *E. coli* genes (e.g., *glpP*, *glpK*, and *glpD*) involved in this pathway into *C. glutamicum* led to the production of glutamic acid from glycerol (Rittmann et al., 2008).

The other way to utilize biomass resources for chemical production via fermentation is to express and secrete the carbohydrate-degrading enzymes. Recently, recombinant *C. glutamicum* strains secreting endoglucanases derived from various microorganisms were constructed and produced glutamic acid from β-glucan upon the addition of Tween 40 (Tsuchidate et al., 2011). In addition, a recombinant *C. glutamicum* strain expressing α-amylase from *Streptococcus bovis* on the cell surface was constructed and successfully produced glutamic acid from starch (Yao et al., 2009).

17.3 GLUTAMIC ACID AS A BUILDING BLOCK

Glutamic acid can be used to produce a variety of useful chemicals (Figure 17.4). The utilization of glutamic acid produced by microbiological fermentation and as a by-product in biofuel production will lead to a reduction in the use of fossil resources. Several studies have been conducted on the microbiological production of chemicals from glutamic acid. This section reviews the microbiological production of chemicals, such as γ-aminobutyric acid (GABA), poly(γ-glutamic acid) (PGA), and 5-aminolevulinic acid (ALA). In addition, the possibility of producing other chemicals from glutamic acid is also discussed.

17.3.1 Production of Chemicals from Glutamic Acid Using Microorganisms

Among the chemicals that can be produced from glutamic acid, the microbiological production of GABA, PGA, and ALA have been investigated. GABA is originally produced by microorganisms, especially lactic acid bacteria (LAB) in fermented foods. GABA is also known as a neurotransmitter in humans and can be used in pharmaceuticals and foods. Recently, GABA was determined to be a building block for the polymer nylon 4 (polyamide 4). PGA is produced by *Bacillus* sp. and increases the tolerance of the microorganism toward environmental changes. PGA can be used both industrially and medically as drug carriers, biological adhesives, water-absorbing polymers and metal-absorbing materials. ALA is a precursor to porphyrin biosynthesis from tRNA-charged glutamic acid (known as C5 pathway) that is used as a pharmaceutical and biodegradable herbicide.

17.3.1.1 Production of γ-Aminobutyric Acid from Glutamic Acid by Microorganisms GABA can be synthesized chemically but alternative production strategies are necessary to reduce the utilization of fossil resources. GABA is synthesized from glutamic acid by glutamate decarboxylase (GAD) and produced by some LAB in fermented foods. Therefore, purified GAD should be able to convert glutamic acid to GABA. The direct conversion of glutamic acid and sugars to GABA by natural and recombinant microorganisms has been performed. The microbial production of GABA from glutamic acid and sugars, and polymer production from GABA is reviewed in this section.

Plokhov et al. (2000) reported the production of GABA using recombinant *E. coli* expressing the *gadA* gene, which encodes GAD. In this study, recombinant *E. coli*

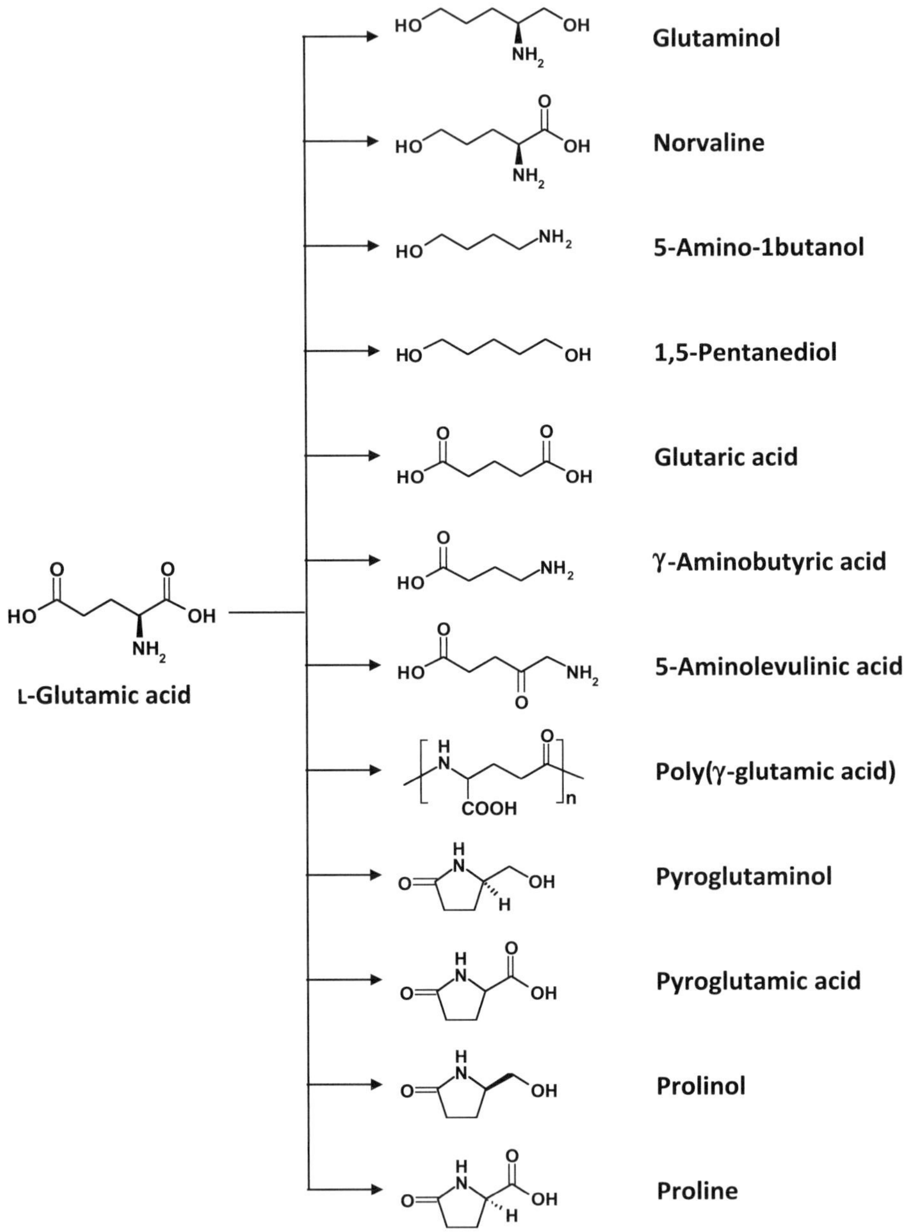

FIGURE 17.4 Derivatives of glutamic acid that can be used as possible targets for bio-based productions. From US DOE Report (Werpy and Petersen, 2004).

was used as whole-cell biocatalyst. As a result, heat-treated recombinant *E. coli* converted glutamic acid to GABA with a productivity of 23–25 g/g of wet cells. Le Vo et al. (2012) demonstrated that the overexpression of *E. coli gadB* and *gadC* genes encoding GAD and the glutamic acid/GABA antiporter, respectively, in the *E. coli* mutant strain of *gabT* encoding GABA aminotransferase yielded GABA from approximately 90% MSG.

GABA is produced by some LAB such as *Lactobacillus paracasei*, *Lactobacillus brevis*, *Lactobacillus plantarum,* and *Streptococcus thermophilus*. For example, enhanced GABA production in *Bifidobacterium longum* was achieved by overexpressing GAD from rice (Park et al., 2005). GABA production by growing and the resting cells of *L. brevis* were also reported (Zhang et al., 2012b). The overexpression of the *gadB* gene from *S. thermophilus* in a non-GABA-producing *S. thermophilus* produced GABA (Somkuti et al., 2012). In addition, the overexpression of *E. coli* GAD in *L. brevis* resulted in enhanced GABA production (Kim et al., 2007). Aside from LAB, Kato et al. (2002) demonstrated that the overexpression of the *gadA* gene from *Aspergillus oryzae* converted glutamic acid into GABA.

Recently, the utilization of a glutamic-acid-producing microorganism *C. glutamicum* was examined for GABA production. Sugars without the addition of glutamic acid can produce GABA using a glutamic acid-producing microorganism. Shi and Li (2011) reported that the expression of GAD genes from *L. brevis* in *C. glutamicum* produced GABA from glucose without any triggering treatments. In addition, Takahashi et al. (2012) reported that the *C. glutamicum* recombinant strain expressing *E. coli gadB* gene could produce GABA from glucose without requiring the addition of Tween 40.

GABA is a building block in the synthesis of nylon 4 (Figure 17.5). Studies on nylon 4 production from biologically produced GABA have been conducted. To synthesize nylon 4, GABA has to be converted into the nylon 4 precursor, 2-pyrrolidone. Yamano et al. (2012) reported the production of 2-pyrrolidone from biologically produced glutamic acid using *E. coli*. The authors prepared a mixture of *E. coli*, glutamic acid, and water, resulting in the release of 303.7 g GABA into water from 560 g of glutamic acid 14 times every 24 hour. The resulting GABA (without purification) was converted to 2-pyrrolidone by removing the water, heating the mixture to 200–240°C, and distilling the product under reduced pressure. Moreover, Park et al. (2013) reported the synthesis of nylon 4 from biologically produced GABA using an *E. coli* recombinant strain. GABA was produced from glutamic acid using *E. coli* expressing the *Lactococcus lactis gadB* gene encoding GAD as a

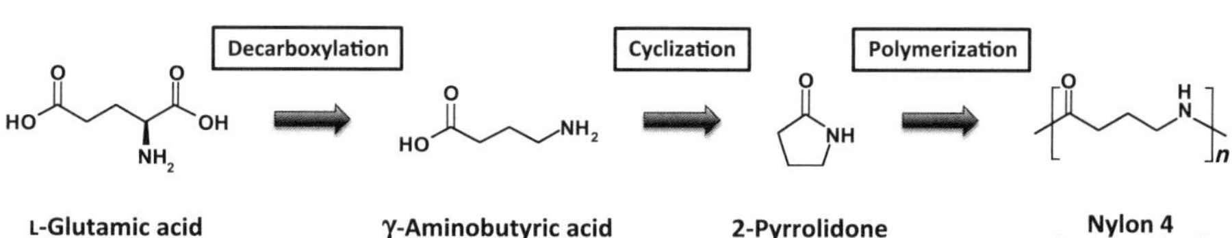

FIGURE 17.5 Synthesis of nylon 4 from glutamic acid via GABA and 2-pyrrolidone.

whole-cell biocatalyst. The resulting GABA was converted to 2-pyrrolidone in a reaction using Al_2O_3 as a catalyst in organic solvent, and nylon 4 could be synthesized by bulk polymerization from obtained 2-pyrrolidone. The molecular weight of nylon 4 was 200,000–300,000, which varied based on the polymerization conditions used. The chemical properties of nylon 4 synthesized from 2-pyrrolidone converted from biologically produced GABA were similar to those nylon 4 chemically synthesized from 2-pyrrolidone. The biologically-produced 2-pyrrolidone has the potential to be used as an alternative raw material to fossil-resources-based materials in order to produce nylon 4.

In addition to nylon 4, the microbial production of polyhydroxyalkanoates (PHA) from GABA has also been studied. Valentin et al., reported the introduction of metabolic reactions to supply 4-hydroxybutyryl-CoA to the recombinant *E. coli* strain expressing *Ralstonia eutropha phaA*, *phaB*, and *phaC* genes, which encode the enzymes that convert 3-hydroxybutyryl-CoA to poly(3-hydroxybutyrate-co-4-hydroxybutyrate) during the conversion of GABA to PHA. To introduce metabolic reactions that produce 4-hydroxybutyryl-CoA, the following four genes were introduced: acetyl-CoA:4-hydroxybutyrate CoA transferase from *Clostridium kluyveri* (*orfZ*), 4-hydroxybutyrate dehydrogenase from *R. eutropha* (*gbd*), glutamate:succinic semialdehyde transaminase from *E. coli* (*gabT*), and GAD from *Arabidopsis thaliana* or *E. coli* (*gad*) (Valentin et al., 2000). Using glutamic acid as a carbon source, the resulting recombinant strain produced PHA, which included 0.7 mol% 4-hydroxybutyrate units. Moreover, PHA could be produced from glutamic acid and 2-oxoglutarate, and the 4-hydroxybutyrate units increased to 1.2 mol%.

17.3.1.2 Production of Poly(γ-Glutamic Acid) by Microorganisms PGA is generally obtained by polymerizing D- and/or L-glutamic acids with the linkage between α-amino and γ-carboxyl groups of two glutamic acid molecules. The synthesis of PGA is ribosome-independent and produced mainly by *Bacillus* sp., including *B. licheniformis*, *B. subtilis*, and *B. anthracis*. The molecular weight of PGA produced by *Bacillus* sp. is in the range of 10^5–10^6. Initially during PGA biosynthesis (Figure 17.6), D- and L-glutamic acids are synthesized. D-Glutamic acid is produced by D-glutamic acid–pyruvic acid aminotransferase and glutamate racemase, and the resulting glutamic acid is polymerized by PGA synthetase encoded by *pgsBCA* genes.

To improve PGA production, culture conditions, such as nutrition and aeration, were examined. PGA production by *Bacillus* sp. can be categorized into the following two groups: glutamic acid-dependent and glutamic acid-independent PGA production. For example, PGA production by *B. licheniformis* ATCC 9945A is glutamic acid-dependent (Thorne et al., 1954). ^{13}C-nuclear magnetic resonance spectroscopy analysis of other glutamic acid-dependent PGA-producing species revealed that carbon sources (e.g., glucose) are mainly used for cell growth and energy supply, such as ATP, whereas glutamic acid is used in PGA biosynthesis (Yao et al., 2010). On the other hand, PGA production by some strains of *B. subtilis* and *B. licheniformis* do not require the addition of glutamic acid but is achieved from nitrogen and carbon sources (Ito et al., 1996; Cao et al., 2011). Moreover, the requirement of metal ions

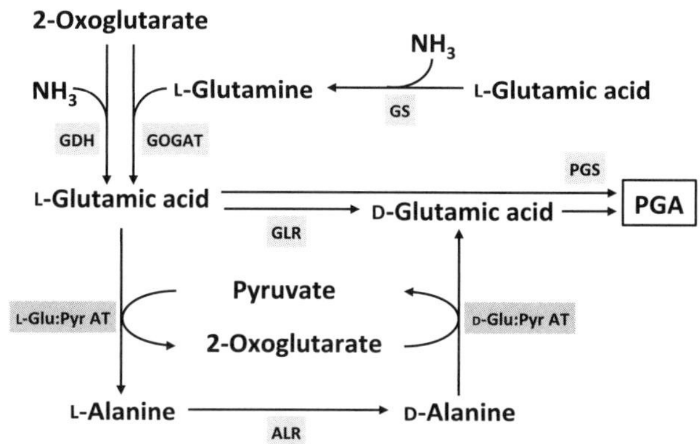

FIGURE 17.6 PGA biosynthetic pathway in *Bacillus* sp. GS, glutamine synthase; GOGAT, glutamate:2-oxoglutarate aminotransferase; GDH, glutamate dehydrogenase; L-Glu:Pyr AT, L-glutamate:pyruvate aminotransferase; D-Glu:Pyr AT, D-glutamate:pyruvate aminotransferase; GLR, glutamate racemase; ALR, alanine racemase; PGS, PGA synthetase.

and aeration for efficient PGA production has been investigated (for review, see Shih and Van (2001) and Bajaj and Singhal (2011)).

Metabolic engineering has been used to enhance PGA production by microorganisms. *B. subtilis* DB430 strain does not produce PGA even though it possesses genes related to PGA biosynthesis, such as *ywsC-ywtAB*. An efficient synthetic expression control sequence was introduced within the upstream region of PGA biosynthetic genes in the genome of this strain and PGA production was successfully achieved (Yeh et al., 2010). Yamashiro et al. (2011) showed that the contribution of PgsE protein with PGA production in *B. subtilis*. Furthermore, the induction of PgsE stimulated PGA production in the presence of zinc ions. PGA production by recombinant microorganisms other than *Bacillus* sp. has also been examined. For example, the genes responsible for PGA biosynthesis in *B. subtilis* IFO 3336 (i.e., *pgsBCA*) were expressed in *E. coli*, and PGA production by the recombinant strain was observed (Ashiuchi et al., 1999). In addition, the enhanced expression of the gene-encoding glutamate racemase, which is involved in D-glutamic acid biosynthesis, was reported to enhance PGA production (Kada et al., 2004). Similarly, PGA production was also achieved by using a recombinant *E. coli* strain expressing the *pgsBCA* gene placed under a strong promoter from another *B. subtilis* strain isolated from food (Jiang et al., 2006). Recently, PGA was produced by untreated cane molasses and MSG waste liquor by *B. subtilis* NX-2 (Zhang et al., 2012a). Metabolic flux analysis of PGA-producing *B. subtilis* has been used to develop culture conditions for improving PGA production (Wu et al., 2008; Zhang et al., 2011).

17.3.1.3 Microbiological Production of 5-Aminolevulinic Acid from Glutamic Acid Currently, ALA production has been achieved using photosynthetic bacteria and algae. There are two known biosynthetic pathways for ALA production in various organisms (Sasaki et al., 2002) (Figure 17.7). In some bacteria and

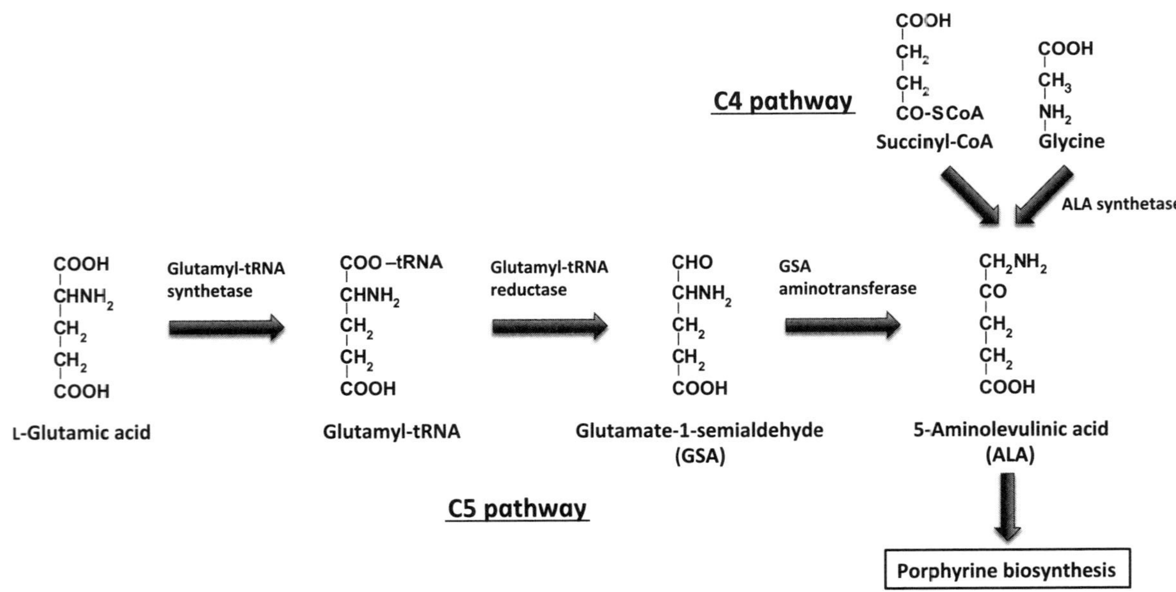

FIGURE 17.7 Biosynthesis pathways of aminolevulinic acid (ALA).

eukaryotes, ALA is synthesized from glycine and succinyl-CoA by ALA synthase via the C4 pathway. In most bacteria, plastids in plants and algae, ALA is synthesized from glutamic acid via glutamyl-tRNA and glutamate-1-semialdehyde via the C5 pathway. Recently, ALA production by recombinant *E. coli* strains was reported. Kang et al. (2011) found that the rate-limiting step in ALA biosynthesis is the reaction of glutamyl-tRNA reductase, which converts glutamyl-tRNA to glutamate-1-semialdehyde and is encoded by the *hemA* gene. Therefore, the overexpression of *hemA* was performed and ALA production slightly increased. Because the stability of the HemA protein is known to be low under conditions with accumulated heme, the mutated *hemA* gene of *Salmonella arizona* was introduced into the *E. coli* strain to increase the stability of mutated HemA protein and a significant increase in ALA production was observed. In addition, the overexpression of the *E. coli rhtA* gene encoding the threonine/homoserine exporter further enhanced the production of ALA. Finally, approximately 4 g/L of ALA was produced from glucose without the addition of cofactors. The enhancement of glutamic acid production in this ALA-producing *E. coli* strain may improve ALA production.

17.3.2 Production of Other Chemicals from Glutamic Acid

Glutamic acid can become an important raw material for the synthesis of other useful chemicals as reported by the US DOE (Werpy and Petersen, 2004) (Figure 17.4). Scott and colleagues reported that glutamic acid chemically converted to *N*-methylpyrrolidone, *N*-vinylpyrrolidone, acrylonitrile, and succinonitrile (Lammens et al., 2010, 2011a, 2011b) (Figure 17.8). Similarly, adiponitrile can also be synthesized from glutamic acid (Dai et al., 2012). *N*-methylpyrrolidone and *N*-vinylpyrrolidone is produced from 2-pyrrolidone, which is synthesized from GABA and can be utilized as raw material to produce polymers. Acrylonitrile, succinonitrile, and adiponitrile have also been utilized to produce a number of chemicals and polymers. For example, succinonitrile is a precursor of 1,4-diaminobutane, which is utilized in the production of polyamides. Additional studies on the chemical and/or biological conversion of glutamic acid to other chemicals are required, which can potentially lead to a decrease in the demand of fossil resources and prevent global warming.

17.4 ECONOMIC OUTLOOK

Currently, the global market for glutamate is over 2.5 million tons per year. Since the production levels of glutamic acid by using microorganisms such as *C. glutamicum* is very high, the production cost of glutamic acid is considered to be low. However, due to the decrease in pH levels in culture during the glutamic acid production by microorganisms, neutralization of culture by the addition of alkaline such as ammonia, which is also utilized as a nitrogen source, is necessary. The resulting glutamic acid becomes a glutamic acid salt, specifically MSG. As mentioned in the US DOE report, the development of low-cost glutamic acid fermentation processes

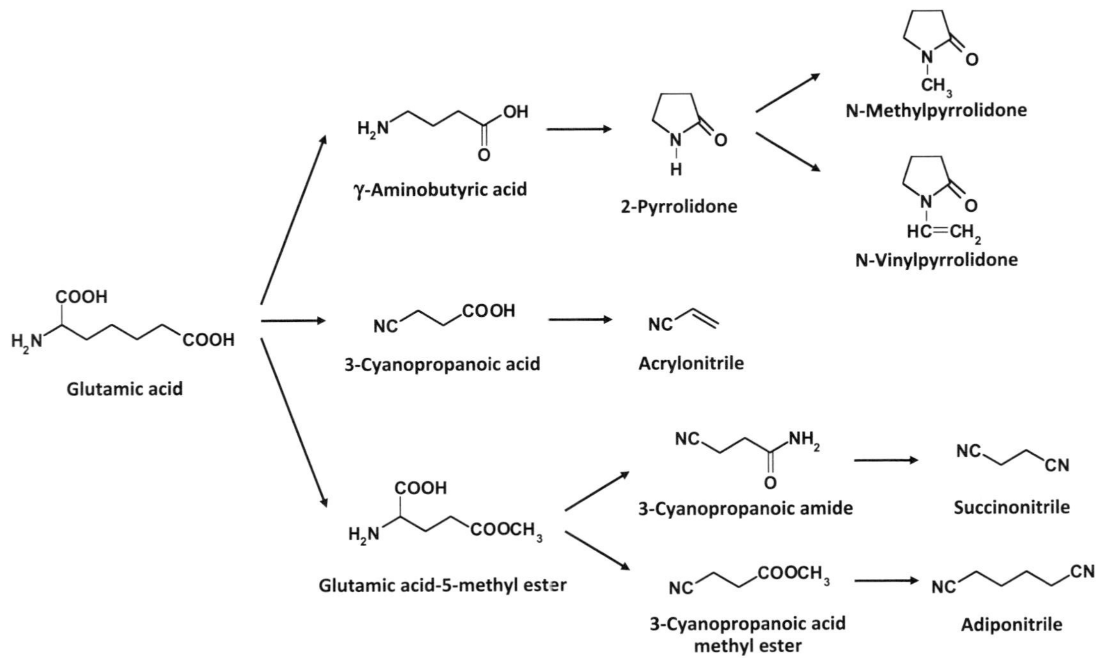

FIGURE 17.8 Synthesis of several chemicals from biologically produced glutamic acid. From Lammens et al. (2011a, 2011b), and Dai et al. (2012).

to obtain the free acid is required (Werpy and Petersen, 2004). These processes will lead to the reduction of costs for the purification and conversion of MSG to the free acid.

17.5 FUTURE PROSPECTS

Currently, the fermentation by microorganisms is predominantly used to produce glutamic acid instead of chemical synthesis, which results in the production of both D- and L-glutamic acid isomers. Although lower amounts of glutamic acid from renewable resources are produced, molecular biology, metabolic engineering, and genomic approaches can be used to enhance glutamic acid production from biomass resources.

As summarized in this chapter, glutamic acid can be used as a building block for producing several chemicals. However, the only examples of microbiological processes for producing chemicals from glutamic acid or sugars via glutamic acid are GABA, PGA, and ALA. Further developments of microbiological processes for producing other chemicals from glutamic acid are essential. To achieve this goal, the discovery of novel enzymes and metabolic reactions is required.

LIST OF ABBREVIATIONS

ALA	5-aminolevulinic acid
DOE	Department of Energy
ICDH	isocitrate dehydrogenase
GABA	γ-aminobutyric acid
GAD	glutamate decarboxylase
GDH	glutamate dehydrogenase
LAB	lactic acid bacteria
MSG	monosodium glutamic acid
ODHC	2-oxoglutarate dehydrogenase complex
PC	pyruvate carboxylase
PDHC	pyruvate dehydrogenase complex
PGA	poly(γ-glutamic acid)
PHA	polyhydroxyalkanoate
PPC	phosphoenolpyruvate carboxylase

REFERENCES

Asakura Y, Kimura E, Usuda Y, Kawahara Y, Matsui K, Osumi T, Nakamatsu T. (2007). Altered metabolic flux due to deletion of *odhA* causes L-glutamate overproduction in *Corynebacterium glutamicum*. *Appl Environ Microbiol*, 73, 1308–1319.

Ashiuchi M, Soda K, Misono H. (1999). A poly-γ-glutamate synthetic system of *Bacillus subtilis* IFO 3336: gene cloning and biochemical analysis of poly-γ-glutamate produced by *Escherichia coli* clone cells. *Biochem Biophys Res Commun*, 263, 6–12.

Bajaj I, Singhal R. (2011). Poly (glutamic acid)–an emerging biopolymer of commercial interest. *Bioresour Technol*, 102, 5551–5561.

Becker J, Zelder O, Hafner S, Schroder H, Wittmann C. (2011). From zero to hero-design-based systems metabolic engineering of *Corynebacterium glutamicum* for L-lysine production. *Metab Eng*, 13, 159–168.

Blombach B, Schreiner ME, Bartek T, Oldiges M, Eikmanns BJ. (2008). *Corynebacterium glutamicum* tailored for high-yield L-valine production. *Appl Microbiol Biotechnol*, 79, 471–479.

Blombach B, Riester T, Wieschalka S, Ziert C, Youn JW, Wendisch VF, Eikmanns BJ. (2011). *Corynebacterium glutamicum* tailored for efficient isobutanol production. *Appl Environ Microbiol*, 77, 3300–3310.

Boulahya KA, Guedon E, Delaunay S, Schultz C, Boudrant J, Bott M, Goergen JL. (2010). OdhI dephosphorylation kinetics during different glutamate production processes involving *Corynebacterium glutamicum*. *Appl Microbiol Biotechnol*, 87, 1867–1874.

Cao M, Geng W, Liu L, Song C, Xie H, Guo W, Jin Y, Wang S. (2011). Glutamic acid independent production of poly-γ-glutamic acid by *Bacillus amyloliquefaciens* LL3 and cloning of *pgsBCA* genes. *Bioresour Technol*, 102, 4251–4257.

Dai JJ, Huang YB, Fang C, Guo QX, Fu Y. (2012). Electrochemical synthesis of adiponitrile from the renewable raw material glutamic acid. *ChemSusChem*, 5, 617–620.

Gopinath V, Meiswinkel TM, Wendisch VF, Nampoothiri KM. (2011). Amino acid production from rice straw and wheat bran hydrolysates by recombinant pentose-utilizing *Corynebacterium glutamicum*. *Appl Microbiol Biotechnol*, 92, 985–996.

Hasegawa T, Hashimoto K, Kawasaki H, Nakamatsu T. (2008). Changes in enzyme activities at the pyruvate node in glutamate-overproducing *Corynebacterium glutamicum*. *J Biosci Bioeng*, 105, 12–19.

Hasegawa S, Uematsu K, Natsuma Y, Suda M, Hiraga K, Jojima T, Inui M, Yukawa H. (2012). Improvement of the redox balance increases L-valine production by *Corynebacterium glutamicum* under oxygen deprivation conditions. *Appl Environ Microbiol*, 78, 865–875.

Hasegawa S, Suda M, Uematsu K, Natsuma Y, Hiraga K, Jojima T, Inui M, Yukawa H. (2013). Engineering of *Corynebacterium glutamicum* for high-yield L-valine production under oxygen deprivation conditions. *Appl Environ Microbiol*, 79, 1250–1257.

Hashimoto K, Kawasaki H, Akazawa K, Nakamura J, Asakura Y, Kudo T, Sakuradani E, Shimizu S, Nakamatsu T. (2006). Changes in composition and content of mycolic acids in glutamate-overproducing *Corynebacterium glutamicum*. *Biosci Biotechnol Biochem*, 70, 22–30.

Hashimoto K, Nakamura K, Kuroda T, Yabe I, Nakamatsu T, Kawasaki H. (2010). The protein encoded by NCgl1221 in *Corynebacterium glutamicum* functions as a mechanosensitive channel. *Biosci Biotechnol Biochem*, 74, 2546–2549.

Hashimoto K, Murata J, Konishi T, Yabe I, Nakamatsu T, Kawasaki H. (2012). Glutamate is excreted across the cytoplasmic membrane through the NCgl1221 channel of *Corynebacterium glutamicum* by passive diffusion. *Biosci Biotechnol Biochem*, 76, 1422–1424.

Hirasawa T, Wachi M, Nagai K. (2000). A mutation in the *Corynebacterium glutamicum ltsA* gene causes susceptibility to lysozyme, temperature-sensitive growth, and L-glutamate production. *J Bacteriol*, 182, 2696–2701.

Hirasawa T, Wachi M, Nagai K. (2001). L-glutamate production by lysozyme-sensitive *Corynebacterium glutamicum ltsA* mutant strains. *BMC Biotechnol*, 1, 9.

Hoffelder M, Raasch K, van Ooyen J, Eggeling L. (2010). The E2 domain of OdhA of *Corynebacterium glutamicum* has succinyltransferase activity dependent on lipoyl residues of the acetyltransferase AceF. *J Bacteriol*, 192, 5203–5211.

Hoischen C, Krämer R. (1990). Membrane alteration is necessary but not sufficient for effective glutamate secretion in *Corynebacterium glutamicum*. *J Bacteriol*, 172, 3409–3416.

Ikeda M, Nakagawa S. (2003). The *Corynebacterium glutamicum* genome: features and impacts on biotechnological processes. *Appl Microbiol Biotechnol*, 62, 99–109.

Ito Y, Tanaka T, Ohmachi T, Asada Y. (1996). Glutamic acid independent production of poly (γ-glutamic acid) by *Bacillus subtilis* TAM-4. *Biosci Biotechnol Biochem*, 60, 1239–1242.

Jiang H, Shang L, Yoon SH, Lee SY, Yu Z. (2006). Optimal production of poly-γ-glutamic acid by metabolically engineered *Escherichia coli*. *Biotechnol Lett*, 28, 1241–1246.

Kada S, Nanamiya H, Kawamura F, Horinouchi S. (2004). Glr, a glutamate racemase, supplies D-glutamate to both peptidoglycan synthesis and poly-γ-glutamate production in γ-PGA-producing *Bacillus subtilis*. *FEMS Microbiol Lett*, 236, 13–20.

Kalinowski J, Bathe B, Bartels D, Bischoff N, Bott M, Burkovski A, Dusch N, Eggeling L, Eikmanns BJ, Gaigalat L, et al. (2003). The complete *Corynebacterium glutamicum* ATCC 13032 genome sequence and its impact on the production of L-aspartate-derived amino acids and vitamins. *J Biotechnol*, 104, 5–25.

Kang Z, Wang Y, Gu P, Wang Q, Qi Q. (2011). Engineering *Escherichia coli* for efficient production of 5-aminolevulinic acid from glucose. *Metab Eng*, 13, 492–498.

Kato Y, Furukawa K, Hara S. (2002). Cloning and nucleotide sequence of the glutamate decarboxylase-encoding gene *gadA* from *Aspergillus oryzae*. *Biosci Biotechnol Biochem*, 66, 2600–2605.

Kawaguchi H, Vertès AA, Okino S, Inui M, Yukawa H. (2006). Engineering of a xylose metabolic pathway in *Corynebacterium glutamicum*. *Appl Environ Microbiol*, 72, 3418–3428.

Kawaguchi H, Sasaki M, Vertès AA, Inui M, Yukawa H. (2008). Engineering of an L-arabinose metabolic pathway in *Corynebacterium glutamicum*. *Appl Microbiol Biotechnol*, 77, 1053–1062.

Kawahara Y, Takahashi-Fuke K, Shimizu E, Nakamatsu T, Nakamori S. (1997). Relationship between the glutamate production and the activity of 2-oxoglutarate dehydrogenase in *Brevibacterium lactofermentum*. *Biosci Biotechnol Biochem*, 61, 1109–1112.

Kikuchi Y, Date M, Yokoyama K, Umezawa Y, Matsui H. (2003). Secretion of active-form *Streptoverticillium mobaraense* transglutaminase by *Corynebacterium glutamicum*: processing of the pro-transglutaminase by a cosecreted subtilisin-like protease from *Streptomyces albogriseolus*. *Appl Environ Microbiol*, 69, 358–366.

Kim S-H, Shin B-H, Kim Y-H, Nam S-W, Jeon S-J. (2007). Cloning and expression of a full-length glutamate decarboxylase gene from *Lactobacillus brevis* BH2. *Biotechnol Bioproc Eng*, 12, 707–712.

Kim J, Hirasawa T, Sato Y, Nagahisa K, Furusawa C, Shimizu H. (2009). Effect of *odhA* overexpression and *odhA* antisense RNA expression on Tween-40-triggered glutamate production by *Corynebacterium glutamicum*. *Appl Microbiol Biotechnol*, 81, 1097–1106.

Kim J, Fukuda H, Hirasawa T, Nagahisa K, Nagai K, Wachi M, Shimizu H. (2010). Requirement of de novo synthesis of the OdhI protein in penicillin-induced glutamate production by *Corynebacterium glutamicum*. *Appl Microbiol Biotechnol*, 86, 911–920.

Kim J, Hirasawa T, Saito M, Furusawa C, Shimizu H. (2011). Investigation of phosphorylation status of OdhI protein during penicillin- and Tween 40-triggered glutamate overproduction by *Corynebacterium glutamicum*. *Appl Microbiol Biotechnol*, 91, 143–151.

Kinoshita S, Udaka S, Shimono M. (1957). Studies on the amino acid fermentation. Part 1. Production of L-glutamic acid by various microorganisms. *J Gen Appl Microbiol*, 50, 193–205.

Krawczyk S, Raasch K, Schultz C, Hoffelder M, Eggeling L, Bott M. (2010). The FHA domain of OdhI interacts with the carboxyterminal 2-oxoglutarate dehydrogenase domain of OdhA in *Corynebacterium glutamicum*. *FEBS Lett*, 584, 1463–1468.

Lammens TM, Franssen MC, Scott EL, Sanders JP. (2010). Synthesis of biobased *N*-methylpyrrolidone by one-pot cyclization and methylation of γ-aminobutyric acid. *Green Chem*, 12, 1430–1436.

Lammens TM, Le Notre J, Franssen MC, Scott EL, Sanders JP. (2011a). Synthesis of biobased succinonitrile from glutamic acid and glutamine. *ChemSusChem*, 4, 785–791.

Lammens TM, Potting J, Sanders JP, De Boer IJ. (2011b). Environmental comparison of biobased chemicals from glutamic acid with their petrochemical equivalents. *Environ Sci Technol*, 45, 8521–8528.

Le Vo TD, Kim TW, Hong SH. (2012). Effects of glutamate decarboxylase and gamma-aminobutyric acid (GABA) transporter on the bioconversion of GABA in engineered *Escherichia coli*. *Bioprocess Biosyst Eng*, 35, 645–650.

Nakamura J, Hirano S, Ito H, Wachi M. (2007). Mutations of the *Corynebacterium glutamicum* NCgl1221 gene, encoding a mechanosensitive channel homolog, induce L-glutamic acid production. *Appl Environ Microbiol*, 73, 4491–4498.

Nakayama K, Kitada S, Kinoshita S. (1961). Studies on lysine fermentation I. The control mechanism on lysine accumulation by homoserine and threonine. *J Gen Appl Microbiol*, 7, 145–154.

Nara T, Samejima H, Kinoshita S. (1964). Effect of penicillin on amino acid fermentation. *Agric Biol Chem*, 28, 120–124.

Niebisch A, Kabus A, Schultz C, Weil B, Bott M. (2006). Corynebacterial protein kinase G controls 2-oxoglutarate dehydrogenase activity via the phosphorylation status of the OdhI protein. *J Biol Chem*, 281, 12300–12307.

O'Regan M, Thierbach G, Bachmann B, Villeval D, Lepage P, Viret JF, Lemoine Y. (1989). Cloning and nucleotide sequence of the phosphoenolpyruvate carboxylase-coding gene of *Corynebacterium glutamicum* ATCC13032. *Gene*, 77, 237–251.

Ohnishi J, Mitsuhashi S, Hayashi M, Ando S, Yokoi H, Ochiai K, Ikeda M. (2002). A novel methodology employing *Corynebacterium glutamicum* genome information to generate a new L-lysine-producing mutant. *Appl Microbiol Biotechnol*, 58, 217–223.

Okino S, Inui M, Yukawa H. (2005). Production of organic acids by *Corynebacterium glutamicum* under oxygen deprivation. *Appl Microbiol Biotechnol*, 68, 475–480.

Park KB, Ji GE, Park MS, Oh SH. (2005). Expression of rice glutamate decarboxylase in *Bifidobacterium longum* enhances γ-aminobutyric acid production. *Biotechnol Lett*, 27, 1681–1684.

Park SJ, Kim EY, Noh W, Oh YH, Kim HY, Song BK, Cho KM, Hong SH, Lee SH, Jegal J. (2013). Synthesis of nylon 4 from gamma-aminobutyrate (GABA) produced by recombinant *Escherichia coli*. *Bioprocess Biosyst Eng*, 36, 885–892.

Peters-Wendisch PG, Kreutzer C, Kalinowski J, Patek M, Sahm H, Eikmanns BJ. (1998). Pyruvate carboxylase from *Corynebacterium glutamicum*: characterization, expression and inactivation of the *pyc* gene. *Microbiology*, 144, 915–927.

Peters-Wendisch PG, Schiel B, Wendisch VF, Katsoulidis E, Möckel B, Sahm H, Eikmanns BJ. (2001). Pyruvate carboxylase is a major bottleneck for glutamate and lysine production by *Corynebacterium glutamicum*. *J Mol Microbiol Biotechnol*, 3, 295–300.

Plokhov AU, Gusyatiner MM, Yampolskaya TA, Kaluzhsky VE, Sukhareva BS, Schluga AA. (2000). Preparation of γ-aminobutyric acid using *E. coli* cells with high activity of glutamate decarboxylase. *Appl Biochem Biotechnol*, 88, 257–265.

Radmacher E, Stansen KC, Besra GS, Alderwick LJ, Maughan WN, Hollweg G, Sahm H, Wendisch VF, Eggeling L. (2005). Ethambutol, a cell wall inhibitor of *Mycobacterium tuberculosis*, elicits L-glutamate efflux of *Corynebacterium glutamicum*. *Microbiology*, 151, 1359–1368.

Rittmann D, Lindner SN, Wendisch VF. (2008). Engineering of a glycerol utilization pathway for amino acid production by *Corynebacterium glutamicum*. *Appl Environ Microbiol*, 74, 6216–6222.

Sano K, Shiio I. (1970). Microbial production of L-lysine III. Production by mutants resistant to S-(2-aminoethyl)-L-cysteine. *J Gen Appl Microbiol*, 16, 373–391.

Sasaki K, Watanabe M, Tanaka T. (2002). Biosynthesis, biotechnological production and applications of 5-aminolevulinic acid. *Appl Microbiol Biotechnol*, 58, 23–29.

Sato H, Orishimo K, Shirai T, Hirasawa T, Nagahisa K, Shimizu H, Wachi M. (2008). Distinct roles of two anaplerotic pathways in glutamate production induced by biotin limitation in *Corynebacterium glutamicum*. *J Biosci Bioeng*, 106, 51–58.

Schneider J, Niermann K, Wendisch VF. (2011). Production of the amino acids L-glutamate, L-lysine, L-ornithine and L-arginine from arabinose by recombinant *Corynebacterium glutamicum*. *J Biotechnol*, 154, 191–198.

Schultz C, Niebisch A, Gebel L, Bott M. (2007). Glutamate production by *Corynebacterium glutamicum*: dependence on the oxoglutarate dehydrogenase inhibitor protein OdhI and protein kinase PknG. *Appl Microbiol Biotechnol*, 76, 691–700.

Schultz C, Niebisch A, Schwaiger A, Viets U, Metzger S, Bramkamp M, Bott M. (2009). Genetic and biochemical analysis of the serine/threonine protein kinases PknA, PknB, PknG and PknL of *Corynebacterium glutamicum*: evidence for non-essentiality and for phosphorylation of OdhI and FtsZ by multiple kinases. *Mol Microbiol*, 74, 724–741.

Shi F, Li Y. (2011). Synthesis of γ-aminobutyric acid by expressing *Lactobacillus brevis*-derived glutamate decarboxylase in the *Corynebacterium glutamicum* strain ATCC 13032. *Biotechnol Lett*, 33, 2469–2474.

Shigu H, Terui G. (1971). Studies on process of glutamic acid fermentation at the enzyme level. I. On the change of α-ketoglutaric acid dehydrogenase in the course of culture. *J Ferm Technol*, 49, 400–405.

Shih IL, Van YT. (2001). The production of poly-(γ-glutamic acid) from microorganisms and its various applications. *Bioresour Technol*, 79, 207–225.

Shiio I, Nakamori S. (1970). Microbial production of L-threonine. Part II. Production by α-amino-β-hydroxyvaleric acid resistant mutants of glutamate producing bacteria. *Agric Biol Chem*, 34, 448–456.

Shiio I, Otsuka SI, Takahashi M. (1962). Effect of biotin on the bacterial formation of glutamic acid. I. Glutamate formation and cellular premeability of amino acids. *J Biochem*, 51, 56–62.

Shimizu H, Tanaka H, Nakato A, Nagahisa K, Kimura E, Shioya S. (2003). Effects of the changes in enzyme activities on metabolic flux redistribution around the 2-oxoglutarate branch in glutamate production by *Corynebacterium glutamicum*. *Bioprocess Biosyst Eng*, 25, 291–298.

Shirai T, Nakato A, Izutani N, Nagahisa K, Shioya S, Kimura E, Kawarabayasi Y, Yamagishi A, Gojobori T, Shimizu H. (2005). Comparative study of flux redistribution of metabolic pathway in glutamate production by two coryneform bacteria. *Metab Eng*, 7, 59–69.

Shirai T, Fujimura K, Furusawa C, Nagahisa K, Shioya S, Shimizu H. (2007). Study on roles of anaplerotic pathways in glutamate overproduction of *Corynebacterium glutamicum* by metabolic flux analysis. *Microb Cell Fact*, 6, 19.

Somkuti GA, Renye JA, Jr., Steinberg DH. (2012). Molecular analysis of the glutamate decarboxylase locus in *Streptococcus thermophilus* ST110. *J Ind Microbiol Biotechnol*, 39, 957–963.

Takahashi C, Shirakawa J, Tsuchidate T, Okai N, Hatada K, Nakayama H, Tateno T, Ogino C, Kondo A. (2012). Robust production of gamma-amino butyric acid using recombinant *Corynebacterium glutamicum* expressing glutamate decarboxylase from *Escherichia coli*. *Enzyme Microb Technol*, 51, 171–176.

Takinami K, Yoshii H, Tsuri H, Okada H. (1965). Biochemical effects of fatty acid and its derivatives on L-glutamic acid fermentation. Part III. Biotin-Tween 60 relationship in the accumulation of L-glutamic acid and the growth of *Brevibacterium lactofermentum*. *Agric Biol Chem*, 29, 351–359.

Thorne CB, Gomez CG, Noyes HE, Housewright RD. (1954). Production of glutamyl polypeptide by *Bacillus subtilis*. *J Bacteriol*, 68, 307–315.

Tsuchida T, Yoshinaga F, Kubota K, Momose H. (1975). Production of L-valine by 2-thiazolealanine resistant mutants derived from glutamic acid producing bacteria. *Agric Biol Chem*, 39, 1319–1322.

Tsuchidate T, Tateno T, Okai N, Tanaka T, Ogino C, Kondo A. (2011). Glutamate production from β-glucan using endoglucanase-secreting *Corynebacterium glutamicum*. *Appl Microbiol Biotechnol*, 90, 895–901.

Udaka S. (1960). Screening method for microorganisms accumulating metabolites and its use in the isolation of *Micrococcus glutamicus*. *J Bacteriol*, 79, 754–755.

Valentin HE, Reiser S, Gruys KJ. (2000). Poly(3-hydroxybutyrate-*co*-4-hydroxybutyrate) formation from γ-aminobutyrate and glutamate. *Biotechnol Bioeng*, 67, 291–299.

Wendisch VF, Bott M, Eikmanns BJ. (2006). Metabolic engineering of *Escherichia coli* and *Corynebacterium glutamicum* for biotechnological production of organic acids and amino acids. *Curr Opin Microbiol*, 9, 268–274.

Werpy T, Petersen G. (2004). Top value added chemicals from biomass: I. Results of screening for potential candidates from sugars and synthesis gas. Oak Ridge: US Department of Energy.

Wu Q, Xu H, Shi N, Yao J, Li S, Ouyang P. (2008). Improvement of poly(gamma-glutamic acid) biosynthesis and redistribution of metabolic flux with the presence of different additives in *Bacillus subtilis* CGMCC 0833. *Appl Microbiol Biotechnol*, 79, 527–535.

Yamano N, Kawasaki N, Takeda S, Nakayama A. (2012). Production of 2-pyrrolidone from biobased glutamate by using *Escherichia coli*. *J Polym Environ*, 21, 528–533.

Yamashiro D, Yoshioka M, Ashiuchi M. (2011). *Bacillus subtilis pgsE* (Formerly *ywtC*) stimulates poly-γ-glutamate production in the presence of zinc. *Biotechnol Bioeng*, 108, 226–230.

Yao W, Chu C, Deng X, Zhang Y, Liu M, Zheng P, Sun Z. (2009). Display of α-amylase on the surface of *Corynebacterium glutamicum* cells by using NCgl1221 as the anchoring protein, and production of glutamate from starch. *Arch Microbiol*, 191, 751–759.

Yao J, Xu H, Shi N, Cao X, Feng X, Li S, Ouyang P. (2010). Analysis of carbon metabolism and improvement of γ-polyglutamic acid production from *Bacillus subtilis* NX-2. *Appl Biochem Biotechnol*, 160, 2332–2341.

Yeh CM, Wang JP, Lo SC, Chan WC, Lin MY. (2010). Chromosomal integration of a synthetic expression control sequence achieves poly-γ-glutamate production in a *Bacillus subtilis* strain. *Biotechnol Prog*, 26, 1001–1007.

Yukawa H, Omumasaba CA, Nonaka H, Kós P, Okai N, Suzuki N, Suda M, Tsuge Y, Watanabe J, Ikeda Y, et al. (2007). Comparative analysis of the *Corynebacterium glutamicum* group and complete genome sequence of strain R. *Microbiology*, 153, 1042–1058.

Zhang D, Xu Z, Xu H, Feng X, Li S, Cai H, Wei Y, Ouyang P. (2011). Improvement of poly(γ-glutamic acid) biosynthesis and quantitative metabolic flux analysis of a two-stage strategy for agitation speed control in the culture of *Bacillus subtilis* NX-2. *Biotechnol Bioproc Eng*, 16, 1144–1151.

Zhang D, Feng X, Zhou Z, Zhang Y, Xu H. (2012a). Economical production of poly(γ-glutamic acid) using untreated cane molasses and monosodium glutamate waste liquor by *Bacillus subtilis* NX-2. *Bioresour Technol*, 114, 583–588.

Zhang Y, Song L, Gao Q, Yu SM, Li L, Gao NF. (2012b). The two-step biotransformation of monosodium glutamate to GABA by *Lactobacillus brevis* growing and resting cells. *Appl Microbiol Biotechnol*, 94, 1619–1627.

CHAPTER 18

Recent Advances for Microbial Production of Xylitol

YONG-CHEOL PARK

Department of Bio and Fermentation Convergence Technology, Kookmin University, Seoul, Korea

SUN-KI KIM and JIN-HO SEO

Department of Agricultural Biotechnology and Center for Food and Bioconvergence, Seoul National University, Seoul, Korea

Abstract

Xylitol is a natural sugar alcohol widely used as a sugar substitute in foods, toothpastes, and mouthwashes and also can serve as a platform chemical for polymer

Bioprocessing of Renewable Resources to Commodity Bioproducts, First Edition.
Edited by Virendra S. Bisaria and Akihiko Kondo.

synthesis. For mass production of xylitol, chemical processes of xylose hydrogenation were commercialized. But the sustainability issue has encouraged biological, especially microbial production of xylitol. From the 1990s, many research groups have identified and developed new microbial strains and processes for efficient conversion of xylose to xylitol. In the 2000s, metabolic engineering approaches based on genetic modulation have been applied for xylitol production with the theoretical yield. In this chapter, general principles for biological production of xylitol and major factors affecting microbial xylitol production are introduced. Recent development of recombinant microorganisms and their xylitol production processes are described, including technical approaches of genetic engineering, metabolic engineering, and cofactor engineering.

18.1 INTRODUCTION

Xylitol ((2*R*,4*S*)-pentane-1,2,3,4,5-pentol, $C_5H_{12}O_5$, CAS No. 87-99-0) is a natural sweetener with a number of advantageous properties. As xylitol has 37.4 kJ/mol of enthalpy for solution and extreme cooling and caries preventative effects, xylitol has been commercially used in chewing gums, candies, toothpastes, and mouthwashes. Xylitol is also safe for diabetes since it does not require insulin for metabolic regulation (Schiweck and Ziesenitz, 1996). As a platform chemical, xylitol has been recently chosen as one of the top 12 value-added chemicals produced from biomass (Werpy and Petersen, 2004). It can be converted to various compounds for other rare sugar production (Granström et al., 2007) and polymer synthesis such as xylaric acid and glycols, directly polymerized for production of unsaturated polyester resin (Werpy and Petersen, 2004). In 2007, the annual production of xylitol was estimated between 20,000 and 40,000 tons per year with a market value of 40–80 million euros (Granström et al., 2007).

For production of xylitol, chemical and biological processes have been developed using xylose as a starting compound (Toivari et al., 2007). For chemical processes, xylose is hydrogenated to xylitol by nickel, ruthenium, and rhodium catalysts at high temperatures and pressures (da Silva and Chandel, 2012). For the chemical reaction, highly pure xylose should be prepared in order to reduce the purification cost, and a series of purification steps be required. Regarding the sustainability issue, biological production of xylitol has gained an interest from academia and industries. For biological processes, microbial conversion of xylose to xylitol has been developed using *Candida*, *Saccharomyces*, *Pichia*, and *Escherichia* species. Enzymatic processes using xylose reductase catalyzing xylose hydrogenation at the expense of a reducing power were also devised.

18.2 GENERAL PRINCIPLES FOR BIOLOGICAL PRODUCTION OF XYLITOL

In a chemical process, an aldehyde group of C_1 in xylose is converted to an alcohol group in xylitol in the presence of metal catalysts using equimolar hydrogen gas

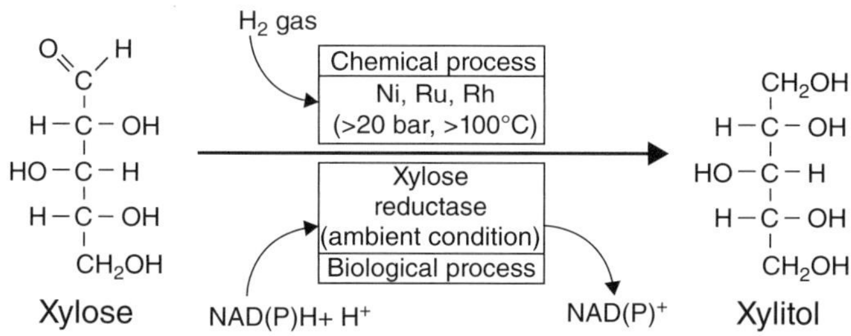

FIGURE 18.1 Processes for xylitol production from xylose.

(Figure 18.1). In the same manner, materials for biological processes consist of xylose reductase as a biological catalyst and NADPH or NADH as a hydrogen donor. Xylose reductase (EC 1.1.1.307, XR) is an oxidoreductase catalyzing an irreversible reaction of xylose hydrogenation to xylitol. XR can use NADPH and NADH as a cofactor and prefers NADPH to NADH generally. For example, a wild type of XR from *Pichia stipitis* possesses about 70% higher xylose reductase activity toward NADPH than NADH. A mutant XR (R276H) of *P. stipitis* has a 2.5-fold higher activity toward NADH than NADPH (Lee et al., 2012). For efficient production of xylitol, NAD(P)H should be sufficiently supplied to the XR reaction. Regeneration of NAD(P)H is one of the important factors affecting the productivity of xylitol production.

The xylose metabolism in eukaryotesis illustrated in Figure 18.2. Xylose is delivered into the cells by specific or nonspecific sugar transporters. *Candida* species can take up extracellular xylose by two sugar transport systems: a carrier-mediated facilitated diffusion transporter in nonstarvation conditions and a proton symporter in starvation conditions (Winkelhausen and Kuzmanova, 1998). *Candida shehatae* CBS 2779 showed various K_m values from 1 to 125 mM for xylose transport. *Saccharomyces cerevisiae* does not have xylose-specific transporters but can take up xylose by both high- and low-affinity glucose transporters (Lee et al., 2002). Intracellular xylose is reduced to xylitol by XR in the presence of NAD(P)H, followed by conversion to xylulose by NAD^+-dependent xylitol dehydrogenase (EC 1.1.1.9, XDH) with a cofactor NAD^+. On the other hand, bacteria have xylose isomerase (EC 5.3.1.5), which are able to convert xylose to xylulose without cofactors. In both prokaryotic and eukaryotic cells, xylulose is then phosphorylated to xylulose-5-phosphate by xylulokinase (EC 2.7.1.17, XK) with ATP. Xylulose-5-phosphate is further metabolized through the pentose phosphate pathway and glycolysis for ATP synthesis and NAD(P)(H) regeneration. Because xylose is utilized as a carbon source for both xylitol production and the central cellular metabolism, a high yield of xylitol is limited. To solve this problem, a system of dual carbon sources has been developed in which a mixture of xylose and glucose is used: xylose for xylitol production and glucose for the central metabolism. As shown in the carbon metabolism (Figure 18.2), glucose is metabolized to produce ATP and NAD(P)H, and regenerate $NAD(P)^+$ through the central carbon metabolism. In some cases, more improvement of xylitol yield reaching the theoretical yield of 1.0 mol/mol could be achieved in the absence of

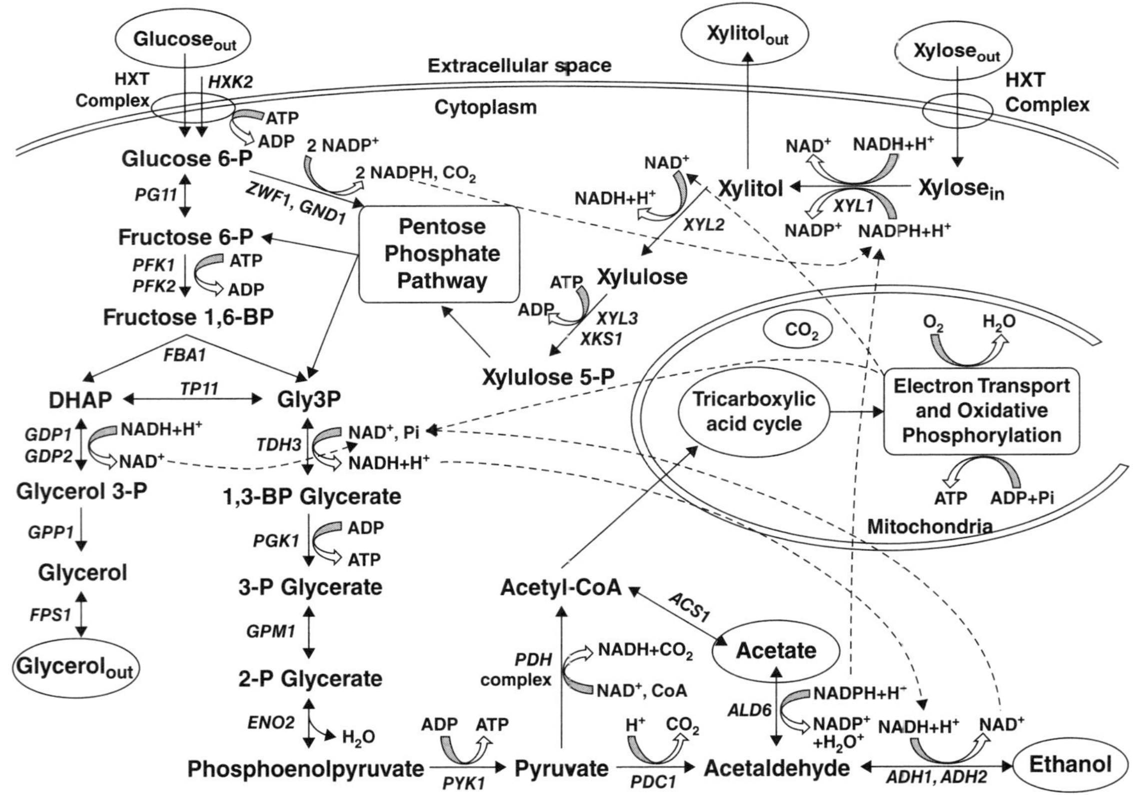

FIGURE 18.2 Xylose and glucose metabolic network. The names of metabolites, genes coding for metabolic enzymes, and so on were indicated as follows: *XYL1*, xylose reductase; *XYL2*, xylitol dehydrogenase; *XYL3*, *XKS1*, xylulokinase; *HXT*, hexose transporter; *HXK2*, hexokinase; *ZWF1*, glucose 6-phosphate dehydrogenase; *GND1*, 6-phosphogluconate dehydrogenase; *PGI1*, phosphoglucoisomerase; *PFK1*, *PKF2*, phosphofructokinase; *FBA1*, fructose 1,6-bisphosphate aldolase; *TPI1*, triose phosphate isomerase; *TDH3*, glyceraldehyde 3-phosphate dehydrogenase; *PGK1*, 3-phosphoglycerate kinase; *GPM1*, phosphoglycerate mutase; *ENO2*, enolase; *PYK1*, pyruvate kinase; *PDH*, pyruvate dehydrogenase; *PDC1*, pyruvate decarboxylase; *ADH1*, *ADH2*, alcohol dehydrogenase; *ACS1*, acetyl-CoA synthetase; *ALD6*, aldehyde dehydrogenase; *GPD1*, *GPD2*, glycerol 3-phosphate dehydrogenase; *GPP1*, glycerol 3-phosphatase; *FPS1*, glycerol efflux protein; *DHAP*, dihydroxyacetone phosphate; Gly3P, glyceraldehyde-3-phosphate; P, phosphate; BP, bisphosphate; in, intracellular; out, extracellular. Solid and dotted single lines and double line indicate the flows of metabolites and

XDH, where xylose is converted only to xylitol, and glucose co-substrate is utilized for NAD(P)H cofactor supply and maintenance (Lee et al., 2000; Chung et al., 2002; Jin et al., 2005; Ko et al., 2006; Kwon et al., 2006; Kim et al., 2011).

18.3 MICROBIAL PRODUCTION OF XYLITOL

In the xylose metabolism, both XR activity and NAD(P)H supply play an important role in xylitol production. High XR activity in wild-type strains or XR overexpressing recombinant strains is critical for efficient xylitol production. Among several microorganisms containing their own xylose reductase, a genus of *Candida* has been regarded as a powerful xylitol producer: *C. tropicalis*, *C. parapsilosis*, *C. guilliermondii*, *C. mogii*, *C. shehatae*, and so on. (Winkelhausen and Kuzmanova, 1998). *Saccharomyces cerevisiae* and *C. tropicalis* were genetically engineered to express the heterologous XR gene or to disrupt the endogenous XDH gene, of which information is presented in section 18.4. Microbial production of xylitol from xylose is affected by various environmental parameters such as substrate, aeration, culture temperature, and medium acidity. Among them, carbon sources and aeration are especially critical for cell growth and xylitol production. Also various fermentation techniques were applied for maximizing the xylitol-producing performance.

18.3.1 Carbon Sources

Xylose is used as the main carbon source for xylitol production. Before xylose is hydrogenated inside the microbial cells by endogenous or heterologous XR, extracellular xylose should be transported into the cells. Because yeasts take up xylose by the facilitated diffusion or active transport mechanism, xylose concentration is a limiting factor to control xylose transport and xylitol production. It is known that high xylose concentrations over 100 g/L induce xylitol production in yeasts (Winkelhausen and Kuzmanova, 1998). However, an increase in initial xylose concentration generally reduces specific growth rate and xylitol productivity, the extent of which depends on the yeast strains. Examples of xylitol production by *Candida* species using various concentrations of xylose are listed in Table 18.1. As mentioned in section 18.2, xylose is consumed for both cell growth and xylitol production. In addition, xylose is more expensive than other sugars such as glucose and sucrose. Development of a cost-effective process requires cheap carbon sources used for cell growth and energy metabolism. Among various monosaccharides and disaccharides (maltose and sucrose), glucose is recognized as a good co-substrate in xylitol production. Even though glucose inhibits the xylose metabolism through the glucose-repression mechanism and the competitive sugar transport, such inhibition can be relieved by controlling the glucose concentration in culture broth.

18.3.2 Aeration

As shown in Figure 18.2, a sole carbon source of xylose is metabolized into xylitol, ATP, NAD(P)H, and some metabolites through the pentose phosphate pathway,

TABLE 18.1 Kinetic Parameters for Microbial Production of Xylitol by *Candida* Species

Host strain	Culture type	Initial xylose concentration (g/L)	Dry cell mass (g/L)	Xylitol concentration (g/L)	Xylitol productivity (g/L·h)	Xylitol yield (g/g)	Reference
Candida tropicalis ATCC 13803	Batch	126	11.7	102	5.06	0.81	Kim et al. (1999)
	Batch	108	22	86	3.86	0.80	Choi et al. (2000)
	Batch	150	24	116	2.60	0.78	Kim et al. (2004)
	Fed-batch	100	10.4	187	3.90	0.75	Kim et al. (2002a)
	Fed-batch	150	17	237	2.00	0.89	Kim and Oh (2003)
	Cell-recycle	100	26	189	4.94	0.82	Choi et al. (2000)
C. parapsilosis KFCC 10875	Batch	140	–	103	3.43	0.74	Oh et al. (1998)
		170		125	3.57	0.74	
		220		159	3.06	0.72	
		270		189	2.22	0.70	
C. guilliermondii NRC 5578	Batch	110	–	68.7	0.29	0.60	Meyrial et al. (1991)
		150		110.3	0.46	0.70	
		200		151.7	0.52	0.71	
		300		221	0.54	0.75	

glycolysis, and electron transport system. Because of the fate of xylose, the xylose flux should be balanced to meet the demand of cell growth and xylitol production. Xylose is fully oxidized in aerobic conditions without xylitol accumulation, however, most xylose-fermentable yeasts failed to grow on xylose under strictly anaerobic environments. It is believed that xylose conversion to both cell mass and xylitol occurs simultaneously only under oxygen limitation, owing to the fact that oxygen is required to synthesize unsaturated fatty acids and ergosterol necessary for sugar transport (Kuriyama and Kobayashi, 1993). To increase xylitol yield, the xylose flux to cell mass has to be minimized, but sufficient maintenance energy and NADPH should be provided by controlling the oxygen supply rate. Concerning the fermentative behavior of the xylose-metabolizing yeasts on oxygen availability, three states can be distinguished (Laplace et al., 1991; Winkelhausen and Kuzmanova, 1998). (1) Under anaerobic or microaerobic conditions, NADH is unable to be fully oxidized to NAD^+ through the electron transfer system. These conditions may keep the NAD^+/NADH ratio and ATP levels low, resulting in the restriction of xylose flux to cell growth by the lack of cofactors necessary for xylitol dehydrogenase and xylulokinase. Consequently, xylitol accumulated inside the cells is secreted into culture medium. (2) Increasing the oxygen transfer rate permits the improvement of xylose fermentation. It is believed that oxygen is consumed as a terminal electron acceptor, thus relieving the imbalance of the two initial steps of anaerobic xylose metabolism. The inverse relationship between the degree of aeration and xylitol production can be found in some yeasts. Under these environments, the main metabolic product is ethanol. (3) In high aeration conditions, the xylose flux via pyruvate favors the tricarboxylic acid cycle rather than ethanol fermentation, resulting in an increase in cell mass. It is generally agreed that an oxygen supply rate is a key parameter which determines whether xylose will be fermented or respired.

To develop an effective process, it is very important to determine the oxygen supply conditions that will balance xylose utilization both for cell growth and xylitol production. Xylitol-producing yeasts require very small amounts of oxygen, which are specific for yeast strains (Winkelhausen and Kuzmanova, 1998). To increase the xylitol yield and productivity, the fermentation stage can be divided into the cell growth stage without xylitol accumulation and the xylitol production stage without cell growth by controlling oxygen supply. In a fed-batch culture, for example, an initial aeration rate was fixed at 1 vvm and 500 rpm of agitation speed where a volumetric oxygen transfer coefficient (k_La) was 1.06/min. After the batch stage, a dissolved oxygen (DO) level was maintained at less than 1% for efficient conversion of xylose to xylitol (Kim et al., 2002a). Various aeration conditions affecting xylitol production are summarized in Table 18.2.

18.3.3 Optimization of Fermentation Strategies

Typical batch production of xylitol using xylose and glucose (co-substrate) is displayed in Figure 18.3 (Kim et al., 1999). In the presence of both xylose and glucose, *Candida* species consumes xylose after glucose utilization because it is a Crabtree-positive microorganism. In general, the assimilation of xylose starts when glucose is

TABLE 18.2 Effects of Aeration on Xylitol Production from Xylose by Yeasts

Yeast	Xylose concentration (g/L)	Agitation rate (rev/min)	Aeration rate (vvm)	Oxygen transfer rate (mmol/L·h)	$k_L a$ (L/h)	Xylitol concentration (g/L)	Xylitol yield (g/g)	Xylitol productivity (g/L·h)	Reference
Candida boidinii	130	150		10		15.7	0.38	0.06	Vandeska et al.
NRRL	130	150		14		58.9	0.48	0.24	(1995)
Y-17213	130	150		18		50.2	0.45	0.21	
	130	150		24		36.3	0.34	0.22	
	130	150		30		23.7	0.23	0.17	
C. guilliermondii	300			1.2			0.50	0.27	Nolleau et al.
FTI 5578	300			2.2			0.66	0.43	(1995)
	300			2.7			0.55	0.41	
	300			4.2			0.32	0.41	
C. parapsilosis	100			0.2			0.57	0.10	Nolleau et al.
ATCC 28474	100			0.4			0.65	0.14	(1995)
	100			1.0			0.59	0.23	
	100			2.2			0.50	0.28	
	100			3.2			0.45	0.36	
C. parapsilosis	50	250	0.08		4.8	22.0	0.61	0.26	Furlan et al.
ATCC 28474	50	250	0.30		16.8	32.5	0.65	0.32	(1994)
	50	250	0.90		35.4	30.4	0.44	0.15	
	50	500	0.60		99.6	5.0	0.10	0.02	
Pachysolen	50	250	0.08		4.8	13.5	0.27	0.06	Furlan et al.
tannophilus	50	250	0.90		35.4	11.0	0.22	0.13	(1994)
ATCC 32691									
Debaryomyces	50		1.0	163		7.7	1.00	0.372	Gírio et al.
hansenii	50		1.0	253		1.6	0.13	0.126	(1994)
DTIA-77	50		1.0	775.8		nr	nr	0.033	

nr, not reported.

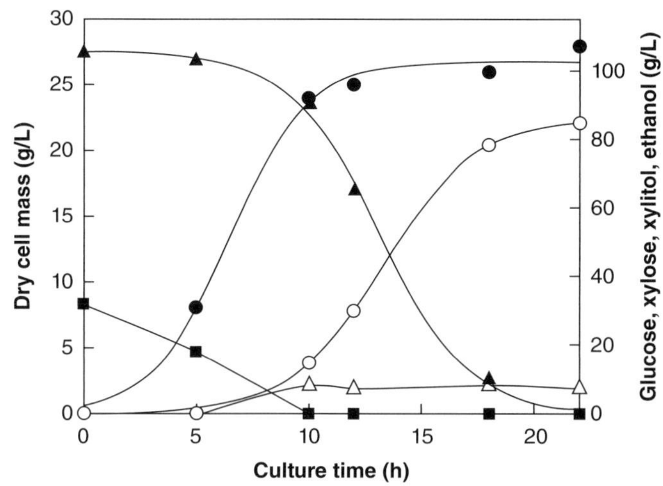

FIGURE 18.3 Results of the two-substrate batch culture by *Candida tropicalis* ATCC 13803 in a 3.5-L fermentor containing 1 L of fermentation medium (10 g/L yeast extract, 20 g/L bactopeptone, 105 g/L xylose, and 32 g/L glucose) at 30°C, pH 6, 500 rpm of agitation speed, and 1 vvm of aeration rate.● , dry cell mass; ■, glucose; ▲, xylose; ○, xylitol; △, ethanol

still present in the medium, indicating that a threshold of glucose repression exists. During 20 hours of batch culture using a mixture of 105 g/L xylose and 32 g/L glucose, 85 g/L of final xylitol concentration, 0.81 g xylitol/g xylose of xylitol yield (equivalent to 90% of the theoretical xylitol yield), and 3.85 g/L·h of volumetric xylitol productivity were obtained (Figure 18.3). Meanwhile, ethanol is a main by-product produced at high concentrations of glucose and xylose. Because high ethanol concentrations reduce cell growth and xylitol production, an optimal concentration of glucose and xylose in the culture broth should be determined (Kim et al., 1999).

To increase the final xylitol concentration and productivity, fed-batch processes have been devised by changing feeding strategies such as feed rate and sugar composition in the feeding solution. Xylitol productivity is related to xylose consumption rate which strongly depends on xylose reductase activity and xylose transport rate. Xylose is transported into yeast cells by the active transport mechanism, where xylose transport depends on extracellular xylose concentrations (Lee et al., 2002). Considering the transport mechanism and substrate inhibition, xylose concentrations should be controlled around optimum values in the fed-batch operation stage. For example, a mathematical modeling was adopted to determine a time-dependent xylose feeding rate and then to maintain the optimal xylose concentration in the broth (Kim et al., 2002a):

$$F = \frac{q_{\mathrm{xyl}}}{(S_{\mathrm{f}} - S_{\mathrm{xyl}})} VX = \frac{q_{\mathrm{xyl}}}{(S_{\mathrm{f}} - S_{\mathrm{xyl}})} V_0 X_0 \mathrm{e}^{\mu t}$$

where F is the feed rate of xylose solution, q_{xyl} is the specific xylose consumption rate, S_{f} is the xylose concentration in the feeding solution, S_{xyl} is the xylose concentration in the culture medium, V and V_0 are the actual and initial volumes of medium, respectively, X and X_0 are the actual and initial dry cell masses, respectively, μ is the specific growth rate at the xylitol production stage, and t is the fermentation time. The equation of xylose feed rate can be derived from the mass balance equations for a fed-batch fermentation (Bailey and Ollis, 1986). F is simply the function of fermentation time because S_f, V_0, X_0, S_{xyl}, q_{xyl}, and μ are constant. Before applying the mathematical approach, some factors are optimized by preliminary experiments: xylose concentration (S_{xyl}) in the culture medium is controlled at 100 g/L, and the specific growth rate (μ) during the xylose conversion phase is maintained at less than 0.02/h by controlling the oxygen supply rate (less than 1% DO). The flow rate can be adjusted in the feedback control mode by using q_{xyl} calculated in each sampling term to maintain the xylose concentration more accurately. Figure 18.4 shows the profiles for an optimized fed-batch fermentation. By controlling the aeration rate and xylose feed rate, the optimum conditions of xylose concentration and specific growth rate were well maintained. A step-wise feeding of a highly concentrated xylose solution and a shift of DO content less than 1% led to a dramatic increase in xylitol production. Finally, 184 g/L of final xylitol concentration, 0.75 g/g xylitol yield, and 3.9 g/L·h of xylitol productivity were obtained.

For the continuous production of xylitol with high yields, techniques of whole cell immobilization and cell-recycle fermentation have been used. *Candida tropicalis* cells were immobilized into calcium alginate, polyacrylamide, a cylindrical porous ceramic, and a nonwoven polyester fabric. At a condition to minimize the loss of the immobilized yeast cells, 86.6 g/L xylitol was produced with a xylitol productivity of

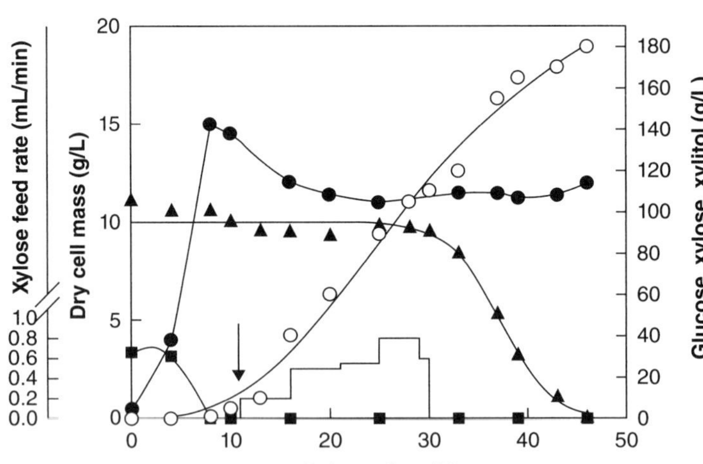

FIGURE 18.4 Profiles of fed-batch fermentation of *Candida tropicalis* ATCC 13803 in a 3.5-L fermentor containing 1 L of YP medium (10 g/L yeast extract, 20 g/L bactopeptone), in which 105 g/L xylose and 32 g/L glucose were added initially. Fermentation environments were fixed at 30°C, pH 6, 500 rpm, and 1 vvm. ●, dry cell mass; ■, glucose; ▲, xylose; ○, xylitol; solid line, feeding of a concentrated xylose solution; arrow, shift of dissolved oxygen content.

1.35 g/L·h and a xylitol yield of 0.60 g/g (Yahashi et al., 1996). In this case, glucose feeding in the presence of xylose improved xylitol productivity, which was similar to the fermentation results using free cells. Even though high productivity could be achieved, cell immobilization has intrinsic problems: lower mass transfer rate from a bulk flow to the cells and the leakage of the cells into the medium. A cell-recycle fermentation system with hollow fiber membrane was designed to maintain high concentrations of cells in a bioreactor for efficient conversion of xylose to xylitol (Choi et al., 2000). A diagram of atypical cell-recycle system is illustrated in Figure 18.5(a). For example, the culture broth was filtered through a hollow fiber membrane with 100,000 molecular weight cut off, and the cells were recycled back to the fermentor. Fresh medium containing 750 g/L xylose, 200 g/L glucose, and 100 g/L

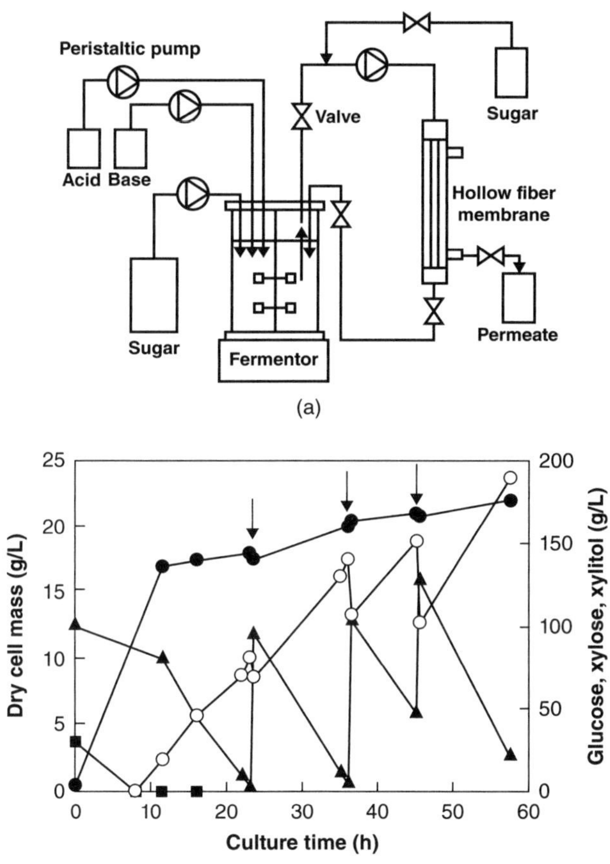

FIGURE 18.5 (a) A diagram of cell-recycle operation system (Bae et al., 2004) and (b) a cell-recycle fermentation of *Candida tropicalis* ATCC 13803 in YP medium with initial concentration of 30 g/L glucose and 100 g/L xylose (Choi et al., 2000). The yeast cells were recycled with a hollow fiber membrane with 100,000 molecular weight cut-off. Feeding solution was composed of 750 g/L xylose, 200 g/L glucose, and 100 g/L yeast extract. ●, dry cell mass; ■, glucose; ▲, xylose; ○, xylitol; arrow, addition of feed solution.

yeast extract was added into the culture broth to adjust xylose concentration from 10 to 100 g/L after transition to the cell-recycle fermentation mode (Choi et al., 2000). As shown in Figure 18.5(b), a cell-recycle fermentation with *C. tropicalis* ATCC 13803 at 30°C and pH 6.0 resulted in 189 g/L xylitol concentration and 4.94 g/L·h productivity, which were 2.2 and 1.3 times higher than those of the batch culture with the same initial concentrations of 100 g/L xylose and 20 g/L glucose. Xylitol yields in batch and cell-recycle fermentations reached 0.8 g/g xylose.

18.4 XYLITOL PRODUCTION BY GENETICALLY ENGINEERED MICROORGANISMS

Natural xylitol-producing microorganisms can metabolize xylose for xylitol production as well as cell growth and maintenance, which causes the reduction of xylitol yield. To improve the xylitol yield and productivity, recombinant yeasts and bacteria have been developed by genetic engineering and metabolic engineering.

18.4.1 Construction of Xylitol-Producing Recombinant *Saccharomyces cerevisiae*

An ethanol-fermenting yeast, *S. cerevisiae*, is a GRAS (Generally Recognized as Safe) microorganism useful for food and drug manufacturing. Even though it has a broad spectrum of sugar utilization, it cannot metabolize xylose because of the absence of xylose reductase (XR) and xylitol dehydrogenase (XDH). The absence of XR gives us an idea to develop xylitol production strategies with the theoretical yield. For the purpose of xylitol production in *S. cerevisiae*, the gene coding for XR was introduced into *S. cerevisiae* by genetic engineering. The *XYL1* gene encoding XR was amplified from the chromosome of *P. stipitis* by a polymerase chain reaction and overexpressed in *S. cerevisiae* (Lee et al., 2000). The *P. stipitis XYL1* gene does not have introns (Hallborn et al., 1991), and its encoded XR protein can be expressed actively in *S. cerevisiae* (Chung et al., 2002). Because the recombinant *S. cerevisiae* expressing the *XYL1* gene does not have metabolic enzymes for full oxidation of xylitol, the recombinant strain requires additional sugars for cell growth and viability maintenance. A cheap carbon source of glucose is generally used as a co-substrate. Figure 18.6 shows typical batch-type production of xylitol using two substrates of xylose and glucose by recombinant *S. cerevisiae* EH13.15 harboring the *XYL1* gene episomally (Lee et al., 2000). As *S. cerevisiae* prefers glucose to xylose, glucose is consumed faster than xylose. At a low concentration of glucose, xylose is transported into the *S. cerevisiae* cells via high-affinity glucose transporters (Lee et al., 2002) and metabolized simultaneously to xylitol. After complete consumption of glucose, however, no more xylose is converted to xylitol, which is due to the fact that NAD(P)H necessary for the XR reaction is not supplied enough (Rizzi et al., 1988; Jeun et al., 2003). A fed-batch fermentation strategy was designed to feed a highly concentrated glucose solution and to add xylose intermittently for elevating xylose concentration to a set point of 30 g/L. The profiles of cell growth, sugar concentrations, and xylitol

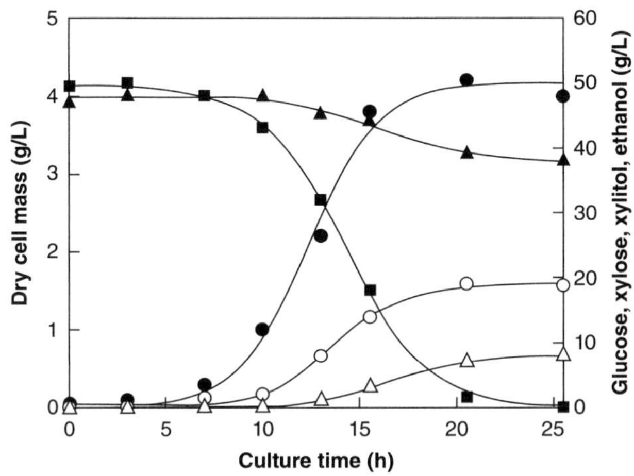

FIGURE 18.6 Batch fermentation profiles of *Saccharomyces cerevisiae* EH13.15:pY2XR at 30°C (Lee et al., 2000). •, dry cell mass; ■, glucose; ▲, xylose; ○, xylitol; △, ethanol.

production in the fed-batch fermentation are illustrated in Figure 18.7. Maintenance of glucose concentration at around 0.35 g/L in culture broth provided 105.2 g/L xylitol concentration, 1.69 g/L·h xylitol productivity, and 0.95 g/g xylitol yield (Lee et al., 2000).

Wild-type strains of *S. cerevisiae* have many nonspecific aldose reductases and are able to produce a small amount of xylitol without XR expression. Among several aldose reductases, the GRE3 protein was expressed in recombinant *S. cerevisiae*, and

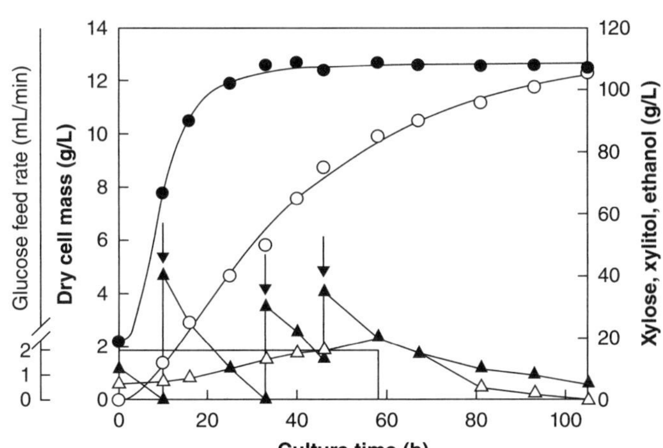

FIGURE 18.7 Fed-batch fermentation profiles of *Saccharomyces cerevisiae* EH13:pY2XR at 30°C. Glucose was maintained at 0.35 g/L throughout the fermentation (Lee et al., 2000). •, dry cell mass; ▲, xylose; ○, xylitol; △, ethanol; solid line, feeding of a concentrated glucose solution; arrow, addition of a concentrated xylose solution.

its expression was compared with a *XYL1*-expressing strain with respect to xylitol production (Kim et al., 2002b). All kinetic parameters of cell growth, glucose and xylose consumption, and xylitol yield were similar between the two recombinant *S. cerevisiae* strains. But, the xylitol productivity of the *GRE3* system was 50% of that of the *XYL1* system in both batch and fed-batch fermentations, which is ascribed to the cofactor dependence (Kim et al., 2002b). The *P. stipitis* XR can use both NADPH and NADH with a ratio of 3:2 (Rizzi et al., 1988), whereas the *S. cerevisiae* GRE3 protein is highly specific for NADPH (Kim et al., 2002b). Plasmid instability has been an inherent obstacle for expression of foreign genes in *S. cerevisiae* (Romanos et al., 1992). According to this plasmid instability, XR expressed in recombinant *S. cerevisiae* harboring episomal plasmid pY2XR showed a 60% decrease in its specific activity during the fed-batch culture (Chung et al., 2002). The YIp-based vectors deliver the target gene into one copy of the chromosomal *URA3* or *TRP1* gene in *S. cerevisiae*. To increase the copies of the integrated target gene, a repetitive chromosomal DNA of the δ-sequence was developed, which is able to insert the cloned gene up to 100 copies (Kim et al., 2001). The *XYL1* gene was cloned into plasmid pδ-neo containing the δ-sequence, and over 98% of its *S. cerevisiae* transformants possessed the cloned gene throughout 60 hours of the fed-batch fermentation (Chung et al., 2002). The integrated system provided a 70% enhancement in final xylitol concentration relative to the episomal system. Kinetic parameters for xylitol production by recombinant *S. cerevisiae* are summarized in Table 18.3.

18.4.2 Cofactor Engineering for Xylitol Production in Recombinant *Saccharomyces cerevisiae*

Cofactors such as NADP(H), NAD(H), vitamins, and coenzymes play many roles in the control of energy balance, enzyme reactions, protein structure, and so on. As a proton donor, xylose reductase (XR) requires 1 mole of NAD(P)H for the conversion of 1 mole of xylose to xylitol (Rizzi et al., 1988). As mentioned above, most xylose reductases prefer NADPH to NADH. To supply NADPH sufficiently, genes coding for metabolic enzymes directly or indirectly engaged in NADPH regeneration were overexpressed in recombinant *S. cerevisiae* expressing the *XYL1* gene.

NADPH is mainly regenerated from the oxidative part of the pentose phosphate pathway (Figure 18.2) in which glucose-6-phosphate dehydrogenase (G6PDH) and 6-phosphogluconate dehydrogenase produce 2 moles of NADPH at the expense of 1 mole of glucose via glucose-6-phosphate. G6PDH encoded by the *ZWF1* gene of *S. cerevisiae* was overexpressed in a xylitol-producing recombinant *S. cerevisiae* (Kwon et al., 2006). An elevation of G6PDH activity increased both final cell mass and xylitol productivity and concentration by about 24%, relative to those of the control strain. Even though a considerable change of the NADPH and $NADP^+$ pools could not be observed inside the yeast cells, it was speculated that rapid regeneration of NADPH by the elevated G6PDH activity enhanced the NADPH-dependent xylose conversion to xylitol (Kwon et al., 2006).

Glcose-6-phosphate is a key metabolite in the sugar metabolism and located at a fork to the glycolysis and pentose phosphate pathway. Reduction of the carbon flux

TABLE 18.3 Kinetic Parameters for Microbial Production of Xylitol by Recombinant *Saccharomyces cerevisiae*

Host strain	Genetic manipulation	Culture type	Initial carbon source	Feeding solution	Xylitol concentration (g/L)	Xylitol productivity (g/L·h)	Xylitol yield (g/g, %)	Reference
EH13.15	Episomal expression of *XYL1*	Batch	47 g/L xylose 50 g/L glucose	–	8.03	0.71	90	Lee et al. (2000)
EH13.15	Episomal expression of *XYL1*	Glucose-limited fed-batch	47 g/L xylose 50 g/L glucose	xylose, glucose (composition not given)	105.2	1.69	95	Lee et al. (2000)
BJ3505	Episomal expression of *GRE3*	Fed-batch	45 g/L xylose 20 g/L glucose	200 g/L ethanol	22.9	0.50	~100	Kim et al. (2002b)
BJ3505	Episomal expression of *XYL1*	Glucose-limited fed-batch	100 g/L xylose 20 g/L glucose	600 g/L glucose	46.0	0.7	60	Chung et al. (2002)
BJ3505	Chromosomal δ-integration of *XYL1*	Glucose-limited fed-batch	100 g/L xylose 20 g/L glucose	600 g/L glucose	78.1	1.1	90	Chung et al. (2002)
BJ3505	Chromosomal δ-integration of *XYL1*	Cell-recycle	80 g/L xylose 20 g/L glucose	600 g/L xylose 600 g/L glucose	67.1	1.23	~100	Bae et al. (2004)
BJ3505	Chromosomal δ-integration of *XYL1*	Cell-recycle	80 g/L xylose 20 g/L glucose	600 g/L xylose 600 g/L glucose 100 g/L yeast extract	116	2.34	~100	Bae et al. (2004)

to the glycolysis is hypothesized to increase the flux toward the pentose phosphate pathway where NADPH is produced. This strategy was applied to xylitol production. The activity of phosphoglucose isomerase (PGI) present in the glycolysis was reduced by the promoter replacement technology (Oh et al., 2007). In this *S. cerevisiae* system, the promoter region located at the upstream of the *PGI1* gene was replaced with the *ADH1* promoter with low transcription ability. Contrary to the expectation, attenuation of PGI activity did not influence xylitol production in both batch and fed-batch fermentations. However, simultaneous modulation of the *ZWF1* overexpression and *PGI1* attenuation not only increased specific xylitol productivity by a 1.9-fold but also reduced the amount of glucose consumed in a glucose-limited fed-batch cultivation (Oh et al., 2007).

Transhydrogenase (EC 1.6.1.1) transfers the reducing equivalent between NADP(H) and NAD(H) and controls the redox balance in bacterial cells. To investigate the effects of redox potential perturbation on xylitol production, a gene coding for a bacterial pyridine nucleotide transhydrogenase from *Azotobacter vinelandii* was introduced into a xylitol-producing recombinant *S. cerevisiae* (Jeun et al., 2003). Even though the bacterial transhydrogenase was expressed actively in the recombinant *S. cerevisiae*, its expression did not affect xylose consumption and xylitol production. The reason was suggested that the bacterial transhydrogenase expressed in the recombinant *S. cerevisiae* cells is favorable for transfer of a proton in NADPH to NAD^+, which reduces the availability of NADPH for xylose conversion to xylitol (Jeun et al., 2003).

Expression of acetaldehyde dehydrogenase 6 and acetyl-CoA synthetase was also tried to increase the NAD(P)H pools. Acetaldehyde dehydrogenase 6 converts acetaldehyde to acetate and produces NADPH concomitantly. Acetyl-CoA synthetase makes acetyl-CoA with coenzyme A, ATP, and acetate produced by acetaldehyde dehydrogenase 6. A recombinant *S. cerevisiae* strain overexpressing acetyl-CoA synthetase produced an 11% higher concentration of xylitol than the recombinant strain overexpressing acetaldehyde dehydrogenase 6. Although acetyl-CoA synthetase does not produce NADPH directly, overexpression of acetyl-CoA synthetase reduced by-product (ethanol and acetate) formation more than acetaldehyde dehydrogenase 6, which might be beneficial for high xylitol yield. Thus, acetyl-CoA synthetase was more effective than acetaldehyde dehydrogenase 6 in xylitol production (Oh et al., 2012).

18.4.3 Other Recombinant Microorganisms for Xylitol Production

Candida tropicalis is one of the high-xylitol-producing microorganisms and uses xylose as a carbon source for both xylitol production and energy metabolism. To increase the conversion of xylose to xylitol, two copies of the *XYL2* gene coding for xylitol dehydrogenase (XDH) in the diploid yeast *C. tropicalis* were destroyed by the *URA*-blasting method (Ko et al., 2006). The mutant could not grow in a mineral medium with a sole carbon source of xylose because of the absence of the *XYL2* gene. Using glycerol as a co-substrate for cell growth and NADPH regeneration, the *C. tropicalis XYL2* mutant produced xylitol with a volumetric productivity of 3.23 g/L·h and a xylitol yield of 98% (Ko et al., 2006). Expression of xylose reductase in *C. tropicalis* is significantly repressed in the presence of glucose (Ikeuchi et al., 2000), so glucose addition should be controlled tightly. To release this repression, the codon-optimized xylose reductase (XR) gene from *Neurospora crassa* was expressed under the control of the constitutive glycerol 3-phosphate dehydrogenase promoter from *C. tropicalis* in the *XYL2*-deficient *C. tropicalis* mutant (Ko et al., 2006). Batch fermentation of this recombinant *C. tropicalis* system using xylose and glucose resulted in about 70% higher xylose production than the control strain without constitutive expression of the *N. crassa* XR gene (Ko et al., 2006). To increase the NADPH level, two NADPH-synthesizing enzymes in the pentose phosphate pathway, glucose 6-phosphate dehydrogenase and 6-phosphogluconate dehydrogenase, were overexpressed in recombinant *C. tropicalis* (Ahmad et al., 2012). Similar to the recombinant *S. cerevisiae* expressing the *ZWF1* gene (Kwon et al., 2006), xylitol productivity increased by 21%, relative to the control strain without overexpression of the two genes (Ko et al., 2006; Ahmad et al., 2012).

Pichia stipitis (newly named as *Scheffersomyces stipitis*) is a xylose-assimilating budding yeast and has became a model microorganism for cellulosic bioethanol production. To produce xylitol from xylose, the *XYL3* gene coding for xylulokinase, the third enzyme in the yeast xylose metabolism (Figure 18.2), was disrupted in *P. stipitis*. In oxygen-limited conditions, a few amount of xylose was used for cell metabolism and xylitol production. In aerobic conditions, the *P. stipitis* FPL-YS30 mutant deficient in the *XYL3* gene consumed 100 g/L pure xylose completely to produce 26 g/L xylitol (Jin et al., 2005) and was also able to produce 13 g/L xylitol from 21 g/L xylose in corn stover hydrolysates treated with vapor-phase diethyl oxalate (Rodrigues et al., 2011). The *P. stipitis* mutant PXM-4 without xylitol dehydrogenase (XDH) activity was screened by chemical mutagenesis using ethyl methanesulfonate (Kim et al., 2000). Gluconate was chosen as a co-substrate because it exerts neither inhibition of xylose transport nor repression of xylose reductase expression in *P. stipitis* (Kim et al., 2011). A two-stage fed-batch fermentation was designed to consist of the cell growth stage using glucose and the xylitol production stage from xylose and gluconate. Xylitol concentration reached 44.8 g/L in 4 days of fermentation with 100% xylitol yield and 0.42 g/L·h overall xylitol productivity (Kim et al., 2011).

Escherichia coli is a popular microorganism for industrial production of chemicals and proteins. In the xylose metabolism, *E. coli* possesses xylose isomerase and xylulokinase encoded by the *xylA* and *xylB* genes, respectively. Xylose isomerase converts xylose to xylulose. Xylulokinase phosphorylates xylulose to xylulose 5-phoshphate, which is metabolized further in the pentose phosphate pathway. *Escherichia coli* is also engineered to express a xylose reductase gene for xylitol production. NADPH-dependent xylose reductases from *C. boidinii*, *C. tenuis*, *P. stipitis*, and *S. cerevisiae* were expressed functionally in recombinant *E. coli* independent on cyclic AMP and deficient in the xylulokinase gene (*xylB*) (Cirino et al., 2006). Recombinant *E. coli* expressing the *C. boidinii* xylose reductase gene showed the highest xylitol production ability in a batch culture with xylose and glucose. In a fed-batch fermentation by controlling a glucose concentration in culture broth, 38 g/L xylitol was produced from 43.5 g/L xylose and 27 g/L glucose (Cirino et al., 2006). Considering the xylose metabolism in *E. coli*, the *xylA* and *xylB* genes are two candidates of gene deletion for xylitol production. Individual disruption of *xylA* and *xylB* showed that the *xylB* deletion yielded more than two times higher xylitol concentration than the *xylA* deletion. The reason was suggested that an unusual compound of xylitol-phosphate produced by *E. coli* xylulokinase encoded by *xylB* might exert a toxic effect on the *E. coli* cell metabolism (Akinterinwa and Cirino, 2009). The replacement of *xylB* with the *P. stipitis XYL3* gene coding for xylulokinase relieved the toxic effect of xylitol-phosphate and hence resulted in an improvement of xylitol concentration by recombinant *E. coli* strains co-expressing xylose reductase. L-Arabinose is a five carbon sugar present in hemicellulose hydrolysates. It is converted to L-arabitol during the xylose reductase reaction. A new pathway from L-arabinose to xylitol was constructed by expression of L-arabinose isomerase (converting L-arabinose to L-ribulose), D-psicose 3-epimerase (converting L-ribulose to L-xylulose), and L-xylulose reductase (converting L-xylulose to xylitol) in *E. coli* (Sakakibara et al., 2009). In a fermentation using L-arabinose and glycerol co-substrate, this recombinant *E. coli* strain produced 14.5 g/L xylitol with a yield of 95% in 30 hours.

Other GRAS bacteria such as *Corynebacterium glutamicum* and *Lactococcus lactis* were genetically engineered for xylitol production. *C. glutamicum* is generally used for commercial production of amino acids. Because it does not have the xylose isomerase gene, the *P. stipitis XYL1* gene was cloned into an IPTG-inducible expression vector pEKEx2 and expressed in *C. glutamicum*. The recombinant *C. glutamicum* strain expressing *XYL1* produced 34 g/L xylitol with over 97% yield in a bioconversion process controlling a low level of glucose (Kim et al., 2010). *Lactococcus lactis* is a widely used lactic acid bacterium in dairy and food industries. Two genes coding for xylose reductase from *P. stipitis* and xylose-H^+symporterXylT from *Lactobacillus brevis* were cloned into a *L. lactis* expression vector pNZ8032, of which expression was under the control of the nicin-inducible (*nisA*) promoter (Nyyssölä et al., 2005). A biotransformation process of the nongrowing recombinant *L. lactis* strain expressing the two genes was designed to use 160 g/L xylose under microaerobic and glucose-limited conditions. Finally, 74 g/L xylitol with a yield of almost 1.0 mol/mol was obtained in 40 hours of process.

18.5 ECONOMIC OUTLOOK

Xylitol is a specialty chemical with the effects of extreme cooling, caries prevention and safety for diabetes, which are beneficial for manufacturing foods and mouth-care products. As a platform chemical, it also can be converted to various compounds for polymer synthesis and directly polymerized for production of unsaturated polyester resin (Werpy and Petersen, 2004). Danisco (currently part of Dupont Co.) is the largest xylitol producer, and many companies in China and South America (Brazil) manufacture xylitol commercially. The global market of xylitol is estimated to be more than 125,000 tons per year, with $4.5–5.5/kg for bulk purchase and $20/kg in supermarket (da Silva and Chandel, 2012). Xylitol has a 12% share of the total polyols market with growth estimated to increase threefold (da Silva and Chandel, 2012). The xylitol market continues to increase worldwide due to an enhancement in health consciousness and chewing gum market, and new application development as a chemical feedstock.

18.6 FUTURE PROSPECTS

For biological production of xylitol, many microbial strains have been developed by new strains screening, chemical and physical mutation, and genetic and metabolic engineering. Wild-type strains or mutants of *Candida* species can consume xylose fast and produce high concentrations of xylitol with around 83% (g/g) yield (equivalent to 0.82 mol/mol). Genetically engineered microbial strains can convert xylose to xylitol efficiently with the theoretical yield (~100%). Depending on the price and purity of xylose, we can choose the wild-type or genetically engineered microbial strains. Hemicellulose, a main source of xylose, can be separated from cellulosic biomass by simple acid treatment. As the solubilized hemicellulose consists of xylose, arabinose, mannose, and galactose, various compositions lead to a low yield of xylitol based on total sugar and production of impurities such as arabitol, mannitol, and galactitol. To solve the yield and impurity problems, recovery processes to make high pure xylose or remove the impurities from crude xylitol solution should be developed. Robust microbial strains able to convert xylose to xylitol and consume other sugars for viability maintenance are also applicable for utilization of crude xylose solution. Currently, xylitol is commercially manufactured by chemical reactions because of its cost-competitiveness. Since microbial processes have many benefits to meet the sustainability issue, these processes will be dominant in future.

ACKNOWLEDGMENTS

This work was financially supported by the Advanced Biomass R&D Center (ABC) of Korea Grant (2011-0031359) and the National Research Foundation of Korea (NRF) Grant (NRF-2013R1A2A2A01004806) funded by the Korean Government (MSIP).

NOMENCLATURE

GRAS	generally recognized as safe
XR	xylose reductase
XDH	xylitol dehydrogenase
XK	xylulokinase
G6PDH	glucose-6-phosphate dehydrogenase
PGI	phosphoglucose isomerase
*XYL*1	XR gene of *Pichia stipitis*
*XYL*2	XDH gene of *Candida tropicalis*
*XYL*3	XK gene of *Pichia stipitis*
*ZWF*1	G6PDH gene of *Saccharomyces cerevisiae*
ADH1	alcohol dehydrogenase gene of *Saccharomyces cerevisiae*
xylA	xylose isomerase gene of *Escherichia coli*
xylB	XK gene of *Escherichia coli*

REFERENCES

Ahmad I, Shim W, Jeon W, Yoon B, Kim JH. (2012). Enhancement of xylitol production in *Candida tropicalis* by co-expression of two genes involved in pentose phosphate pathway. *Bioprocess Biosys Eng*, 35, 199–204.

Akinterinwa O, Cirino PC. (2009). Heterologous expression of D-xylulokinase from *Pichia stipitis* enables high levels of xylitol production by engineered *Escherichia coli* growing on xylose. *Metab Eng*, 11, 48–55.

Bae SM, Park YC, Lee TH, Kweon DH, Choi JH, Kim SK, Ryu YW, Seo JH. (2004). Production of xylitol by recombinant *Saccharomyces cerevisiae* containing xylose reductase gene in repeated fed-batch and cell-recycle fermentations. *Enzyme Microb Tech*, 35, 545–549.

Bailey JE, Ollis DF. (1986). *Biochemical Engineering Fundamentals*, 2nd ed. New York: McGraw-Hill, Inc.

Choi JH, Moon KH, Ryu YW, Seo JH. (2000). Production of xylitol in cell recycle fermentations of *Candida tropicalis*. *Biotechnol Lett*, 22, 1625–1628.

Chung YS, Kim MD, Lee WJ, Ryu YW, Kim JH, Seo JH. (2002). Stable expression of xylose reductase gene enhances xylitol production in recombinant *Saccharomyces cerevisiae*. *Enzyme Microb Tech*, 30, 809–816.

Cirino PC, Chin JW, Ingram LO. (2006). Engineering *Escherichia coli* for xylitol production from glucose-xylose mixtures. *Biotechnol Bioeng*, 95, 1167–1176.

da Silva SS, Chandel AK. (2012). *D-Xylitol: Fermentative Production, Application and Commercialization*. Heidelberg: Springer.

Furlan SA, Bouilloud P, de Castro HF. (1994). Influence of oxygen on ethanol and xylitol production by xylose fermenting yeasts. *Process Biochem*, 29, 657–662.

Gírio FM, Roseiro JC, Sá-Machado P, Duarte-Reis AR, Amaral-Collaço MT. (1994). Effect of oxygen transfer rate on levels of key enzymes of xylose metabolism in *Debaryomyces hansenii*. *Enzyme Microb Tech*, 16, 1074–1078.

Granström T, Izumori K, Leisola M. (2007). A rare sugar xylitol. Part II: biotechnological production and future applications of xylitol. *Appl Microbiol Biotechnol*, 74, 273–276.

Hallborn J, Walfridsson M, Airaksinen U, Ojamo H, Hahn-Hägerdal B, Penttilä M, Keräsnen S. (1991). Xylitol production by recombinant *Saccharomyces cerevisiae*. *Nat Biotechnol*, 9, 1090–1095.

Ikeuchi T, Kiritani R, Azuma M, Ooshima H. (2000). Effect of D-glucose on induction of xylose reductase and xylitol dehydrogenase in *Candida tropicalis* in the presence of NaCl. *J Basic Microbiol*, 40, 167–175.

Jeun YS, Kim MD, Park YC, Lee TH, Yoo MS, Ryu YW, Seo JH. (2003). Expression of *Azotobacter vinelandii* soluble transhydrogenase perturbs xylose reductase-mediated conversion of xylose to xylitol by recombinant *Saccharomyces cerevisiae*. *J Mol Catal B-Enzym*, 26, 251–256.

Jin YS, Cruz J, Jeffries TW. (2005). Xylitol production by a *Pichia stipitis* D-xylulokinase mutant. *Appl Microbiol Biotechnol*, 68, 42–45.

Kim TB, Oh DK. (2003). Xylitol production by *Candida tropicalis* in a chemically defined medium. *Biotechnol Lett*, 25, 2085–2088.

Kim JH, Ryu YW, Seo JH. (1999). Analysis and optimization of a two-substrate fermentation for xylitol production using *Candida tropicalis*. *J Ind Microbiol Biotechnol*, 22, 181–186.

Kim MS, Kim C, Seo JH, Ryu YW. (2000). Enhancement of xylitol yield by xylitol dehydrogenase defective mutant of *Pichia stipitis*. *Korean J Biotechnol Bioeng*, 15, 113–119.

Kim MD, Rhee SK, Seo JH. (2001). Enhanced production of anticoagulant hirudin in recombinant *Saccharomyces cerevisiae* by chromosomal δ-integration. *J Biotechnol*, 85, 41–48.

Kim JH, Han KC, Koh YH, Ryu YW, Seo JH. (2002a). Optimization of fed-batch fermentation for xylitol production by *Candida tropicalis*. *J Ind Microbiol Biotechnol*, 29, 16–19.

Kim MD, Jeun YS, Kim SG, Ryu YW, Seo JH. (2002b). Comparison of xylitol production in recombinant *Saccharomyces cerevisiae* strains harboring *XYL1* gene of *Pichia stipitis* and *GRE3* gene of *S. cerevisiae*. *Enzyme Microb Tech*, 31, 862–866.

Kim TB, Lee YJ, Kim P, Kim CS, Oh DK. (2004). Increased xylitol production rate during long-term cell recycle fermentation of *Candida tropicalis*. *Biotechnol Lett*, 26, 623–627.

Kim SH, Yun JY, Kim SG, Seo JH, Park JB. (2010). Production of xylitol from D-xylose and glucose with recombinant *Corynebacterium glutamicum*. *Enzyme Microb Tech*, 46, 366–371.

Kim MS, Chung YS, Seo JH, Jo DH, Park YH, Ryu YW. (2011). High-yield production of xylitol from xylose by a xylitol dehydrogenase defective mutant of *Pichia stipitis*. *J Microbiol Biotechnol*, 11, 564–569.

Ko BS, Kim J, Kim JH. (2006). Production of xylitol from D-xylose by a xylitol dehydrogenase Gene-disrupted mutant of *Candida tropicalis*. *Appl Environ Microbiol*, 72, 4207–4213.

Kuriyama H, Kobayashi H. (1993). Effects of oxygen supply on yeast growth and metabolism in continuous fermentation. *J Ferment Bioeng*, 75, 364–367.

Kwon DH, Kim MD, Lee TH, Oh YJ, Ryu YW, Seo JH. (2006). Elevation of glucose 6-phosphate dehydrogenase activity increases xylitol production in recombinant *Saccharomyces cerevisiae*. *J Mol Catal B-Enzym*, 43, 86–89.

Laplace JM, Delgenes JP, Moletta R, Navarro JM. (1991). Alcoholic fermentation of glucose and xylose by *Pichia stipitis*, *Candida shehatae*, *Saccharomyces cerevisiae* and *Zymomonas mobilis*: oxygen requirement as a key factor. *Appl Microbiol Biotechnol*, 36, 158–162.

Lee WJ, Ryu YW, Seo JH. (2000). Characterization of two-substrate fermentation processes for xylitol production using recombinant *Saccharomyces cerevisiae* containing xylose reductase gene. *Process Biochem*, 35, 1199–1203.

Lee WJ, Kim MD, Ryu YW, Bisson L, Seo JH. (2002). Kinetic studies on glucose and xylose transport in *Saccharomyces cerevisiae*. *Appl Microbiol Biotechnol*, 60, 186–191.

Lee SH, Kodaki T, Park YC, Seo JH. (2012). Effects of NADH-preferring xylose reductase expression on ethanol production from xylose in xylose-metabolizing recombinant *Saccharomyces cerevisiae*. *J Biotechnol*, 158, 184–191.

Meyrial V, Delgenes JP, Moletta R, Navarro JM. (1991). Xylitol production from D-xylose by *Candida guillermondii*: fermentation behaviour. *Biotechnol Lett*, 13, 281–286.

Nolleau V, Preziosi-Belloy L, Navarro JM. (1995). The reduction of xylose to xylitol by *Candida guilliermondii* and *Candida parapsilosis*: incidence of oxygen and pH. *Biotechnol Lett*, 17, 417–422.

Nyyssölä A, Pihlajaniemi A, Palva A, von Weymarn N, Leisola M. (2005). Production of xylitol from d-xylose by recombinant *Lactococcus lactis*. *J Biotechnol*, 118, 55–66.

Oh DK, Kim SY, Kim JH. (1998). Increase of xylitol production rate by controlling redox potential in *Candida parapsilosis*. *Biotechnol Bioeng*, 58, 440–444.

Oh YJ, Lee TH, Lee SH, Oh EJ, Ryu YW, Kim MD, Seo JH. (2007). Dual modulation of glucose 6-phosphate metabolism to increase NADPH-dependent xylitol production in recombinant *Saccharomyces cerevisiae*. *J Mol Catal B-Enzym*, 47, 37–42.

Oh EJ, Bae YH, Kim KH, Park YC, Seo JH. (2012). Effects of overexpression of acetaldehyde dehydrogenase 6 and acetyl-CoA synthetase 1 on xylitol production in recombinant *Saccharomyces cerevisiae*. *Biocatal Agric Biotech*, 1, 15–19.

Rizzi M, Erlemann P, Bui-Thanh N-A, Dellweg H. (1988). Xylose fermentation by yeasts. *Appl Microbiol Biotechnol*, 29, 148–154.

Rodrigues R, Kenealy W, Jeffries T. (2011). Xylitol production from DEO hydrolysate of corn stover by *Pichia stipitis* YS-30. *J Ind Microbiol Biotechnol*, 38, 1649–1655.

Romanos MA, Scorer CA, Clare JJ. (1992). Foreign gene expression in yeast: a review. *Yeast*, 8, 423–488.

Sakakibara Y, Saha BC, Taylor P. (2009). Microbial production of xylitol from L-arabinose by metabolically engineered *Escherichia coli*. *J Biosci Bioeng*, 107, 506–511.

Schiweck H, Ziesenitz SC. 1996. Physiological properties of polyols in comparison with easily metabolisable saccharides. In: Grenby TH, editor. *Advances in Sweeteners*. Glasgow, UK: Blackie Academic & Professional, pp. 56–84.

Toivari MH, Ruohonen L, Miasnikov AN, Richard P, Penttilä M. (2007). Metabolic engineering of *Saccharomyces cerevisiae* for conversion of D-glucose to xylitol and other five-carbon sugars and sugar alcohols. *Appl Environ Microbiol*, 73, 5471–5476.

Vandeska E, Kuzmanova S, Jeffries TW. (1995). Xylitol formation and key enzyme activities in *Candida boidinii* under different oxygen transfer rates. *J Ferment Bioeng*, 80, 513–516.

Werpy T, Petersen G. 2004. Top Value Added Chemicals from Biomass: Volume 1—Results of Screening for Potential Candidates from Sugars and Synthesis Gas. Oak Ridge: US Department of Energy.

Winkelhausen E, Kuzmanova S. (1998). Microbial conversion of D-xylose to xylitol. *J Ferment Bioeng*, 86, 1–14.

Yahashi Y, Hatsu M, Horitsu H, Kawai K, Suzuki T, Takamizawa K. (1996). D-glucose feeding for improvement of xylitol productivity from D-xylose using *Candida tropicalis* immobilized on a non-woven fabric. *Biotechnol Lett*, 18, 1395–1400.

CHAPTER 19

First and Second Generation Production of Bio-Adipic Acid

JOZEF BERNHARD JOHANN HENRI VAN DUUREN

Biochemical Engineering Institute, Technische Universität Braunschweig, Braunschweig, Germany
Systembiotechnologie, Universität des Saarlandes, Saarbrücken, Germany

CHRISTOPH WITTMANN

Systembiotechnologie, Universität des Saarlandes, Saarbrücken, Germany

Abstract

Adipic acid is an important platform chemical with a turnover of 2.5 million tons per year. It is mainly applied as an intermediate reactant for the production of nylon-6,6, a world-leading polymer in the textile and the automobile industry. Due to a relatively high energy demand and greenhouse gas emission, inherently linked to the traditional petrochemical production, great potential exists for more sustainable bio-based alternatives. In pioneering developments, sugars, fats, and alkanes are used as novel feedstock for the production of bio-adipic acid. Recent approaches that recruit small

Bioprocessing of Renewable Resources to Commodity Bioproducts, First Edition.
Edited by Virendra S. Bisaria and Akihiko Kondo.

aromatic compounds from lignin-based materials seem even more promising. Based on life cycle assessment, such aromatics from lignin reduce energy demand and carbon dioxide equivalent emission significantly, in comparison to the conventional petrochemical process. Lignin-rich streams are available in huge amounts as waste streams of low value that accumulate during processing of lignocellulosic biomass. Recent development of superior producing microorganisms and integrated bioprocesses are enabling technologies toward future industrial production of bio-adipic acid, which is highlighted in this chapter.

19.1 INTRODUCTION

Since the development of nylon-6,6, one of the most important industrial polymers worldwide, almost 100 years ago, adipic acid has been an important chemical, as it is the main precursor for this product. Still today, about 85% of the total market of about 2.5 million tons adipic acid per year is used for the production of nylon-6,6 in a copolymerization process with hexamethylenediamine. In addition, a range of other applications for adipic acid have evolved during the past decades involving polyurethane (5%) and adipate esters (4%) in the automotive and the textile industry as two important examples. Additionally, adipic acid is used in the food, pharmaceutical, and agrochemical market. From early on, adipic acid has been produced from petrochemicals. Today, the market leaders for the production of adipic acid are Invista (21%) and Ascend (13%) in the United States, Rodia (18%), as part of the Solvay group in Belgium, and BASF (9%) in Germany. The major production route involves nitric acid oxidation of cyclohexanone and cyclohexanol at 250°C and 55 bars, which are both obtained from naphtha-based benzene (Draths and Frost, 1993). The two conventional feedstock, benzene and nitric acid, are derived from oil and natural gas, respectively. Alternative petrochemical routes are the carbonylation of butadiene and the oxidation of cyclohexane with hydrogen peroxide. In the latter case, four moles hydrogen peroxide are consumed per mole adipic acid. The prices for butadiene and hydrogen peroxide are, however, high (Sato et al., 1998). Furthermore, petrochemical production for adipic acid requires an enormous amount of energy and causes significant greenhouse gas emissions (Vink et al., 2003). The cumulative energy demand (CED) for petrochemical adipic acid is as high as 100 GJ/ton, which includes raw material and process energy input (Patel, 2003). Moreover, the nitric acid oxidation directly yields nitrous oxide (N_2O) in addition to adipic acid (Shimizu et al., 2000). This is of major environmental concern, because N_2O represents a greenhouse gas, 310 times more severe as carbon dioxide (CO_2).

In order to become less dependent on oil and to reduce the ecological footprint of adipic acid production, the use of biomass as alternative feedstock is promising. However, such bio-based routes to a novel bio-based adipic acid (bio-adipic acid) have to compete with conventional processes with regard to the economic efficiency. In this regard, the low price of adipic acid about €1500/ton (ICIS Pricing) and the fully depreciated petrochemical plants are challenging for new bio-based processes, as this requests for high titers, yields, and productivities. Consequently, low-cost

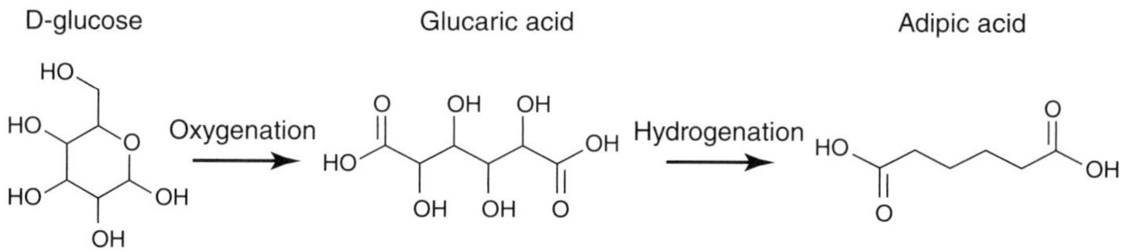

FIGURE 19.1 Rennovia process for production of bio-adipic acid from glucose via two chemical catalytic steps (Boussie et al., 2010; de Guzman, 2010; Dapsens et al., 2012).

feedstocks have to be used in an efficient and integrated way in order to enable good earnings and to reduce ecological footprints (Hatti-Kaul et al., 2007).

As demonstrated by several early studies, the achievable purity of bio-adipic acid meets the requirements for the fabrication of nylon-6,6 fibers, and resins (Schmidt and Knackmuss, 1984; Mizuno et al., 1988; Draths and Frost, 1993). Thus, future commercial bio-adipic acid might work as a drop-in product. This makes bio-based, chemical or biochemical alternatives particularly attractive: The product is compatible with existing processes downstream of the value chain to the industrial product, so that a large market can be addressed and growth is foreseen. First attempts have already demonstrated production of bio-adipic acid from glucose by a purely chemical process. Shortly, glucose as starting material is converted to bio-adipic acid by two chemical steps using metal catalysts (Boussie et al., 2010; de Guzman, 2010; Dapsens et al., 2012) (Figure 19.1).

First, glucose is oxygenated to the intermediate glucaric acid, which is subsequently hydrogenated to bio-adipic acid. Both conversion steps, however, require high temperature and pressure and the use of supporting solvents. In addition to chemical production, bio-adipic acid can be formed through occurring biosynthetic routes from benzene, found naturally in different microorganisms (Cheng et al., 2002). In particular, this is a good starting point toward *de novo* biosynthetic processes, as described below in more detail. Meanwhile, the use of metabolically engineered microorganisms even allows production of bio-adipic acid from glucose under nature-friendly conditions (Figure 19.2).

In recent years, corresponding biotechnological processes from renewable feedstocks have been developed. Today, both routes—chemical conversion and biochemical conversion of glucose—are about to be commercialized. Beyond glucose, second generation production is on the horizon which involves non-food raw materials such as fats, alkanes, pentose sugars, as well as small aromatics from lignin (Figure 19.3).

Particularly, lignin is an interesting resource, because it is a cheap and widely available starting material (Tuck et al., 2012). Moreover, the chemical structures of the major aromatics are, in contrast to other potential substrates, similar to the desired products, which allows straightforward bioconversion. Huge amounts of lignin accumulate as side-product from lignocellulosic bioethanol. Lignin waste displays up to 25% of the initial resource (de Wild et al., 2010). Until now, it is mainly burned for the generation of electricity. In recent years, thermochemical and lignolytic enzyme

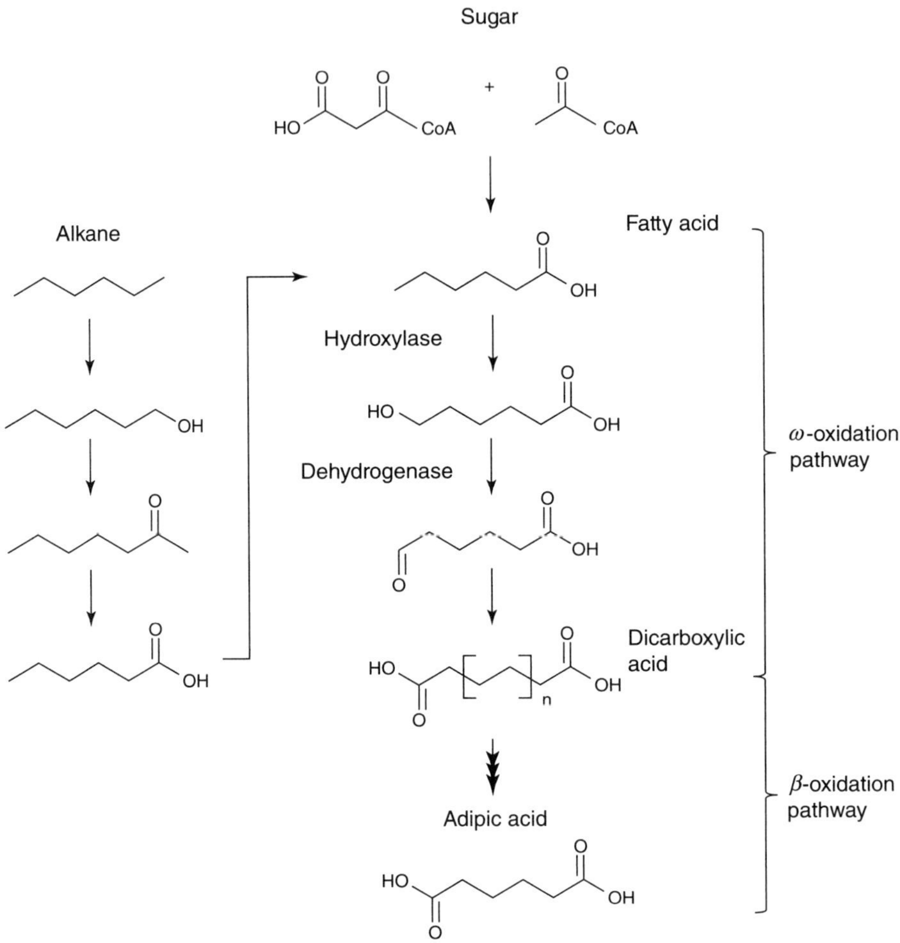

FIGURE 19.2 Verdezyne process for production of bio-adipic acid from sugars, fats, and alkanes with *Candia tropicalis* via the ω-oxidation pathway (Picataggio et al., 1992; Picataggio and Beardslee, 2012). When alkanes or fatty acids are used as feedstock, these typically long-chain substrates are depending on the number of carbons, repeatedly shortened via the β-oxidation pathway to obtain bio-adipic acid.

technologies have been developed to depolymerize lignin to valuable products such as small aromatic compounds (Bu et al., 2011; Ma et al., 2012). So far, it is difficult to directly produce commercial chemicals from lignin at sufficient purity. For a validation protocol the complex structure of lignin needs to be understood to predict the mechanisms of depolymerization. An additional complicating factor is the variety on aromatic chemical composition of different plant species. These problems can be overcome by application of fermentation, since microorganisms can use crude/impure compounds as substrate.

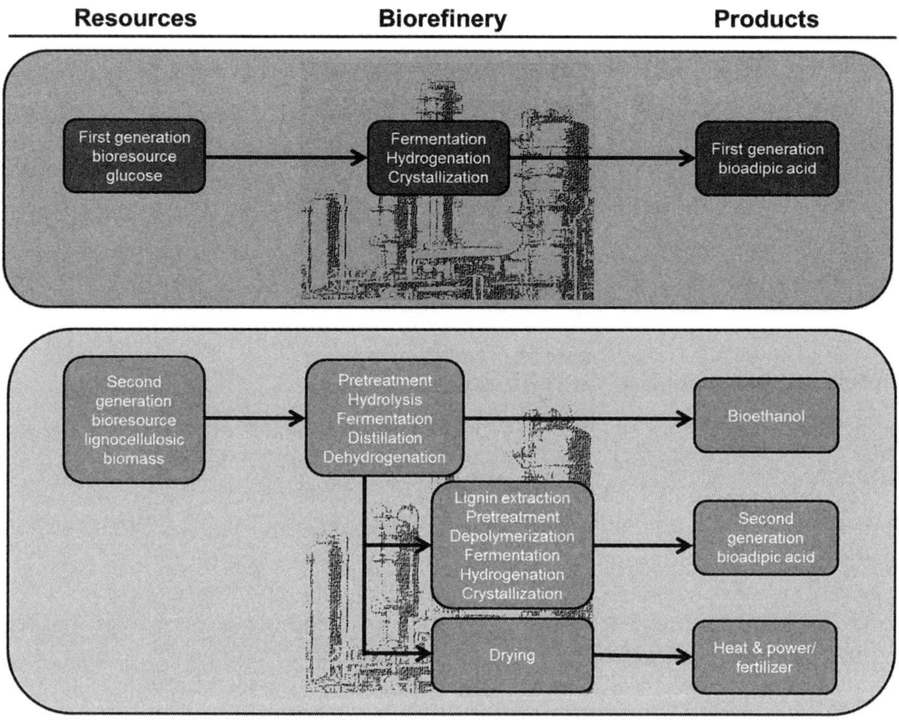

FIGURE 19.3 Biorefinery concepts that integrate the first and second generation bio-adipic acid. (*For a color version, see the color plate section.*)

19.2 PRODUCTION OF BIO-ADIPIC ACID

19.2.1 Natural Formation by Microorganisms

Bio-adipic acid is formed by different bacteria (Cheng et al., 2002). As example, it results from naturally occurring degradation of cyclohexane, cyclohexanol, and cyclohexanone (Figure 19.4).

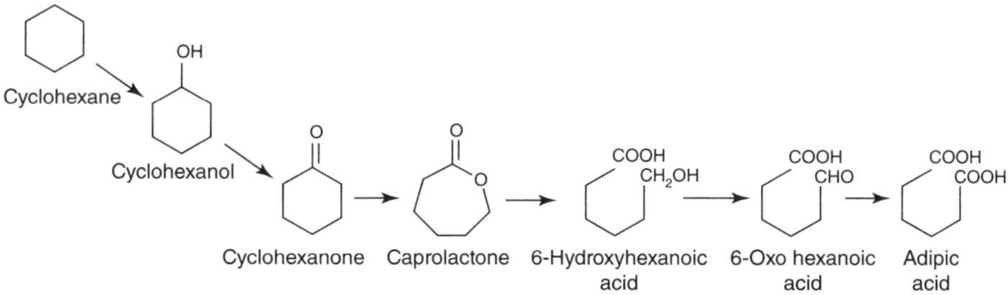

FIGURE 19.4 Natural degradation of cyclohexane to adipic acid by microorganisms (Cheng et al., 2002).

Relevant metabolic genes and enzymes from different species have been analyzed and tested for activity in recent years (Polen et al., 2012). The corresponding pathways from *Acinetobacter* sp., *Arthrobacter* sp., or *Rodococcus* sp. have been found functional in *Escherichia coli* as heterologous host (Cheng et al., 2000; Brzostowicz et al., 2003). These mutants indeed accumulated bio-adipic acid from cyclohexanol, although at low amount and rate, far below a commercially attractive level. As recently reviewed, no natural metabolic pathway, discovered so far, allows efficient biotechnological production (Polen et al., 2012) so that new concepts from systems metabolic engineering and synthetic pathway engineering are needed for sustainable production of bio-adipic acid (Becker and Wittmann, 2012; Wittmann and Lee, 2012).

19.2.2 First Generation Bio-Adipic Acid

19.2.2.1 Biotechnological De-Novo Biosynthesis From Glucose Glucose, the major traditional raw material in industrial biotechnology, can be used as renewable feedstock for bio-adipic acid. However, the structure of the sugar substrate is different from that of the product, so that the biotechnological conversion is complex, involves a cascade of enzymatic steps, and requires efficient cofactor regeneration. Particularly, *Candida tropicalis* has been engineered into an excellent whole-cell biocatalyst for direct production of bio-adipic acid from glucose (Picataggio et al., 1992; Picataggio and Beardslee, 2012) (Figure 19.2). The yeast is successfully applied in different biotechnological processes and able to grow at low pH. Such a characteristic is very important for the development of a continuous process. *cis,cis*-Muconic acid precipitates at a pH below its second pKa value of 3.03, which enables cell and medium recycling (Leong, 2007). The genome of *C. tropicalis* has been sequenced and largely annotated, which provides a rich source of knowledge for its targeted genetic modification (Butler et al., 2009). Important improvements of the yeast's performance have been achieved through metabolic engineering of its fatty acid metabolism. This involved stepwise reduction of the activity of the β-oxidation pathway by subsequent deletion of the four genes that encode for both isozymes of acetyl-CoA oxidase, catalyzing the first reaction of the β-oxidation pathway. As consequence, glucose is converted into bio-adipic acid via acetyl-CoA through fatty acid synthesis and the ω-oxidation pathway. This reaction sequence forms fatty acids as intermediates from glucose, which are then further processed into the final product. Hereby, a hydroxyl group is introduced at the ω-carbon of the fatty acid and then oxidized into a carboxyl group. The hydroxyl group is catalyzed by two successive oxidation steps, which involve alcohol dehydrogenase and aldehyde dehydrogenase. Among other modifications, debottlenecking of ω-hydroxylation by amplifying the genes encoding for cytochrome P450 monooxygenase and NADPH-cytochrome reductase was found crucial for efficient production. An attractive molar yield of about 40% from glucose has been claimed in certain embodiments (Picataggio and Beardslee, 2012). A production process has been developed on basis of these designer mutants by the company Verdezyne. Hereby, the use of fatty acids as pathway intermediates opens the possibility to widen the substrate spectrum of this process toward fats, vegetable oils, and alkanes, as non-food raw materials. Depending on the number of carbons in these typically long-chain substrates, this requires the β-oxidation pathway, of a partially β-oxidation blocked *C. tropicalis* strain, to repeatedly shorten the carbon backbone and to accumulate bio-adipic acid. At present, this route competes with the chemical process from glucose, operated by Rennovia (Figure 19.1) (Boussie et al., 2010; de Guzman, 2010; Dapsens et al., 2012). The chemical alternative from glucose claims a molar yield of 66% for the first step to glucaric acid and of 89% for the second step to bio-adipic acid (Boussie et al., 2010).

19.2.2.2 Combined Biotechnological–Chemical Synthesis From Glucose An interesting concept toward bio-adipic acids comprises a combined strategy that links initial biochemical conversion of glucose into the intermediate *cis,cis*-muconic acid with subsequent chemical reduction into bio-adipic acid. In the first step, *cis,cis*-muconic acid is formed under aerobic conditions at pH 6.3 and 36°C from glucose by heterologous *E. coli* (Draths and Frost, 1993; Niu et al., 2002). Given interesting process advantages, including lower pH values, a similar heterologous strain was developed with *Saccharomyces cerevisiae* (Curran et al., 2013; Weber et al., 2012). Acid resistance may also be possible for *E. coli*, *P. putida*, and *Bacillus subtilis* through heterologous gene expression and directed evolution (Guazzaroni et al., 2013). This novel route is based on a synthetic pathway that links aromatic amino acid biosynthesis with catalytic steps for aromatization of the intermediate 3-dehydroshikimic acid and its further conversion via protocatechuate and catechol into the final product (Figure 19.5).

The heterologous genes that were introduced into in *E. coli* and *S. cerevisiae* comprise 3-dehydroshikimic acid dehydratase, protocatechuic acid decarboxylase, and catechol 1,2-dioxygenase. In *E. coli*, the first two genes were expressed from *Klebsiella pneumoniae* strain A170-40 and the last gene from *Acinetobacter calcoaceticus*. For this strategy, a molar yield of 22% *cis,cis*-muconic acid from glucose was reached (Table 19.1). The intermediate accumulated to about 37 g/L and could be easily hydrogenated at a yield of 90% after addition of catalytic 10% platinum on carbon at a low elevated pressure of 3.5 bars within 3 hours. The three heterologous genes integrated into *S. cerevisiae* are genes from *Podospora anserine*, *Enterobacter cloacae*, and *Candida albicans*, respectively. Together with other modifications of the metabolism enabled the production of *cis,cis*-muconic acid with *S. cerevisiae* at a molar yield of 0.5% *cis,cis*-muconic acid and the intermediate accumulated to about 0.14 g/L. Another new artificial biosynthetic route has been described for *E. coli* that uses besides glucose also glycerol as substrate. Anthranilic acid, the first intermediate in the blocked tryptophan biosynthetic branch, was subsequently converted to catechol and *cis,cis*-muconic acid (Sun et al., 2013). Genes for anthranilate 1,2-dioxygenase and catechol 1,2-dioxygenase were coexpressed from *P. aeruginosa* and *P. putida* KT2440, respectively. A maximum titer of 0.39 g/L of *cis,cis*-muconic acid and a molar yield of 2.2% was described for flask experiments. In addition, an alternative strategy via glucaric acid as intermediate has been proposed, which is based on three heterologous expressed genes in *E. coli* encoding myo-inositol-1-phosphate synthase, myo-inositol oxygenase, and uronate dehydrogenase from *S. cerevisiae*, mice and *Pseudomonas syringae*, respectively (Moon et al., 2009). At present, the

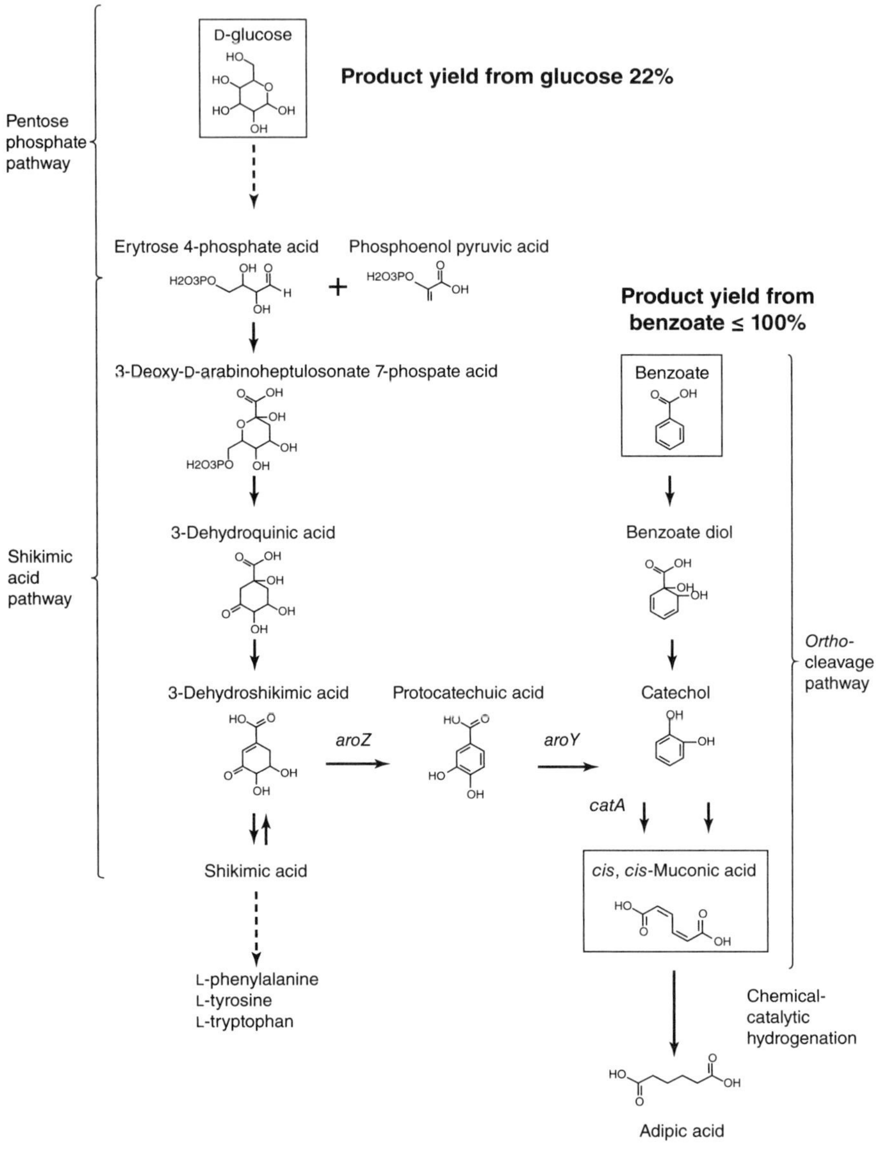

FIGURE 19.5 Combined biotechnological–chemical synthesis of bio-adipic acid from glucose. Via heterologous expressed genes in *Escherichia coli* and *Saccharomyces cerevisiae* glucose was converted to *cis,cis*-muconic acid at a molar yield of 22% and 0.5%, respectively, via heterologous expression of genes [*aroZ* (3-dehydroshikimic acid dehydratase), *aroY* (protocatechuic acid decarboxylase), and *catA* (catechol 1,2-dioxygenase)] (Draths and Frost, 1993; Niu et al., 2002, Curran *et al.*, 2013). *Arthrobacter* sp. and *Pseudomonas* sp. are able to convert benzoate to *cis,cis*-muconic acid at a molar yield close to 100% (Schmidt and Knackmuss, 1984; Mizuno et al., 1988; Chua and Hsieh, 1990; Bang and Choi, 1995; Choi et al., 1997; van Duuren et al., 2012). Subsequently, *cis,cis*-muconic acid is converted to bio-adipic acid by chemical catalytic hydrogenation.

TABLE 19.1 Biotechnological Production of *Cis*,*Cis*-Muconic Acid

Process	Strain	Substrate	Maximum specific productivity (g/g-dcw·h)	Maximum volumetric productivity (g/L·h)	Final concentration (g/L)	Molar yield %
pH-stat fed-batch	*Pseudomonas putida* KT2440-JD1	Benzoate	0.60	0.8	18.5	≤100
Cell-recycle system[a]	*P. putida* BM014		0.14	5.5	12.0	73
DO-stat fed-batch	*P. putida* BM014		0.21	2.2	32.0	≤100
Fed-batch	*Arthrobacter* sp.		nd	1.1	44.0	96
Fed-batch	*Pseudomonas* sp. B13		0.24	0.8	7.0	91
Fed-batch	*Pseudomonas* sp.	Toluene	nd	nd	45.0	nd
Shake flask cultures	*E. coli* BW25113	ΔtrpD::kan Glucose and Glycerol	nd	3×10^{-3}	0.39	2.2
DO-stat fed-batch	*Escherichia coli* WN1/pWN2.248	Glucose	0.05	1.3	36.8	22
Shake flask culture	*Saccharomyces cerevisiae*		nd	8.3×10^{-3}	141×10^{-3}	0.4

The data given comprise maximum specific productivity, maximum volumetric productivity, final concentration, and molar yield and are taken from previous work (Schmidt and Knackmuss, 1984; Mizuno et al., 1988; Chua and Hsieh, 1990; Bang and Choi, 1995; Choi et al., 1997; Niu et al., 2002; van Duuren et al., 2012; Curran et al., 2013; Sun at al., 2013)

nd, not determined.

[a]Steady values.

production of bio-adipic acid via glucaric acid seems not efficient enough for a commercial process. An alternative synthetic pathway for the production of this compound is proposed via glucodialdose (Lippow et al., 2010).

19.2.3 Second Generation Bio-Adipic Acid

Beyond sugar-based production, non-food raw materials appear particularly promising for bio-adipic acid. A first attempt into this direction is the extension of the substrate spectrum for recombinant *C. tropicalis* to fats and oils (see 19.2.2.1). As mentioned, the chemical structure of these substrates differs substantially from that of the target product, which requires a comprehensive pathway set for production. In this regard low-value aromatic compounds from lignin seem interesting alternative feedstocks. Naturally, the β-ketoadipate pathway serves for the convergent conversion of such aromatic compounds derived from lignin and other plant materials (Harwood and Parales, 1996). Additionally to this pathway, phenyl acetic acid and homogentisic acid are central aromatic metabolites (Jiménez et al., 2002). The ability to convert these compounds can be used for the degradation of some aromatic pollutants. The β-ketoadipate pathway is widely distributed among the microbial world and is present, for example, in *Pseudomonas putida*, *Acinetobacter calcoaceticus*, *Agrobacterium tumefaciens*, and *Rodococcus erythropolis* (Harwood and Parales, 1996). These exhibit differences in gene organization and the enzymatic set. Depending on the nature of starting aromatic molecule, the pathway converges at catechol or at protocatechuic acid as central metabolite (Figure 19.6). As example, quinic acid, vanillic acid, 4-coumaric acid, ferulic acid, and coniferyl alcohol are converted to protocatechuic acid. Unsubstituted aromatic compounds such as naphthalene, benzene, anthranilic acid, toluene, and phenol are metabolized via the catechol branch. In the latter cases, this gives direct access to *cis,cis*-muconic acid, an intermediate directly downstream in the β-ketoadipate pathway (Figure 19.6). The aromatic compounds that are not converted via catechol can be used as feedstock for growth and maintenance.

This available biocatalytic potential has been exploited since many years by selecting classical mutants with beneficial phenotypes. Different microorganisms accumulate *cis,cis*-muconic acid from benzoate or toluene, as model substrates for small aromatic compounds, examples being *Pseudomonas* sp. and *Arthrobacter* sp. (Schmidt and Knackmuss, 1984; Mizuno et al., 1988; Chua and Hsieh, 1990; Bang and Choi, 1995; Choi et al., 1997; van Duuren et al., 2012) (Table 19.1). As explained above, the intermediate can be easily hydrogenated to yield bio-adipic acid. Overall, surprisingly high titers for *cis,cis*-muconic acid of more than 40 g/L have been reported. Hereby, *cis,cis*-muconic acid was accumulated at a molar yield close to 100%. Such a high yield is enabled by the fact that the conversion is independent from the central metabolism. For pure benzoate and toluene, the process was successfully scaled to 500 and 1500 L, respectively (Table 19.1). Sophisticated bioprocess development further improved the process performance. A precise process control is rather important to avoid the accumulation of catechol, because it is toxic for the cells (Schreiber et al., 1980; Schmidt and Knackmuss, 1984; Benndorf et al., 2001; Benndorf et al., 2006).

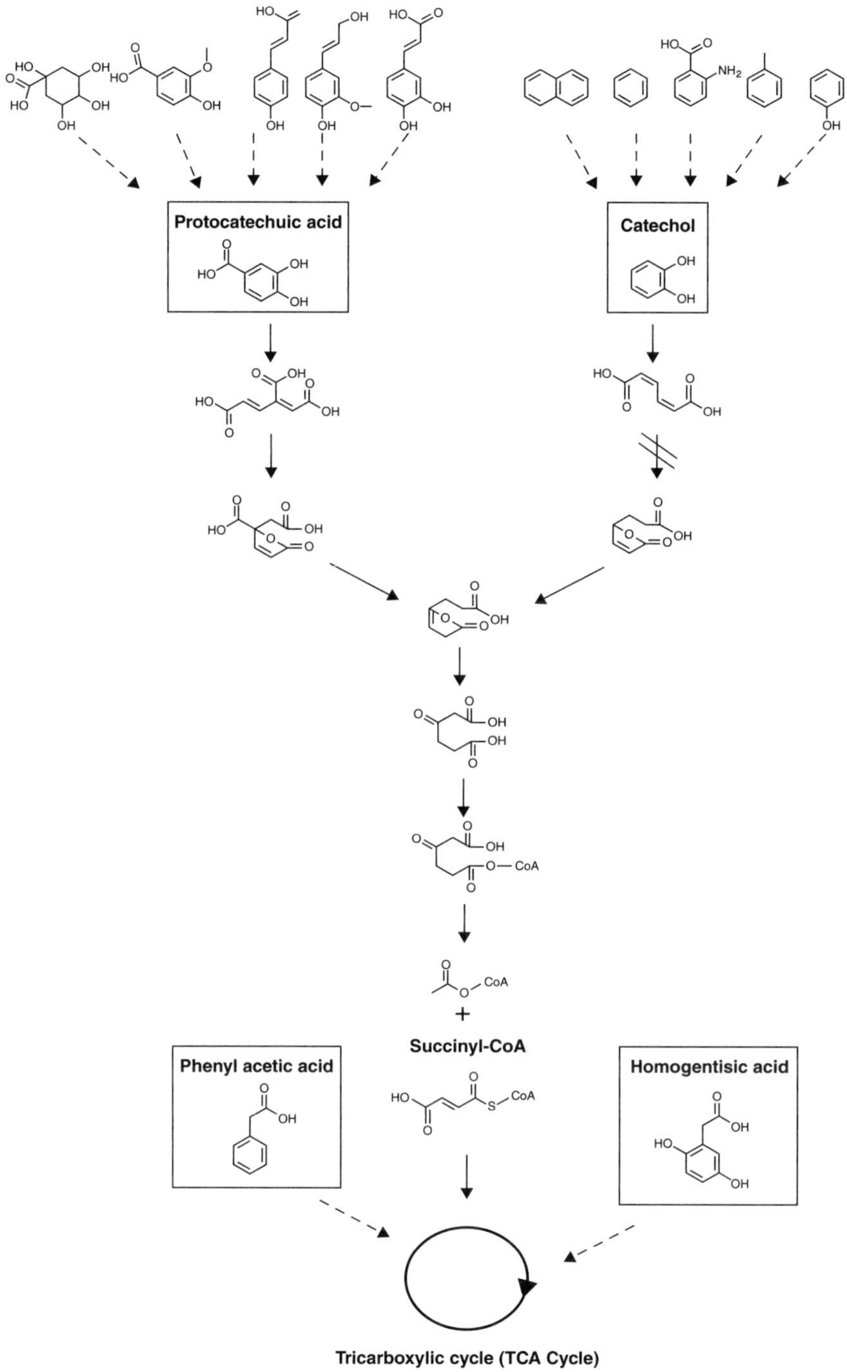

FIGURE 19.6 The two branches of the β-ketoadipate pathway, which comprise catechol and protocatechuic acid as central metabolites (Harwood and Parales, 1996). To obtain *cis,cis*-muconic acid its further conversion has to be blocked. Additionally, other small aromatic compounds are converted via different pathways of which phenyl acetic acid and homogentisic acid are central metabolites (Jiménez et al., 2002).

This is complicated by the fact that the conversion of catechol to *cis,cis*-muconic acid is considered rate-limiting (Kim et al., 1988).

For efficient accumulation of *cis,cis*-muconic acid, a pH-stat fed-batch and a dissolved oxygen (DO)-stat fed-batch approach were developed to avoid the accumulation of toxic intermediate metabolites (Bang and Choi, 1995; Niu et al., 2002; van Duuren et al., 2012). The pH strategy directly couples the microbial activity to the feed rate. When benzoate is converted to *cis,cis*-muconic acid, the medium acidifies in stoichiometric amounts. This enables to couple the feeding of benzoate directly to the formation of *cis,cis*-muconic acid via the pH control. Accordingly, the feed rate is automatically set optimal for production, even when the production is unexpectedly reduced. The pH-stat fed-batch process was found reliable, given the uncontrolled catechol accumulation, when glucose was fed on basis of the dissolved oxygen level. During the pH-stat fed-batch process, a specific production rate of 0.6 g/g·h was obtained, which is higher than values of other process (Table 19.1). For the future, rational metabolic engineering strategies, available in the meantime, promise a next level of producing strains. It might be possible to reach a final concentration of up to 80 g/L of *cis,cis*-muconic acid (van Duuren et al., 2012).

19.3 ECOLOGICAL FOOTPRINT OF BIO-ADIPIC ACID

To evaluate the ecological footprint of bio-adipic acid and to compare it to that of conventional petrochemistry, an extended life cycle assessment (LCA) was recently performed (van Duuren et al., 2011). Given that the kind of feedstock for production is determining the most costly portion of the ecological footprint, the assessment was taken further by additionally incorporating the glucose-based process for comparison (glucose 1 Figure 19.7). This focused on the investigation of the combined biotechnological–chemical process from glucose to *cis,cis*-muconic acid by recombinant *E. coli*, where sufficient data were available to support a valid analysis (Niu et al., 2002) (Figure 19.7). To increase the efficiency of the process five recycle times were considered. To evaluate the relevance of cell and medium recycling for the LCA, by conducting a fermentation at low pH, the energy demand and CO_2eq emission for medium preparation were divided five times (glucose 2 in Figure 19.7). To cover the increased biomass concentration the quantity of added glucose and ammonium sulphate remained unchanged per ton product as under fed-batch conditions. Low pH causes stress for microorganisms, which negative effect was assumed to be compensated during the process by high cell density. Furthermore, the impact on the downstream process was expected to be negligible. To recycle cells after micro-filtration an extra cross-flow filtration step is applied. Besides the extra effort, this technique is also expected to improve the conditions for hydrogenation and evaporation due to a higher final concentration and the use of solvents with a lower boiling point.

As basis for the assessment, the ISO14040 series for environmental management systems were followed to a great extent for the CED and the CO_2 equivalent (CO_2eq) emission (CO_2 and N_2O measured separately) (ISO 14040:2006). The assessment

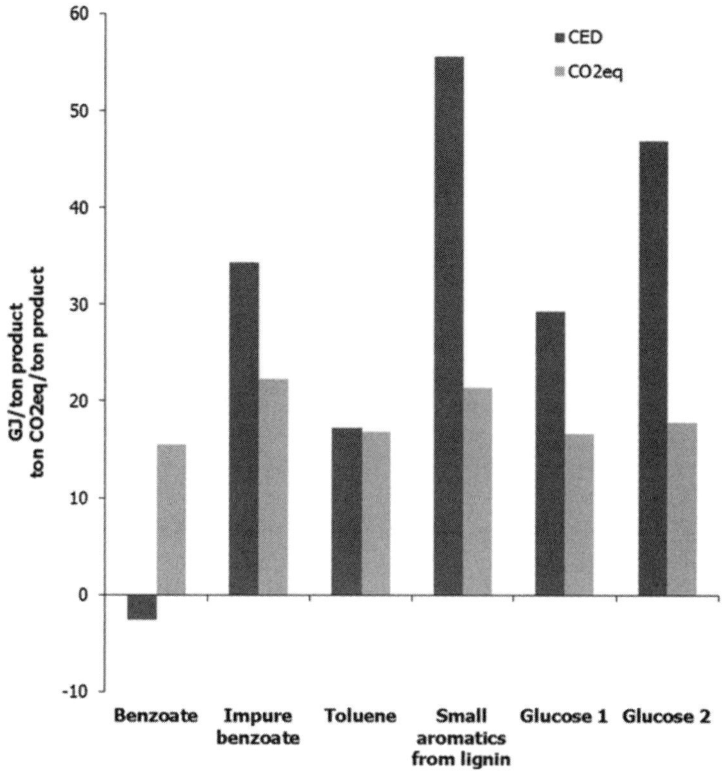

FIGURE 19.7 Reduction of the cumulative energy demand (CED) and CO_2 equivalent (CO_2eq) emission per ton bio-adipic acid for the integrated biotechnological process using benzoate (62.4 GJ/ton, 6.8 ton/ton), impure benzoate (25.4 GJ/ton, no CO_2eq emission), toluene (56.4 GJ/ton, 7.2 ton/ton), small aromatics from lignin (5.1 GJ/ton, 1.2 ton/ton) and glucose (5.1 GJ/ton, 1.0 ton/ton) as feedstock in relation to the petrochemical process (104.24 GJ/ton, 25.9 ton/ton). To compare the LCA outcome for aromatic compounds and glucose as feedstock, a final concentration of 37 g/L was assumed, given this concentration was reached by *Escherichia coli* with glucose as feedstock (Niu et al., 2002). The medium and the process were kept similar. Glucose 1 represents a process without cell and medium recycling, Glucose 2 represents a process in which the biomass and medium are recycled 5 times.

was started at the point of fossil fuel delivery as naphtha products and natural gas. Furthermore, the assessment was defined till the factory gate for nylon-6,6 production, which is essentially the delivery point of pure bio-adipic acid. For CO_2 emission relating to the formation of benzoate and toluene, the assessment was taken beyond the factory gate to the point of incineration. Because impure benzoate represents a waste product from the BTEX naphtha cracker, solely its intrinsic energy value was taken into account. Phenol, as model substrate, represents impure small aromatic compounds from lignin. For its formation derived from wheat, the theoretical energy demand was based on the associated agricultural system and the dry weight of biomass

(Brehmer, 2008; Brehmer et al., 2009). Via pyrolysis a molar yield of 40% small aromatics can be generated from lignocellulosic biomass and directly form lignin, which are converted biotechnologically to *cis,cis*-muconic acid (Bu et al., 2011; Ma et al., 2012). The energy required for this process step is provided by the combustion of pyrolysis gases. Lignin is considered as a waste product from the production of the second generation raw materials cellulose and hemicellulose, because it is not the primary product for which plants are grown and processed (Tuck et al., 2012). The physicochemical formation of lignin from lignocellulose was therefore not eligible for the LCA given these process steps are calculated for the primary product formation. The production of glucose was based on the saccharide content of whole sugar beets. A similar approach for glucose was selected, as was done for lignin to calculate the theoretical energy demand and greenhouse gas emission. However, additionally 2 GJ/ton was encountered for the process. Furthermore, a molar yield of 96% and 22% *cis,cis*-muconic acid was considered, when small aromatic compounds or glucose were used as feedstock, respectively (Bang and Choi, 1995; Choi et al., 1997; Schmidt and Knackmuss, 1984; Mizuno et al., 1988; Chua and Hsieh, 1990; Niu et al., 2002; van Duuren et al., 2012). For the chemical catalytic hydrogenation step of *cis,cis*-muconic acid to bio-adipic acid, a slightly increased pressure (3.5 bar) and a yield of 90% was taken into account. The CO_2 emission was simulated by multiplying the energy use with the emission factor of heating oil as defined by the National Greenhouse Accounts Factors of the Department of Climate Change and Energy Efficiency of the Australian Government. The N_2O emission related to the use of fertilizers was included in the LCA. Namely, by agricultural soil management and nitric acid formation for the production of the fertilizer NPK-CAN, N_2O is produced (Nevison, 2002; Brehmer et al., 2008; Neuffer et al., 2010). Possibly in the future, the negative effect of nitric acid production can be reduced by the application of *Rhizobium* symbiosis to non-*papilionaceous* plants for the fixation of nitrogen (Ivanov et al., 2012).

Without medium reuse, 27 L of water are required for 1 kg product at a titer of 37 g/L *cis,cis*-muconic acid. The environmental impact of two times autoclaving 1 ton of water equals 1.4 GJ and 95 kg CO_2eq, respectively. Only one the two autoclavation steps was incorporated in the LCA to compensate for the intrinsic energy value of biomass that can be further used in closed loop processes. For this calculation, clear water treatment and waste water treatment was also taken into account, but it encompasses only less than one percent of the calculated energy demand (Vairavamoorthy, 2011). Furthermore, pH regulation with HCl (4.2 GJ/ton, 0.2 ton/ton) was included at a feed level of 0.58 ton/ton. To offset the ammonium use, it is interesting to feed the protein-rich coproduct slurry of the glucose production during the fermentation step.

Bio-adipic acid from lignin, revealed a 53% reduction in the total energy demand as compared to classical adipic acid from the petrochemical route (Figure 19.7). Moreover, the greenhouse gas emission was reduced even by 83%. In addition, the glucose-based process exhibited a better ecological footprint than that from traditional petrochemistry, although it did not reach the excellent performance of the lignin-based process. However, when cell and medium recycling is applied with glucose as substrate the environmental impact is competitive.

Glucose as feedstock for production of *cis,cis*-muconic acid was added to model the present leading industrial business concepts. However, since the processes applied by Verdezyne and Rennovia have a different setup, they cannot directly be represented by this LCA. Furthermore, the ecological and commercial competitiveness of these two processes is mainly dependent on undefined parameters of the processes, for example, product yield, product rate, and final concentration. In view of the good properties of glucose as feedstock, the efficiency of the process is expected to be competitive compared to the conventional petrochemical process. To give an impression of how the yield and product concentration of *cis,cis*-muconic acid influence the environmental impact, a molar yield of 25–100% and a concentration of 25–100 g/L were taken into account in Figure 19.8. When glucose is used for production, as well as for growth and maintenance, the theoretical molar yield of 100% is unattainable. At the same yield and titer for glucose or impure small aromatic compounds as feedstock, the environmental impact has a fairly uniform development. The process of Verdezyne does not apply any chemical catalytic step, because it converts glucose, fats, and alkanes directly to bio-adipic acid (Picataggio et al., 1992; Picataggio and Beardslee, 2012) (Figure 19.2). This possibly results in significantly less energy demand for the process given the chemical catalytic hydrogenation step is the most energy demanding step. The process of Rennovia applies an extra chemical catalytic step concerning the oxidation of glucose to glucaric acid (Boussie et al., 2010; de Guzman, 2010; Dapsens et al., 2012) (Figure 19.1). Although chemical catalytic steps cost a lot of energy, they also might be more efficient compared to a fermentative conversion given the conditions can be optimally chosen. By cell and medium recycling the energy demand for the process was reduced up to 60% and 50% for the substrates glucose and lignin, respectively, and the CO_2eq emission was reduced by a maximum of 10% for both substrates at 25 g/L and 25% molar yield. At 100 g/L the decrease of energy demand was only 10% and CO_2eq emission remained the same for both substrates. The reduction in environmental impact increased exponentially with a decreasing titer. With a concentration of 25 g/L the decrease in energy demand was additionally negatively affected by an increasing molar yield in an exponential manner, leading to a reduction at 100% of only 40% GJ/ton for both substrates.

The great advantage of applying plant biomass as feedstock is that it is formed by photosynthesis and therefore low energy values for growth have to be taken into account. Even with a molar yield of 22% *cis,cis*-muconic acid from glucose, it is still possible to save energy as compared to the petrochemical production process of adipic acid (Figure 19.7). In comparison to second generation bioprocesses, first generation processes are in the majority, because organisms can easily use these carbon sources for growth and maintenance. By growing the species under limited conditions and/or by applying metabolic engineering, it becomes possible to accumulate metabolic intermediates including *cis,cis*-muconic acid (Polen et al., 2012). When the second generation feedstock lignin is used, the CED and CO_2eq emission are rather similar as for glucose (Figure 19.8). However, in particular because the conversion of small aromatics can lead to a high molar yield, the application of this second generation feedstock is very interesting. Therefore, based on this assessment, developed depolymerization techniques for lignin should be applied in industry as soon as possible to open the gate for the most sustainable bio-adipic acid.

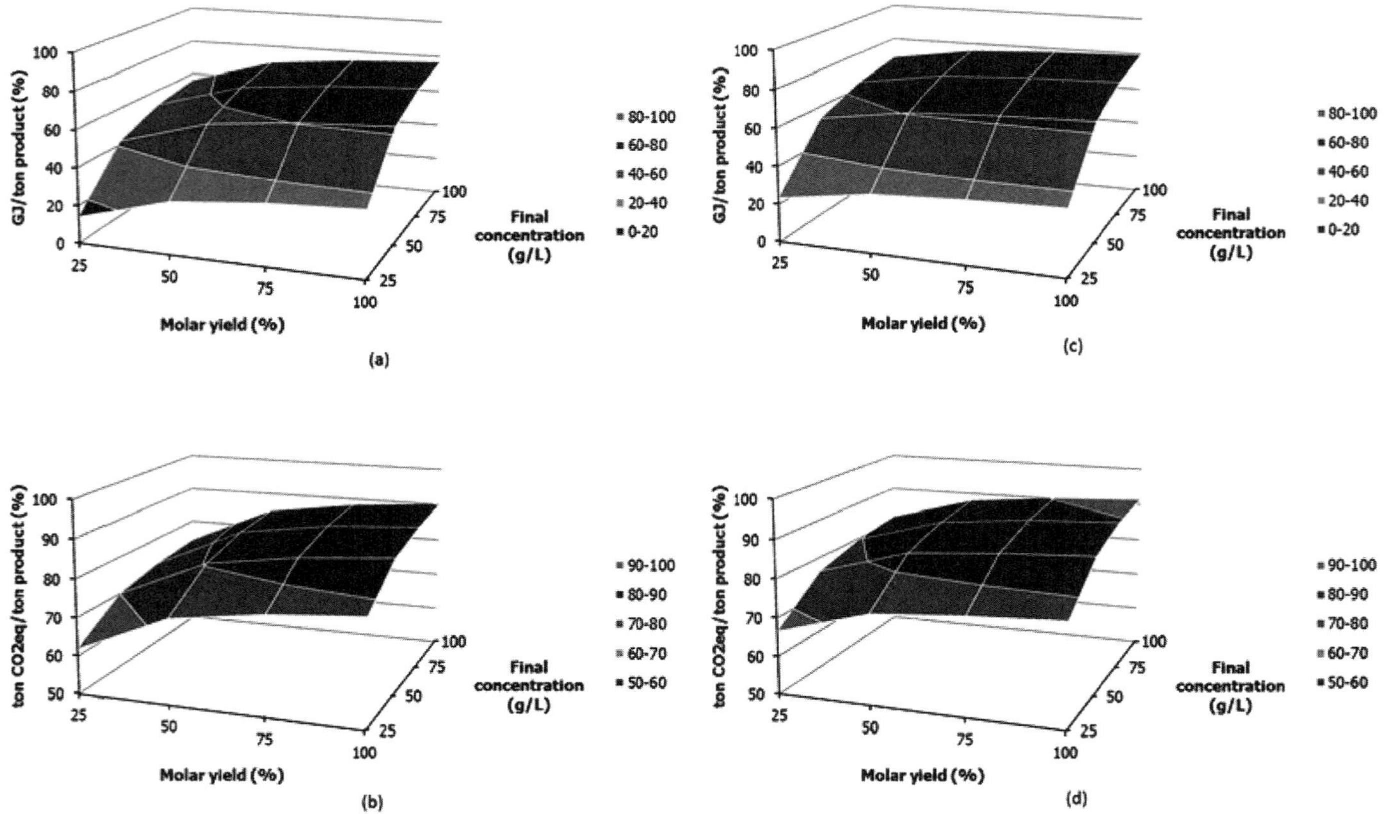

FIGURE 19.8 Percentage of cumulative energy demand (a, c) and CO_2 equivalent emission (b, d) savings per ton bio-adipic acid at a molar yield of 25–100% and a concentration of 25–100 g/L of *cis*,*cis*-muconic acid for the integrated biotechnological process using glucose (a, b) and small aromatics from lignin (c, d) as feedstock compared to the petrochemical process.

19.4 ECONOMIC OUTLOOK

For an economic comparison of the various feedstocks to be used in an integrated biotechnological process, the costs per ton bio-adipic acid can be calculated. Table 19.2 lists the variable costs for a process with a final concentration of 37 g/L *cis,cis*-muconic acid in the fermentation broth. The variable costs encounter feedstocks, process, energy, and greenhouse gas emission costs. To estimate the price of lignin from wheat, €4/GJ is applied and a yield of 40% is taken into account (Reith et al., 2009). The intrinsic energy level of various second generation biomass resources does differ between 9.5 and 17.5 GJ/ton as defined by the European Biomass Industry Association. For wheat in particular, 14.5 GJ/ton was considered. When glucose is used as feedstock, the volatile price and relative low molar yield are points that require attention. Additionally, this substrate competes with the food supply, which should be avoided given the price-raising effect. Process costs in Table 19.2 include medium (trace elements, phosphate, ammonium, HCl) (197 Euro/ton), glucose as growth and maintenance substrate and the use of water (106 Euro/ton) based on Alibaba products, Indexmundi, and the price of water supply and sewage (BMU, 2011), respectively. When five recycle times were incorporated with glucose as substrate the cost were for medium formation 29 Euro/ton and for water use 21 Euro/ton. The medium formation was particularly reduced in price since pH regulation with HCl is not necessary at low pH. When glucose is converted to *cis,cis*-muconic acid, the process cost is lower compared to when small aromatic compounds are used, since no additional carbon source for growth and maintenance must be added. The energy costs and greenhouse gas emission are related to each step of the process. Given feedstocks are purchased, it is assumed that all costs are covered concerning the production, including the CO_2 emission costs. For 1 GJ of energy, €19.57 was calculated based on the price for heating oil (37.3 MJ/L) indicated by Europe's Energy Portal. Furthermore, for 1 ton CO_2eq emission €7.00 was calculated, which price

TABLE 19.2 Variable Cost Price of the Production of Bio-Adipic Acid from Different Feedstocks Via an Integrated Biotechnological Process With *Cis,Cis*-Muconic Acid as Intermediate Metabolite at 37 g/L in the Fermentation Broth

			Variable cost price bio-adipic acid Euro/ton				
Substrate	Euro/ton	Molar yield %	Feedstock	Process	Energy	Ghg	Total
Benzoate	1188	0.86	1149	466	761	19	2394
Impure benzoate	nd						
Toluene	951	0.86	694	466	761	19	1939
Small aromatics	145	0.86	108	466	761	19	1353
Glucose 1	322	0.20	2005	303	761	19	3087
Glucose 2	322	0.20	2005	50	468	12	2535

Ghg, greenhouse gas; nd, not determined. Glucose 1 represents a process without cell and medium recycling, Glucose 2 represents a process in which the biomass and medium are recycled 5 times.

was determined at Boerse.de. These two costs cover the energy use and greenhouse gas emission during the process, respectively. The process from small aromatic compounds is competitive in price compared to glucose given the low economic value of lignin and the high molar yield of *cis,cis*-muconic acid. It is important to note that the variable costs are not equal to the selling price. Besides profit, the fixed costs associated with labor, investment, housing, cleaning, and transportation as well as other costs, for example, for licensing, sale, logistics, marketing, and R&D should also be encountered. These costs should be handled with care, as dependent on individual cases. Figure 19.9 displays the impact of changing yield and titer. A low molar yield causes an increased demand for feedstock, an indirect extra demand for energy, and an extra emission of greenhouse gas. These effects increase the costs, that is, more money is spent on feedstock, energy, and CO_2-certificates to cover greenhouse gas emission. In order to allow economically attractive production (variable costs below the selling price of 1500 Euro/ton), at least for glucose, a yield of +/− 60% has to be achieved. Taking into account the effect of five recycle times the variable costs with glucose were reduced so that a yield of 40% is required. With cell and biomass recycling for glucose as well as lignin the variable costs decreased from 258 to 773 Euro/ton. The reduction in price increased exponentially with a decreasing titer. Based on the calculation, the variable cost price with lignin is under the most relevant circumstances under €1500. Given the favorable conditions for lignin, the use of this substrate can lead to a very successful second generation process for the production of bio-adipic acid.

19.5 FUTURE PROSPECTS

Petrochemical production of adipic acid has a high, unfavorable ecological footprint (Vink et al., 2003). As a result, there is a great potential for sustainable bio-based alternatives. In recent years, this had led to substantial developments toward bio-adipic acid, which already have reached pilot scale operations by the companies Verdezyne and Rennovia. DSM, a large Dutch biotech company, has also indicated to add the compound to its portfolio. The future use of lignin-based raw materials promises an even higher potential. Raw materials for second-generation biofuels consist of 25% lignin, which displays a huge waste stream (de Wild et al., 2010; Tuck et al., 2012). To indicate the potential significance of this waste stream, 70% of oil is used for transport fuels. First second-generation lignocellulosic biorefineries at commercial scale (e.g. by ZeaChem, Süd-Chemie), recently been put into operation, cannot make efficient use of their lignin wastes. A possible use of this raw material for bio-adipic acid is therefore expected highly attractive. Additionally by providing an added value to this significant waste stream, the use of biomass becomes more profitable as a whole. This route might also supply products, other than bio-adipic acid. As first example, terephthalic acid is produced from *cis,cis*-muconic acid by Amyris in the United States. An extended use of *cis,cis*-muconic acid is currently limited by the rather high price, but markets could grow, when the price would be lowered. In particular, the second generation integrated-biotechnological process

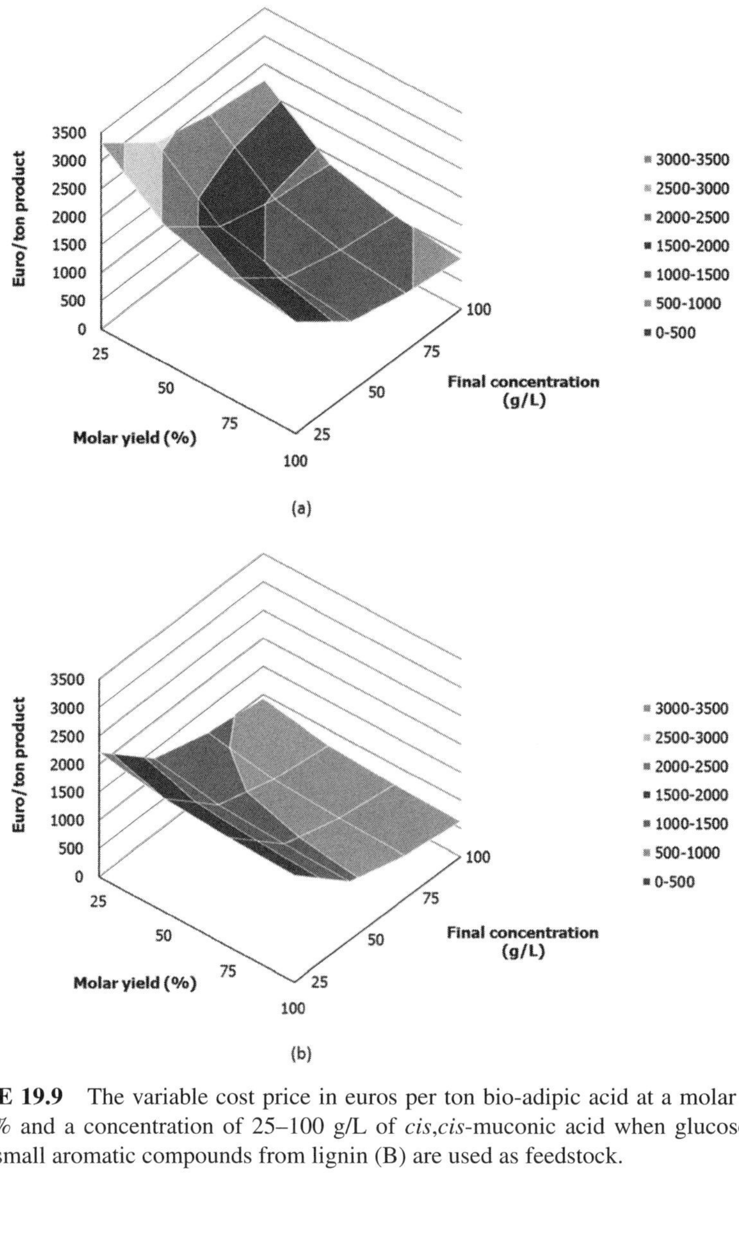

FIGURE 19.9 The variable cost price in euros per ton bio-adipic acid at a molar yield of 25–100% and a concentration of 25–100 g/L of *cis,cis*-muconic acid when glucose (A) or impure small aromatic compounds from lignin (B) are used as feedstock.

for bio-adipic acid, which uses lignin as substrate, can be seen as a trendsetter for the production of bio-based bulk chemicals. Important tasks remaining, involve the validation on pilot scale how lignin can best be depolymerized for industrial integrated-biotechnological processes. Moreover, to ensure a low environmental impact under such circumstances, it is important to establish a high molar yield and final concentration with the integrated biotechnological process.

REFERENCES

Bang SG, Choi CY. (1995). DO-stat fed-batch production of *cis, cis*-muconic acid from benzoic-acid by *Pseudomonas putida* Bm014. *J Ferment Bioeng*, 79, 381–383.

Becker J, Wittmann C. (2012). Systems and synthetic metabolic engineering for amino acid production—the heartbeat of industrial strain development. *Curr Opin Biotechnol*, 23, 718–726.

Benndorf D, Loffhagen N, Babel W. (2001). Protein synthesis patterns in *Acinetobacter calcoaceticus* induced by phenol and catechol show specificities of responses to chemostress. *FEMS Microbiol Lett*, 200, 247–252.

Benndorf D, Thiersch M, Loffhagen N, Kunath C, Harms H. (2006). *Pseudomonas putida* KT2440 responds specifically to chlorophenoxy herbicides and their initial metabolites. *Proteomics*, 6, 3319–3329.

BMU, Division WA I 1. (2011). Federal Ministry for the Environment, Nature Conservation and Nuclear Safety (BMU).

Boussie TR, Dias EL, Fresco ZM, Murphy VJ, Shoemaker J, Archer R, Jiang H, Rennovia, Inc, Assignee. (2010). Production of adipic acid and derivatives from carbohydrate-containing materials. US patent 2010/0317823 A1.

Brehmer B. (2008). Chemical biorefinery perspectives. Available at: http://de.scribd.com/doc/56588067/Chemical-Bio-Refinery-Perspectives.

Brehmer B, Struik PC, Sanders J. (2008). Using an energetic and exergetic life cycle analysis to assess the best applications of legumes within a biobased economy. *Biomas Bioenerg*, 32, 1175–1186.

Brehmer B, Boom R, Sanders J. (2009). Maximum fossil fuel feedstock replacement potential of petrochemicals via biorefineries. *Chem Eng Res Des*, 87, 1103–1119.

Brzostowicz PC, Walters DM, Thomas SM, Nagarajan V, Rouvière PE. (2003). mRNA differential display in a microbial enrichment culture: simultaneous identification of three cyclohexanonemonooxygenases from three species. *Appl Environ Microbiol*, 69, 334–342.

Bu Q, Lei H, Ren S, Wanga L, Holladay J, Qin Zhang A, Tang J, Ruan R. (2011). Phenol and phenolics from lignocellulosic biomass by catalytic microwave pyrolysis. *Bioresour Technol*, 102, 7004–7007.

Butler G, Rasmussen MD, Lin MF, Santos MA, Sakthikumar S, Munro CA, Rheinbay E, Grabherr M, Forche A, Reedy JL, et al. (2009). Evolution of pathogenicity and sexual reproduction in eight Candida genomes. *Nature*, 459, 657–662.

Cheng Q, Thomas SM, Kostichka K, Valentine JR, Nagarajan V. (2000). Genetic analysis of a gene cluster for cyclohexanol oxidation in *Acinetobacter* sp. Strain SE19 by in vitro transposition. *J Bacteriol*, 182, 4744–4751.

Cheng Q, Thomas SM, Rouviere P. (2002). Biological conversion of cyclic alkanes and cyclic alcohols into dicarboxylic acids: biochemical and molecular basis. *Appl Microbiol Biotechnol*, 58, 704–711.

Choi WJ, Lee EY, Cho MH, Choi CY. (1997). Enhanced production of *cis, cis*-muconate in a cell-recycle bioreactor. *J Ferment Bioeng*, 84, 70–76.

Chua JW, Hsieh J. (1990).Oxidative bioconversion of toluene to 1,3-butadiene-1,4-dicarboxylic acid (*cis, cis*-muconic acid). *World J Microb Biot*, 6, 127–143.

Curran KA, Leavitt JM, Karim AS, Alper HS. (2013). Metabolic engineering of muconic acid production in *Saccharomyces cerevisiae*. *Metab Eng*, 15, 55–66.

Dapsens PY, Mondelli C, Pérez-Ramírez J. (2012). Biobased chemicals from conception toward industrial reality: lessons learned and to be learned. *ACS Catal*, 2, 1487–1499.

de Guzman D. (2010). *Introducing Rennovia*. ICIS green Chemicals.

de Wild PJ, van der Laan RR, Wilbering RWA. (2010). Thermolysis of lignin for value-added products. XV Meeting of the International Humic Substances Society Tenerife, Canary Islands, Spain, 27 June – 2 July 2010.

Draths KM, Frost JW. (1993). Environmentally compatible synthesis of adipic acid from D-Glucose. *J Am Chem Soc*, 116, 399–400.

Guazzaroni ME, Morgante V, Mirete S, González-Pastor JE. (2013). Novel acid resistance genes from the metagenome of the Tinto River, an extremely acidic environment. *Environ Microbiol*, 15, 1088–1102.

Hatti-Kaul R, Törnvall U, Gustafsson L, Börjesson P. (2007). Industrial biotechnology for the production of bio-based chemicals-a cradle-to-grave perspective. *Trends Biotechnol*, 25, 119–124.

Harwood CS, Parales RE. (1996). The beta-ketoadipate pathway and the biology of self-identity. *Annu Rev Microbiol*, 50, 553–590.

Ivanov S, Fedorova EE, Limpens E, De Mita S, Genre A, Bonfante P, Bisseling T. (2012). *Rhizobium*-legume symbiosis shares an exocytotic pathway required for arbuscule formation. *Proc Natl Acad Sci USA*, 109, 8316–8321.

Jiménez JI, Miñambres B, García E, Díaz JL. (2002). Genomic analysis of the aromatic catabolic pathways from *Pseudomonas putida* KT2440. *Environ Microbiol*, 4, 824–841.

Kim BJ, Choi WJ, Lee EY, Cho CY. (1988). Enhancement of *cis*, *cis*-muconate productivity by overexpressing of catechol 1,2-dioxygenase in *Pseudomonas putida* BCM114. *Biotechnol Bioprocess Eng*, 3, 112–114.

Leong YK. (2007). Particle bridging in dispersions by small charged molecules: chain length and rigidity, architecture and functional groups spatial position. *Phys Chem Chem Phys*, 9, 5608–5618.

Lippow SM, Moon TS, Basu S, Yoon SH, Li X, Chapman BA, Robison K, Lipovsek D, Prather KL. (2010). Engineering enzyme specificity using computational design of a defined-sequence library. *Chem Biol*, 17, 1306–1315.

Ma Z, Troussard E, van Bokhoven JA. (2012). Controlling the selectivity to chemicals from lignin via catalytic fast pyrolysis. *Appl Catal A Gen*, 423–424, 130–136.

Mizuno S, Yoshikawa N, Schi M, Mikawa T, Imada Y. (1988). Microbial-production of *cis*, *cis*-muconic acid from benzoic-acid. *Appl Microbiol Biotechnol*, 28, 20–25.

Moon TS, Yoon SH, Lanza AM, Roy-Mayhew JD, Prather KL. (2009). Production of glucaric acid from a synthetic pathway in recombinant *Escherichia coli*. *Appl Environ Microbiol*, 75, 589–595.

Neuffer B, Frank N, Desai M. (2010). Available and emerging technologies for reducing greenhouse gas emissions from the nitric acid production industry. EPA United States Environmental Protection Agency.

Nevison CD. (2002). Indirect N_2O Emissions from Agriculture. Background Papers—IPCC Expert Meetings on Good Practice Guidance and Uncertainty Management in National Greenhouse Gas Inventories, 381–397.

Niu W, Draths KM, Frost JW. (2002). Benzene-free synthesis of adipic acid. *Biotechnol Prog*, 18, 201–211.

Patel M. (2003). Cumulative energy demand (CED) and cumulative CO_2 emissions for products of the organic chemical industry. *Energy*, 28, 721–740.

Picataggio S, Rohrer T, Deanda K, Lanning D, Reynolds R, Mielenz J, Eirich LD. (1992). Metabolic engineering of *Candida tropicalis* for the production of long-chain dicarboxylic acids. *Biotechnology (N Y)*, 10, 894–898.

Picataggio S, Beardslee T, Verdezyne, Inc, Assignee. (2012). Biological methods for preparing adipic acid. US Patent 2012/0021474 A1.

Polen T, Spelberg M, Bott M. (2012). Toward biotechnological production of adipic acid and precursors from biorenewables. *J Biotechnol*, 167(2), 75–84.

Reith JH, van Ree R, Capote Campos R, Bakker RR, de Wild PJ, Monot F, Estrine B, Bridgwater AV, Agostini A. (2009). Lignocellulosic feedstock biorefinery for co-production of chemicals, transportation fuels, electricity and heat. International Workshop on Biorefinery, Madrid, Spain, 22 June 2009.

Sato K, Aoki M, Noyori RA. (1998). "Green" route to adipic acid: direct oxidation of cyclohexenes with 30 percent hydrogen peroxide. *Science*, 281, 1646–1647.

Schmidt E, Knackmuss HJ. (1984). Production of *cis*, *cis*-Muconate from benzoate and 2-fluoro-*cis*, *cis*-muconate from 3-fluorobenzoate by 3-chlorobenzoate degrading bacteria. *Appl Microbiol Biotechnol*, 20, 351–355.

Schreiber A, Hellwig M, Dorn E, Reineke W, Knackmuss HJ. (1980). Critical reactions in fluorobenzoic acid degradation by *Pseudomonas* sp. B13. *Appl Environ Microbiol*, 39, 58–67.

Shimizu A, Tanaka K, Fujimori M. (2000). Abatement technologies for N_2O emissions in the adipic acid industry. *Chemosphere Glob Change Sci*, 2, 425–434.

Sun X, Lin Y, Huang Q, Yuan Q, Yan Y. (2013). A novel muconic acid biosynthesis approach by shunting tryptophan biosynthesis via anthranilate. *Appl Environ Microbiol*, 79, 4024–4030.

Tuck CO, Pérez E, Horváth IT, Sheldon RA, Poliakoff M. (2012). Valorization of biomass: deriving more value from waste. *Science*, 337, 695–699.

Vairavamoorthy K. 2011. The urban challenge. The water, energy and food security nexus. "The Water Energy and Food Security Nexus – Solutions for the Green Economy", 16–18 November, 2011, Bonn, Germany.

van Duuren JB, Brehmer B, Mars AE, Eggink G, Dos Santos VA, Sanders JP. (2011). A limited LCA of bio-adipic acid: manufacturing the nylon-6,6 precursor adipic acid using the benzoic acid degradation pathway from different feedstocks. *Biotechnol Bioeng*, 108, 1298–1306.

van Duuren JB, Wijte D, Karge B, dos Santos VA, Yang Y, Mars AE, Eggink G. (2012). pH-stat fed-batch process to enhance the production of *cis*, *cis*-muconate frombenzoate by *Pseudomonas putida* KT2440-JD1. *Biotechnol Prog*, 28, 85–92.

Vink ETH, Rabago KR, Glassner DA, Gruber PR. (2003). Applications of life cycle assessment to NatureWorks™ polylactide (PLA) production. *Polym Degrad Stab*, 80, 403–419.

Weber C, Brückner C, Weinreb S, Lehr C, Essl C, Boles E. (2012). Biosynthesis of *cis,cis*-muconic acid and its aromatic precursors, catechol and protocatechuic acid, from renewable feedstocks by Saccharomyces cerevisiae. *Appl Environ Microbiol*, 78, 8421–8430.

Wittmann C, Lee SY. (2012). *Systems Metabolic Engineering*. Springer.

INDEX

Bioprocessing of Renewable Resources to Commodity Bioproducts, First Edition.
Edited by Virendra S. Bisaria and Akihiko Kondo.